Thymic Hormones and Lymphokines

Basic Chemistry and Clinical Applications

GWUMC Department of Biochemistry Annual Spring Symposia

Series Editors:

Allan L. Goldstein, Ajit Kumar, and George V. Vahouny

The George Washington University Medical Center

DIETARY FIBER IN HEALTH AND DISEASE
Edited by George V. Vahouny and David Kritchevsky

EUKARYOTIC GENE EXPRESSION
Edited by Ajit Kumar

THYMIC HORMONES AND LYMPHOKINES
Basic Chemistry and Clinical Applications
Edited by Allan L. Goldstein

Thymic Hormones and Lymphokines

Basic Chemistry and Clinical Applications

Edited by

Allan L. Goldstein

The George Washington University
School of Medicine and Health Sciences
Washington, D.C.

SPRING SYMPOSIUM SERIES

PLENUM PRESS • NEW YORK AND LONDON

Library of Congress Cataloging in Publication Data

Main entry under title:

Thymic hormones and lymphokines.

(GWUMC Department of Biochemistry annual spring symposia)
Bibliography: p.
Includes index.
1. Thymic hormones—Congresses. 2. Thymic hormones—Therapeutic use—Congresses. 3. Lymphokines—Congresses. 4. Lymphokines—Therapeutic use—Congresses. I. Goldstein, Allan L. II. Series. [DNLM: 1. Thymus Hormones—congresses. 2. Lymphokines—congresses. WK 400 T5485 1983]
QP572.T46T48 1984 612′.43 84-9995

ISBN-13: 978-1-4684-4747-7 e-ISBN-13: 978-1-4684-4745-3
DOI:10.1007/ 978-1-4684-4745-3

Softcover reprint of the hardcover 1st edition 1984

A Division of Plenum Publishing Corporation
233 Spring Street, New York, N.Y. 10013

Preface

The Third Annual Symposium on Health Sciences attracted over 400 scientists from 15 countries. It was held at the National Academy of Sciences in Washington, D.C. The theme of this meeting was "Thymic Hormones and Lymphokines: Basic Chemistry and Clinical Applications."

The meeting emphasized the state of the art with regard to chemistry, mechanism of action, and clinical applications of thymic hormones and lymphokines. In addition to the five sessions, the chairmen of each session were asked to prepare a synthesis and overview of each of the sessions. Some of the chairmen used this time to summarize the new material presented while others addressed important areas of conflict and terminology. The chairmen of the plenary sessions prepared short summary papers which have been incorporated into this volume at the end of each major section. We also invited several of the scientists who took part in the poster presentations to submit short papers.

In addition to the scientific sessions, there were two awards and lectures. One, the Abraham White Distinguished Science Award, was given to Dr. Robert C. Gallo, who gave the keynote lecture. He was recognized for his pioneering research and scientific contributions to our understanding of the regulation of hematopoietic cell differentiation and the role of viruses in the development of human leukemias. The second award, a Distinguished Public Service Award, was given to Senator John Glenn from Ohio in recognition of his leadership and outstanding achievements in the United States Senate, and for his insightful support for basic science biomedical research and graduate education.

Considerable progress was reported in the chemical and biological characterization of thymic hormones and lymphokines. This included the sequencing and synthesis of several of the key thymic hormones, the role of the endocrine thymus in modulating neuroendocrine function, the purification of several lymphokines, and the characterization of monoclonal antibodies and lymphokine receptors.

An important aspect of the symposium was the presentation of several clinical studies. These clinical reports suggest that thymic hormones and lymphokines may play a key role in the treatment of a number of diseases, including cancer, autoimmune diseases, and infectious diseases.

The symposium provided a forum for new ideas on the mechanism of action, pharmacology, and clinical applications of the immune system peptides that were both provocative and intellectually exciting. It is clear from the high quality of most of the presentations that the field of thymic hormones and lymphokines has come of age and that studies using chemically defined molecules, rather than crude extracts, are now possible.

The availability of synthetic thymic hormones, purified lymphokines, and antibodies to these peptides enables both researchers and clinicians to acquire definitive information on the immunology, pharmacology, and clinical applications of thymic hormones and lymphokines. Clinicians will be particularly interested in the reports of clinical responses and diagnostic applications of these biological response modifiers in diseases ranging from the acquired immune deficiency syndrome (AIDS) to rheumatoid arthritis. This volume will also be of keen interest to biochemists, pharmacologists, immunologists, and other basic scientists interested in acquiring information on the state of the art of purification procedures for, and the chemical characterization (including sequencing), pharmacokinetics, immunopharmacology, and immunobiology of thymic hormones and lymphokines.

Allan L. Goldstein

Contents

Opening Remarks

Human T-Cell Growth Factor, Growth of Human Neoplastic T Cells, and Human T-Cell Leukemia-Lymphoma Virus

ROBERT C. GALLO, SURESH K. ARYA, STEPHAN G. LINDNER, FLOSSIE WONG-STAAL, and MANGALASSERIL G. SARNGADHARAN

1. INTRODUCTION

Identification of conditions permitting growth of different human hematopoietic cell strains was a goal for a long time in many laboratories including ours. To this end we screened a large number of sources including conditioned media from freshly initiated cultures of a variety of tissue and cell types for factors that would support growth of specific hematopoietic cell types in suspension culture. These attempts led to the initial description of an activity in the media from short-term cultures of human peripheral blood lymphocytes treated with phytohemagglutinin (PHA) that allowed the specific growth of T lymphocytes from human peripheral blood or bone marrow (Morgan *et al.*, 1976; Ruscetti *et al.*, 1977). Extensive characterization of the resultant cell population revealed that these were functional T lymphocytes (Ruscetti *et al.*, 1977), and the term "T-cell growth factor" (TCGF) was assigned to the activity. Preliminary analyses quickly revealed that the T-cell response is a two-step process: an initial activation of the lymphocytes by lectin (or an antigen)

ROBERT C. GALLO, SURESH K. ARYA, STEPHAN G. LINDNER, FLOSSIE WONG-STAAL, and MANGALASSERIL G. SARNGADHARAN • Laboratory of Tumor Cell Biology, Division of Cancer Treatment, National Cancer Institute, National Institutes of Health, Bethesda, Maryland 20205.

which makes the cells synthesize receptors for TCGF and subsequently bind TCGF resulting in the mitogenic activity. The effect was highly specific and the only cells that grew out of the mixed population of cells that we started with were cells with morphological and functional characteristics of mature T cells and only after preliminary activation with PHA.

The findings in the human system were soon adapted to the murine system (Gillis and Smith, 1977; Rosenberg *et al.*, 1978) and this resulted in some key advances in our understanding of the early cell biology of TCGF production and response. Soon it became apparent that a small number of macrophage-monocytes are essential for TCGF production. These cells release another factor called lymphocyte-activating factor (LAF) or interleukin-1, which in some fashion interacts with the lymphocytes and leads to the production of TCGF (Larsson *et al.*, 1980).

The conditioned media used as sources for TCGF were very rich sources of a variety of other lymphokines (Oppenheim *et al.*, 1979). A number of reports appeared in the literature describing many biological effects of such conditioned media and ascribed them to TCGF. However, subsequent studies using partially purified, lectin-free fractions of TCGF showed that TCGF does not have such activities as B-cell growth factor (Swain *et al.*, 1981; Paul *et al.*, 1981) and T-cell replacing factor (Takatsu *et al.*, 1980). Also, many of the initial observations were made with media still containing the lectin used to induce the lymphocytes to produce TCGF in the first place.

Availability of partially purified, lectin-free preparations of TCGF (Mier and Gallo, 1980; Lotze and Rosenberg, 1981) presented an enormous opportunity to maintain and study lines of proliferating T-cell blasts that retained specific functional characteristics. The value of such systems for studies of T-cell biology and immunology was immediately appreciated. The system was used most effectively to generate and maintain antigen-specific human and other animal cytotoxic T lymphocytes. Cultured T cells mediating alloantigen-directed cytotoxicity retained this property remarkably constantly after several months in culture (Gillis and Smith, 1977; Kasakura, 1977; Strausser and Rosenberg, 1978; Kurnick *et al.*, 1979). T-cell lines with other reactivities have also been developed. Schrier and Tees (1980) demonstrated that erythrocyte antigen-specific helper murine T cells could be grown in long-term culture. Such cloned helper T cells could selectively reconstitute nude mice to produce antibodies against the antigen that the cells were activated with *in vitro* (Tees and Schrier, 1980).

The availability of a system for the long-term growth of pure populations of human T cells also made it possible to grow large quantities of such cells for biochemical studies. Most importantly, it became possible to grow neoplastic T cells and to characterize them for various biological properties and particularly to use these systems to study the involvement of retroviruses in human malignancies.

2. PURIFICATION OF TCGF

Human TCGF is present usually in small amounts in the culture conditioned media of lectin-stimulated lymphocyte cultures. Higher levels are expressed in certain continuous cell lines, e.g., a leukemic cell line, JURKAT (Gillis and Watson, 1980), when induced with PHA and phorbol myristate acetate (PMA). Certain other cell lines, e.g., a gibbon leukemia cell line, MLA 144 (Rabin *et al.*, 1981), and a human lymphoma cell line, HUT 102 (Gootenberg *et al.*, 1981), constitutively produce TCGF, but both these lines also produce retroviruses (Kawakami *et al.*, 1972; Poiesz *et al.*, 1980a). We describe here the purification of TCGF from the conditioned media of PHA-treated human normal peripheral blood lymphocytes (PBL). Because of the enormous volumes of conditioned media that need to be processed to obtain significant amounts of pure TCGF, the initial step is a major concentration procedure. This is routinely achieved by diafiltration using a Pellicon Cassette system (Millipore) employing the polysulfone membrane filter PTGC with a 10,000-dalton cutoff limit. A concentration of 40 liters of medium to 1.5–2 liters can be achieved in less than 3 hr.

2.1 Anion-Exchange Chromatography on DEAE-Sepharose

This step is extremely useful when the conditioned media also contain serum proteins, because it effectively fractionates TCGF from serum proteins. When TCGF is produced under serum-free and albumin-free conditions, this anion-exchange step is not very critical. Twentyfold concentrated conditioned medium (100 ml) from PHA-stimulated human PBL is dialyzed extensively against 10 mM Tris-HCl (pH 7.8) containing 0.1 mM phenylmethylsulfonyl fluoride (PMSF) and 0.1% polyethylene glycol (PEG), applied to a 100-ml column of DEAE-Sepharose (Pharmacia) equilibrated with the above buffer and washed with two to three bed volumes of the buffer to remove unadsorbed proteins. The column is then developed with a 500-ml 0–0.15 M NaCl gradient in the equilibration buffer. Fractions of 5 ml are collected, and their absorbance measured at 280 nm. After sterile filtration through 0.22-μm filters, aliquots are removed for assay of TCGF activity in a [^{3}H]thymidine incorporation assay (Sarngadharan *et al.*, 1984) using a TCGF-dependent mouse cytotoxic T-cell line (Gillis *et al.*, 1978). A typical chromatographic profile obtained is shown in Fig. 1. Most of the TCGF activity elutes at about 0.05 M NaCl, although low levels of TCGF activity can be seen occasionally at a slightly higher salt concentration. Most of the protein remains bound to the column and only begins to elute when the column is washed with 0.15 M NaCl at the end of the gradient.

The ion-exchange chromatography described above can be scaled up. Concentrates of up to 40 liters conditioned media have been successfully fractionated on a single column of DEAE-Sepharose. In addition to separating all the albumin from TCGF, this procedure also removes all the PHA present in the conditioned medium. This is an extremely important practical point. In most biological experiments, the absolute purity of TCGF is not critical, but the continued presence of PHA will complicate the analysis of the direct effects of TCGF such as selective growth of antigen-primed T cells. For purification of TCGF induced from normal PBL cultures in media containing no serum or protein additives (Cellular Products, Inc.), the above step of anion-exchange chromatography can be eliminated. Concentrated media are then processed directly through the next step.

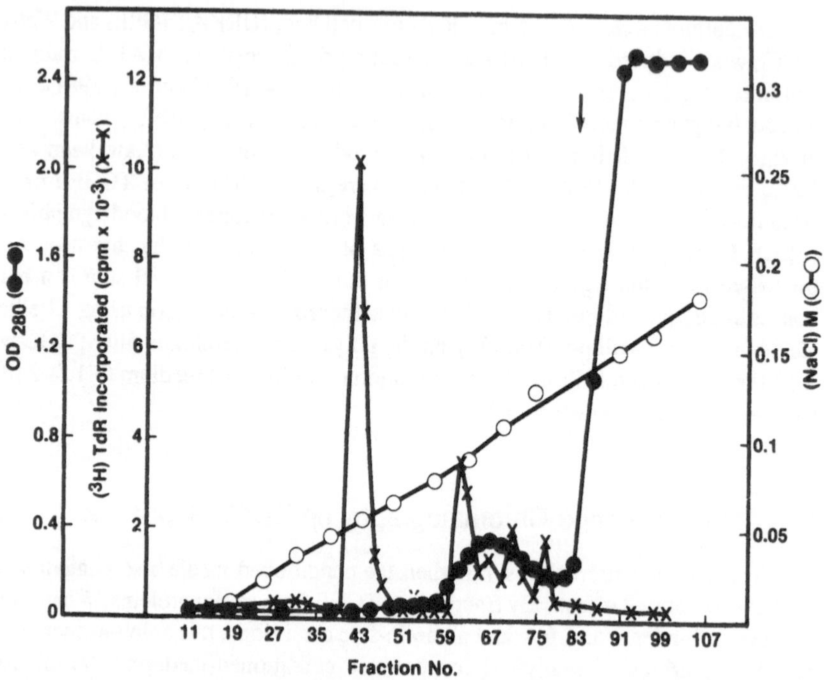

FIGURE 1. DEAE-Sepharose chromatography of TCGF in PHA-stimulated human PBL-conditioned media. Concentrated conditioned media were dialyzed against Tris-HCl (pH 7.9) containing 0.1% PEG and 0.1 mM PMSF and applied to a 100-ml column of DEAE-Sepharose equilibrated with the same buffer. After extensive washing to remove unadsorbed proteins, the column was developed with a 0–0.15 M NaCl gradient in the buffer. At the end of the gradient (arrow), the column was further washed with 0.15 M NaCl. Fractions of 5 ml were collected and assayed for TCGF activity (X), absorbance at 280 nm (●), and NaCl concentration (by conductance) (○).

2.2. Adsorption and Elution of TCGF from Controlled-Pore Glass

Media (40 liters) from TCGF induced from normal PBL cultures were concentrated to 1.4 liters using the Pellicon Cassette filtration. To the concentrate taken in a roller bottle, 120 g of Controlled-Pore glass (CPG-10, 75 Å pure, 80–100 mesh, ElectroNucleonics, Inc.) was added and the contents mixed overnight in the cold using a roller apparatus. The medium was removed by centrifugation and the glass beads were packed into a column and washed successively with PBS (500 ml) and 10 mM Tris-HCl, pH 7.8 (700 ml). TCGF was eluted with Tris-HCl (pH 8.2) containing 0.1% PEG and 0.1 mM PMSF and 1 M tetramethylammonium chloride (TMAC). Generally, the first two bed volumes of the eluate contain most of the TCGF activity. As fractions containing TMAC cannot be directly assayed for TCGF activity because of cytotoxicity, we collect up to four bed volumes of the eluate. The fractions are pooled to correspond to the initial two bed volumes and the next successive bed volumes. They are separately dialyzed against 10 to 20 volumes of 10 mM Tris-HCl buffer (pH 7.8) containing 0.1% PEG and 0.1 mM

PMSF, the buffer being changed at least three times. Dialyzed pools are assayed for TCGF activity and the active pools are combined and used in the next step.

2.3. High-Performance Liquid Chromatography (HPLC)-I

Active fractions from the CPG step were acidified to pH 2 with 0.1% trifluoroacetic acid (TFA) and filtered through a 0.22-μm filter. The sample was pumped through a column of C_{18}-silica (2.5 × 30 cm) (Waters Associates) at a flow rate of 5 ml/min. The column was successively washed at the same flow rate with 10, 45, 50, and 65% aqueous acetonitrile containing 0.1% TFA, each elution step being continued until the absorbance at 214 nm reached a steady low background. The eluted fractions are assayed for TCGF activity in a [^{3}H]thymidine incorporation assay with a starting dilution of 1 : 80 and subsequent serial dilutions in the assay medium. TCGF is eluted in the 50–65% acetonitrile fraction.

2.4. HPLC-II

The 50–65% acetonitrile fraction from HPLC-I is diluted with an equal volume of 0.1% aqueous TFA containing 0.2% PEG and loaded onto a 0.38 × 30-cm column of μBondapak C_{18} (Waters Associates) at a flow rate of 1 ml/min. The column is washed successively at 32.5 and 45% aqueous acetonitrile containing 0.1% TFA. After achieving a stable baseline, a linear 2-hr gradient between 45 and 65% acetonitrile is initiated. The elution of proteins is monitored by 214-nm absorbance and fractions are collected manually. Active fractions are pooled and if necessary rechromatographed as above, but using a 1-hr 0–60% linear gradient of acetonitrile for elution. Figure 2 (bottom panel) shows the profile of a representative experiment. The TCGF activity was recovered in the fraction shown by the shaded area. For comparison, a sample of immunoaffinity-purified TCGF from JURKAT medium (Smith, 1983) was chromatographed on the same column and its elution profile is shown in the upper panel of Fig. 2. The small difference in the elution characteristics of the two TCGFs probably implies a slight difference in the hydrophobicities of the two proteins, possibly because of differences in posttranslational modification.

The amino-terminal 15 amino acids of the PBL TCGF were determined and compared with the sequence of the JURKAT TCGF. They were identical, including the residue at position 3 where the threonine was modified by glycosylation (Table I). Additional sequencing of the TCGF molecule was performed using peptides generated from *S*-carboxamide methylated JURKAT TCGF by treatment with Lys-C endopeptidase (Boehringer) (Copeland *et al.*, Chapter 17, this volume). Knowledge of the amino acid sequence of TCGF was extremely useful in the successful cloning of the human gene for TCGF from lectin-stimulated normal human PBL as described below.

The fraction from HPLC-II was found to contain homogeneous TCGF. Amino acid sequence determined by the micro sequence procedure using Edman degradation provided a single amino acid residue in the first cycle.

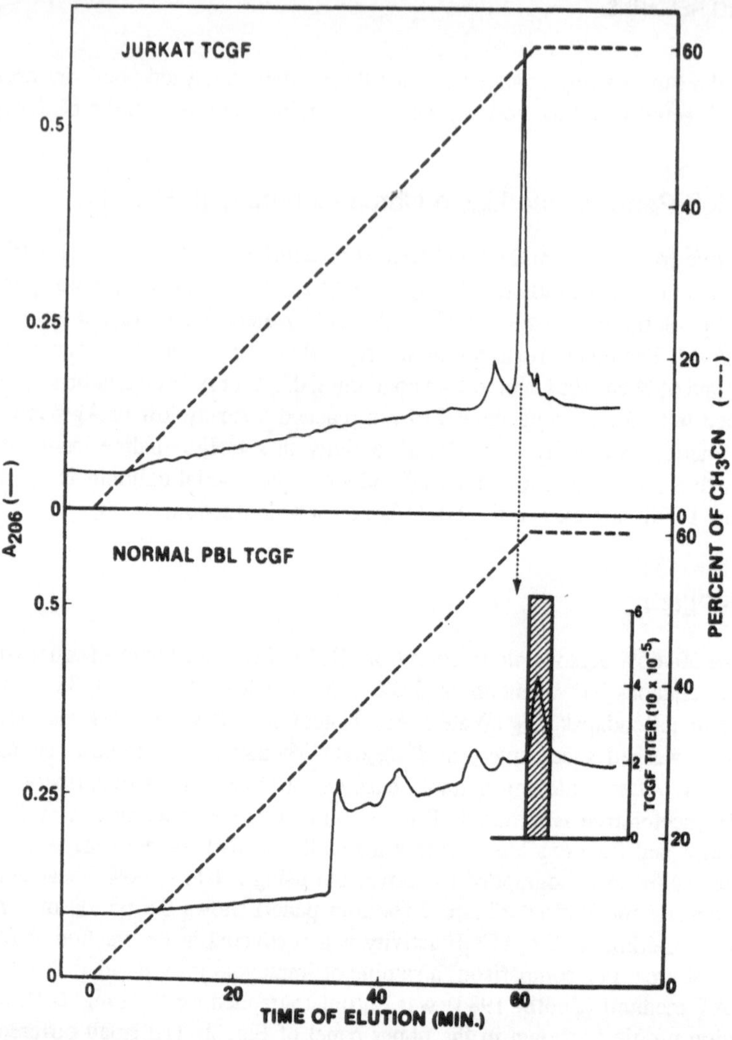

FIGURE 2. Reverse-phase high-performance liquid chromatography of TCGF. The samples were injected into a μBondapak C_{18} column (Waters Associates) and developed with a 60-min, 0–60% aqueous acetonitrile gradient containing 0.05% TFA for 60 min at a flow rate of 1 ml/min. Absorbance of the effluent was monitored at 206 nm and fractions were collected manually to correspond to the absorbance peaks. TCGF activity in the fractions was measured at serial dilutions starting with an initial dilution of 1 : 80 in the assay medium using [^{3}H]thymidine incorporation by a TCGF-dependent cytotoxic mouse T-cell line as the assay. The sample run in the upper panel is immunoaffinity-purified JURKAT TCGF (Smith, 1983) and that in the bottom panel is TCGF fractionated from lectin-stimulated normal human PBL as described in the text.

TABLE I. Amino-Terminal Amino Acid Sequence of T-Cell Growth Factor from Normal Human Peripheral Blood Lymphocytes and from JURKAT Cells[a]

Normal PBL	Ala-Pro-Thr[b]-Ser-Ser-Ser-Thr-Lys-Lys-Thr-Gln-Leu-Gln-Leu-Glu
JURKAT	Ala-Pro-Thr[b]-Ser-Ser-Ser-Thr-Lys-Lys-Thr-Gln-Leu-Gln-Leu-Glu

[a] Sequence determined by Copeland *et al.* (Chapter 17, this volume).
[b] Threonine is modified by glycosylation (see Robb and Lin, Chapter 22, this volume).

3. MOLECULAR CLONING OF THE HUMAN GENE FOR TCGF

We prepared a pool of tetradecameric oligonucleotides complementary to the first 14 nucleotides of the codon corresponding to a pentapeptide Lys-Phe-Tyr-Met-Pro selected from the amino acid sequence of TCGF. These nucleotides were used as probes to screen a cDNA library prepared from PHA-stimulated normal human PBL mRNA. Several cDNA clones were obtained that hybridized to these probes. When inserted into eukaryotic expression vectors, these clones directed the synthesis of biologically active TCGF. The nucleotide sequence of these clones was determined and was consistent with the amino acid sequence information on TCGF (Clark *et al.*, 1984). The nucleotide sequence of the cDNA clone from human normal lymphocytes was essentially identical with that from JURKAT TCGF published by Taniguchi *et al.* (1983).

4. STRUCTURAL ORGANIZATION OF THE TCGF GENE

The structural organization of the TCGF gene in normal PBL was examined by the hybridization of cellular DNA to radiolabeled cloned TCGF cDNA using the standard Southern blot procedure. Data obtained with a number of restriction endonucleases are shown in Fig. 3. Enzymes, Bam HI, Kpn I, Pst I, and Pvu II do not cleave with TCGF cDNA, but each of them generates one specific DNA

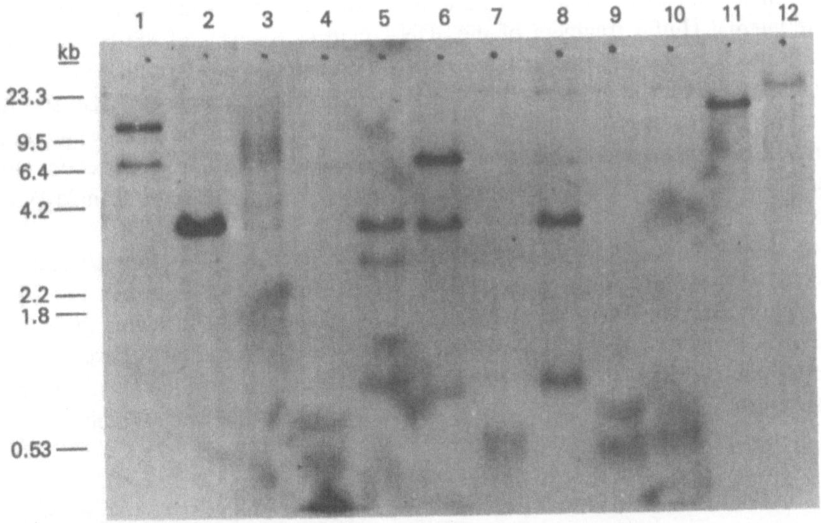

FIGURE 3. Nick-translated cloned TCGF cDNA insert was hybridized using the Southern procedure with high-molecular-weight DNA from normal lymphocytes digested with Bgl II (lane 1), Eco RI (lane 2), Xba I (lane 3), Hinf I (lane 4), Bgl II plus Eco RI (lane 5), Bgl II plus Xba I (lane 6), Bgl II plus Hinf I (lane 7), Eco RI plus Xba I (lane 8), Eco RI plus Hinf I (lane 9), Xba I plus Hinf I (lane 10), Bam HI (lane 11), and Kpn I (lane 12).

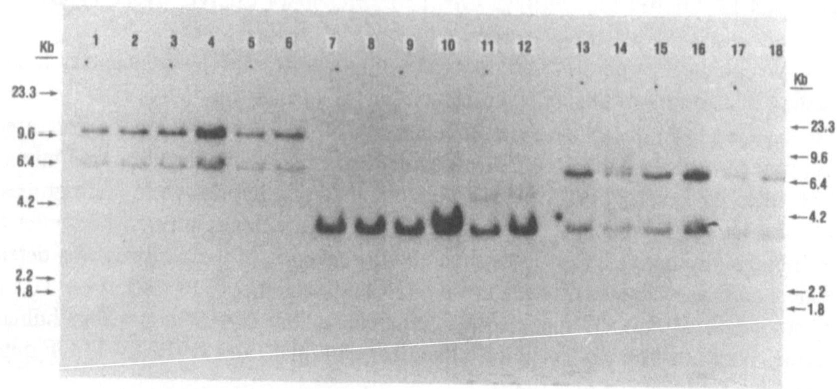

FIGURE 4. Hybridization of nick-translated cloned TCGF cDNA insert with DNAs from normal lymphocytes, JURKAT cells, lymphocytes from a patient with acute lymphocytic leukemia (ALL), HUT 102 cells, HUT 78 cells, and MT-2 cells, digested with Bgl II (lanes 1–6, respectively), digested with Eco RI (lanes 7–12, respectively), and digested with Xba I (lanes 13–18, respectively).

fragment larger than 10 Kb. Eco RI, which also does not cleave within the TCGF cDNA, generated two comigrating TCGF-specific DNA fragments (3.8–4.0 Kb), one of which can be distinguished because it has an internal Bgl II site. Xba I, which cleaves the cDNA clone once, gives at least three TCGF-specific fragments (6.8, 3.8, and 1.04 Kb), and Hinf I, which cleaves the TCGF cDNA twice, generates two major fragments (about 0.72 and 0.51 Kb) neither of which comigrates with the internal Hinf I fragment of the cDNA clones. Analysis of the Hinf I digest indicates that the genomic and the cDNA TCGF sequences are not colinear and that the TCGF gene must have at least one intron. The patterns of TCGF-specific fragments from the Xba I and Eco RI digests are also consistent with the presence of introns and exons. The exact number and arrangement of introns and exons have not been determined but the data presented strongly suggest there is a single TCGF gene.

The organization of the TCGF sequences in cellular DNAs from several different sources was compared by cleaving the DNAs with Bgl II, Eco RI, and Xba I (Fig. 4). The six DNAs shown in Fig. 4 and 12 other DNAs including those from normal, uninfected and HTLV-infected leukemic cells (not shown) gave identical restriction patterns, indicating the absence of detectable polymorphism or rearrangement of the TCGF sequences in the limited number of cells examined.

5. EXPRESSION OF THE HUMAN TCGF GENE

TCGF is synthesized by normal human T cells upon stimulation with lectins or antigens, while a T-cell line (JURKAT) derived from a human leukemia patient produces TCGF after stimulation with lectin plus PMA. There are, however, some

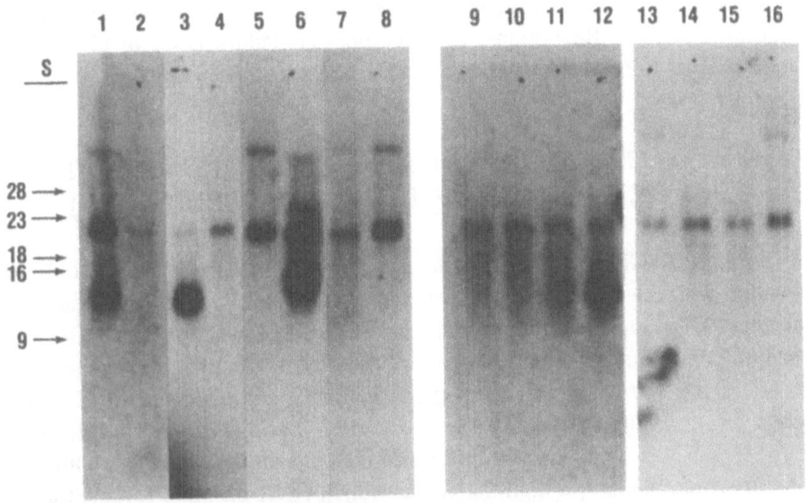

FIGURE 5. Hybridization of nick-translated cloned TCGF cDNA to RNA from PHA-stimulated lymphocytes (lane 1), unstimulated lymphocytes (lane 2), PHA plus PMA-stimulated JURKAT cells (lane 3), unstimulated JURKAT cells (lane 4), 6G1 cells (lane 5), MLA 144 cells (lane 6), Molt-4 cells (lane 7), HL-60 cells (lane 8), JURKAT cells, unstimulated and stimulated with PMA, PHA, and PHA plus PMA (lanes 9–12, respectively), Daudi cells (lane 13), Raji cells (lane 14), 3A cells (lane 15), and SD cells (lane 16). TCGF-specific RNA appears as a 900-nucleotide band (1100 nucleotides for MLA 144 cells). The 2300-nucleotide band resulted from hybridization of a trace-contaminating sequence in the probe to an abundant mRNA species (data not shown), which is constitutively expressed in all cell types examined. Hybridization to this mRNA provided an internal control for the experiment.

neoplastic T cells, such as a gibbon cell line, MLA 144 (Rabin *et al.*, 1981), and a human cell line, HUT 102 (Gootenberg *et al.*, 1981), that produce TCGF constitutively. To understand how TCGF synthesis is regulated in these cell lines, cloned PBL TCGF cDNA was hybridized to mRNA using the Northern technique. Cellular mRNA was fractionated by electrophoresis, through denaturing agarose gels, transferred to Gene Screen (New England Nuclear), and hybridized with ^{32}P-labeled cloned cDNA. A TCGF-specific mRNA of approximately 900 nucleotides (11 S) was readily detected in the RNA isolated from PHA-stimulated PBL and JURKAT cells stimulated with both PHA and PMA (Fig. 5). TCGF-specific mRNA was also seen in MLA-144, but its size was slightly larger (~1100 nucleotides long). In all positive cases, only one species of TCGF mRNA was found even under relaxed hybridization conditions. There is a complete correlation between the mRNA expression in the cells and the secretion of TCGF into the culture medium. A specific TCGF mRNA could not be found in unstimulated PBL, unstimulated JURKAT cells, another gibbon T-cell line 6G1 or human cells Molt-4, CCRF-CEM (immature T cells), HL-60 (myeloid cells), Daudi (B cells), Raji (B cells), as well as the trophoblast lines 3A and SD (Fig. 5). In the case of JURKAT cells, PHA stimulation alone causes production of low levels of TCGF mRNA, and low but detectable levels of TCGF. Both these levels increase upon the combined stimulation with PHA and PMA. PMA alone, however, does not induce TCGF or TCGF mRNA synthesis. Therefore, the regulation of TCGF gene expression is at the mRNA level.

6. GROWTH OF MATURE HUMAN T CELLS AND ISOLATION OF HUMAN T-CELL LEUKEMIA-LYMPHOMA VIRUS (HTLV)

Purified TCGF supports growth of T cells from normal people only after activation with a lectin or antigen, whereas T cells from patients with adult T-cell leukemia (ATL) are directly responsive to TCGF. Several of these became established cell lines and grew independent of the requirement for exogenous TCGF, and released a retrovirus that we named "human T-cell leukemia virus" (HTLV) (Poiesz *et al.*, 1980a, 1981; Popovic *et al.*, 1983; Sarin *et al.*, 1983). Morphologically, HTLV is a typical type-C virus and contains all the functional components of a retrovirus, namely, a high-molecular-weight RNA genome, reverse transcriptase, and core *gag* proteins homologous to other retroviral proteins (Poiesz *et al.*, 1980a; Reitz *et al.*, 1981; Rho *et al.*, 1981; Kalyanaraman *et al.*, 1981, 1984; Oroszlan *et al.*, 1982; Copeland *et al.*, 1983). By nucleic acid hybridization studies (Reitz *et al.*, 1981; Gallo *et al.*, 1982), and by immunological analysis of proteins (Kalyanaraman *et al.*, 1981, 1984; Robert-Guroff *et al.*, 1981), HTLV has been identified as a unique exogenously acquired human retrovirus and with the exception of a distant evolutionary relationship in the amino acid sequence of the *gag* proteins with bovine leukemia virus (Oroszlan *et al.*, 1982), it was substantially unrelated to all known animal retroviruses. Since the initial isolation of HTLV, several additional isolates have been made (Popovic *et al.*, 1983; Yoshida *et al.*, 1982; Haynes *et al.*, 1983; Sarin *et al.*, 1983), many of them in our own laboratory. In immunological cross-reactivities of their structural proteins (Popovic *et al.*, 1982, 1983; Sarin *et al.*, 1983) and in nucleic acid homology (Reitz *et al.*, 1983), all these isolates appeared extremely related to the original isolate. There is one isolate, however, derived from T-cell cultures of a patient with a T-cell variant of hairy cell leukemia (Saxon *et al.*, 1978), that exhibited substantial differences in immunological properties (Kalyanaraman *et al.*, 1982c) and in nucleic acid homology (Reitz *et al.*, 1983) from the rest of the isolates. The majority of the isolates that are totally cross-reactive with the initial isolate are classified as HTLV-I and the isolate from hairy cell leukemia is classified as HTLV-II (Kalyanaraman *et al.*, 1982). The provirus for HTLV-I and that for HTLV-II have been molecularly cloned (Seiki *et al.*, 1982; Manzari *et al.*, 1983a; Gelmann *et al.*, 1984), and the differences between the two subtypes of HTLV have been confirmed also in hybridization experiments employing the cloned DNA probes (Gelmann *et al.*, 1984).

ATL, characterized by HTLV infection, shows a pattern of geographic clustering in certain parts of the world, notably southern Japan, Caribbean islands, and regions of South America and Africa (Uchiyama *et al.*, 1977; Catovsky *et al.*, 1982; Blayney *et al.*, 1983a). Multiple cases have been identified also in the southeastern United States (Gallo *et al.*, 1983; Blayney *et al.*, 1983a). HTLV-I disease is usually a fulminant acute leukemia often associated with lytic bone lesions,

TABLE II. Comparison of Properties of HTLV-Positive Human Neoplastic T Cells with Normal Uninfected and HTLV-Infected Human Cord Blood T Cells

Property	HTLV-positive primary neoplastic T cell	HTLV-transformed T cell	Mitogen-stimulated normal T cell
In vitro growth	Temporary or immortal	Immortal	Temporary
Requirement for exogenous TCGF	Minimal or none	None	Yes
TCGF receptor (TAC)[a]	+ + +	+ + +	+
Transferrin receptor[a]	+ + +	+ + +	+
S-Ig,[b] EBNA,[c] TdT[d]	−	−	−
Cell phenotype			
Inducer/helper (OKT4, Leu3)	Most or all	Most or all	Most or all
Suppressor/cytotoxic (OKT8, Leu2)[a]	Few or none	Few or none	Few or none
Cell morphology			
Presence of multinucleated giant cells	+	+	−
Presence of lobulated nuclei	+	+	−
HLA "modification"			
Expression of additional HLA antigens	+	+	−
Expression of HLA-Dr[a]	+	+	−
Expression of HT-3 sequences[e]	+ + +	+ + +	+
HTLV p19, p24, and RT expression	+	+	−
Type-C virus particles (EM)	+	+	−

[a] Determined by cell sorter using monoclonal antibodies.
[b] S-IgG, cell surface immunoglobulins.
[c] EBNA, Epstein–Barr nuclear antigens.
[d] TdT, terminal deoxynucleotidyl transferase.
[e] By molecular hybridization using cloned HT-3 DNA (Manzari *et al.*, 1983b).

hypercalcemia, and a predisposition for leukemic involvement of the skin. Many patients may have visceral organ enlargement and lymphadenopathy (Gallo *et al.*, 1983; Blayney *et al.*, 1983b). The most striking finding is that they all have antibodies to HTLV proteins (Kalyanaraman *et al.*, 1982a,b; Robert-Guroff *et al.*, 1982; Gallo *et al.*, 1983). Seroepidemiological surveys have indicated widespread occurrence of serum antibodies to HTLV (4–16%) in the normal populations in areas where HTLV disease is endemic (Robert-Guroff *et al.*, 1983; Gallo *et al.*, 1983).

The mode of natural transmission of HTLV among populations appears to require prolonged and intimate contact with HTLV-positive patients. Naturally, the largest single group of seropositive individuals have been close family contacts of virus-positive leukemia patients. It is not uncommon to find multiple antibody-positive households in which a member is a virus-positive leukemia patient.

Biological transmission can also be demonstrated *in vitro*. HTLV isolated from several T-cell lines established in culture from T-cell leukemic patients from many parts of the world is able to infect fresh umbilical cord blood T cells and, in most instances, the resulting lymphocyte cultures replicate *in vitro* without added TCGF (Popovic *et al.*, 1983; Markham *et al.*, 1983). Target cells for infection are also found among human bone marrow leukocytes (Markham *et al.*, unpublished results). This infection and transformation are accomplished by cocultivation of lethally irradiated or mitomycin C-treated HTLV-producing T cells from cell cultures initiated from leukemic patients. The immortalized cord blood T lymphocytes exhibit in general the same properties as the fresh or cultured leukemic T cells. For instance, the transformed cultures have the same type of multinucleated cells and similar frequency of cells with convoluted nuclei as the primary HTLV-positive leukemic cells. The virally transformed T lymphocytes also resemble primary virus-producing leukemic T-cell cultures in several other functional and surface characteristics (Table II) (Popovic *et al.*, 1983; Markham *et al.*, 1983). HTLV can also infect adult peripheral blood T lymphocytes at a lower incidence (Markham *et al.*, unpublished observation). Once established, they share most of the properties described for HTLV-transformed cord blood T cells.

7. HTLV INFECTION AND THE EXPRESSION OF TCGF RECEPTOR

Resting normal T cells do not express receptors for TCGF and those cells do not respond to TCGF. So it was a novel finding that T cells from certain patients with leukemias and lymphomas involving mature T cells could be grown in direct response to exogenously added TCGF (Poiesz *et al.*, 1980b). This immediately provided a means to selectively grow human neoplastic T cells and to establish permanent T-cell lines from these patients (Poiesz *et al.*, 1980b). Many such cells began expressing HTLV as indicated in an earlier section. Unlike normal human PBL, these cell lines had the capacity to absorb TCGF activity when incubated with solutions containing TCGF (Gootenberg *et al.*, 1981), indicating that they possess surface receptors for TCGF. This was confirmed subsequently by direct binding studies using metabolically labeled TCGF. The latter studies also demonstrated TCGF receptors on lectin- or alloantigen-stimulated normal T lymphocytes (Robb *et al.*, 1981).

The status of the TCGF receptor on human neoplastic cells has been studied extensively using a monoclonal antibody to an activated T-lymphoid cell surface antigen (TAC) believed to be the TCGF receptor (Uchiyama *et al.*, 1981; Leonard *et al.*, 1982). Although immature leukemic T lymphoblasts are normally negative for TCGF receptor, some of them (e.g., JURKAT and HSB-2) express surface determinants recognized by anti-TAC upon induction with PMA and PHA (Greene *et al.*, 1983). These determinants on HSB-2 and JURKAT could be immunoprecipitated using anti-TAC following induction. It is known that both JURKAT (Gillis and Watson, 1980) and HSB-2 (Stadler and Oppenheim, 1982) can be induced with PHA and PMA to produce high levels of TCGF. Also, it has been shown that PMA alone neither induced TCGF secretion nor expression of TCGF mRNA in JURKAT cells (see earlier). In contrast, induction with PMA alone resulted in well-preserved expression of TCGF receptor (Greene *et al.*, 1983). Therefore, the genes encoding the proteins TCGF and TCGF receptor are not coordinately expressed.

The clinical entity ATL (Uchiyama *et al.*, 1977; Catovsky *et al.*, 1982; Blayney *et al.*, 1983b; Bunn *et al.*, 1983) and the Sézary syndrome (Broder *et al.*, 1976; Lutzner *et al.*, 1975; Broder and Bunn, 1980) represent lymphoproliferative disorders involving mature T cells. Of these, ATL has an etiological association with the type-C virus, HTLV (Poiesz *et al.*, 1980a; Yoshida *et al.*, 1982). Comparisons of the surface phenotypes of ATL and Sézary cells using the monoclonal antibodies to TAC revealed that ATL cells were uniformly positive for TCGF receptors whereas Sézary cells were mostly negative (Waldmann *et al.*, 1983). Studies of [^{3}H]-anti-TAC binding have revealed that ATL cells express at least 5–10 times more TAC molecules per cell than PHA-stimulated normal human PBL (J. M. Depper *et al.*, unpublished data). TAC antigen is also expressed in human cord blood cell lines transformed by HTLV. Thus, anti-TAC binding may provide a valuable tool in the diagnosis of HTLV-associated T-cell malignancies.

Many HTLV-producing primary neoplastic T-cell lines secrete TCGF (Gootenberg *et al.*, 1981). The combined expression of TCGF and TCGF receptor could provide a mechanism for the malignant growth of these cells. However, some of the HTLV-transformed cord blood T cells do not produce detectable TCGF and they grow independent of exogenous TCGF. These cells also do not express TCGF mRNA (S. Arya *et al.*, unpublished results). Therefore, alternative mechanisms of malignant cell growth should be operating at least in these *in vitro* systems. These transformed cord blood T-cell cultures produce a large number of lymphokines (P. Markham *et al.*, unpublished) and study of some of these may yet provide information on alternative mechanisms for T-cell growth.

REFERENCES

Blayney, D. W., Blattner, W. A., Robert-Guroff, M., Jaffe, E. S., Fisher, R. I., Bunn, P. A., Patton, M. G., Rarick, H. R., and Gallo, R. C., 1983a, The human T-cell leukemia/lymphoma virus (HTLV) in the southeastern United States, *J. Am. Med. Assoc.* **250**:1048–1052.

Blayney, D. W., Jaffe, E. S., Fisher, R. I., Schechter, G. P., Cossman, J., Robert-Guroff, M., Kalyanaraman, V. S., Blattner, W. A., and Gallo, R. C., 1983b, The human T-cell leukemia/lymphoma virus, lymphoma, lytic bone lesions, and hypercalcemia, *Ann. Intern. Med.* **98**:144–151.

Broder, S., and Bunn, P. A., 1980, Cutaneous T-cell lymphomas, *Semin. Oncol.* **7**:310–331.

Broder, S., Edelson, R., Lutzner, M., Nelson, D., MacDermott, R., Durm, M., Goldman, C., Meade, B., and Waldmann, T. A., 1976, The Sézary syndrome, *J. Clin. Invest.* **58**:1297–1306.

Bunn, P. A., Schechter, G. P., Jaffe, E., Blayney, D., Young, R. C., Matthews, M. J., Blattner, W., Broder, S., Robert-Guroff, M., and Gallo, R. C., 1983, Clinical course of retrovirus-associated adult T-cell lymphoma in the United States, *N. Engl. J. Med.* **309**:257–264.

Catovsky, D., Greaves, M. F., Rose, M., Galton, D. A. G., Goolden, A. W. G., McCluskey, D. R., White, J. M., Lampert, I., Bourikas, G., Ireland, R., Bridges, J. M., Blattner, W. A., and Gallo, R. C., 1982, Adult T-cell lymphoma-leukemia in blacks from the West Indies, *Lancet* **1**:639–643.

Clark, S. C., Arya, S. K., Wong-Staal, F., Matsumoto-Kobayashi, M., Kay, R. M., Kaufman, R. J., Brown, E. L., Shoemaker, C., Copeland, T., Oroszlan, S., Smith, K., Sarngadharan, M. G., Lindner, S. G., and Gallo, R. C., 1984, Human T-cell growth factor: Partial amino acid sequence, cDNA cloning, and organization and expression in normal and leukemic cells, *Proc. Natl. Acad. Sci. USA* **81**:2543–2547.

Copeland, T. D., Oroszlan, S. D., Kalyanaraman, V. S., Sarngadharan, M. G., and Gallo, R. C., 1983, Complete amino acid sequence of human T-cell leukemia virus structural protein p15, *FEBS Lett.* **162:**390–395.

Gallo, R. C., Mann, D., Broder, S., Ruscetti, F. W., Maeda, M., Kalyanaraman, V. S., Robert-Guroff, M., and Reitz, M. S., 1982, Human T-cell leukemia-lymphoma virus (HTLV) is in T- but not in B-lymphocytes from a patient with cutaneous T-cell lymphoma, *Proc. Natl. Acad. Sci. USA* **79:**5680–5683.

Gallo, R. C., Kalyanaraman, V. S., Sarngadharan, M. G., Sliski, A., Vonderheid, E. C., Maeda, M., Nakao, Y., Yamada, K., Ito, Y., Gutensohn, N., Murphy, S., Bunn, P. A., Catovsky, D., Greaves, M. F., Blayney, D. W., Blattner, W., Jarrett, W. F. H., zur Hausen, H., Seligman, M., Brouet, J. C., Haynes, B. F., Jegasothy, B. V., Jaffe, E., Cossman, J., Broder, S., Fisher, R. I., Golde, D. W., and Robert-Guroff, M., 1983, Association of the human type-C retrovirus with a subset of adult T-cell cancers, *Cancer Res.* **43:**3892–3899.

Gelmann, E. P., Franchini, G., Manzari, V., Wong-Staal, F., and Gallo, R. C., 1984, Molecular cloning of a new unique human T-leukemia virus (HTLV-II_{MO}), *Proc. Natl. Acad. Sci. USA* **81:**993–997.

Gillis, S., and Smith, K. A., 1977, Long-term culture of tumor-specific cytotoxic T-cells, *Nature (London)* **268:**154–155.

Gillis, S., and Watson, J., 1980, Biochemical and biological characterization of lymphocyte regulatory molecules. V. Identification of an interleukin-2 producing human leukemia T-cell line, *J. Exp. Med.* **152:**1709–1719.

Gillis, S., Ferm, M. M., Ou, W., and Smith, K. A., 1978, T-cell growth factor: Parameters for production and a quantitative microassay for activity, *J. Immunol.* **120:**2027–2032.

Gootenberg, J. E., Ruscetti, F. W., Mier, J. W., Gazdar, A., and Gallo, R. C., 1981, Human cutaneous T-cell lymphoma and leukemia cell lines produce and respond to T-cell growth factor, *J. Exp. Med.* **154:**1403–1418.

Greene, W. C., Wong-Staal, F. Y., Depper, J. M., Leonard, W. J., Gallo, R. C., and Waldmann, T. A., 1983, Acute lymphocytic leukemia T-cells can be induced to express membrane receptors for T-cell growth factor (TCGF), *Clin. Res.* **31:**344–344A.

Haynes, B., Miller, S., Palker, T., Moore, J., Dunn, P., Bolognesi, D., and Metzgar, R., 1983, Identification of human T-cell leukemia virus in a Japanese patient with adult T-cell leukemia and cutaneous lymphomatous vesiculitis, *Proc. Natl. Acad. Sci. USA* **80:**2054–2058.

Kalyanaraman, V. S., Sarngadharan, M. G., Poiesz, B. J., Ruscetti, F. W., and Gallo, R. C., 1981, Immunological properties of a type-C retrovirus isolated from human T-lymphoma cells and comparison to other mammalian retroviruses, *J. Virol.* **38:**906–915.

Kalyanaraman, V. S., Sarngadharan, M. G., Bunn, P. A., Minna, J. D., and Gallo, R. C., 1982a, Antibodies in human sera reactive against an internal structural protein (p24) of human T-cell lymphoma virus, *Nature (London)* **294:**271–273.

Kalyanaraman, V. S., Sarngadharan, M. G., Nakao, Y., Ito, Y., Aoki, T., and Gallo, R. C., 1982b, Natural antibodies to the structural core protein (p24) of the human T-cell leukemia (lymphoma) retrovirus found in sera of leukemia patients in Japan, *Proc. Natl. Acad. Sci. USA* **79:** 1653–1657.

Kalyanaraman, V. S., Sarngadharan, M. G., Robert-Guroff, M., Miyoshi, I., Blayney, D., Golde, D., and Gallo, R. C., 1982c, A new subtype of human T-cell leukemia virus (HTLV-II) associated with a T-cell variant of hairy cell leukemia, *Science* **218:**571–573.

Kalyanaraman, V. S., Jarvis-Morar, M., Sarngadharan, M. G., and Gallo, R. C., 1984, Immunological characterization of the low molecular weight *gag* gene proteins p19 and p15 of human T-cell leukemia-lymphoma virus (HTLV), and demonstration of human natural antibodies to them. *Virology* **132:**61–70.

Kasakura, S., 1977, Cytotoxic lymphocytes induced by soluble factors derived from the medium of leukocyte cultures, *J. Immunol.* **118:**43–47.

Kawakami, T. G., Huff, S. D., Buckley, P. M., Dungworth, D. C., Snyder, J. D., and Gilden, R. V., 1972, Isolation and characterization of a C-type virus associated with lymphosarcoma, *Nature New Biol.* **235:**170–171.

Kurnick, J. T., Grovnik, K. D., Kimura, J. A., Lindblom, J. B., Skoog, V. T., Sjoberg, O., and Wigzell, H., 1979, Long-term growth *in vitro* of human T-blasts with maintenance of specificity and function, *J. Immunol.* **122:**1255–1260.

Larsson, E. L., Iscove, N. N., and Coutinho, A., 1980, The distinct factors are required for induction of T-cell growth, *Nature (London)* **283:**664–666.

Leonard, W. J., Depper, J. M., Uchiyama, T., Smith, K. A., Waldmann, T. A., and Greene, W. C., 1982, A monoclonal antibody that appears to recognize the receptor for human T-cell growth factor, *Nature (London)* **300:**267–269.

Lotze, M. T., and Rosenberg, S. A., 1981, *In vitro* growth of cytotoxic human lymphocytes. III. Preparation of lectin-free T-cell growth factor and an analysis of its activity, *J. Immunol.* **126:**2215–2220.

Lutzner, M., Edelson, R., Sckein, P., Green, I., Kirkpatrick, C., and Ahmed, A., 1975, Cutaneous T-cell lymphomas: The Sézary syndrome, mycosis fungoides, and related disorders, *Ann. Intern. Med.* **83:**534–552.

Manzari, V., Wong-Staal, F., Franchini, G., Columbini, S., Gelmann, E. P., Oroszlan, S., Staal, S., and Gallo, R. C., 1983a, Human T-cell leukemia-lymphoma virus (HTLV): Cloning of an integrated defective provirus and flanking cellular sequences, *Proc. Natl. Acad. Sci. USA* **80:**1574–1578.

Manzari, V., Gallo, R. C., Franchini, G., Westin, E., Ceccherini-Nelli, L., Popovic, M., and Wong-Staal, F., 1983b, Abundant transcription of a cellular gene in T-cells infected with human T-cell leukemia-lymphoma virus (HTLV), *Proc. Natl. Acad. Sci. USA* **80:**11–15.

Markham, P. D., Salahuddin, S. Z., Popovic, M., Sarin, P., and Gallo, R. C., 1983, Infection and transformation of fresh human umbilical cord blood cells by multiple sources of human T-cell leukemia-lymphoma virus (HTLV), *Int. J. Cancer* **31:**413–420.

Mier, J. W., and Gallo, R. C., 1980, Purification and some properties of human T-cell growth factor from phytohemagglutinin-stimulated lymphocyte conditioned media, *Proc. Natl. Acad. Sci. USA* **77:**6134–6138.

Morgan, D. A., Ruscetti, F. W., and Gallo, R. C., 1976, Selective *in vitro* growth of T-lymphocytes from normal human bone marrows, *Science* **193:**1007–1008.

Oppenheim, J. J., Mizel, S. B., and Meltzer, M. S., 1979, Biological effects of lymphocyte and macrophage derived mitogenic "amplification" factors, in: *Biology of the Lymphokines* (S. Cohen, E. Pick, and J. J. Oppenheim, eds.), pp. 291–328, Academic Press, New York.

Oroszlan, S., Sarngadharan, M. G., Copeland, T. D., Kalyanaraman, V. S., Gilden, R. V., and Gallo, R. C., 1982, Primary structure analysis of the major internal protein p24 of human type-C T-cell leukemia virus, *Proc. Natl. Acad. Sci. USA* **79:**1291–1294.

Paul, W. E., Sredni, B., and Schwartz, R. H., 1981, Long-term growth and cloning of nontransformed lymphocytes, *Nature (London)* **294:**697–699.

Poiesz, B. J., Ruscetti, F. W., Gazdar, A. F., Bunn, P. A., Minna, J. D., and Gallo, R. C., 1980a, Isolation of type-C retrovirus particles from cultured and fresh lymphocytes from a patient with cutaneous T-cell lymphoma, *Proc. Natl. Acad. Sci. USA* **77:**7415–7419.

Poiesz, B. J., Ruscetti, F. W., Mier, J. W., Woods, A. M., and Gallo, R. C., 1980b, T-cell lines established from human T-lymphocyte neoplasias by direct response to T-cell growth factor, *Proc. Natl. Acad. Sci. USA* **77:**6815–6819.

Poiesz, B. J., Ruscetti, F. W., Reitz, M. S., Kalyanaraman, V. S., and Gallo, R. C., 1981, Isolation of a new type-C retrovirus (HTLV) in primary uncultured cells of a patient with Sézary T-cell leukemia, *Nature (London)* **294:**268–271.

Popovic, M., Reitz, M. S., Sarngadharan, M. G., Robert-Guroff, M., Kalyanaraman, V. S., Nakao, Y., Miyoshi, I., Minowada, J., Yoshida, M., Ito, Y., and Gallo, R. C., 1982, The virus of Japanese adult T-cell leukemia is a member of the human T-cell leukemia virus group, *Nature (London)* **30:**63–66.

Popovic, M., Sarin, P. S., Robert-Guroff, M., Kalyanaraman, V. S., Mann, D., Minowada, J., and Gallo, R. C., 1983, Isolation and transmission of human retrovirus (HTLV), *Science* **219:**856–859.

Rabin, H., Hopkins, R. F., Ruscetti, F. W., Newbauer, R. H., Brown, R. L., and Kawakami, T. G., 1981, Spontaneous release of a factor with T-cell growth factor activity from a continuous line of primate tumor T-cells, *J. Immunol.* **127:**1852–1857.

Reitz, M. S., Poiesz, B. J., Ruscetti, F. W., and Gallo, R. C., 1981, Characterization and distribution of nucleic acid sequences of a novel type-C retrovirus isolated from neoplastic human T-lymphocytes, *Proc. Natl. Acad. Sci. USA* **78:**1887–1891.

Reitz, M. S., Popovic, M., Haynes, B. F., Clark, S. C., and Gallo, R. C., 1983, Relatedness by nucleic acid hybridization of new isolates of human T-cell leukemia-lymphoma virus (HTLV) and demonstration of provirus in uncultured leukemic blood cells, *Virology* **126:**688–692.

Rho, H. M., Poiesz, B. J., Ruscetti, F. W., and Gallo, R. C., 1981, Characterization of the new reverse transcriptase from a new retrovirus (HTLV) produced by a human cutaneous T-cell lymphoma cell line, *Virology* **112:**355–359.

Robb, R. J., Munck, A., and Smith, K. A., 1981, T-cell growth factor receptors: Quantitation, specificity, and biological relevance, *J. Exp. Med.* **154:**1455–1474.

Robert-Guroff, M., Ruscetti, F. W., Posner, L. E., Poiesz, B. J., and Gallo, R. C., 1981, Detection of the human T-cell lymphoma virus p19 in cells of some patients with cutaneous T-cell lymphoma and leukemia using a monoclonal antibody, *J. Exp. Med.* **154:**1957–1964.

Robert-Guroff, M., Nakao, Y., Notake, K., Ito, Y., Sliski, A., and Gallo, R. C., 1982, Natural antibodies to human retrovirus HTLV in a cluster of Japanese patients with adult T-cell leukemia, *Science* **215:**975–978.

Robert-Guroff, M., Kalyanaraman, V. S., Blattner, W. A., Popovic, M., Sarngadharan, M. G., Maeda, M., Blayney, D., Catovsky, D., Bunn, P. A., Shibata, A., Nakao, Y., Ito, Y., Aoki, T., and Gallo, R. C., 1983, Evidence for human T-cell lymphoma-leukemia virus infection of family members of human T-cell leukemia-lymphoma patients, *J. Exp. Med.* **157:**248–258.

Rosenberg, S. A., Schwartz, S., and Spiess, P. J., 1978, *In vitro* growth of murine T-cells. II. Growth of *in vitro* sensitized cells cytotoxic for alloantigens, *J. Immunol.* **121:**1951–1955.

Ruscetti, F. W., Morgan, D. A., and Gallo, R. C., 1977, Functional and morphological characterization of human T-cells continuously grown *in vitro*, *J. Immunol.* **119:**131–138.

Sarin, P. S., Aoki, T., Shibata, A., Ohnishi, Y., Aoyagi, Y., Miyakoshi, H., Emura, I., Kalyanaraman, V. S., Robert-Guroff, M., Popovic, M., Sarngadharan, M. G., Nowell, P. C., and Gallo, R. C., 1983, High incidence of human type-C retrovirus (HTLV) in family members of an HTLV-positive Japanese T-cell leukemia patient, *Proc. Natl. Acad. Sci. USA* **80:**2370–2374.

Sarngadharan, M. G., Ting, R. C., and Gallo, R. C., 1984, Methods for production and purification of human T-cell growth factor, in: *Cell Culture Methods in Molecular and Cell Biology*, Vol. 4 (D. Barnes, D. Sirbasku, and G. Sato, eds.).

Saxon, A., Stevens, R. H., and Golde, D. W., 1978, T-lymphocyte variant of hairy cell leukemia, *Ann. Intern. Med.* **88:**323–326.

Schrier, M. H., and Tees, R., 1980, Clonal induction of helper T-cells: Conversion of specific into non-specific signals, *Int. Arch. Allergy Appl. Immunol.* **61:**227–237.

Seiki, M., Hattori, S., and Yoshida, M., 1982, Human adult T-cell leukemia virus: Molecular cloning of the provirus DNA and the unique terminal structure, *Proc. Natl. Acad. Sci. USA* **79:**6899–6902.

Smith, K. A., 1983, T-cell growth factor, a lymphocytotrophic hormone, in: *Genetics of the Immune Response* (E. Moller and G. Moller, eds.), pp. 151–185, Plenum Press, New York.

Stadler, B. M., and Oppenheim, J. J., 1982, Human interleukin-2, biological studies using purified IL-2 and monoclonal anti-IL-2 antibodies, in: *Lymphokines* (S.B. Mizel, ed.), pp. 117–135, Academic Press, New York.

Strausser, J. L., and Rosenberg, S. A., 1978, *In vitro* growth of human cytotoxic lymphocytes: Growth of cells sensitized *in vitro* to alloantigens, *J. Immunol.* **121:**1491–1495.

Swain, S. L., Dennert, G., Warner, J. F., and Dutton, D. W., 1981, Culture supernatants of a stimulated T-cell line have helper activity that acts synergistically with interleukin-2 in the response of T-cell to antigen, *Proc. Natl. Acad. Sci. USA* **78:**2517–2521.

Takatsu, K., Tanaka, K., Tominaga, A., Kumahara, Y., and Hamaoka, T., 1980, Antigen-induced T-cell replacing factor (TRF). III. Establishment of T-cell hybrid clone continuously releasing TRF, *J. Immunol.* **125:**2646–2653.

Taniguchi, T., Matsui, H., Fujita, T., Takaoka, C., Kashima, M., Yoshimoto, R., and Hamuro, J., 1983, Structure and expression of a cloned cDNA for human interleukin-2, *Nature (London)* **302:**305–310.

Tees, R., and Schrier, M. H., 1980, Selective reconstitution of nude mice with long-term cultures and cloned specific helper T-cells, *Nature (London)* **283:**780–782.

Uchiyama, T., Yodoi, J., Sagawa, K., Takatsuki, K., and Uchino, H., 1977, Adult T-cell leukemia: Clinical and hematological features of 16 cases, *Blood* **50:**481–492.

Uchiyama, T., Broder, S., and Waldmann, T. A., 1981, A monoclonal antibody (anti-Tac) reactive with activated and functionally mature human T-cells. I. Production of anti-Tac monoclonal antibody and distribution of Tac (+) cells, *J. Immunol.* **126:**1393–1397.

Waldmann, T., Broder, S., Greene, W., Sarin, P., Goldman, C., Frost, K., Sharrow, S., Depper, J., Leonard, W., Uchiyama, T., and Gallo, R., 1983, A comparison of the function and phenotype of Sézary T-cells with human T-cell leukemia-lymphoma virus (HTLV) associated adult T-cell leukemia cells, *Clin. Res.* **31:**547A.

Yoshida, M., Miyoshi, I., and Hinuma, Y., 1982, Isolation and characterization of retroviruses from cell lines from human adult T-cell leukemia and its implication in the disease, *Proc. Natl. Acad. Sci. USA* **79:**2031–2035.

I

Thymic Hormones

1

Thymosins

Isolation, Structural Studies, and Biological Activities

TERESA L. K. LOW and ALLAN L. GOLDSTEIN

1. INTRODUCTION

The studies in the early 1960s by Miller (Miller, 1961), and Good (Good *et al.*, 1962) have established that the thymus is necessary for the normal development of the immune response. For the past 20 years, a large research effort has been directed toward the isolation and identification of the thymic factors or hormones responsible for the physiological functions of the thymus gland (Low and Goldstein, 1978). Our previous studies demonstrated that a partially purified bovine thymic preparation termed thymosin fraction 5 could partially or fully reconstitute immune functions in immunodeficient diseases, autoimmune diseases, and cancer.

As shown in Fig. 1, the endocrine thymus produces thymosin and other thymic factors to maintain normal immune balance. The lack of adequate production and utilization of these thymic factors due to genetic, chemical, viral, or radiation damage causes an immune imbalance and may contribute to the etiology of many diseases.

Our ongoing studies have provided evidence that thymosin fraction 5 consists of a family of biologically active polypeptide components with hormonelike activities. Several of the components in fraction 5 have been purified to homogeneity and amino acid sequences determined, such as α_1, β_1, and β_4; others have been partially purified. It appears that these peptides act at different sites and on different subsets of T cells and contribute to the normal maintenance of immune function and balance.

In this chapter, we will summarize the results of our ongoing studies on the isolation, chemical and biological characterization of thymosin polypeptides. The clinical studies with thymosins are presented elsewhere in this volume.

TERESA L. K. LOW and ALLAN L. GOLDSTEIN • Department of Biochemistry, The George Washington University School of Medicine and Health Sciences, Washington, D.C. 20037.

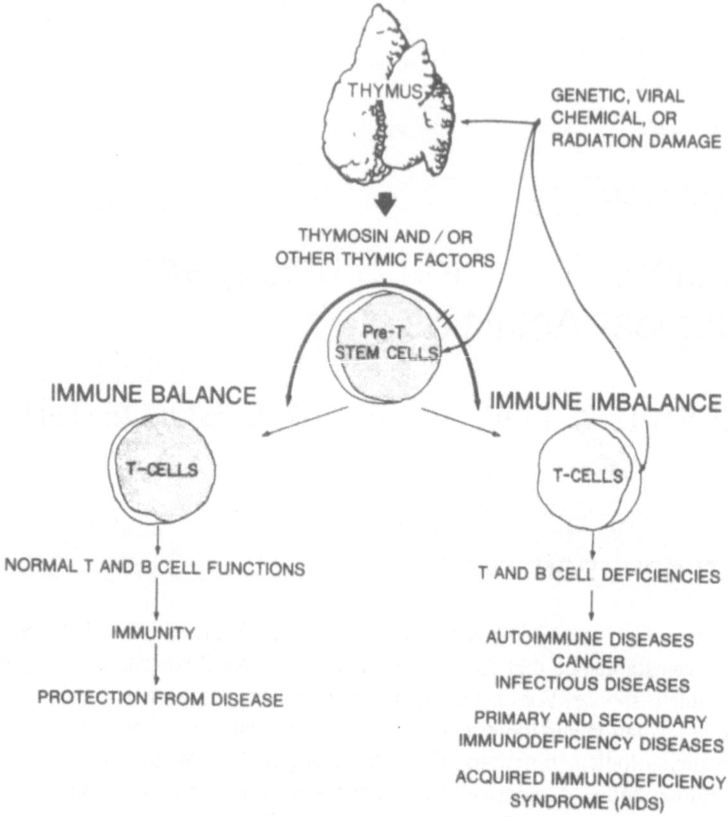

FIGURE 1. Role of the endocrine thymus in immune regulation.

2. ISOLATION AND STRUCTURAL STUDIES OF THYMOSINS

2.1. Preparation of Thymosin Fractions 5 and 5A

Thymosin fraction 5 and fraction 5A are prepared from calf thymus as described previously (Low *et al.*, 1979). The preparations are obtained by homogenization of thymus tissue, followed by 80°C heat denaturation, acetone precipitation, and fractionation with ammonium sulfate. The 25–50% ammonium sulfate precipitate is further subjected to ultrafiltration in an Amicon DC-2 hollow-fiber system to yield fraction 5. The 50–95% ammonium sulfate cut is also collected and processed through DC-2 to yield fraction 5A.

2.2. Nomenclature of Thymosin Polypeptides

On an isoelectric focusing gel, fraction 5 is shown to consist of 30 or more individual polypeptide components. We have proposed a nomenclature system for thymosin peptides based on this isoelectric focusing pattern. As shown in Fig. 2,

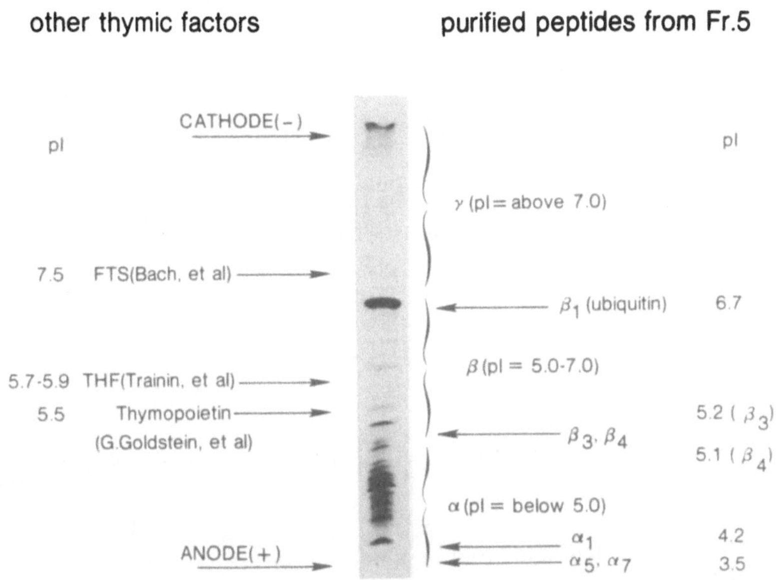

FIGURE 2. Isoelectric focusing of thymosin fraction 5 in LKB PAG plate (pH 3.5–9.5). Purified thymosin peptides from the α, β, and γ regions are identified. The isoelectric points of several other well-characterized thymic factors are illustrated for comparison.

the separated protein bands are divided into three regions. The α region consists of polypeptides with isoelectric points below 5.0; the β region, 5.0–7.0; and the γ region, above 7.0. Subscript numbers α_1, α_2, β_1, β_2, etc. are used to identify the polypeptides from each region in the order that they are isolated. For comparative purposes, some other well-defined thymic factors such as FTS (Bach *et al.*, 1977), thymopoietin (Schlesinger and Goldstein, 1975), and THF (Kook *et al.*, 1975) are also shown in Fig. 2 at appropriate positions according to their reported isoelectric points.

On the same isoelectric focusing gel system, thymosin fraction 5A appears to contain mostly the β-region peptides and very few peptides from the α region (data not shown). It appears that fraction 5A is a better source for isolation of β -region peptides, such as β_1 (Low *et al.*, 1979) and β_4 (Low and Goldstein, 1982).

2.3. Purification and Chemical Characterization of the Component Polypeptides in Thymosin Fraction 5

2.3.1. Thymosin α_1

The first thymosin polypeptide isolated from fraction 5 has been termed thymosin α_1 (Low *et al.*, 1979). The isolation procedure for thymosin α_1 from fraction 5 includes column chromatography on ion-exchangers CM-cellulose and DEAE-cellulose and gel filtration on Sephadex G-75. Thymosin α_1 migrates as a single band on analytical polyacrylamide gels at pH 8.3 and 2.9 and as a major band with

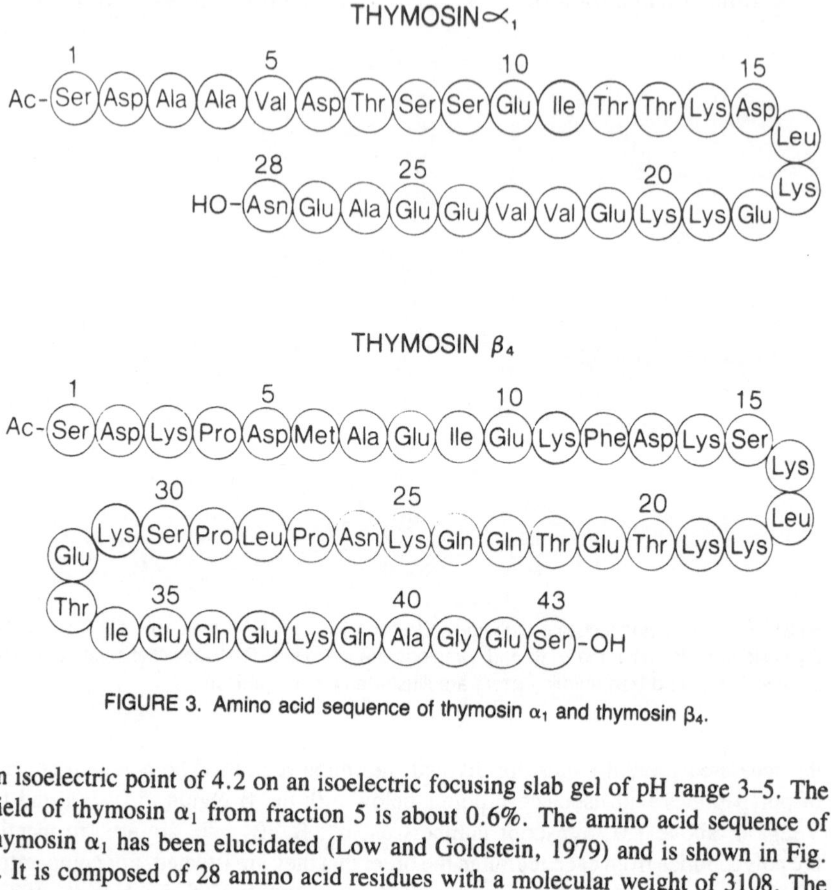

FIGURE 3. Amino acid sequence of thymosin α_1 and thymosin β_4.

an isoelectric point of 4.2 on an isoelectric focusing slab gel of pH range 3–5. The yield of thymosin α_1 from fraction 5 is about 0.6%. The amino acid sequence of thymosin α_1 has been elucidated (Low and Goldstein, 1979) and is shown in Fig. 3. It is composed of 28 amino acid residues with a molecular weight of 3108. The amino terminus of thymosin α_1 is acetylated. Using parameters for the prediction of protein conformation described by Chou and Fasman (1974), it would appear that the C-terminal half of the molecule has a high helical potential.

2.3.2. Thymosin β_4

Thymosin β_4 is the first of the biologically active polypeptides from the β region of thymosin fraction 5 to be completely characterized (Low and Goldstein, 1982). Thymosin β_4 is composed of 43 amino acid residues with an acetyl group at the N-terminal end. It has a molecular weight of 4963 and an isoelectric point of 5.1. The complete amino acid sequence of thymosin β_4 is shown in Fig. 3.

A computer search for possible sequence homology between the sequence of thymosin β_4 and other published protein sequences has been conducted at the National Biomedical Research Foundation, Washington, D.C. The results do not indicate a statistically significant relationship of thymosin β_4 to any other sequenced protein currently stored in the data bank. However, there is an interesting internal duplication between residues 18–30 and 31–43, as illustrated in Fig. 4. Using data for prediction of protein conformation, described by Chou and Fasman (1974), two

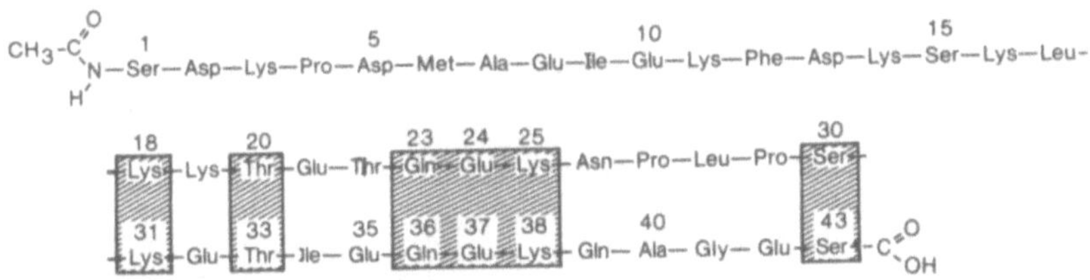

FIGURE 4. Regions of internal duplication in thymosin β_4. Residues 31–43 are aligned with 18–30, and six identities are identified by the shaded areas.

regions were identified to contain high helical potential. As shown in Fig. 5, residues 4–12 and 32–40 were identified to be helical regions. No region suggestive of β-sheet structure was found.

2.3.3. Thymosin α_7

Thymosin α_7 was isolated from thymosin fraction 5 by ion-exchange chromatography on CM-cellulose and gel filtration on Sephadex G-75 (Low and Goldstein, unpublished). It is highly acidic with an isoelectric point around 3.5. Thymosin α_7 is a partially purified preparation and can be resolved into several components by high-performance liquid chromatography (HPLC) on a reverse-phase column (Low and Goldstein, unpublished). Purification of these components is now in progress.

2.3.4. Polypeptide β_1

Polypeptide β_1 is the predominant band on the isoelectric focusing gel of fraction 5 (see Fig. 2). It was isolated from thymosin fraction 5A as described (Low *et al.*, 1979). It has a molecular weight of 8451 and an isoelectric point of 6.7. The amino acid sequence of β_1 has been determined (Low and Goldstein, 1979) and has revealed that this molecule is identical to ubiquitin and a portion of protein A24, a nuclear chromosomal protein. Polypeptide β_1 does not show biological activity in our bioassay systems, indicating that it is not an important molecule for T-cell maturation. Ubiquitin has been found to be identical to APF-1, the polypeptide cofactor of ATP-dependent protein degradation (Wilkinson *et al.*, 1980). This defines an important biological role of the molecule which is unrelated to the specific function of the thymus. Furthermore, its role in protein degradation explains the ubiquitous nature of this molecule (Schlesinger *et al.*, 1975).

2.3.5. Other Peptides

Several peptides isolated from thymosin fraction 5 have structures closely related to thymosin β_4. Thymosin β_3 has an isoelectric point of 5.2 and a molecular weight of approximately 5500. Preliminary studies (Low and Goldstein, unpublished) indicate that thymosin β_3 shares an identical sequence to β_4 through most

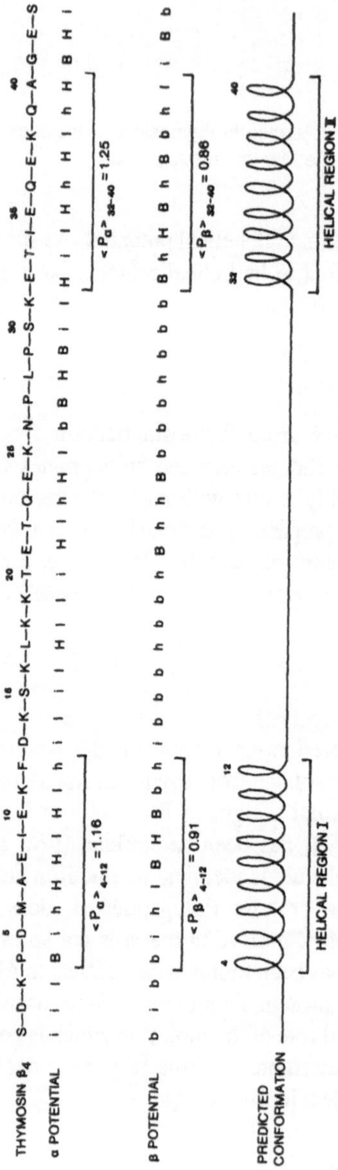

FIGURE 5. Prediction of thymosin β_4 conformation. Two regions (residues 4–12 and 32–40) were identified to contain high helical potential.

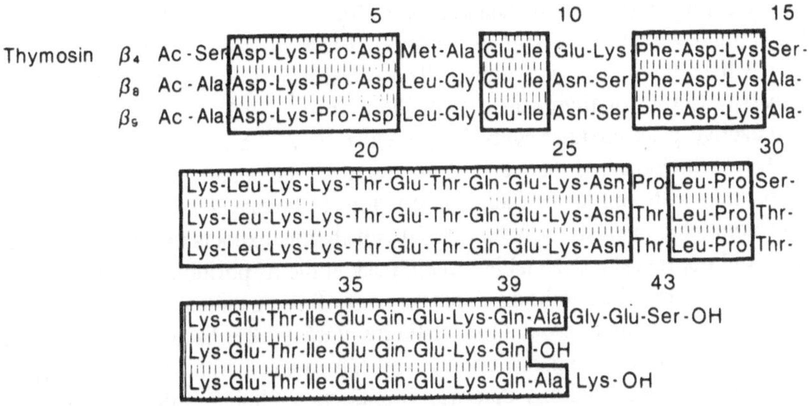

FIGURE 6. Comparison of the amino acid sequences of thymosin β_4, β_8 and β_9.

of its N-terminal portion and differs only at the C-terminal end. However, the complete sequence of this molecule has not been determined.

Recently, Hannappel *et al.* (1982) reported the isolation of thymosin β_8 from thymosin fraction 5 and thymosin β_9 from fresh-frozen calf thymus. As shown in Fig. 6, a large portion of the amino acid sequences of thymosin β_8 and β_9 were found to be homologous to thymosin β_4. Of the 38 amino acid residues in thymosin β_8, 31 are identical to the corresponding residues in thymosin β_4. Thymosin β_9 is identical to thymosin β_8 except for the presence of an additional dipeptide, -Ala-Lys-OH, at the C terminus.

Previously, we have isolated from thymosin fraction 5A a peptide termed thymosin β_7, which was closely associated with polypeptide β_1 (Low *et al.*, 1979). Structural studies have revealed that thymosin β_7 is identical to the sequence of thymosin β_9 (Low and Goldstein, unpublished). Thus, thymosin β_7 is present both in thymosin fraction 5A and in fresh-frozen thymus.

Using a combination of column chromatography, gel filtration, and HPLC, a polypeptide that is active in mixed lymphocyte reaction (MLR) has been isolated from thymosin fraction 5 (Low *et al.*, 1983b). Structural studies on this polypeptide are now in progress.

2.4. Thymosin α_1 from Other Species

To evaluate the species variation of thymosin polypeptides, we have prepared fraction 5 from thymic tissues of human, pig, sheep, chinchilla, and mouse. The human thymic tissue was excised during open heart surgery and from selected autopsies.

Thymosin α_1 from several species, was prepared from fraction 5, using a modification of the extraction and fractionation procedures developed for the isolation of bovine α_1. Human (Low and Goldstein, 1978), porcine, and ovine (Hu, 1980) thymosin α_1 have also been partially sequenced. From the results obtained, they appear to have an identical sequence to bovine α_1. Most recently, thymosin fraction 5 preparations were analyzed by HPLC on a μBondapak C_{18} (Waters) reverse-phase column (Low *et al.*, 1983a). Fractions were collected around the region where bovine α_1 was eluted. These fractions were analyzed by radioimmunoassay (RIA) for thymosin α_1 (McClure *et al.*, 1982). In each case, the immunoreactive peak coelutes with the protein peak at the retention time of bovine α_1. The results indicated that each of the fraction 5 preparations examined consists of a peptide with similar or identical structure to bovine α_1. On the other hand, chromatographic analysis of fresh rat and calf thymic extracts using similar HPLC system did not reveal a detectable protein peak at the α_1 position. Our observations suggest that α_1 may be synthesized in a precursor form in animal tissues, supporting the report of Hannappel *et al.* (1982). Freire *et al.* (1981) have translated mRNA from fresh calf thymus and carried out synthesis in a cell-free wheat germ system. A peptide of 16,000 daltons that was immunoprecipitable with antisera against thymosin fractions was analyzed and found to contain tryptic peptides corresponding to fragments of thymosin α_1. Thus, the 16,000-dalton peptide appears to be a precursor of thymosin α_1. The precursor is probably processed by proteolysis outside of the thymus to produce thymosin α_1. The enzymatic cleavage of an α_1 precursor is apparently not random, as only one form of α_1 is detected in the HPLC elution profile by RIA. Furthermore, our recent studies on the isolation of thymosin α_1 from human blood (Chen *et al.*, Chapter 16, this volume) suggest that the immunoreactive blood α_1 has physicochemical properties similar to thymosin α_1.

2.5. Thymosin β_4 from Other Species

Porcine, ovine, murine, and human thymosin fraction 5A preparations were analyzed by HPLC on a reverse-phase column. Porcine and ovine fraction 5A each contain a predominant peak at retention time similar to synthetic thymosin β_4. Amino acid analysis, as well as tryptic peptide mapping of these peaks, indicate that they have identical sequence to bovine thymosin β_4.

On the other hand, human and murine fraction 5A preparations gave more complex HPLC elution profile in the β_4 region. Although peaks were present at the retention time of synthetic β_4 in both preparations, they are not predominant peaks. Our results indicate that human and murine fraction 5A contain thymosin β_4 in much lower concentrations in comparison to porcine, ovine, and bovine fraction 5A. Furthermore, murine and human fraction 5A contain several other peptides with similar retention time as β_4.

3. SYNTHETIC ANALOGS OF THYMOSIN α_1 AND β_4

3.1. Thymosin α_1

Thymosin α_1 has been synthesized by several laboratories, including Wang *et al.* (1978), Birr and Stollenwerk (1979), and Abiko *et al.* (1980) using solution synthesis procedures, and Folkers *et al.* (1980), Wang *et al.* (1981), Wong and Merrifield (1980), and Colombo (1981) using solid-phase procedures.

The synthetic material prepared by Wang *et al.* (1978) migrated as a single band on acrylamide gel isoelectric focusing (pH 3.5–9.5) and on high-voltage silica gel thin-layer electrophoresis (pH 1.9 and 5.6), and is indistinguishable from the natural thymosin α_1. As shown in Fig. 7, the synthetic polypeptide migrates to the same position as the natural thymosin α_1 preparation on LKB PAG plate with pH range 3.5–9.5. The synthetic material has been tested in our laboratories for biological activities, and has been found to be similar to the natural α_1 in a TdT assay (Hu *et al.*, 1981) and a MIF assay (Thurman *et al.*, 1981).

Thymosin α_1 has also been synthesized utilizing recombinant DNA procedures (Wetzel *et al.*, 1980). In this important new development, the gene for thymosin α_1 was synthesized. The gene was inserted under *lac* operon control into the plasmid and expressed as part of a β-galactosidase chimeric protein. The plasmid is then cloned in a strain of *E. coli*. The structure of the N^{α}-desacetyl thymosin α_1 was confirmed by sequence analysis. The molecule was found to be as active as the

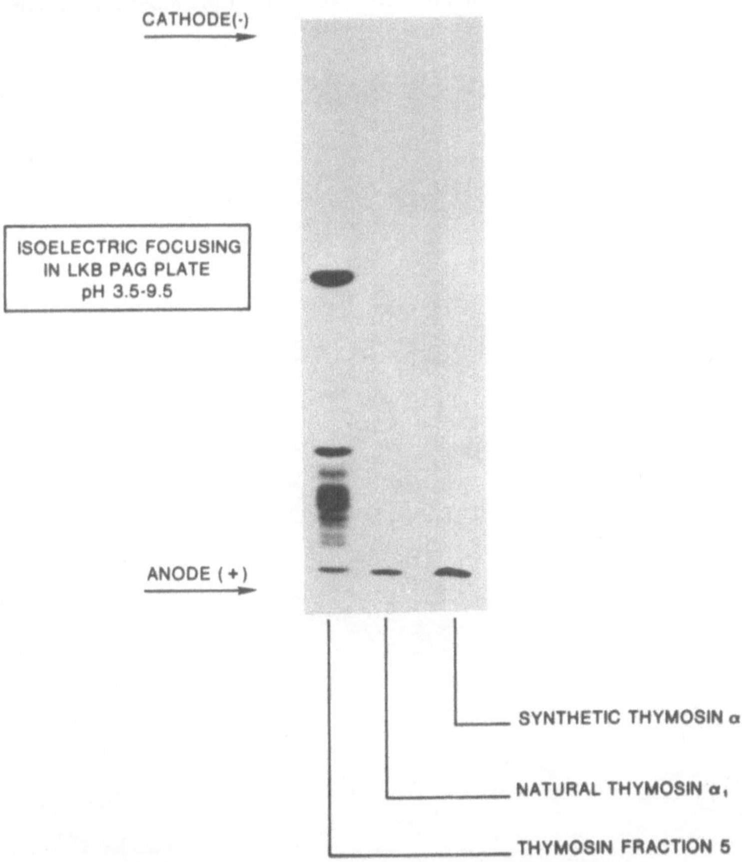

FIGURE 7. Isoelectric focusing gel of thymosin fraction 5, natural thymosin α_1, and synthetic thymosin α_1 on LKB PAG plate at pH 3.5–9.5.

chemically synthesized α_1 (Wang *et al.*, 1978) in a MIF assay (Thurman *et al.*, 1981) and a TdT assay (Hu *et al.*, 1981). A method for amino-acetylation of the N^{α} -desacetyl thymosin α_1 has been reported by Kido *et al.* (1981) using a trans-acetylase associated with a wheat germ ribosomal preparation.

Most recently, Ciardelli *et al.* (1982) have found that several synthetic C-terminal fragments (spanning positions 17–28) of α_1 possess activity in the aza-thioprine E-rosette inhibition assay. They suggested that a basic–acidic–lipophilic sequence character is a possible essential feature of a molecular signal for T-cell differentiation.

3.2. Thymosin β_4

The chemical synthesis of thymosin β_4 using a solid-phase procedure has been accomplished (Low *et al.*, 1983c). The synthetic material was shown to be identical to the natural thymosin β_4 by tryptic peptide mapping and amino acid compositional analyses. As shown in Fig. 8, on an LKB PAG plate at pH 3.5–9.5, the synthetic

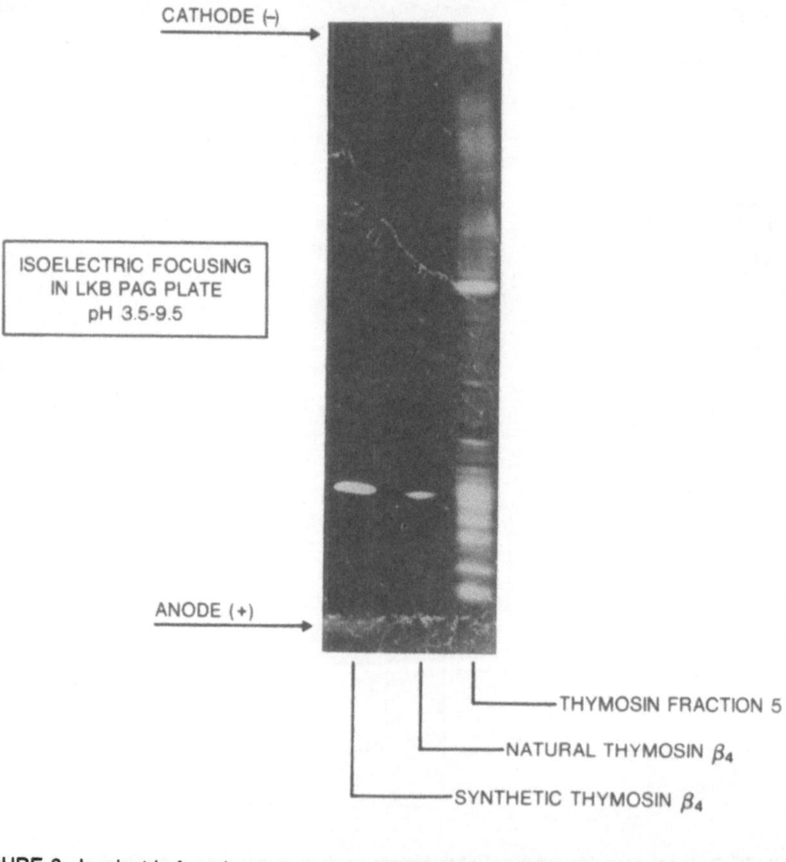

FIGURE 8. Isoelectric focusing gel of thymosin fraction 5, natural thymosin β_4, and synthetic thymosin β_4.

thymosin β_4 moves to a similar position as the natural material with a pH of 5.1. Biologically, synthetic thymosin β_4 was found to be as active as the natural compound in a TdT assay and a MIF assay. Thus, the proposed structure of this peptide hormone was confirmed by a chemical synthesis.

4. BIOLOGICAL PROPERTIES OF THYMOSINS

The biological properties of thymosin fraction 5, α_1, α_7, β_3, and β_4 are summarized in Table I.

TABLE I. Major Biological Properties of Thymosin Fraction 5 and Its Component Polypeptides

Thymic preparations	Principal investigators	Chemical properties	Biological effects
Thymosin fraction 5	Hooper, A. Goldstein *et al.*	Family of heat-stable, acidic polypeptides, MW 1000–15,000	Induces T-cell differentiation and enhances immunological function in animal models and humans; increases ACTH, β-endorphin, and glucocorticoid release
Thymosin α_1	Low, A. Goldstein *et al.*	Polypeptide of 28 residues, MW 3108, pI 4.2, sequence determined	Induces enhancement of MIF, interferon, and lymphotoxin production, modulates TdT activity, increases viral, fungal, and tumor immunity
Thymosin β_4	Low, A. Goldstein *et al.*	Polypeptide of 43 residues, MW 4963, pI 5.1, sequence determined	Induces TdT *in vivo* and *in vitro* in bone marrow cells from normal and athymic mice; *in vivo* induction of TdT in thymocytes of immunosuppressed mice; stimulates release of LH-RH and LH
Thymosin β_8	Hannappel, Horecker *et al.*	Polypeptide of 39 residues, sequence determined, homologous to β_4	No reported biological activities
Thymosin β_9	Hannappel, Horecker *et al.*	Polypeptide of 41 residues, sequence determined, homologous to β_4	No reported biological activities
Thymosin α_7	Low, A. Goldstein *et al.*	Acidic polypeptide, MW 2500, pI 3.5	*In vitro* enhancement of suppressor T cells; expression of Lyt 1,2,3-positive cells

In addition, thymosin α_1 was found to enhance T-cell functions in aged mice (Frasca *et al.*, 1982), and enhance resistance to *Candida* (Neta and Salvin, 1983; Bistoni *et al.*, 1982) and other infectious agents (Ishitsuka *et al.*, Chapter 40, this volume).

Biological studies (Hu, 1980) on human, porcine, ovine, and murine fraction 5 indicated that they possess similar activities as bovine fraction 5 in an MLR assay, a mouse mitogen assay, a MIF assay, and a human E-rosette assay. No biological activity has been reported for thymosin β_8 or β_9.

5. PERSPECTIVES

Our ongoing studies on the chemistry and biology of thymosins have documented a family of polypeptide components in thymosin fraction 5, that appears to be acting on different subsets of T cells or T-cell precursors to influence their maturation and function.

As shown in Fig. 9, thymosin α_1 appears to act at both early and late stages of thymocyte maturation. Thymosin α_7 converts immature T cells to cells with suppressor function. Thymosin β_3 and β_4 appear to act at very early stages of the maturation sequence on TdT-negative bone marrow stem cells to form TdT-positive prothymocytes. From the various thymic preparations presented in this chapter and elsewhere in this volume, it is evident that a great multiplicity of apparently unique thymic factors exist which are capable of promoting thymus-dependent immune

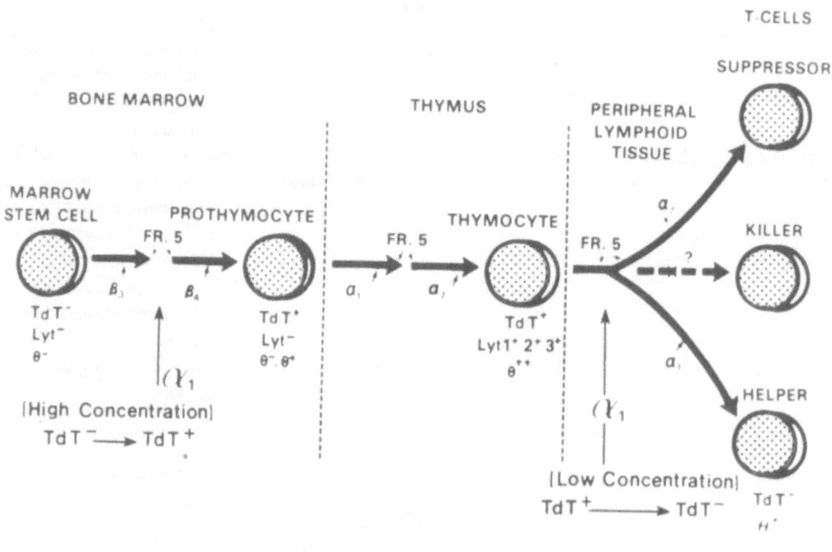

FIGURE 9. Proposed role of thymosin peptides in T-cell maturation. Thymosin β_3, and β_4 promote early stem cell differentiation to the prothymocyte stage. Thymosin α_1 promotes both early and late steps of T-cell differentiation. Thymosin α_7 is associated with the generation of functionally mature suppressor T cells and α_1 with the generation of helper T cells.

response. Although the precise relationships among the various hormonelike factors are still not defined, a number of generalizations can be made regarding their biological properties:

1. In murine systems, thymic factors can induce the appearance of T-cell surface antigens from precursor cells and can enhance the production of a number of lymphokines, including MIF, lymphotoxin, α- and γ-interferon, and TCGF.
2. Thymic factors tend to increase the proportions and absolute numbers of human E-rosette-forming cells and functional T cells *in vitro*. The magnitude of the effects correlates inversely with initial T-cell levels.
3. Thymic factors can affect the expression and modulation of TdT in T cells and T-cell precursors.
4. Thymic preparations exhibit a wide range of immune restorative properties *in vivo* in animals and in humans.

The most important contribution of thymic hormone research resides in its application to the clinical management of thymus-dependent diseases. Our ongoing clinical trials of thymosin have suggested that thymosin has a major role in restoring immune responsiveness and augmenting specific lymphocyte activities in children with hypothymic function and in patients with secondary T-cell deficiencies resulting from a variety of disorders, including cancer and autoimmune disease.

Most of the clinical trials with thymosin have been carried out using thymosin fraction 5. The complete elucidation of the structure of thymosin α_1, thymosin β_4, and other thymic factors and successful chemical synthesis of these molecules render the clinical trials of purified thymic polypeptide hormones feasible.

Furthermore, our studies on the interrelationships of thymus and central nervous systems (Hall *et al.*, Chapter 29, this volume), as well as reproductive system (Rebar, Chapter 30, this volume) will provide possible new approaches for treatment of many endocrine disorders with thymic peptides. The acquisition of a full understanding of the biological function of thymosin peptides will make possible the treatment of different diseases with different components or a combination of them. This approach should prove to be extremely useful in the area of immunotherapy and immunopharmacology.

ACKNOWLEDGMENTS. This study was supported in part by grants or gifts from the NIH (AI-17710 and CA-24974), Hoffmann–La Roche Inc., and Alpha One Biomedicals Inc.

REFERENCES

Abiko, T., Onodera, I., and Sekino, H., 1980, Synthesis and immunological effects of thymosin α_1 and its fragments on inhibitory factor in minimal changes nephrotic syndrome, *Chem. Pharm. Bull.* **28**:3542.

Bach, J.-F., Dardenne, M., and Pléau, J. M., 1977, Biochemical characterization of a serum thymic factor, *Nature (London)* **266**:55.

Birr, C., and Stollenwerk, U., 1979, Synthesis of thymosin α_1, a polypeptide of the thymus, *Angew. Chem. Int. Ed. Engl.* **18**:394.

Bistoni, F., Marconi, P., Frati, L., Bonmassar, E., and Garaci, E., 1982, Increase of mouse resistance to *Candida albicans* infection by thymosin α_1, *Infect. Immun.* **36**:609.

Chou, P. Y., and Fasman, G. D., 1974, Conformational parameters for amino acids in helical, β-sheet and random coil regions calculated from proteins, *Biochemistry* **13**:211.

Ciardelli, T. L., Incefy, G. S., and Birr, C., 1982, Activity of synthetic thymosin α_1 C-terminal peptides in the azathioprine E-rosette inhibition assay, *Biochemistry* **21**:4233.

Colombo, R., 1981, Solid-phase peptide synthesis without acidolysis: The synthesis of thymosin α_1 on a new benzhydrylamine resin, *J. Chem. Soc. Chem. Commun.* **1981**:1012.

Folkers, K., Leban, J., Sakura, N., Rampold, G., Lundanes, E., Dahmen, J., Lebek, M., Ohta, J., and Bowers, C. Y., 1980, Current advances on biologically active synthetic peptides, in: *Polypeptide Hormones* (R. F. Beers and E. G. Bassett, eds.), p. 149, Raven Press, New York.

Frasca, D., Garavini, M., and Doria, G., 1982, Recovery of T-cell functions in aged mice injected with synthetic thymosin α_1, *Cell. Immunol.* **72**:384.

Freire, M., Hannappel, E., Rey, M., Freire, J. M., Kido, H., and Horecker, B. L., 1981, Purification of thymus mRNA coding for a 16,000-dalton polypeptide containing the thymosin α_1 sequence, *Proc. Natl. Acad. Sci. USA* **78**:192.

Good, R. A., Dalmasso, A. P., Martinez, G., Archer, O. K., Pierce, J. C., and Papermaster, B. W., 1962, The role of the thymus in development of immunologic capacity in rabbits and mice, *J. Exp. Med.* **116**:773.

Hannappel, E., Davoust, S., and Horecker, B. L., 1982, Thymosins β_8 and β_9: Two new peptides isolated from calf thymus homologous to thymosin β_4, *Proc. Natl. Acad. Sci. USA* **79**:1708.

Hu, S. K., 1980, Studies on thymosin: (1) Regulation of terminal deoxynucleotidyl transferase activity; (2) Evaluation of chemical and biological relationships among species, Ph.D. dissertation, The University of Texas Medical Branch, Galveston.

Hu, S. K., Low, T. L. K., and Goldstein, A. L., 1981, Modulation of terminal deoxynucleotidyl transferase activity by thymosin, *Mol. Cell. Biochem.* **41**:49.

Kido, H., Vita, A., Hannappel, E., and Horecker, B. L., 1981, Aminoterminal acetylation of synthetic N^{α}-desacetyl thymosin α_1, *Arch. Biochem. Biophys.* **208**:101.

Kook, A. I., Yakir, Y., and Trainin, N., 1975, Isolation and partial chemical characterization of THF, a thymus hormone involved in immune maturation of lymphoid cells, *Cell. Immunol.* **19**:151.

Low, T. L. K., and Goldstein, A. L., 1978, Structure and function of thymosin and other thymic factors, in: *The Year in Hematology* (R. Silber, J. Lobue, and A. S. Gordon, eds.), pp. 281–319, Plenum Press, New York.

Low, T. L. K., and Goldstein, A. L., 1979, The chemistry and biology of thymosin. II. Amino acid sequence analysis of thymosin α_1 and polypeptide β_1, *J. Biol. Chem.* **254**:987.

Low, T. L. K., and Goldstein, A. L., 1982, Chemical characterization of thymosin β_4, *J. Biol. Chem.* **257**:1000.

Low, T. L. K., Thurman, G. B., McAdoo, M., McClure, J., Rossio, J., Naylor, P. H., and Goldstein, A. L., 1979, The chemistry and biology of thymosin. I. Isolation, characterization and biological activities of thymosin α_1 and polypeptide β_1 from calf thymus, *J. Biol. Chem.* **254**:981.

Low, T. L. K., McClure, J. E., Naylor, P. H., Spangelo, B. L., and Goldstein, A. L., 1983a, Isolation of thymosin α_1 from thymosin fraction 5 of different species by high performance liquid chromatography, *J. Chromatogr.* **266**:533.

Low, T. L. K., Seals, C., and Goldstein, A. L., 1983b, Isolation of a thymic polypeptide active in mixed lymphocyte reaction (MLR), *Fed. Proc.* **42**:1873.

Low, T. L. K., Wang, S. S., and Goldstein, A. L., 1983c, Solid-phase synthesis of thymosin β_4: Chemical and biological characterization of the synthetic peptide, *Biochemistry* **22**:733.

McClure, J. E., Lameris, N., Wara, D. W., and Goldstein, A. L., 1982, Immunochemical studies on thymosin: Radioimmunoassay of thymosin α_1, *J. Immunol.* **128**:368.

Miller, J. F. A. P., 1961, Immunological function of the thymus, *Lancet* **2**:748.

Neta, R., and Salvin, S. B., 1983, Resistance and susceptibility to infection in inbred murine strains, *Cell. Immunol.* **75**:173.

Schlesinger, D. H., and Goldstein, G., 1975, The amino acid sequence of thymopoietin II, *Cell* **5**:361.

Schlesinger, D. H., Goldstein, G., and Niall, H. D., 1975, The complete amino acid sequence of ubiquitin, and adenylate cyclase stimulating polypeptide probably universal in living cells, *Biochemistry* **14**:2214.

Thurman, G. B., Low, T. L. K., Rossio, J. L., and Goldstein, A. L., 1981, Specific and non-specific macrophage migration inhibition, in: *Lymphokines and Thymic Hormones: Their Potential in Cancer Therapeutics* (A. L. Goldstein and M. Chirigos, eds.), p. 145, Raven Press, New York.

Wang, S. S., Kulesha, I. D., and Winter, D. P., 1978, Synthesis of thymosin α_1, *J. Am. Chem. Soc.* **101**:253.

Wang, S. S., Makosfke, R., Bach, A., and Merrifield, R. B., 1981, Automated solid phase synthesis of thymosin α_1, *Int. J. Pept. Protein Res.* **15**:1.

Wetzel, R., Heyneker, H. L., Goeddel, D. V., Jhurani, P., Shapiro, J., Crea, R., Low, T. L. K., McClure, J. E., and Goldstein, A. L., 1980, Production of biologically active N^{α}-desacetylthymosin α_1 in *E. coli* through expression of a chemically synthesized gene, *Biochemistry* **19**:6096.

Wilkinson, K. D., Urban, M. K., and Haas, A. L., 1980, Ubiquitin is the ATP-dependent proteolysis factor 1 of rabbit reticulocytes, *J. Biol. Chem.* **255**:7529.

Wong, T. W., and Merrifield, R. B., 1980, Solid-phase synthesis of thymosin α_1 using tert-butyloxycarbonylaminoacyl-4(oxymethyl)phenylacetamidomethyl-resin, *Biochemistry* **19**:3233.

2

Thymulin

New Biochemical Aspects

M. DARDENNE, W. SAVINO, L. GASTINEL, and J.-F. BACH

1. INTRODUCTION

The thymus gland produces numerous polypeptide hormones that can induce T-cell markers and functions in immature cells and, consequently, are considered to be physiologically involved in T-cell differentiation (Bach, 1979). One of these hormones, thymulin, formerly called FTS (for facteur thymique sérique), was initially isolated from porcine serum (Dardenne *et al.*, 1977) but also purified from human serum. Its amino acid sequence was determined ⟨Glu-Ala-Lys-Ser-Gln-Gly-Gly-Ser-Asn⟩ (Pléau *et al.*, 1977) and the synthetic peptide was shown to be fully biologically active (Bach *et al.*, 1977).

More than 40 analogs of thymulin have been synthesized and immunologically evaluated (in bioassays and receptor assays), permitting the localization of the molecule's activity on the seven terminal amino acids (Bach, 1983). Some analogs have been shown to bind to the receptor and to inhibit thymulin activity, behaving like antihormones (Pléau *et al.*, 1979).

Direct evidence for the presence of thymulin in the thymus has been obtained by different approaches including its isolation from thymic extracts (Dardenne *et al.*, 1980) and the demonstration of its exclusive presence in thymic epithelial cells by immunochemistry, using xenoantisera (Monier *et al.*, 1980; Jambon *et al.*, 1981) and specific monoclonal antibodies (Savino *et al.*, 1982; Auger *et al.*, 1982) which were selected on their ability to inhibit *in vitro* and *in vivo* the biological activity of the hormone (Dardenne *et al.*, 1982a).

Recent studies in our laboratory suggested that thymulin might exist in two forms: one biologically active and the other one biologically inactive, for the same

M. DARDENNE, W. SAVINO, L. GASTINEL, and J.-F. BACH • INSERM U-25, Hôpital Necker, 75015 Paris, France.

rigorously controlled amino acid sequence (Dardenne *et al.*, 1982b). The presence of a transition metal in the biologically active form was suggested by the following argument: thymulin used in its synthetic or natural form lost its biological activity in a rosette assay after treatment with a metal ion chelating agent. This activity was restored by the addition of zinc salts or, to a lesser extent, other metals. This activation was secondary to the binding of the metal to the peptide. A metal-to-peptide molar ratio of 1 : 1 provided the best activation. The binding affinity was calculated to be around 10^{-6} M in equilibrium chromatography (Gastinel *et al.*, 1984).

In this report, we shall review the main data on the biochemistry of this peptide, emphasizing the new findings concerning the presence of zinc in the molecule. Before getting to this specific point, we think it is worthwhile to summarize the principal data previously known about the chemistry of this hormone.

2. CHEMISTRY OF THYMULIN: CLASSICAL DATA

2.1. Sequence Studies

Using the rosette theta and/or azathioprine conversion assay, we have characterized a serum factor capable of inducing T-cell markers on T-cell precursors. This serum factor (thymulin) was absent in the serum of nude or thymectomized (Tx) mice and reappeared after thymus grafting. Chemical analysis showed that it was a peptide of molecular weight 847 (Dardenne *et al.*, 1977).

As a result of amino acid analysis and sequence studies (Pléau *et al.*, 1977) on the intact peptide and on the peptide treated with proteolytic enzymes by Edman's method, the amino acid sequence proposed for thymulin was the following: ⟨Glu-Ala-Lys-Ser-Gln-Gly-Gly-Ser-Asn-OH⟩.

There is no apparent species specificity, as the amino acid analysis of calf and human thymulin was identical to that of porcine thymulin (Lacovara and Utermohlen, 1983). This sequence did not show any homology with that of the other thymic peptides that have also been sequenced (thymopoietin, thymosin α_1). It cannot be excluded, however, that peptides not chemically related to thymulin may serve as cleavage factors for a thymulin precursor, as is known to be the case for growth hormone which induces the release of small peptides, the somatedines, that mediate most of its biological activities.

2.2. Synthesis

On the basis of this sequence, a peptide has been synthesized according to two methods: using solid-phase synthesis (Merrifield's technique) by the Merck peptide group (Strachan *et al.*, 1979) and classical solution methods by Bricas *et al.*, (1977) and P. Lefrancier (Choay's peptide group). The synthetic material showed full biological activity and chromatographically displayed characteristics identical to those of natural thymulin in several chromatography systems.

2.3. Analog Studies

Many thymulin analogs have been synthesized in our laboratory (Blanot *et al.*, 1979; Martinez *et al.*, 1981). They were generally evaluated in a mouse or human rosette assay, and less frequently in other bioassays including the induction of suppressor cells (Kaiserlian and Dardenne, 1982) or of NK cell activity (Bardos and Bach, 1983). The analogs were further analyzed in a radioimmunoassay and a receptor assay.

These receptor studies lead us to the concept of competitive inhibitors, i.e., analogs which bind to thymulin receptors but exhibit effects antagonistic to thymulin (Pléau *et al.*, 1979).

3. PRESENCE OF ZINC IN THYMULIN

3.1. Direct Demonstration of Zinc in the Molecule

The fortuitous preparation of inactive or unstable lots of thymulin suggested that the peptide could exist in two forms: one biologically active and the other one inactive (Dardenne *et al.*, 1982b). Data have recently been collected showing that the active form contains a metal, probably zinc, whereas the inactive form lacks metal. We demonstrated that thymulin used in its synthetic or natural form lost its biological activity in the rosette assay after treatment with the metal ion chelating agent, Chelex 100. This activity was restored by the addition of zinc salts and, to a lesser extent, by certain other metal salts, notably aluminum and gallium. The specificity of the effect was assessed by the absence of biological effects in the assay of zinc used alone. The interaction between zinc and thymulin was directly shown by gel chromatography of a mixture of Chelex 100-treated [^{3}H]-FTS and $^{65}Zn^{2+}$ on Bio-Gel P-2. The [^{3}H]-FTS and bound $^{65}Zn^{2+}$ were precisely coeluted with the peak of thymulin biological activity. The presence of zinc in active lots of synthetic thymulin has been confirmed by atomic absorption spectrometry.

The relationship between zinc and biological activity of thymulin was further investigated by the study of the consequences of zinc deprivation in mice and humans on the serum levels of the hormone: previous studies showed that zinc-deficient animals presented low levels of active thymulin (Iwata *et al.*, 1979). More recently, we demonstrated that in fact thymulin was present in serum in its inactive form (zinc-deprived) and that we could restore the biological activity of the hormone by the *in vitro* addition of $ZnCl_2$ (Dardenne *et al.*, 1984). Similar results were obtained in children with nephrotic syndrome characterized by zinc deficiency (Bensman *et al.*, 1983).

3.2. Binding Affinity Studies

Gel filtration studies in ligand equilibrium showed that the nonapeptide could strongly bind one Zn^{2+} ion at pH 7.5 with an affinity of about $5 \pm 2 \times 10^{-7}$ M. This binding site was relatively specific for Zn^{2+} although Al^{3+}, Cu^{2+}, and Mn^{2+}

were found to be good competitors for Zn^{2+} thymulin binding. Furthermore, we observed a good correlation between the metal binding affinity to thymulin and the capacity of the metal to reactivate the Chelex-treated peptide.

Studies using structural analogs of thymulin strongly suggested that the metal binding site is dependent on the presence of the asparagin β-COOH group in position 9 (Gastinel *et al.*, 1984).

However, a better understanding of the coordination geometry of metal ions and thymulin will probably be derived from physicochemical methods such as magnetic circular dichroism and nuclear magnetic resonance.

3.3. Evidence for Zinc-Dependent Antibody Binding Sites in the Thymulin Molecule

In our laboratory, we recently produced antithymulin monoclonal antibodies, for which we used two different antigen sources. In one case, synthetic thymulin coupled to bovine serum albumin was used as the immunogen (Dardenne *et al.*, 1982a), while in the other, trypsinized cultured human epithelial cells were injected (Berrih *et al.*, 1983). Both types of monoclonals were screened for their ability to inhibit the activity of synthetic or natural thymulin in the rosette assay routinely used to measure thymulin biological activity, and to bind specifically to thymic epithelial cells. Interestingly, we observed that the anti-synthetic thymulin monoclonal antibodies recognized a relatively small subpopulation of thymic epithelial cells (Savino *et al.*, 1982), whereas the thymulin monoclonal antibody raised against epithelial cells was able to detect not only those cells recognized by the anti-synthetic product but a further group of epithelial cells as well. These results suggest the existence of differences either in the affinity of these antibodies for thymulin, or in the epitope recognized.

Another point of interest concerning the recognition of thymulin by these monoclonals is that the hormone is recognized only if zinc is present in the molecule, a conclusion based on immunofluorescence as well as biological activity (rosette) assays (Savino *et al.*, 1984). Thus, for example, when synthetic or natural (serum) thymulin was subjected to a chelation procedure, it was no longer detectable by antithymulin monoclonal antibodies. However, when we added $ZnCl_2$ to the chelated molecule, it was recognized by the antibodies. Similar results were also obtained using supernatants from thymic epithelial cell cultures as another source of natural thymulin.

Further evidence for this zinc-dependent binding site(s) of the antibodies to thymulin is that sera from mice or humans, known to have zinc deficiencies, were not able to inhibit the fluorescence on thymic sections incubated with the antithymulin monoclonals. However, these zinc-deficient sera inhibited the fluorescence if they were preincubated with $ZnCl_2$ (Savino *et al.*, 1984).

These data suggest that the presence of zinc in the thymulin molecule determines a spatial configuration which, in addition to being necessary for the biological activity of the hormone, yields new antigenic determinants that can be specifically recognized by monoclonal antibodies.

4. CONCLUSION

Finally, thymulin appears to be a naturally occurring metallopeptide. The importance of the presence of zinc for the biological activity of this hormone strongly indicates the future use of the peptide coupled to zinc for clinical use and radioimmunological studies performed on biological samples.

REFERENCES

Auger, G., Monier, J. C., Dardenne, M., Pléau, J. M., and Bach, J.-F., 1982, Identification of FTS (facteur thymique sérique) on thymus ultrathin sections using monoclonal antibodies, *Immunol. Lett.* **5**:213–216.

Bach, J.-F., 1979, Thymic hormones, *Int. J. Immunopharmacol.* **1**:277–285.

Bach, J.-F., 1983, Thymulin (FTS-Zn), in: *Clinics in Immunology and Allergy,* Vol. 3, pp. 133–156.

Bach, J.-F., Dardenne, M., Pléau, J. M., and Rosa, J., 1977, Biochemical characterization of a serum thymic factor, *Nature (London)* **226**:55–56.

Bardos, P., and Bach, J.-F., 1983, Modulation of mouse natural killer cell activity by the serum thymic factor (FTS), *Scand. J. Immunol.* **75**:321–325.

Bensman, A., Dardenne, M., Morgant, G., Vasmant, D., and Bach, J.-F., 1983, Decrease of biological activity of serum thymic factor in children with nephrotic syndrome, *VIth Int. Symp. Paediatr. Neurol.* p. 59.

Berrih, S., Savino, W., Azoulay, M., and Dardenne, M., 1983, Cultured human epithelial thymic cells as a source of antigen for the production of anti-thymulin monoclonal antibody, *Joint Congr. Eur. Soc. (ETCS–Eur. Reticuloendothelial. Soc. (EURES),* p. 3.

Blanot, D., Martinez, J., Auger, G., and Bricas, E., 1979, Synthesis of analogs of the serum thymic nonapeptide "facteur thymique sérique" (FTS) (Part I), *Int. J. Pept. Protein Res.* **14**:41–56.

Bricas, E., Martinez, J., Blanot, D., Auger, G., Dardenne, M., Pléau, J. M., and Bach, J.-F., 1977, The serum thymic factor and its synthesis, in: *Proceedings , 5th International Peptide Symposium* (M. Goodman and J. Meienhofer, eds.), pp. 564–567, Wiley, New York.

Dardenne, M., Pléau, J. M., Man, N. K., and Bach, J.-F., 1977, Structural study of circulating thymic factor: A peptide isolated from pig serum. I. Isolation and purification, *J. Biol. Chem.* **252**:8040–8044.

Dardenne, M., Pléau, J. M., and Bach, J.-F., 1980, Characterization of facteur thymique sérique (FTS) in the thymus. II. Direct demonstration of the presence of FTS in thymosin fraction V, *Clin. Exp. Immunol.* **42**:477–482.

Dardenne, M., Pléau, J. M., Savino, W., and Bach, J.-F., 1982a, Monoclonal antibody against the serum thymic factor (FTS), *Immunol. Lett.* **4**:61–69.

Dardenne, M., Pléau, J. M., Nabarra, B., Lefrancier, P., Derrien, M., and Choay, J., 1982b, Contribution of zinc and other metals to the biological activity of the serum thymic factor (FTS), *Proc. Natl. Acad. Sci. USA* **79**:5370–5375.

Dardenne, M., Savino, W., Wade, S., Kaiserlian, D., Lemonnier, D., and Bach, J.-F., 1984, *In vivo* and *in vitro* studies of thymulin in marginally zinc-deficient mice, *Eur. J. Immunol.* (in press).

Gastinel, L. N., Pléau, J. M., Dardenne, M., and Bach, J.-F., 1984, Characterization of zinc binding sites on the nonapeptide thymulin, *Biochim. Biophys. Acta* **797**:147–155.

Iwata, T., Incefy, G. S., Tanaka, T., Fernandes, G., Menendez-Botet, C. J., Pitt, K., and Good, R. A., 1979, Circulating thymic hormone levels in zinc deficiency, *Cell. Immunol.* **471**:100–105.

Jambon, B., Montagne, P., Bene, M. C., Brayer, M. P., Faure, G., and Bach, J.-F., 1981, Immunohistologic localization of "facteur thymique sérique" (FTS) in human thymic epithelium, *J. Immunol.* **127**:2055–2059.

Kaiserlian, D., and Dardenne, M., 1982, Studies on the mechanisms of the inhibitory effects of serum thymic factor on murine allograft immunity, *Cell. Immunol.* **66**:360–371.

Lacovara, J., and Utermohlen, V., 1983, Isolation and assay of a human plasma factor affecting human thymus-derived lymphocytes, *Clin. Immunol. Immunopathol.* **27**:428–432.

Martinez, J., Blanot, D., Auger, G., Sasaki, A., and Bricas, E., 1980, Synthesis of analogs of the serum thymic nonapeptide "facteur thymique sérique" (FTS), Part II, *Int. J. Pept. Protein Res.* **16**:267–279.

Monier, J. C., Dardenne, M., Pléau, J. M., Deschaux, P., and Bach, J.-F., 1980, Characterization of facteur thymique sérique (FTS) in the thymus. I. Fixation of anti-FTS antibodies on thymic reticulo-epithelial cells, *Clin. Exp. Immunol.* **42**:470–476.

Pléau, J. M., Dardenne, M., Blouquit, Y., and Bach, J.-F., 1977, Structural study of circulating thymic factor: A peptide isolated from pig serum. II. Amino acid sequence, *J. Biol. Chem.* **252**:8045–8047.

Pléau, J. M., Dardenne, M., Blanot, D., Bricas, E., and Bach, J.-F., 1979, Antagonistic analogue of serum thymic factor (FTS) interacting with the FTS cellular receptor, *Immunol. Lett.* **12**:179–182.

Savino, W., Dardenne, M., Papiernik, M., and Bach, J. F., 1982, Thymic hormone-containing cells: Characterization and localization of serum thymic factor in young mouse thymus studied by monoclonal antibodies, *J. Exp. Med.* **156**:628–634.

Savino, W., Dardenne, M., Berrih, S., and Bach, J. F., 1984, Evidence for zinc dependent antibody binding sites in the molecule of thymulin, a thymic hormone: A study using antithymulin monoclonal antibodies, submitted for publication.

Strachan, R. G., Paleveda, W. J., Bergstrand, S. J., Nutt, R. F., Holly, F. W., and Veber, D. I., 1979, Synthesis of a proposed thymic factor, *J. Med. Chem.* **22**:586–588.

3

Use of Monoclonal Antibodies to Identify Antigens of Human Endocrine Thymic Epithelium

BARTON F. HAYNES

1. INTRODUCTION

The thymus plays a central role in the differentiation of T lymphocytes (Cantor and Weissman, 1976). Lymphocytes in various stages of maturation are found throughout the thymic cortex and medulla; however, only a minority of thymic lymphocytes mature completely and migrate to peripheral lymphoid organs (Matsuyama *et al.*, 1966; McPhee *et al.*, 1979). Thymic epithelial cells are in intimate contact with thymic lymphocytes, are thought to promote normal intrathymic maturation, and contain thymic hormones such as thymopoietin and thymosin α_1 (Bach and Goldstein, 1980).

In murine systems, cells of the nonlymphoid portion of the thymus are thought to educate maturing thymocytes with regard to major histocompatibility (MHC) antigen allotypes which immunocompetent T cells will recognize in the mediation of phases of the normal immune response (Zinkernagel *et al.*, 1978). Moreover, Zielinski *et al.* (1982) have suggested that thymic epithelium is programmed to induce preleukemic changes in thymocyte differentiation in leukemia-susceptible mice. Thus, study of the human thymic microenvironment and its role in normal and aberrant intrathymic T-cell maturation is essential toward understanding mechanisms of genesis of autoimmune and T-cell lymphoproliferative states.

While the surface antigen characteristics of human T lymphocytes and human thymocytes are relatively well characterized (Reinherz and Schlossman, 1980; Haynes, 1981), surface antigens expressed by cells comprising the nonlymphoid component of the thymic microenvironment have not been well characterized.

BARTON F. HAYNES • Department of Medicine, Division of Rheumatic and Genetic Diseases, Duke University School of Medicine, Durham, North Carolina 27710.

Both human (Janossy *et al.*, 1980; Bhan *et al.*, 1980) and murine (Scollay *et al.*, 1980; Jenkinson *et al.*, 1981) cells in the nonlymphoid thymic microenvironment have been shown to express MHC antigens. However, demonstration of expression of non-MHC-encoded antigens on thymic epithelium has been limited to the demonstration of expression of the human Thy-1 antigen on subcapsular cortical thymic epithelial cells (Ritter *et al.*, 1981).

Thus, this chapter will review our recent work on the study of cell surface antigens on the neuroendocrine portion of the human thymus, i.e., those thymic epithelial cells which contain thymotrophic hormones.

2. THYMIC EPITHELIAL SURFACE ANTIGENS AND THYMIC HORMONE DISTRIBUTION IN NORMAL THYMUS

Eisenbarth *et al.* (1979, 1981, 1982) have recently characterized a murine monoclonal antibody (A2B5) which reacts with a complex neuronal G_Q ganglioside expressed on the cell surface of neurons, neural crest-derived cells, and peptide-secreting endocrine cells. Tetanus toxin (TT), which binds to G_D and G_T gangliosides, also binds to this neuroendocrine family of cell types (Eisenbarth *et al.*, 1981). These observations, coupled with previous studies demonstrating that a portion of thymic epithelium in fowl is neural crest derived (Le Douarin and Jotereau, 1975), led us to determine if antibody A2B5 and/or TT bound to cells in human and rodent thymus (Haynes *et al.*, 1983a).

Monoclonal antibody A2B5 showed a similar pattern of reactivity with all human, rat, and mouse thymic sections tested (Fig. 1A). Thymic medullary areas reacted strongly with A2B5 and demonstrated a dense reticular network of epithelial cells (Fig. 1B). These A2B5$^+$ processes went up to and surrounded Hassall's corpuscles, but the central areas of Hassall's corpuscles were unreactive with A2B5. Rat and mouse thymic medulla also showed a similar pattern of reactivity with A2B5. In the thymic cortex, many areas were completely nonreactive with A2B5, only a few scattered A2B5$^+$ cells being seen. However, epithelial cells in the subcapsular cortex area of normal human thymus stained strongly with A2B5 in a pattern similar to that seen in the thymic medulla (Fig. 1A).

To correlate A2B5 reactivity with the endocrine function of the thymus, labeling of thymic epithelium with rabbit antithymopoietin (Goldstein, 1976) or rabbit antithymosin α_1 (Hirokawa *et al.*, 1982) followed by A2B5 was performed. We found essentially a 1 : 1 correlation of A2B5 reactivity with bright antithymopoietin staining or antithymosin α_1 (not shown).

Because all neuroendocrine cell types we have studied which bound A2B5 also bound TT, utilizing indirect immunofluorescence with monoclonal anti-TT antibodies we studied TT binding to thymic tissue. We found that TT bound to sections of rat and human thymus in a pattern identical to A2B5.

In addition to reactivity of A2B5 and TT, we have delineated the intrathymic location of the human Thy-1 antigen, and the thymic hormones thymosin α_1,

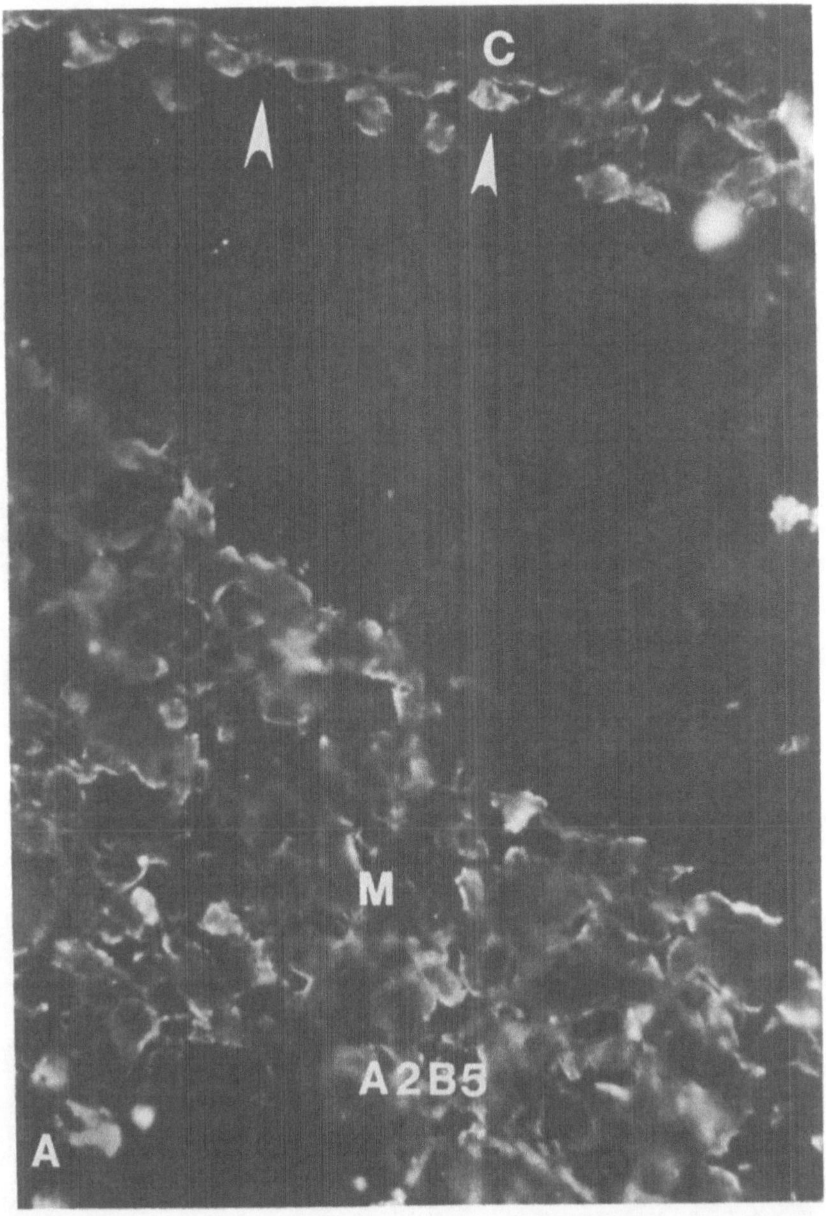

FIGURE 1. Reactivity pattern of antibody A2B5 with normal human thymus. (A) Fluorescent micrograph demonstrating that A2B5-positive cells are distributed in two areas—the subcapsular cortex (arrowheads) and medulla (M). The thymic cortex and the capsule (C) overlying the subcapsular cortex are unreactive with A2B5. (B) In the thymic medulla, A2B5-reactive cells have long dendritic processes that encircle Hassall's bodies (H). ×400 (Haynes *et al.* 1983b; Haynes and Eisenbarth, 1983) .

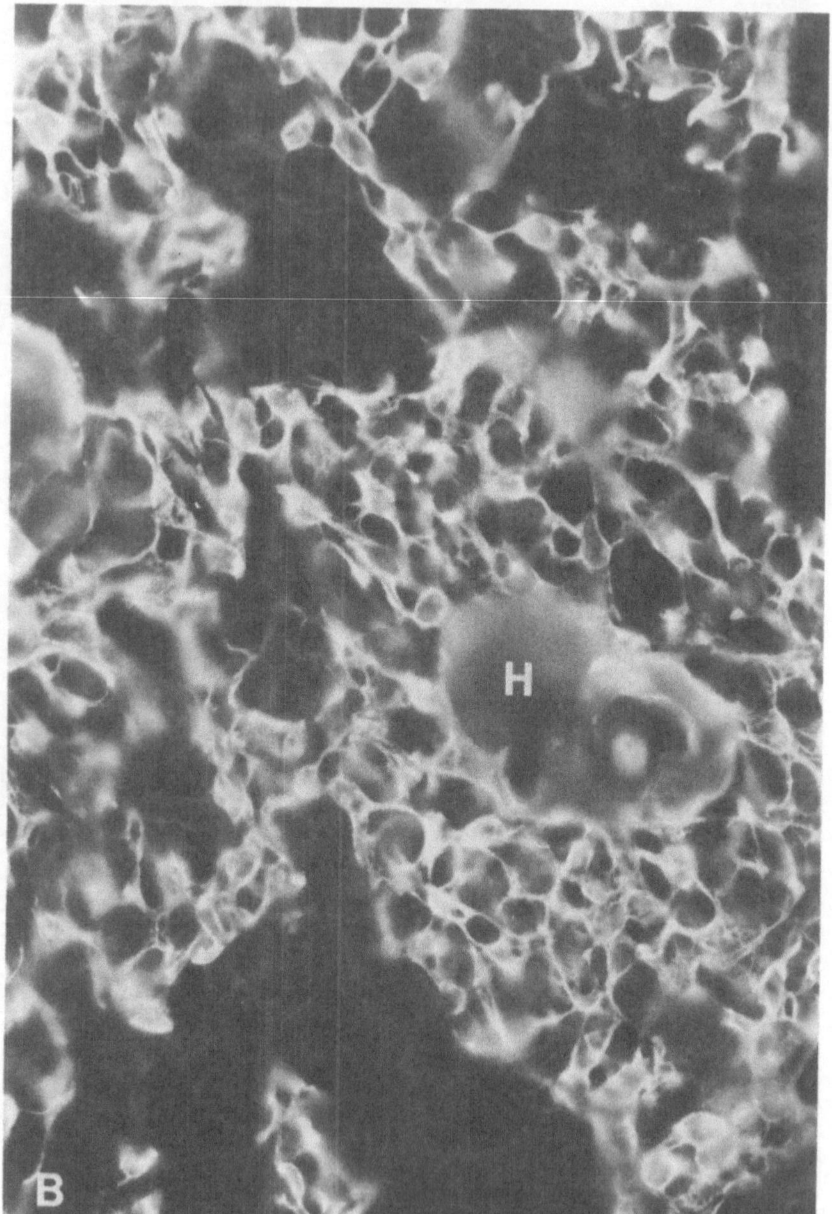

FIGURE 1. *(Continued)*

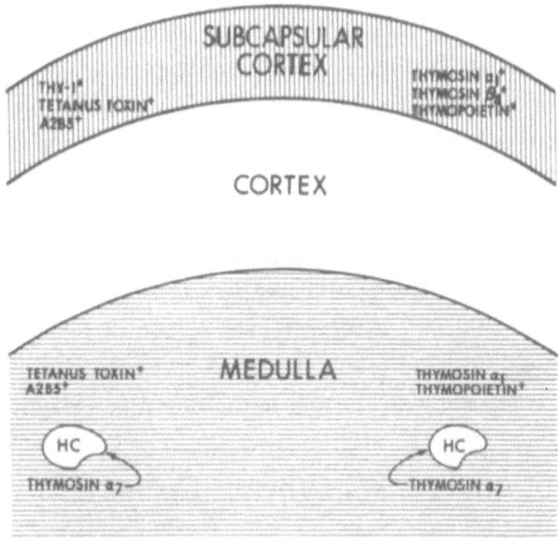

FIGURE 2. Schematic representation of reactivity patterns of anti-thymic hormone antibodies in normal human thymus (Haynes and Eisenbarth, 1983).

thymosin β_4, thymosin α_7, and thymopoietin (Haynes *et al.*, 1983b; Haynes and A. Goldstein, unpublished observations).

Figure 2 summarizes the reactivity patterns of A2B5, TT, anti-Thy-1, antithymosin α_1, antithymopoietin, and antithymosin β_4 in studies of four normal thymuses. As shown by A2B5 reactivity in Fig. 1A and summarized in Fig. 2, the epithelial component of the human thymus is present in two definable compartments—the subcapsular cortical and medullary thymic epithelium. Thymic lymphocytes teased from normal thymic tissue did not react with any of the rabbit antithymic hormone antisera, or anti-Thy-1. The normal subcapsular cortical thymic epithelial region differed from the medullary region by reactivity with anti-Thy-1 and antithymosin β_4. Otherwise, the subcapsular cortical and medullary thymic epithelia were identical in their reactivity with A2B5, TT, antithymosin α_1, and antithymopoietin. In contrast to the other thymic hormones, thymosin α_7 localized only to Hassall's bodies in the thymic medulla (Fig. 2).

3. A HUMAN THYMIC EPITHELIAL ANTIGEN ACQUIRED DURING ONTOGENY IS RECOGNIZED BY A MONOCLONAL ANTIBODY AGAINST HUMAN T-CELL LEUKEMIA VIRUS (HTLV) p19 STRUCTURAL PROTEIN

A novel human retrovirus, HTLV has recently been isolated from a variety of adult T-cell leukemias and lymphomas that have so far generally exhibited a particular geographic distribution in the southern United States, West Indies, and Japan (Poiesz *et al.*, 1980, 1981). Specific antibodies to HTLV proteins have been found in virus-positive cases, often in family members, and much less frequently in the general population (Posner *et al.*, 1981; Kalyanaraman *et al.*, 1981, 1982; Robert-Guroff *et al.*, 1982). To further study the incidence of HTLV in large patient populations, a monoclonal antibody to the HTLV structural protein has been used to examine T cells of patients and normal donors for evidence of HTLV expression (Robert-Guroff *et al.*, 1981).

The monoclonal anti-p19 immunoprecipitates a 19,000-dalton structural protein of HTLV. The antigen is located intracellularly in fixed HTLV -infected T cells, and is not detected in HTLV-negative malignant or normal T cells of any type (Robert-Guroff *et al.*, 1981). During a recent screen of anti-p19 reactivity with a large number of normal human tissues, it was observed that anti-p19 reacted strongly with the epithelial component of normal human thymus (Haynes *et al.*, 1983c).

3.1. Reactivity of Anti-p19 with Thymic Epithelium

As previously reported, anti-p19 did not react with antigens of normal thymocytes. However, when anti-p19 was assayed for reactivity on frozen thymic 4-μm tissue sections from a 30-month-old normal child, the entire subcapsular cortex and medullary thymic epithelium (as defined by A2B5 and TT reactivity) was anti-p19^{+} (Haynes *et al.*, 1983c) (Fig. 3). However, in contrast to A2B5 or TT reactivity,

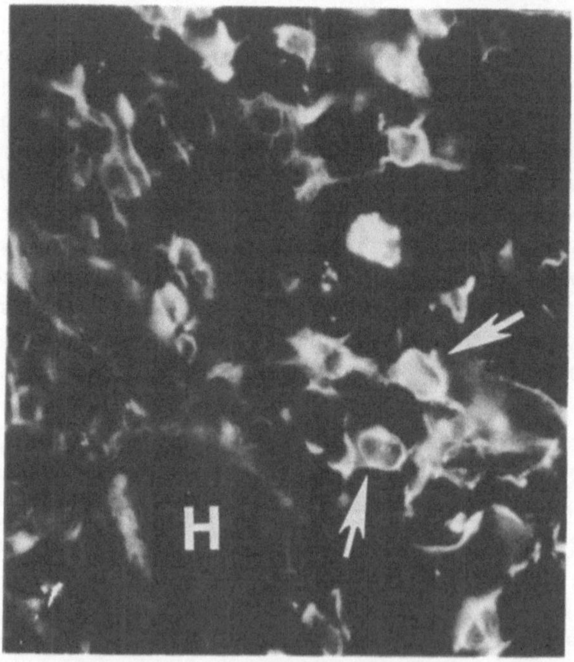

FIGURE 3. Fluorescent micrograph of normal thymic medulla stained with anti-p19 and fluoresceinated goat anti-mouse IgG. Arrows show p19^{+} medullary thymic epithelial cells. H, Hassall's body. ×400.

anti-p19 reactivity was variable from thymus to thymus in the amount of thymic epithelium labeled. Ontogenic study of normal thymic tissue from 8 weeks of fetal gestation through 60 years of age demonstrated that the reactive antigen was acquired during thymic ontogeny. Whereas the entire thymic epithelial component of all fetal and adult thymuses tested was A2B5$^+$, no anti-p19$^+$ epithelium was present until 15 weeks of gestation and these anti-p19$^+$ cells were limited to only a few areas of the subcapsular cortex. By birth, the entire subcapsular cortical thymic epithelial region was anti-p19$^+$, whereas the A2B5$^+$ medullary component of thymic epithelium was anti-p19$^-$. By 30 months of age, the entire A2B5$^+$ neuroendocrine component of thymic epithelium was anti-p19$^+$.

3.2. All Anti-p19-Reactive Thymic Epithelial Cells Contain Thymopoietin and Thymosin α_1

Anti-p19 reacted with both subcapsular cortical and medullary thymic epithelium in all thymic tissue from donors over the age of 30 months. As this reactivity pattern was identical to the pattern seen with A2B5 and TT, and as all A2B5$^+$ and TT$^+$ thymic epithelial cells contain thymopoietin and thymosin α_1 (Haynes *et al.*, 1983a), we wanted to confirm that anti-p19$^+$ epithelial cells as well contained thymopoietin and thymosin α_1. Using double indirect immunofluorescence with rabbit antithymosin α_1 or antithymopoietin developed with rhodamine-conjugated goat anti-mouse IgG, we demonstrated that all anti-p19$^+$ thymic epithelial cells contained these two thymotrophic hormones (Haynes *et al.*, 1983c).

3.3. Reactivity of Anti-p19 in Abnormal Thymic Tissue

Because anti-p19 reactivity is acquired during thymic ontogeny, we looked to see if the antigen recognized was present in abnormal thymic tissue from patients with thymic dysplasia (severe combined immunodeficiency disease and Nezelof syndrome), thymic epithelial malignancies (thymoma), and thymus lymphoid malignancy (HTLV$^-$ T-cell acute lymphoblastoid leukemia). As expected, we found both reactive and nonreactive dysplastic and malignant thymic epithelial tissues.

3.4. Assay for the Presence of HTLV Proteins or HTLV DNA in Normal or Myasthenia Gravis Thymic Epithelium

Because p19 is an HTLV structural protein, the question arose whether normal thymic epithelium was infected with, and thus a reservoir for HTLV. To study this question, a goat anti-p24 antiserum which reacts with a cytoplasmic antigen in HTLV-infected T cells and immunoprecipitates the 24,000-dalton HTLV internal core protein (Kalyanaraman *et al.*, 1981) was used. It did not react by indirect immunofluorescence with any thymic tissue previously shown to have strongly anti-p19-positive thymic epithelium. Finally, DNA from normal and myasthenia gravis purified thymic epithelium, normal whole thymus, normal thymocytes, and DNA from an HTLV-infected cell line were extracted, digested with Kpnl, and probed

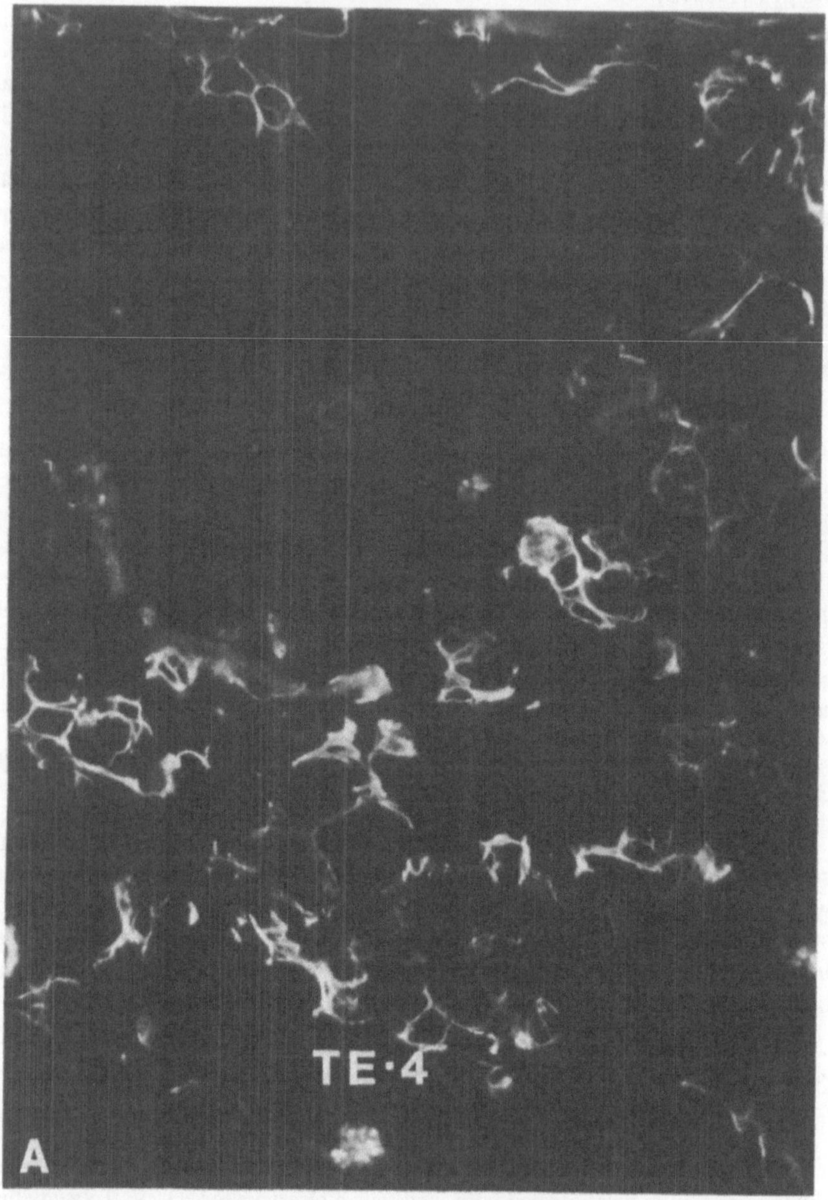

FIGURE 4. Reactivity patterns of anti-human thymic stromal antibodies TE-4 and TE-7. (A) TE-4 reacts with the same subcapsular cortical and medullary epithelial cells as do antibodies A2B5 and anti-p19. (B) TE-7 reacts with nonendocrine stroma that is $p19^-$, $A2B5^-$, and consists of the thymus fibrous capsule, intralobular septae, and vessels. ×400 (Haynes *et al.*, 1984).

FIGURE 4. *(Continued)*

via hybridization with the 5′ (450N) and 3′ (500N) sequence of HTLV. DNA from the cell line infected with HTLV showed distinct bands of hybridization with the HTLV probes, whereas the normal thymic epithelium, whole thymus, and thymocyte DNA did not (Haynes *et al.*, 1983c).

4. NEW ANTIGENS OF THYMIC STROMAL ELEMENTS DEFINED BY MONOCLONAL ANTIBODIES TE-4, TE-7, AND TE-8

We have recently produced murine monoclonal antibodies raised against purified thymic epithelial components that selectively define either the endocrine thymic epithelium (TE-4), nonneuroendocrine thymic stroma (fibrous capsules, fibrous stroma, and vessels), (TE-7), or Hassall's bodies (TE -8).

Using double labeling techniques for indirect immunofluorescence described above in Section 2, we showed (1) the TE-4 and TE-7 subsets of thymic stroma are mutually exclusive (Fig. 4); (2) TE-4$^+$ endocrine stroma contains thymosin α_1 and is p19$^+$ and A2B5$^+$; (3) TE-7$^+$ nonneuroendocrine thymic stroma does not contain thymosin α_1 and is p19$^-$ and A2B5$^-$; and (4) TE-8 selectively binds to cells within Hassall's bodies (Fig. 5).

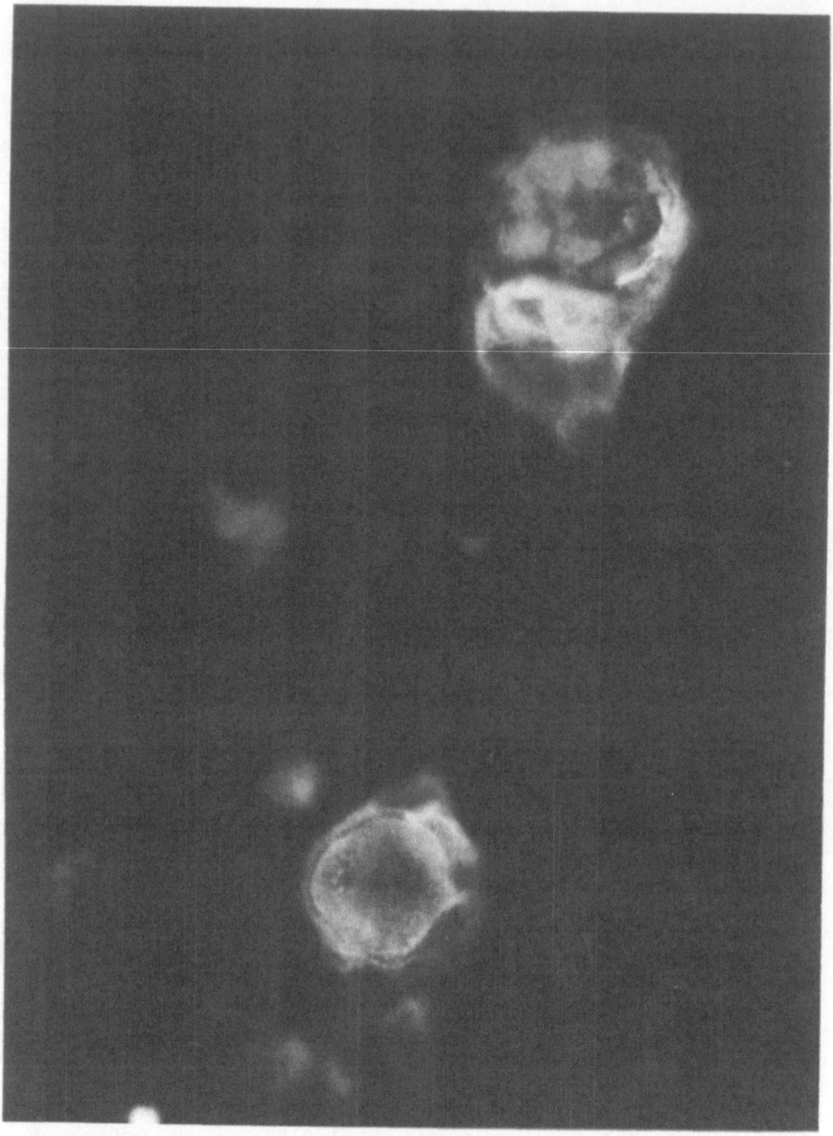

FIGURE 5. Antibody TE-8 reacts selectively with human Hassall's bodies.

Regarding the tissue distribution of antigens TE-4, TE -7, and TE-8, TE-4 was found on no other normal human tissue except the basal layer of squamous epithelium. TE-7 was distributed widely on the fibrous stroma of virtually all tissues tested. TE-8 selectively identified the granulosa cell layer of squamous epithelium, and as well was found on other types of epithelium in various tissues.

5. DISCUSSION

Recent studies by Hirokawa *et al.* (1982), Ritter *et al.* (1981), and Goldstein *et al.* (1981) have suggested that normal human thymic epithelium exists in two regions—the subcapsular cortex and the medulla. Ritter *et al.* (1981) demonstrated that human subcapsular cortical thymic epithelium expressed Thy-1 antigen while medullary thymic epithelium did not. Hirokawa *et al.* (1982) Goldstein *et al.* (1981) demonstrated that the thymosin α_1-containing thymic epithelium is arranged in two areas—the subcapsular cortex and the medulla. In contrast, thymosin β_4, another thymus-derived polypeptide, is present only in the subcapsular cortical epithelial region and not in the medullary epithelial region. Additional evidence for antigenically distinct regions of the endocrine thymic epithelium comes from the observation of Hirokawa *et al.* (1982) that during aging, the medullary thymosin α_1-containing epithelial cells atrophy while the subcapsular cortical epithelial cells do not. Our studies confirm the localization of thymosin α_1 in the subcapsular cortex and medullary thymic epithelium and demonstrate a intrathymic localization for thymopoietin. Moreover, we show that those similar to other neuroendocrine tissues, thymic epithelial cells bind TT and monoclonal antibody A2B5 (Haynes *et al.*, 1983a,b).

The observations regarding anti-HTLV p19 and thymic epithelium have given rise to several hypotheses regarding the nature of the reactive antigen on thymic epithelium: (1) Thymic epithelium may be infected during ontogeny with HTLV. Strong evidence against this is the observations that anti-p19$^+$ thymic epithelium is HTLV p24$^-$, that no HTLV proviral DNA sequences can be detected with HTLV cDNA probes by Southern blot hybridization techniques, and that all thymic tissues tested from donors older than 8 weeks of gestation were anti-p19$^+$. Moreover, epidemiologic data strongly suggest that HTLV-associated T-cell leukemia is horizontally passed as an infectious disease and not by vertical transmission (Robert-Guroff *et al.*, 1981, 1982). Molecular hybridization studies have verified this and proven that the infection is postzygotic (Reitz *et al.*, 1981). The observation that all normal, myasthenic, and T ALL (HTLV$^-$) thymuses tested were anti-p19$^+$ also mitigates against p19 expression on thymic epithelium reflecting HTLV infection. (2) Anti-p19 recognizes a structural protein of HTLV and also may bind to a "cross-reactive" host-derived and therefore non-HTLV-associated antigen on thymic epithelium. This possibility is suggested by the specificity data and by the lack of both HTLV proviral sequences and HTLV p24 antigen in anti-p19$^+$ thymic epithelium. (3) Monoclonal anti-p19 may recognize a 19,000-dalton host-encoded protein which is induced by HTLV infection of cells, contaminates HTLV preparations, and copurifies with virally encoded HTLV p19. Studies are under way to characterize the reactive thymic epithelial antigen. If this antigen has a molecular weight of 19,000, this possibility must be considered more carefully. (4) Finally, HTLV p19, although a viral structural protein, may be encoded by the host rather than the virus. Thus, monoclonal anti-p19 would specifically react not only with HTLV p19 but also with a host protein induced to be expressed by HTLV infection in malignant T cells which is also normally expressed on thymic epithelium. This host protein would be incorporated into the HTLV virion structure during viral morphogenesis. It is important to note that HTLV-infected patients with Japanese and U.S. adult T-cell leukemia make anti-p19 antibodies that compete with monoclonal anti-p19 for binding to HTLV (Robert-Guroff *et al.*, 1982).

This latter observation directly implies that HTLV-infected patients with T-cell leukemia have circulating anti-p19 serum antibodies that react with their own endocrine thymic epithelium. That HTLV-induced leukemia patients have antibodies against thymic epithelium can be directly demonstrated in radioimmunoprecipitation assays using patient serum and ^{125}I-labeled solubilized thymic epithelial membranes (T. J. Palker and B. F. Haynes, unpublished observations).

The notion that HTLV-associated leukemia patients have antibodies against thymic epithelium may have relevance to the pathogenesis of the acquired immune deficiency syndrome (AIDS). Recently, AIDS has been linked to HTLV infection in some patients (Essex *et al.*, 1983; Gallo *et al.*, 1983). Moreover, AIDS thymuses have been shown to be severely atrophic, fibrotic, and to contain a plasma cell infiltrate (Elie *et al.*, 1983). Thus, one possible mechanism of T-cell deficiency in AIDS is HTLV-induced production of anti-HTLV p19 antibodies that also react with normal thymic epithelium; in AIDS, antithymic epithelial antibodies could effect thymic epithelial dysfunction leading to lack of formation of Hassall's bodies and as well T-cell deficiency.

6. SUMMARY

Given the critical role the endocrine thymic epithelium plays in T-cell maturation, study of these cells may be central to our understanding to T-cell leukemogenesis and abnormal T-cell reactivity in autoimmune diseases. Thus, in this chapter our initial series of studies has been summarized regarding the surface antigen characteristics of neural crest-derived thymic epithelial cells. Isolation and purification of these cells coupled with the development of relevant *in vitro* functional assays of their endocrine and thymotrophic functions should provide new insight into the mechanisms of both normal and aberrant intrathymic T-cell maturation.

ACKNOWLEDGMENTS. The author thanks Ms. Joyce Lowery for expert secretarial and editorial assistance. Supported by NIH Grants CA-28936 and RCDA CA-00695.

REFERENCES

Bach, J.-F., and Goldstein, G., 1980, Newer concepts of thymic hormones. *Thymus* **2**:1–4.

Bhan, A. K., Reinherz, E. L., Poppema, S., McCluskey, R. T., and Schlossman, S. F., 1980, Location of T cell and major histocompatibility complex antigens in the human thymus, *J. Exp. Med.* **152**:771–782.

Cantor, H., and Weissman, I., 1976, Development and function of subpopulations of thymocytes and T lymphocytes, *Prog. Allergy* **20:**1–64.

Eisenbarth, G. S., Walsh, F. S., and Nirenberg, M., 1979, Monoclonal antibody to a plasma membrane antigen of neurons, *Proc. Natl. Acad. Sci. USA* **76:**4913–4917.

Eisenbarth, G. S., Shimizu, K., Conn, M., Mittler, B., and Wells, S., 1981, Monoclonal antibody F12A2B5: Reaction with a plasma membrane antigen of vertebrate neurons and peptide-secreting endocrine cells, in: *Monoclonal Antibodies to Neural Antigens* (R. McKay, M. Raff, and L. Reichardt, eds.), Vol. 2, pp. 209–218, Cold Spring Harbor Laboratory, New York.

Eisenbarth, G. S., Shimizu, K., Bowring, M. A., and Wells, S., 1982, Expression of receptors for tetanus toxin and monoclonal antibody A2B5 by pancreatic islet cells, *Proc. Natl. Acad. Sci. USA* **79:**5066–5070.

Elie, R., LaRoche, A. C., Arnoux, E., Guerin, J. M., Pierre, G., Malebranche, R., Seemayer, T., Dupuy, T. M., Russo, P., and Lapp, W. S., 1983, Thymic dysplasia in acquired immunodeficiency syndrome, *N. Engl. J. Med.* **308:**841.

Essex, M., McLane, M. F., Lee, T. H., Falk, L., Howe, C. W. S., and Mullins, J. I., 1983, Antibodies to cell membrane antigens associated with human T cell leukemia virus in patients with AIDS, *Science* **220:**859–862.

Gallo, R. C., Saran, P. S., Gelmann, E. P., Robert-Guroff, M., Richardson, E., Kalyanaraman, V. S., Mann, D., Sidhu, G. D., Stahl, R. E., Zolla-Pazner, S., Leibowitch, J., and Popovic, M., 1983, Isolation of human T-cell leukemia virus in acquired immune deficiency syndrome (AIDS), *Science* **220:**865–867.

Goldstein, A. L., Low, T. L. K., Thurman, G. B., Zatz, M. M., Hall, N., Chen, J., Hu, S. K., Naylor, P. B., and McClure, J. E., 1981, Current status of thymosin and other hormones of the thymus gland, *Horm. Res.* **37:**369–415.

Goldstein, G., 1976, Radioimmunoassay for thymopoietin, *J. Immunol.* **117:**690–692.

Haynes, B. F., 1981, Human T lymphocyte antigens as defined by monoclonal antibodies, *Immunol. Rev.* **57:**127–161.

Haynes, B. F. and Eisenbarth, G. S. (eds.), 1983, *Use of Monoclonal Antibodies to Identify Cell Surface Antigens of Human Neuroendocrine Thymic Epithelium in Monoclonal Antibodies*, pp. 47–65, Academic Press, New York.

Haynes, B. F., Shimizu, K., and Eisenbarth, G. S., 1983a, Identification of human and rodent thymic epithelium using tetanus toxin and monoclonal antibody A2B5, *J. Clin. Invest.* **71:**9–14.

Haynes, B. F., Warren, R. W., Buckley, R. H., McClure, J. E., Goldstein, A. L., Henderson, F. W., Hensley, L. L., and Eisenbarth, G. S., 1983b, Demonstration of abnormalities in expression of thymic epithelial surface antigens in severe cellular immunodeficiency diseases, *J. Immunol.* **130:**1182–1188.

Haynes, B. F., Robert-Guroff, M., Metzgar, R. S., Frauchni, G., Kalyanaraman, V. S., Palker, T. J., and Gallo, R. C., 1983c, Monoclonal antibody against human T-cell leukemia virus p19 defines a human thymic epithelial antigen acquired during ontogeny, *J. Exp. Med.* **157:**907–920.

Haynes, B. F., Scearce, R. M., Lobach, D. F., and Hensley, L. L., 1984, Phenotypic characterization and ontogeny of mesodermal-derived and endocrine epithelial components of the human thymic microenvironment, *J. Exp. Med.* (in press).

Hirokawa, K., McClure, J. E., and Goldstein, A. L., 1982, Age-related changes in localization of thymosin in the human thymus, *Thymus* **4:**19–29.

Janossy, G., Thomas, J., Bollum, F., Granger, S., Pizzolo, G., Bradstock, K., Wong, L., McMichael, A., Ganeshaguru, K., and Hoffbrane, A. V., 1980, The human thymic microenvironment: An immunohistologic study, *J. Immunol.* **125:**202–212.

Jenkinson, E. J., van Ewik, W., and Owen, J. J. T., 1981, Major histocompatibility complex antigen expression on the epithelium of the development thymus in normal and nude mice, *J. Exp. Med.* **153:**280–292.

Kalyanaraman, V. S., Sarngadharan, M. G., Bunn, P. A., Minna, J. D., and Gallo, R. C., 1981, Immunologic properties of a type C retrovirus isolated from cultured human T -lymphoma cells and comparison to other mammalian retroviruses, *J. Virol.* **38:**906.

Kalyanaraman, V. S., Sarngadharan, M. G., Nakao, Y., Ito, Y., Aoki, T., and Gallo, R. C., 1982,

Natural antibodies to the structural core protein (p24) of the human T cell leukemia (lymphoma) virus found in sera of leukemia patients in Japan, *Proc. Natl. Acad. Sci. USA* **79:**1653.

Le Douarin, N. M., and Jotereau, F. V., 1975, Tracing of cells of the avian thymus through embryonic life in interspecific chimeras, *J. Exp. Med.* **142:**17–40.

McPhee, D., Pye, J., and Shortman, K., 1979, The differentiation of T lymphocytes. V. Evidence for intrathymic death of most thymocytes, *Thymus* **1:**151–162.

Matsuyama, M., Wiadroski, M., and Metcalf, D., 1966, Autoradiographic analysis of lymphopoiesis and lymphocyte migration in mice bearing multiple thymus grafts, *J. Exp. Med.* **123:**559–576.

Poiesz, B. J., Ruscetti, F. W., Gazdar, A. F., Bunn, P. A., Minna, J. D., and Gallo, R. C., 1980, Detection and isolation of type C retrovirus particles from fresh and cultured lymphocytes of a patient with cutaneous T cell lymphoma, *Proc. Natl. Acad. Sci. USA* **77:**7415.

Poiesz, B. J., Ruscetti, F. W., Reitz, M. S., Kalyanaraman, V. S., and Gallo, R. C., 1981, Isolation of a new type C retrovirus (HTLV) in primary uncultured cells of a patient with Sézary T-cell leukemia, *Nature (London)* **294:**268.

Posner, L. E., Robert-Guroff, M., Kalyanaraman, V. S., Poiesz, B. J., Ruscetti, F. W., Fossieck, B., Bunn, P. A., Minna, J. D., and Gallo, R. C., 1981, Natural antibodies to the human T cell lymphoma virus in patients with cutaneous T cell lymphomas, *J. Exp. Med.* **154:**333.

Reinherz, E. L., and Schlossman, S. F., 1980, The differentiation and function of human T lymphocytes, *Cell* **19:**821–827.

Reitz, M. S., Poiesz, B. J., Ruscetti, F. W., and Gallo, R. C., 1981, Characterization and distribution of nucleic acid sequences of a novel type C retrovirus isolated from neoplastic human T lympocytes, *Proc. Natl. Acad. Sci. USA* **78:**1887.

Ritter, M. A., Sauvage, C. A., and Cotmore, S. F., 1981, The human thymus microenvironment: *In vivo* identification of thymic nurse cells and other antigenically distinct subpopulations of epithelial cells, *Immunology* **44:**439–446.

Robert-Guroff, M., Ruscetti, F. W., Posner, L. E., Poiesz, B. J., and Gallo, R. C., 1981, Detection of the human T cell lymphoma virus p19 in cells of some patients with cutaneous T cell lymphoma and leukemia using a monoclonal antibody, *J. Exp. Med.* **154:**1957.

Robert-Guroff, M., Fahey, K. A., Maeda, M., Ito, Y., and Gallo, R. C., 1982, Identification of HTLV p19 specific natural human antibodies by competition with monoclonal antibodies, *Virology* **122:**297–305.

Scollay, R., Jacobs, S., Jeraber, L., Butcher, E., and Weissman, I., 1980, T cell maturation: Thymocyte and thymus migrant subpopulations defined with monoclonal antibodies to MHC region antigens, *J. Immunol.* **124:**2845–2853.

Zielinski, C. C., Waksal, S. D., and Datta, S. K., 1982, Thymic epithelium is programmed to induce preleukemic changes in retrovirus expression and thymocyte differentiation in leukemia susceptible mice: Studies on bone marrow and thymic chimeras, *J. Immunol.* **129:**882–889.

Zinkernagel, R. M., Callahan, A., Althage, A., Cooper, S., Klein, P., and Klein, J., 1978, On the thymus in the differentiation of "H-2 self-recognition" by T cells: Evidence for dual recognition?, *J. Exp. Med.* **147:**882–896.

4

Abnormal Levels of Thymosin α_1 and the Destruction of the Thymus Gland in the Acquired Immune Deficiency Syndrome

ARTHUR E. DAVIS, Jr.

1. INTRODUCTION

The acquired immune deficiency syndrome (AIDS) apparently represents a progressive and irreversible loss of cell-mediated immunity (Fauci, 1983). This may be in part a result of an organ-specific immune complex attack on the thymus gland. This disease has been manifested clinically by fever, malaise, anorexia, progressive weight loss, night sweats, diarrhea, lymphadenopathy, and finally the development of serious opportunistic infections and/or Kaposi's sarcoma. The mortality is at least over 50% and may approach 100%. Originally, AIDS was thought to be limited to the male homosexual population, but now has been found in i.v. drug users, Haitians, and close contacts in these groups (Harris *et al.*, 1983). Unfortunately, AIDS has now been identified in recipients of blood and blood products, such as hemophiliacs, and may indicate wide dissemination (Lederman *et al.*, 1983). The epidemiology indicates that AIDS is an infectious disease that is transmitted by body fluids and is most probably a virus (Gallo *et al.*, 1983).

The laboratory findings are variable, depending on what stage of the disease the testing is performed. There is a progressive decrease in the total lymphocyte count with a marked decrease in the helper T cells resulting in a reversal of the helper/suppressor ratio. Viral serological studies are often positive for cytomegalovirus, herpes, hepatitis B, Epstein–Barr, and others. There is a dramatic reduction

ARTHUR E. DAVIS, Jr. • Roche Biomedical Laboratories, Inc., Burlington, North Carolina 27215.

in mitogen response and there is often anergy to skin testing. The immunoglobulins are usually elevated with a polyclonal hypergammaglobulinemia, and circulating immune complexes are often found (Rubinstein *et al.*, 1983). Thymosin α_1 and β_2-microglobulin are often elevated, especially in the early stages. Many of these laboratory abnormalities reflect functions of the immune system that are thought to be related to the thymus gland.

2. MATERIALS AND METHODS

Because of this possibility, a systematic study of thymus glands was initiated late in 1982. Normal controls were established by the study of autopsy material from a wide variety of sources. The ages ranged from premature infants to adults up to 84 years old. These studies reaffirmed the known histology of the thymus gland and confirmed that adults in the eighth decade have normal-appearing Hassall's corpuscles and thymic epithelial cells. Thymus glands were obtained from patients who died of AIDS and included male homosexuals, i.v. drug users, and Haitians.

3. RESULTS

The study of the AIDS thymus glands revealed marked histological abnormalities with destruction of the normal architecture. There appeared to be a systematic destruction of Hassall's corpuscles and alteration in the appearance of the thymic epithelial cells. The process appears to begin with a progressive infiltrate of plasma cells and lymphocytes that are producing polyclonal IgG (Fig. 1). Hassall's corpuscles undergo cytolytic and degenerative changes and finally become hyalinized or disappear. Most of the thymic epithelial cells become spindled, have pyknotic nuclei, and appear inactive. There is depletion of thymocytes, areas of patchy fibrosis, and loss of corticomedullary demarcation (Fig. 2). Immunoperoxidase stains reveal a heavy deposition of polyclonal IgG in the areas of former Hassall's corpuscles and on the surface of the spindled thymic epithelial cells. This is contrasted to the normal aging process where the changes are independent of an active infiltrate and represent passive atrophy. The pathological process appears to

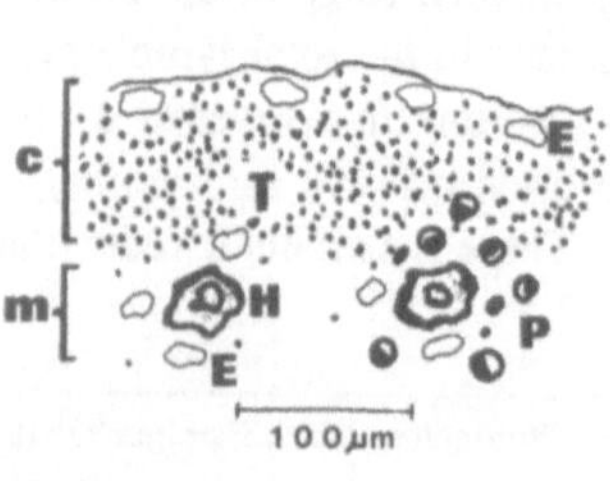

FIGURE 1. Thymus gland in early AIDS. C, cortex; M, medulla; H, Hassall's corpuscles; E, epithelial cells; T, thymocytes; P, plasma cell and lymphocyte infiltrate.

FIGURE 2. Thymus gland in late AIDS. Note loss of corticomedullary demarcation, patchy fibrosis (F), spindled epithelial cells (E), and plasma cell and lymphocytic infiltrate (P).

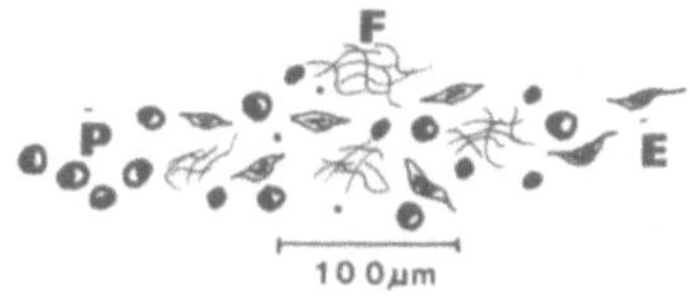

be complement and neutrophil independent, at least in the later stages, and is consistent with an organ-specific immune complex attack on the thymus gland.

4. DISCUSSION

A comparable histological model to AIDS is found in autoimmune thyroiditis (Hashimoto's disease). Again there is a heavy infiltration of plasma cells and lymphocytes that are making antibodies against at least six different thyroid components. This has been shown to represent an organ-specific local immune complex disease and probably not local T-cell cytotoxicity, complement- or neutrophil-mediated process. Passive transfer of this thyroiditis with serum has been effected (Yashida, 1969). In the early stages, there is an elevation of thyroid hormones due to cellular damage (Yashida, 1982). Most cases recover. In others, the epithelial cell damage continues, leading to thyroid insufficiency, and supplementary thyroid hormone must be given. Thus, this is an antibody-mediated disease resulting from antibody combining with antigen with a resultant cytotoxic or cytolytic effect on the target cells. Autoimmune lymphocytic thyroiditis is thought to be a virus-induced process. A retrovirus human T-cell leukemia virus (HTLV) has recently been isolated from some AIDS patients (Gallo *et al.*, 1983). HTLV contains an intracellular structural protein (p19) that strongly cross-reacts with the epithelial cells of the thymus (Haynes *et al.*, 1983). In the immune response to a viral infection, is the thymus the harmed innocent bystander? The basic histological changes in the thymus glands from patients with AIDS are remarkably similar to the histological changes in thyroid glands from patients with lymphocytic thyroiditis. The basic defect in lymphocytic thyroiditis is within the suppressor T cell.

Some of the abnormal laboratory values found in AIDS patients reflect an autoimmune process. There are often very great elevations of the immunoglobulins, especially polyclonal IgG. Circulating immune complexes are frequently found. The differentiation of the circulating T lymphocytes is thymus dependent and there are profound changes in these lymphocytes in AIDS. The β_2-microglobulin elevation is a reflection of cellular death, especially lymphocytes. Most important, the elevation of thymosin α_1 is often found in the early stages and may well reflect thymic epithelial cell damage. Thymosin α_1 levels were determined in three of the autopsy cases studied and were found to be significantly elevated. Thymosin α_1 levels are elevated in many active homosexuals, in many patients with AIDS, and in many patients with serious disease (Hersh *et al.*, 1983). Thymosin α_1 may be an excellent surrogate laboratory test for AIDS, especially in the early stages.

5. SUMMARY

Histological study of the thymus glands from patients who have died of AIDS reveals changes consistent with an organ-specific immune complex attack by polyclonal immunoglobulins directed against the thymic epithelial cells and Hassall's corpuscles. This results in marked architectural alteration. There is eventual disappearance of the corpuscles. The epithelial cells become spindled with pyknotic nuclei. There is thymocyte depletion and patchy fibrosis. Thymosin α_1 levels appear to be elevated, especially in the earlier stages and may reflect epithelial cell damage.

ACKNOWLEDGMENTS. The author would like to give special thanks to Thomas A. Seemayer and other pathologists who have allowed me to review the microscopic slides of thymus glands from patients who have died of AIDS.

REFERENCES

Fauci, A. S., 1983, The syndrome of Kaposi's sarcoma and opportunistic infections: An epidemiologically restricted disorder of immunoregulation, *Ann. Inter. Med.* **96:**777–779.

Fauci, A. S., 1983, The acquired immune deficiency syndrome: The ever broadening clinical spectrum, *J. Am. Med. Assoc.* **249:**2375–2376.

Gallo, R. C., Sarin, P. S., Gelmann, E. P., Robert-Guroff, M., Richardson, E., Kalyanaraman, V. S., Mann, D., Sidhu, G. D., Stahl, R. E., Zolla-Pazner, S., Leibowitch, J., and Popovic, M., 1983, Isolation of human T-cell leukemia virus in acquired immune deficiency syndrome (AIDS), *Science* **220:**865–867.

Harris, C., Small, C. B., Klein, R. S., Friedland, G. H., Moll. B., Emeson, E. E., Spigland, I., and Steigbigel, N. H., 1983, Immunodeficiency in female sexual partners of men with the acquired immunodeficiency syndrome, *N. Engl. J. Med.* **308:**1181–1184.

Haynes, B. F., Robert-Guroff, M., Metzgar, R. S., Frauchim, G., Kalyanaraman, V. S., Palker, T. J., and Gallo, R. C., 1983, Monoclonal antibody against human T-cell leukemia virus p19 defines a human thymic epithelial antigen acquired during ontogeny, *J. Exp. Med.* **157:**907–920.

Hersh, E. M., Rios, A., Mansell, W. A., Newell, G. R., McClure, J. E., and Goldstein, A. L., 1983, Elevated serum thymosin α_1 levels associated with evidence of immune dysregulation in male homosexuals with a history of infectious disease or Kaposi's sarcoma, *N. Engl. J. Med.* **308:**46–46.

Lederman, M. M., Ratnoff, O. D., Scillian, J. J., Jones, P. K., and Schacter, B., 1983, Impaired cell-mediated immunity in patients with classic hemophilia, *N. Engl. J. Med.* **308:**79–83.

Rubinstein, A., Sicklick, M., Gupta, A., Bernstein, L., Klein, N., Rubinstein, E., Spigland, I., Fruchter, L., Litman, N., Lee, H., and Hollander, M., 1983, Acquired immunodeficiency with reversed T_4/T_8 ratios in infants born to promiscuous and drug-addicted mothers, *J. Am. Med. Assoc.* **249:**2350–2356.

Yashida, K., 1969, Transfer of experimental thyroiditis with serum from thymectomized donors, *J. Exp. Med.* **130:**263–268.

Yashida, K., 1982, Serum free thyroxine and triiodothyronine concentrations in subacute thyroiditis, *J. Clin. Endocrinol. Metab.* **55:**185–188.

5

Thymic Epithelial Injury in the Acquired Immune Deficiency Syndrome

THOMAS A. SEEMAYER, A. CLAUDE LAROCHE, PIERRE RUSSO, HY GOLDMAN, RODOLPHE MALEBRANCE, EMMANUEL ARNOUX, JEAN-MICHEL GUÉRIN, GÉRARD PIERRE, JEAN-MARIE DUPUY, JOHN G. GARTNER, WAYNE S. LAPP, and ROBERT ELIE

1. INTRODUCTION

An outbreak of opportunistic infections and/or Kaposi's sarcoma (Gottlieb *et al.*, 1981; Masur *et al.*, 1981; Siegal *et al.*, 1981; Hymes *et al.*, 1981) was described in 1981 among homosexual men in California and New York. Since these early reports a similar syndrome has been described in diverse segments of society throughout North America and Europe. The acquired immune deficiency syndrome (AIDS), as it has come to be known, is currently regarded as one major medical issue confronting the Western world.

As presently delineated, four groups of individuals are principally involved: homosexual males, intravenous drug users, Haitians, and hemophiliacs. Among

THOMAS A. SEEMAYER and PIERRE RUSSO • Department of Pathology, Montreal Children's Hospital and McGill University–Montreal Children's Hospital Research Institute, Montreal, Quebec, Canada. A. CLAUDE LAROCHE, RODOLPHE MALEBRANCE, EMMANUEL ARNOUX, JEAN-MICHEL GUÉRIN, GÉRARD PIERRE, and ROBERT ELIE • Groupe de recherche sur les Maladies Immunitaires en Haiti, Port-au-Prince, Haiti. HY GOLDMAN • Department of Pediatrics, McGill University Faculty of Medicine, Montreal, Quebec, Canada. JEAN-MARIE DUPUY • Immunology Research Center, Institut Armand-Frappier, Université du Québec, Laval-des-Rapides, Quebec, Canada. JOHN G. GARTNER and WAYNE S. LAPP • Department of Physiology, McGill University Faculty of Medicine, Montreal, Quebec, Canada.

such diversity resides the common denominator of profound suppression of cellular immunity as manifest by anergy, lymphopenia, impaired T-cell function *in vitro*, and reduction of peripheral blood T-helper cells with relative sparing of T-suppressor/cytotoxic cells and humoral immunity (Gottlieb *et al.*, 1981; Masur *et al.*, 1981; Siegal et al., 1981; Stahl *et al.*, 1982; Kornfeld *et al.*, 1982).

The nearly singular involvement of the cellular immune system, coupled with the recently described abnormal levels of thymosin α_1 (Hersh *et al.*, 1983) in male homosexuals, have led us to question whether critical events might be initiated in the thymus of AIDS patients. Indeed, in a brief note (O'Reilly, 1982) the thymus of two infants with presumed AIDS was described as "hypoplastic"; in a second report (Fliegel and Naeim, 1983), Hassall's corpuscles were reduced in number in nine male homosexuals with AIDS. We have tested our hypothesis in a preliminary histological study of six Haitian patients with AIDS. Our findings suggest that the thymus may well be a site of cellular injury as manifest by a loss of at least one population of epithelium, patchy fibrosis, and an ingress of plasma cells into a tissue normally devoid of such cells. The description of these findings is the subject of this report. A brief notation (Elie *et al.*, 1983) and a more complete documentation (Seemayer *et al.*, 1984) of these findings have been previously described.

2. MATERIALS AND METHODS

Six heterosexual adult Haitians with AIDS from Port-au-Prince, Haiti, were studied. The patients, four females and two males, were 29–49 years of age. The thymus was fixed in 10% formalin, embedded in paraffin, step-sectioned, stained with hematoxylin–phloxine–saffron, and examined by conventional light microscopy.

Control thymuses (prepared and examined in like fashion) were provided by: (1) five age-matched hospitalized Haitian patients, (2) 10 age- and sex-matched Montreal patients following sudden death or a brief illness, and (3) 20 middle-aged–elderly Montreal patients following either a chronic illness or a prolonged hospitalization. None of the controls had an immune deficiency condition, although many were emaciated, profoundly stressed, and had succumbed to infection.

3. RESULTS

3.1. AIDS Patients

The thymus from each patient was involuted, devoid of definitive cortex and medulla, depleted of thymocytes, and composed of a central condensed core of epithelium (Figs. 1, 2). The latter was represented by round to spindle-shaped cells and occasional rosettelike structures. Hassall's corpuscles and solid aggregates of "plump" polygonal epithelial cells were not identified. Variable degrees of fibrosis

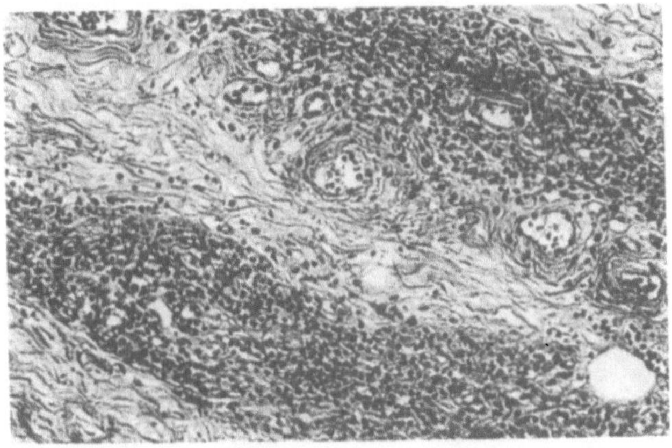

FIGURE 1. Photomicrograph of thymus from 35-year-old female with AIDS depicting involution and architectural effacement. Hematoxylin–phloxine–saffron; ×100.

and plasma cell infiltration were noted within and around the loose fibroareolar tissue of the gland (Fig. 3).

3.2. Control Patients

Thymuses from the five age-matched Haitian hospitalized patients and 10 age- and sex-matched Montreal controls demonstrated abundant Hassall's corpuscles (Fig. 4) and variable degrees of stress involution manifest by thymocyte depletion and cortically distributed debris-laden histiocytes.

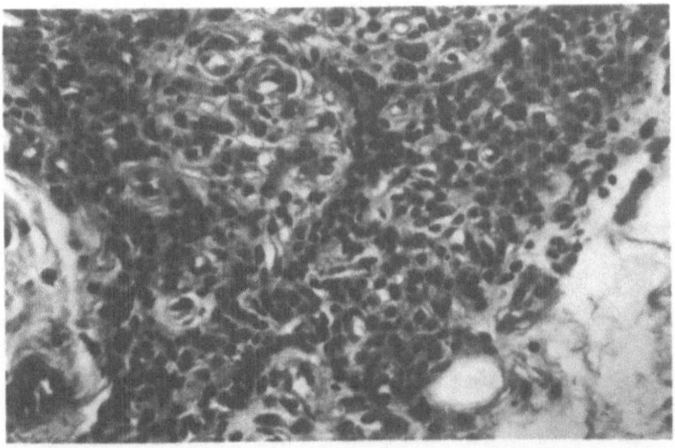

FIGURE 2. Photomicrograph of thymus from 40-year-old female with AIDS depicting involution, prominent vascularization, lack of definitive cortex and medulla, thymocyte depletion, and an absence of Hassall's corpuscles. Hematoxylin–phloxine–saffron; ×200.

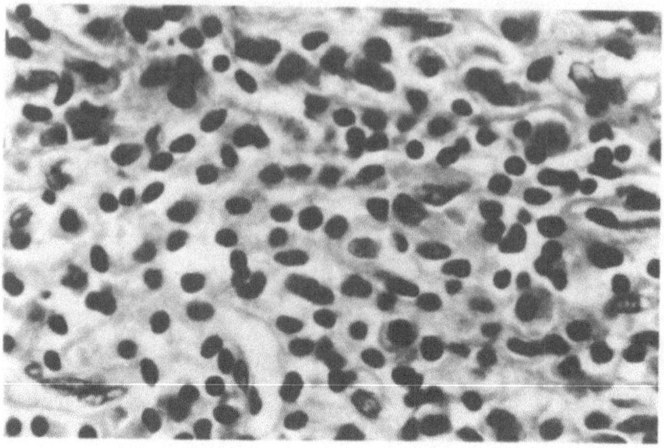

FIGURE 3. High-power photomicrograph of same thymus as Fig. 1 illustrating plasma cell infiltration within and contiguous to gland. Hematoxylin–phloxine–saffron; ×400.

Thymuses from 12 middle-aged–elderly chronically ill Montreal patients demonstrated involution, architectural effacement, and an absence of Hassall's corpuscles. Nevertheless, in eight such patients, variable degrees of architectural preservation were noted and Hassall's corpuscles and/or plump aggregates of medullary epithelium were identified. In none was there evidence of patchy fibrosis or plasma cell infiltration as described for AIDS patients.

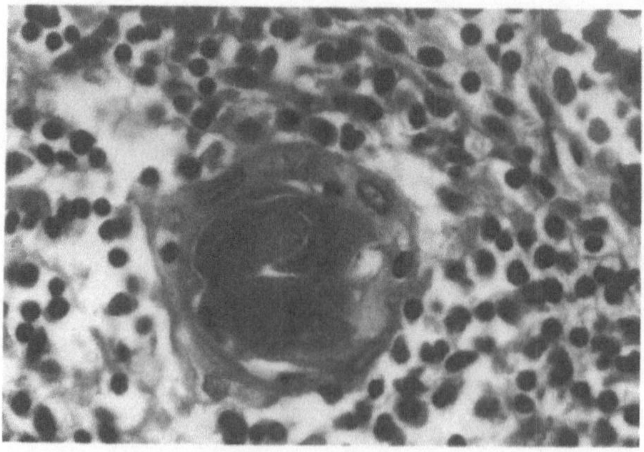

FIGURE 4. Photomicrograph from Haitian control patient, a 55-year-old male who died following a cerebral hemorrhage, illustrating well-developed Hassall's corpuscle. Hematoxylin–phloxine–saffron; ×400.

4. DISCUSSION

Although the thymus plays a pivotal role in the ontogeny of the immune system, its function in adult life remains poorly understood. Much has been written about thymic neoplasms and their clinical associations, yet little is known about the thymus in many pathological states. Several thymic alterations are relatively well characterized and will be described as they are judged to be pertinent to this report.

The first, *stress involution,* is associated with diverse forms of acute and chronic stress and is believed to be mediated by adrenal corticosteroids acting on corticosteroid-sensitive thymocytes. Under stressful conditions, the thymus involutes and becomes depleted of cortical thymocytes and replete with cortical histiocytes, the latter laden with nuclear and cellular debris. Hassall's corpuscles are unaffected, and often appear quite prominent owing to thymocyte depletion.

A second alteration, *physiological involution,* represents the morphological expression of thymic aging. Over the course of years (beginning in the second decade), the gland incurs a sustained reduction in size, number of thymocytes, and Hassall's corpuscles. As this proceeds, fatty replacement of parenchyma occurs; the net effect is to convert a lymphoid organ into a lipomatous mass containing variable amounts of thymic epithelial remnants. It is to be noted that this process appears to be age-programmed, i.e., physiological; hence, neither fibrosis nor invasion of the gland by cells normally excluded from the thymic microenvironment is a feature of physiological involution.

A third form of involution is that sustained as a result of a graft-versus-host (GVH) reaction. In both mice (Seemayer *et al.*, 1977, 1978) and man (Seemayer and Bolande, 1980), the thymus incurs a form of involution in GVH reactions characterized by weight and size reduction, architectural effacement, and epithelial injury manifest by the loss of Hassall's corpuscles. In such glands it is not uncommon to observe variable numbers of plasmacytoid lymphocytes and even fibrosis. This thymic epithelial injury is not without immunological consequences, as a major defect in thymocyte function (Seddik *et al.*, 1980) has been demonstrated in animals sustaining GVH reactions.

The thymic involution described in the six Haitian patients with AIDS is judged to be different from that seen in response to stress and/or aging. In AIDS the thymus demonstrates complete architectural effacement and loss of Hassall's corpuscles, patchy fibrosis, and variable degrees of plasma cell infiltration. The extent of involution observed clearly supersedes that sustained by stress and inanition and antedates that incurred with aging. Moreover, the fibrosis within and contiguous to the gland, coupled with the ingress of plasma cells, suggest that such thymuses are sites of cellular injury during the course of the illness. As such, the features are not too dissimilar from those inflicted during GVH reactions. The thymic changes cannot be ascribed solely to transfusion-related GVH reactions (as might occur in an immunodeficient patient on receipt of exogenous T lymphocytes), as only three of the six AIDS patients received blood or blood products prior to or during their illness. Possibly the thymic alterations stem from a GVH-like mechanism initiated directly or indirectly by a thymotropic transmissible agent. In this regard, sequences of human T-cell leukemia virus (HTLV) have been identified (Gelmann *et al.*, 1983; Gallo *et al.*, 1983) in peripheral blood lymphocytes of several AIDS patients. Moreover, studies of normal human thymus (Haynes *et al.*, 1983) have demonstrated a neuroendocrine population of thymic epithelium which reacts strongly with an HTLV-associated antigen. These interesting, as yet unresolved findings, when viewed in the context of the stated thymic alterations, dictate that further studies of AIDS patients include thorough histological, immunological, and virological examination of the thymus.

5. SUMMARY

Thymuses from six adult Haitian patients with AIDS were examined histologically and compared with controls represented by Haitian and Montreal patients succumbing from a variety of conditions other than AIDS. Control thymuses exhibited a spectrum of findings which ranged from normal histology to severe physiological and/or stress involution. Nonetheless, in nearly half of the middle-aged–elderly chronically ill control patients, the thymus demonstrated variable degrees of architectural preservation and scattered Hassall's corpuscles. In contrast, thymuses from AIDS patients exhibited profound involution, architectural effacement, thymocyte depletion, patchy fibrosis, variable degrees of plasma cell infiltration, and, above all, an absence of Hassall's corpuscles. Thus, the thymic changes in AIDS antedate those associated with aging (physiological involution) and supersede those incurred with stress (stress involution). The precocious loss of Hassall's corpuscles, one of a population of thymic epithelial cells, coupled with the observed fibrosis and plasma cell infiltrates, suggest that the thymus sustains injury during the course of the illness. These features collectively suggest that the thymus may be a site of important, as yet cryptic events in AIDS. Future studies of AIDS should include histological, immunological, and virological study of the thymus.

ACKNOWLEDGMENTS. The authors thank L. Pegorari and N. Ranger for excellent technical and secretarial assistance. This work was supported in part by the McGill University–Montreal Children's Hospital Research Institute and the Laboratoire d'Investigations Biologiques, Port-au-Prince, Haiti. T.A.S. is a McPherson, Fraser, Monat University Associate.

REFERENCES

Elie, R., Laroche, A. C., Arnoux, E., Guérin, J.-M., Pierre, G., Malebranche, R., Seemayer, T. A., Dupuy, J.-M., Russo, P., and Lapp, W. S., 1983, Thymic dysplasia in acquired immunodeficiency syndrome, *N. Engl. J. Med.* **308**:841–842.

Fliegel, S., and Naeim, F., 1983, Immunopathologic findings in male homosexual patients with acquired immunodeficiency syndrome, *Lab. Invest.* **48**:25A.

Gallo, R. C., Sarin, P. S., Gelmann, E. P., Robert-Guroff, M., Richardson, E., Kalyanaraman, V. S., Mann, D., Sidhu, G. D., Stahl, R. E., Zolla-Pazner, S., Leibowitch, J., and Popovic, M., 1983, Isolation of human T-cell leukemia virus in acquired immune deficiency syndrome (AIDS), *Science* **220**:865–867.

Gelmann, E. P., Popovic, M., Blayney, D., Masur, H., Sidhu, G., Stahl, R. E., and Gallo, R. C., 1983, Proviral DNA of a retrovirus, human T-cell leukemia virus, in two patients with AIDS, *Science* **220:**862–864.

Gottlieb, M. S., Schroff, R., Schanker, H. M., Weisman, J. D., Fan, P. T., Wolf, R. A., and Saxon, A., 1981, *Pneumocystis carinii* pneumonia and mucosal candidiasis in previously healthy homosexual men, *N. Engl. J. Med.* **305:**1425–1431.

Haynes, B. F., Robert-Guroff, M., Metzgar, R. S., Frauchim, G., Kalyanaraman, V. S., Palker, T. J., and Gallo, R. C., 1983, Monoclonal antibody against human T-cell leukemia virus p19 defines a human thymic epithelial antigen acquired during ontogeny, *J. Exp. Med.* **157:**907–920.

Hersh, E. M., Reuben, J. M., Rios, A., *et al.*, 1983, Elevated serum thymosin α_1 levels associated with evidence of immune dysregulation in male homosexuals with a history of infectious disease or Kaposi's sarcoma, *N. Engl. J. Med.* **308:**45–46.

Hymes, K. B., Greene, J. B., Marcus, A., William, D. C., Cheung, T., Prose, N. S., Ballard, H., and Lauberstein, L. J., 1981, Kaposi's sarcoma in homosexual men—A report of eight cases, *Lancet* **2:**598–600.

Kornfeld, H., Vande Stouwe, R. A., Lange, M., Reddy, M. M., and Greico, M. H., 1982, T-lymphocyte subpopulations in homosexual men, *N. Engl. J. Med.* **307:**729–731.

Masur, H., Michelis, M. A., Greene, J. B., Oronato, I., Vande Stouwe, R. A., Holzman, R. S., Wormser, G., Brettman, L., Lange, M., Murray, H. W., and Cunningham-Rundles, S., 1981, An outbreak of community-acquired *Pneumocystis carinii* pneumonia, *N. Engl. J. Med.* **305:**1431–1438.

O'Reilly, R., 1982, Unexplained immunodeficiency and opportunistic infections in infants—New York, New Jersey, California, *U.S. Morbidity and Mortality Weekly Report* **31:**665–667.

Seddik, M., Seemayer, T. A., and Lapp, W. S., 1980, T-cell functional defect associated with thymic epithelial injury induced by a graft-versus-host reaction, *Transplantation* **29:**61–66.

Seemayer, T. A., and Bolande, R. P., 1980, Thymic involution mimicking thymic dysplasia, *Arch. Pathol. Lab. Med.* **104:**141–145.

Seemayer, T. A., Lapp, W. S., and Bolande, R. P., 1977, Thymic involution in murine graft-versus-host reaction, *Am. J. Pathol.* **88:**119–134.

Seemayer, T. A., Lapp, W. S., and Bolande, R. P., 1978, Thymic epithelial injury in graft-versus-host reactions following adrenalectomy, *Am. J. Pathol.* **93:**325–338.

Seemayer, T. A., Laroche, A. C., Russo, P., Malebranche, R., Arnoux, E., Guérin, J.-M., Pierre, G., Dupuy, J.-M., Gartner, J. G., Lapp, W. S., Spira, T. J., and Elie, R., 1984, Precocious thymic involution manifest by epithelial injury in the acquired immune deficiency syndrome, *Hum. Pathol.* in press.

Siegal, F. P., Lopez, C., Hammer, G. S., Brown, A. E., Kornfeld, S. J., Gold, J., Hassette, J., Hirschman, S. Z., Cunningham-Rundles, C., Adelsberg, B. R., Parkham, D. M., Siegal, M., Cunningham-Rundles, S., and Armstrong, D., 1981, Severe acquired immunodeficiency in male homosexuals, manifested by chronic perianal herpes simplex lesions, *N. Engl. J. Med.* **305:**1439–1444.

Stahl, R. E., Friedman-Kien, A., Dubin, R., Marmor, M., and Zolla-Pazner, S., 1982, Immunologic abnormalities in homosexual men, *Am. J. Med.* **73:**171–178.

6

Increased Thymosin Levels Associated with Acquired Immune Deficiency Syndrome

PAUL H. NAYLOR, MICHAEL R. ERDOS, and ALLAN L. GOLDSTEIN

1. INTRODUCTION

Acquired immune deficiency syndrome (AIDS) has elicited great concern in its progression to epidemic proportions worldwide (Fig. 1). Initially identified in young male homosexuals, AIDS has recently appeared in other groups including hemophiliacs, transfusion recipients, Haitian immigrants, black Africans, and drug addicts in the USA, as well as Europe, Canada, and Haiti (CDC, 1982a,b; West, 1983).

Preliminary symptoms of AIDS are nonspecific, including: fever, weight loss, lymphadenopathy, inverse helper to suppressor cell ratios, depression of T-cell-mediated immune parameters, skin test anergy, elevated levels of immunoglobulins, and lymphopenia (Durack, 1981; Friedman-Kien, 1981; Levine, 1982). The thymus gland is an integral part of the T-cell-mediated immune system. As reported previously and confirmed in these proceedings, the thymus gland undergoes histological alterations as a result of AIDS infection (Elie *et al.*, 1983). Lymphocytes and plasma cells infiltrate the thymic tissue destroying the thymic epithelium and causing the disappearance of Hassall's corpuscles. Abnormal levels of thymosin α_1 are related to the progressive destruction of thymic tissues (Hersh *et al.*, 1983; Naylor and Goldstein, 1983, 1984; Reuben *et al.*, 1983). As the immune system deteriorates, susceptibility develops toward opportunistic infections such as *Pneumocystis carinii* pneumonia and about 40% of the patients present with a rare type of cancer termed Kaposi's sarcoma (CDC, 1982b; Durack, 1981; Friedman-Kien, 1981; Gottlieb *et al.*, 1981; Levine, 1982).

PAUL H. NAYLOR, MICHAEL R. ERDOS, and ALLAN L. GOLDSTEIN • Department of Biochemistry, The George Washington University School of Medicine and Health Sciences, Washington, D.C. 20037.

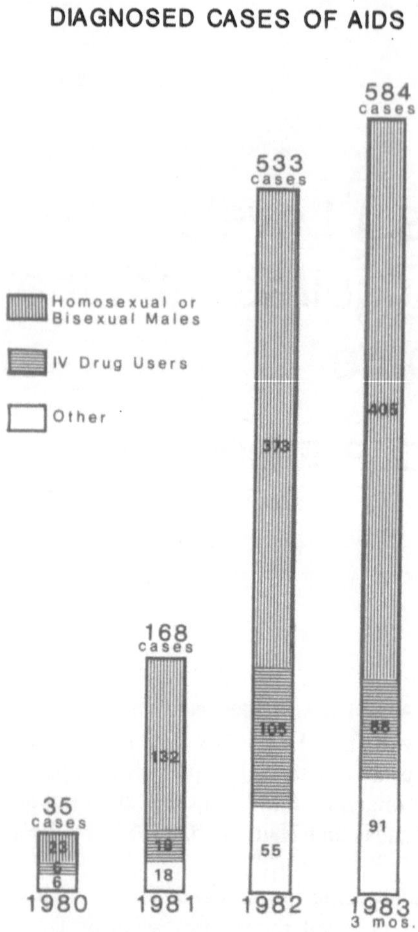

FIGURE 1. Epidemic incidence of AIDS in the U.S. (March 1983) as reported by the Centers for Disease Control.

AIDS appeared as a progressive crippling of the body's thymus-dependent immune system resulting in an exceedingly high mortality rate of 90% in patients having the disease for more than 2 years. The thymus secretes a family of polypeptides, called thymosins, which modulate the immune system (see Fig. 1 of Chapter 1). At least one thymosin peptide (thymosin α_1) is present in serum and can be measured by RIA (McClure *et al.*, 1981).

2. MATERIALS AND METHODS

2.1. Blood Samples

2.1.1. Normal Donor Sera

Blood samples were drawn daily from volunteer donors in the Washington, D.C., area by the Washington Red Cross. Freshly drawn blood was refrigerated immediately and delivered twice a day. Upon arrival, the individual samples were separated and the serum aliquoted and frozen at −70°C. The samples were generously provided by Drs. Paul McCurdy and Fred Darr (Red Cross, Washington, D.C.).

2.1.2. Patient Sera

Frozen serum from patients well-documented with AIDS was provided by Dr. Friedman-Kien (New York University Medical Center, New York, N.Y.).

2.1.3. High-Risk Homosexual Sera

Frozen serum samples were obtained from nonhospitalized individuals in Houston, Texas, and Denmark. Each set of samples included age-matched normal serum obtained in the same region. The serum from Houston was supplied by Dr. Evan Hersh (M. D. Anderson Tumor Institute, Houston). The serum from Denmark was provided by Dr. Robert Biggar (NIH, Bethesda, Md.).

2.2. Thymosin α_1 Assay

Thymosin α_1 levels in serum were determined using an RIA technique employing some modifications of the previously reported assay (McClure *et al.*, 1981). An aliquot of serum (100 μl) was incubated with ^{125}I-labeled *N*-Ac(Tyr^1)-thymosin α_1 and an antiserum raised in rabbits. The antigen in the system was KLH-conjugated *N*-Ac-thymosin α_1. After an incubation period of 24 hr, the antibody–antigen complexes were precipitated by an additional 18-hr incubation with goat anti-rabbit immunoglobulin.

3. RESULTS

A clear definition of normal levels of thymosin α_1 in the blood was obtained by assaying 150 male and female Red Cross blood donors. Significant differences in thymosin α_1 levels were not found to be associated with the age (aged 20–50 years) or the sex of the donors (in preparation). In the three experiments, only one donor (a 46-year-old male) had levels of thymosin α_1 above two standard deviations of the mean (Fig. 2). Concurrently, thymosin α_1 levels were assayed in normal individuals' sera received from various researchers over the past year and the results have been similar. These results demonstrate that healthy males and females rarely express levels of thymosin α_1 above two standard deviations of the mean.

In a study of 94 patients afflicted with AIDS or Kaposi's sarcoma, thymosin α_1 levels were shown to be elevated (Fig. 3). Sixty percent (44/72) of Kaposi's sarcoma patients expressed levels ranging from two standard deviations above the mean to three times the mean levels of the hormone population; 54% (12/22) of AIDS patients showed similar elevated levels.

Investigation of thymosin α_1 levels among populations at risk of contracting AIDS showed remarkably similar results (Biggar *et al.*, 1983; Hersh *et al.*, 1983; Naylor and Goldstein, 1983, 1984; Reuben *et al.*, 1983). Two populations composed

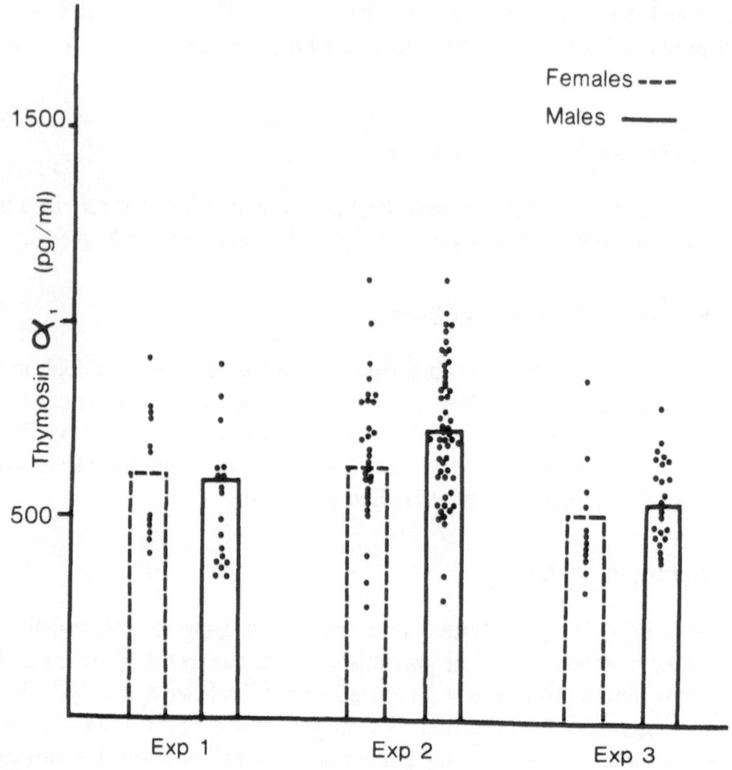

FIGURE 2. Thymosin α_1 levels in serum from normal male and female Red Cross volunteer donors measured by RIA.

of nonhospitalized homosexuals were used as "at risk" individuals in these studies. One set of sera from ambulatory homosexuals who visited a clinic in the Houston area showed 75% of this "at risk" population having elevated serum α_1 levels. In the second set—nonhospitalized homosexuals from Denmark—33% (25/75) of the individuals assayed expressed significantly elevated serum α_1 levels (Fig. 4). Helper (OKT4) T-cell to suppressor (OKT8) T-cell ratios were also investigated in this Danish population (Biggar *et al.*, 1983). No correlation was found between elevated α_1 levels and low or inverse helper to suppressor cell ratios (Fig. 5).

4. DISCUSSION

Thymic hormones influence the immune system by modulation of T-cell number and function. Significant deviations in thymic hormone levels should be considered indicative of a serious impairment of thymic function. As primary immune deficiencies have generally shown lower or normal thymosin α_1 levels, the expres-

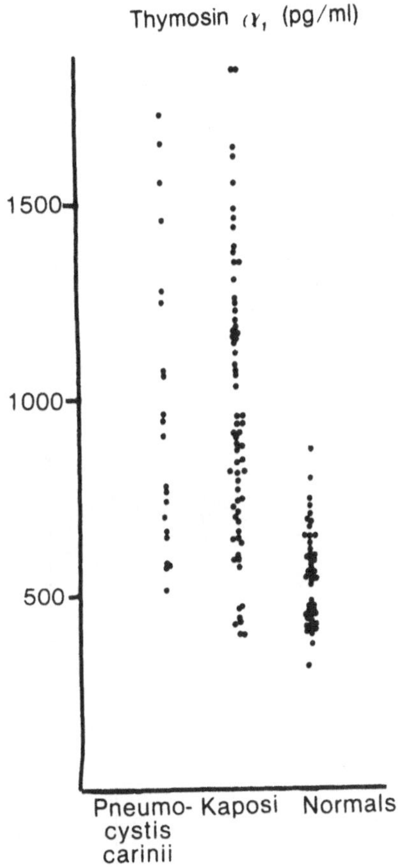

FIGURE 3. Thymosin α_1 levels in serum from AIDS patients with either Kaposi's sarcoma or *pneumocystis carinii* measured by RIA.

sion of significantly elevated levels in AIDS patients and populations at high risk of AIDS is very unusual. There are several possible mechanisms by which such high thymosin α_1 levels may occur. The end organ failure hypothesis suggests that a subpopulation of lymphocytes fails to respond to thymosin α_1 and the faulty feedback mechanism results in continued production of thymosin α_1. Accumulation of α_1 in this manner would result in elevated levels of biologically active hormone. A viral infection hypothesis suggests the thymosin α_1 produced may not be biologically active. A virus-infected thymic epithelium, or possibly a T-cell subpopulation could result in production of a modified peptide hormone. A cross-reactive pathogen which is transmissible may also be responsible for altering the normal structure and function of the thymosin α_1 peptide. It is similarly possible that elevated hormone levels may result from an activation of the authentic α_1 gene by an infectious agent in lymphocytes or other cell types. Non-thymus-mediated thymosin α_1 production could result from infection of a nonthymic tissue stimulating the production of thymosin α_1, either directly or by stimulation of production of a thymosin-releasing factor.

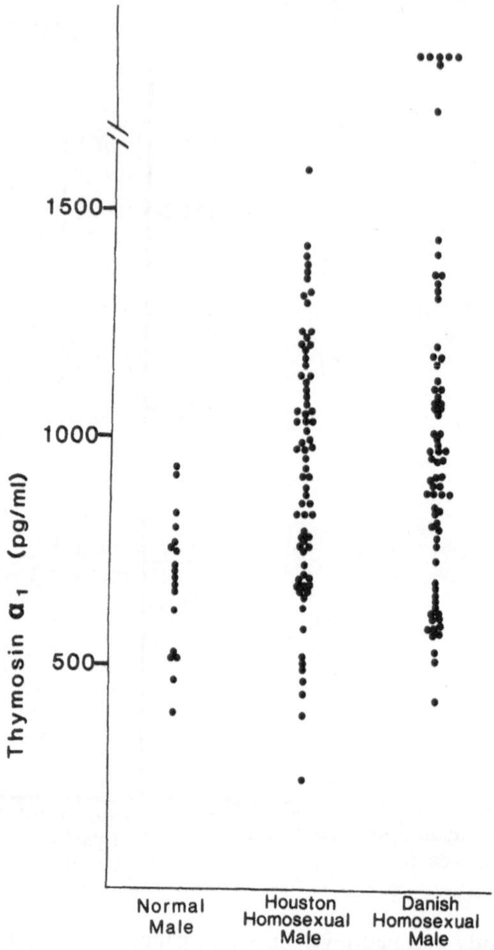

FIGURE 4. Thymosin α_1 levels in serum from "at risk" homosexuals from Denmark and USA (Houston).

Regardless of the mechanism, it is clear that homosexuals are at risk for AIDS and they have a different thymosin α_1 profile than heterosexuals. Coupled with the observations that thymic structure is abnormal in AIDS victims and that AIDS patients have elevated thymosin α_1, we suggest that elevated thymosin α_1 levels in serum may be early warning signal for AIDS. Continuing investigations will demonstrate whether or not elevated thymosin α_1 levels will be important to physicians as an early marker for AIDS. Hemophiliacs afflicted with AIDS presumably contract the disease either from direct transfusion or from the use of concentrated clotting factor. Thymosin α_1 was found in two of seven separate lots of lyophilized Factor VIII concentrates in amounts similar to serum. Factor IX concentrates measured showed large amounts of thymosin α_1 (Kessler *et al.*, 1983). It appears that AIDS is contracted by a bloodborne transmissible agent and may be preserved through the purification and concentration of blood products.

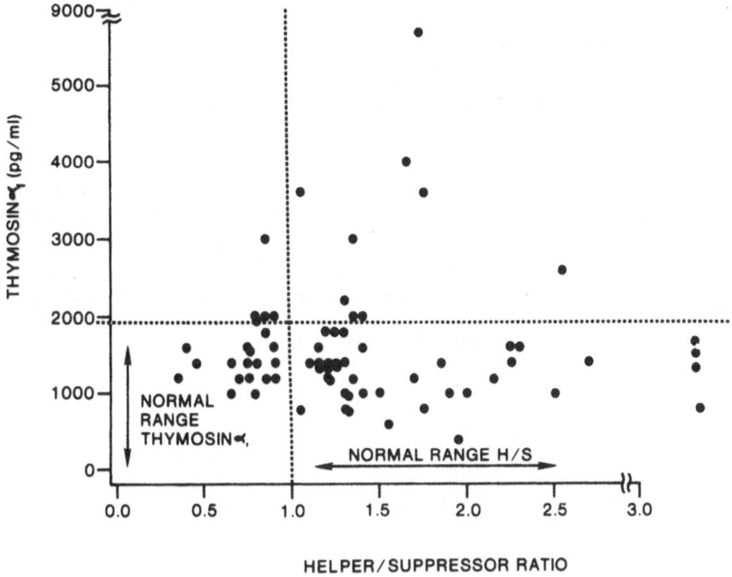

FIGURE 5. Elevated thymosin α_1 levels in serum as compared with helper/suppressor ratios in a group of Danish homosexual men.

REFERENCES

Biggar, R. J., Naylor, P. H., Goldstein, A. L., Melbye, M., Ebheson, P., Mann, D. L., and Strong, D. M., 1983, Thymosin α_1 levels and helper : suppressor ratios in homosexual males, *N. Engl. J. Med.* **309**:49.

Center for Disease Control Task Force on Kaposi's Sarcomas and Opportunistic Infections, 1982a, Epidemiologic aspects of the current outbreak of Kaposi's sarcoma and opportunistic infections, *N. Engl. J. Med.* **306**:248–252.

Center for Disease Control, 1982b, Update on acquired immune deficiency syndrome (AIDS), *U.S. Morbidity and Mortality Weekly Report* **31**:507–514.

Durack, D. T., 1981, Opportunistic infections and Kaposi's sarcoma in homosexual men, *N. Engl. J. Med.* **305**:1465–1467.

Elie, R., LaRoche, A. C., Arnoux, E., Guerin, J., Pierre, G., Malebranche, R., Seemayer, T. A., Dupuy, J., Russo, P., and Lapp, N., 1983, Thymic dysplasia in acquired immunodeficiency syndrome, *N. Engl. J. Med.* **305**:1425–1431.

Friedman-Kien, A. E., 1981, Disseminated Kaposi's sarcoma syndrome in young homosexual men, *J. Am. Acad. Dermatol.* **5**:468–471.

Gottlieb, M. S., Schroff, R., Shanker, N. M., Weisman, D. O., Fan, P. T., Wolf, R. A., and Saxon, A., 1981, *Pneumocystis carinii* pneumonia and mucosal candidiasis in previously healthy homosexual men: Evidence of a newly acquired cellular immunodeficiency, *N. Engl. J. Med.* **305**:1425–1431.

Hersh, E. M., Reuben, J. M., Rios, A., Mansell, P. N. A., Newell, G. R., McClure, J. E., and Goldstein, A. L., 1983, Elevated serum thymosin α_1 levels in male homosexuals with evidence of immune dysregulation in male homosexuals with a history of infectious diseases or Kaposi's sarcoma, *N. Engl. J. Med.* **308**:45–46.

Kessler, C. M., Schulof, R. S., Goldstein, A. L., Naylor, P. H., Luban, N., Kelleher, J. F., and Reaman, G. H., 1983, Abnormal T-lymphocyte subpopulations associated with transfusions of blood-derived products, *Lancet* **1**:991–992.

Levine, A. S., 1982, The epidemic of acquired immune dysfunction in homosexual men and its sequel—Opportunistic infections, Kaposi's sarcoma, and other malignancies in update and interpretation, *Cancer Treatment Rep.* **66**:1391–1395.

McClure, J. E., Lameris, N., Wara, D. W., and Goldstein, A. L., 1981, Immunochemical studies on thymosin: Radioimmunuoassay of thymosin α_1, *J. Immunol.*, **128**:368–375.

Naylor, P. H., and Goldstein, A. L., 1983, Abnormally elevated thymosin α_1 levels in acquired immunodeficiency syndrome (AIDS), *Clin. Lett.* **4**:126–128.

Naylor, P. H., and Goldstein, A. L., 1984, Elevated serum thymosin α_1 as an early marker for acquired immunodeficiency syndrome (AIDS), in: *AIDS: The Epidemic of Kaposi Sarcoma and Opportunistic Infections* (A. E. Friedman-Kien, ed.), pp. 173–180, Masson, New York.

Reuben, J. M., Hersh, E. M., Mansell, P. N., Newell, G., Rios, A., Rossen, R., Goldstein, A. L., and McClure, J. E., 1983, Immunological characteristics of homosexual males, *Cancer Res.* **43**:897–904.

West, S., 1983, One step behind a killer, *Science 83*, **4**:36–45.

7

Thymosin β_4

Distribution and Biosynthesis in Vertebrate Cells and Tissues

BERNARD L. HORECKER

1. INTRODUCTION

Thymosin β_4 is one of several peptides that have been isolated from thymosin fraction 5, a preparation from calf thymus whose properties have been discussed by Low and Goldstein (Chapter 1, this volume). Following the demonstration that thymosin fraction 5 was a mixture of many peptides, ranging in molecular weight from 1000 to 14,000 and in isoelectric point from 4.0 to 7.0 (Goldstein *et al.*, 1977; Low *et al.*, 1979) and the elegant isolation and structural analysis of several of these peptides by Goldstein and his co-workers (Goldstein *et al.*, 1977; Low and Goldstein, 1979; Low *et al.*, 1981), we became interested in the mechanism of their biochemical synthesis.

Our strategy was to translate calf thymus mRNA in an appropriate *in vitro* system (e.g., the wheat germ extract) and identify the biological precursors of the thymic peptides by immunoprecipitation and by characterization of peptide fragments containing amino acid sequences characteristic of the peptides isolated from thymosin fraction 5.

When this procedure was applied to studies of the biosynthesis of thymosin α_1, using an anti-thymosin fraction 6 serum provided by John E. McClure of The George Washington University School of Medicine and Allied Health Sciences, we succeeded in identifying a larger polypeptide, with a molecular weight of approximately 16,000, containing a 15-amino acid sequence present in thymosin α_1 (Freire *et al.*, 1978, 1981). However, our efforts to synthesize the cDNA corresponding to the thymosin α_1 mRNA and thereby deduce the sequence of the thymosin α_1 precursor were unsuccessful. We therefore concentrated our efforts on thymosin β_4, whose isolation and sequence had just been reported by Low *et al.* (1981).

BERNARD L. HORECKER • Roche Institute of Molecular Biology, Roche Research Center, Nutley, New Jersey 07110.

1. Frozen tissue (10-1500 mg) homogenized in 10 ml 6 M guanidinium chloride.
2. Diluted with 10 ml 0.2 M pyridine/1 M HCOOH, pH 2.8.
3. Centrifuged to remove lipids.
4. Desalted on Waters C18 Sep-Pak Cartridge.
   Peptides eluted with 5 ml 20% propanol in pyridine/HCOOH.
5. Eluate lyophilized and analyzed by HPLC (RP18 column).

FIGURE 1. Outline of procedure for isolation of peptides from fresh-frozen calf thymus. Tissue frozen in liquid nitrogen at the slaughterhouse was processed in small batches as indicated (from Hannappel *et al.*, 1982a).

2. ISOLATION OF PEPTIDES RELATED TO THYMOSIN β_4

Unlike thymosin α_1, which had been chemically synthesized (Wang *et al.*, 1979), thymosin β_4 was not available in sufficient quantities for addition as carrier or characterization of peptide fragments. We therefore undertook its isolation both from fraction 5 and from fresh calf thymus, using, in the latter case, a procedure designed to minimize the possibility of proteolytic modification (Hannappel *et al.*, 1982a). The procedure is briefly outlined in Fig. 1. The frozen tissue was cut into small segments at the slaughterhouse, quick-frozen in liquid nitrogen, and stored in liquid nitrogen until processed. To inactivate proteases, the still-frozen tissue was homogenized in 6 M guanidinium chloride, using a Polytron homogenizer. The homogenate was diluted with an equal volume of pyridine–formic acid, pH 2.8, and immediately desalted by forcing it through disposable SepPak filters, which absorbed the peptide but not the guanidinium salts. After washing the filters with the pyridine–formic acid buffer, the peptides were eluted with the same buffer containing 20% 1-propanol. For analytical purposes, the SepPak eluates were simply diluted with buffer and subjected to reverse-phase HPLC on RP18 columns. The entire analytical procedure required less than 2 hr. For the separation of larger quantities, it was necessary to concentrate the SepPak elutes by lyophilization before chromatography.

2.1. Peptides in Fresh Calf Thymus

When the procedure outlined in Fig. 1 was applied to fresh calf thymus, a major peptide peak emerged at a concentration of 1-propanol of about 24–28% (peak 2, Fig. 2). This region contained only a few other peaks; most of the material emerged as larger peptides (peaks 4–8), none of which was identified. Thymosin α_1, which would have emerged at 58 min (arrow), could not be detected. The peptide in peak 3 was found to be similar in composition to thymosin β_4, but lacked

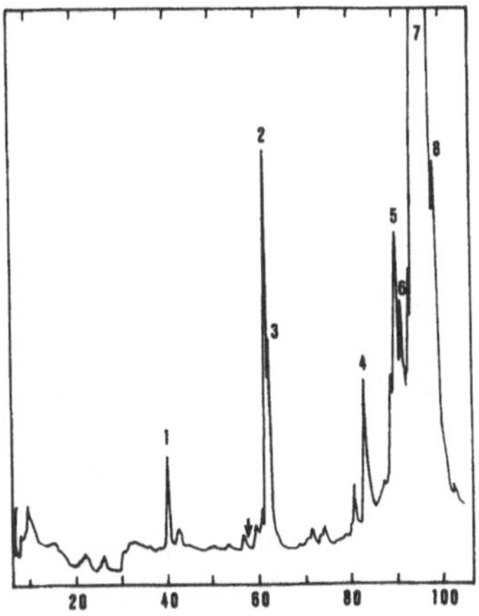

FIGURE 2. Reverse-phase HPLC of a desalted guanidinium chloride extract of fresh-frozen calf thymus. Elution was with a stepwise gradient of 0–40% 1-propanol, increasing in 4% increments every 10 min. In the experiment shown, the SepPaks were eluted with pyridine–formic acid containing 40% propanol, which accounts for the presence of the larger polypeptides (peaks 4–8). These polypeptides were not identified. (From Hannappel *et al.*, 1982a.)

methionine (see Fig. 5). This peptide could be separated from thymosin β_4 by oxidizing the mixture with H_2O_2 (Hannappel *et al.*, 1982a), which converted the methionyl residue in thymosin β_4 to methionyl sulfoxide; on rechromatography under the same conditions, the oxidized thymosin β_4 emerged 9 min earlier, well separated from the new peptide, designated thymosin β_9, whose elution position was not altered. Thymosin β_9 was found to contain 41 amino acids, 31 of which were identical to those found in thymosin β_4 (Hannappel *et al.*, 1982b). Thymosin β_9 was not detected in preparations of fraction 5; in its place we found a peptide containing the first 39 amino acid residues but lacking the terminal dipeptide, Ala-Lys. This modified form of thymosin β_9 was designated thymosin β_8. This finding, and the absence of thymosin α_1 in fresh calf thymus, supported the suggestion that fraction 5 may contain proteolytically modified forms of native thymic peptides.

2.2. Peptides in Rat and Mouse Tissues

When the same isolation procedure was applied to rat and mouse tissues, we were surprised to find that thymosin β_4 was present not only in the thymus, but also in substantial quantities in other tissues, including brain, spleen, and peritoneal macrophages (Fig. 3). Smaller quantities were found to be present in lung, liver, and kidney (Hannappel *et al.*, 1982c). Tissues from athymic (*nu/nu*) mice were found to contain higher concentrations of thymosin β_4 than tissues from heterozygous littermates. These observations indicated that tissues other than the thymus gland might be capable of synthesizing thymosin β_4.

In rat and mouse tissues, thymosin β_9 could not be detected. Instead, we found another peptide, designated thymosin β_{10}, that emerged approximately 5 min later than thymosin β_4 under the conditions employed. Thymosin β_{10} contained methionine at position 6, and showed 85% homology with thymosin β_9 (Erickson-Viitanen *et al.* 1983b) (see Fig. 5).

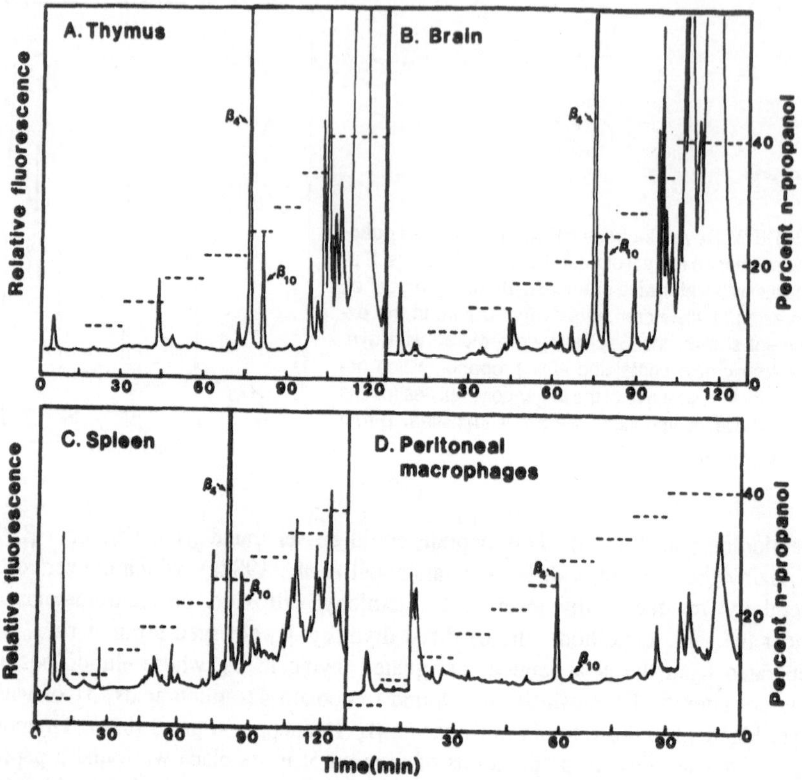

FIGURE 3. HPLC analysis of peptides in rat tissues. Quantities of tissues extracted were thymus, 0.15 g; brain 1.0 g; spleen, 1.8 g; peritoneal macrophages, 3 × 10^7 cells (from Hannappel *et al.*, 1982c).

2.3. Peptides in Tissues of Other Vertebrates

Thymosin β_4 was found to be present in tissues of all vertebrate classes except bony fish, which contained yet another related peptide, designated thymosin β_{11} (Figs. 4 and 5, Table I). In the spleens of all mammals examined, the content of thymosin β_4 ranged from 60 to 100 μg/g wet tissue, corresponding to nearly 0.1% of the total protein in this tissue. Similar high concentrations were found in the bursa of Fabricius of the chicken and in livers of a lizard and a toad, and somewhat similar concentrations of thymosin β_{11} were present in fish liver (Erickson-Viitanen *et al.*, 1983a).

2.4. Structures of β-Thymosins and Phylogenetic Distribution

A high degree of conservation was observed in the structures of the peptides present in vertebrate species ranging from fish to mammals (Fig. 5). Their chain length varied between 41 and 43 amino acid residues and 29 of these residues were invariant in all of the peptides examined.

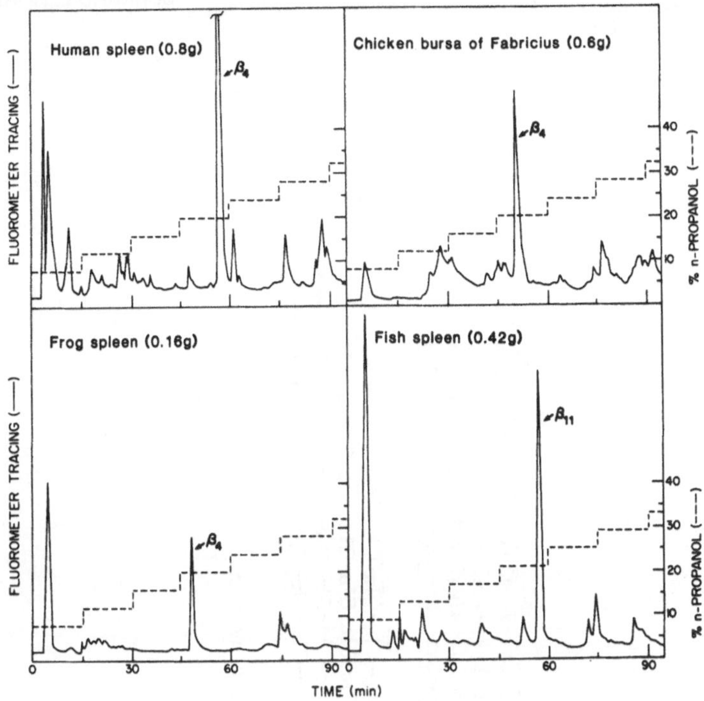

FIGURE 4. HPLC analysis of peptides in extracts of human, frog, and fish spleen and the chicken bursa of Fabricius. In these experiments the SepPaks were eluted with 20%, instead of 40%, 1-propanol, which accounts for the absence of the large peptides seen in Figs. 2 and 3. (From Erickson-Viitanen *et al.*, 1983a.)

TABLE I. Tissue Content of Thymosin β_4-like Peptides in Vertebrate Classes

	μg/g Wet weight tissue			
	β_4	β_9	β_{10}	β_{11}
Mammals (spleen)				
Calf	60	24	—	—
Cat	97	—	9	—
Man	68	—	11	—
Mouse	98	—	12	—
Rabbit	88	—	75	—
Rat	100	—	14	—
Birds (bursa of Fabricius)				
Chicken	83	—	—	—
Reptiles (liver)				
Gecko	27	—	—	—
Amphibians (liver)				
Frog	67	—	—	—
Bony fishes (liver)				
Oscar	—	—	—	35
Trout	—	—	—	24

```
                  5                      10                  15                20
β4   AcSer-Asp-Lys-Pro-Asp-Met-Ala-Glu-Ile-Glu-Lys-Phe-Asp-Lys-Ser-Lys-Leu-Lys-Lys-Thr-
β9   AcAla-Asp-Lys-Pro-Asp-Leu-Gly-Glu-Ile-Asn-Ser-Phe-Asp-Lys-Ala-Lys-Leu-Lys-Lys-Thr-
β10   xAla-Asp-Lys-Pro-Asp-Met-Gly-Glu-Ile-Ala-Ser-Phe-Asp-Lys-Ala-Lys-Leu-Lys-Lys-Thr-
β11   xSer-Asp-Lys-Pro-Asp-Leu-Glu-Glu-Val-Ala-Ser-Phe-Asp-Lys-Thr-Lys-Leu-Lys-Lys-Thr-

                     25                  30              35                  40
β4   Glu-Thr-Gln-Glu-Lys-Asn-Pro-Leu-Pro-Ser-Lys-Glu-Thr-Ile-Glu-Gln-Glu-Lys-Gln-Ala-Gly-Glu-SerOH
β9   Glu-Thr-Gln-Glu-Lys-Asn-Thr-Leu-Pro-Thr-Lys-Glu-Thr-Ile-Glu-Gln-Glu-Lys-Gln-Ala-LysOH
β10  Glu-Thr-Gln-Glu-Lys-Asn-Thr-Leu-Pro-Thr-Lys-Glu-Thr-Ile-Glu-Gln-Glu-Lys-Ser-Glu-Ile-SerOH
β11  Glu-Thr-Gln-Glu-Lys-Asn-Pro-Leu-Pro-Thr-Lys-Glu-Thr-Ile-Glu-Gln-Glu-Lys-Gln-Ala-SerOH
```

FIGURE 5. Amino acid sequences of β-thymosins.

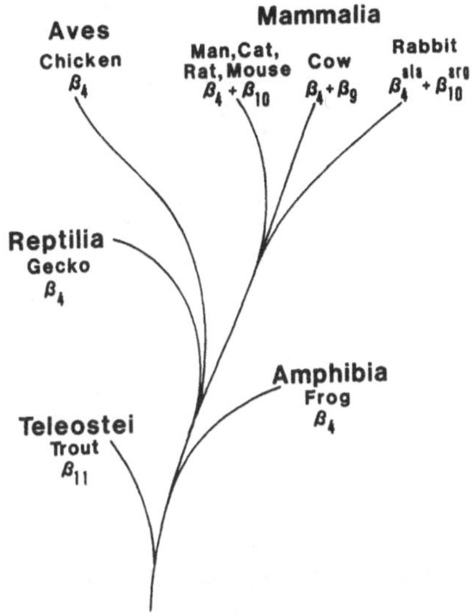

FIGURE 6. Phylogenetic distribution of β-thymosins. (Constructed from analyses reported by Erickson Viitanen *et al.*, 1983b, and Ruggieri *et al.*, 1983.)

The phylogenetic distributions of the β-thymosins is illustrated in Fig. 6. Thymosin β_4, the most widely distributed member of this family of peptides, has been found in all vertebrate classes except bony fish, which contain thymosin β_{11}. In mammals, but not in lower vertebrates, a second homologous peptide occurs together with thymosin β_4. This second peptide is thymosin β_9 in the calf (related forms have not yet been examined) or thymosin β_{10} in other species of mammals.

A notable exception to the above generalization is the rabbit, which contains variant forms of both thymosin β_4 and thymosin β_{10} (Ruggieri *et al.*, 1983). The variant form of thymosin β_4, designated thymosin β_4^{Ala}, contains alanine in place of serine in position 1. The variant of thymosin β_{10}, designated thymosin β_{10}^{Arg}, contains an additional arginine residue inserted between residues 38 and 39 of thymosin β_{10}. Preliminary evidence (Erickson-Viitanen, unpublished) suggests that thymosin β_{10}^{Arg} may also be present as a third β-thymosin in tissues of other mammals.

The quantities of thymosin β_{10} in mammalian tissues are generally 10–30% of those of thymosin β_4, except in rabbit spleen where thymosin β_{10}^{Arg} and thymosin β_4^{Ala} are present in almost equal quantities (Ruggieri *et al.*, 1983).

3. BIOSYNTHESIS OF THYMOSIN β_4

3.1. Synthesis in Isolated Peritoneal Macrophages and in Established Cell Lines

In isolated peritoneal macrophages and adherent spleen cells, thymosin β_4 is rapidly synthesized from radioactive methionine added to the culture medium (Xu *et al.*, 1982). The radioactive peptide, isolated from the cell pellet after extraction with 6 M guanidinium chloride and separation by HPLC, was identified by its

TABLE II. Synthesis of β_4 in Rat and Mouse Cells

	cpm/dish	Peak height
Cell lines		
Rat myoblast (L6)	105,000	48
Rat neuroblastoma (B104)	61,500	31
Mouse thymoma (BW5147)	22,000	14
Rat pheochromocytoma (PC12)	18,000	8
Mouse erythroleukemia (DS19)	—	(no β_4)
Mouse myeloma (P3X63)	—	(no β_4)
Mouse myeloma (NS1)	—	(no β_4)
Primary cultures		
Rat peritoneal macrophages	95,000	19
Thymocytes, nonattached	—	(no β_4)
Thymocytes	1,000	2.5

coelution with authentic thymosin β_4 and by the identification of a radioactive heptapeptide (residues 6–12) after partial acid hydrolysis.

Thymosin β_4 was also actively synthesized by a variety of established cell lines, including myoblasts and macrophages (Goodall *et al.*, 1983). The rates of synthesis were correlated with the content of thymosin β_4 in these cell lines (Table II). Both rates of synthesis and content appeared to be related to the ability of these cells to adhere and their motility. This prompted the suggestion that thymosin β_4 might function as a component of the cytoskeleton. It was not synthesized by several tumor cell lines, and was not detected in these cells, but relatively high concentrations and rates of synthesis were observed with a neuroblastoma cell line.

3.2. Cell-Free Synthesis of Thymosin β_4

These experiments were similar to those described earlier for thymosin α_1 and were expected to yield evidence for a larger precursor polypeptide from which thymosin β_4 would be formed by proteolytic processing. The protocol for the experiment (Filipowicz and Horecker, 1983) is outlined in Fig. 7. Rat spleen was chosen as the source of mRNA because of its convenience and because rat spleen had been found to be a particularly rich source of thymosin β_4 (see Table I).

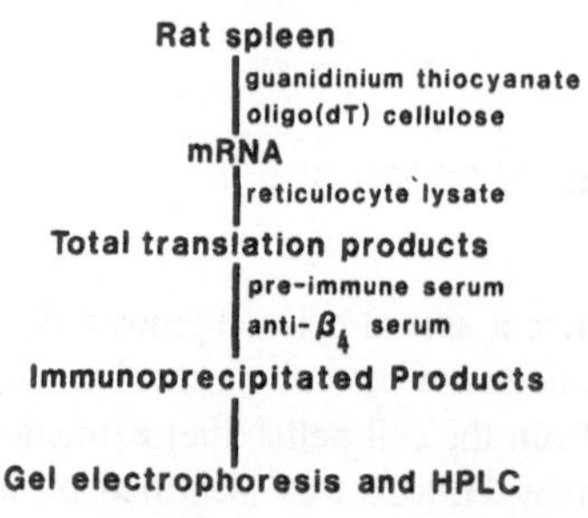

FIGURE 7. Protocol for studies of the cell-free synthesis of thymosin β_4.

The cell-free synthesis product, immunoprecipitated with a specific rabbit antithymosin β_4 γ-globulin provided by Dr. J. I. Morgan of this Institute, was found to be identical to thymosin β_4, with no indication for the formation of a larger precursor polypeptide (Fig. 8). Thymosin β_4 labeled with [^{35}S]methionine formed from rat spleen mRNA in a protein-synthesizing system from rabbit reticulocytes was identified by: (1) its identical behavior in SDS–PAGE and its absence in the antibody precipitate when an excess of authentic thymosin β_4 was added (Fig. 8); (2) its coelution with authentic thymosin β_4 in reverse-phase HPLC (Fig. 9); and (3) the identity of methionine-containing peptides generated from authentic thymosin β_4 by mild acid hydrolysis (residues 6–12, see Fig. 5) or digestion with chymotrypsin (residues 1–12). This last evidence confirms that the cell-free synthesis product possesses the acetylated NH_2 terminus.

In order to exclude the possibility of specific processing of a larger polypeptide precursor by proteinases in the rabbit reticulocyte system, the rat spleen mRNA was also translated in two nonmammalian systems, a wheat germ extract (Fig. 8) and a yeast protein-synthesizing system. It was considered that these would be

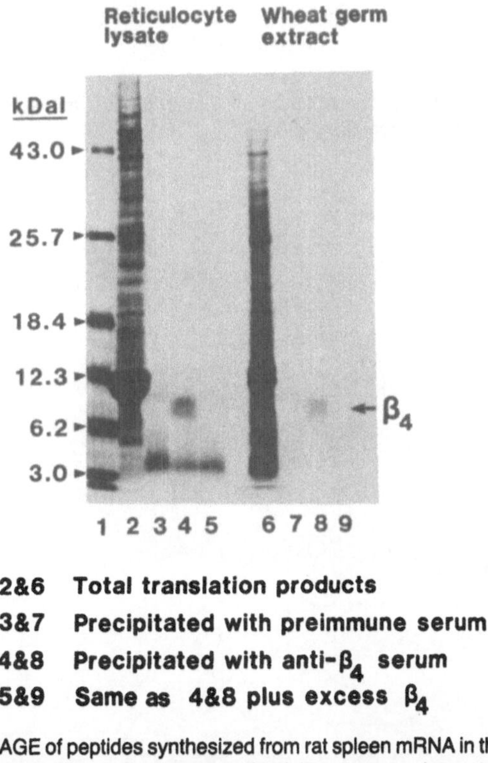

FIGURE 8. SDS–PAGE of peptides synthesized from rat spleen mRNA in the reticulocyte (lanes 1–5) and wheat germ (lanes 6–9) protein-synthesizing systems. Lane 1, ^{14}C-labeled protein markers. Lanes 2 and 6, total translation products. Lanes 3 and 7, peptides precipitated with preimmune serum. Lanes 4 and 8, peptides precipitated with anti-β_4 serum. Lanes 5 and 9, same as 4 and 8 except that nonradioactive thymosin β_4 was added before addition of antibody. (From Filipowicz and Horecker, 1983.)

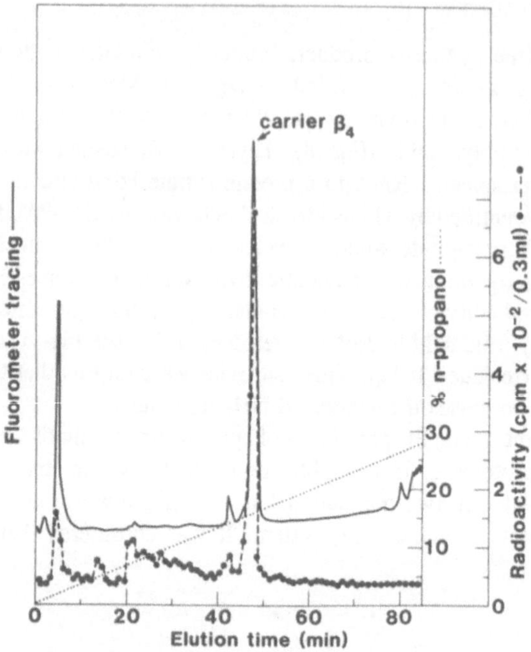

FIGURE 9. Coelution of immunoprecipitated translation product and authentic thymosin β_4 in HPLC. (From Filipowicz and Horecker, 1983.)

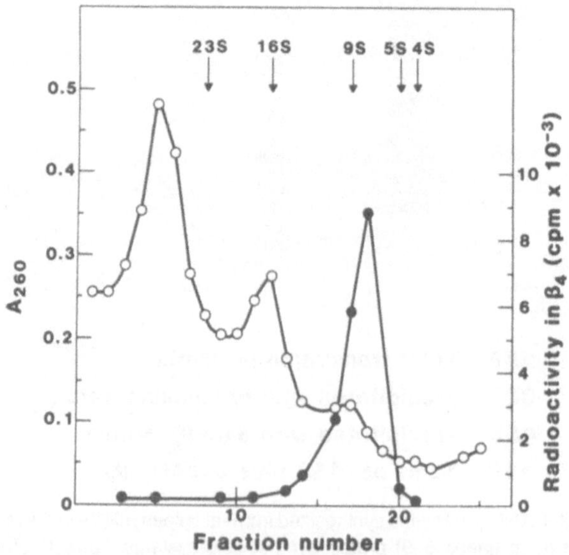

FIGURE 10. Sucrose density gradient analysis of rat spleen poly(A) RNA. RNA purified on an oligo(dT) cellulose column was sedimented through a 15–40% sucrose gradient and the samples, collected from the bottom of the tube, were analyzed in the reticulocyte lysate translation system. (From Filipowicz and Horecker, 1983.)

unlikely to contain specific processing enzymes, as thymosin β_4 has been found only in vertebrate tissues. The immunoprecipitated product formed with these cell-free protein-synthesizing systems was also characterized as thymosin β_4, although with the wheat germ system the NH_2 terminus was only partially acetylated.

3.3. Characterization of the mRNA Coding for Thymosin β_4

The synthesis of a peptide containing as few as 43 amino acid residues without formation of a larger precursor is a rare event in eukaryotes. It has been demonstrated only for the protamines, polypeptides containing 31 amino acid residues present in fish sperm. Protamine mRNA has been shown to be small, sedimenting in the 4–6 S RNA region. The mRNA coding for thymosin β_4 belongs to a somewhat larger size class, sedimenting with the 7–8 S RNA fraction (Fig. 10). The relatively small size of the thymosin β_4 mRNA is consistent with the chain length of the thymosin β_4 molecule.

4. CONCLUSIONS

The evidence presented here suggests that thymosin β_4 is unlikely to function solely or principally as a thymic hormone. Its ubiquitous distribution in vertebrate tissues and its high concentrations, approaching 0.1% of the total protein, suggest that it plays some more general role, possibly as a component of the cytoskeletal system. Six members of the β-thymosin family of peptides have now been identified. Thymosin β_4 is found in vertebrate classes from amphibians to mammals. Mammals contain a second related peptide, thymosin β_9 in calf and thymosin β_{10} in other species. One exception is the rabbit, which contains a variant of thymosin β_4, designated thymosin β_4^{Ala}, and also a variant of thymosin β_{10}, designated thymosin β_{10}^{Arg}. The related peptide found in bony fish, thymosin β_{11}, may be the most primitive form.

Thymosin β_4 and thymosin β_{10} are synthesized by a variety of cells and do not necessarily originate in the thymus gland. They are found in high concentration in tissues of athymic (*nu/nu*) mice. Thymosin β_4 is synthesized directly from a relatively small mRNA, without formation of a larger polypeptide precursor. This mechanism of synthesis, similar to that reported for protamines, is consistent with a function as a structural element, and is not shared by any of the known polypeptide hormones.

Although the function of the β-thymosin family of peptides remains unknown, their wide distribution, high tissue concentrations, and remarkable conservation of primary structure suggest that they play a vital role in the physiology of the cell.

ACKNOWLEDGMENTS. The author gratefully acknowledges the hard work and major contributions of the following colleagues: Ewald Hannappel developed the iso-electric focusing and HPLC procedures for the isolation of peptides from thymosin fraction 5 and from fresh calf thymus and was the first to isolate thymosins β_8 and β_9 and establish their structures. Gen-jun Xu, Ewald Hannappel, James Hempstead, and Gregory Goodall showed that thymosin β_4 was widely distributed in rat and mouse tissues and cells and that a variety of cell types were capable of synthesizing thymosin β_4. Susan Erickson-Viitanen, with Silverio Ruggieri and Paolo Natalini, established the wide distribution of thymosin β_4 in the vertebrate kingdom and established the structures of the new variants, thymosins β_{10}, β_{11}, β_4^{Ala}, and β_{10}^{Arg}, and their distributions in vertebrate species and classes. Aleksandra Wodnar Filipowicz demonstrated that thymosin β_4 was synthesized directly from a relatively small mRNA species.

REFERENCES

Erickson-Viitanen, S., Ruggieri, S., Natalini, P., and Horecker, B. L., 1983a, Distribution of Thymosin β_4 in vertebrate classes, *Arch. Biochem. Biophys.* **221**:570–576.

Erickson-Viitanen, S., Ruggieri, S., Natalini, P., and Horecker, B. L., 1983b, Thymosin β_{10}, a new analogue of Thymosin β_4 in mammalian tissues, *Arch. Biochem. Biophys. 225*:407–413.

lipowicz, A. W., and Horecker, B. L., 1983, *In vitro* synthesis of Thymosin β_4 encoded by rat spleen mRNA, *Proc. Natl. Acad. Sci. USA* **80**:1811–1815.

Freire, M., Crivellaro, O., Isaacs, C., and Horecker, B. L., 1978, Translation of mRNA from calf thymus in the wheat germ system: Evidence for a precursor of Thymosin α_1, *Proc. Natl. Acad. Sci. USA* **75**:6007–6011.

Freire, M., Hannappel, E., Rey, M., Freire, J. M., Kido, H., and Horecker, B. L., 1981, Purification of thymus mRNA coding for a 16,000 dalton polypeptide containing the thymosin α_1 sequence, *Proc. Natl. Acad. Sci. USA* **78**:192–195.

Goldstein, A. L., Low, T. L. K., McAdoo, M., McClure, J., Thurman, J. B., Rossio, J., Lai, C.-Y., Chang, D., Wang, S. S., Harvey, C., Ramel, A. H., and Meienhofer, Jr., 1977, Thymosin α_1: Isolation and sequence analysis of an immunologically active thymic polypeptide, *Proc. Natl. Acad. Sci. USA* **74**:725–729.

Goodall, G. J., Morgan, J. I., and Horecker, B. L., 1983, Thymosin β_4 in cultured mammalian cell lines, *Arch. Biochem. Biophys.* **221**:598–601.

Hannappel, E., Davoust, S., and Horecker, B. L., 1982a, Isolation of peptides from calf thymus, *Biochem. Biophys. Res. Communs.* **104**:266–271.

Hannappel, E., Davoust, S., and Horecker, B. L., 1982b, Thymosins β_8 and β_9, two new peptides isolated from calf thymus homologous to Thymosin β_4, *Proc. Natl. Acad. Sci. USA* **79**:1708–1711.

Hannappel, E., Xu, G.-J., Morgan, J., Hempstead, J., and Horecker, B. L., 1982c, Thymosin β_4—a ubiquitous peptide in rat and mouse tissues, *Proc. Natl. Acad. Sci.* **79**:2172–2175.

Low, T. L. K., and Goldstein, A. L., 1979, The chemistry and biology of Thymosin II, *J. Biol. Chem.* **254**:987–995.

Low, T. L. K., Thurman, G. B., McAdoo, M., McClure, J., Rossio, J. L., Naylor, P. H., and Goldstein, A. L., 1979, The chemistry and biology of Thymosin I, *J. Biol. Chem.* **254**:981–986.

Low, T. L. K., Hu, S.-K., and Goldstein, A. L., 1981, Complete amino acid sequence of bovine thymosin β_4: A thymic hormone that induces terminal deoxynucleotidyl transferase activity in thymocyte populations, *Proc. Natl. Acad. Sci. USA* **78**:1162–1166.

Ruggieri, S., Erickson-Viitanen, S., and Horecker, B. L., 1983, Thymosin β_{10}^{Arg}, a major variant of Thymosin β_{10} in rabbit tissues, *Arch. Biochem. Biophys.* **226**:388–392.

Wang, S. S., Kuleska, I. D., and Winter, D. P., 1979, Synthesis of Thymosin α_1, *J. Am. Chem. Soc.* **101**:253–254.

Xu, G.-J., Hannappel, E., Morgan, J., Hempstead, J., and Horecker, B. L., 1982, Synthesis of "Thymosin" β_4 by peritoneal macrophages and adherent spleen cells, *Proc. Natl. Acad. Sci. USA* **79**:4006–4009.

8

The Finding and Significance of Spermidine and Spermine in Thymic Tissue and Extracts

KARL FOLKERS and HONG-MING SHIEH

1. INTRODUCTION

Several years ago, I (K.F.) listened to a lecture by Allan Goldstein on the thymic hormones, and he inspired me to initiate research in this field. I am pleased to thank and acknowledge Allan for his inspiration and leadership in the complicated research on the thymus and the immune system. After my initiation of efforts to isolate thymic hormones, I was also significantly influenced by the research of Nathan Trainin and his group in this field.

To isolate a thymic hormone, one needs a biological assay to guide the multiple fractionations. Kook and Trainin (1974) had reported on the control of cellular cAMP levels which was observed by fractions of their THF, and they believed that THF is a hormone with an action mediated by cAMP. On this basis, we chose the stimulation of cAMP as one *in vitro* assay which could guide our chemical fractionation to a thymic hormone.

2. GLUTATHIONE

We extracted bovine thymus and extensively fractionated the extracts as guided by a cAMP assay. The isolation steps led us to a peptide which was unexpectedly identified as glutathione (Folkers *et al.*, 1980a) which is a very well-known tripeptide:

KARL FOLKERS and HONG-MING SHIEH • Institute for Biomedical Research, The University of Texas, Austin, Texas 78712.

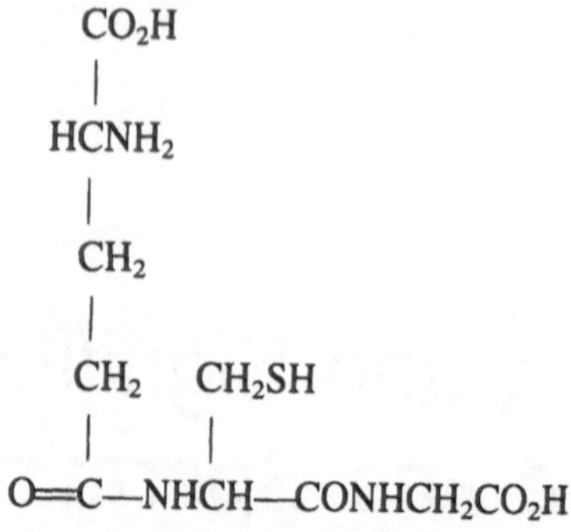

The isolated and purchased glutathiones showed unambiguous stimulation of cAMP at 1, 10, 20, 40, and 100 μg/ml with a significance of $p < 0.05$–0.01. Glutathione showed no activity in the mixed lymphocyte assay, but did show activity in the T-rosette assay.

It was credible that glutathione might be present in Allan Goldstein's fraction 5, and in the fraction which Kook, Yakir, and Trainin had designated as "CTO" (Rotter and Trainin, 1979). The possible presence of glutathione in fraction 5 and "CTO" seemed important, because both of these crude materials had been clinically investigated.

Glutathione is known to be a coenzyme. It was proposed that glutathione may function in transport of amino acids across membranes (Miura and Metzler, 1976). Such transport and the reported essentiality of L-alanine (Rotter *et al.*, 1979) for human leukocytes to respond to mitogenic and allogeneic stimulation indicated that glutathione might be functional in the complex immune systems.

3. A NEW ASSAY BASED ON LYMPHOCYTE STIMULATION (THE "PL" ASSAY)

After the unexpected isolation of glutathione as based on an assay lacking sufficient specificity, we sought a new *in vitro* method to guide our continuing fractionation. The use of spleen cells (T and B lymphocytes) from neonatally thymectomized mice appeared fundamental for our purpose. The measurement of the incorporation of [^{3}H]thymidine into DNA allowed quantitative measurements. The use of Con A for preliminary mitogenic stimulation of the spleen cells appeared to be undesirable, because the known chemical heterogeneity of Con A could lessen quantitation. If a thymic hormone directly stimulated the incorporation of [^{3}H]thymidine into DNA by spleen cells, this method might provide an excellent quantitative assay after standardization of all the steps. There was a background in this field for this method, as exemplified by the following four citations: Trainin *et al.* (1975), Houck *et al.* (1971), Caspary and Hughes (1972), and Kiger *et al.* (1977).

We found that direct stimulation of the incorporation of [^{3}H]thymidine into DNA by the immunoincompetent lymphocytes of spleen cells from neonatally thymectomized mice actually provided a reliable, quantitative, *in vitro* assay which allowed convincing standardization. We used this assay to monitor hundreds of fractions and achieved isolations of thymones A and B (Folkers *et al.*, 1980b,c) and a high purification of thymone C (Folkers *et al.*, 1981).

4. ASSAY OF KEY SAMPLES

We assayed some key materials in this field:

1. Thymosin fraction 5 (kindly provided by Drs. Goldstein and Low) was inactive at levels of 1–250 μg/ml.
2. Synthetic thymosin α_1 was inactive at levels of 0.001–50 μg/ml.
3. Synthetic FTS (thymulin) was inactive at levels from 10^{-6} to 50 μg/ml.
4. Glutathione was inactive at levels of 0.3–30 nM.
5. Of particular interest, "CTO," which we prepared as described in a publication by Rotter and Trainin (1979), did show activity at 10 μg/ml ($p < 0.05$) and at 50 μg/ml ($p < 0.001$). Our assay was published in 1982 (Stepien *et al.*).

5. THYMONE A

Bovine thymus was our source tissue and we used the assay for the proliferation of lymphocytes. Diverse sequences and series of steps were explored during the fractionation toward the isolation of one or more active factors or hormones.

After eight steps of fractionation, we isolated a preparation of thymone A (Folkers *et al.*, 1980b). This preparation was identified by single spots in three electrophoretic systems. The preparation was active at a range of 10–100 ng, and it showed activity in the cAMP assay at 1 μg. Trypsin destroyed the activity of the preparation, and amino acid analysis revealed up to 14 amino acids: Asp, Glu, Gly, Ala, Val, Ile, Leu, Pro, Ser, Thr, Met, Lys, Arg, His. This analysis indicated a total of 68–71 amino acids which would have a molecular weight of 7291–7677. The molecular weight estimated using Bio-Gel P-6 was about 6000.

We concluded that thymone A may be a functional hormone, or it may be a prohormone which has activity. The chemical and biological properties of thymone A differentiated it from other peptides, purified or isolated, as described by other investigators in this field.

6. THYMONE B

During the multiple fractionations which led to thymone A, an activity in other fractions, which were widely separated from those containing thymone A, was increasingly observed. These fractions ultimately yielded a different substance in essentially pure form which was designated thymone B (Folkers *et al.*, 1980c).

Thymone B was present in lower concentrations than was thymone A, and was correspondingly more difficult to isolate in a pure state, and the paucity of the substance greatly restricted the initial chemical and biological determination of its properties. Thymone B showed essentially single spots by TLC in three solvent systems and a single spot by electrophoresis in one system. This preparation of thymone B appeared to be essentially pure.

Thymone B showed activity at 100 ng and at 100 μg in the assay for the proliferation of lymphocytes, and showed activity in an assay for cGMP, but not in an assay for cAMP. Trypsin destroyed the activity of thymone B. The preparation contained up to 13 amino acids: Asp, Glu, Gly, Ala, Val, Ile, Leu, Ser, Pro, Thr, His, Lys, Arg. The molecular weight appeared to be considerably less than the value of about 6000 for thymone A.

Repetition of our chemical fractionations and bioassays which revealed thymones A and B also revealed the presence of another biologically active entity which was designated thymone C.

7. THYMONE C

Thymone C was obtained as a highly purified fraction after about 10 steps of fractionation. This preparation of thymone C showed activity in the assay for cGMP and for cAMP. The preparation was active in the mixed lymphocyte assay.

By this time, other investigators had described fractions of biologically active entities from thymic tissue, lymphocytes, and other tissues, including Baker *et al.* (1980) on a T-cell growth factor; Dabrowski *et al.* (1980) on a calf thymic hormone; Lenfant *et al.* (1980) on a spleen-derived immunosuppressive peptide; and Söder and Ernström (1980) on a thymocyte-specific growth factor.

The chemical nature of thymone C can be related to or differentiated from these cited and uncited studies by other investigators when thymone C is finally characterized.

8. SPERMINE AND SPERMIDINE

During these fractionations which yielded thymones A, B, and C, some fractions were obtained which were strongly inhibitory and showed a dose–response relationship. An inhibitory activity appeared to conflict with the stimulatory activity which had been observed in the assay on the proliferation of lymphocytes. This conflict between inhibition and stimulation led to the design of fractionations to isolate the inhibitory substance. Ultimately, the inhibitory substance was found to be at least dual and we identified spermine and spermidine, apparently for the first time from thymic tissue.

Spermine and spermidine are well-known polyamines. Pure and authentic spermine was active in the assay at 10 μg/ml. Spermidine was active at 1 μg/ml. N^1-acetylspermidine [$CH_3CONH(CH_2)_3NH(CH_2)_4NH_2$] was active at 10 μg/ml and

N^8-acetylspermidine [$NH_2(CH_2)_3NH(CH_2)_4NHCOCH_3$] was active at 1 μg/ml. Putrescine [$NH_2(CH_2)_4NH_2$], agmatine [$NH_2C(NH)NH(CH_2)_4NH_2$], and acetylputrescine [$CH_3CONH(CH_2)_4NH_2$] were inactive.

The conclusive organic chemical evidence for the presence of spermine and spermidine in thymic extracts raises new questions on the possible importance of these polyamines to mechanisms of immunoregulation. Mitogen activation of polyamine synthesis in human T lymphocytes has been described. Polyamine synthesis is believed to occur early in lymphocyte activation. Investigators in this field may not often or ever have considered the participation of polyamines and their conjugates. I wonder if there is a relationship between these amines and conjugates and the studies by Djerassi, and by Blazsek *et al.*, according to their presentations in this symposium.

9. POLYAMINES IN THYMOSIN FRACTION 5

Fraction 5 of Allan Goldstein and his co-workers has been clinically studied rather extensively, and we thought it might be important to know whether spermine and spermidine were constituents of fraction 5. We had observed during the development of our *in vitro* assay that levels of fraction 5 up to 250 μg/ml did not stimulate lymphocyte proliferation, but at a level of 500 μg/ml, fraction 5 was inhibitory, and such inhibition indicated the possible presence of spermine and spermidine in fraction 5. By a new analysis, we found that fraction 5 contained 0.013% of spermidine and 0.11% of spermine (Kubiak *et al.*, 1982). We had to devise an assay for these polyamines, and we based the assay on the tri- and tetradansyl derivatives of spermidine and spermine, respectively.

Glutathionyl-spermidine occurs in *E. coli* (Tabor and Tabor, 1975) and there is a diverse literature on the probable existence of other peptidic derivatives of these polyamines. Consequently, we subjected fraction 5 to acid hydrolysis after receiving more gifts of this fraction from Dr. Allan Goldstein and Dr. Teresa Low. We found that 50% of the spermidine was apparently present as a "covalently linked conjugate."

10. TWO PERSPECTIVES

At this stage in our research, two perspectives became apparent which could be of clinical importance. These are:

1. The presence of spermine and spermidine and a "covalently linked conjugate" of spermidine in fraction 5 may *diminish overriding stimulating effects* of this fraction on T cells in immunocompetence in patients.
2. A fraction free of polyamines and their conjugates should be tested in patients to enhance immunocompetence.

11. SYNTHESIS OF MODEL PEPTIDYL-POLYAMINES

According to our search of the literature, synthesis of peptidic derivatives of spermine and spermidine seems never to have been achieved. Consequently, we are now synthesizing peptidyl-polyamines.

I thought we should select one of the thymic peptides of known sequence rather than just any peptide to link to spermine. Bach's FTS, or thymulin:

pGln-Ala-Lys-Ser,Gln,Gly,Gly,Ser,Asn-OH

which is a known nonapeptide, appeared to be ideal for our first synthesis, and we have apparently successfully synthesized the nonapeptidyl-spermine having the structure:

pGln-Ala-Lys-Ser,Gln,Gly,Gly,Ser,Asn-$NH(CH_2)_3NH(CH_2)_4NH(CH_2)_3NH_2$

Next, we chose thymosin β_4, which has 43 amino acids, because it is considerably larger than FTS and because of the ubiquitous occurrence of thymosin β_4, as reported by Hannappel *et al.* (1982).

Presently, we are conducting synthetic reactions to achieve the covalent bond between the C-terminal of thymosin β_4 and one amino group of spermine.

We are also testing for an effect of the nonapeptidyl derivative of spermine in the assay for the proliferation of lymphocytes. Before the test, we considered that the peptide would more likely be inhibitory than stimulatory, and such seems to be the case, but I do not yet have final data.

In conclusion, I was seeking stimulatory thymic substances at the start of this research. Our finding the inhibitory polyamines was unexpected. Spermine and spermidine and their naturally occurring conjugates in the thymus may be essential to immunoregulation.

12. SUMMARY

Bovine thymus was fractionated toward isolation of thymic hormones. Guidance by a cAMP assay unexpectedly led to the isolation of glutathione. The presence of glutathione in thymic extracts, its known role as a coenzyme, and its function in transport of amino acids across membranes indicate that glutathione could be functional in immune systems. Using a new *in vitro* "PL" assay based on the incorporation of [^{3}H]thymidine into DNA by spleen cells from neonatally thymectomized mice, led to the isolation of thymones A and B and the high purification of thymone C. The PL assay not only revealed these stimulatory thymones, but potent inhibitory substances which were isolated and identified as spermine and spermidine. These inhibitory substances in the thymus may have growth-controlling roles in mechanisms of immunoregulation. The presence of both stimulatory and inhibitory substances in the thymus indicates possible interrelated growth-promoting and growth-limiting mechanisms of immunoregulation.

ACKNOWLEDGMENT. Appreciation is expressed to the Robert A. Welch Foundation for their partial support of this research.

REFERENCES

Baker, P. E., Brooks, P. L., and Smith, K. A., 1980, T-cell growth factor therapy in athymic mice, *Fourth International Congress of Immunology*, Abstract 17.2.02.

Caspary, E. A., and Hughes, D., 1972, Liquid scintillation counting techniques in lymphocyte transformation *in vitro* measured by tritiated thymidine uptake, *J. Immunol. Methods* **1**:263–272.

Dabrowski, M. P., Dabrowska, B. K., Babiuch, L., and Brzoski, W. J., 1980, Calf thymic hormone (TFX) driven maturation of human T-lymphocytes, *Fourth International Congress of Immunology*, Abstract 17.2.05.

Folkers, K., Dahmen, J., Ohta, M., Stepien, H., Leban, J., Sakura, N., Lundanes, E., Rampold, G., Patt, Y., and Goldman, R., 1980a, Isolation of glutathione from bovine thymus and its significance to research relevant to immune systems, *Biochem. Biophys. Res. Commun.* **97**:590–594.

Folkers, K., Kubiak, T., Stepien, H., and Sakura, N., 1980b, Isolation, partial chemical and biological characterization of thymone B, *Biochem. Biophys. Res. Commun.* **97**:601–606.

Folkers, K., Sakura, N., Kubiak, T., and Stepien, H., 1980c, Isolation of thymone A from bovine thymus, partial chemical and biological characterization, *Biochem. Biophys. Res. Commun.* **97**:595–600.

Folkers, K., Stepien, H., Kubiak, T., Sakura, N., and Sakagami, M., 1981, The finding and partial purification and characterization of thymone C, *Biochem. Biophys. Res. Commun.* **98**:115–121.

Hannappel, E., Davoust, T., and Horecker, B. L., 1982, Thymosins β_8 and β_9: Two new peptides isolated from calf thymus homologous to thymosin β_4, *Proc. Natl. Acad. Sci. USA* **79**:1708–1711.

Houck, J., Irasquin, H., and Leikin, S., 1971, Lymphocyte DNA synthesis inhibition, *Science* **173**:1139–1141.

Kiger, N., Florentin, J., Lorans, G., and Mathe, G., 1977, Further purification and chemical characterization of the lymphocyte-inhibiting-factor extracted from thymus (LIFT), *Immunology* **33**:438–448.

Kook, A. I., and Trainin, N., 1974, Hormone-like activity of a thymus humoral factor on the induction of immune competence in lymphoid cells, *J. Exp. Med.* **139**:193–207.

Kubiak, T., Stepien, H., Blonska, B., and Folkers, K., 1982, The finding and significance of spermidine and spermine in fraction 5, *Biochem. Biophys. Res. Commun.* **108**:1482–1487.

Lenfant, M., Duchange, M., and Millerioux-Di Giusto, L., 1980, Activity of a spleen derived immunosuppressive peptide (SDIP) in plaque forming cells assay, *Fourth International Congress of Immunology, Abstract 17.2.19.*

Miura, R., and Metzler, D. E., 1976, Interaction of a 5-*trans*-vinylcarboxylic acid analogue of pyridoxal 5′-phosphate with apoaspartate aminotransferase: Covalent labeling of the enzyme, *Biochemistry* **15**:283–290.

Rotter, V., and Trainin, N., 1979, Role of thymic hormone (THF) and of a thymic hormone plasma recirculating factor (TPRF) in the modulation of human lymphocyte response to PHA and Con A, *J. Immunol.* **122**:414–420.

Rotter, V., Yakir, Y., and Trainin, N., 1979, Role of L-alanine in the response of human lymphocytes to PHA and Con A, *J. Immunol.* **123**:1726–1731.

Söder, O., and Ernström, U., 1980, Characterization of a thymocyte specific growth factor from calf thymus, *Fourth International Congress of Immunology*, Abstract 3.3.26.

Stepien, H., Sakura, N., Dahmen, J., Lundanes, E., Rampold, G., and Folkers, K., 1982, Stimulation of lymphocyte proliferation to monitor fractionation of thymus extracts in *Res. Commun. Chem. Pathol. Pharmacol.* **37**(3):403–412.

Tabor, H. and Tabor, C. W., 1975, Isolation, characterization, and turnover of glutathionylspermidine from *Escherichia coli* in *J. Biol. Chem.* **250**(7):2648–2654.

Trainin, N., Small, M., Zipori, D., Umiel, T., Kook, A. I., and Rotter, V., 1975, Characteristics of THF, a thymic hormone, in: *The Biological Activity of Thymic Hormones* (D. W. van Bekkum, ed.), pp. 117–144, Kooyker Scientific Publications, Rotterdam.

Synthetic Small Thymic Peptides

An Immunoregulatory Concept

CHRISTIAN BIRR

1. INTRODUCTION

In 1976 we had started to collect human thymic tissue for the isolation and structural elucidation of immunoregulatory proteins. Specialized in protein chemistry and peptide synthesis (Birr, 1978, 1980), however, we immediately initiated the total synthesis of thymosin α_1 when the sequence of this polypeptide consisting of 28 amino acid residues was published in 1977 (Goldstein *et al.*). Our methodological know-how in the synthesis of polypeptides rich in trifunctional amino acids furnished the total synthesis in solution of thymosin α_1 within 2 years (Birr and Stollenwerk, 1979) including one less successful trial for the synthesis of the polypeptide.

2. SYNTHETIC THYMOSIN α_1

By the utilization of the Ddz-amino acids (Birr, 1972) we have performed the total synthesis of thymosin α_1 (Fig. 1) in a stepwise manner starting from the C-terminus of the polypeptide chain. Related to the synthesis strategy with Ddz-amino acids, most of the trifunctional amino acids were protected on their side functions by the tertiary butyl protecting group. Five small-sized peptide fragments were synthesized by the repetitive use of excess Ddz-amino acid mixed anhydrides (Birr *et al.*, 1979a). These intermediate fragments I–V again were combined to each other in a straight stepwise manner, furnishing the medium-sized polypeptide fragments VI, VII, and VIII. As can be seen in Fig. 1, the fully protected sequence IX of thymosin α_1 was obtained in 52% yield condensing fragment I with fragment

CHRISTIAN BIRR • Max-Planck-Institut für Medizinische Forschung, and Organogen, Medizinisch-Molekularbiologische Forschungsgesellschaft m.b.H., D-6900 Heidelberg, Federal Republic of Germany. This chapter is dedicated to the memory of Ulrich Stollenwerk and Erhard Gross, excellent scientists and close friends.

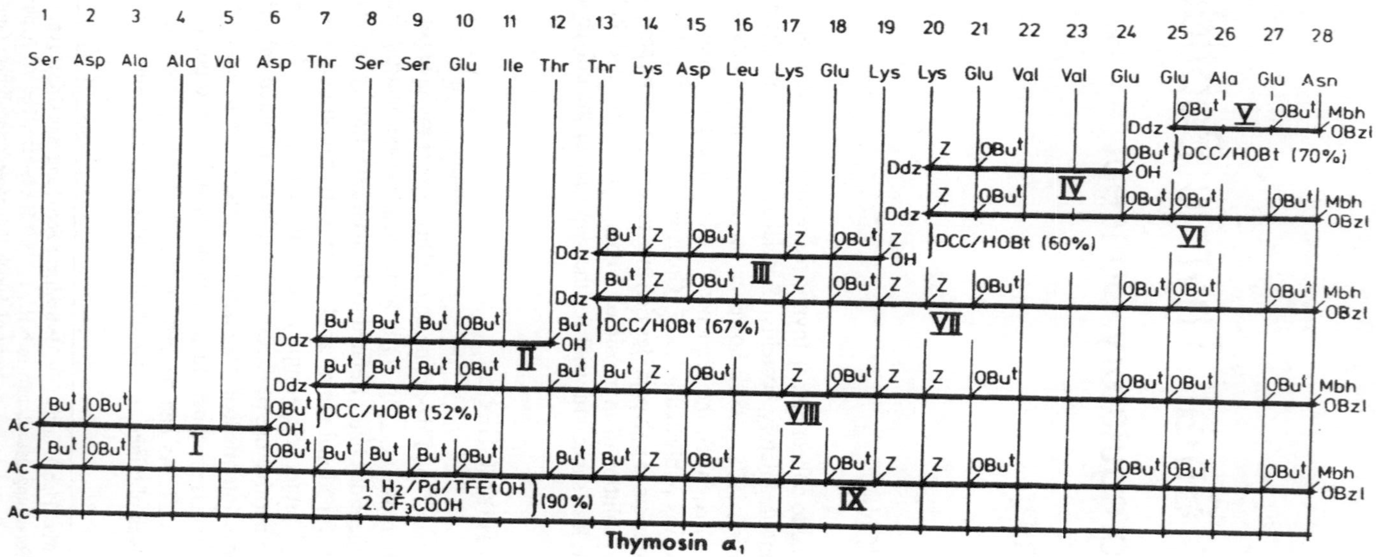

FIGURE 1. Total synthesis of thymosin α_1. The individual fragments I–V were prepared from excess Ddz-amino acid anhydrides in solution beginning C-terminally in each case. The fragments were combined in the same stepwise manner utilizing dicyclohexylcarbodiimide/1-hydroxybenzotriazole (DCC/HOBt) for coupling.

VIII by the aid of dicyclohexylcarbodiimide/1-hydroxybenzotriazole (König and Geiger, 1970), which was the method of choice for all fragment couplings throughout the total synthesis. In three well-defined steps the fully protected form of thymosin α_1 was converted into the free immunoregulatory polypeptide with 90% overall yield. In this respect we would like to acknowledge the total synthesis of Su-Sun Wang and others of thymosin α_1, which was published a few months earlier (Wang *et al.*, 1979) indicating a final yield of thymosin α_1 of 9.2%. Furthermore, our strategy of the synthesis of thymosin α_1 resulted in five molecular elements of the parent compound ready for detailed studies of the immunoregulatory properties of the individual portions in thymosin α_1. From physicochemical data (Birr *et al.*, 1979b) and from structural calculations (Chou and Fasmann, 1978) we estimated that the C-terminal portion of thymosin α_1 (residues 17–28) has a helical structure, whereas the N-terminal part 1–16 is random. A closer inspection of the C-terminus shows that the helix contains an amphiphilic character with the orientation of lipophilic amino acids mainly on one side of the helical wheel and the more polar residues opposite to them. This together with the flexible random N-terminal portion renders a close interaction of thymosin α_1 with polar structural elements in a lipophilic membrane environment on the cell surface. From this point of view the question is fascinating how the individual structural elements from the helical area and from the random part of thymosin α_1 contribute to the immunoregulatory property of the thymic polypeptide.

Already in the beginning of our study for the total synthesis of thymosin α_1 we had collected some information about possible immunoregulatory properties of small portions of thymosin α_1, because our first trial of the synthesis (12% end-product yield) ended in a very impure material containing at least five sections of thymosin α_1. Immunologically, this composition of small portions of thymosin α_1 showed already a surprising biological activity in the α-amanitine-inhibited E-rosette assay (Birr *et al.*, 1979b). In Table I the biological activities in this assay are

TABLE I. Biological Activity of Thymosin α_1[a]

Thymosin α_1				Natural	
Synthetic pure		Synthetic impure[b]		thymic proteins[c]	
[c][d]	%	[c]	%	[c]	%
5.00	33	**1.00**	**55**	6.25	10
2.50	57	0.50	39	3.12	67
1.00	64	0.25	48	**1.25**	**100**
0.50	**100**	0.13	48	0.62	62
0.25	39	0.06	3	0.31	6

[a] Comparison of synthetic pure thymosin α_1, an impure preparation containing five additional sections of the parent molecule, and a natural protein mixture from calf thymus in the E-rosette assay with α-amanitine inhibition on human peripheral T lymphocytes. (For the assay see Birr *et al.*, 1981b.)
[b] Contains synthetic fragments.
[c] According to H.-G. Manke similar to TP-1 (Serono).
[d] Concentration μg/200 μl cell culture.

compared between synthetic pure thymosin α_1, the earlier impure preparation consisting of five molecular sections and thymosin α_1, and a thymic polypeptide mixture isolated from organ tissue. Surprisingly enough, 1 μg/200 μl of the impure synthetic preparation restored the E-rosette response to 55%. Pure thymosin α_1 restores the full E-rosette response at 0.5 μg/cell culture. From this early result we were able to deduce how those small peptides might exhibit immunoregulatory properties. In our opinion the biological activity of thymic peptides in this modified E-rosette assay is a most interesting and provocative biological phenomenon. As described in more detail elsewhere (Birr *et al.*, 1981b), in this assay peripheral T-lymphocyte populations are preincubated with α-amanitine. This mushroom toxin is a most potent RNA polymerase B inhibitor active at 10^{-9} M concentrations in cell cultures. In T lymphocytes this inhibitor temporarily blocks protein biosynthesis, resulting in a less differentiated state of the cells, as determined by the disappearance of cell surface markers like the E-receptor and others. After 13 hr incubation with the inhibitor, the E-rosette count of the T lymphocytes drops from 65% to 25% on the average. The cells then are briefly washed and stimulated with thymosin α_1 or small thymic peptides for 1 hr at 20°C. Addition of sheep red blood cells restores the original number (or close to it) of E-rosettes, indicating the restart of the ribosomal protein biosynthesis of the E-receptor by the action of thymic peptides on the cell surface. It is an open question how the temporary blockade by α-amanitine of the protein synthesis is released by thymic peptides. From earlier investigations it is known (Vaisius and Wieland, 1982) that α-amanitine forms a ternary complex with polymerase B and the RNA dinucleotide, locked in this state of elongation by the inhibitor. From other studies it is known (Vaisius and Horgen, 1980) that inorganic bivalent cations can disintegrate this inhibiting complex. It is our working hypothesis that small thymic peptides of polycationic nature might follow the same mechanistic routes. From our point of view this scientific phenomenon will be further elucidated by the detection of general signal sequences in small thymic peptides.

3. SYNTHETIC ELEMENTS AND ANALOGS OF THYMOSIN α_1

Based on our experiences in the total synthesis of thymosin α_1 we have repeated the synthesis of small intermediate fragments I–V (see Fig. 1) to collect further information on the immunoregulatory properties of these individual portions of thymosin α_1. In this series of studies we have first modified fragment I by exchange of acetyl serine for free lysine in position 1. In addition we have constructed analogs of fragments IV and V with γ-carboxylglutamyl residues in position 24 and 25, respectively. In this way we were not only in a position to study the influence of the individual thymic peptides with altered ionic charges in the N- and C-terminal portions but also to synthesize a thymosin α_1 analog containing a pair of γ-carboxylglutamyl residues in position 24 and 25 (Krueck, 1984). Our guiding idea was to construct a thymosin α_1 variant with calcium complexing properties. Indeed, Gla^{24},Gla^{25}-thymosin α_1 in the presence of Ca^{2+} changes its biological activity compared to the parent molecule in the mixed lymphocyte culture and also in the α-aminitine-inhibited E-rosette assay (Table II). Furthermore, it is worth mentioning that we can reconstitute thymosin α_1 by decarboxylation of the $Gla^{24,25}$ analog.

Most interesting are the immunological activities of the individual sections I–V of thymosin α_1 and of the analog mentioned (Birr *et al.*, 1981a,b; Ciardelli *et al.*, 1981). Though conventional immunological assays like the E-rosette assay and the mixed lymphocyte culture do not yield very consistent data, we have observed the common trend that N-terminal portions of thymosin α_1 show an increased stimulatory activity in the E-rosette assay, whereas the C-terminal portions show a drastically increased stimulatory activity in the mixed lymphocyte culture. Most active in our E-rosette assay, however, was fragment I (1–6) in which the position of acetyl serine was exchanged for free lysine. This finding of an increase in the immunostimulatory property by exchange of a noncharged residue (acetyl serine) for a bivalent cationic lysine (vide supra) forced us to initiate a whole study on this influence of N-terminal lysyl residues in small thymic peptides.

Thymosin α_1 in its sequence contains four lysine residues in positon 14, 17, 19, and 20, located in the C-terminal half of the polypeptide chain. We like to recall that this portion of the molecule mainly is constituted from helix-forming amino acids. On the other hand, the C-terminal fragments IV and V of thymosin α_1 showed strong stimulatory properties in the mixed lymphocyte culture. We therefore subdivided the C-terminal region spanning from positions 17 to 28 into 11 overlapping peptides to collect further information on a molecular signal hypothesis related to ionic charges for a possibly structure-dependent regulation of the immune balance expressed in T-lymphocyte subset proportions. As described elsewhere (Ciardelli *et al.*, 1982), we have synthesized the fragments as presented in Fig. 2 starting from the fragment 20–24 of thymosin-α_1. N- and C-terminally, this central sequence was extended in a stepwise fashion by mixed anhydride

TABLE II. Biological Activity of Gla^{24},Gla^{25}-Thymosin α_1

Synthetic thymic polypeptides	Assay I[a]		Assay II[b]	
	%	[c][c]	%	[c]
$Gla^{24,25}$-thymosin α_1	21	5.0	66	0.5
$Gla^{24,25}$-thymosin α_1 (Ca^{2+})	38	0.5	95	2.5
Thymosin α_1	41	0.5	76	0.5
Thymosin α_1[d]	26	0.5	60	0.5

[a] Excess human peripheral mixed lymphocyte response (%) to Gla^{24},Gla^{25}-thymosin α_1 with and without Ca^{2+} and compared to the parent compound. (Blank responses are set to 0%; for the assay see Birr *et al.*, 1981b.)

[b] E-rosette assay with α-amanitine inhibition (0.2 μg/200 μl) on human peripheral T lymphocytes in the presence of Gla^{24},Gla^{25}-thymosin α_1 with and without Ca^{2+} and compared to the parent compound. [Relative scale; inhibited E-rosette number (blank response) is set to 0%; the normal E-rosette count without inhibitor is set to 100%.]

[c] Sample concentration μg/200 μl at maximum response.

[d] Foreign source.

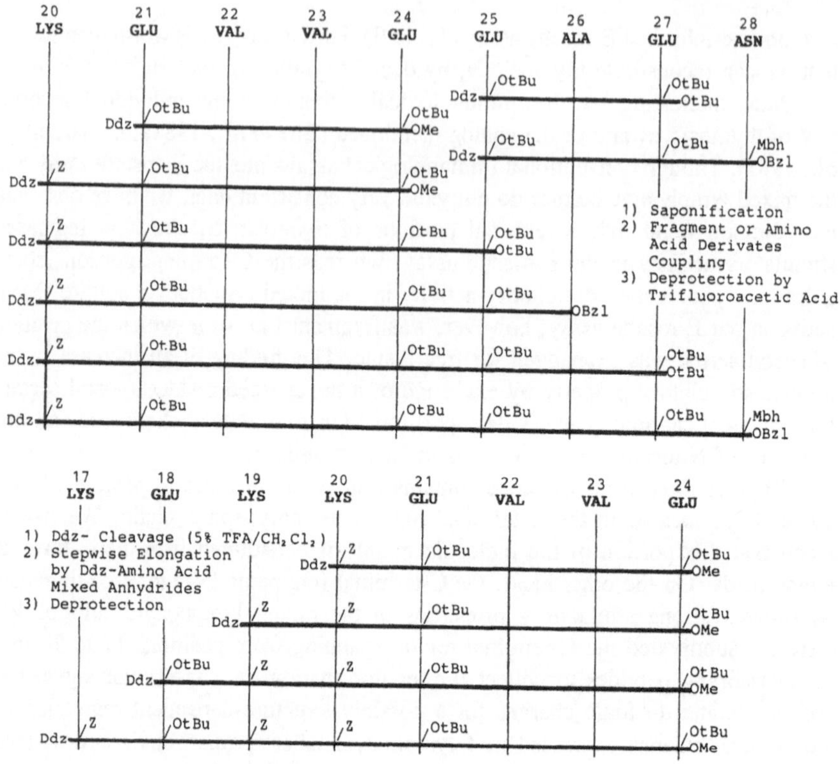

FIGURE 2. Synthesis of 11 small thymic peptides overlapping the C-terminal region 17–28 of thymosin α_1 (Ciardelli *et al.*, 1982). The Ddz-protective group used throughout the synthesis is 3,5-dimethoxyphenyl-2,2-propyloxycarbonyl (Birr, 1972).

synthesis with Ddz-amino acids. In this way we obtained a series of synthetic small thymic peptides mainly showing a lysyl residue at the N-terminus (Fig. 2). These structures together with the lysyl analog of fragment I, Lys-Asp-Ala-Ala-Val-Asp, and thymosin α_1 were comparatively investigated for immunological *in vitro* activities in three assay systems, namely the azathioprine E-rosette inhibition assay, the α-amanitine-inhibited mixed lymphocyte culture, and the α-amanitine E-rosette inhibition assay (Ciardelli *et al.*, 1981, 1982; Birr *et al.*, 1981b). The results of these assays are presented in Table III.

4. BIOLOGICAL ACTIVITIES OF THYMIC PEPTIDES

Six of the investigated structures showed activities in the azathioprine E-rosette inhibition assay similar or even better than thymosin α_1, which is active in this assay at about 10^{-6} M. It may be of significance that five of the eight structures

TABLE III. Biological Activities of Peptide Sections from Thymosin α_1[a]

Position	Sequence	E_{Az} [M]	MLC_{Am} (%)	E_{Am} (%)
17–24	KEKKEVVE	10^{-7}	58	37
18–24	EKKEVVE	10^{-5}	60	53
19–24	KKEVVE	IA	64	15
20–24	KEVVE	10^{-5}	56	31
21–24	EVVE	IA	49	35
20–25	KEVVEE	10^{-6}	28	42
20–26	KEVVEEA	IA	90	12
20–27	KEVVEEAE	10^{-6}	34	—
20–28	KEVVEEAEN	10^{-5}	51	15
25–27	EAE	IA	94	35
25–28	EAEN	IA	88	42
1′–6	KDAAVD	IA	28	66
Thy α_1	Ac-SDAAVDTSSEITTKDL- -KEKKEVVEEAEN	10^{-6}	100	100

[a] Biological activities of structural elements of thymosin α_1 as compared in three assay systems. E_{Az} [M], azathioprine E-rosette inhibition assay; activity is expressed as the lowest molar concentration [M] of peptide inducing in immature mouse spleen cells the rosetting ability sensitive to azathioprine inhibition. MLC_{Am} (%), mixed lymphocyte culture from human peripheral T lymphocytes; stimulator cells are blocked with mitomycin C, responder cells are preincubated with α-amanitine. Results are excess stimulation (%) over the blank MLC_{Am} response effected by thymic peptides; the effect of thymosin α_1 is taken as 100% (Birr *et al.*, 1981b). E_{Am} (%), E-rosette assay with α-amanitine inhibition on human peripheral T lymphocytes. Results are percent restoration to normal of the rosetting ability with sheep red blood cells in a T-cell culture preincubated with α-amanitine. Full restoration of the normal E-rosette number is effected by thymosin α_1 (100%); the blank response is set for 0% (Birr *et al.*, 1981b; Ciardelli *et al.*, 1982).

beginning N-terminally with lysine exhibit activity while of the four peptides beginning N-terminally with glutamic acid only one was active. The occurrence of a basic–acidic–lipophilic sequence of amino acids in thymus-derived peptides and its relationship (Birr *et al.*, 1981a,b) to activity in E-rosette assay systems have been discussed by us and subsequently by R. Geiger (personal communication, 1980). The active pentapeptide of the thymopoietins I and II (regions 32–36, TP-5) (Goldstein *et al.*, 1979), Arg-Lys-Asp-Val-Tyr, incorporates such a relationship and does show activity in certain E-rosette assay systems (Nash *et al.*, 1981; Abiko *et al.*, 1980a–c) including one involving azathioprine inhibition (Verhagen *et al.*, 1980). Longer portions of thymopoietin containing the entire pentapeptide showed activity, while those containing only a part of the pentapeptide sequence possessed no activity (Abiko *et al.*, 1979a,b). Removal of one or more of the N-terminal basic residues resulted in inactive peptides, whereas substitution of residues afforded varying results depending on the assays employed (Abiko *et al.*, 1980a–c). However, in no case was a peptide found to be active that did not contain a basic–basic–acidic–lipophilic–lipophilic amino acid sequence. In our study, seven of the peptides beginning N-terminally with lysine contain at least one basic–acidic–lipophilic sequence. As stated, five of these peptides were active in the assay. Of the five fragments beginning N-terminally with glutamic acid, three were inactive and did not possess a basic–acidic–lipophilic sequence. The single active N-terminal glutamic acid fragment (18–24) did include such a sequence. The failure of the fragments (19–24) and (20–26) containing one or a pair of lysines N-terminally to show activity is difficult to interpret in particular with respect to the finding of others via an N-terminal basic–basic sequence character. The requirements for activity in this assay system are certainly more subtle than this simple relationship. A correct spatial electronic presentation, essential for binding in a receptor and eventually in the release mechanism for the α-amanitine blockade, is dependent on both sequence and conformation, the latter being subject to other allosteric influences.

At first appearance, FTS does not seem to contain such a sequence relationship at all, yet it is highly active in the azathioprine E-rosette assay. Recently, however, a pentapeptide sequence essential for activity in this assay has been reported for FTS (Imaizumi *et al.*, 1981). This peptide fragment 3–7 of the FTS sequence Lys-Ser-Gln^5-Gly-Gly does begin N-terminally with lysine followed by hydrophilic and then more lipophilic amino acid residues. It is of additional significance that an FTS analog containing glutamic acid substituted for glutamine in position 5 retains full biological activity while substitution by the lipophilic amio acid norvaline at the 5 position results in an inactive analog (Blanot *et al.*, 1979). Therefore, FTS and particularly the Glu-5 analog may, in fact, be related to the previously mentioned thymus-derived peptides through sequence similarities. All these comparisons evidently support the suggestion that a basic–acidic–lipophilic sequence of amino acids is of significance in the interaction of small thymus-related peptides with T cells. It is a most essential question as to whether specific subsets of peripheral T lymphocytes are stimulated by those signal molecules.

5. DISCUSSION AND HYPOTHESES

On comparison of the data from the azathioprine E-rosette inhibition with those from the mixed lymphocyte cultures (see Table III), it is obvious that sequences inactive in the first assay, surprisingly enough, show significant activity in the second one. This latter investigation was performed on human peripheral T-cell populations. We can assume that all responder populations in this assay are maintained in an immature state of differentiation by the inhibition of protein biosynthesis by α-amanitine. Addition of those peptides, which were found inactive (IA) to restore spleen cell sensitivity to azathioprine (first assay), does sensitize cells in the mixed lymphocyte culture for cellular stimulation by lymphocytes. On the other hand, the same peptides alone, in the absence of cellular antigens, are scarcely active to stimulate α-amanitine-inhibited T lymphocytes in the E-rosette assay (fifth column, Table III, entries 3, 5, 7, 10, and 11). For the mixed lymphocyte culture response this means, in conclusion, that these active peptides stimulate one specific subgroup possibly of the mitomycine-inhibited stimulator cells in the mixed lymphocyte culture, most probably nonlytic T-helper cells (Baker *et al.*, 1979; von Boehmer *et al.*, 1979), to become responsive to the triggering allogeneic lymphocytes.

The opposite arguments can be used to explain the action of peptides found as active as thymosin α_1 in the azathioprine E-rosette inhibition assay. It is obvious from the results of the mixed lymphocyte culture (see Table III, fourth column) that these peptides which do contain the basic–acidic–lipophilic sequence character seem to function as a soluble growth factor for a cytolytic subset of T lymphocytes in the responding cells, stimulated by the cellular antigens in the mixed lymphocyte culture. These growth factor-like activities result in general proliferation, as also can be seen in the E-rosette assay with released α-amanitine inhibition (Table III, fifth column). This proliferation in the mixed lymphocyte culture is accompanied by the destructive action of cytolytic T lymphocytes. It effects in summary less than 50% increase of the counts as measured by [^{3}H]-TdR incorporation in the mixed lymphocyte culture (Table III).

6. THE IMMUNOREGULATORY CONCEPT

The biological activity of peptides representing C-terminal structural elements of thymosin α_1 in three independent assay systems, performed with different blood samples at different times (Table III, indicates that under the assay conditions described (Birr *et al.*, 1981b), one specific T-lymphocyte subset is responding to the molecular signal peptide added. Those structural elements of thymosin α_1 that are active in the azathioprine E-rosette assay seem to stimulate the mixed lymphocyte culture response of suppressive cytolytic T-lymphocyte subsets. Sequences inactive in the azathioprine E-rosette assay do stimulate another T-lymphocyte subset, most probably the nonlytic T-helper lymphocytes. In this sequence of speculations, our results from the α-amanitine-inhibited E-rosette assay (Table III) should be judged as control experiments. We recall that in this assay, human peripheral T-lymphocyte populations are first preincubated with α-amanitine inhibiting the polymerase, before the cellular protein biosynthesis is restarted by the thymic peptide added.

Our finding of a variety of different immunodulatory active structural elements located along the sequence of thymosin α_1 suggests an immunoregulatory concept based on small thymic peptides.

The occurence of thymosin α_1 in the standardized preparation named thymosin fraction 5 (Hooper *et al.*, 1975) is already the result of an enzymatic degradation of a precursor molecule extended at the C-terminus by at least five amino acids (Seipke, Hoechst AG, workshop communication, 1982). There are pharmaceutical drugs on the market composed of low-molecular-weight peptides resulting from random enzymatic processing of thymic proteins, which we found as active as thymosin fraction 5 in the mixed lymphocyte culture and our E-rosette assay (Birr, 1983). From embryonic calf thymus we have isolated a thymosin α_1 precursor utilizing the guanidinium salt extraction and high-pressure liquid chromatography technique of Hannappel *et al.* (1982).* The material eluted precisely at the position indicated in the literature flanked by thymosin β_4/β_9 and β_4(ox.) (Fig. 3). From an amino acid analysis we calculated the molecular weight of this precursor containing one unit of thymosin α_1 to be 8360, which is exactly half of the size published by Freire *et al.* (1981). This finding together with the detection of cysteine sulfonic acid in the total hydrolysate after performic acid oxidation and of 15 equivalents of arginine suggests a dimeric form of the precursor, probably linked by a connecting

* The excellent advise and technical help of E. Hannappel are gratefully acknowledged.

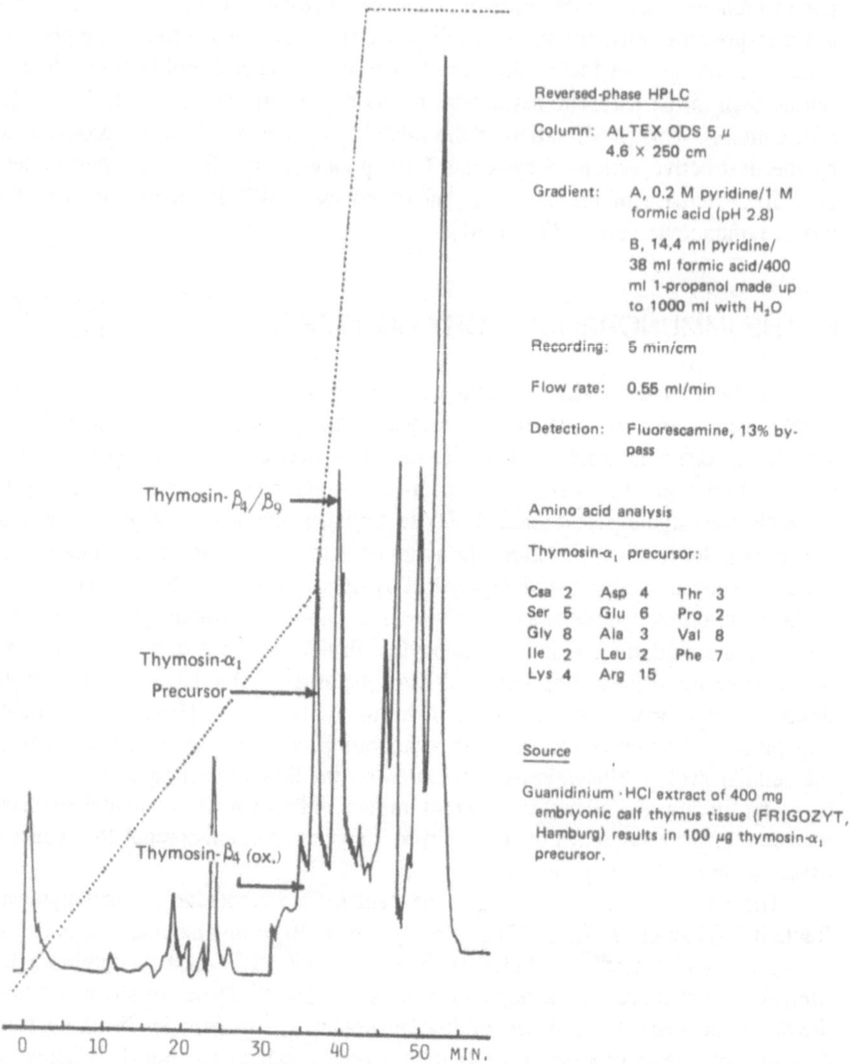

FIGURE 3. HPLC identification of thymosin α_1 precursor in the guanidinium hydrochloride extract of embryonic calf thymus. (Preparative and HPLC conditions are as published by Hannappel *et al.*, 1982.) Amino acid analysis of the precursor peak was after sulfoxidation (performic acid, 4 hr 80°C) and total hydrolysis (6 N HCl, 24 hr 120°C). Molecular weight of the precursor/thymosin α_1 unit, 8360.

peptide prone to enzymatic processing as with the C-peptide in the generation of insulin. The isolation of this posttranslational precursor together with the finding that most of the immunoregulatory proteins isolated from thymic tissue are acetylated at the N-terminus suggest a thymic processing mechanism for the natural generation of immunoregulatory signal molecules. It is our working hypothesis that thymosin

α_1 and several other thymic polypeptides are molecular storage or carrier forms for a variety of signal peptides, which are released from those intermediates by enzymatic processing. In our view, thymic polypeptides are released from the thymus and are selectively degraded further into small feedback signal peptides on the cell surface of T-lymphocyte subsets by enzymes specific for cellular differentiation (Fig. 4). This generation of peptidic signals regulates the specific subset, or stimulates another subclone for proliferation. Both events would stabilize the immune balance by the exchange of molecular signals generated enzymatically on the level of a cellular interaction.

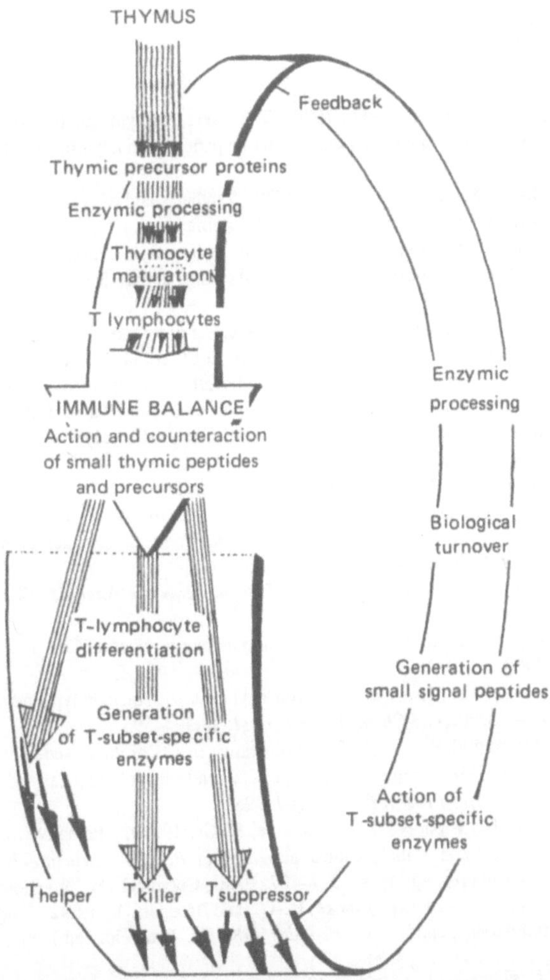

FIGURE 4. Generation of immunoregulatory signal peptides from thymic precursor proteins and polypeptides like thymosin α_1 in an enzymatic processing cascade of events connecting by molecular feedback the primary immune organs and the secondaries' response.

This hypothesis is in best agreement with our finding of many immunologically active small peptides comprising portions of thymosin α_1. As known from investigations of Freire *et al.* (1978, 1981), the enzymatic digest of thymosin α_1 and its precursor results in fragments similar in size and molecular character to our segments I–V, utilized in the total synthesis and investigated for individual immunological activities. Currently, studies in our laboratory are in progress to verify this concept by the aid of an Ortho Cytofluorograph and subset-specific Orthomune antibodies.

We are confident that several of those peptides investigated by us are of great immunotherapeutic potential. Some of them belong to the group of immunopharmacological molecules similar to TP-5, whereas others unquestionably have different immunoregulatory properties.

REFERENCES

Abiko, T., Onodera, I., and Sekino, H., 1979a, The effect of a synthetic thymosin-α_1 fragment on the inhibition of E-rosette formation by the serum of a patient with nephrotic syndrome, *Chem. Pharm. Bull.* **27**:3171–3180.

Abiko, T., Kumikawa, M., and Sekino, H., 1979b, Synthesis and effects of two peptide fragments of thymopoietin II on E-rosette forming cells in the uremic state, *Chem. Pharm. Bull.* **27**:2233–2242.

Abiko, T., Onodera, I., and Sekino, H., 1980a, Synthesis and immunological effects of thymosin-α_1 and its fragments on inhibitory factor in minimal change nephrotic syndrome, *Chem. Pharm. Bull.* **28**:3542–3550.

Abiko, T., Sekino, H., and Higuchi, H., 1980b, The effect of thymosin-α_1 fragments on T-lymphocyte transformation in the uremic state, *Chem. Pharm. Bull.* **28**:3411–3418.

Abiko, T., Onodera, I., and Sekino, H., 1980c, The effect of thymopoietin II fragments and their analogs on E-rosette forming cells in the uremic state, *Chem. Pharm. Bull.* **28**:2507–2513.

Baker, P. E., Gillis, S., and Smith, K. A., 1979, Monoclonal cytolytic T-cell lines, *J. Exp. Med.* **149**:273–278.

Birr, C., 1972, Der αα-Dimethyl-3.5.-dimethoxybenzyloxycarbonyl (Ddz)-Rest, eine photo- und säurelabile Stickstoffschutzgruppe für die Peptidchemie, *Liebigs Ann. Chem.* **763**:162–172.

Birr, C., 1978, Reactivity and structure concepts in organic chemistry, *Aspects of the Merrifield Peptide Synthesis*, Vol. 8, Springer Verlag, Berlin.

Birr, C. (ed), 1980, *Methods in Peptides and Protein Sequence Analysis*, Elsevier/North-Holland, Amsterdam.

Birr, C., 1983, unpublished investigations on Thymus Mulli, Pharm. Co., Dr. Kurt Mulli Nachf., Neuenburg, FRG.

Birr, C., and Stollenwerk, U., 1979, Synthese von Thymosin-α_1, einem Polypeptid des Thymus, *Angew. Chem.* **91**:422–423; *Angew. Chem. Int. Ed. Engl.* **18**:394–395.

Birr, C., Nassal, M., and Pipkorn, R., 1979, Preparative merits of the mixed anhydride (MA) method in the excess use of Ddz-amino acids in the peptide synthesis of biologically active new antamanide analogues, *Int. J. Pept. Protein Res.* **13**:287–295.

Birr, C., Stollenwerk, U., Brodner, O., and Manke, H.-G., 1979b, Alternative routes in the synthesis of thymosin-α_1 and some of its biological activities, in: *Peptides: Structure–Biology–Function* (E. Gross and J. Meienhofer, eds.), pp. 397–407, Pierce Chemical Co., Rockford, Ill.

Birr, C., Stollenwerk, U., Werner, I., Manke, H.-G., and Brodner, O., 1981a, Synthesis and properties of immunostimulating peptides, in: *Peptides 1980* (K. Brunfeldt, ed.), pp. 420–427, Scriptor, Copenhagen.

Birr, C., Krueck, I., Stollenwerk, U., Ciardelli, T. L., and Brodner, O., 1981b, Synthesis and properties of immunostimulating peptides. III. Potentiated immunomodulation by varied sequences of thymosin-α_1, in: *Peptides: Synthesis–Structure–Function* (D. Rich and E. Gross, eds.), pp. 545–548, Pierce Chemical Co., Rockford, Ill.

Blanot, D., Martinez, J., Sasaki, A., Anger, G., Bricas, E., Dardenne, M., and Bach, J. F., 1979, Synthetic analogs of serum thymic factor and their biological activities, in: *Peptides: Structure–Biology–Function* (E. Gross and J. Meienhofer, eds.), pp. 551–554, Pierce Chemical Co., Rockford, Ill.

Chou, P. Y., and Fasmann, G. D., 1978, Prediction of the secondary structure of proteins from their amino acid sequence, *Adv. Enzymol.* **47**:45–115.

Ciardelli, T. L., Krueck, I., Birr, C., and Brodner, O., 1981, Synthesis and properties of immunostimulating peptides. II, in: *Peptides: Synthesis–Structure–Function* (D. Rich and E. Gross eds.), pp. 541–544, Pierce Chemical Co., Rockford, Ill.

Ciardelli, T. L., Incefy, G. S., and Birr, C., 1982, Activity of synthetic thymosin-α_1 C-terminal peptides in the azathioprine E-rosette inhibition assay, *Biochemistry* **21**:4233–4237.

Freire, M., Crivellaro, O., Isaacs, C., Moschera, J., and Horecker, B. L., 1978, Translation of mRNA from calf thymus in the wheat germ system: Evidence for a precursor of thymosin-α_1, *Proc. Natl. Acad. Sci. USA* **75**:6007–6011.

Freire, M., Hannappel, E., Rey, M., Freire, J. M., Kido, H., and Horecker, B. L., 1981, Purification of thymus mRNA coding for a 16,000-dalton polypeptide containing the thymosin-α_1 sequence, *Proc. Natl. Acad. Sci. USA* **78**:192–195.

Goldstein, A. L., Low, T. L. K., McAdoo, M., McClure, J., Thurman, G. B., Rossio, J., Lai, C.-Y., Chang, D., Wang, S.-S., Harvey, C., Ramel, A. H., and Meienhofer, J., 1977, Thymosin-α_1: Isolation and sequence analysis of an immunologically active thymic polypeptide, *Proc. Natl. Acad. Sci. USA* **74**:725–729.

Goldstein, G., Scheid, M., Boyse, E. A., Schlesinger, D. H., and Van Wauwe, J., 1979, A synthetic pentapeptide with biological activity characteristic of the thymic hormone thymopoietin, *Science* **204**:1309–1310.

Hannappel, E., Davoust, S., and Horecker, B. L., 1982, Isolation of peptides from calf thymus, *Biochem. Biophys. Res. Commun.* **104**:266–271.

Hooper, J. A., McDaniel, M. C., Thurman, G. B., Cohen, G. H., Schulof, R. S., and Goldstein, A. L., 1975, Purification and properties of bovine thymosin, *Ann. N.Y. Acad. Sci.* **249**:125–138.

Imaizumi, A., Gyotoku, S., Terada, S., and Kimoto, E., 1981, Structural requirement for the biological activity of serum thymic factor, *FEBS Lett.* **128**:108–111.

König, W., and Geiger, R., 1970, Eine neue Methode zur Synthese von Peptiden: Aktivierung der Carboxylgruppe mit Dicyclohexylcarbodiimid unter Zusatz von 1-Hydroxy-benzotriazol, *Chem. Ber.* **103**:788–795.

Krueck, I., 1984, Doctoral thesis, University of Heidelberg, FRG.

Nash, L., Good, R. A., Hatzfield, A., Goldstein, G., and Incefy, G. S., 1981, In vitro differentiation of two surface markers for immature T cells by the synthetic pentapeptide thymopoietin, *J. Immunol.* **126**:150–156.

Vaisius, A. C., and Horgen, P. A., 1980, The effects of several divalent cations on the activation or inhibition of RNA polymerases II, *Arch. Biochem. Biophys.* **203**:553–564.

Vaisius, A. C., and Wieland, T., 1982, Formation of a single phosphodiester bond by RNA polymerase B from calf thymus is not inhibited by α-amanitine, *Biochemistry* **21**:3097–3101.

Verhagen, H., DeCock, W., DeCree, J., and Goldstein, G., 1980, Comparison of the in vitro effects of thymopoietin pentapeptide and levamisole on peripheral E-rosette forming cells, *Thymus* **1**:195–202.

von Boehmer, H., Hengartner, H., Nabholz, M., Leonhardt, W., Schreier, M. H., and Haas, W., 1979, Firm specificity of a continuously growing killer cell clone specific for H-Y antigen, *Eur. J. Immunol.* **9**:592–594.

Wang, S.-S., Kulesha, J. D., and Winter, D. P., 1979, Synthesis of thymosin-α_1, *J. Am. Chem. Soc.* **101**:253–257.

Human Peripheral Blood Lymphocytes Bear Markers for Thymosins (α_1, α_7, β_4)

MARINOS C. DALAKAS, DAVID L. MADDEN, AURELLA KREZLEWICZ, JOHN L. SEVER, and ALLAN L. GOLDSTEIN

1. INTRODUCTION

Within a partially purified calf thymic extract termed thymosin fraction 5 (Goldstein *et al.*, 1972), there exist several active polypeptides with distinct immunobiological action (Trivers and Goldstein, 1980; Low and Goldstein, 1982). Among those, thymosin α_1 affects the early stages of T-lymphocyte maturation and induces helper T cells; thymosin α_7 affects the induction of suppressor T-cell development (Trivers and Goldstein, 1980); and thymosin β_4 affects the maturation of prothymocytes and has a MIF-like activity (Low and Goldstein, 1982).

Using specific antibodies, these thymic polypeptides have been localized immunocytochemically in the thymic epithelial cells of the human thymus (Dalakas *et al.*, 1981a,b; Hirokawa *et al.*, 1982) and have been detected with radioimmunoassay in the normal human serum (McClure *et al.*, 1982; Naylor *et al.*, 1983). Although these thymic peptides influence the transformation of precursor to more mature cells, probably through an intrathymic association, it is not known whether they also exert an immunoregulatory role in the periphery and interact with mature circulating lymphocytes recognizing antigenic determinants on their cell surface.

MARINOS C. DALAKAS, DAVID L. MADDEN, AURELLA KREZLEWICZ, and JOHN L. SEVER • National Institute of Neurological and Communicative Disorders and Stroke, National Institutes of Health, Bethesda, Maryland 20205. ALLAN L. GOLDSTEIN • Department of Biochemistry, The George Washington University School of Medicine and Health Sciences, Washington, D.C. 20037.

In the present study we report that antibodies against thymosins α_1, α_7, and β_4 immunoreact with a subset of circulating peripheral blood lymphocytes (PBL).

2. MATERIALS AND METHODS

2.1. Staining of PBL for Thymosin α_1

Venous blood was collected in heparinized tubes from 15 normal volunteers and 16 patients with myasthenia gravis (MG). PBL were separated on a Ficoll–Hypaque density gradient and T- and B-cell-enriched populations were prepared by rosetting with sheep red blood cells as previously described (Dalakas *et al.*, 1983). The obtained T-cell fraction contained less than 2% IgM-bearing cells and less than 2% esterase-positive cells; the B-cell fraction contained 60–70% IgM-bearing cells and 30–40% esterase-positive cells. Esterase-positive cells could be eliminated by their adherence to polystyrene culture flasks in the presence of media enriched with fetal calf serum at 37°C. Cells were incubated with one of the following antisera: 1 : 10 dilution of antibody against synthetic thymosin α_1 raised in rabbits, normal rabbit serum, rabbit anti-human albumin, and antithymosin α_1 previously absorbed with thymosin α_1 on an immunosorbent column. The cells were subsequently stained with $F(ab')_2$ fraction of goat anti-rabbit IgG (Cappel Lab), mounted on slides, and observed with a Zeiss epi-illumination fluorescence and bright-field microscope. A minimum of 400–500 cells were counted.

2.2 Staining of PBL for Thymosin α_1 and β_4

PBL from 36 individuals of different ages (from 18 to 80) were separated as described above. A portion of PBL was incubated with rabbit antisera against thymosin β_4 and α_7 and with normal rabbit serum. Reactive cells were stained with FITC-conjugated sheep anti-rabbit IgG ($F(ab')_2$ fragment (Accurate Co., Hicksville, N.Y.) and analyzed in a flow cytofluorograph FC 200/4800 A (Ortho, Raritan, N.J.). A total of 10,000 cells were examined and their background fluorescence (defined by the nonspecific reaction of PBL with the normal rabbit serum) was subtracted from antibody-specific fluorescence. Results were expressed as the percent of positive cells of the total number of counted cells. The other portion of the separated PBL was simultaneously stained with the following monoclonal antibodies that identify surface membrane markers: OKT_3, OKT_4, OKT_8, Ia_1, OKM_1, OKT_{11} (Ortho) and anti-IgM (BRL, Bethesda, Md.).

Immunoreactivity of PBL for thymosin α_7, β_4, and normal rabbit serum was evaluated on (1) total PBL, (2) T- and B-cell-enriched population, and (3) PBL depleted of adherent cells, obtained as described above for thymosin α_1.

Antibodies to thymosins α_1 and β_4 were prepared and characterized as previously described (McClure *et al.*, 1982; Naylor *et al.*, 1983). Antibodies to α_7 were similarly obtained (McClure and Goldstein, unpublished data).

3. RESULTS

3.1. Detection of Antithymosin α_1-Reactive PBL

Only cells with a homogeneous rim pattern or cells with a clearly membrane-associated patchy fluorescence that resembled capping were counted. As previously reported (Dalakas *et al.*, 1983), cells (mostly macrophages or monocytes) with irregular and scattered spots of staining were not counted because such nonspecific staining was also seen when incubation with antithymosin α_1 was omitted.

As shown in Fig. 1, antibodies to thymosin α_1 immunoreacted specifically with T cells ($T\alpha_1$). Twenty determinations on fifteen normal volunteers revealed that $1.5 \pm 0.3\%$ of the total T cells were stained for thymosin α_1. In B-cell-enriched fractions, $0.1 \pm 0.1\%$ of the cells were positive (Fig. 1). Elimination of adherent cells did not alter the number of thymosin α_1-positive cells. Control antisera did not reveal cells with specific staining.

$T\alpha_1$ levels were increased (above the 3 S.D. from the mean) in 11 patients with MG (Fig. 1) who had active disease but were within the normal range in five patients with asymptomatic disease; from 11 other patients with autoimmune diseases, only one with inactive lupus had slightly increased $T\alpha_1$ (Fig. 1). In patients with active MG, the number of $T\alpha_1$ cells fell within the normal range a few weeks after thymectomy (Fig. 2) and this was associated with improvement of the neurological function (Dalakas *et al.*, 1983).

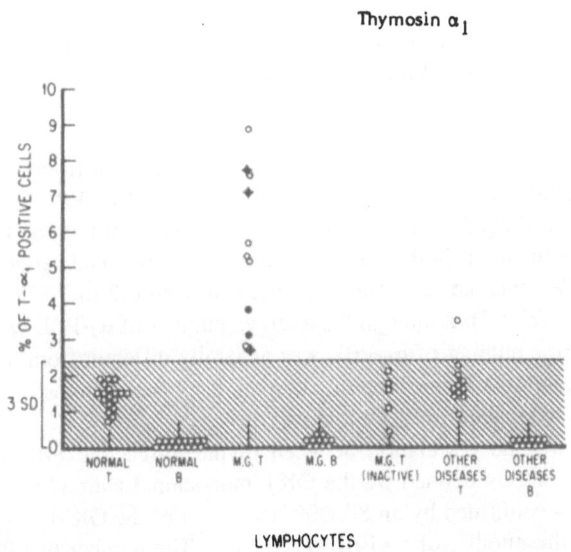

FIGURE 1. Levels of thymosin α_1-positive T cells (T-α_1) in T- and B-cell fractions of normals, myasthenia gravis (M.G.) patients, and other immune diseases; ● represents two elderly patients and ⊕ three previously thymectomized patients, all with symptomatic disease.

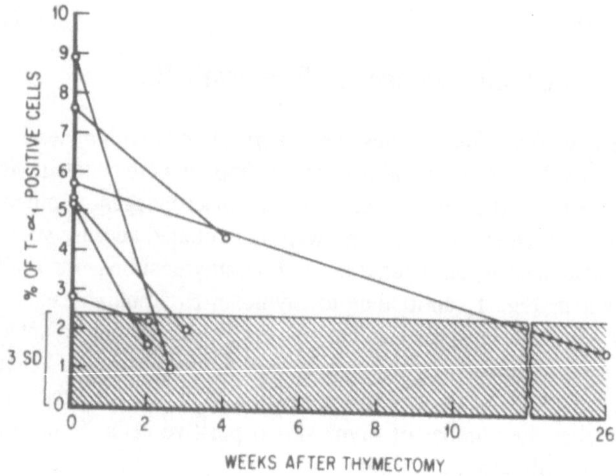

FIGURE 2. Levels of $T\alpha_1$ cells in six MG patients before and after thymectomy.

3.2. Detection of Antithymosin α_7 and β_4-Reactive PBL

A number of PBL also immunoreacted with antisera to α_7 (α_7-PBL) and β_4 (β_4-PBL) (Figs. 3, 4). The absolute number of α_7-PBL and β_4-PBL was obtained by superimposing the histogram of their reactivity to that of cells stained nonspecifically with normal rabbit serum and subtracting the difference (Figs. 3, 4). The staining pattern of α_7-PBL was somewhat different from that of β_4-PBL (Figs. 3, 4).

As shown in Table I, the number of α_7-PBL ranged from 1.02 to 31.81% of the total and averaged 5.21%, whereas the number of β_4-PBL ranged from 0.5 to 19.58% and averaged 5.47%. There was a fluctuation of both α_7-PBL and β_4-PBL on serial examinations in the same person with an individual variation ranging from 0.2 to 12.4% (average 3.3%) for β_4-PBL and from 0.2 to 3.7% (average 1.67%) for α_7-PBL (Table I). Although the average number of α_7-PBL was similar to that of β_4-PBL, the number of α_7-PBL was generally different from that of β_4-PBL in the same individual; this difference ranged in all the examined subjects from 0.1 to 22% and averaged 2.64% (Table I).

There was no correlation between thymosin-reactive PBL with any of the lymphocyte subsets defined by the OKT monoclonal antibodies. Removal of adherent cells—confirmed by an 80–90% reduction of the OKM_1-positive cells—did not change the number of α_7-PBL or β_4-PBL. The number of PBL reactive with antithymosin α_7 and β_4 was essentially unchanged in both the total (unseparated) PBL and their T-cell-enriched fraction, suggesting that most of the reactive cells were probably T cells. The stained cells within the B-cell-enriched fraction, gave

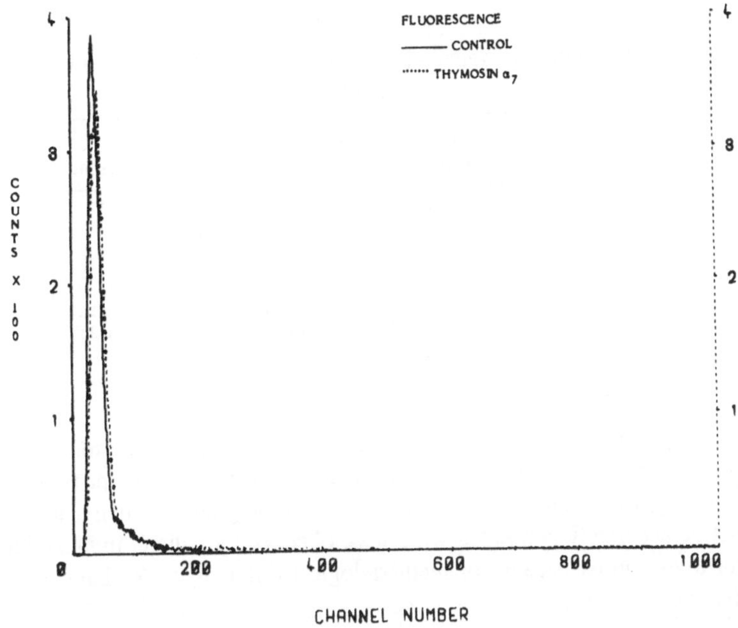

FIGURE 3. Histogram of specific fluorescence for thymosin α_7-positive PBL (- - - - -) superimposed on the histogram obtained for the nonspecifically stained cells incubated with normal rabbit serum (——).

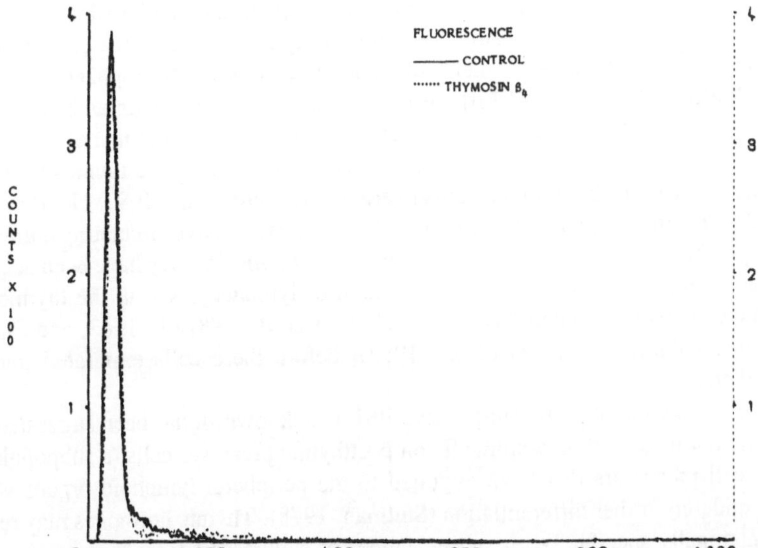

FIGURE 4. Histogram of specific fluorescence for thymosin β_4-positive PBL (- - - -) superimposed on the histogram obtained for the nonspecifically stained cells incubated with normal rabbit serum (——).

TABLE I. Immunoreactive PBL with Antibodies to Thymosins α_7 and β_4

Antithymosin	Number of reactive PBL (36 individuals)[a]	Individual variation (serial studies in 9 persons)
α_7	1.02–31.81% (5.21%)	0.2–3.7% (1.67%)
β_4	0.5–19.58% (5.47%)	0.2–12.4% (3.3%)

[a] Differences between α_7-PBL and β_4-PBL in the same person, ranged from 0.1 to 22% and averaged 2.64%.

bizarrely shaped histograms from which it was impossible to determine whether any of the B cells had specifically immunoreacted for thymosin α_7 or β_4.

Obviously, the results obtained with the cytofluorograph for thymosins α_7 and β_4 cannot be accurately correlated with those obtained by indirect immunofluorescence for thymosin α_1 because of methodological differences. We have not compared the two techniques.

4. DISCUSSION

We have demonstrated that antibodies to thymic hormones thymosin α_1, α_7, and β_4 recognize antigenic determinants in a small subset of PBL. By indirect immunofluorescence, thymosin α_1 was uniformly distributed on the cell surface of a small subset of human T cells but not in B cells or macrophages; with the cytofluorograph, a subset of PBL specifically immunoreacted with antithymosin α_7 and β_4. The findings suggest that circulating PBL carry thymic markers.

The source of thymosins on the PBL is uncertain. Because thymosins α_1 and β_4 are detected in the normal human serum (McClure *et al.*, 1982; Naylor *et al.*, 1983), PBL may bind thymosins derived from their respective circulating hormones, via specific cell surface receptors. Alternatively, thymosins may have been acquired in the thymus by an intrathymic association of lymphocytes with the thymic epithelial cells [which contain thymosins (Dalakas *et al.*, 1981a,b, 1983, and Chapter 11, this volume; Hirokawa *et al.*, 1982)] before these cells emigrated into the circulation.

The function of thymosin-positive PBL is unknown. It has been suggested that thymic hormones act predominantly on postthymic precursor cells, a subpopulation of T-cell precursors that have migrated to the peripheral lymphoid organs where they undergo further differentiation (Stutman, 1978). Thymic hormones may recruit a functionally specific T-cell subset (helper or suppressor) from those precursors, depending on the stage of maturation of the precursor cells and the activity of the hormone (Goldstein *et al.*, 1978). Because thymosin α_1 recruits helper cells and α_7 suppressors (Trivers and Goldstein, 1980), it is possible that α_1 or α_7-positive PBL may represent a subset of cells with a helper or suppressor function, respectively. Thymosin β_4 affects the early stages of T-cell differentiation and inhibits the migration of macrophages (Trivers *et al.*, 1980); β_4-PBL may therefore be a subset of cells that participate in these functions. Whether thymosin-positive PBL play a role in the immune regulation and whether their levels change in several immune or neuroendocrine conditions remain to be determined.

Thymosin α_1-positive T cells have been found increased in patients with MG and their levels correlated with symptoms (Dalakas *et al.*, 1983), suggesting that $T\alpha_1$ cells may represent an immunologically competent subpopulation of T cells which could play a role in the immunopathogenesis of MG. This is compatible with the observation that in the thymus gland of MG patients, there is proliferation of epithelial cells which appear to be active and hypersecretory for thymosin α_1 (Dalakas *et al.*, 1980).

REFERENCES

Dalakas, M. C., Engel, W. K., McClure, J. E., and Goldstein, A. L., 1980, Thymosin α_1 in myasthenia gravis (MG), *N. Engl. J. Med.* **19:**1092–1093.

Dalakas, M. C., Engel, W. K., McClure, J. E., Goldstein, A. L., and Askanas, V., 1981a, Immunocytochemical localization of thymosin α_1 in thymic epithelial cells of normal and myasthenia gravis patients and in thymic cultures, *J. Neurol. Sci.* **50:**239–247.

Dalakas, M. C., Engel, W. K., McClure, J. E., Goldstein, A. L., and Askanas, V., 1981b, Identification of human thymic epithelial cells with antibodies to thymosin α_1 in myasthenia gravis, *Ann. N.Y. Acad. Sci.* **377:**477–485.

Dalakas, M. C., Rose, J. W., Paul, J., Engel, W. K., McClure, J. E., and Goldstein, A. L., 1983, Increased circulation of T lymphocytes bearing surface thymosin α_1 in patients with myasthenia gravis: Effect of thymectomy, *Neurology* **33:**144–149.

Goldstein, A. L., Guhar, A., Zatz, M. M., Hardy, A., and White, A., 1972, Purification and biological activity of thymosin, a hormone of the thymosin gland, *Proc. Natl. Acad. Sci. USA* **69:**1800–1803.

Goldstein, A. L., Thurman, G. B., Low, T. L. K., Rossio, J. L., and Trivers, G. E., 1978, Hormonal influences on the reticuloendothelial system—Current status of the role of the thymosin in the regulation and modulation of immunity, *J. Reticuloendothel. Soc.* **23:**253–266.

Hirokawa, K., McClure, J. E., and Goldstein, A. L., 1982, Age-related changes in localization of thymosin in the human thymus, *Thymus* **4:**19–29.

Low, T. L. K., and Goldstein, A. L., 1982, Chemical characterization of thymosin β_4, *J. Biol. Chem.* **257:**1000–1006.

McClure, J. E., Lameris, N., Wara, D., and Goldstein, A. L., 1982, Immunochemical studies on thymosin: Radioimmunoassay of thymosin α_1, *J. Immunol.* **128:**368–375.

Naylor, P. H., McClure, J. E., Spangelo, B. L., Low, T. L. K., and Goldstein, A. L., 1983, Immunocytochemical studies of thymosin: Radioimmunoassay of thymosin β_4, *Immunopharmacology* (in press).

Stutman, O., 1978, Intrathymic and extrathymic T cell maturation, *Immunol. Rev.* **42:**138.

Trivers, G. E., and Goldstein, A. L., 1980, The endocrine thymus—A role for thymosin and other thymic factors in cancer immunity and therapy, in: *Canadian Biological Review* (J. J. Marchalonis, M. G. Hanna, and M. J. Fidler, eds.), pp. 49–131, Dekker, New York.

Thymosin β_4 Is Present in a Subset of Oligodendrocytes in the Normal Human Brain

MARINOS C. DALAKAS, RAYMOND HUBBARD, GUY CUNNINGHAM, BRUCE TRAPP, JOHN L. SEVER, and ALLAN L. GOLDSTEIN

1. INTRODUCTION

Thymosin β_4, a polypeptide extracted from the original thymosin fraction 5 (Hopper *et al.*, 1975), has been characterized, sequenced, and synthesized (Low and Goldstein, 1981, 1982). Thymosin β_4 affects the early stages of T-cell differentiation, has a MIF-like activity (Low and Goldstein, 1981), and may play a role in cellular immune mechanisms (Hannappel *et al.*, 1982). It has been found in a variety of rat and mouse tissue extracts including spleen, thymus, brain, liver, and peritoneal macrophages (Hannappel *et al.*, 1982), but its cellular distribution and presence in human tissues are unknown.

The presence of thymosin β_4 in tissue extracts of brain as well as lymphoid organs prompted us to examine whether thymosin β_4, a peptide of thymic origin, is a common antigen present in a specific subpopulation of cells in the brain and lymphoid tissues. Such information may help us to understand how the brain, thought to represent an immunologically "privileged" site, becomes a target of several autoimmune mechanisms.

The immuocytochemical localization of thymosin β_4 in normal human and monkey brain, liver, spleen, thymus, skin, and cultured peritoneal macrophages is the purpose of this report.

MARINOS C. DALAKAS, RAYMOND HUBBARD, GUY CUNNINGHAM, BRUCE TRAPP, and JOHN L. SEVER • National Institute of Neurological and Communicative Disorders and Stroke, National Institutes of Health, Bethesda, Maryland 20205. ALLAN L. GOLDSTEIN • Department of Biochemistry, The George Washington University School of Medicine and Health Sciences, Bethesda, Maryland 20205.

2. MATERIAL AND METHODS

We studied the following tissues: (1) normal human thymuses from patients undergoing cardiac surgery and human thymuses from patients with myasthenia gravis undergoing therapeutic thymectomy; (2) normal human skin from patients undergoing above-knee amputation for sarcomas; (3) lymph nodes, liver, spleen, and lung from normal rhesus monkeys; (4) normal human, monkey, and rat brain; and (5) rhesus monkey peritoneal macrophages cultured for 1–4 days.

Tissues were fixed in formalin and embedded in paraffin. Six-micrometer-thick sections were cut on glass slides, deparaffinized, and stained immunocytochemically by the PAP method (Sternberger, 1979). In brief, the sections were incubated with the following solutions: (1) 3% normal goat serum in 0.5 M Tris (pH 7.6); (2) rabbit anti-synthetic thymosin β_4 from 1 : 50 to 1 : 10000 in the same Tris buffer containing 1% normal goat serum; (3) goat anti-rabbit IgG 1 : 20 in Tris buffer; (4) rabbit PAP diluted 1 : 80 in Tris buffer with 1% normal goat serum; and (5) 0.01% H_2O_2 and 0.05% 3,3′-diaminobenzidine HCl in 0.1 M Tris buffer. Selected sections were counterstained with hematoxylin and treated with 2% osmium tetroxide. The sections were dehydrated and mounted. For control, tissues were incubated with antithymosin α_1 and α_7 prepared identically to that of antithymosin β_4 (McClure *et al.*, 1982) or with normal rabbit serum instead of the primary antibody (step No. 2).

Antiserum to synthetic thymosin β_4 was induced in rabbit and characterized as previously described (Naylor *et al.*, 1983).

3. RESULTS

In the thymus, anti-β_4 immunoreacted with cells known as interdigitating reticular cells (IDC) (Fig. 1a) found predominantly in the medulla and corticomedullary junction and occasionally in the inner cortex (Kaiserling *et al.*, 1974). These β_4-positive IDC were distinctly stained and formed a configuration with a group of surrounded thymocytes resembling "rosettes" (Figs. 1b,c); their processes engulfing the surrounded thymocytes were also immunoreacted and could be visualized at higher magnification (Figs. 1b,c). The cellular distribution of thymosin β_4 was different from that previously described for thymosin α_1 (Dalakas *et al.*, 1981a,b) which was found in epithelial cells predominantly around Hassall's corpuscles or forming a large network of long rows.

In the skin, antithymosin β_4 strongly stained the Langerhans cells and their processes (Fig. 1d). Peritoneal macrophages immunoreacted with antithymosin β_4 (Fig. 1e) but the number of positive cells and the intensity of staining varied each day from 20% (first day in culture) to almost 100% (4th day). In the lymph node and spleen, both macrophages and IDC were stained. In the liver, some of the Kupffer cells as well as the dendritic cells around the portal triads immunoreacted with antithymosin β_4.

In the brain, certain cells and their processes in both gray and white matter were specifically stained (Fig. 2a); many of these β_4-positive cells appeared next

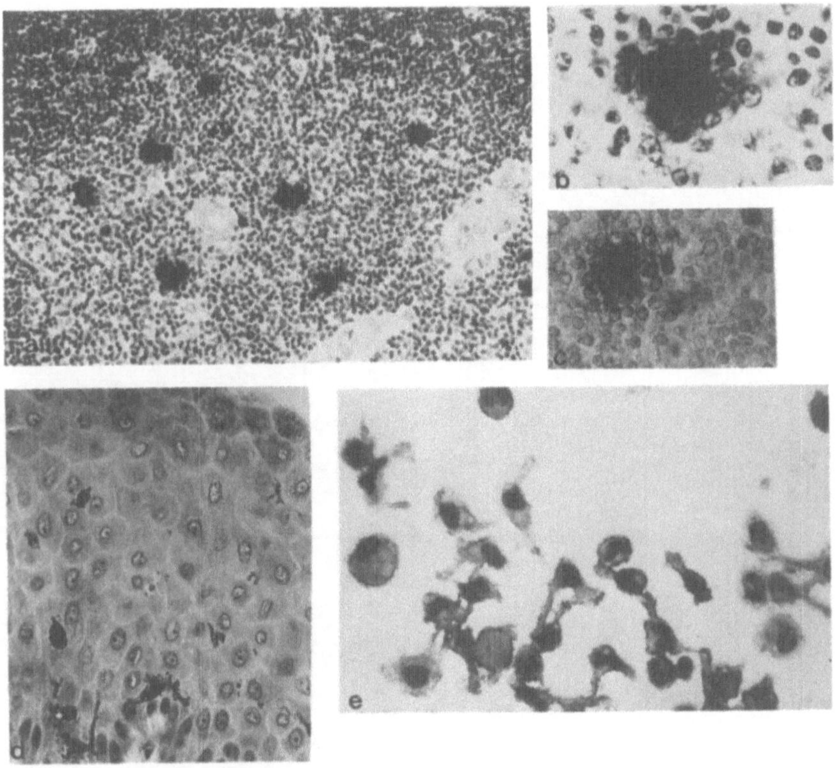

FIGURE 1. (a) Interdigitating cells in the medulla and corticomedullary junction stained specifically with antithymosin β_4; (b,c) interdigitating cells form rosettes with engulfing thymocytes; (d) Langerhans cells of the skin immunoreact specifically with antithymosin β_4; (e) peritoneal macrophages and their processes (4th day in culture), demonstrate membrane staining with anti-β_4.

to axons with their processes terminating around them. This staining pattern was different from that obtained when serial sections were stained with anti-GFAP (Fig. 2b), suggesting that the thymosin β_4-stained cells were not astrocytes. With higher magnification, the thymosin β_4-positive cells were identified as oligodendrocytes because (1) they appeared to extend their stained processes around or along axons and their myelin sheaths (Figs. 2d–g) and (2) in longitudinal sections of the white matter (corpus callosum, optic nerve, and medulla), they were among those interfascicular round cells, typical of oligodendrocytes, known to form rows along the long tracts (Fig. 2c). The staining was cytoplasmic and the strongly stained processes could be traced to the neighboring long tracts of myelinated fibers in both white (Figs. 2d–f) and gray matter (Figs. 2g, h). Some of the satellite oligodendrocytes and their processes next to neurons were also thymosin β_4-positive (Fig. 2h). At a certain level of focus, two or three stained processes could be seen terminating on

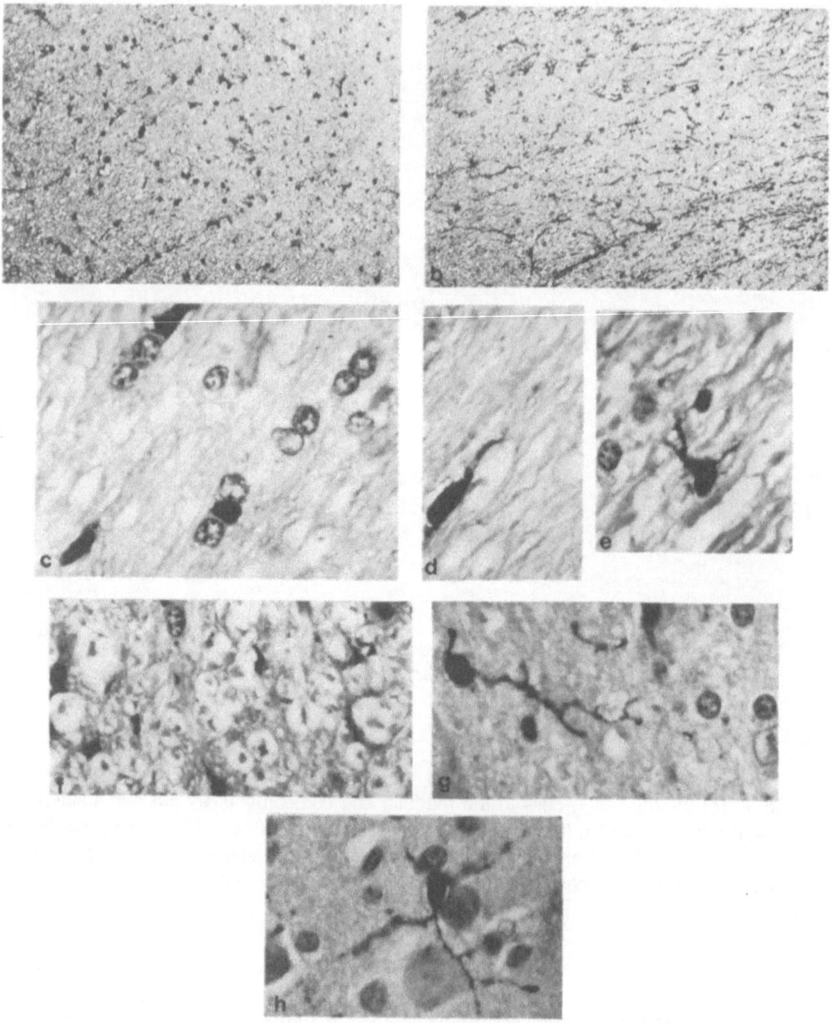

FIGURE 2. (a) Transverse section of spinal cord (edge of gray matter, bottom left, and white matter). Several cells and their processes immunoreact with antithymosin β_4. Many of the positive cells and their processes appear next to axons. (b) Section serial to a stained with anti-GFAP, demonstrates that the pattern of staining for astrocytes is different from that obtained with antithymosin β_4. (c) Longitudinal section of corpus callosum shows that some of the interfascicular oligodendrocytes stain specifically for thymosin β_4. (d,e) Processes of oligodendrocytes surround axons (d) or terminate on them (e). (f) Oligodendrocytes and their processes surround at least one axon (transverse section of medulla). At different levels of focusing, these processes can be clearly seen terminating at different axons. Stained dark lines and dots represent transverse sections of oligodendrocytic processes. (g,h) Transverse sections of gray matter. Satellite oligodendrocytes and their processes surrounding axons (PAP, counterstained with hematoxylin).

the surrounding sheaths (Figs. 2d–f), but at different levels of focusing few other processes could be seen in the same section. Not all but only a subset of oligodendrocytes in gray and white matter were positive with antithymosin β_4.

No species differences were noted in the staining pattern of oligodendrocytes in rat, monkey, or human brain. With rabbit antisera against the two other small polypeptides from the original thymosin fraction 5, thymosin α_7 and α_1, which have been produced similarly to the antiserum against thymosin β_4 (McClure *et al.*, 1982), there was a complete absence of reaction with the brain tissue; normal rabbit serum also did not react.

4. DISCUSSION

We have demonstrated using paraffin-embedded sections and PAP immunocytochemistry that thymosin β_4 is present in the IDC of the thymus, Langerhans cells of the skin, peritoneal macrophages, and in a subset of brain and spinal cord oligodendrocytes and their processes. The stained cells were identified as (1) IDC in the thymus, because they had all the morphological and topographical characteristics described for IDC by Kaiserling *et al.* including processes engulfing thymocytes, a known property of these cells (Kaiserling *et al.*, 1974). When immunostaining was performed on frozen sections, these cells were ATPase positive, a known histochemical reaction for IDC (Müller-Hermelink, 1974); (2) Langerhans cells in the skin, because of their star-shaped configuration and location in the deeper portions of the germinative layer of the epidermis; and (3) oligodendrocytes, because of their round or ovoid cell body with their many slender processes branching before surrounding axons or terminating along myelin sheaths. Both interfascicular oligodendrocytes along the long tracts as well as some of the satellite oligodendrocytes with their processes next to neuronal perikarya were specifically stained with antithymosin β_4. Their staining pattern resembled that obtained with anti-MBP antiserum in the myelin-forming oligodendrocytes of newborn rats (Sternberger *et al.*, 1978). The myelin sheaths, which have antigenic properties differing from those of the parent cell (Rumsby, 1979), were not stained with antithymosin β_4.

The findings suggest that IDC, Langerhans cells, and macrophages—all Ia^+ cells of thymic origin which bind and present antigens to T cells (Katz *et al.*, 1979; Shelly and Juhlin, 1976; Unanue, 1980; Janossy *et al.*, 1980; Lambert *et al.*, 1980)—share a common antigen of thymic origin with a subset of oligodendrocytes. Ia^+ cells of glial type are present in the human brain (Natali *et al.*, 1981) and represent approximately 1% of the total cells in the CNS (Nixon *et al.*, 1982) with a fraction of those deriving from the bone marrow (Nixon *et al.*, 1982). Because only a subset of oligodendrocytes in our study were thymosin β_4 positive, it is tempting to speculate that all thymosin β_4-positive cells (IDC, Langerhans, macrophages, and a subset of oligodendrocytes) are Ia^+ cells that share common antigens (and perhaps similar immune properties) and may belong to the same mobile ancillary system for T-cell reactions. As has been shown for the Langerhans cells (Katz *et al.*, 1979), these cells may have been generated from the same precursor bone marrow cell and circulated afterwards to specialized organs. Thymosin β_4, which affects the maturation and differentiation of bone marrow stem cells (Low and Goldstein, 1981), may prove to play a role in the transformation and migration of these cells.

The presence of a common antigen of thymic origin on a subset of oligodendrocytes and the antigen-presenting immunocompetent cells of the lymphoid organs, may be of importance in understanding the immunological mechanisms of demyelination and the suggested primary role of activated lymphocytes and specific macrophages in myelin destruction and oligodendrocyte depletion (Preneas and Graham, 1981). The specific function of the thymosin β_4-positive subset of oligodendrocytes and their role in the immune surveillance of the CNS remain to be determined.

REFERENCES

Dalakas, M. C., Engel, W. K., McClure, J. E., Goldstein, A. L., and Askanas, V., 1981a, Immunocytochemical localization of thymosin α_1 in thymic epithelial cells of normal and myasthenia gravis patients and in thymic cultures, *J. Neurol. Sci.* **50:**239–247.

Dalakas, M. C., Engel, W. K., McClure, J. E., Goldstein, A. L., and Askanas, V., 1981b, Identification of human thymic epithelial cells with antibodies to thymosin α_1 in myasthenia gravis, *Ann. N.Y. Acad. Sci.* **377:**477–485.

Hannappel, E., Xu, G.-j, Morgan, J., Hempstead, J., and Horecker, B., 1982, Thymosin β_4: An ubiquitous peptide in rat and mouse tissues, *Proc. Natl. Acad. Sci. USA* **79:**2172–2175.

Hooper, J. A., McDaniel, M. C., Thurman, G. B., Cohen, G. H., Schulof, R. S., and Goldstein, A. L., 1975, Purification and properties of bovine thymosin, *Ann. N.Y. Acad. Sci.* **249:**125–144.

Janossy, G., Tidman, N., Selby, W. S., Thomas, J. A., Granger, S., Kung, P. G., and Goldstein, G., 1980, Human T lymphocytes of inducer and suppressor type occupy different microenvironments, *Nature (London)* **288:**81–88.

Kaiserling, E., Stein, H., and Müller-Hermelink, H. K., 1974, Interdigitating reticulum cells in the human thymus, *Cell Tissue Res.* **155:**47–55.

Katz, S. I., Tomaki, K., and Sachs, D. H., 1979, Epidermal Langerhans cells are derived from cells originating in bone marrow, *Nature (London)* **282:**324–326.

Lambert, I. A., Pizzolo, G., Thomas, A., and Janossy, G., 1980, Immunocytochemical characterization of cells involved in dermatopathic lymphadenopathy, *J. Pathol.* **131:**145–156.

Low, T. L. K., and Goldstein, A. L., 1981, Complete amino acid sequence of bovine thymosin β_4:A thymic hormone that induces terminal deoxynucleotidyl transferase activity in thymocyte populations, *Proc. Natl. Acad. Sci. USA* **78:**1162–1166.

Low, T. L. K., and Goldstein, A. L., 1982, Chemical characterization of thymosin β_4, *J. Biol. Chem.* **257:**1000–1006.

McClure, J. E., Lameris, N., Wara, D. W., and Goldstein, A. L., 1982, Immunochemical studies of thymosin: Radioimmunoassay of thymosin α_1, *J. Immunol.* **128:**368–375.

Müller-Hermelink, H. K., 1974, Characterization of the B-cell and T-cell regions of human lymphatic tissue through enzyme histochemical demonstration of ATPase and 5′-nucleotidase activities, *Virchows Arch. B* **16:**371–378.

Natali, P. G., DeMartino, C., Quaranta, V., Nicotia, M. R., Frezza, F., Pellegrinno, M. A., and Ferrane, S., 1981, Expression of Ia-like antigens in normal human nonlymphoid tissues, *Transplantation* **31:**75.

Naylor, P. H., McClure, J. E., Spangelo, B. L., Low, T. L. K., and Goldstein, A. L., 1983, Immunochemical studies on thymosin: Radioimmunoassay of thymosin β_4, *Immunopharmacology* (in press).

Nixon, D. F., Ting, J. P. Y., and Frelinger, A., 1982, Ia antigens on non-lymphoid tissues: Their origins and functions, *Immunology Today* **3:**339–342.

Prineas, J. W., and Graham, B. A., 1981, Multiple sclerosis: Capping of surface immunoglobulin G on macrophages engaged in myelin breakdown, *Ann. Neurol.* **10:**149–158.

Rumsby, M. G., 1979, Oligodendrocyte–myelin sheath interrelationships, in: *Search for the Cause of Multiple Sclerosis and Other Chronic Diseases of the CNS* (A. Boese, ed.), First National Symposium of Hertie Foundation, Frankfurt/Main, pp. 49–63.

Shelly, W. B., and Juhlin, L., 1976, Langerhans cells form a reticulo-epithelial trap for external contact antigens, *Nature (London)* **461:**46–47.

Sternberger, L. A., 1979, *Immunocytochemistry*, 2nd ed., Wiley, New York.

Sternberger, N. H., Itoyama, Y., Kies, M. W., and Webster, H. deF., 1978, Immunocytochemical method to identify basic protein in myelin-forming oligodendrocytes of newborn rat CNS, *J. Neurocytol.* **7:**251–263.

Unanue, E. R., 1980, Cooperation between mononuclear phagocytes and lymphocytes in immunity, *N. Engl. J. Med.* **303:**977–985.

The Synthesis and Physical Properties of Thymosin α_1 and Its Analogs Prepared by Fragment Condensation

ARTHUR M. FELIX, EDGAR P. HEIMER, CHING-TSO WANG, THEODORE J. LAMBROS, JOSEPH SWISTOK, MUSHTAQ AHMAD, MARTIN ROSZKOWSKI, ARNOLD TRZECIAK, DIETER GILLESSEN, VOLDEMAR TOOME, BOGDA WEGRZYNSKI, ROSS PITCHER, and JOHANNES MEIENHOFER

1. INTRODUCTION

The isolation of biologically important peptides from the thymus gland has been studied extensively in the last few years. Several thymic peptides have been shown to play certain roles in T-cell maturation (White, 1980; Trainin *et al.*, 1980a,b; Goldstein and Lau, 1980; Bach and Goldstein, 1980). Thymosin α_1, a highly acidic $N\alpha$-acetyl octacosapeptide, isolated from calf thymus gland (Goldstein *et al.*, 1977) and characterized by sequence analysis (Low and Goldstein, 1979), has been reported to exhibit biological activities involved in the development of thymus-de-

ARTHUR M. FELIX, EDGAR P. HEIMER, CHING-TSO WANG, THEODORE J. LAMBROS, JOSEPH SWISTOK, MUSHTAQ AHMAD, MARTIN ROSZKOWSKI, VOLDEMAR TOOME, BOGDA WEGRZYNSKI, ROSS PITCHER, and JOHANNES MEIENHOFER • Bio-Organic Chemistry Department, Roche Research Center, Hoffman–La Roche Inc., Nutley, New Jersey 07110. ARNOLD TRZECIAK and DIETER GILLESSEN • F. Hoffmann–La Roche and Co., Ltd., Basel, Switzerland.

pendent lymphocytes (T cells) (Goldstein *et al.*, 1977). Thymosin α_1 has been synthesized by classical procedures in solution (Wang *et al.*, 1979; Birr and Stollenwerk, 1979) and by solid-phase methods (Wong and Merrifield, 1980; Wang *et al.*, 1980; Colombo, 1981).

The need for larger quantities of thymosin α_1 (100 g scale) for clinical evaluation required a solution synthesis which could be readily scaled up. A more efficient synthesis of thymosin α_1 was therefore designed using fragments with *tert*-butyl side chain protecting groups. Coupling sites and conditions were chosen to minimize racemization and provide the flexibility to prepare analogs of thymosin α_1 (e.g., desacetyl-thymosin α_1). An efficient purification procedure was required to maximize the overall yield.

2. RESULTS AND DISCUSSION

2.1. Synthesis of Thymosin α_1

The novel synthesis of thymosin α_1 required the preparation of seven fragments (Fig. 1; I–VII) which were prepared by stepwise chain elongation using either the mixed anhydride or the *N*-hydroxysuccinimide ester method. Assemblage of the intermediate fragments led to the preparation of two key protected intermediates (**8, 13**) as shown in Fig. 1. All the protected and deprotected peptide intermediates were characterized by IR, NMR, TLC, elemental analysis, optical rotation, and amino acid analysis.

Alternate coupling conditions of **9** and **13** were investigated to optimize the yield and purity of the fully protected 28-peptide hormone, **14.** The final coupling was best achieved in trifluoroethanol (0°C for 1 hr and 25°C for 21 hr) with 1.8 equiv. of **13** using DCC (5.6 equiv.) and HOBt (10.1 equiv.).

2.2. Purification of Thymosin α_1

Following the deprotection of the crude protected 28-peptide hormone, **14,** preparative HPLC of the resultant thymosin α_1 was performed in two stages. Crude thymosin α_1 was partially purified (~ 8.5 g product per run) on a Jobin–Yvon Chromatospac Prep. 100 instrument (8 × 100 cm) using C_8 reverse-phase silica gel. Final purification (~ 2.5 g product per run) was carried out on the Whatman Partisil-10 Magnum 20 ODS-3 system (2 × 100 cm) using C_{18} reverse-phase silica gel with simultaneous discontinuous monitoring of each fraction with fluoropa which permitted pooling of appropriate fractions (Fig. 2).

The overall yield for the final four stages of thymosin α_1 synthesis was ~ 30% and the product was shown to be homogeneous and in agreement with expected values (TLC, HVE, IEF, amino acid analysis, peptide mapping, and ^{1}H NMR). Thymosin α_1 prepared by this procedure (nearly 100 g) was obtained in microcrystalline form and released for clinical studies.

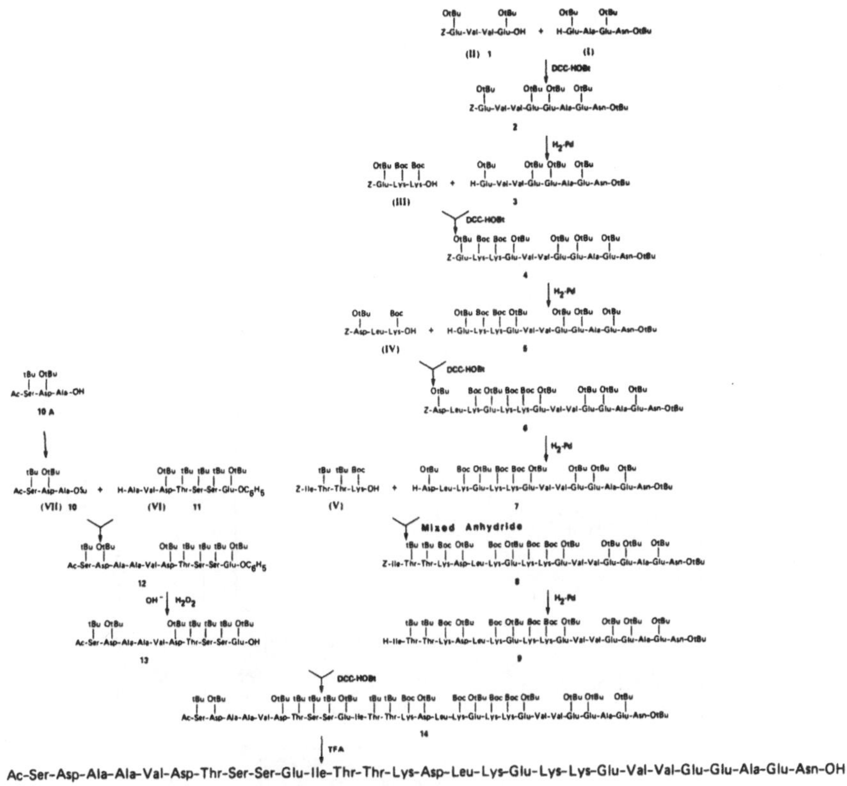

THYMOSIN α_1

FIGURE 1. Synthesis of thymosin α_1.

2.3. Analogs and Stereoisomers of Thymosin α_1

A series of COOH-terminal analogs of thymosin α_1 was prepared directly from the *tert*-butyl protected fragments (Fig. 1). These analogs were purified by ion-exchange or preparative HPLC and shown to be homogeneous and in agreement with expected values (Table I). Preliminary results in the *in vivo* immunosuppressed mouse bioassay (Ishitsuka *et al.*, 1983) indicate that the biological activity of thymosin α_1 resides in the COOH terminus of the molecule increasing with increasing chain length. Circular dichroism studies were carried out on the fragments (Fig. 3) and it was observed that the onset of secondary structure coincided with biological activity.

The three minor stereoisomer contaminants from the thymosin α_1 synthesis were isolated by preparative HPLC (Fig. 2) and identified by a series of analytical measurements. Tryptic digestion, amino acid analysis, ^1H NMR spectroscopy, and glass capillary chromatography were used to identify each of the stereoisomers:

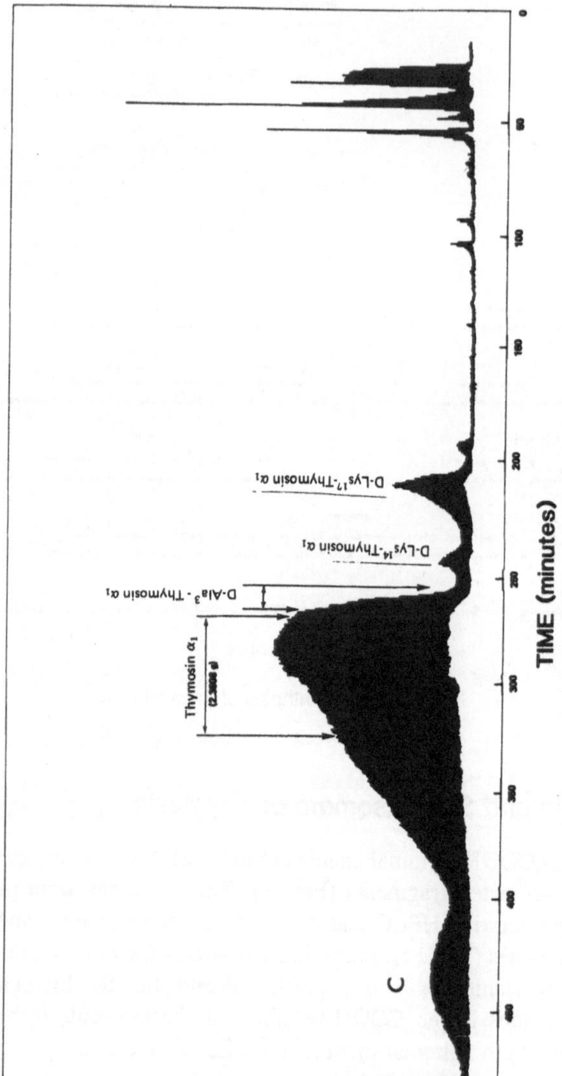

FIGURE 2. Preparative HPLC purification.

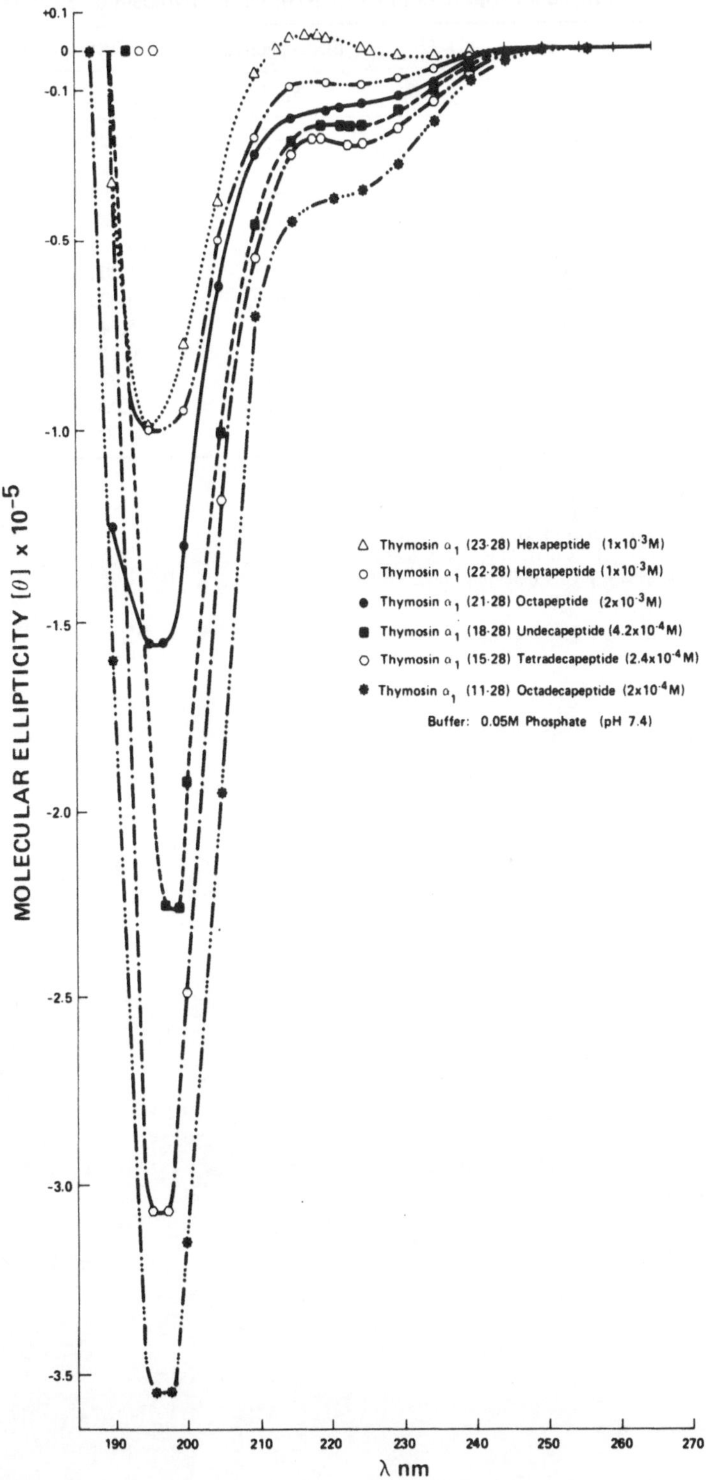

FIGURE 3. Circular dichroism spectra of thymosin α_1 H-peptide-OH fragments.

TABLE I. Physical Properties of COOH-Terminal Thymosin α_1 Peptides

Peptide	$[\alpha]_D^{25}$[a]	TLC R_f[b]	TLC R_f[c]	Electrophoresis R_{Glu}[d]	Circular dichroism $[\theta]_{220-225\ nm}$[e]	Circular dichroism $[\theta]^{MRW}_{220-225\ nm}$[f]
Di-	+7.25°	0.53	—	—	+1,528	+764
Tri-	−20.47°	0.47	—	—	—	—
Tetra-	−37.25°	0.47	—	—	+2,799	+670
Hexa-	−67.77°	0.50	—	—	+3,987	+665
Hepta-	−75.59°	0.58	—	—	−9,409	−1344
Octa-	−88.51°	0.55	0.90	0.92	−15,200	−1900
Nona-	−88.50°	0.35	0.88	0.63	−14,000	−1556
Deca-	−71.73°	0.21	0.80	0.31	−15,949	−1595
Undeca-	−88.55°	0.29	0.73	0.47	−20,219	−1838
Tetradeca-	−86.98°	0.14	0.80	0.34	−24,672	−1762
Octadeca-	−86.73°	—	0.75	0.14	−39,000	−2167
Thymosin α_1	−92.33°	—	0.82	0.56	−80,000	−2857

[a] Specific rotation: concn., 1% + 0.2%; in 0.1 M HCl.
[b] Solvent system: *n*-BuOH : AcOH : EtOAc : H_2O, 1–1–1–1.
[c] Solvent system: *n*-BuOH : Pyr : AcOH : H_2O, 10–3–15–12.
[d] Electrophoresis: Pyr : AcOH : H_2O, 2.3–0.6–97 (pH 5.6; 1000 V; 90min).
[e] Molecular ellipticity: pH 7.4 (0.05 M phosphate buffer).
[f] Mean residue ellipticity: pH 7.4 (0.05 M phosphate buffer).

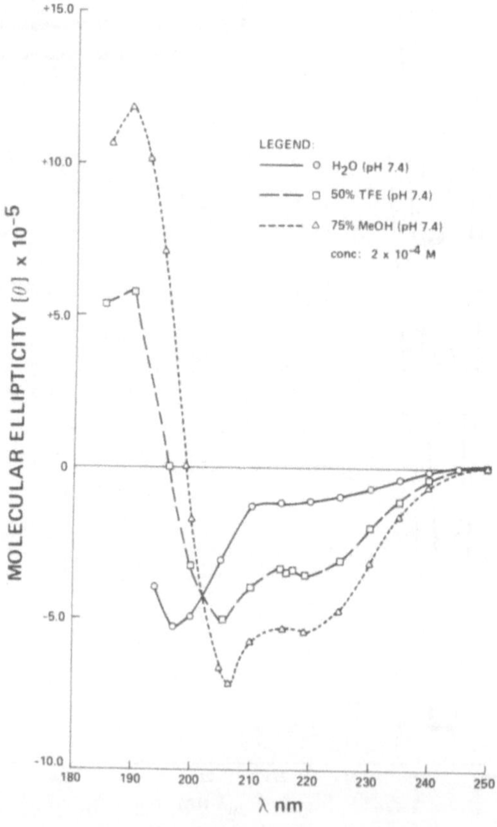

FIGURE 4. Circular dichroism of thymosin α_1.

TABLE II. Thymosin α_1 and Related Analogs: Circular Dichroism in 75% MeOH (pH 7.4)

Compound	First cotton effect (n → II*)				Second cotton effect (II → II*)			
	λ	[θ]	$[\theta]_{MRW}$	% helix	λ	[θ]	$[\theta]_{MRW}$	% helix
Thymosin α_1	219	−545,000	−19,464	63.9	206–207	−715,000	−25,536	71.6
D-Ala^3-thymosin α_1	217	−440,000	−15,714	49.7	206–207	−600,000	−21,429	55.1
Des-Ac-thymosin α_1	216	−436,000	−15,571	49.1	206	−596,000	−21,286	54.6
D-Lys^{14}-thymosin α_1	219–220	−307,500	−10,982	31.8	205	−470,000	−16,786	36.5
D-Lys^{17}-thymosin α_1	220	−255,000	−9,107	24.7	203–204	−435,000	−15,536	31.5
Thymosin α_1 (11–28)	217	−154,000	−8,556	22.6	204–205	−288,000	−16,000	33.3
Thymosin α_1 (15–28)	217	−104,000	−7,429	18.3	—	—	—	
Thymosin α_1 (1–14)	215–217	−75,000	−5,357	10.4	—	—	—	
Thymosin α_1 (21–28)	222–223	−24,000	−3,000	1.5	—	—	—	

D-Ala[3]-thymosin α_1, D-Lys[14]-thymosin α_1, and D-Lys[17]-thymosin α_1. Circular dichroism studies were carried out on thymosin α_1 (Fig. 4) in three solvents (H_2O, 50% TFE, and 75% MeOH) and compared with the stereoisomers in 75% MeOH (Table II). It was observed that the amount of secondary structure was decreased by the presence of D-amino acid residues. This decrease in percent helicity was most pronounced when the D-amino acid residue was present in the COOH-terminal region of thymosin α_1 (i.e., D-Lys[17]-thymosin α_1). ^{13}C NMR spectra were run on all the thymosin α_1 peptide fragments in D_2O. This series of ^{13}C NMR spectra permitted complete assignments for all 129 carbon atoms present in thymosin α_1.

REFERENCES

Bach, J.-F., and Goldstein, G., 1980, Newer concepts of thymic hormones, *Thymus* **2**:1–4.

Birr, C., and Stollenwerk, U., 1979, Synthesis of thymosin α_1, a polypeptide of the thymus, *Angew. Chem. Int. Ed. Engl.* **18**:394–395.

Colombo, R., 1981, Solid-phase peptide synthesis without acidolysis: The synthesis of thymosin α_1 on a new benzhydrylamine resin, *J. Chem. Soc. Chem. Commun.* **1981**:1012–1013.

Goldstein, A. L., Low, T. L. K., McAdoo, M., McClure, J. E., Thurman, G. B., Rossio, J. L., Lai, C.-Y., Chang, D., Wang, S.-S., Harvey, C., Ramel, A. H., and Meienhofer, J., 1977, Thymosin α_1: Isolation and sequence analysis of an immunologically active thymic polypeptide, *Proc. Natl. Acad. Sci. USA* **74**:725–729.

Goldstein, G., and Lau, C., 1980, Thymopoietin and immunoregulation, in: *Polypeptide Hormones* (R.F. Beers, Jr., and E.G. Bassett, eds.), pp. 459–465, Raven Press, New York.

Ishitsuka, H., Umeda, Y., Nakamura, J., and Yagi, 1983, Protective activity of thymosin against opportunistic infections in animal models, *Cancer Immunol. Immunother.* **14**:145–150.

Low, T. L. K., and Goldstein, A. L., 1979, The chemistry and biology of thymosin, *J. Biol. Chem.* **254**:987–993.

Trainin, N., Umiel, T., Klein, B., and Kleir, I., 1980a, Role of thymus humoral factor, a thymic hormone, in the physiology of the thymus, in: *Polypeptide Hormones* (R.F. Beers, Jr., and E.G. Bassett, eds.), pp. 467–488, Raven Press, New York.

Trainin, N., Zaisov, R., Yakir, Y., and Rotter, V., 1980b, Thymic hormones: Characterization and perspectives, in: *The Immune System: Functions and Therapy of Dysfunction* (G. Doria and A. Eshkol, eds.), pp. 159–169, Academic Press, New York.

Wang, S.-S., Kulesha, I. D., and Winter, D. P., 1979, Synthesis of thymosin α_1, *J. Am. Chem. Soc.* **101**:253–254.

Wang, S.-S., Makofske, R., Bach, A., and Merrifield, R. B., 1980, Automated solid phase synthesis of thymosin α_1, *Int. J. Pept. Protein Res.* **15**:1–4.

White, A., 1980, Chemistry and biological actions of products with thymic hormone-like activity, in: *Biochemical Actions of Hormones*, Vol. VII, (G. Litwack, ed.) pp. 1–46, Academic Press, New York.

Wong, T. W., and Merrifield, R. B., 1980, Solid phase synthesis of thymosin α_1 using tert-butyloxycarbonyl-aminoacyl-4-(oxymethyl)-phenylacetamidomethyl-resin, *Biochemistry* **19**:3233–3238.

13

Immunocytochemical Localization of Thymosin and Thymopoietin in Human, Rat, and Murine Thymus

HOWARD R. HIGLEY and GEOFFERY ROWDEN

1. INTRODUCTION

Several distinct protein preparations isolated from the thymus have been shown to be involved in the differentiation of T lymphocytes. Thymopoietin, the thymosin complex, and facteur thymique sérique (FTS), referred to collectively as thymic hormones, have been localized in the nonlymphoid stromal network of the thymus by immunocytochemical techniques (Mandi *et al.*, 1979; Teodorczyk *et al.*, 1975; Monier *et al.*, 1980; van den Tweel *et al.*, 1979). However, as the thymic nonlymphoid cell population is a highly heterogeneous one, the specific cell type or types of thymic hormone origin have not been unequivocally described. Several forms of epithelial cells and macrophages have been identified in the different zones of the thymus and all are potential candidates for thymic hormone production (Duijvestijn and Hoefsmit, 1981; von Gaudecker and Müller-Hermelink, 1980). There also seems to be a difference in the distribution of cells immunoreactive for various thymic hormone preparations within the thymuses and other organs of different species (Mandi *et al.*, 1979; Teodorczyk *et al.*, 1975; Monier *et al.*, 1980; van den Tweel *et al.*, 1979). Finally, in all cases except one (Schmitt *et al.*, 1982), there has been no ultrastructural immunocytochemical investigation of the synthetic apparatus or secretory mechanism in cells containing thymic hormones. The objectives of the present investigation were: (1) to compare the immunocytochemical staining patterns using antisera to thymopoietin and thymosins α_1, α_7, and β_4, to those produced with antisera against keratin, lysozyme (or peanut agglutinin), and

HOWARD R. HIGLEY and GEOFFERY ROWDEN • Department of Pathology, Loyola University Medical School, Maywood, Illinois 60153.

S-100; known markers for epithelial cells (Battifora *et al.*, 1980), macrophages (Mendelsohn *et al.*, 1980), and dendritic cells (Rowden *et al.*, 1982), respectively; (2) to note differences in the distribution of thymic hormone-positive cell populations in human, mouse, and rat thymus and in human epidermis; and (3) to achieve an electron microscopic identification of the subcellular components associated with thymic hormone localization.

2. MATERIALS AND METHODS

Human thymic tissue was obtained at autopsy from an infant who had died of a congenital heart defect. Thymus was fixed in 10% buffered formalin for 8 hr, dehydrated in ethanol, and embedded in paraffin. Five-micrometer sections were mounted on glued slides, deparaffinized in xylene, and rehydrated to phosphate-buffered saline (PBS).

Human epidermis was obtained at autopsy and hooded rat and C57BL mouse thymuses were obtained from animals of from 1 to 3 weeks of age. Tissue was mounted in OCT and frozen at −20°C in a cryostat. Ten-micrometer sections were cut, well air-dried, and prior to immunocytochemical staining washed in cold PBS for 30 min, then fixed in cold 95% ethanol for 5 min and washed again in PBS.

2.1. Light Microscopic Immunocytochemistry

1. Paraffin sections were treated with 0.125% trypsin in $CaCl_2$ buffer for 10 min to better expose antigenic sites (frozen sections were not subjected to this step). PBS wash.

2. Endogenous peroxidase activity was eliminated by a 10-min incubation with 3% H_2O_2 (sections for immunofluorescence were not treated in this way). PBS wash.

3. A 15-min pretreatment with a 1 : 10 dilution of normal goat serum served to reduce nonspecific background staining.

4. Primary antiserum diluted 1 : 20–1 : 50 with PBS was applied to the sections for 1 hr at room temperature in a humid chamber. PBS wash. [Rabbit antithymopoietin was kindly provided by Dr. G. Goldstein. Rabbit antithymosin 6, 5, α_1, α_7, and β_4 were prepared by Dr. J. McClure and were generous gifts of Dr. A. Goldstein. Antiserum against human lysozyme (muramidase) was purchased from Immulok Inc., Carpenteria, Calif. Antiserum against human callus keratin and bovine S-100 was prepared by Dr. K. Sheikh. Antiserum against peanut agglutinin and PNA were purchased from Vector Laboratories, Burlingame, Calif.]

5. Sections were incubated with goat anti-rabbit IgG–peroxidase conjugate (Dako Corp., Santa Barbara, Calif.) or with fluorescein-conjugated GAR (Cappel Laboratories, Cochranville, Penn.) for 30 min at room temperature. PBS wash.

6. Tissues stained by the indirect immunofluorescence technique were viewed by ultraviolet illumination with appropriate barrier/exciter filters. Immunoperoxidase-labeled sections were developed by a 10-min treatment with aminoethyl carbazol–H_2O_2 or diaminobenzidine–H_2O_2 substrates.

7. Control reactions consisted of deletion of the primary antisera or substitution with normal rabbit sera and absorption of the primary antisera with thymosin 6 or thymopoietin overnight at 4°C before use. None of the control reactions produced any immunoreactivity above background.

2.2. Electron Microscopic Immunocytochemistry

Mouse thymus was used for immuno-EM staining and was fixed for 8 hr in 10% buffered formalin and then washed overnight in PBS. Tissue was then mounted in agar and 50-μm slices were cut on a tissue chopper.

1. Tissue chopper slices were incubated with thymic hormone antisera overnight at 4°C, then washed for 3 hr in three changes of PBS.

2. Tissues were then incubated overnight at 4°C with goat anti-rabbit IgG Fab fragment–peroxidase conjugate (Cappel), then washed three times in PBS.

3. Immunoreacted tissue was stained with diaminobenzidine–H_2O_2, osmicated, dehydrated, and embedded in Epon 812. Thin sections were cut and grids examined without heavy metal staining.

3. RESULTS

3.1. Light Microscopy of Human Thymus

Immunocytochemical analysis of parallel sections of human thymus revealed a different distribution of thymosin 6- and thymopoietin-positive cells (Fig. 1, 2). Thymosin 6 staining was restricted to cortical nonlymphoid cells. These cells were both punctate and dendritic in form. Staining with antikeratin and lysozyme showed these cells to be both epithelial cells and macrophages. There were few thymosin 6-containing cells in the medulla of this formalin-fixed specimen of infant thymus.

In contrast, thymopoietin-containing cells were largely restricted to the medulla with a prominent band of reactive cells in a subcapsular location (Fig. 2). Peripheral lamellae of Hassall's corpuscles contained thymopoietin and the majority of reactive cells also stained with antikeratin. No lysozyme-positive cells were seen to stain for thymopoietin nor did the occasional thymopoietin-positive, keratin-negative cell stain with S-100. Thymosin 6, 5, α_1, α_7, and β_4 were not detectable in conventionally fixed, paraffin-embedded human thymus.

3.2. Light Microscopy of Mouse Thymus

Thymosin α_1, α_7, and β_4 antisera did stain cortical epithelial cells in frozen sections of mouse thymus fixed in ethanol (Fig. 3). Interpretation of parallel sections suggested all three peptides were present in the same cortical epithelial cells in the mouse, but that not all keratin-positive cortical epithelia contained thymic hormones. Medullary epithelial cells were weakly positive with all three antisera. Macrophages (PNA positive) did not stain.

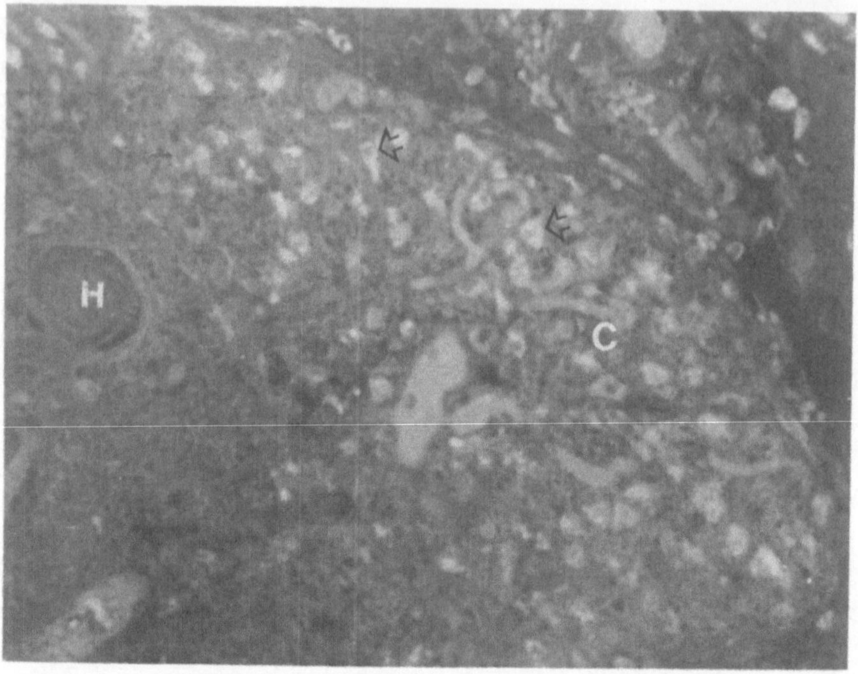

FIGURE 1. Paraffin section of human thymus immunostained for thymosin 6. Note restriction of fluorescence to cells (arrowheads) of the cortex (C) and the absence of reaction in the medullary zone. H, Hassall's corpuscle.

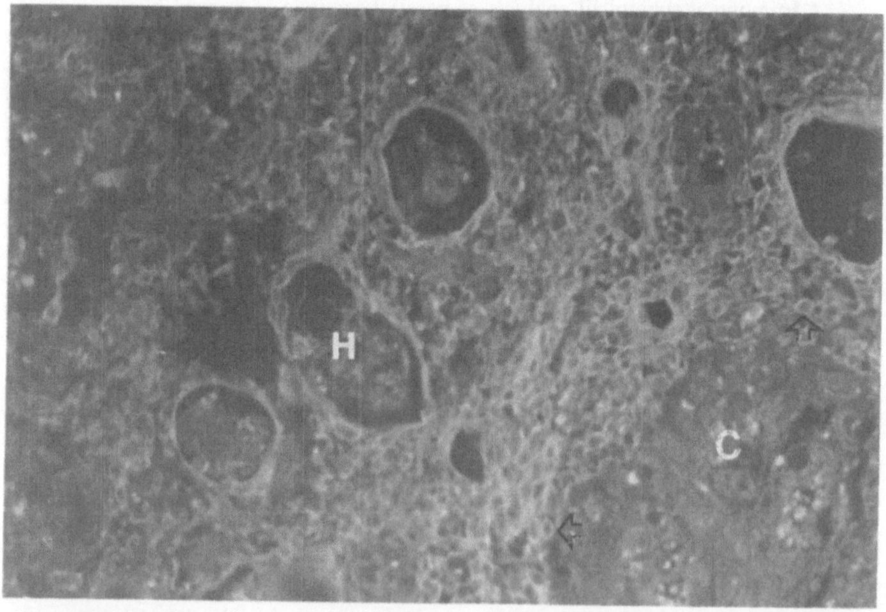

FIGURE 2. Paraffin section of human thymus immunostained for thymopoietin. Note fluorescence of medullary epithelial network (arrowheads), including the peripheral lamellae of Hassall's corpuscles (H). Cells in the cortical zone are unstained.

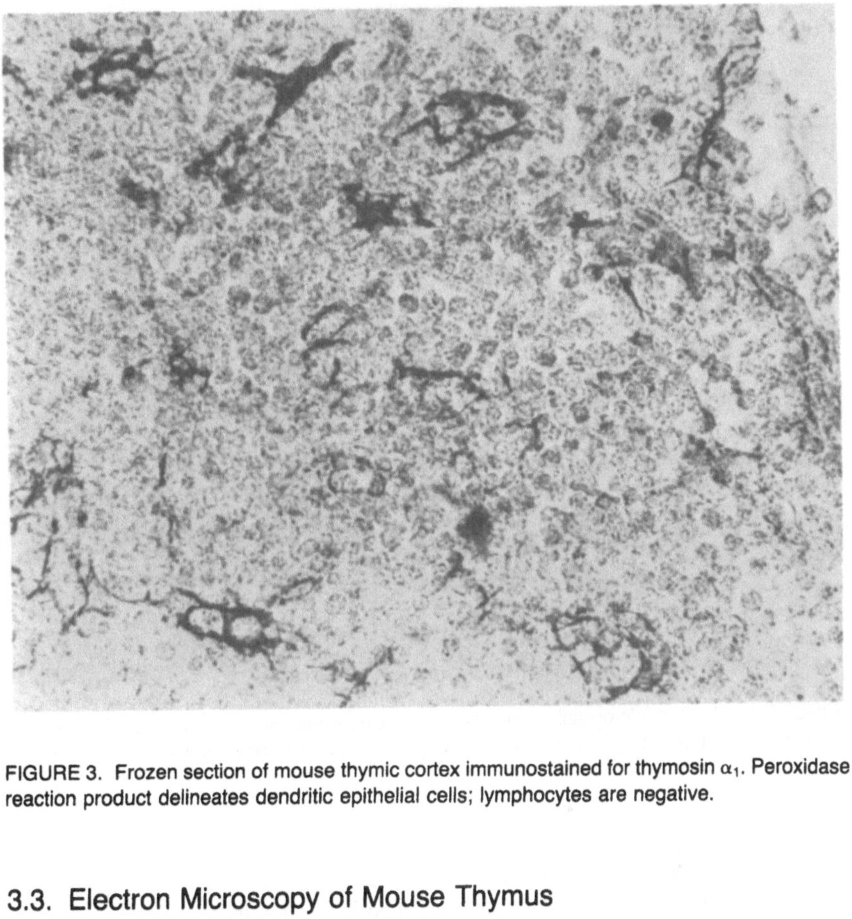

FIGURE 3. Frozen section of mouse thymic cortex immunostained for thymosin α_1. Peroxidase reaction product delineates dendritic epithelial cells; lymphocytes are negative.

3.3. Electron Microscopy of Mouse Thymus

Ultrastructural visualization of immunoreacted mouse thymic tissue chopper sections confirmed the light microscopic observations. Thymosin 6 antisera delineated cortical epithelial cells containing desmosomes and tonofilaments, and could be detected in macrophage cytoplasm as well (Fig. 4). Reaction product was associated with epithelial cell cytoplasm and specific cortical granules. No thymosin 6-positive medullary epithelial cells were noted.

Thymopoietin cytoplasmic reactivity was restricted to medullary and subcapsular forms (Fig. 5). However, cystic epithelial cells and the central cells of Hassall's corpuscles were not immunoreactive. No specific granules, as those seen in cortical epithelial cells, were seen to contain thymopoietin. Antisera to thymosin α_1, α_7, and β_4 were poorly reactive with formalin fixed tissue used for immuno-EM.

3.4. Light Microscopy of Rat Thymus and Human Epidermis

Thymosin α_1 and β_4 antisera stained thymic cortical epithelial cells in the rat; however, unlike the mouse, thymosin α_7 was found only in rat medullary epithelium. α_1 and β_4 peptides and thymopoietin were present in human epidermis. The dis-

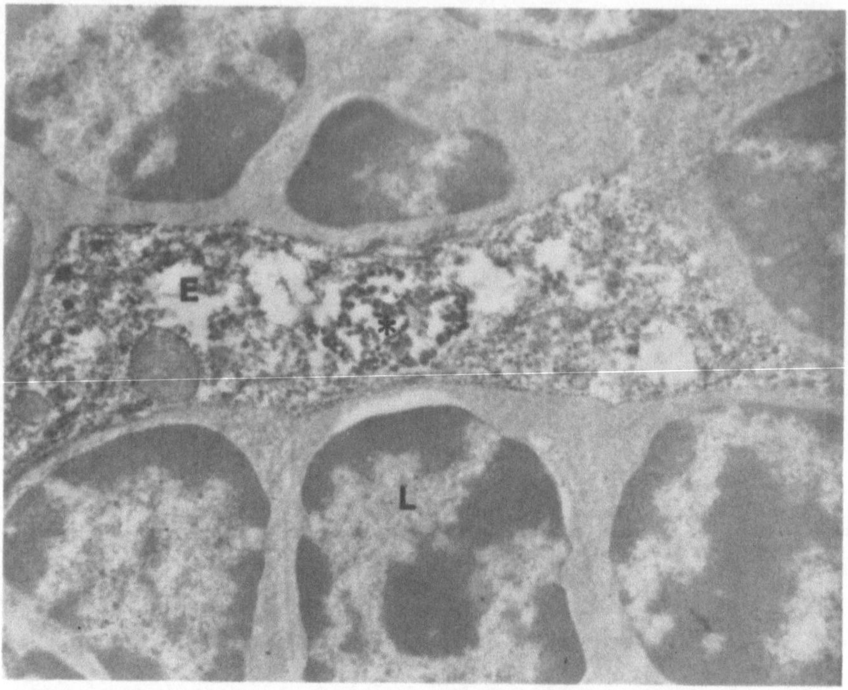

FIGURE 4. Electron micrograph of mouse thymic cortical epithelial cell process immunostained for thymosin 6. Reaction product is seen in cytoplasm (E) and cortical granule (*). No lymphocyte (L) staining is seen.

tribution was the reverse of that seen for callus keratin. Thymic factors were localized in basal keratinocytes that contained little immunoreactive keratin.

4. DISCUSSION

Thymosin is a complex collection of polypeptides, and antiserum to it stained both cortical epithelial cells and macrophages. Thymosin 6-positive macrophages and epithelia were found in other organs and lymphoid tissues (Hirokawa *et al.*, 1982, and personal observations), indicating that not all components in this preparation are thymus specific. Purified thymosin subfractions α_1, α_7, and β_4 seem to be exclusively epithelial cell products, but only α_7 seems to be thymus specific. As shown in this study, α_1 and β_4 are cross-reactive with basal cells of the epidermis, and staining of epithelium in other organs has been reported (A. Goldstein, personal communication). The restriction of α_1, α_7, and β_4 to cortical epithelium seems to be unique to the mouse thymus. α_1 and β_4 have been demonstrated in both the cortex and the medulla in frozen sections of human thymus (Hirokawa *et al.*, 1982; Haynes *et al.*, 1983), with α_7 found only in medullary epithelium (A. Goldstein, personal communication). As shown here, the staining pattern is different yet in

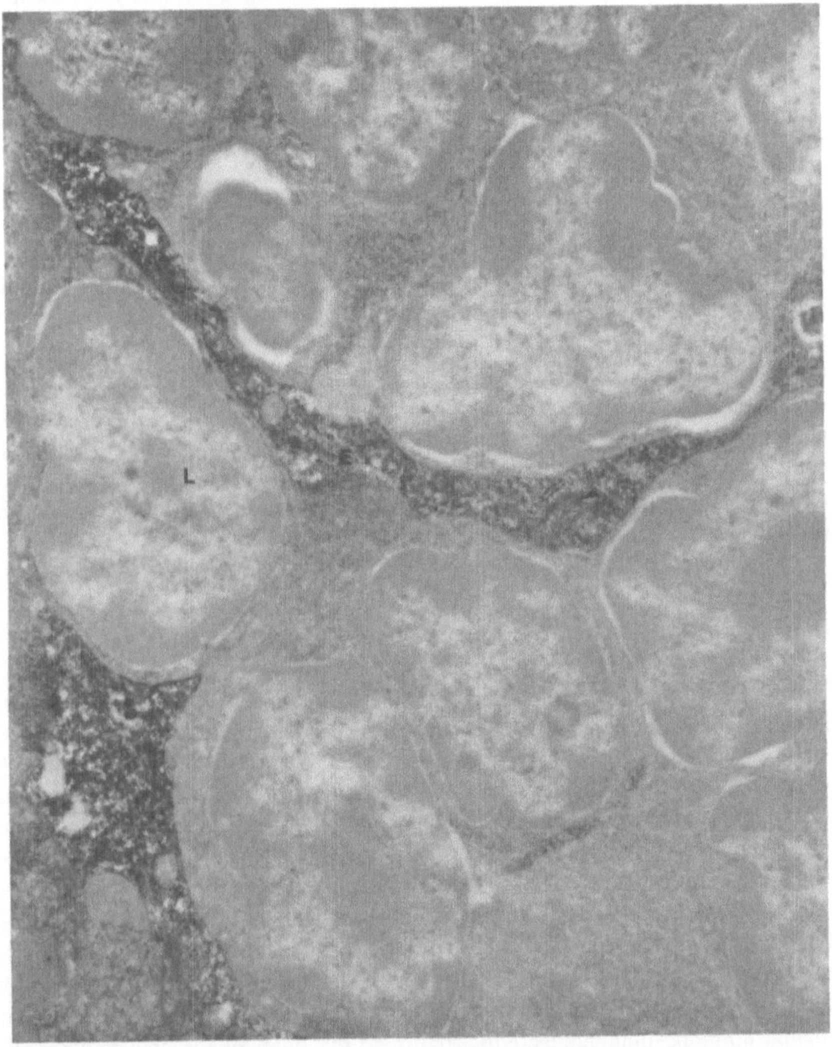

FIGURE 5. Electron micrograph of mouse thymic subcapsular epithelial cells (E) immunostained for thymopoietin. Arrowheads indicate tonofilaments. Note absence of reaction product in surrounding lymphocytes (L).

the rat, with α_1 and β_4 found in the cortex and α_7 found in the medulla indicating a species specificity in the distribution of these polypeptides.

With the exception of basal keratinocyte staining for thymopoietin (Chu *et al.*, 1982; Haynes, Chapter 3, this volume), this thymic hormone seems to be unique to the thymus. But the present study illustrates that not all thymic epithelia contain this factor and an as yet unidentified thymopoietin-positive medullary hypertrophic cell may not be epithelial.

The only previous electron microscopic demonstration of thymic hormone-containing structures was of the serum thymic factor FTS in cortical epithelial cell granules (Schmitt *et al.*, 1982). However, that observation has been complicated by the report of FTS staining of nonthymic epithelial structures and its possible association with intermediate filaments (Kato *et al.*, 1981). Furthermore, a subsequent report of FTS staining indicated that most of the immunoreactive cells were present in the medulla (Savino *et al.*, 1982). The formalin-resistant component of thymosin 6 was shown here to be present in both cortical epithelial cell granules and cytoplasm. Thymopoietin's ultrastructural localization was exclusively cytoplasmic, possibly associated with filamentous material. Whether these observations indicate localization of hormone in both cellular compartments or represents a high-resolution diffusion artifact remains to be proven. Although both hormones exhibit definite plasma levels demonstrated by radioimmunoassay (Pahwa *et al.*, 1980), a conventional method of secretion of an intermediate filament-associated protein is difficult to envision. Therefore, the morphological mechanism of local induction of T-cell differentiation by thymic hormones is still problematic.

5. SUMMARY

This study examined thymic hormone distribution and species specificity by immunocytochemical analysis. The thymosins and thymopoietin appeared to be largely products of different thymic nonlymphoid cells. Thymosin α_1, α_7, and β_4 were exclusively associated with epithelial cells but only α_7 was thymus specific. The majority of thymopoietin-containing cells were thymic epithelia, but epidermal basal keratinocytes contained this hormone as well. Thymosin fractions 5 and 6 contained macrophage products that could be found in other lymphoid tissues. There were differences in cortical versus medullary locations of thymic hormone-containing cells in man, rat, and mouse. These differences did not appear to be attributable to the variety of developmental stages examined or tissue preparative techniques employed. An ultrastructual examination of organelles associated with thymic hormone immunoreactivity confirmed the light microscopic findings, but raised some questions about how thymic hormones are secreted and interact with their target lymphocytes.

REFERENCES

Battifora, H., Sun, T., Bahu, R., and Sambasiva, R., 1980, Use of antikeratin antiserum as a diagnostic tool: Thymoma vs lymphoma, *Hum. Pathol.* **11**:653.

Chu, A., Goldstein, G., Berger, C., Paterson, J., and Edelson, R., 1982. Identification of thymopoietin-like substance in epidermal cells from normal human skin, *J. Invest. Dermatol.* **78**:330 (abstract).

Duijvestijn, A., and Hoefsmit, E., 1981, Ultrastructure of the rat thymus, the microenvironment of T-lymphocyte maturation, *Cell Tissue Res.* **218**:279.

Haynes, B., Warren, R., Buckley, R., McClure, J., Goldstein, A., Henderson, F., Hensley, L., and

Eisenbarth, G., 1983, Demonstration of abnormalities in expression of thymic epithelial surface antigens in severe cellular immunodeficiency diseases, *J. Immunol.* **130:**1182.

Hirokawa, K., McClure, J., and Goldstein, A., 1982, Age related changes in the localization of thymosin in human thymus, *Thymus* **4:**19.

Kato, K., Ikeyama, S., Takaoki, M., Shino, A., Takeuchi, M., and Kakinuma, A., 1981, Epithelial cell components immunoreact with anti-serum thymic factor (FTS) antibodies: Possible association with intermediate-sized filaments, *Cell* **24:**885.

Mandi, B., Holub, M., Rossmann, P., Csaba, B., Glant, T., and Olivetti, E., 1979, Detection of thymosin 5 in calf and mouse thymus and in nude mouse dysgenetic thymus, *Folia Biol. (Prague)* **25:**49.

Mendelsohn, G., Eggleston, J., and Man, R., 1980, Relationship of lysozyme to histiocytic differentiation in malignant histiocytosis, *Caner* **45:**273.

Monier, J., Dardenne, M., Pléau, M., Schmitt, D., Deschaux, P., and Bach, J., 1980, Characterization of facteur thymique serique (FTS) in the thymus, *Clin. Exp. Immunol.* **42:**420.

Pahwa, R., Ikehara, S., Pahwa, S., and Good, R., 1980, Thymic function in man, *Thymus* **1:**27.

Rowden, G., Sheikh, K., Misra, B., Higley, H., Relfson, M. and Connelly, E., 1982, Investigations of glial protein S-100 and GFA in human dendritic cells in the epidermis and lymph nodes, *J. Invest. Dermatol.* **78**(4):354 (abstract).

Savino, W., Dardenne, M., Papiernik, M., and Bach, J., 1982, Thymic hormone containing cells: Characterization and localization of serum thymic factor in young mouse thymus studied by monoclonal antibodies, *J. Exp. Med.* **156:**628.

Schmitt, D., Monier, J., Dardenne, M., Pléau, M., and Bach, J., 1982, Location of FTS in the thymus of normal and autoimmune mice, *Thymus* **4:**21.

Teodorczyk, J., Potworowski, E., and Silviculus, A., 1975, Cellular localization and antigenic species specificity of thymic factors, *Nature (London)* **258:**617.

van den Tweel, J., Taylor, C., McClure, J., and Goldstein, A., 1979, Detection of thymosin in thymic epithelial cells by an immunoperoxidase method, *Proc. Int. Germinal Center Conf. Kiel.*

von Gaudecker, B., and Müller-Hermelink, H., 1980, Ontogeny and organization of the stationary non-lymphoid cells in the human thymus, *Cell Tissue Res.* **207:**287.

14

Thymus- and Spleen-Derived Immunosuppressive Peptides

I. BLAZSEK, G. MATHÉ, M. LENFANT, and N. KIGER

The hematoimmunological machinery appears to be regulated by soluble factors acting in two opposite directions, those of subpopulation size and function amplification and reduction (suppression), respectively (Table I).

In 1972, we prepared two immunosuppressive factors, one from the spleen (SISF) and one from the thymus (TISF) (Mathé, 1972). We have also obtained a suppressive factor produced by histamine-induced suppressor cells (Susuki and Huchet, 1982) (Table I). Table II compares the factor's susceptibility to hydrolytic enzymes to that of FTS according to Dardenne *et al.* (1977), Bach *et al.* (1978), and Pléau *et al.* (1977).

Figures 1 and 2 show the effect of SISF on the *in vivo* GvH mortalities induced by a semiallogeneic bone marrow and spleen cell graft or after bone marrow graft incompatible for the minor histocompatibility antigens, respectively. We have reported the severity of the GvH complicity of such a graft (Mathé *et al.*, 1979; Rappaport *et al.*, 1979).

Figure 3 indicates the preparation modality we have used to obtain thymic factor TISF-ICIG 1991. Figure 4 shows the effect of the subfraction T4-P2 on the spontaneous proliferation of different cell types, measured by [^{3}H]-TdR incorporation. The thymocytes and bone marrow cells were most sensitive targets. Figure 5 illustrates the effect of 30 subfractions prepared by isoelectric focusing of the T4-P2 fraction on Ampholine gradient on DNA synthesis of thymocytes, bone marrow, spleen, and transformed cells. Its action is significant on the thymocytes and Yac cells.

I. BLAZSEK, G. MATHÉ, M. LENFANT, and N. KIGER • Institut de Cancérologie et d'Immunogénétique (INSERM U-50), Hôpital Paul-Brousse, 94804 Villejuif, France.

TABLE I. Main Factors Known as Regulators of the Hematoimmunological Machinery in the Suppressive Direction

	Thymus	Spleen	Granulocyte	Peripheral blood system	Cells	Added substance	References[a]
Thymic immunosuppressive factor (TISF-ICIG 1991)	+						1, 2
Thymosin α_7	+						3
Splenic immunosuppressive factor (SISF-ICIG 1992)		+					1, 4–11
Bovine spleen chalone		+					12
Lymphocyte growth-inhibiting splenic factor		+					13
Spleen immunosuppressive extract		+					14
Granulocyte chalone			+				15
Thymic inhibitory factor	+						16
Tuftsin agonist tripeptide				+			17
Tuftsin agonist pentapeptide				(Synthetic)			18
Serum suppressive factor of the primary response				+			19
Serum immunosuppressive factor				+			20
α-Globulin-like peptides				+			21
Suppressive immunoglubulin binding factor (IBF)				+			22
					Spleen cells	Con A	23
Histamine-induced immunosuppressor factor (HIIF-ICIG 2039)					Lymphocytes	Histamine	24
Lymphocyte DNA synthesis inhibitor (IDS)					T lymphocytes		25

[a] (1) Mathé, 1972; (2) Kiger *et al.*, 1977; (3) Goldstein *et al.*, 1981; (4) Garcia-Giralt *et al.*, 1978; (5–10) Lenfant *et al.*, 1976, 1978, 1980, 1981a, 1981b, 1982; (11) Millerioux *et al.*, 1981; (12) Grundboeck, 1976; (13) Nordling *et al.*, 1979; (14) Ward and Munro, 1979; (15) Maurer and Laerum, 1976; (16) Blazsek and Gaàl, 1978; (17) Conference on Tuftsin, 1983; (18) Nozaki *et al.*, 1977; (19) Lee and Paraskevas, 1977; (20) Woorting-Hawking and Michael, 1977; (21) Wang *et al.*, 1977; (22) Friedman *et al.*, 1977; (23) Tadakuma *et al.*, 1976; (24) Suzuki and Huchet, 1982; (25) Namba *et al.*, 1978.

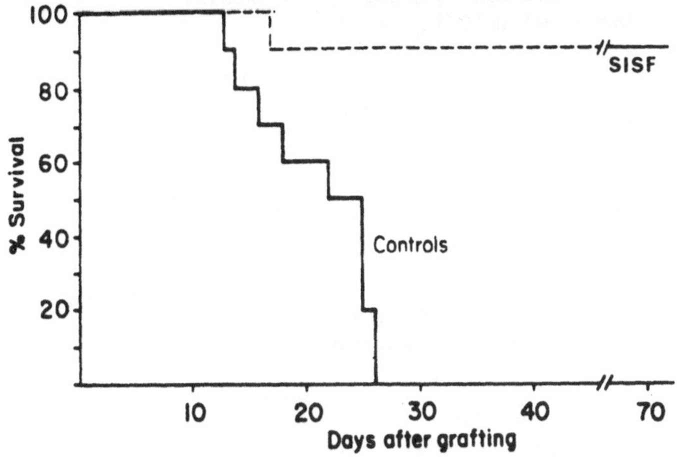

FIGURE 1. The effect of SISF on the *in vivo* GvH mortalities induced by a semiallogeneic bone marrow and spleen cell graft.

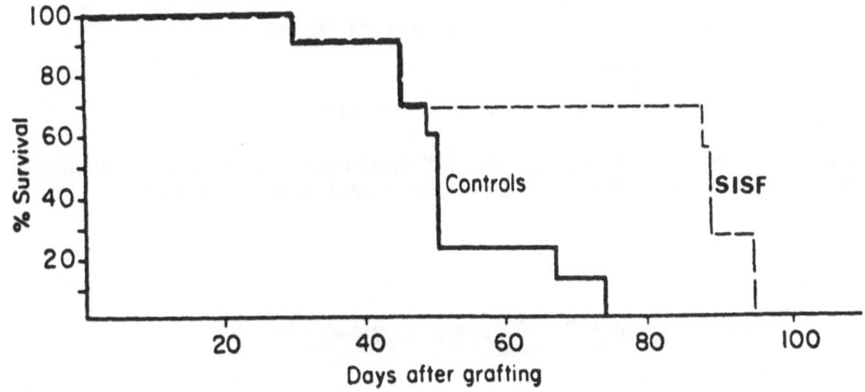

FIGURE 2. The effect of SISF on bone marrow graft incompatible for the minor histocompatibility antigens.

TABLE II. Comparison of SISF and FTS Behavior

	SISF	FTS[a]
Enzymatic susceptibility		
Pronase	+	+
Trypsin	+	+
DNase	−	−
RNase	−	−
Carboxypeptidase	+	+
Leucine aminopeptidase	−	ND
Pyroglutamate aminopeptidase	+	+
Electrophoretic mobility, m (pH 6.5)	$0 < m < 0.1$	0.1

[a] Data from Dardenne *et al.* (1977), Pléau *et al.* (1977), Bach *et al.* (1978).

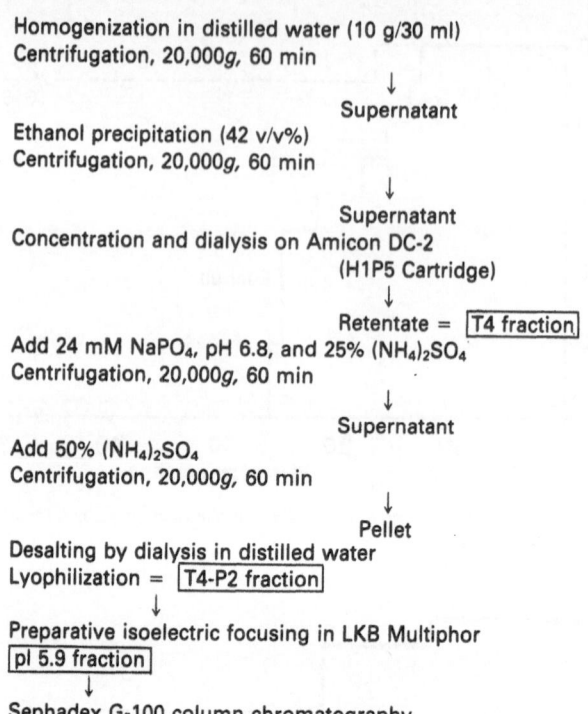

Homogenization in distilled water (10 g/30 ml)
Centrifugation, 20,000*g*, 60 min

↓

Supernatant

Ethanol precipitation (42 v/v%)
Centrifugation, 20,000*g*, 60 min

↓

Supernatant

Concentration and dialysis on Amicon DC-2
(H1P5 Cartridge)

↓

Retentate = **T4 fraction**

Add 24 mM $NaPO_4$, pH 6.8, and 25% $(NH_4)_2SO_4$
Centrifugation, 20,000*g*, 60 min

↓

Supernatant

Add 50% $(NH_4)_2SO_4$
Centrifugation, 20,000*g*, 60 min

↓

Pellet

Desalting by dialysis in distilled water
Lyophilization = **T4-P2 fraction**

↓

Preparative isoelectric focusing in LKB Multiphor
pI 5.9 fraction

↓

Sephadex G-100 column chromatography

FIGURE 3. Extraction and purification of a calf thymic protein fraction. The marked fractions are active in inhibiting the DNA synthesis of lymphocytes *in vitro* and/or *in vivo*.

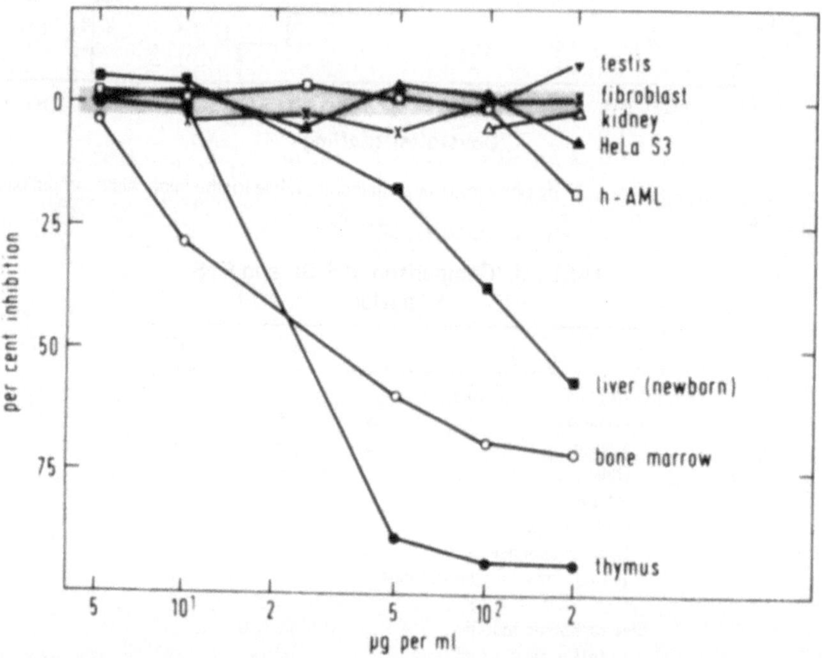

FIGURE 4. The effect of the subfraction T4-P2 on the spontaneous proliferation of different cell types, measured by [^{3}H]-TdR incorporation.

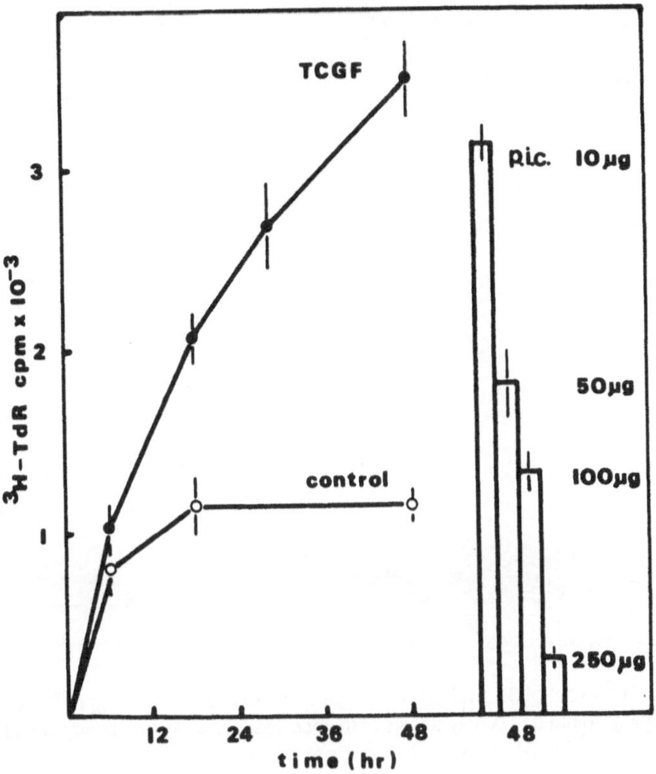

FIGURE 7. Concentration-dependent inhibitory effect of pi c subfraction on [^{3}H]-TdR incorporation of cytolytic T cells.

TABLE III. Effect of Time and Dose Administration of Thymic Fractions on Mouse IgM Response against a Thymus-Dependent Antigen (SRBC)

Expt.	Treatment	Dose (µg protein i.v.)	Day of injection	PFC/spleen[a]	p value[b]
1	—	—	—	93,600 ± 1,000	—
2	—	—	—	31,800 ± 5,300	—
1	Fraction o	480	0	80,500 ± 3,000	NS
1	(control)	480	3	86,000 ± 4,600	NS
2		200	3	25,800 ± 2,000	NS
1	Fraction b	1000	0	85,500 ± 4,700	NS
1		1000	3	56,000 ± 5,200	$0.02 < p < 0.05$
2		600	3	16,700 ± 3,900	$0.02 < p < 0.05$
1	Fraction c	480	0	135,800 ± 12,300	NS
1		480	3	30,400 ± 2,300	$p < 0.001$
2		200	3	16,000 ± 2,000	$0.01 < p < 0.02$

[a] Mean ± S.E.; eight mice per group.
[b] Student's *t* test.

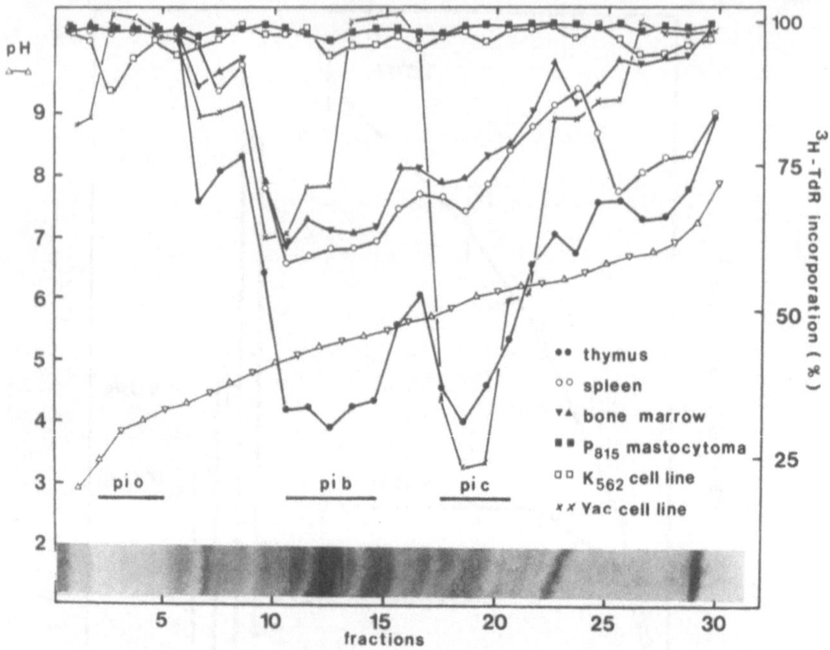

FIGURE 5. The effect of 30 subfractions prepared by isoelectric focusing of the T4-P2 fraction on Ampholine gradient on DNA synthesis of thymocytes, bone marrow, spleen, and transformed cells.

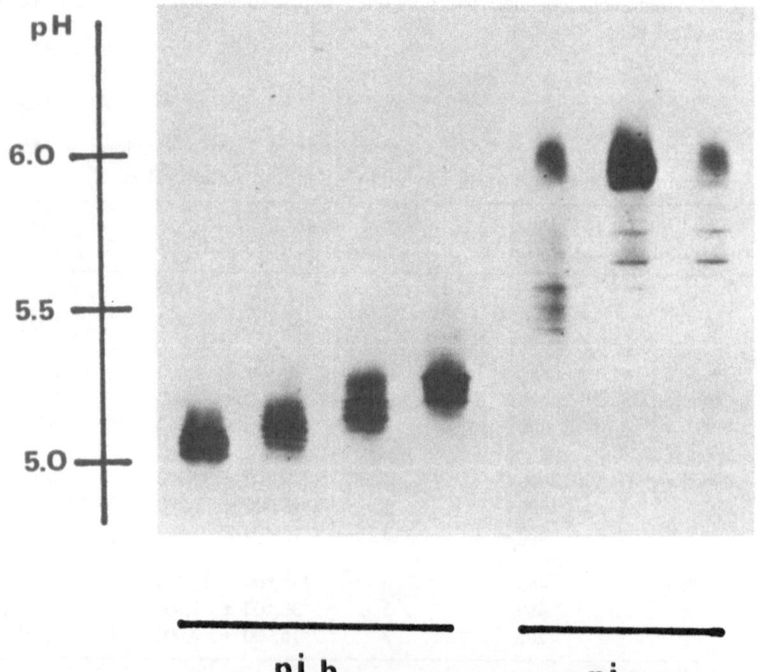

FIGURE 6. Isoelectric focusing analysis of the two subfractions pi b and pi c.

TABLE IV. Effect of Preincubation with the Fraction Purified by Preparative Isoelectric Focusing on the Capacity of Spleen Cells to Induce *In Vivo* a GvH Reaction Measured by Splenomegaly of the Recipient F_1 Mice

Spleen cell donors	F_1	C57BL/6			
In vitro pretreatment by thymic fraction	—	—	Fraction o (control)	Fraction b	Fraction c
Mean relative spleen weight ± S.E.	352 ± 35	785 ± 80	745 ± 75	479 ± 89	510 ± 42
No. of mice per group	9	6	8	9	10
p value (vs. group 2)[a]	$p < 0.001$	—	NS	$p = 0.05$	$0.01 < p < 0.02$
Spleen index[b]	—	2.0	2.1	1.36	1.4

[a] Student's *t* test.
[b] Spleen index = relative spleen weight (experimental group)/relative spleen weight (first column).

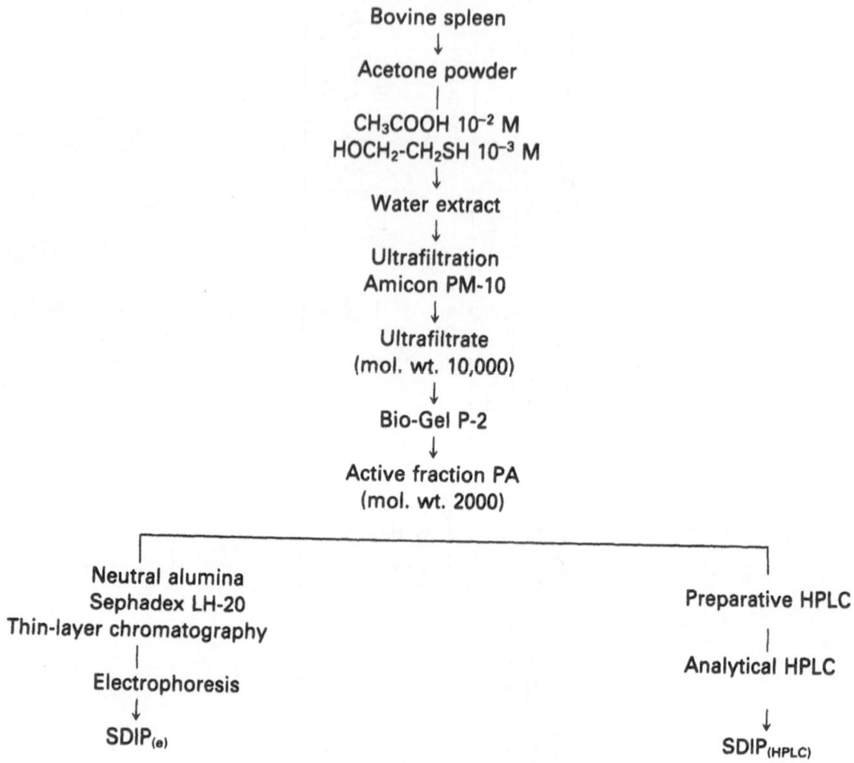

FIGURE 8. Purification of SDIP.

In Fig. 6 is shown the result of the isoelectric focusing analysis of the two subfractions pi b and pi c; it is seen that both contain several polypeptides in the pH range of β-thymosin polypeptides (Goldstein *et al.*, 1981). Figure 7 shows the concentration-dependent inhibitory effect of pi c subfraction on [^{3}H]-TdR incorporation by cytolytic T cells.

Table III indicates the effect of pi b and pi c subfractions on IgM antibody response against SRBC (a thymus-dependent antigen); the fractions are not active on the response against TNP–LPS (a thymus-independent antigen). Table IV shows the effect of pi b and pi c in the GvH-induced splenomegaly test: both, but especially pi c, reduce significantly the splenic effect.

Figure 8 lists the steps in the purification of SDIP.

REFERENCES

Bach, J.-F., Bach, A.-M., Blanot, D., Bricas, E., Charreire, J., Dardenne, M., Fournier, C., and Pléau, J. M., 1978, Facteur thymique sêrique (FTS), *Bull. Inst. Pasteur* **76:**325.

Blazsek, I., and Gaal, D., 1978, Endogenous thymic factors regulating cell proliferation and analysis of their mechanism of action, *Cell Tissue Kinet.* **11:**265.

Dardenne, M., Pléau, J. M., Man, N. K., and Bach, J.-F., 1977, Structural study of circulating thymic factor: A peptide isolated from pig serum, *J. Biol. Chem.* **252:**8040.

Fridman, W. M., Guimezanes, A., and Gisler, R. H., 1977, Characterization of suppressive immunoglobulin binding factor (IBF). I. Production of IBF by a O positive lymphone (L 5178-Y), *J. Immunol.* **119:**1266.

Garcia-Giralt, E., Lenfant, M., Privat De Garilhe, M., Mayadoux, E., and Motta, R., 1978, Purification of immunosuppressive factors extracted from bovine spleen. II. Biological activity, *Cell Tissue Kinet.* **11:**465.

Goldstein, A. L., Low, T. L. K., Thurman, G. B., Zatz, M. M., Hale, N., Chen, J. Hu, S. K., Naylor, P. B., and McClure, G. E., et al., 1981, Current status of thymosin and other hormones of the thymus gland, *Recent Prog. Horm. Res.* **37:**369.

Grundboeck Jusko, J., 1976, Chemical characteristics of chalones isolated from bovine spleen, *Acta Biochim. Pol.* **23:**165.

Kiger, N., Florentin, I., Lorans, G., and Mathe, G., 1977, Further purification and chemical characterization of the lymphocyte inhibitory factor extracted from thymus, *Immunology* **33:**439.

Lee, S. T., and Paraskevas, F., 1977, Soluble factors in mouse immune sera. III. Origin, Nature and target of small molecular weight suppressive factors isolated from serum during primary responses, *Cell. Immunol.* **32:**171.

Lenfant, M., and Millerioux-Di Giusto, L., 1982, New insight into lymphocytic chalone research: The 'facteur thymique serique' (FTS) might be responsible for part of the immunosuppressive activity detected in the 'chalone fraction,' *Int. Arch. Allergy Appl. Immunol.* **68:**387.

Lenfant, M., Privat De Garilhe, M., Garcia-Giralt, E., and Tempete, C., 1976, Further purification of bovine spleen inhibitors of lymphocyte DNA synthesis, *Biochim. Biophys. Acta* **451:**106.

Lenfant, M., Garcia-Giralt, E., Thomas, M., and Di Giusto, L., 1978, Purification of immunosuppressive factors extracted from bovine spleen (lymphoid chalone). I, *Cell Tissue Kinet.* **11:**455.

Lenfant, M., Garcia-Giralt, E., Di Giusto, L., and Thomas, M., 1980, The purification of an immunosuppressive factor extracted from bovine spleen. III. Purification process, *Mol. Immunol.* **17:**119.

Lenfant, M., Millerioux-Di Giusto, L., Masson, A., and Gasc, J. C., 1981a, Shortened purification procedure of a spleen derived immunosuppressive peptide, *J. Chromatogr.* **20b:**177.

Lenfant, M., Millerioux, L., Duchange, N., and Tanzer, J., 1981b, De la recherche d'une chalone lymphocytaire du facteur thymique sérique, *C. R. Acad. Sci.* **292:**403.

Mathé, G., 1972, Lymphocyte inhibitors fulfilling the definition of chalones and immunosuppression, *Rev. Eur. Etudes Clin. Biol.* **17:**548.

Mathé, G., Pritchard, L. L., and Halle-Pannenko, O., 1979, Mismatching for minor transplantation antigens in bone marrow transplantation: Consequence for the development and control of severe graft-versus-host disease, *Transplant. Proc.* **11:**235.

Maurer, H. R., and Laerum, O. D., 1976, Granulocyte chalone testing, a critical review, in: *Chalones* (J. C. Houck, ed.), p. 331, North-Holland, Amsterdam.

Millerioux, L., Duchange, N., Masson, A., and Lenfant, M., 1981, Relation between a spleen derived immunosuppressive peptide and the immunoglobulin binding factor, *Cell. Immunol.* **44:**249.

Namba, Y., Jegasothy, B. V., and Waksman, B. H., 1977, Regulatory substances produced by lymphocytes. V. Production of inhibitor DNA synthesis (IDS) by proliferating T lymphocytes, *J. Immunol.* **118:**1379.

New York Academy of Sciences Conference on Tuftsin, 1983, in press.

Nordland, K., Sandberg, G., and Thyberg, J., 1979, Further studies on splenic material inhibiting the growth of lymphocytes in vitro, *Int. Arch. Allergy Appl. Immunol.* **60:**195.

Nozaki *et al.*, 1977, *Bull. Chem. Soc. Jpn.* **50:** 422. Cited in Serva, *Fine Chemicals for the Scientist*, 1983/84, p. 357.

Pléau, J. M., Dardenne, M., Blouquit, Y., and Bach, J.-F., 1977, Structural study of circulating thymic factor: A peptide isolated from pig serum. II. Amino-acid sequence, *J. Biol. Chem.* **252:**8045.

Rappaport, H., Khalil, A., Halle-Pannenko, O., Pritchard, L. L., Dantchev, D., and Mathe, G., 1979, Histopathologic sequence of events in adult mice undergoing lethal graft-versus-host reaction developed across H-2 and/or non H-2 and/or H-2 histocompatibility barriers, *Am. J. Pathol.* **96:**121.

Susuki, S., and Huchet, R., 1982, Properties of histamine-induced suppressor factor in the regulation of lymphocyte response to PHA in mice, *Cell. Immunol.* **68:**349.

Tadakuma, T., Kuhner, A. L., Rich, R. R., David, J. R., and Pierce, C. W., 1976, Biological expressions of lymphocyte activation. V. Characterization of a soluble immune response suppressor (SIRS) produced by concanavalin A-activated spleen cells, *J. Immunol.* **117**:323.

Voorting-Hawkins, M., and Michael, G., 1977, Isolation and characterization of immunoregulatory factors from normal human serum. I. Preliminary biochemical and biological characterization of immunosuppressive factors, *J. Immunol.* **118**:505.

Wang, B. F., Badger, A. M., Nimberg, R. R., Cooperband, S. R., Schmid, K., and Mannick, J. A., 1977, Suppression of tumor specific cell mediated cytotoxicity by immunoregulatory alpha globulin-like peptides from cancer patients, *Cancer Res.* **37**:3022.

Ward, K., and Munro, A. J., 1979, Non-specific immunosuppressive effects of mouse spleen extracts in vitro, *Immunology* **37**:61.

15

Production of a Thymosin α_1-like Material by T-Cell Lymphomas

MARION M. ZATZ, PAUL H. NAYLOR, ALLAN L. GOLDSTEIN, JOHN E. McCLURE, and BARTON F. HAYNES

1. INTRODUCTION

Thymosin α_1 is a peptide (MW 3108) originally isolated from calf thymosin fraction 5, a partially purified preparation of bovine thymus (Low and Goldstein, 1979). It is believed to be one of the thymic hormones secreted by the thymic epithelium (Kater *et al.*, 1979; Dalakas *et al.*, 1981; Hirokawa *et al.*, 1982; Haynes *et al.*, 1983) and has known biological properties in inducing differentiation and maturation of T lymphocytes in experimental animal models as well as in man (Zatz *et al.*, 1982).

The thymic microenvironment has long been suspected as one of the critical factors in the spontaneous development of thymic lymphomas of leukemia-prone AKR mice (Metcalf *et al.*, 1966; Zielinski *et al.*, 1981). Metcalf first noted in 1956 that a circulating lymphocytosis-stimulating factor appeared in the serum of leukemic AKR mice. We therefore sought to examine the serum thymosin α_1 levels in AKR mice bearing spontaneously arising or transplanted lymphomas. We took advantage of the existence of two well-characterized sublines of an AKR tumor,

MARION M. ZATZ, PAUL H. NAYLOR, and ALLAN L. GOLDSTEIN • Department of Biochemistry, The George Washington University School of Medicine and Health Sciences, Washington, D.C. 20037. JOHN E. McCLURE • Allergy–Immunology Service, Department of Pediatrics, Texas Children's Hospital, Houston, Texas 77030. BARTON F. HAYNES • Department of Medicine, Division of Rheumatic and Genetic Diseases, Duke University School of Medicine, Durham, North Carolina 27710.

AKTB-lt and AKTB-lb, with distinct T-cell or B-cell characteristics, respectively (Zatz *et al.*, 1981; Mathieson *et al.*, 1982), to determine whether thymosin α_1 might be associated with subclasses of murine lymphomas.

2. RESULTS

2.1. Immunoreactive Thymosin α_1 in Sera of Leukemic AKR Mice

In our initial studies (Table I), we observed that 6- to 10-month-old male leukemic AKR mice had three- to sixfold elevated immunoreactive serum thymosin α_1 (IRα_1) levels, as measured by radioimmunoassay (RIA) (McClure *et al.*, 1982), in contrast to serum levels observed in young or old AKR mice.

When 10^5 thymic lymphoma cells were injected i.v. into young AKR mice, the IRα_1 levels increased four- to sixfold as the tumor grew in the mice. These results are shown in two separate experiments in Fig. 1. Thus, at the early time points (5–7 days after passage), no significant difference between medium-injected control and tumor-injected mice could be noted, although the serum levels of the tumor-injected mice in fact were consistently below normal, at this early time point. By 11–14 days after passage, when the tumor was palpable in peripheral lymphoid organs, IRα_1 levels were noticeably elevated, and by 16–22 days, levels of 4000–5000 pg/ml were present in the serum. Mice usually are killed by the transplanted lymphoma under these conditions at 18–25 days following injection.

As these results could be explained either by production of IRα_1 by tumor cells, or by activation of thymic epithelial secretory activity, the following experiment was performed to distinguish between these possibilities: 10^4 AKTB-lt or AKTB-lb lymphoma cells were injected i.v. into normal or thymectomized (tx) young AKR mice; completeness of thymectomy was verified in all animals at autopsy. As shown in Fig. 2, the AKTB-lt T-cell tumor resulted in an initial decrease in IRα_1 levels at day 8, followed by a fivefold elevation at day 14. This elevation was indistinguishable from that seen in tx recipients. Injection of the AKTB-lb B-cell tumor caused no significant change in serum IRα_1 levels at any time point (note: no tx recipients of AKTB-lt were alive at day 17 for comparison). Thus, the elevation of thymosin α_1 was directly associated with growth of the T-cell tumor,

TABLE I. Thymosin α_1 in Serum of Normal and Leukemic AKR Mice

Group	No.	α_1 (pg/ml)[a]
Normal, 2 months old	13	677 ± 160
Normal, 6–10 months old	20	395 ± 70
Leukemic, 6–10 months old	13	5386 ± 979*

[a] Mean ± S.E.
* Significant at $p < 0.001$ by Student's t test compared to normal groups.

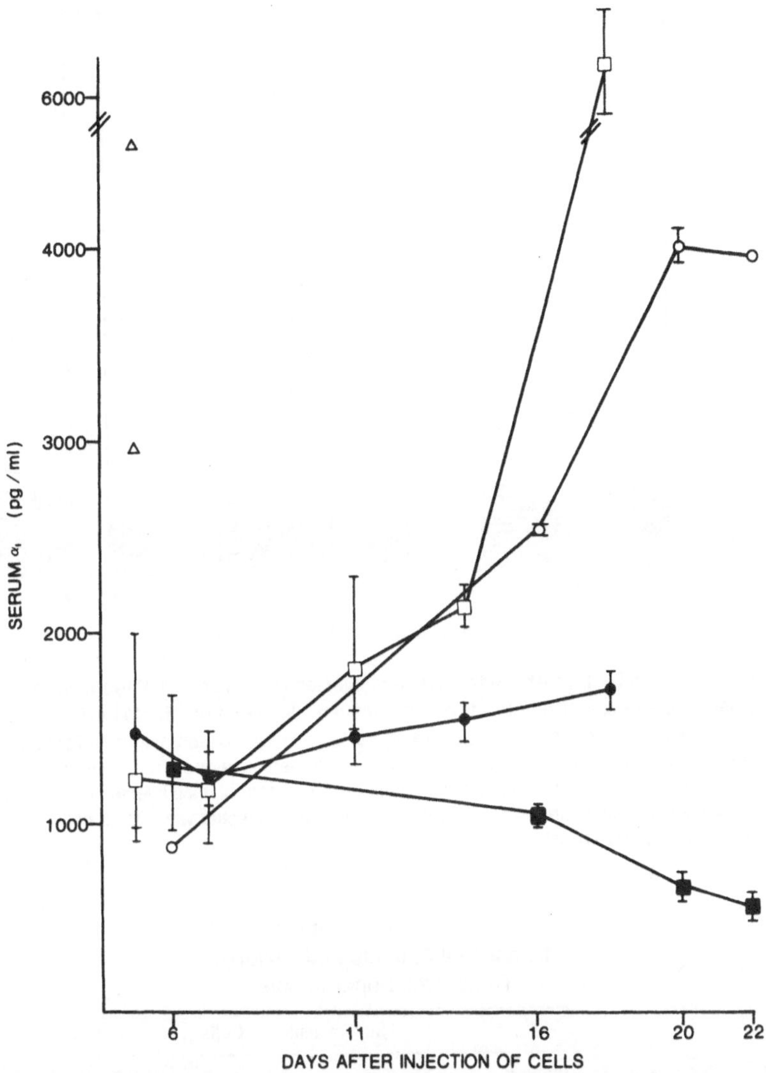

FIGURE 1. Two separate experiments are presented, in which 10^5 thymic lymphoma cells (open symbols) or medium (black symbols) were injected i.v. into groups of three mice. Mice were bled and killed at the times indicated to determine serum α_1 levels by RIA. Means ± S.E. are shown. Expt. 1 = squares; Expt. 2 = circles; △ indicates serum levels of the mice in which the spontaneous thymic lymphomas arose.

and was not mediated by the thymus. The pattern of growth and elevation of IRα_1 seen with this serially passaged *in vivo* T-cell tumor line mimics that shown in Fig. 1 with the passage of two spontaneously arising thymomas. It should also be noted that the B-cell line causes splenic enlargement comparable to that seen with the T-cell line and kills the mice at approximately the same time.

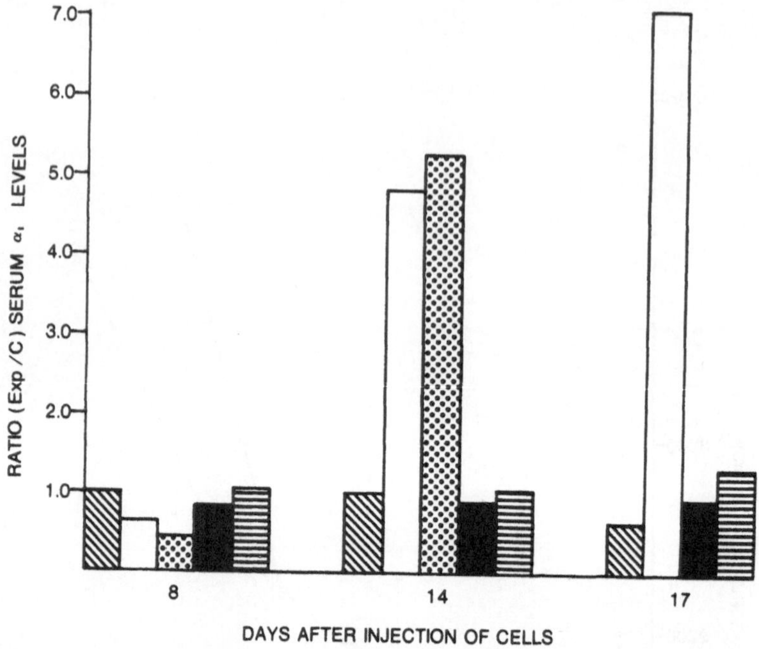

FIGURE 2. 10^4 AKTB-1t or AKTB-1b cells were injected i.v. into groups of five to seven normal or thymectomized mice at day 0. Mice were sequentially bled from the tail vein on the days shown. For ease of presentation, results are presented as a ratio of serum levels in experimental/control mice where control = normal mice injected with medium alone. The experimental groups are as follows: ▧ = tx mice injected with medium; □ = normal mice injected with TB-1t cells; ⛆ = tx mice injected with TB-1t cells; ■ = normal mice injected with TB-1b cells; ▤ = tx mice injected with TB-1b cells.

TABLE II. $IR\alpha_1$ Content (pg/ml) of Tumor Cell Extracts and Cultured Tumor Cell Supernatants[a]

	Supernatant	Cells
Expt. 1		
Normal spleen	280	
Lymphoma[b]	590	
AKTB-lt	510	
AKTB-lb	<160[c]	
Expt. 2		
AKTB-lt		1240
AKTB-lb		<160

[a] 10^6 cells were either extracted (Expt. 2) or cultured overnight (Expt. 1). After overnight culture, the cell-free supernatants were analyzed for $IR\alpha_1$ content by RIA.

[b] Enlarged spleen from an AKR mouse with spontaneously arising T-cell lymphoma was used.

[c] <160 = nondetectable.

2.2. IRα_1 in Cells and Supernatants of T-Cell Lymphomas

The elevation of IRα_1, while not dependent upon an intact thymus, could still be an indirect result of T-lymphoma growth. In order to determine whether IRα_1 was directly associated with the tumor cells, tumor cells suspensions were prepared from spleen, thoroughly washed, and disrupted by ultrasonic shock: nonsolubilized debris was removed by ultracentrifugation. Short-term tissue culture supernatants also were prepared and evaluated for IRα_1 content. Results are shown in Table II and demonstrate that the IRα_1 is associated directly with the T-lymphoma cells, but not the B-lymphoma cells, and detectable amounts are released into the culture medium during overnight incubation at 37°C. Preliminary further analyses of two additional AKR T-cell lymphomas (TB-6, AKX-3), as well as one additional B-cell tumor (AKX1-b), show high levels of IRα_1 associated with the T-cell lymphomas and low IRα_1 levels associated with the B-cell lymphoma (data not shown).

2.3. Elevated Thymosin α_1 in Sera of Human Lymphoma Patients

These results in the murine system led us to examine sera from 21 patients with a variety of lymphomas and leukemias. The data (Fig. 3) reveal that some patients with T-cell lymphomas, including one individual positive for the human T-cell leukemia virus (HTLV$^+$, JATL), also have markedly elevated serum α_1 levels. With the exception of one patient with common acute lymphoid leukemia (CALL), none of the non-T-cell-lymphoma patients exhibited serum α_1 levels above the normal range. Thymosin β_4 levels did not correlate with thymosin α_1 levels, and were significantly elevated in only two patients (Nos. 1 and 2).

Studies are now in progress to determine whether: (1) the IRα_1, associated with murine and human T-cell lymphomas, is in fact identical in its primary amino acid sequence to the sequence of thymosin α_1 isolated from thymus and (2) this tumor-associated thymosin α_1 has those biological activities associated with native thymosin α_1 isolated from bovine thymus thymosin fraction 5.

3. DISCUSSION

These studies are of importance in several respects. (1) It has long been suspected that the thymus may be the primary, but not the only source, of thymic hormones. Thus, thymectomy results in an eventual, but not immediate, drop of serum thymosin α_1 levels, and levels are diminished but not absent in athymic nude mice. There is some evidence that the CNS may be one source of IRα_1 (Hall *et al.*, 1982) and although thymic epithelial cells can be strongly stained with heterologous antibody to α_1, epithelial cells in other organs also can be weakly labeled (Dalakas *et al.*, 1981). It should be noted that the antiserum to thymosin α_1 used in the RIA does not cross-react with α-fetal proteins, carcinoembryonic antigen, or a spectrum of other well-defined thymic and nonthymic hormones (McClure *et al.*, 1982).

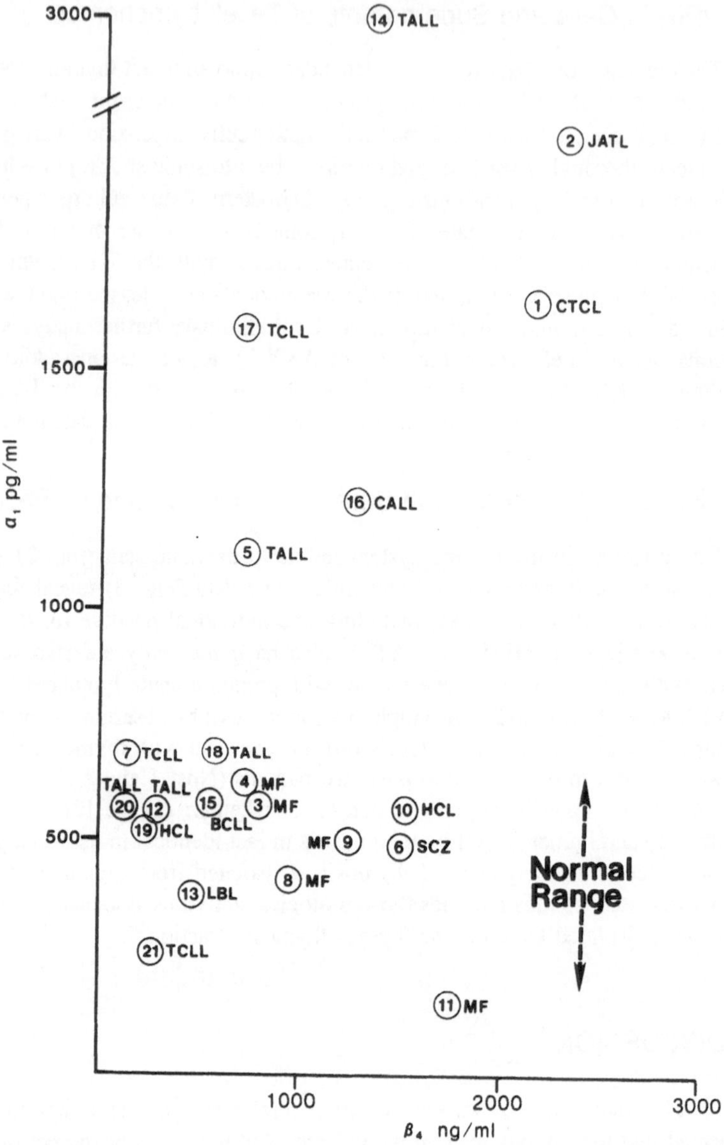

FIGURE 3. Sera from 21 patients and a panel of 70 normals (obtained from the Red Cross) were evaluated by RIA for $IR\alpha_1$ and β_4 content. The individual patients are numbered and their diagnoses are abbreviated as follows: BCLL = B-cell lymphoma and leukemia; CALL = common acute lymphocytic leukemia; CTCL = cutaneous T-cell leukemia; HCL = hairy cell leukemia; JATL = Japanese adult T-cell leukemia ($HTLV^+$); LBL = lymphoblastic lymphoma; MF = mycoses fungoides; SCZ = Sezary cell leukemia; TALL = T-cell acute lymphocytic leukemia; TCLL = T-cell lymphoma and leukemia.

The biochemical analyses now in progress will enable us to determine whether the tumor sources of IRα_1 produce a peptide biochemically and functionally identical to the thymic peptide, or an inactive and/or cross-reactive product. Recent studies have established that a number of hormone and hormonelike factors, termed "tissue hormones," including insulin, ACTH, glucagon, gastrin, and bombesin, are produced by tissue far removed from the endocrine glands normally associated with these molecules (Roth *et al.*, 1982). In addition, there is a considerable body of evidence documenting the production of biologically active hormones by dedifferentiated tumor cells (Moody *et al.*, 1981; Radcliffe *et al.*, 1981; Pavelic *et al.*, 1981). A thymopoietin-like material, termed "thymopoietin III," has been isolated from calf spleen (Audhya *et al.*, 1981). Also, it recently has been reported that a leukemic cell line produces a variant of the lymphokine T-cell growth factor (TCGF) which is biochemically distinct from normal TCGF (Gootenberg *et al.*, 1982).

A second important aspect of this work is in the potential for detection and classification of human tumors. In at least one case of a patient with thymic lymphoma, elevated serum IRα_1 levels decreased to the normal range following thymectomy (unpublished data). Tissue extracts of the thymoma contained twice as much IRα_1 as did thymus from an age- and sex-matched normal individual. In a recent study of an immunosuppressed homosexual population with acquired immune deficiency syndrome (AIDS), shown to be at high risk for development of rare malignancies such as undifferentiated Hodgkin's and Kaposi's sarcoma (Gottlieb *et al.*, 1981; Siegal *et al.*, 1981; Durack, 1981), elevated IRα_1 levels were consistently found (Hersh *et al.*, 1983; Naylor and Goldstein, 1983). It is of particular interest to note that evidence of HTLV (Robert-Guroff *et al.*, 1983; Popovic *et al.*, 1983) has also been found in high frequency in the AIDS population (Essex *et al.*, 1983; Gelmann *et al.*, 1983; Gallo *et al.*, 1983; Barre-Sinoussi *et al.*, 1983).

The association of elevated serum thymosin α_1 in certain tumor-bearing individuals may provide a valuable diagnostic tool to monitor the growth and regression of the tumor as well as clues to the biology and possible viral etiology of these lymphomas.

ACKNOWLEDGMENTS. This work was supported in part by NIH Grants CA-29943 and CA-24974 and by funds from Hoffmann–La Roche Inc.

REFERENCES

Audhya, T., Schlesinger, D. H., and Goldstein, G., 1981, Complete amino acid sequence of bovine thymopoietins I, II and III: Closely homologous peptides, *Biochemistry* **30**:6195–6200.

Barre-Sinoussi, F., Chermann, J. C., Rey, F., Nugeyre, M. T., Chameret, S., Gruest, J., Dauguet, C., Axler-Blin, C., Vezinet-Brun, F., Rouzoux, C., Rozenbaum, W., and Montagnier, L., 1983, Isolation of a T-lymphotropic retrovirus from a patient at risk for acquired immunodeficiency syndrome AIDS, *Science* **220**:868–871.

Dalakas, M. C., Engel, W. K., McClure, J. E., Goldstein, A. L., and Askanas, V., 1981, Immunocytochemical localization of thymosin α_1 in thymic epithelial cells of normal and myasthenia gravis patients and in thymic culture, *J. Neurol. Sci.* **50**:239–247.

Durack, D. T., 1981, Opportunistic infections and Kaposi's sarcoma in homosexual men, *N. Engl. J. Med.* **305**:1439–1464.

Essex, M., McLane, M. F., Lee, T. H., Falk, L., Howe, C. W. S., Mullins, J. I., Cabradilla, C., and Francis, D. P., 1983, Antibodies associated with human T-cell leukemia virus in patients with AIDS, *Science* **220:**859–862.

Gallo, R. C., Sarin, P. S., Gelmann, E. P., Robert-Guroff, M., Richardson, E., Kalyanaraman, V. S., Mann, D., Sidhu, G. D., Stahl, R., Zolla-Pazner, S., Leibowitch, J., and Popovic, M., 1983, Isolation of human T-cell leukemia virus in acquired immune deficiency syndrome (AIDS), *Science* **220:**865–867.

Gelmann, E. P., Popovic, M., Blayney, D., Masur, H., Sidhu, G., Stahl, R., and Gallo, R. C., 1983, Proviral DNA of a retrovirus, human T-cell leukemia virus in two patients with AIDS, *Science* **220:**862–865.

Gootenberg, J. E., Ruscetti, F. W., and Gallo, R. C., 1982, A biochemical variant of human T-cell growth factor produced by a cutaneous T-cell lymphoma cell line, *J. Immunol.* **129:**1499–1505.

Gottleib, M. S., Schroff, R., Schanker, H. M., Weissman, J. D., Fan, P. T., Wolf, R. A., and Saxon, A., 1981, Pneumocystis carinii pneumonia and mucosal candidasis in previously healthy homosexual men: Evidence of a new acquired cellular immunodeficiency, *N. Engl. J. Med.* **305:**1425–1430.

Hall, N. R., McGillis, J. P., Spangelo, B., Palaszynski, E., Moody, T., and Goldstein, A. L., 1982, Evidence for a neuroendocrine-thymus axis mediated by thymosin polypeptides, in: *Current Concepts in Human Immunology and Cancer Immunomodulation* (B. Seron *et al.*, eds.), pp. 653–660, Elsevier Biomedical Press, New York.

Haynes, B. F., Warren, R. W., Buckley, R. H., McClure, J. E., Goldstein, A. L., Henderson, F. W., Hensley, L. L., and Eisenbarth, G. S., 1983, Demonstration of abnormalities in expression of thymic epithelial surface antigens in severe cellular immunodeficiency diseases, *J. Immunol.* **130:**1182–1188.

Hersh, E. M., Reuben, J. M., Rios, A., Mansell, P. W. A., Newell, G. R., McClure, J. E., and Goldstein, A. L., 1983, Elevated serum thymosin α_1 levels associated with evidence of immune dysregulation in male homosexuals with a history of infectious diseases or Kaposi's sarcoma, *N. Engl. J. Med.* **308:**45–46.

Hirokawa, K., McClure, J. E., and Goldstein, A. L., 1982, Age-related changes in localization of thymosin in the human thymus, *Thymus* **4:**19–29.

Kater, L., Oosterom, R., McClure, J., and Goldstein, A. L., 1979, Presence of thymosin-like factors in human thymic epithelium conditioned medium, *Int. J. Immunopharmacol.* **1:**273–284.

Low, T. L. K., and Goldstein, A. L., 1979, The chemistry and biology of thymosin, *J. Biol. Chem.* **254:**987–995.

McClure, J. E., Lameris, D. W., Wara, D. W., and Goldstein, A. L., 1982, Immunochemical studies on thymosin: Radioimmunoassay of thymosin α_1, *J. Immunol.* **128:**368–375.

Mathieson, B. J., Zatz, M. M., Sharrow, S. O., Asofsky, R., Logan, W., and Kanellopoulos -Langevin, C., 1982, Separation and characterization of two component tumor lines within the AKR lymphoma, AKTB-1, by fluorescence activated cell sorting and flow microfluorometry analysis, *J. Immunol.* **128:**1832–1838.

Metcalf, D., Wiadrowski, M., and Bradley, R., 1966, Analysis of the role of thymus in Leukemia development with the use of thymectomy and thymus grafts, *Natl. Cancer Inst. Monograph* **22:**571–583.

Metcalf, D., 1956, The origin of the plasma lymphocytosis stimulating factor, *Br. J. Cancer* **10:**442–457.

Moody, T. W., Pert, C. B., Gazder, A. F., Carney, D. N., and Minna, J. D., 1981, High levels of intracellular bombesin characterize human small cell lung carcinoma, *Science* **219:**1246–1248.

Naylor, P. H., and Goldstein, A. L., 1984, Elevated serum thymosin as an early marker for acquired immunodeficiency syndrome (AIDS), in: *AIDS: The Epidemic of Kaposi's Sarcoma and Opportunistic Infections* (A. E. Friedman-Kien and L. J. Lauberstein, eds.), pp. 173–180, Masson, New York.

Pavelic, K., Ferle-Vidovic, A., Osmak, M., and Vuk-Pavlovic, S., 1981, Synthesis of immunoreactive insulin in vitro by aplastic mammary carcinoma preconditioned in diabetic mice, *J. Natl. Cancer Inst.* **67:**687–688.

Popovic, M., Sarin, P. S., Robert-Guroff, M., Kalyanaraman, V. S., Mann, D., Minowada, J., and Gallo, R. C., 1983, Isolation and transmission of human retrovirus (human T-cell leukemia virus), *Science* **219:**856–859.

Radcliffe, J. G., Knight, R. A., Besser, G. M., Landon, J., and Stansfeld, A. G., 1981, Tumor and plasma ACTH concentrations in patients with and without the ectopic ACTH syndrome, *Clin. Endocrinol.* **1**:27–44.

Robert-Guroff, M., Kalyanaraman, V. S., Blattner, W. A., Popovic, M. C., Sarngadharan, M. M., Blayney, D., Catovsky, D., Bunn, P. A., Shibata, A., Yoshinobu, N., Ito, Y., Aoki, T., and Gallo, R. C., 1983, Evidence for human T-cell lymphoma-leukemia virus infection of family members of human T-cell lymphoma-leukemia virus positive T-cell leukemia-lymphoma patients, *J. Exp. Med.* **157**:248–258.

Roth, J., LeRoith, D., Shiloach, J., Rosenzweig, J. L., Lesniak, M. A., and Havrankova, J., 1982, The evolutionary origins of hormones, neurotransmitters, and other extracellular chemical messengers, *N. Engl. J. Med.* **306**:523–527.

Siegal, F. P., Lopez, C., Hammer, G. S., Brown, A. E., Kornfeld, S. J., Gold, J., Hassett, J., Hirschman, S. Z., Cunningham-Rundles, S., and Armstrong, D., 1981, Severe acquired immunodeficiency in male homosexuals manifested by chronic perianal ulcerative herpes simplex lesions, *N. Engl. J. Med.* **305**:1439–1444.

Zatz, M. M., Mathieson, B. J., Kanellopoulos-Langevin, C., and Sharrow, S. O., 1981, Separation and characterization of two component tumor lines within the AKR lymphoma, AKTB-1, by fluorescence-activated cell sorting and flow microfluorometry analysis, *J. Immunol.* **126**:608–613.

Zatz, M. M., Low, T. L. K., and Goldstein, A. L., 1982, Role of thymosin and other thymic hormones in T-cell differentiation, in: *Biological Responses in Cancer*, Vol. 1 (E. Mihich, ed.), pp. 219–247, Plenum Press, New York.

Zielinski, C. C., Waters, D. L., Datta, S. K., and Waksal, S. D., 1981, Analysis of intrathymic differentiation patterns during the course of AKR leukemogenesis, *Cell. Immunol.* **59**:355–366.

16

Purification and Characterization of an Immunoreactive Thymosin α_1 from Human Blood

Evidence for a Thymosin α_1 Carrier Protein

JIEPING CHEN, TERESA L. K. LOW, and ALLAN L. GOLDSTEIN

1. INTRODUCTION

Thymosin α_1 was first isolated from bovine thymus gland (Goldstein *et al.*, 1977; Low *et al.*, 1979). It is composed of 28 amino acid residues with a molecular weight of 3108 and an isoelectric point of 4.2. This peptide can stimulate lymphocytes to produce macrophage inhibitory factor (Thurman *et al.*, 1981), interferon (Huang *et al.*, 1981), and T-cell growth factor (Zatz *et al.*, 1984). Thymosin α_1 also modulates the expression of terminal deoxynucleotidyl transferase (TdT) *in vivo* and *in vitro* (Hu *et al.*, 1981; Goldschneider *et al.*, 1981).

A radioimmunoassay (RIA) for thymosin α_1 was developed to determine the concentrations of thymosin α_1 in human serum (McClure *et al.*, 1982). However, until this study, there was no chemical evidence that the immunoreactive thymosin α_1 (IRα_1) in human blood was similar in structure to bovine thymosin α_1. In the study reported here, an immunoaffinity chromatography has been developed to purify IRα_1 from human blood. Our results indicate that the IRα_1 in human blood is similar in physicochemical properties to bovine thymosin α_1.

JIEPING CHEN, TERESA L. K. LOW, and ALLAN L. GOLDSTEIN • Department of Biochemistry, The George Washington University School of Medicine and Health Sciences, Washington, D.C. 20037.

2. METHODS AND RESULTS

A Cohn fraction IV-1, a fraction of human plasma obtained by fractional ethanol precipitation (Cohn *et al.*, 1950) used as starting material for the isolation of IRα_1, was generously provided by Dr. G. A. Jamieson, American Red Cross, Bethesda, Maryland.

2.1. CM-Cellulose Chromatography

A total of 15 g of Cohn fraction IV-1 was chromatographed on a column (5 × 50 cm) of carboxymethyl-cellulose (CM-52) in 10 mM sodium acetate, pH 5.5. The column was eluted first with the starting buffer followed by 50 mM sodium phosphate buffer, pH 8.0, containing 0.5 M NaCl.

The IRα_1 was measured by an RIA for thymosin α_1 (McClure *et al.*, 1982). The concentration of IRα_1 in Cohn fraction IV-1 is 3.54 ng/mg (McClure *et al.*, 1982). As shown in Table I, following ion-exchange chromatography on CM-52, the concentration of IRα_1 in the void peak (CM-A) is 10-fold higher than in the starting material, Cohn fraction IV-1.

All CM-52 subfractions were examined by analytical electrofocusing on LKB PAG plate using a previously described method (Low *et al.*, 1979). Only the CM-A fraction was found to contain a band at thymosin α_1 region (pI 4.2).

2.2. Immunoaffinity Chromatography

One milliliter of ^{125}I-(Tyr^1)-thymosin α_1 (200,000 cpm) and 320 mg of CM-A were chromatographed on a column (2.2 × 29 cm) of CNBr-activated Sepharose 4B coupled with γ-globulin of antithymosin α_1 antiserum in 10 mM sodium phosphate buffer containing 0.15 M NaCl, pH 7.5. The column was eluted first with starting buffer followed by 20 mM glycine buffer containing 0.5 NaCl, pH 2.8. The affinity chromatography was monitored by gamma counting for labeled thymosin α_1, by UV absorption (at 280 or 235 nm), and by RIA for thymosin α_1. In Figs. 1 and 2, peak 2 represents the specific binding. In Fig. 1, it comprises 83.2%

TABLE I. IRα_1 in CM Subfractions of Cohn Fraction IV-1[a]

CM subfractions	Weight (g)	IRα_1 concn (ng/mg)	Total amount of IRα_1 recovered in subfraction (μg)
CM-A	0.25	39.5	9.8
Others	13.85	0.62	8.6

[a] A total of 15 g of Cohn fraction was fractionated on a CM-cellulose column.

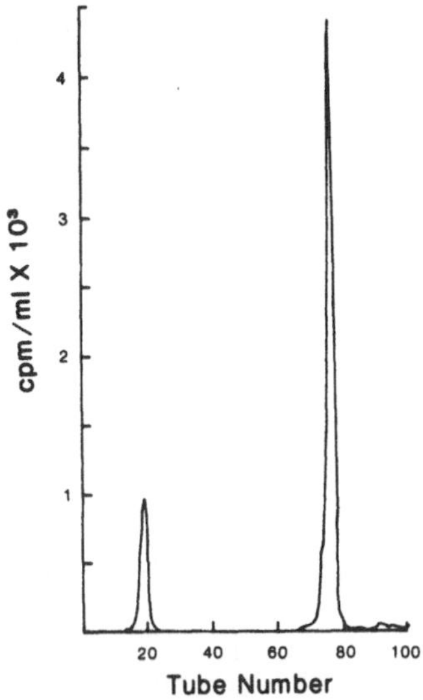

FIGURE 1. Immunoaffinity chromatography of labeled thymosin α_1 on CNBr-activated Sepharose 4B–antithymosin α_1 column (2.2 × 29 cm). One milliliter of labeled thymosin α_1 (10,000 cpm/50 μl) was loaded onto the column. The column was eluted first with 190 ml starting buffer (0.01 M sodium phosphate, pH 7.5) and then with glycine buffer (0.02 M glycine-HCl, pH 2.8, containing 0.5 M NaCl).

of the sum of cpm of peak 1 and peak 2. In Fig. 2, there is a very high and sharp peak of IRα_1 activity. This peak represents the specific binding of IRα_1 with the column.

2.3. Ultrafiltration

Eighty grams of Cohn fraction IV-1 and 1 ml of labeled thymosin α_1 (200,000 cpm) in 10 mM acetate buffer containing 0.15 M NaCl, pH 6.0, were fractionated by ultrafiltration using an Amicon DC-2 hollow-fiber system with a molecular weight cutoff at 10,000. The experiment was also performed under the same condition with buffer containing 2 M guanidine-HCl.

The results are shown in Table II. When acetate buffer was used, 43.5 and 56.5% of the labeled thymosin α_1 were in the retentate and ultrafiltrate, respectively. In contrast, when the acetate buffer contained 2 M guanidine-HCl, 9.4% of the labeled thymosin α_1 was in the retentate and 90.6% in the ultrafiltrate.

2.4. Preparative Electrofocusing

Preparative flat-bed electrofocusing was used to fractionate CM-A and to determine the pI of IRα_1. An amount of 5.5 ml of Ampholine (LKB), pH 3.5–5.0, was dissolved in distilled water and made up to 100 ml. Five grams of LKB Ultrodex was added into 97 ml diluted Ampholine solution. The suspension was stirred gently and poured immediately onto the tray. After evaporating the suspension with a gentle stream of air using a hair dryer, 80 mg of CM-A in 3 ml diluted Ampholine solution was loaded on the plate 3 cm from the cathode side. Electrofocusing was

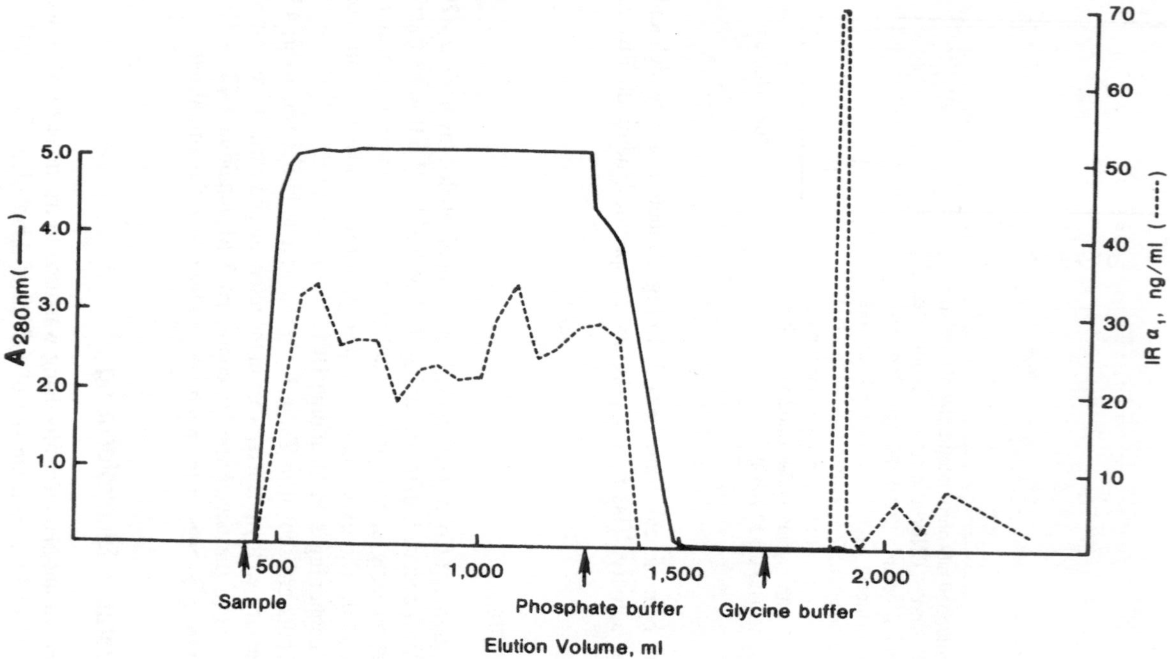

FIGURE 2. Purification of IRα_1 from CM-A by immunoaffinity chromatography on CNBr-activated Sepharose 4B coupled with anti-thymosin α_1. Sodium phosphate buffer (0.01 M, pH 7.5) and glycine buffer (0.02 M, pH 2.8 containing 0.5 M NaCl) were used to elute nonspecific proteins and IRα_1, respectively.

TABLE II. Ultrafiltration of Cohn Fraction IV-1 Using an Amicon DC-2 Hollow-Fiber System[a]

	Acetate buffer		Acetate buffer containing 2 M guanidine-HCl	
	cpm	Percentage	cpm	Percentage
Retentate	76,200	43.5	17,675	9.4
Ultrafiltrate	98,900	56.5	170,500	90.6

[a] Twenty grams of human Cohn fraction IV-1 was dissolved in 100 ml of 0.01 M sodium acetate, pH 6.0, with or without 2 M quanidine-HCl. One milliliter of ^{125}I-(Tyr^1)-thymosin α_1 (2×10^5 cpm/ml) was added to the sample solution and incubated for 1 hr at room temperature.

performed for 16 hr at 4°C with maximal voltage of 1000 V and maximal power of 8 W. The separated zones were collected by sectioning the gel bed with the fractionating grid. The proteins in each section were eluted with distilled water twice, 2 ml each time per section. For each fraction, pH determination, RIA for thymosin α_1, and optical density determination (at 280 nm) were performed.

The results are shown in Fig. 3. One small peak was observed at pH 4.2, and a larger peak at pH 4.4. The percentage of IRα_1 in the first peak was 18.1%, and in the second peak, 81.9%.

2.5. High-Performance Liquid Chromatography (HPLC)

The immunoaffinity chromatography-purified fraction (1.35 mg) was analyzed by HPLC as described previously (Low *et al.*, 1983). A gradient of 0–50% acetonitrile in 0.05% trifluoroacetic acid was used to elute the peptides. About 10 fractions were collected at 1-min intervals around the region where synthetic thymosin α_1 is eluted (26.4 min). The fractions were lyophilized and analyzed by RIA for thymosin α_1.

As shown in Fig. 4, a single peak was found. It had a retention time at 26.4 min, which is the same as that of synthetic and natural thymosin α_1.

2.6. Molecular Weight Determination

Molecular weight of the IRα_1 was determined on a calibrated Sephadex G-50 column (1.5×86 cm) in 2 M urea, 10 mM acetate buffer, pH 6.0, according to the method of Fish *et al.* (1969) with modification of the buffer. Ten milligrams of the desalted ultrafiltrate fraction of Cohn fraction IV-1 in 1 ml of acetate buffer was loaded onto the column. The eluate was monitored by RIA for thymosin α_1.

The results are shown in Fig. 5. IRα_1 has a molecular weight of 3200, which is very close to that of thymosin α_1 (3108).

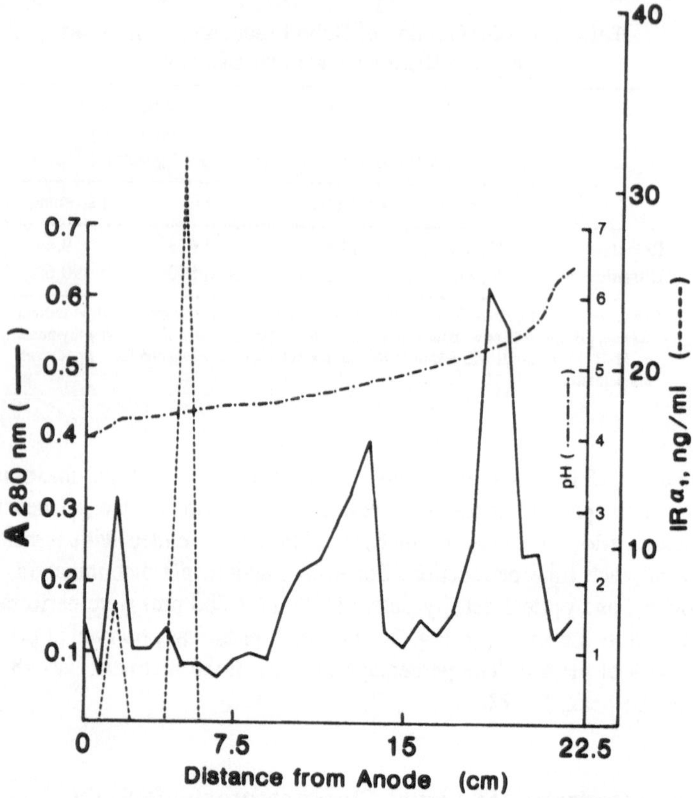

FIGURE 3. Electrofocusing of CM-A. After electrophoresis, the gel was cut into 30 sections (0.75 cm each). They were then eluted with 0.01 M sodium phosphate buffer, pH 7.4, and analyzed by RIA. Two IRα_1 peaks (pH 4.2 and 4.4) were found.

3. DISCUSSION

In the present study, we have found that Cohn fraction IV-1 isolated from human blood contains a peptide that cross-reacts with an antibody prepared to synthetic thymosin α_1 and this peptide has several of the physicochemical properties of thymosin α_1 isolated from calf thymic tissue. In addition, ultrafiltration experiments suggest that thymosin α_1 is bound to a carrier protein with a molecular weight over 10,000. Approximately 82% of the IRα_1 is in bound form and the remaining 18% is in free form. Further studies are necessary to isolate and identify this high-molecular-weight carrier protein.

Using the denaturing agent guanidine-HCl or urea, we have found that IRα_1 can be dissociated from the carrier protein without irreversibly denaturing the IRα_1 molecules. This has allowed us to isolate IRα_1 from blood using ultrafiltration.

The RIA for thymosin α_1 has been used for the first time in tandem with

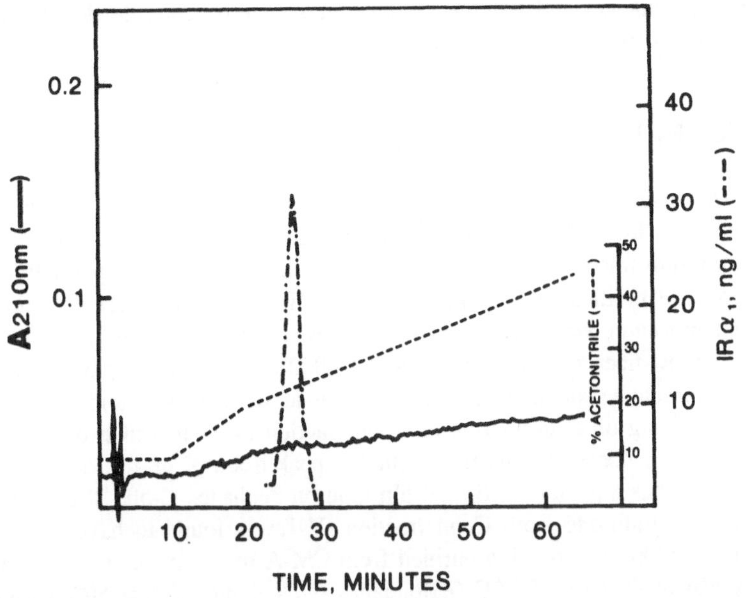

FIGURE 4. HPLC analysis of IRα_1. The IRα_1 fraction purified by affinity chromatography was analyzed by HPLC (μBondapak C_{18} reverse-phase column with 0–50% acetonitrile in 0.05% trifluoroacetic acid as solvents). A single immunoreactive peak similar in retention time as synthetic thymosin α_1 was detected.

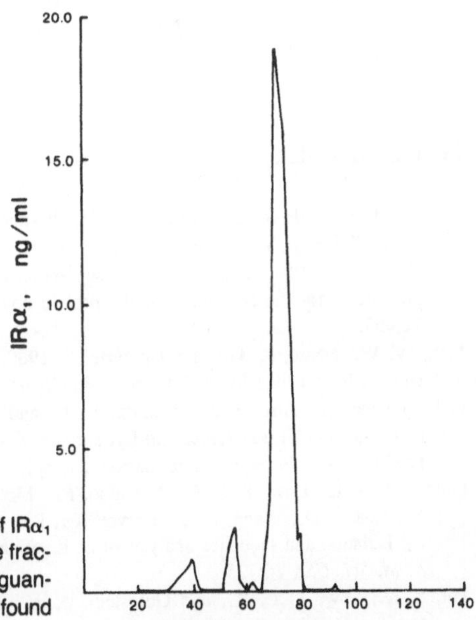

FIGURE 5. Molecular weight determination of IRα_1 on a calibrated Sephadex G-50 column. The fractions were analyzed by RIA. IRα_1 in the 2 M guanidine ultrafiltrate from Cohn fraction IV-1 was found to have a molecular weight of 3200.

electrofocusing gel and column chromatographic procedures to detect the trace amounts of IRα_1 present in the blood.

4. SUMMARY

Thymosin α_1 isolated from calf thymus has been shown to be an immunologically active thymic hormone (M_r 3108, pI 4.2). In this study, IRα_1 was isolated from a serum fraction (Cohn fraction IV-1) of human blood and partially characterized. The purification was monitored by RIA. An immunoaffinity chromatographic procedure was developed to isolate IRα_1 in blood. Ultrafiltration (Amicon DC-2 hollow-fiber system, MW_r cutoff 10,000) experiments suggested that thymosin α_1 might be bound to a carrier protein under normal physiological conditions. Electrofocusing of a subfraction of Cohn fraction IV-1 showed two IRα_1 peaks. The substance corresponding to the smaller peak has a pI of 4.2, and that of the larger peak has a pI of 4.4. By gel filtration on Sephadex G-50, IRα_1 in the 2 M guanidine ultrafiltrate from Cohn fraction IV-1 was found to have a molecular weight of 3200. The fraction purified from CM-A by immunoaffinity chromatography was analyzed by HPLC (μBondapak C_{18} column, with 0–50% acetonitrile in 0.05% trifluoroacetic acid as solvents) to give a single immunoreactive peak similar in retention time to synthetic thymosin α_1. Our results indicate that IRα_1 isolated from human blood is similar in physicochemical properties to thymosin α_1 isolated from bovine thymus.

ACKNOWLEDGMENTS. These studies were supported in part by grants and gifts from the NCI (CA-24974 and AI-17710), Hoffmann–La Roche Inc., and Alpha One Biomedicals Inc.

REFERENCES

Cohn, E. J., Gurd, F. R. N., Surgenor, D. M., Barnes, B. A., Brown, R. K., Derouaux, G., Gillespie, J. M., Kahnt, F. W., Lever, W. F., Liu, C. H., Mittelman, D., Mouton, R. F., Schmid, K., and Uroma, E., 1950, A system for the separation of the components of human blood: Quantitative procedures for the separation of the protein components of human plasma, *J. Am. Chem. Soc.* **72**:465.

Fish, W. W., Mann, K. G., and Tanford, C., 1969, The estimation of polypeptide chain molecular weights by gel filtration in 6 M guanidine hydrochloride, *J. Biol. Chem.* **244**:4989.

Goldschneider, I., Ahmed, A., Bollum, F. J., and Goldstein, A. L., 1981, Induction of terminal deoxynucleotidyl transferase and Lyt antigens with thymosin: Identification of multiple subsets of prothymocytes in mouse bone marrow and spleen, *Proc. Natl. Acad. Sci. USA* **78**:2469.

Goldstein, A. L., Low, T. L. K., McAdoo, M., McClure, J., Thurman, G. B., Rossio, J., Lai, C.-Y., Chang, D., Wang, S.-S., Harvey, C., Ramel, A. H., and Meienhofer, J., 1977, Thymosin α_1: Isolation and sequence analysis of an immunologically active thymic polypeptide, *Proc. Natl. Acad. Sci. USA* **74**:725.

Hu, S.-K., Low, T. L. K., and Goldstein, A. L., 1981, Modulation of terminal deoxynucleotidyl transferase activity by thymosin, *Mol. Cell. Biochem.* **41**:49.

Huang, K. Y., Kind, P. D., Jagoda, E. M., and Goldstein, A. L., 1981, Thymosin treatment modulates production of interferon, *Int. J. Interferon Res.* **1**:411.

Low, T. L. K., Thurman, G. B., McAdoo, M., McClure, J. E., Rossio, J. L., Naylor, P. H., and Goldstein, A. L., 1979, The chemistry and biology of thymosin. I. Isolation, characterization and biological activities of thymosin α_1 and polypeptide β_1 from calf thymus, *J. Biol. Chem.* **254**:981.

Low, T. L. K., McClure, J. E., Naylor, P. H., Spangelo, B. L., and Goldstein, A. L., 1983, Isolation of thymosin α_1 from thymosin fraction 5 of different species by high performance liquid chromatography, *J. Chromatogr.* **266**:533.

McClure, J. E., Lameris, N., Wara, D. W., and Goldstein, A. L., 1982, Immunochemical studies on thymosin: Radioimmunoassay of thymosin α_1, *J. Immunol.* **128**:368.

Thurman, G. B., Low, T. L. K., Rossio, J. L., and Goldstein, A. L., 1981, Specific and nonspecific macrophage migration inhibition, in: *Lymphokines and Thymic Hormones: Their Potential Utilization in Cancer Therapeutics* (A. L. Goldstein and M. A. Chirigos, eds.), pp. 145–157, Raven Press, New York.

Zatz, M. M., Oliver, J., Samuels, C., Skotnicki, A. B., Sztein, M., and Goldstein, A. L., 1984, Thymosin increases T-cell growth factor production by normal peripheral blood lymphocytes, *Proc. Natl. Acad. Sci. USA* (in press).

Part I Summary

Thymic Hormones

HERMAN FRIEDMAN and JEAN-FRANÇOIS BACH

The study of the role of the thymus in immunity was revived and received widespread attention in the early 1960s based largely on the elegant experimental work of Jacques Miller and associates, using thymectomized newborn mice, and the now classic studies of Robert Good and colleagues with thymectomized chickens and rabbits. It became clear at that time that the many earlier observations in man and experimental animals that the thymus gland may play a role in leukemogenesis and some immunological disorders, especially acquired and/or general immunodeficiencies, could be related to thymic dysfunction. The explosive interest over the last few years throughout the USA and indeed throughout the world concerning the acquired immune deficiency syndrome has again brought the subject of the nature and mechanism of immunoregulation and dysfunction of the thymus and products from the thymus to the forefront of biomedical sciences.

It is apparent that one of the major mechanisms of thymic control and regulation of immune function resides in the elaboration of what now appears to be a large variety of immunoreactive thymus-derived or associated molecules, including an almost bewildering array of peptides and hormones or hormonelike substances. Thus, it is of interest that the intricate role of the thymus has been examined, although sporadically, over the past five decades or so. Nevertheless, it is only within the last 15 years or so that a concerted effort has been made to study thymic products biochemically. This opening session of the Symposium on Thymic Hormones and Lymphokines presented a forum for a wide-ranging discussion of the biochemistry of some of the most studied peptides and biologically active components of the thymus and related substances. In addition, several new observations concerning the devastating effects of acquired immune deficiency syndrome (AIDS) on thymic function were also presented.

HERMAN FRIEDMAN • Department of Medical Microbiology and Immunology, University of South Florida College of Medicine, Tampa, Florida 33612. JEAN-FRANÇOIS BACH • INSERM U-25, Hôpital Necker, 75015 Paris, France.

Over the last few years, much information has been obtained concerning the biological activity of purified thymus-associated factors and molecules, including those present in the serum and various lymphoid and nonlymphoid tissues of both man and animals. Dr. M. Dardenne and colleagues from the Hospital Necker, Paris, have studied in detail polypeptide hormones in human serum which can induce thymic cell markers on immature lymphoid cells and stimulate various immunologically active functions in treated cells. One of these factors, now termed thymulin and previously called facteur thymique sérique (FTS), has been sequenced (Glu-Ala-Tyr-Ser-Gln-Gly-Gly-Ser-Asn) and the synthetic peptide has been found to be fully active biologically. Minimal changes in the C-terminal area result in alteration of biological and antigenic activity. High-affinity-specific receptors have been found on lymphoid cells using ^{3}H-isotopically labeled peptide. The thymic origin of the peptide has been confirmed using polyclonal and monoclonal antibodies in immunofluorescence assays. It has been found that the peptide binds the metal zinc and this appears essential for biological activity. Zinc is also important in the biological activity of the thymus-derived peptide.

Dr. Theresa Low and colleagues, including Dr. Allan Goldstein, reported on extensive chemical analyses of various thymic peptides present in thymosin fraction 5, which has been extensively studied in both experimental animals and man. Clinical trials on the biologically active, partially purified extract are under way and this material has been shown under appropriate conditions to augment or influence a wide variety of immunological functions. New procedures have been utilized to analyze and examine several of the peptides present in thymosin fraction 5, including thymosin α_1, which has been synthesized, thymosin β_4, and thymosin α_7. A number of papers were presented either orally or as poster sessions concerning these factors. Dr. Dalakas and colleagues utilized thymosin β_4 as a marker for receptor sites and showed that this factor is present in or on the dendritic process of thymic cells. Dr. Chen and associates examined thymosin α_1 as an immunoreactive blood peptide and showed that the level of this peptide is altered under various conditions. Dr. Danho and colleagues found that there were variations in thymosin α_1 activity as a function of the extraction procedure, while Dr. Felix and associates showed that this material can be synthesized completely in the laboratory and the various properties of the synthetic material were found to be similar to the native peptide. Dr. Zatz reported that thymosin α_1-like peptides can be produced *in vitro* by T-cell lymphomas growing in culture. Drs. Higley and Rowden utilized the presence of thymosin α_1, α_7, and β_4 as surface markers for peripheral blood leukocytes of man and discussed the distribution of these markers in various cell populations. Dr. Horecker examined thymosin β_4 distribution and biosynthesis in many vertebrate cells and found a basic conservation of this peptide among various animal species.

The concept of thymic hormones as immunoregulatory molecules was described in detail by Dr. Birr, while Dr. Folkers indicated that spermidine and spermine present in thymic tissues and extracts may act as a self-regulator for thymic hormone activity. Dr. Mathe reported the presence of a suppressive thymic factor as an important substance in T-cell deficiency. Several short papers were presented concerning the possible role of the thymus in acquired immune deficiency.

Dr. Haynes reported that thymic hormones are present in thymic tissue, including epithelial cells of the thymus in both normal and AIDS patients. Dr. A. Davis found that thymosin α_1 is distributed in various areas of the thymus and that there is marked thymic involution in AIDS. Dr. Seemayer also reported on the marked thymic involution in AIDS. Although thymosin α_1 was found elevated, thymulin levels are lower in AIDS, as reported by Dr. Dardenne, in concordance with the finding of thymic involution.

It was apparent from the symposium talks as well as the contributed poster papers that much work is being done in the area of the chemistry and biochemistry of thymic hormones and peptides. Nevertheless, there is still some important questions to be answered. Although there are apparently a large number of peptides and factors extractable from the thymus, it still is not clear how these relate to similar immunoregulatory peptides present in the serum and other tissues of individuals. It is clear that after thymectomy, many of these thymic-like peptides or molecules decrease in concentration in the serum and that these factors increase biological activity, especially of T lymphocytes. Further work is certainly necessary to characterize these factors and to determine their role in the immune system. It is not clear whether the synthetic peptides are completely identical to thymic peptides found naturally or whether they are end products of larger precursor molecules. This is still essentially unknown.

The structure of thymic peptides is still to be fully understood as only a few have been analyzed to date. One still does not know the target cells for all or even a few of the thymus-associated peptides. Are they all immature T cells, or are other cells reactive with these factors? Work is now under way in some laboratories to characterize some of the receptors to selected factors, but no one is certain whether similar or different receptors are present on the same cell to the different factors. Analysis of receptors will permit a more rapid expansion of knowledge concerning the immunobiology of thymic hormones. Future approaches which are still necessary include determining the sequence and structure of the various immunoregulatory peptides. This can be approached by preparing monospecific and monoclonal antibodies, not only to the peptides but also to specific membrane receptors. Genetic engineering is also important for future analysis of thymic hormones and peptides. Recombinant DNA technology will permit delineation of the genes responsible for production of these biologically reactive factors. Finally, the judicious use of antibodies to thymic factors in standardized immunoserological techniques will permit analysis of those factors important not only in immunodeficiency disease states associated with thymus dysfunction, but also as a possible predictor of potential immunological disorders, such as those associated with acquired immune deficiency.

II

Lymphokines

Characterization of Immunoaffinity-Purified Human T-Cell Growth Factor from JURKAT Cells

TERRY D. COPELAND, KENDALL A. SMITH, and STEPHEN OROSZLAN

1. INTRODUCTION

T-cell growth factor (TCGF) is a lymphokine that was first identified in conditioned media of mitogen-activated human T-cell culture (Morgan *et al.*, 1976). TCGF also occurs in other mammalian systems and its biological properties were recently reviewed (Smith, 1983). TCGF proved to be essential for the isolation of human T-cell leukemia virus. For a recent review see Gallo and Wong-Staal (1982).

TCGF is a glycoprotein (Robb and Smith, 1981) with a molecular weight of approximately 15,000 (Mier and Gallo, 1980). As an initial approach to understand the lymphokine, including its biological activity, we undertook primary structure studies of human (h)-TCGF. Here we report the results of amino acid sequence analysis of immunoaffinity-purified h-TCGF from JURKAT leukemic cells. In addition, the data indicate the location and the probable nature of the posttranslational modifications(s) of the protein molecule.

TERRY D. COPELAND and STEPHEN OROSZLAN • Laboratory of Molecular Virology and Carcinogenesis, LBI-Basic Research Program, NCI-Frederick Cancer Research Facility, Frederick, Maryland 21701. KENDALL A. SMITH • Department of Medicine, Dartmouth Medical School, Hanover, New Hampshire 03756.

2. MATERIALS AND METHODS

2.1. Protein Purification

TCGF was obtained from phytohemagglutinin (PHA)-stimulated JURKAT cell supernatant. The purification of biologically active h-TCGF by immunoaffinity chromatography utilizing monoclonal antibodies coupled to Affi-Gel 10 has been described elsewhere (Smith *et al.*, 1983). The immunoaffinity-purified protein was subjected to reverse-phase liquid chromatography (RPLC). The protein solution acidified to pH 2 with trifluoracetic acid (TFA) was loaded onto a Waters μBondapak C_{18} column and the column developed with a linear gradient of 0–60% acetonitrile in water containing 0.05% TFA. The peak fractions detected by absorbance at 215 nm were collected and protein was recovered by lyophilization.

2.2. Protein Modification

Prior to enzymatic cleavage, protein was reduced with dithiothreitol and *S*-carboxamidomethylated with iodoacetamide as described elsewhere (Henderson *et al.*, 1981). Modified protein was desalted by RPLC (Copeland *et al.*, 1983). *S*-carboxamidomethyl h-TCGF eluted at ~ 54% acetonitrile and was recovered by lyophilization.

2.3. Protein Fragmentation and Purification of Peptides

Lyophilized *S*-carboxamidomethyl h-TCGF was suspended in 0.1 M sodium bicarbonate, pH 8, containing 30% acetonitrile (Henderson *et al.*, 1981). Lys-C endoproteinase (Boehringer-Mannheim) was added to give an enzyme to substrate ratio of 1 : 50. Digestion occurred under nitrogen at room temperature for 24 hr. Digestion was arrested by freezing and lyophilization. To purify peptides the mixture was dissolved in guanidine hydrochloride, acidified to ~ pH 2 with aqueous TFA, and loaded onto the same RPLC column used for protein isolation. Peptide peaks were collected manually and aliquots taken for amino acid analysis.

2.4. Amino-Terminal Sequence Analysis

Unmodified protein as obtained from RPLC and peptides generated by enzymatic digestion of *S*-carboxamidomethyl protein with Lys-C proteinase were subjected to semiautomated micro Edman degradation (Edman and Begg, 1967) in the presence of Polybrene (Tarr *et al.*, 1978) utilizing established procedures (Copeland *et al.*, 1980).

2.5. Carboxyl-Terminal Analysis

Intact h-TCGF was digested with carboxypeptidase A as described elsewhere (Oroszlan *et al.*, 1978). Amino acids released during digestion at various time points were determined on the amino acid analyzer.

3. RESULTS

3.1. RPLC and Terminal Amino Acid Sequences of Intact h-TCGF

Immunoaffinity-purified protein was subjected to RPLC and the chromatogram obtained is illustrated in Fig. 1A. Protein eluted from the column as a single homogeneous peak at 58.5% acetonitrile. Biological activity was shown to coelute with the protein (Smith *et al.*, 1983). The quantitative data showed that the lymphokine retained complete activity when exposed to pH 2 and to organic solvents. The results of RPLC indicated that TCGF is a hydrophobic protein as it was retained on the hydrophobic support until elution with an acetonitrile concentration of 58.5%.

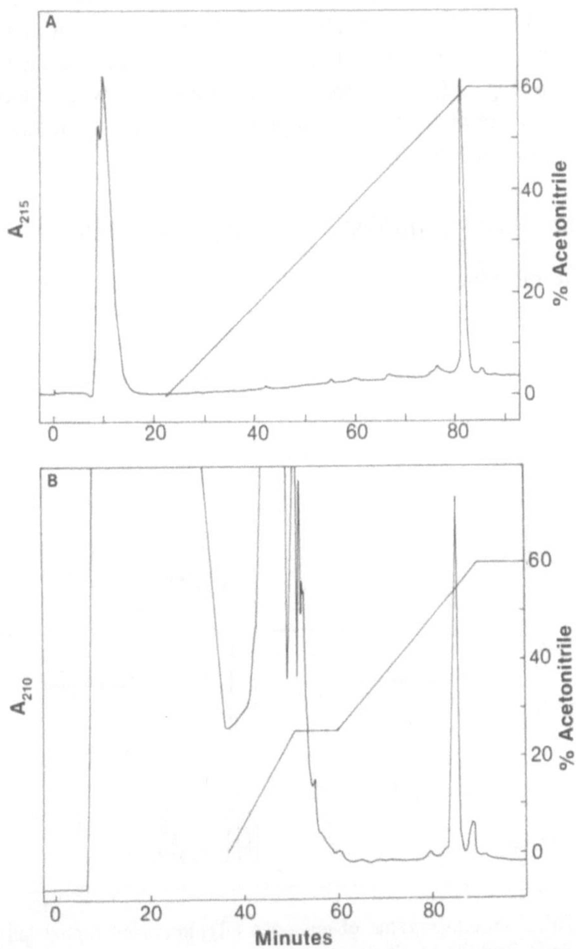

FIGURE 1. Purification by reverse-phase liquid chromatography on μBondapak C_{18} support (see text) of immunoaffinity-purified h-TCGF (A) and *S*-carboxamidomethyl h-TCGF (B).

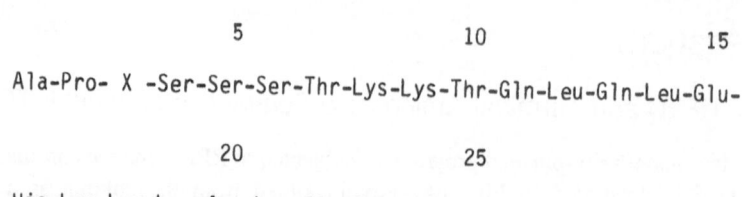

```
                5                   10                  15
Ala-Pro- X -Ser-Ser-Ser-Thr-Lys-Lys-Thr-Gln-Leu-Gln-Leu-Glu-

                20                  25
His-Leu-Leu-Leu-Asp-Leu-Gln-Met-Ile-Leu-
```

FIGURE 2. Amino-terminal amino acid sequence determined by semiautomated Edman degradation of immunoaffinity-purified h-TCGF.

Purified h-TCGF as obtained from RPLC was subjected to semiautomated microsequence analysis and the results for 25 cycles are given in Fig. 2. The presence of a single-chain polypeptide as obtained by the amino-terminal sequence also demonstrates that the h-TCGF preparation is homogeneous. A definitive assignment could not be made at position 3. The chromatogram for this cycle of the degradation is presented in Fig. 3 (protein cycle 3) and the results will be discussed below.

Utilizing carboxypeptidase A digestion the carboxyl-terminal sequence was found to be Leu-Thr-OH.

3.2. Protein Fragmentation and Sequence Analyses of Peptides

Prior to cleaving the protein in order to obtain peptides for further sequence analysis, the cysteines were alkylated. TCGF was reduced with dithiothreitol and

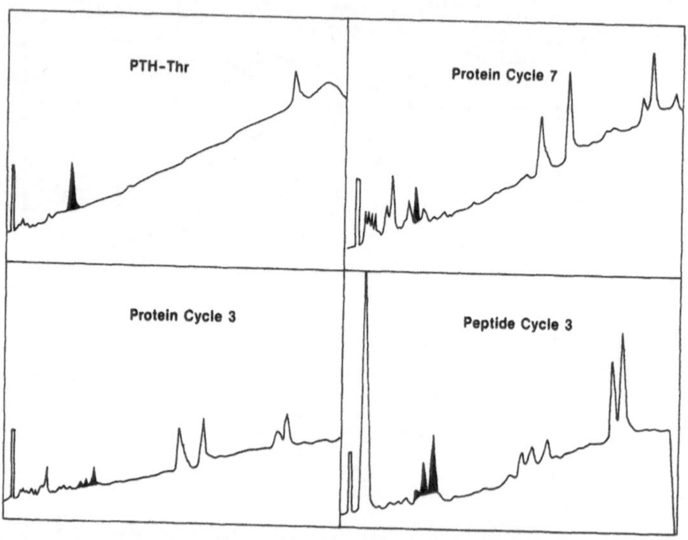

FIGURE 3. RPLC chromatograms of authentic PTH-threonine (upper left panel) and of the results from cycle 3 of the semiautomated degradation of intact protein (lower left panel) and of the amino-terminal peptide (lower right panel). The chromatogram of cycle 7 during the degradation of intact protein is shown in the upper right panel. Peaks from each chromatogram which are discussed in the text are shaded.

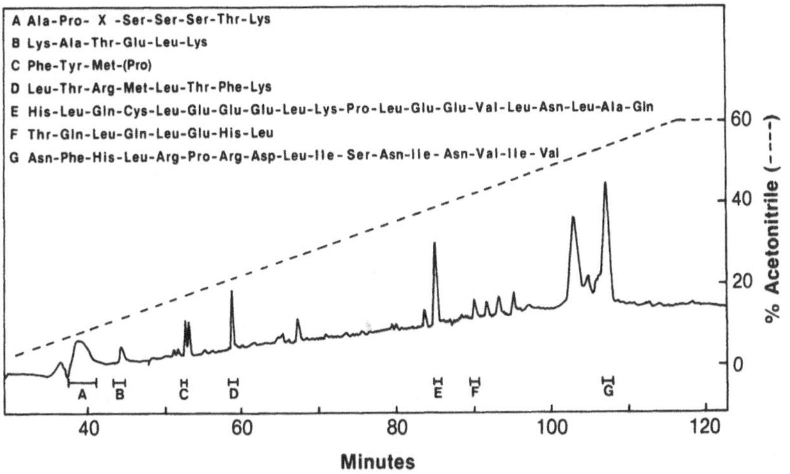

FIGURE 4. Separation by RPLC on μBondapak C_{18} support (see text) of peptides generated by endoproteinase Lys-C digestion of *S*-carboxamidomethyl h-TCGF. Peaks designated by letters proved by Edman degradation to be unique peptides and the sequences obtained are indicated.

S-carboxamidomethylated with iodoacetamide. Modified protein was removed from excess reagents and purified by a single step on RPLC as illustrated in Fig. 1B. *S*-carboxamidomethyl h-TCGF eluted from the column at 54% acetonitrile. The modified protein to be used for cleavage was recovered by lyophilization. Cleavage was accomplished by digestion with Lys-C endoproteinase. The generated peptides were separated and purified by RPLC as shown in Fig. 4. Peptides which gave unique amino-terminal sequences are indicated and the sequence obtained for each peptide is also given in Fig. 4. While this study was in progress, the nucleotide sequence of a cDNA coding for h-TCGF from JURKAT cells was published (Taniguchi *et al.*, 1983). The deduced amino acid sequence of the protein is given in Fig. 5. An identical amino acid sequence was deduced from the nucleotide sequence of TCGF DNA cloned from normal peripheral blood lymphocytes (Clark *et al.*, 1984). The amino acid sequences obtained by analysis of unmodified h-TCGF (Fig. 2) and of peptides obtained from modified h-TCGF (Fig. 4) are compared to the deduced amino acid sequence for h-TCGF in Fig. 5. Those amino acids determined in this study are underlined. No differences were found.

4. DISCUSSION

By amino-terminal and carboxyl-terminal analysis we have defined the ends of the h-TCGF molecule. Indeed, a signal peptide has been cleaved during processing of the precursor, and the biologically active lymphokine begins with alanine. The carboxyl-terminal Leu-Thr protein sequence is also in agreement with the

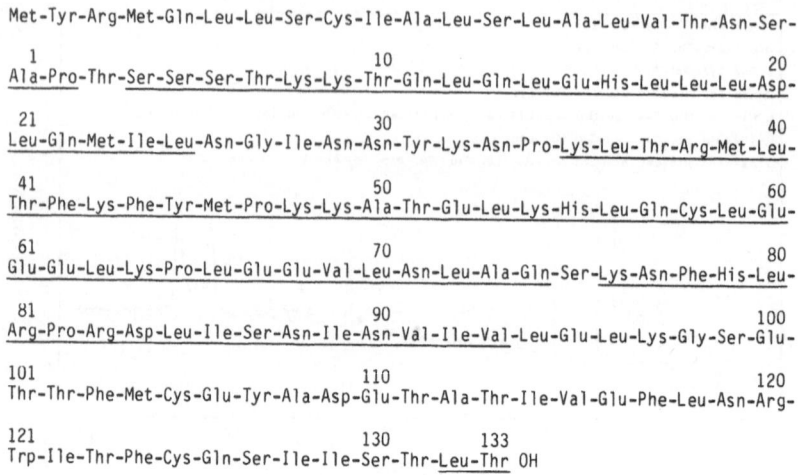

Met-Tyr-Arg-Met-Gln-Leu-Leu-Ser-Cys-Ile-Ala-Leu-Ser-Leu-Ala-Leu-Val-Thr-Asn-Ser-

1 10 20
Ala-Pro-Thr-Ser-Ser-Ser-Thr-Lys-Lys-Thr-Gln-Leu-Gln-Leu-Glu-His-Leu-Leu-Leu-Asp-

21 30 40
Leu-Gln-Met-Ile-Leu-Asn-Gly-Ile-Asn-Asn-Tyr-Lys-Asn-Pro-Lys-Leu-Thr-Arg-Met-Leu-

41 50 60
Thr-Phe-Lys-Phe-Tyr-Met-Pro-Lys-Lys-Ala-Thr-Glu-Leu-Lys-His-Leu-Gln-Cys-Leu-Glu-

61 70 80
Glu-Glu-Leu-Lys-Pro-Leu-Glu-Glu-Val-Leu-Asn-Leu-Ala-Gln-Ser-Lys-Asn-Phe-His-Leu-

81 90 100
Arg-Pro-Arg-Asp-Leu-Ile-Ser-Asn-Ile-Asn-Val-Ile-Val-Leu-Glu-Leu-Lys-Gly-Ser-Glu-

101 110 120
Thr-Thr-Phe-Met-Cys-Glu-Tyr-Ala-Asp-Glu-Thr-Ala-Thr-Ile-Val-Glu-Phe-Leu-Asn-Arg-

121 130 133
Trp-Ile-Thr-Phe-Cys-Gln-Ser-Ile-Ile-Ser-Thr-Leu-Thr OH

FIGURE 5. Amino acid sequence of h-TCGF as deduced from the nucleotide sequence (Taniguchi *et al.*, 1983). The numbering system is for the processed protein as utilized in this report. Amino acids that we have placed in sequence by amino- and carboxyl-terminal sequences of intact protein and amino-terminal sequences of internal peptides are underlined.

prediction from the nucleotide sequence in which a stop codon follows the threonine. Thus, no processing occurs at the carboxyl end.

It was noted above that no amino acid could be assigned at position 3 during the sequence analysis of intact unmodified protein. A similar result was also obtained during analysis of the amino-terminal peptide (see Fig. 3, peptide cycle 3, and Fig. 4, peptide A). The nucleotide sequence predicted position 3 to be threonine. In addition, the amino acid composition of the amino-terminal peptide (peptide A, data not shown) suggests that position 3 should be threonine. Based on the chromatograms illustrated in Fig. 3, an explanation for our inability to identify threonine is given as follows. In the upper left panel of Fig. 3, authentic phenylthiohydantoin (PTH) threonine is depicted. The sample from cycle 3 of the intact protein degradation is shown in the lower left panel. There is no peak with a retention time of authentic PTH-threonine. Rather a triplet of peaks (shaded) is obtained which do not correspond to the PTH derivative of any common amino acid. An authentic threonine assignment was made at position 7. The chromatogram obtained from the analysis of the degradation product obtained at cycle 7 of the intact protein is illustrated in the upper right panel. The shaded peak coelutes with authentic PTH-threonine. Similar results were obtained for cycle 7 during degradation of the amino-terminal peptide (not shown). The lower right panel is the chromatogram obtained for cycle 3 during degradation of the amino-terminal peptide. Again the triplet of peaks was obtained. Thus, by degrading both intact protein and the amino-terminal peptide a reproducible set of peaks was detected at cycle 3 none of which represents PTH-threonine. It is likely that residue 3 is a posttranslationally modified threonine. The RPLC elution pattern for cycle 3 (Fig. 3) and the elution at low acetonitrile concentration of the amino-terminal peptide itself (peptide A, Fig. 4) suggest that

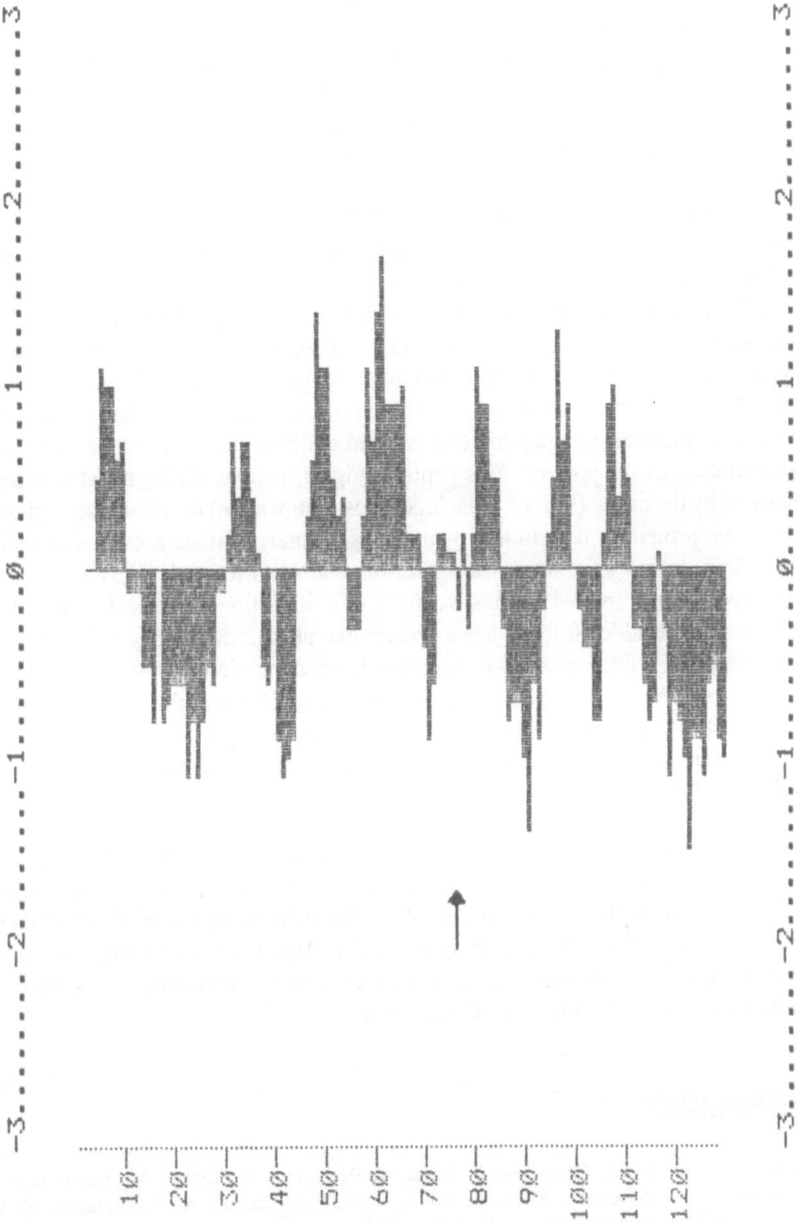

FIGURE 6. An illustration of the h-TCGF amino acid sequence plotted according to its hydrophilic character (Hopp and Woods, 1981). Hydrophilicity is above the line (positive values) while hydrophobicity is below the line (negative values). The arrow indicates the amino end of the carboxyl-terminal peptide.

the modification of threonine (residue 3) is not hydrophobic but rather hydrophilic in nature and may be due to the presence of carbohydrate moiety.

It was noted above that during the RPLC of immunoaffinity-purified h-TCGF, the protein was retained on the column until eluted with 58.5% acetonitrile. Behavior of the protein on the hydrophobic column support implied that h-TCGF is a highly hydrophobic protein. Figure 6 is a plot based on the method of Hopp and Woods (1981) to assess the hydrophilic and hydrophobic character of a protein based on the amino acid sequence. The abscissa of the plot is the residue number reading from the amino terminus to the carboxyl terminus. On the ordinate, hydrophilic values are expressed as positive numbers and hydrophobic values as negative numbers. The apparent high hydrophobicity of h-TCGF is not predicted by this analysis. In order to account for the chromatographic behavior on the hydrophobic RPLC support, it is suggested that h-TCGF may be posttranslationally modified by a hydrophobic moiety. During the purification of the h-TCGF peptides generated by Lys-C endoproteinase, a peptide was isolated which eluted at the same acetonitrile concentration as the protein. This peptide (Fig. 4, peptide G) begins at residue 77 indicated by the arrow (Fig. 6). It includes most, if not all, of the remaining carboxyl-half of the protein as determined by amino acid analysis (data not shown). On the basis of its amino acid sequence alone, this carboxyl-terminal peptide is not the most hydrophobic peptide generated by Lys-C digestion. Peptide F (Fig. 4) extending from residue 10 to 32 has a greater overall hydrophobicity (Fig. 6). However, peptide F eluted at a lower acetonitrile concentration than did the carboxyl-terminal peptide. Hence, we suggest that the proposed hydrophobic moiety resides on the carboxyl-terminal half of the molecule. The possible role of these proposed modifications in the function of h-TCGF is at present unknown and awaits further study.

ACKNOWLEDGMENTS. We wish to acknowledge the expert technical assistance of Young D. Kim. Research sponsored by the National Cancer Institute under Contract NO1-CO-23909 with Litton Bionetics, Inc. The contents of this publication do not necessarily reflect the views or policies of the Department of Health and Human Services, nor does mention of trade names, commercial products, or organizations imply endorsement by the U.S. Government.

REFERENCES

Clark, S. C., Arya, S. K., Wong-Staal, F., Matsumoto-Kobayashi, M., Kay, R. M., Kaufman, R. J., Brown, E. L., Shoemaker, C., Copeland, T., Oroszlan, S., Smith, K., Sarngadharan, M. G., Lindner, S. G., and Gallo, R. C., 1984, Human T-cell growth factor: Partial amino acid sequence, cDNA cloning, and organization and expression in normal and leukemic cells, *Proc. Natl. Acad. Sci. USA* (in press).

Copeland, T. D., Grandgenett, D. P., and Oroszlan, S., 1980, Amino acid sequence analysis of reverse transcriptase subunits from avian myeloblastosis virus, *J. Virol.* **36**:115–119.

Copeland, T. D., Morgan, M. A., and Oroszlan, S., 1983, Complete amino acid sequence of the nucleic acid binding protein of bovine leukemia virus, *FEBS Lett.* **156**:37–40.

Edman, P., and Begg, G., 1967, A protein sequenator, *Eur. J. Biochem.* **1**:80–91.

Gallo, R. C., and Wong-Staal, F., 1982, Retroviruses as etiologic agents of some animal and human leukemias and lymphomas and as tools for elucidating the molecular mechanism of leukemogenesis, *Blood* **60:**545–557.

Henderson, L. E., Copeland, T. D., Sowder, R. C., Smythers, G. W., and Oroszlan, S., 1981, Primary structure of the low-molecular-weight nucleic acid binding proteins of murine leukemia viruses, *J. Biol. Chem.* **256:**8400–8406.

Hopp, T. P., and Woods, K. R., 1981, Prediction of protein antigenic determinants from amino acid sequences, *Proc. Natl. Acad. Sci. USA* **78:**3824–3828.

Mier, J. W., and Gallo, R. C., 1980, Purification and some characteristics of human T-cell growth factor from phytohemagglutinin-stimulated lymphocyte-conditioned media, *Proc. Natl. Acad. Sci.* **77:**6134–6138.

Morgan, D. A., Ruscetti, F. W., and Gallo, R. C., 1976, Selective *in vitro* growth of T lymphocytes from normal human bone marrows, *Science* **193:**1007–1008.

Oroszlan, S., Henderson, L. E., Stephenson, J. R., Copeland, T. D., Long, C. W., Ihle, J. N., and Gilden, R. V., 1978, Amino- and carboxyl-terminal amino acid sequences of proteins coded by *gag* gene of murine leukemia virus, *Proc. Natl. Acad. Sci. USA* **75:**1404–1408.

Robb, R. J., and Smith, K. A., 1981, Heterogeneity of human T-cell growth factor (TCGF) due to variable glycosylation, *Mol. Immunol.* **18:**1087–1094.

Smith, K. A., 1983, T-cell growth factor, a lymphocytotrophic hormone, in: *Genetics of the Immune Response* (G. Moller, ed.), pp. 151–185, Plenum Press, New York.

Smith, K. A., Favata, M. F., and Oroszlan, S., 1983, Production and characterization of monoclonal antibodies to human interleukin-2, *J. Immunol.* **131:**1808–1815.

Taniguchi, T., Matsui, H., Fujita, T., Takaoka, C., Kashima, N., Yoshimoto, R., and Hamuro, J., 1983, Structure and expression of a cloned cDNA for human interleukin-2, *Nature (London)* **302:**305–310.

Tarr, G. E., Beecher, J. F., Bell, M., and McKean, D. J., 1978. Polyquaternary amines prevent peptide loss from sequenators, *Anal. Biochem.* **84:**622–627.

18

The Use of Lymphoid Cells Expanded in IL-2 for the Adoptive Immunotherapy of Murine and Human Tumors

STEVEN A. ROSENBERG, MAURY ROSENSTEIN, ELIZABETH GRIMM, MICHAEL LOTZE, and AMITABHA MAZUMDER

1. BACKGROUND

The adoptive immunotherapy of cancer refers to the transfer, to the tumor-bearing host, of immunologically competent cells capable of mediating responses against tumor. Specific adoptive immunotherapy is a theoretically attractive approach to the treatment of tumors although few examples exist of the effective treatment of established syngeneic solid tumors using this modality. Early reports from Delorme and Alexander (1964) claimed that thoracic duct lymphocytes from immunized rats as well as lymphocytes from immunized xenogeneic animals (Alexander *et al.*, 1966) could mediate the regression of solid methylcholanthrene-induced sarcomas. Borberg *et al.* (1972) treated mice bearing the Meth-A sarcoma with up to 4×10^9 immunized syngeneic lymphocytes and succeeded in causing the regression of established tumors. Using the Meth-A tumor, but an alternative method of immunization, Berendt and North (1980) have demonstrated that the intravenous infusion of sensitized T cells from immune donors could cause complete regression of established tumors growing in a T-deficient host. They also showed that infusion of splenocytes from tumor-bearing donors could inhibit this regression, suggesting that suppressor T cells existed in the tumor-bearing host. Fernandez-Cruz *et al.* (1980) showed that intravenous infusion of immune lymphocytes was capable of

STEVEN A. ROSENBERG, MAURY ROSENSTEIN, ELIZABETH GRIMM, MICHAEL LOTZE, and AMITABHA MAZUMDER • Surgery Branch, Division of Cancer Treatment, National Cancer Institute, National Institutes of Health, Bethesda, Maryland 20205.

curing rats bearing a subcutaneous tumor, and Smith *et al.* (1977) showed that immunized peritoneal exudate cells from syngeneic guinea pigs were capable of curing guinea pigs with established line 10 hepatoma.

In a review of the adoptive immunotherapy of cancer, Rosenberg and Terry (1977) summarized the evidence that the adoptive transfer of immunologically reactive lymphoid cells was capable of mediating the regression of established tumors in a variety of animal tumor models. An analysis of successful adoptive immunotherapy revealed several factors that seemed essential for successful treatment. Three of the most important factors were: (1) obtaining highly sensitized cells with specific antitumor reactivity; (2) obtaining large numbers of sensitized cells (at least 10^8 cells to treat a 1-cm-diameter tumor in a mouse); and (3) use of syngeneic cells for therapy. In the light of recent information, a fourth factor essential for the success of adoptive immunotherapy should be added, i.e., the elimination of suppressor cells in the tumor-bearing host that will prevent the action of adoptively transferred cells.

Early attempts at adoptive immunotherapy of human tumors fell far short of meeting each of the above four criteria. The development of techniques to expand T-lymphoid cells to large numbers *in vitro* using interleukin-2 (IL-2) has, however, provided a means for overcoming many previous problems in the adoptive immunotherapy of human tumors. The potential now exists for isolating from a cancer-bearing human, reactive lymphoid cells that can be cloned and expanded to large numbers for subsequent reinfusion into the autologous host. This approach to the adoptive immunotherapy of human tumors has great appeal though major problems to its successful application must still be overcome.

The generation of large numbers of lymphoid cells in IL-2 presents little difficulty because of the rapid expansion of both murine and human cells in IL-2 *in vitro*. Starting with 10^5 cells, more than 9×10^{13} (or 9 kg of packed cells) can be obtained after 2 months of growth, if all cells in these cultures were saved (Rosenberg, 1982). Though technical problems exist in developing large-scale culture techniques, these problems are surmountable. The recent development of tumor lines that produce IL-2 and cloning of the gene that produces IL-2 provides a means for attaining the amounts of IL-2 necessary for these cell expansions.

The development of techniques for identifying and isolating lymphoid cells with antitumor reactivity is a major obstacle to adoptive immunotherapy. The poor immunogenicity of tumor antigens in both the mouse and the human presents considerable difficulties in attempting to isolate the very rare T-cell clones that may have specific reactivity to tumor antigens.

Previous work from our laboratory (Rosenstein *et al.*, 1981) showed that systemic administration of sensitized lymphoid cells expanded in IL-2 was capable of mediating the rejection of skin allografts with a high degree of immunological specificity. Recent experiments have shown that the adoptive transfer of cytolytic clones with reactivity for antigens on the skin grafts was incapable of mediating skin graft rejection (Kim *et al.*, 1983). However, the adoptive transfer of clones with proliferative specificity for the target antigens on the skin graft was highly effective in mediating skin graft rejection on both irradiated as well as nude mice.

In the present report we will summarize some of our studies attempting to apply this approach to the adoptive immunotherapy of a syngeneic tumor. We will summarize our attempts to find the lymphocyte subpopulations involved in mediating tumor regression in mice and summarize some of our recent experiments attempting to produce lymphoid cells in the human with antitumor reactivity.

2. SENSITIZED LYMPHOID CELLS EXPANDED IN IL-2 CAN MEDIATE THE SUCCESSFUL ADOPTIVE IMMUNOTHERAPY OF A DISSEMINATED FBL-3 LYMPHOMA

The syngeneic FBL-3 lymphoma in C57BL/6 mice will grow progressively in about 25% of C57BL/6 mice following intramuscular injection of FBL-3 tumor. In the remaining animals, tumor will grow for approximately 10 days and then regress. Tumor grows progressively and kills mice uniformly when the tumor is injected intra-footpad into syngeneic mice that have received 500 rads of total body irradiation. Therefore, to immunize mice with FBL-3 lymphoma, we injected irradiated FBL-3 cells on two successive weeks and then waited 4 weeks before harvesting immune spleens. An alternative method of immunization that was often used was to inject live FBL-3 lymphoma intramuscularly followed by tumor growth and regression in those animals that rejected the tumor. Approximately 1 month after tumor rejection, spleens were harvested and used in experiments.

A model for the adoptive immunotherapy of solid and disseminated FBL-3 tumor was developed (Eberlein *et al.*, 1982a,b,c). In this model, mice were given 500 rads of whole body irradiation and approximately 4 hr later injected with 10^7 live FBL-3 tumor cells into the right hind footpad in 0.05 cm^3 of sterile Hanks' balanced salt solution (HBSS).

Five days later, tumors were easily palpable and the footpad diameter measured between 2.5 and 3 mm (the normal mouse footpad measures between 1.5 and 2 mm). If left untreated, both the footpad tumor and its metastases grew and eventually killed the animal by 20 days after the initial tumor injection. Also by day 5, either the right popliteal lymph node or 0.75 ml of blood from tumor-bearing mice could cause progressive tumor growth and death when injected intraperitoneally into a normal mouse pretreated with 500 rads of total body irradiation.

Five days after tumor was induced, when the tumor was clearly measurable in all animals, mice were randomly assigned to treatment groups and injected intravenously in the tail vein with experimental or control cells in 1 cm^3 of sterile HBSS. Each mouse was ear-tagged and measured every second or third day in a blinded fashion without knowledge of the previous treatment or the previous measurements for that mouse.

This model was used to test the effects of lymphoid cells expanded in IL-2 to cure mice with syngeneic disseminated FBL-3 lymphoma. In these experiments, spleens were harvested from immune mice and boosted *in vitro* to the FBL-3 lymphoma. To perform these sensitizations, 6×10^7 of viable responder splenocytes and 10^6 irradiated stimulator cells were placed in upright tissue culture flasks in 20 ml of tissue culture medium. Normal C57BL/6 stimulator cells were irradiated with 2000 rads, and fresh FBL-3 tumor stimulator cells were irradiated with 10,000 rads. Cells were harvested from these *in vitro* sensitization cultures on day 5 and

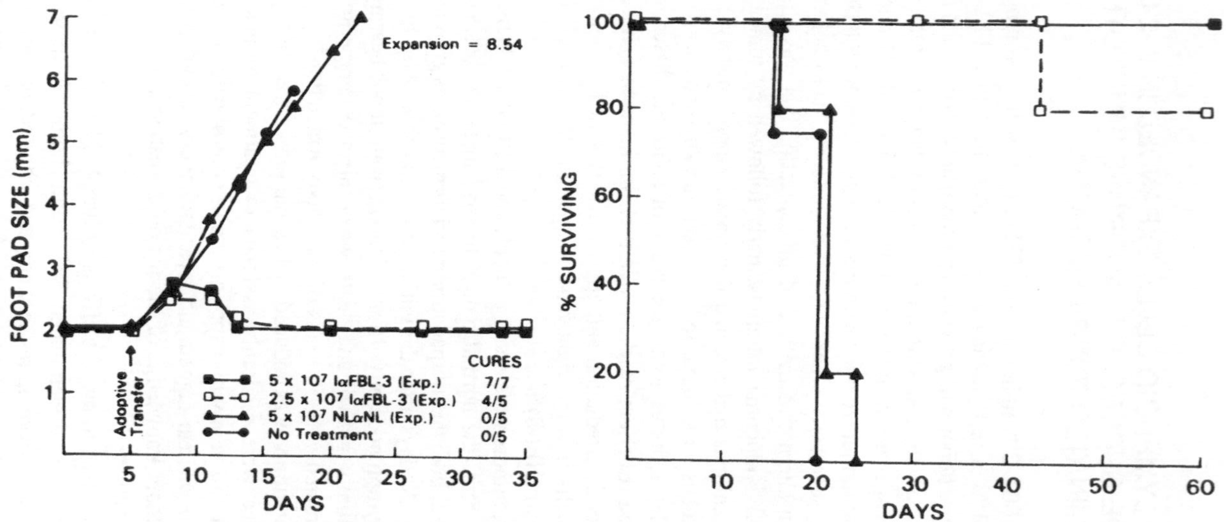

FIGURE 1. Adoptive immunotherapy of a solid disseminated syngeneic murine lymphoma using lymphoid cells expanded in IL-2. The size of the local tumor (left) and the survival of mice (right) are presented. (From Eberlein *et al.*, 1982a.)

tested *in vitro* for cytotoxicity and then placed in IL-2 for cell expansion. In virtually all experiments, control populations of either immune lymphocytes or nonimmune lymphocytes cocultured with normal C57BL/6 irradiated stimulators *in vitro* were also generated and expanded in IL-2.

Figure 1 shows the results of a typical experiment with the adoptive transfer of expanded cells in this disseminated footpad tumor model (Eberlein *et al.*, 1982a). Immune lymphocytes resensitized to FBL-3 tumor for 5 days and then expanded 8.5-fold in lectin-free (LF) IL-2 for 7 days (squares) were capable of curing 11 of 12 mice treated with one intravenous injection of either 5×10^7 or 2.5×10^7 cells ($p < 0.0005$). The only mouse in these groups that died did so on day 42 without evidence of tumor, and this animal received a lower dose of effector cells (2.5×10^7). There was no significant difference either in footpad size (Fig. 1, left) or in survival (Fig. 1, right) between mice receiving 5×10^7 or 2.5×10^7 lymphocytes. Normal lymphocytes expanded in LF IL-2 for 7 days failed to have any impact on either footpad tumor size or survival (triangles).

3. GENERATION OF LYMPHOID CLONES WITH *IN VITRO* REACTIVITY TO THE FBL-3 LYMPHOMA AND THEIR EFFECTS *IN VIVO*

Replica plating techniques that we have previously described (Eberlein *et al.*, 1982c) were used to isolate clones with a high degree of cytolytic or proliferative reactivity to the FBL-3 tumor. An example of the reactivity of several of these cytolytic clones is shown in Fig. 2. Several cytolytic clones with high degrees of lytic specificity for the FBL-3 tumor were adoptively transferred to mice in our tumor model and no increase in survival was seen (data not shown).

Using the replica screening techniques we have described previously for the generation of cytolytic cells, we have now been successful in raising proliferative cell lines and clones to the FBL-3 tumor. These clones have proliferative specificity for the FBL-3 tumor and do not proliferate when stimulated by other syngeneic tumors. To raise these clones with proliferative activity to the FBL-3 tumor, we have taken advantage of two of our previous findings. Cells from immune animals have been depleted of Lyt-2^+ cells using anti-Lyt-2 antibody plus complement. This technique was highly successful for developing proliferative clones in allogeneic systems (Rosenstein *et al.*, 1983). These depleted cells were then placed into *in vitro* sensitization cultures and then immediately cloned by limiting dilution techniques. These clones were screened for proliferative activity in a 2- to 4-day *in vitro* assay (Eberlein *et al.*, 1982c). Mitrotiter wells containing cells with high degrees of proliferative activity in this screening assay were then harvested and expanded *in vitro*.

An example of a proliferative clone (No. 28) that was isolated from the replica plate screening technique and then recloned is shown in Table I. All six of the reclones shown in Table I showed high degrees of proliferation to the FBL-3 tumor. The proliferative specificity of one of these reclones (clone 28-7) is shown in Table II. As can be seen, this clone had a very small degree of lytic activity to the FBL-

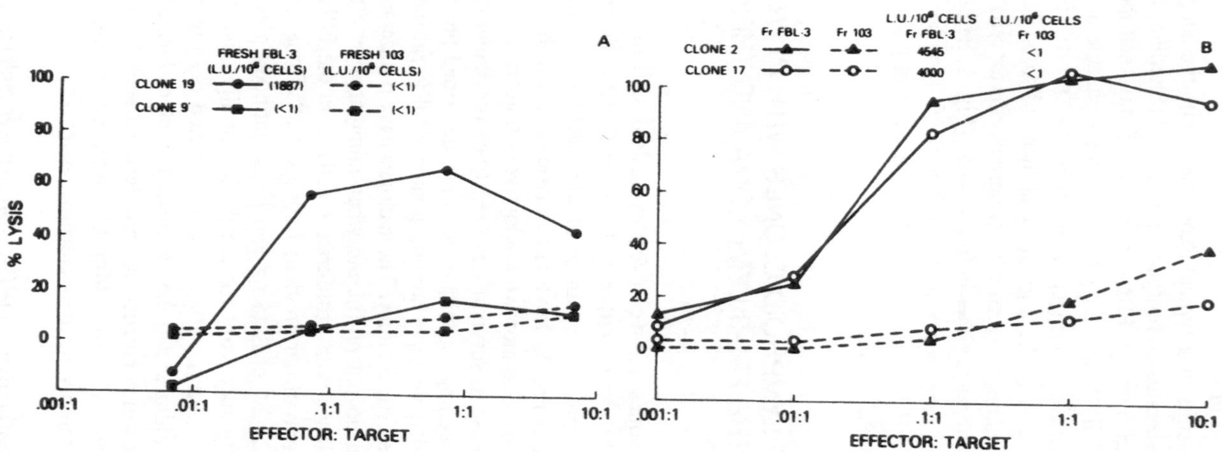

FIGURE 2. Lytic activity in an 18-hr ^{51}Cr-release assay of four clones isolated from an *in vitro* sensitization against the syngeneic FBL-3 lymphoma. Clones were tested for lysis against the syngeneic FBL-3 tumor and also were tested against another syngeneic non-cross-reacting methylcholanthrene induced MCA-103 tumor. Three lytic clones are shown (clones 2, 17, and 19) and one clone selected to be nonlytic is also shown (clone 9′). Note the high levels of specific lysis seen in the three lytic clones. (From Eberlein *et al.*, 1982b.)

TABLE I. Cloning of Proliferative Cells with Reactivity to Syngeneic Lymphoma[a]

100 cells/well	74/96 +
10 cells/well	19/96 + (harvested)
1 cell/well	1/96 + (harvested)

Clone No.	(1) Clone alone	(2) + C57BL/6 splenocytes (CPM)[b]	(3) + FBL-3 tumor	(2) + (3)
None	478	785	2,453	1,500
SR 7	448	332	360,065	216,392
SR 25	201	658	181,844	108,056
SR 35	670	229	129,908	137,811
SS 5	345	479	39,814	57,789
SS 16	130	259	103,214	83,581
SS 53	293	514	48,244	27,904

[a] From Rosenstein and Rosenberg (1983).
[b] Mean of triplicates (coefficient of variance 5–15%); harvested on day 2.

TABLE II. Specificity of Reactivity of Clone 28-7[a]

Target	Cytotoxicity (% lysis) E : T 100 : 1	10 : 1	Proliferative activity (stimulation index)	cpm/cpm
FBL-3	29 ± 2	24 ± 2	19.1	79,170/4139
C57BL/6 lymphocyte	−20 ± 4	−17 ± 2	2.6	
EL-4 tumor (C57BL/6)	78 ± 2	81 ± 5	<1	
P815 tumor (DBA/2)	10 ± 1	7 ± 1	<1	
MCA-106 tumor (C57BL/6)	5 ± 1	2 ± 1	<1	
FBL-3 tissue culture line	1 ± 2	−1 ± 1	<1	
Line 53 (T-cell line from C57BL/6)	−27 ± 19	−31 ± 9	1.1	329/298
Meth-A tumor (BALB/c)	3 ± 1	1 ± 1	<1	

[a] From Rosenstein and Rosenberg (1983).

TABLE III. Mean Survival Time after Adoptive Transfer of Cytotoxic Clones, Proliferative Lines, and Proliferative Clones[a,b]

Experiment	Treatment: None	Treatment: Cytotoxic clone	p^c
A	16.5 ± 0.9 (6)	17.8 ± 1.0 (5)	NS
B	20.4 ± 1.1 (8)	19.6 ± 2.9 (5)	NS
C	17.8 ± 1.8 (5)	19.2 ± 0.5 (5)	NS
		Proliferative line	
A	19.9 ± 0.4 (8)	24 ± 2 (5)	<0.05
B	14.7 ± 1.7 (6)	28.2 ± 10.2 (6)	NS
C	16.7 ± 0.4 (6)	22.4 ± 0.6 (5)	<0.001
		Proliferative clone	
D	16.3 ± 1.2 (6)	25.8 ± 2.4 (5)	<0.01
E	19.4 ± 0.4 (7)	27.3 ± 2.0	<0.01

[a] After Rosenstein and Rosenberg (1983).
[b] Values are days ± S.E.M. Number of mice in parentheses.
[c] *p* value clone or line vs. no treatment.

3 tumor and somewhat greater lytic activity to the EL-4 tumor. Clone 28-7 was not lytic to several other syngeneic or allogeneic murine tumors. This clone exhibited significant proliferative activity only to the FBL-3 tumor and not to other syngeneic or allogeneic tumors (Rosenstein and Rosenberg, 1983).

We have only recently begun testing these proliferative clones for their ability to cure mice in our FBL-3 tumor model. In preliminary experiments, it appears that these proliferative clones are indeed able to prolong the survival of mice bearing the FBL-3 lymphoma. An example of several preliminary experiments is presented in Table III (Rosenstein and Rosenberg, 1983).

4. GENERATION OF LYMPHOID CELLS WITH SPECIFIC ANTITUMOR REACTIVITY FOR AUTOLOGOUS HUMAN TUMOR CELLS

4.1. Human Activated Killer Cells

In the past several years we have studied a unique phenomenon capable of generating human lymphoid cells with significant levels of lytic reactivity for fresh autologous human tumor cells (Lotze and Rosenberg, 1981; Strausser *et al.*, 1981; Grimm *et al.*, 1982, 1983a,b; Mazumder *et al.*, 1982, 1983a,b). Human peripheral blood lymphoid cells (PBL) activated by a variety of techniques develop the ability to lyse fresh autologous tumor cells in a short-term ^{51}Cr-release assay. Three separate techniques for generating these activated killer cells from human PBL have been described. These activated killer cells can be generated by exposure of fresh PBL to IL-2 [lymphokine-activated killer (LAK) cells] by incubation with the lectin phytohemagglutinin [phytohemagglutinin-activated killer (PAK) cells] and by allosensitization with pooled stimulator cells [alloactivated killer (AAK) cells]. These activated killer cells are capable of killing NK-resistant fresh autologous tumor as well as a variety of allogeneic tumor cells. An example of the kinetics of development of LAK cell killing of fresh tumor cells is shown in Fig. 3 (Grimm *et al.*, 1982). Extensive studies of the characteristics of human activated killer cells have been performed and are summarized in Table IV (Mazumder et al., 1983). The biological role of these cells *in vivo* is unknown. We are currently exploring the possible use of these cells in the adoptive immunotherapy of human tumors.

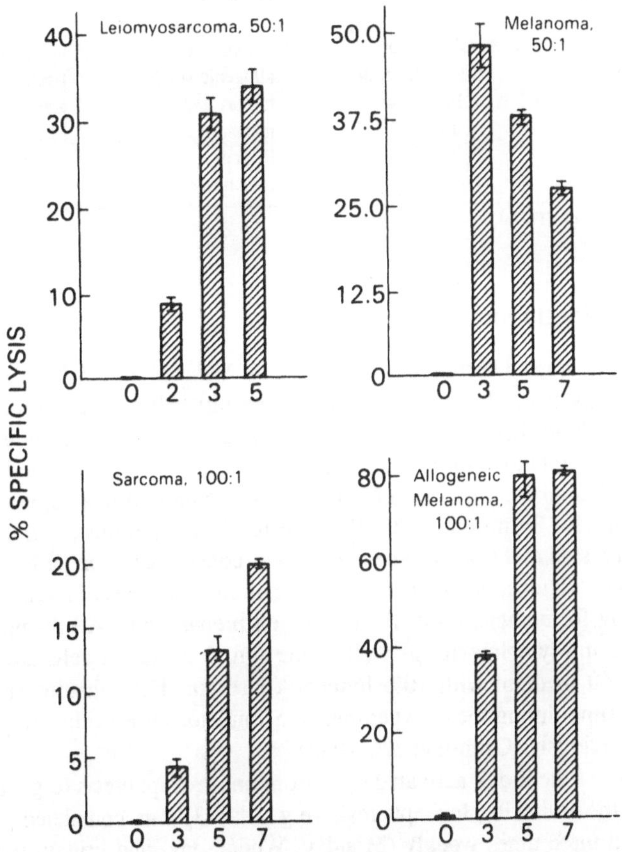

FIGURE 3. The kinetics of generation of lymphokine-activated killer cell activity against three autologous fresh solid tumor cells and against one allogeneic tumor cell (lower right). In all cases, significant killer activity for fresh solid tumor cells was generated by 3–5 days after exposure to lymphokine. (From Grimm *et al.*, 1982.)

TABLE IV. Characteristics of Human Activated Killer Cells[a]

Characteristic	Lectin incubation	Allosensitization	Activation
Precursor location	TDL$^+$, PBL$^+$	TDL$^+$, PBL$^+$	TDL$^+$, PBL$^+$
Precursor phenotype	OKT3$^+$, OKM1$^-$	OKT3$^+$, OKM1$^-$	OKT3$^-$, OKM1$^-$
Radiation sensitivity and adherence of precursor	Radioresistant and nonadherent	Radiosensitive and nonadherent	Radiosensitive and nonadherent
Role of macrophages	Necessary but can be suppressive	Necessary but can be suppressive	Necessary but can be suppressive
Development kinetics	Day 1, 2	Day 2, 3	Day 2, 3
Stimulus	Lectins	Alloantigens	IL-2-containing medium
Effector phenotype	OKT3$^+$, OKM1$^-$	OKT3$^+$, OKM1$^-$	OKT3$^+$, OKM1$^-$
Radiation sensitivity and adherence of effector	Radioresistant and nonadherent	Radioresistant and nonadherent	Radioresistant and nonadherent
Specificity of cytotoxicity	Autologous or allogenic fresh solid tumors and metastases. Not PBL or lymphoblasts	Autologous or allogenic solid tumors and metastases. Not PBL or lymphoblasts	Autologous or allogenic fresh solid tumors and metastases. Not PBL or lymphoblasts

[a] From Mazumder *et al.* (1983a).

4.2. Mixed Lymphocyte Tumor Cultures

We have adapted techniques described by Vose *et al.* (1978) and Vanky *et al.* (1979) for the generation of human autologous mixed lymphocyte tumor cultures. Responder cells in these cultures are PBL separated on Ficoll–hypaque gradients from the peripheral blood of cancer patients. Adherent cells are depleted in a plastic flask by incubation at 37°C for 1 hr. These cells are then further depleted on nylon wool columns for 30 min at 37°C. Because in most experiments PBL and tumor cells are harvested at the same time and it takes considerably longer to perform the tumor cell preparations, the lymphoid cells are often incubated at 37°C at 10^6 cells/ml while waiting for completion of the tumor cell preparations. All manipulations of these responding lymphocytes as well as the stimulator tumor cells are performed in RPMI 1640 medium with 10% human AB serum. Heterologous serum is not used at any time during these experiments. Stimulator tumor cells are prepared as described previously (Grimm *et al.*, 1982).

Infusion of the 2-day activated cells from one leukapheresis (e.g., day 1) was given after the following leukapheresis (e.g., day 3) was completed. This cycle was repeated three times weekly (Monday, Wednesday, and Friday) for a total of 7 to 15 leukaphereses and infusions.

In 37 separate autologous mixed lymphocyte tumor cultures, 19 (51%) positive proliferative responses have been seen. A sample culture exhibiting a modest positive autologous mixed lymphocyte tumor interaction is shown in Table V. On 15 occasions thus far, we have been able to obtain normal cells (either lung, liver,

TABLE V. Human Autologous MLTI[a]

No. of tumor cells/well	Tumor alone (cpm)	Tumor and PBL (cpm)
1×10^5	1511 ± 365	1,482 ± 360
0.5×10^5	3479 ± 751	3,655 ± 469
0.25×10^5	874 ± 279	6,734 ± 1658
0.125×10^5	1401 ± 186	15,946 ± 898
0.06×10^5	218 ± 762	5,324 ± 1101
0	—	702 ± 137

[a] 10^5 PBL responders from a patient with sarcoma stimulated by autologous tumor.

colon, or small bowel) from the same individual from whom tumor cells were isolated. In seven of these experiments, we saw a similar degree of stimulation caused by normal cells as was caused by tumor cells. We are currently attempting to clone lymphoid cells from these cultures in an effort to find cells with specific antitumor reactivity.

5. *IN VIVO* STUDIES IN HUMANS USING ACTIVATED KILLER LYMPHOCYTES

A phase I clinical protocol was conducted to test the feasibility and toxicity of the infusion of large numbers of PHA-activated autologous PBL into humans (Mazumder *et al.*, 1984). Measurements of the toxicity of this procedure, the presence of circulating activated cells on the traffic patterns of infused ^{111}In-labeled PHA-activated PBL, were conducted. Patients were accepted into this protocol if they had evaluable metastases and had failed all standard treatment and had an expected survival of greater than 3 months. Patients were excluded who had undergone in the previous month any other form of therapy for their cancer.

Leukaphereses were performed using either an Aminco Centrifuge II or IBM Model 2997, following standard procedures. An adjustable port optimized for lymphocyte removal distributed the separated fraction to a plastic bag. Each leukapheresis lasted 3 hr and the procedure was repeated three times weekly.

The mononuclear cells were then separated from the buffy coat by Ficoll–Hypaque (LSM-Litton Bionetics, Kensington, Md.) gradient centrifugation, according to standard methods. The cells were washed three times with HBSS and were resuspended in medium consisting of RPMI 1640 (GIBCO, Grand Island, N.Y.) with 10% heat-inactivated human AB serum (KC Biological, Lenexa, Kans.), 100 U/ml of penicillin, and 100 μg/ml of streptomycin (NIH Media Unit). An aliquot of these fresh cells was cryopreserved in 90% human AB serum plus 10% dimethylsulfoxide for future *in vitro* testing.

The mononuclear cells obtained above were placed into 175-cm^2 (750 ml) incubation flasks (Falcon, Oxnard, Calif.) laid supine in 175 ml of medium containing 1.0 μg/ml of PHA (Burroughs-Wellcome, Beckenham, England) at a concentration of 2.5×10^6 cells/ml. The flasks were incubated at 37°C, 5% CO_2, for 40–48 hr.

They were then harvested into sterile 250-ml centrifuge tubes (Falcon), washed three times with HBSS, and finally resuspended in 200 ml HBSS with 5% autologous plasma (obtained at leukapheresis) for infusion.

On the morning of infusion, 2×10^9 PHA-activated cells were harvested as above and resuspended in medium. In some trials, cells were incubated with approximately 1 mCi of ^{111}In oxine (Mediphysics, South Plainfield, N.J.) for 30 min at 37°C with occasional shaking. The labeled cells were washed three times with HBSS and resuspended in 200 ml HBSS with 5% autologous plasma and infused.

Gram stain and culture for aerobes, fungus, and anaerobes were carried out from the activation medium supernatant. The infusion was canceled if these gram stains or cultures were positive. 10^6 cells were initially infused over 5 min, followed by the remainder of the cells over 1 hr if no immediate reactions occurred. Vital signs were obtained preinfusion and every hour until stable. The first four patients had cardiac monitoring during their infusions.

Infusion of the 2-day activated cells from one leukapheresis (e.g., day 1) was given after the following leukapheresis (e.g., day 3) was completed. This cycle was repeated three times weekly (Monday, Wednesday, and Friday) for a total of 7 to 15 leukaphereses and infusions.

5.1. Results of This Treatment Protocol

A total of 10 patients were treated on the protocol (Mazumder *et al.*, 1984) (Table VI). Six patients had pulmonary metastases from sarcoma (two synovial cell sarcoma, one epithelioid sarcoma, one osteosarcoma, one fibrosarcoma, and one undifferentiated sarcoma). Three patients had colorectal cancer metastases (one in the lung, two in the liver), and one patient had melanoma nodules in the skin. All patients had greater than several kilograms of body tumor burden and had failed all standard treatments. No patient demonstrated an antitumor effect of the cell infusions in this protocol. The ages of patients ranged from 22 to 60. The number of infusions given was from 7 to 15 and the total number of cells infused ranged from 1.6 to 17.1×10^{10}. Patients 5, 6, 7, and 8 received 14 or 15 leukaphereses and reinfusions based on the finding of rapidly increasing numbers of circulating activated cells after six to seven infusions in the first four patients.

The number of cells obtained from each leukapheresis and the recovery after 2 days of PHA activation were variable in the patients. In patient 1, for example, the number of cells obtained from each of the nine leukaphereses varied between 7.2×10^9 and 12.5×10^9. The recovery from the 2-day PHA activation culture ranged from 59% to 100%.

The toxicity of the cell reinfusion consisted primarily of fever and chills in all patients. The symptoms were self-limited and generally returned to normal within 18 hr after the infusion. Shaking chills were occasionally seen as well though were easily controlled by meperidine. In each of three patients tested, there was a mean decrease of 30% in the pulmonary D_{CO} between the pulmonary function test performed immediately before and after infusions. However, in all three patients the

TABLE VI. Composite of Patient Characteristics[a]

	Patient no.									
	1	2	3	4	5	6	7	8	9	10
Age	22	37	47	36	39	51	60	42	29	37
Diagnosis	Synovial cell sarcoma	Epithelioid sarcoma	Melanoma	Osteo-sarcoma	Colorectal carcinoma	Colorectal carcinoma	Colorectal carcinoma	Fibro-sarcoma	Undiff-erentiated sarcoma	Synovial cell sarcoma
Evaluable disease	Lung	Lung	Skin	Lung	Liver	Liver	Lung	Lung	Lung	Lung
No. of infusions	9	9	9	9	14	15	15	14	9	7
Total No. of cells infused ($\times 10^{-10}$)	7.2	5.5	2.8	1.9	17.1	7.3	8.4	10.1	3.2	1.6

[a] From Mazumder *et al*. (1983b).

D_{CO} returned to baseline within 24 hr and remained normal on follow-up studies. Thus, it appears that large numbers of PHA-activated autologous cells can be safely obtained and infused into humans (up to 17.1×10^{10} cells over 5 weeks).

In vivo cell distribution studies were performed for the initial and final infusions in patient 9 and for the initial infusion in patient 10, with similar results. In patient 9, from the estimated blood volume, a maximum of 6% of the infused counts was detected in the peripheral blood 1 hr after the infusion of ^{111}In-labeled PBL. Significant radioactivity could be recovered for a prolonged period of time (up to 2 weeks) after the infusion. Greater than 90% of the counts in the blood was in the cellular fraction of the whole blood collected during the first 96 hr and during this time, very few cell-free counts (less than 0.1% of the injected dose) were found in the urine.

The percent of counts remaining in the entire body remained high for a prolonged period of time, with almost 70% of the initial counts remaining for 2 weeks. In the initial infusion, the cells appeared to migrate preferentially to the spleen and liver rather than the lungs with a rapid tripling of the initial counts over these organs, concurrent with a rapid decrease in counts over the lungs and heart. Subsequently, counts in the lungs and heart continued to decline while substantial activity persisted in the liver and spleen. It was also found that while the counts over the normal thigh did not vary significantly from the baseline, a tumor mass in the tumor-bearing thigh appeared to continue to accumulate labeled PBL for up to 8 days after infusion although with counts much lower than those in the liver and spleen.

Rectilinear scanning confirmed this preferential distribution of labeled cells to the liver and spleen in the initial infusion seen in the organ point counting (Fig. 4, top). However, scanning at similar intervals after the final infusion revealed a different pattern of cell trafficking (Fig. 4, bottom). Whole body scanning at 24, 48, and 96 hr after the initial infusion showed far fewer counts in the lungs than

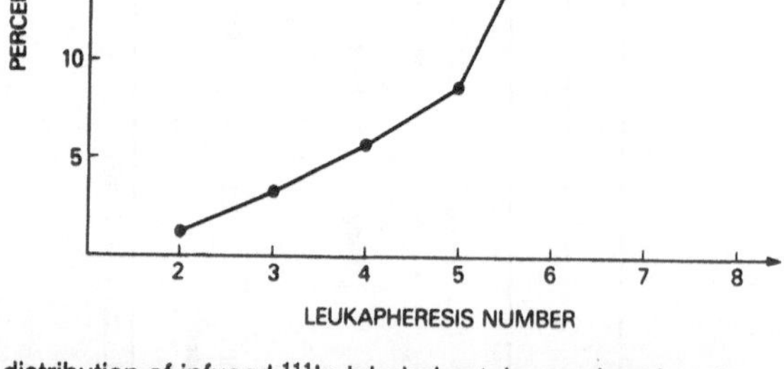

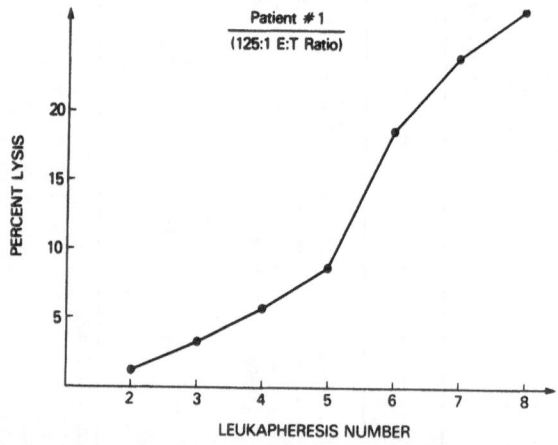

FIGURE 4. The distribution of infused ^{111}In-labeled autologous lymphocytes activated by PHA. Note that in the initial infusion (top), distribution in the first several days is exclusively to liver and spleen. By the final infusion, however, a substantial increase in distribution to the lungs is seen in the first 2 days after infusion. (From Mazumder *et al.*, 1983b.)

in the liver and spleen as compared to the same intervals after the final infusion where much higher uptake in the lungs was seen. Also, the patient's tumor mass (left thigh) accumulated label visible on scans only in the final infusion.

Thus, it appears that PHA-activated cells migrate preferentially to the liver and spleen with some accumulation in the tumor and lungs, especially after repeated infusions.

5.2. *In Vitro* Testing

In vitro tests were carried out in all patients using unstimulated cells obtained directly from the leukaphereses to determine whether there were circulating activated cells (Mazumder *et al.*, 1983b). Background incorporation of [^{3}H]-TdR into unstimulated PBL in medium alone for 2 days increased four- to fivefold in cells from late compared to early leukaphereses. This increase could be further enhanced by the addition of LF IL-2 in the 2-day assay.

The presence of circulating activated cells was further assayed by lysis of fresh human tumor cells by fresh PBL obtained directly from the leukaphereses. Figure 5 shows that the lysis of fresh tumor cells by circulating lymphocytes collected from the successive leukaphereses also increased. In fact, in 9 of 10 patients tested, the lysis of fresh tumor cells (autologous in four cases and allogeneic for the remainder) increased from a mean of 2% initially, to a mean of 31% on successive fresh PBL collections. It should be emphasized that in all of our *in vitro* experiments performed to date with fresh, unstimulated PBL from cancer patients, these protocol patients are the only ones whose fresh PBL consistently show lysis of fresh tumor cells in a short-term ^{51}Cr-release assay. As with the [^{3}H]-TdR uptake, in the patients who had a higher number of leukaphereses, the lysis of tumor rose until about the ninth leukapheresis and then remained at a high level.

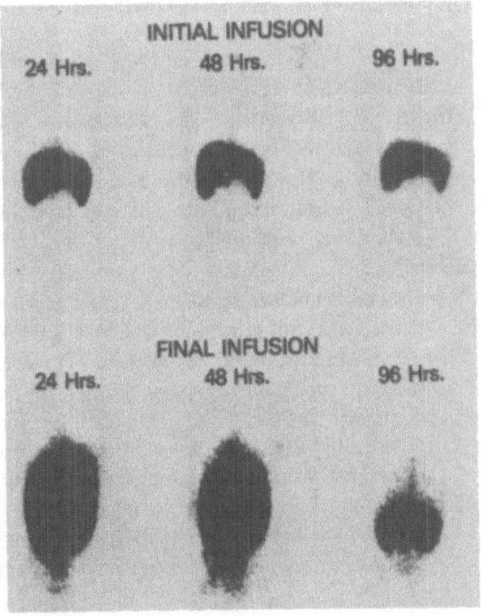

FIGURE 5. Lysis of fresh autologous tumor cells by PBL taken from patients prior to each leukapheresis of PHA-activated killer cells. Note the increase in killer cell activity directed against fresh autologous tumor present in the peripheral blood of this patient that increases with successive leukaphereses. (From Mazumder *et al.*, 1983b.)

Thus, it appears that we can achieve an increase in the number of circulating activated cells, and maintain a high level of activated cells in the circulation with repeated infusions of autologous PHA-activated lymphocytes.

6. SUMMARY

In this brief review we have described our approach to developing an adoptive immunotherapy that may be applicable to the treatment of human tumors. This approach is based on the expansion in IL-2 of appropriately sensitized lymphoid cells reactive against autologous tumor. In several mouse models we have been able to develop cell lines and clones capable of prolonging survival or curing mice with syngeneic local and disseminated tumors. In the human we are attempting to develop cells capable of reacting with fresh autologous tumors using two main approaches. Activated killer cells, which can be simply activated by a variety of techniques, are currently being studied both *in vitro* and *in vivo*. Attempts are also being made to develop specific lines or clones of human lymphoid cells expanded in IL-2 with reactivity against fresh autologous human tumor. It is hoped that the development of such lines may prove to be of value in the adoptive immunotherapy of autologous tumors.

REFERENCES

Alexander, P., Delorme, E. J., and Hall, J. G., 1966, The effect of lymphoid cells from the lymph of specifically immunized sheep on the growth of primary sarcomata in rats, *Lancet* **1:**1186.

Berendt, M. J., and North, R. J., 1980, T-cell-mediated suppression of anti-tumor immunity: An explanation for progressive growth of an immunogenic tumor, *J. Exp. Med.* **151:**69.

Borberg, H., Oettgen, H. F., Choudry, K., and Beattie, E. J., Jr., 1972, Inhibition of established transplants of chemically induced sarcoma in syngeneic mice by lymphocytes from immunized donors, *Int. J. Cancer* **10:**539.

Delorme, E. J., and Alexander, P., 1964, Treatment of primary fibrosarcoma in the rat with immune lymphocytes, *Lancet* **2:**117.

Eberlein, T. J., Rosenstein, M., and Rosenberg, S. A., 1982a, Regression of a disseminated syngeneic solid tumor by systemic transfer of lymphoid cells expanded in IL-2, *J. Exp. Med.* **156:**285.

Eberlein, T. J., Rosenstein, M., Spiess, P. J., and Rosenberg, S. A., 1982b, The generation of long term T-lymphoid cell lines with specific cytotoxic reactivity for a syngeneic murine lymphoma, *J. Natl. Cancer Inst.* **69:**109.

Eberlein, T. J., Rosenstein, M., Spiess, P., Wesley, R., and Rosenberg, S. A., 1982c, Adoptive chemoimmunotherapy of a syngeneic murine lymphoma using long term lymphoid cell lines expanded in T cell growth factor (PP-TCGF), *Cell. Immunol.* **70:**248.

Fernandez-Cruz, E., Woda, B. A., and Feldman, J. D., 1980, Elimination of syngeneic sarcomas in rats by a subset of T-lymphocytes, *J. Exp. Med.* **152:**823.

Grimm, E. A., Mazumder, A., Zhang, H. Z., and Rosenberg, S. A., 1982, The lymphokine activated killer cell phenomenon: Lysis of NK resistant fresh solid tumor cells by IL-2 activated autologous human peripheral blood lymphocytes, *J. Exp. Med.* **155:**1823.

Grimm, E. A., Ramsey, K. M., Mazumder, A., Wilson, D. J., Djeu, J. Y., and Rosenberg, S. A., 1983a, Lymphokine-activated killer cell phenomenon. II. The precursor phenotype is serologically distinct from peripheral T lymphocytes, memory CTL, and NK cells, *J. Exp. Med.* **157:**884.

Grimm, E. A., Robb, R. J., Roth, J. A., Neckers, L. M., Lachman, L. B., Wilson, D. J., and Rosenberg, S. A., 1983b, Lymphokine-activated killer cell (LAK) phenomenon. III. Evidence that IL-2 alone is sufficient for direct activation of PBL into LAK, *J. Exp. Med.* **158:**1356–1361.

Kim, B., Rosenstein, M., and Rosenberg, S. A., 1983, Clonal analysis of the lymphoid cells mediating skin allograft rejection: cloned Ly $1^{+}2^{-}$ proliferative, noncytotoxic long term cell lines mediate graft rejection in vivo, *Transplantation* **36:**525–532.

Lotze, M. T., and Rosenberg, S. A., 1981, In vitro growth of cytotoxic human lymphocytes. III. The preparation of lectin free T cell growth factor (TCGF) and an analysis of its activity, *J. Immunol.* **126:**2215.

Mazumder, A. M., Grimm, E. A., Zhang, H. Z., and Rosenberg, S. A., 1982, Lysis of fresh human solid tumors by autologous lymphocytes activated in vitro with lectins, *Cancer Res.* **42:**913.

Mazumder, A., Grimm, E. A., and Rosenberg, S. A., 1983a, Characterization of the lysis of fresh human solid tumors by autologous lymphocytes activated in vitro with phytohemagglutinin, *J. Immunol.* **130:**958.

Mazumder, A., Eberlein, T. J., Grimm, E. A., Wilson, D. J., Keenan, A. M., Aamodt, R., and Rosenberg, S. A., 1984, Phase I study of the adoptive immunotherapy of human cancer with lectin activated autologous mononuclear cells, *Cancer* **53:**896–905.

Rosenberg, S. A., 1982, Potential use of expanded T-lymphoid cells and T-cell clones for the immunotherapy of cancer, in: *Isolation, Characterization and Utilization of T Lymphocyte Clones*, Academic Press, New York, p. 451.

Rosenberg, S. A., and Terry, W., 1977, Passive immunotherapy of cancer in animals and man, *Adv. Cancer Res.* **25:**323.

Rosenstein, M., and Rosenberg, S. A., 1983, T lymphoid cell clones with lytic and proliferative activity to syngeneic tumor, *J. Natl. Cancer Inst.* (in press).

Rosenstein, M., Eberlein, T., Kemeny, M. M., Sugarbaker, P. H., and Rosenberg, S. A., 1981, In vitro growth of murine T cells. VI. Accelerated skin graft rejection caused by adoptively transferred cells expanded in T cell growth factor, *J. Immunol.* **127:**566.

Rosenstein, M., Eberlein, T., Schwarz, S., and Rosenberg, S. A., 1983, Simplified techniques for the isolation of alloreactive cell lines and clones with specific cytotoxic or proliferative activity, *J. Immunol. Methods* **61:**183–193.

Smith, H. G., Harmel, R. P., Hanna, M. G., Jr., Zwilling, B. S., Zbar, B., and Rapp, H. J., 1977, Regression of established intradermal tumors and lymph node metastases in guinea pigs after systemic transfer of immune cells, *J. Natl. Cancer Inst.* **58:**1315.

Strausser, J. L., Mazumder, A., Grimm, E. A., Lotze, M. T., and Rosenberg, S. A., 1981, Lysis of human solid tumors by autologous cells sensitized in vitro to alloantigens, *J. Immunol.* **127:**266.

Vanky, F. T., Vose, B. M., Fopp, M., and Klein, E., 1979, Human tumor–lymphocyte interaction in vitro. VI. Specificity of primary and secondary autologous lymphocyte-mediated cytotoxicity, *J. Natl. Cancer Inst.* **62:**1407.

Vose, B. M., Vanky, F., Fopp, M., and Klein, E., 1978, In vitro generation of cytotoxicity against autologous human tumor biopsy cells, *Int. J. Cancer* **21:**588.

Biochemical and Biological Properties of Interleukin-3

JAMES N. IHLE, JONATHAN KELLER, EDMUND PALASZYNSKI, TERRY BOWLIN, RICHARD MURAL, RUDOLPH MEDICUS, and HERBERT C. MORSE, III

1. INTRODUCTION

Studies of the sequence of lymphocyte differentiation and the identification of factors associated with differentiation have been important but difficult areas of research. In part the difficulty arises from the complexity of the regulation. In T-cell differentiation a variety of experimental approaches have demonstrated a sequence involving the initial commitment of a bone marrow stem cell population, the maturation of this population in the thymus, and finally functional differentiation which occurs following antigenic stimulation. The last step in differentiation resulting in the generation and expansion of fully functional cytotoxic T cells can be demonstrated to be due to interleukin-2 (IL-2). Conceivably, comparable events may be involved in the differentiation of functional helper T cells capable of producing a variety of lymphokines including IL-2. The factors which regulate the differentiation of precursors of either helper or cytotoxic T cells are not known. The requirement for the thymic microenvironment for this phase of T-cell differentiation has been well established and represents the basis for studies of thymus derived factors capable of promoting T-cell differentiation. Studies directed at identifying factors which regulate the differentiation of "prothymocytes" from bone marrow stem cell populations have been more limited due to the lack of appropriate assays for this stage of differentiation. In order to approach this phase of differentiation we have

JAMES N. IHLE, JONATHAN KELLER, EDMUND PALASZYNSKI, TERRY BOWLIN, RICHARD MURAL, and RUDOLPH MEDICUS • LBI-Basic Research Program, NCI-Frederick Cancer Research Facility, Frederick, Maryland 21701. HERBERT C. MORSE, III • National Institute of Allergy and Infectious Diseases, National Institutes of Health, Bethesda, Maryland 20205.

studied factors which regulate the distribution and ontogeny of cells expressing the enzyme 20-α-hydroxysteroid dehydrogenase (20αSDH) which has been proposed to be acquired early in T-cell differentiation. As summarized in this review, IL-3 is a specific lymphokine which is capable of inducing 20αSDH in stem cell populations and can be shown to induce a lymphocyte with a prothymocyte phenotype. In addition, other phenotypes are induced indicating that the prothymocyte phenotype may be only one progeny of the IL-3 lineage.

2. DISTRIBUTION AND ONTOGENY OF 20αSDH-POSITIVE LYMPHOCYTES

IL-3 was initially identified as a factor in conditioned media of mitogen-stimulated T cells which could induce the expression of the enzyme 20αSDH in *nu/nu* splenic lymphocytes (Ihle *et al.*, 1981b). An interest in such a factor came from a variety of studies suggesting that 20αSDH expression was uniquely associated with T-cell differentiation (Weinstein, 1977, 1981; Weinstein *et al.*, 1977; Pepersack *et al.*, 1980). Some of these data are summarized in Table I which shows the distribution and ontogeny of 20αSDH-positive cells. In normal mice, 20αSDH is found in peripheral lymphoid tissues where it is primarily found in Thy-1-positive lymphocytes. In the thymus, 20αSDH is uniquely associated with Thy-1-positive, PNA-nonagglutinated, hydrocortisone-resistant medullary thymocytes. This popu-

TABLE I. Distribution and Ontogeny of 20 αSDH-Positive Cells

In vivo distribution	
Normal	
Peripheral lymphoid tissues	Thy-1$^+$, Lyt-1$^+$2$^-$ (80%)
	Thy-1$^-$ (20%)
Thymus	Thy-1$^+$, TdT$^-$, PNA$^-$, Con A responsive
	IL-2-responsive medullary thymocyte
Bone marrow	Thy-1$^-$ blast population
Genetically athymic or newborn thymectomized	
Peripheral lymphoid tissues	Not detectable
Bone marrow	Thy-1$^-$ blast population
Ontogeny	
Predominant differentiation 2–5 days after birth concomitant with the differentiation of TdT$^+$ thymocytes	
Cell lines: Distribution	
Present	Il-2-dependent cytotoxic T cells
	Helper T-cell-like lines Thy-1$^+$, Lyt-1$^+$2$^-$
	IL-3-dependent cell lines
Absent	B-cell lines
	Macrophage/myelomonocytic lines
	TdT$^+$ T-cell lines
	Erythroleukemias
	Melanoma, fibroblastic, etc.

lation of thymocytes is capable of responding to Con A and following stimulation can be further expanded *in vitro* with IL-2. Conversely, the PNA-agglutinable, terminal transferase-positive, hydrocortisone-sensitive population of corticol thymocytes is deficient in 20αSDH. In the bone marrow, 20αSDH is primarily associated with a Thy-1-negative blast population. The most provocative evidence relating 20αSDH expression to the T-cell lineage comes from the observation that the peripheral lymphoid tissues of *nu/nu* athymic or newborn thymectomized mice have levels of 20αSDH which are dramatically lower than those observed in normal mice.

A relationship of 20αSDH expression to the T-cell lineage has also been suggested from the ontogeny, during development, of 20αSDH-positive cells. In particular, only low levels of 20αSDH are present in tissues from fetal mice including fetal liver or spleen. The lack of high levels of 20αSDH in fetal liver is important as this is a major embryonic source of non-T-cell hematopoietic cell differentiation. The majority of 20αSDH-positive cells appear between 0 and 5 days after birth with the ontogeny of functional T cells, cellular infiltration of the thymus, and the ontogeny of terminal transferase-positive cells. As noted above, newborn thymectomy eliminates the appearance of 20αSDH-positive peripheral lymphocytes.

The distribution of 20αSDH in a variety of cell lines provides further evidence for a T-cell-lineage restriction. In particular, a wide variety of cell types have no detectable 20αSDH activity including B-cell lines, macrophage/myelomonocytic cell lines, erythroleukemia cell lines, melanoma cell lines, and fibroblastic cell lines. Among the T-cell lines examined, a wide variety of terminal transferase-positive T-cell lymphoma lines have been found to be 20αSDH negative. This result is consistent with the *in vivo* data which demonstrated that terminal transferase-positive, cortical thymocytes lacked 20αSDH. A number of Thy-1-positive, terminal transferase-negative, helperlike T-cell lines have been found to have very low levels of 20αSDH. Among the lines examined, the highest levels of 20αSDH have been found to occur in IL-2-dependent cytotoxic T-cell lines. In addition to the above, the only other cell lines which have been found to express 20αSDH are IL-3-dependent cell lines which as noted below characteristically express 20αSDH.

3. INDUCTION OF 20αSDH EXPRESSION *IN VITRO* WITH IL-3

The presence of 20αSDH in bone marrow Thy-1-negative lymphoblasts as well as in the medullary thymocyte population suggested that the induction of 20αSDH in the T-cell lineage might occur early in differentiation and specifically prior to the stage at which they acquire IL-2 responsiveness. Therefore, we reasoned that a factor capable of inducing 20αSDH in appropriate stem cells might be important in T-cell differentiation and developed an assay based on the induction of expression of 20αSDH in cultures of *nu/nu* splenic lymphocytes. Using this assay a factor was detectable in conditioned media from mitogen-stimulated lymphocytes which could induce 10- to 15-fold increases in 20αSDH activity over 24 hr. Maximal induction occurred rapidly with a peak of induction evident at 24–36

TABLE II. Sources of IL-3

Con A, PHA-stimulated normal splenic lymphocytes	+
Con A, PHA-stimulated *nu/nu* splenic lymphocytes	–
LPS-stimulated normal splenic lymphocytes	–
Antigen-stimulated Thy-1$^+$, Lyt-1$^+$2$^-$ immune specific lymphocytes	+
Mitogen- or antigen-stimulated helper/inducer T cells	+
Conditioned media from Thy-1$^+$ WEHI-3 cells	+
Mitogen, PMA-stimulated P388D1, RAW 264, PU5-18	–

hr suggesting that the factor was directly promoting 20αSDH expression. Because the factor was biochemically distinguishable from other lymphokines and because the assay was designed to measure an effect on T-cell differentiation, the term interleukin-3 was proposed to distinguish it from other factors affecting T-cell differentiation.

The major, if not exclusive source of IL-3 is T cells. As indicated in Table II, IL-3 activity is readily detectable in conditioned media from mitogen-activated normal splenic lymphocytes but is not detectable in conditioned media from comparably treated splenic lymphocytes from *nu/nu* athymic mice (Ihle *et al.*, 1981b). The B-cell mitogen LPS does not induce IL-3 activity in cultures of normal splenic lymphocytes. IL-3 has also been shown to be induced by immunologically specific Thy-1$^+$, Lyt-1$^+$2$^-$ helperlike T cells from immunized animals (Ihle *et al.*, 1981a) and has been shown to be produced by antigen-specific cloned helper/inducer-like T-cell lines (Prystowsky *et al.*, 1982). The only possible exception to the evidence which suggests an exclusive T-cell origin of IL-3 is the constitutive production of IL-3 by the WEHI-3 cell line (Lee *et al.*, 1982) which has been classified as a myelomonocytic cell line (Warner *et al.*, 1969). The ability of the WEHI-3 cell line to produce IL-3, however, clearly distinguishes it from a variety of other myelomonocytic/macrophage cell lines. In addition to IL-3 production, the WEHI-3 cell line is also distinguishable from the other myelomonocytic/macrophage cell lines by the expression of Thy-1. The uniqueness of the WEHI-3 line for Thy-1 expression and IL-3 production suggests that either it is a myelomonocytic line which aberrantly expresses these properties or perhaps the initial characterization of the WEHI-3 line failed to recognize its potential T-cell lineage and relied on less specific criteria of lineage classification.

4. PURIFICATION AND BIOCHEMICAL CHARACTERIZATION OF IL-3

The availability of a highly specific assay requiring only 24 hr and of a cell line which constitutively produced high titers of IL-3 has allowed the development of a purification scheme capable of yielding microgram quantities of homogeneous IL-3. This purification scheme has been described in detail and therefore will not be discussed here (Ihle *et al.*, 1982c, 1983). The criteria which were used to indicate purity of the final preparation included: (1) the presence of a single major protein detectable on SDS–PAGE which migrates with an apparent molecular weight of 28,000; (2) the presence of a single N-terminal sequence of Asp-Thr-His-Arg-Leu-Thr-Arg-Thr-Leu; (3) elution as a single peak from the final chromatography step with a constant biological specific activity across the peak; (4) an ED_{50} (effective dose for 50% of maximal biological activity) of 0.2 ng/ml which corresponds to a molar concentration of approximately 10^{-11} M; and (5) the ability of the iodinated protein to specifically bind in radioreceptor assays to cell lines known to require IL-3 for growth but not to control cell lines. Using this type of preparation of IL-3, a number of the biochemical properties have been examined. The charge of homogeneous IL-3 is heterogeneous and appears to be due to the presence of carbohydrate. By direct analysis, 100 μg of IL-3 contains approximately 12 μg of glucosamine. By amino acid composition, the protein moiety of IL-3 is approximately 15,000 daltons. One possible glycosylation site has been detected at residue 10 in the sequence. Attempts to remove the carbohydrate and assess biological activity have not been possible, because under the conditions employed the activity was lost prior to the removal of the carbohydrate, suggesting denaturation. The amino acid composition is somewhat notable in containing a high percentage of charged residues (≃ 38%). The N-terminal sequence through 29 residues shows no homology to any proteins contained in an atlas of protein structure (Dayhoff, 1976) and in particular shows no regions of homology to the sequence of IL-2. Efforts are currently being directed at cloning the gene for IL-3 to obtain more sequence information for comparison with other lymphokines.

5. BIOLOGICAL ACTIVITY OF HOMOGENEOUS IL-3

Early in studies employing homogeneous IL-3 it became evident that the induction of 20αSDH in an appropriate precursor lymphocyte in *nu/nu* spleen might represent only the initial step in a series of events mediated by IL-3. The spectrum of events and their possible sequence were first appreciated by following the phenotypic changes occurring in cultures of *nu/nu* splenic lymphocytes in IL-3. As shown in Fig. 1, there are a number of changes which occur in such cultures. The earliest event detectable is the induction of 20αSDH activity in cultures which occurs maximally at 1–2 days. The peak of the specific activity of cells for 20αSDH, however, does not occur until approximately 7–8 days, reflecting the loss of non-responding cells and the expansion of early, highly 20αSDH-positive cells. This level of specific activity decreases subsequent to the peak at 1 week and levels off at a specific activity of approximately 1000 pmoles/hr per 10^8 cells. In the cultures there is a dramatic increase in Ly-5-positive and H-11-positive cells, suggesting the acquisition of these markers with time in culture. This was further substantiated by the observation that elimination of Ly-5-positive cells with antibodies and complement does not reduce the initial 20αSDH-inducible population. In addition to Ly-5 and H-11, there is a lower but consistent increase in Thy-1-positive cells and to a lesser degree Lyt-2-positive cells. This population does not persist in the cultures, however, and by 40 days no Thy-1- or Lyt-2-positive cells are detectable. The specificity of response seen for Thy-1, Lyt-2, Ly-5, or H-11 is indicated by

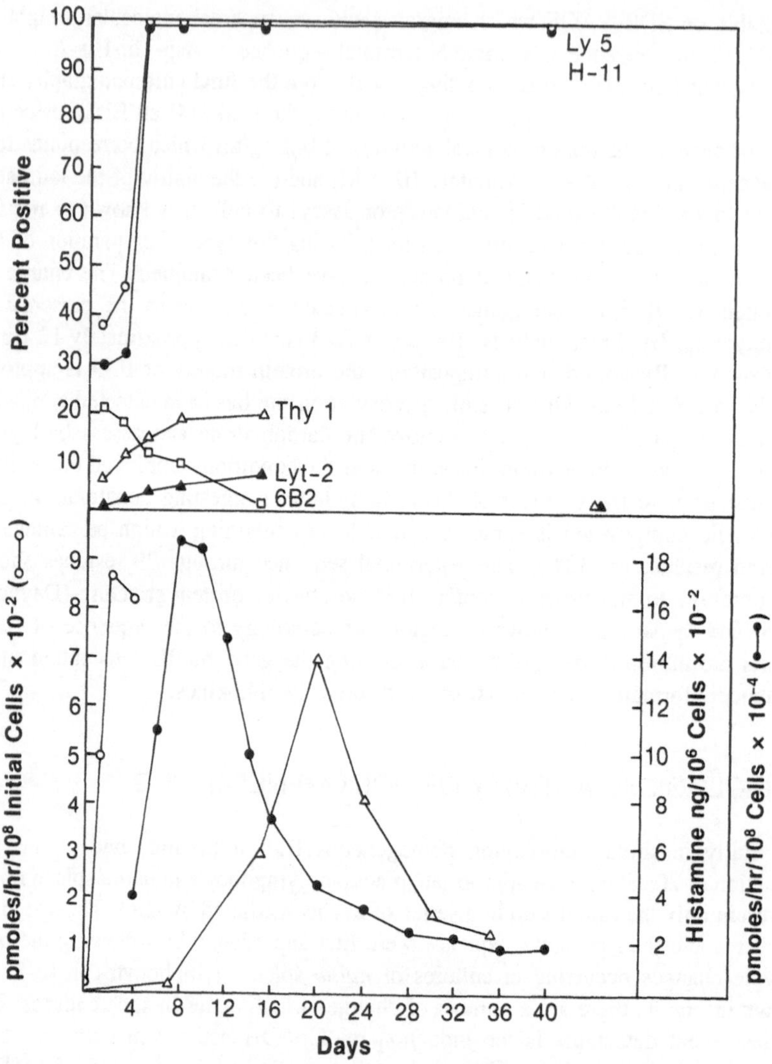

FIGURE 1. Sequence of phenotypic changes occurring in cultures of *nu/nu* splenic lymphocytes in IL-3. Splenic lymphocytes from *nu/nu* mice were cultured with IL-3 for the indicated times and analyzed for the percentage of the cells expressing the markers indicated (top panel) by FACS analysis. The cells were also assayed for 20αSDH which is presented as the total enzyme activity present in early cultures (○) or as the specific activity in long-term cultures (●). Last, the cell-associated histamine was assayed at the indicated times (△).

the observation that a marker associated with B cells (6B2) as well as a macrophage marker (Mac-1) (not shown) decrease with time in culture. Last, the levels of histamine in the cultures increase from about 1 week in culture and peak at approximately 3 weeks and then decline. Consistent with this, granulated, mastlike cells become evident in the cultures at approximately 1 week and increase with time such that after 2–3 weeks the cultures are homogeneous populations of mastlike cells. At various times of culture, cloning attempts have not yielded lines of IL-3-dependent cells of a homogeneous phenotype. Moreover, although the cultures survive for up to 2 months, the rate of proliferation decreases with time and in our experience rarely gives rise to long-term factor-dependent cell lines. These results as well as others have suggested that in such cultures IL-3 initiates an obligatory sequence of differentiation which has as the major progeny a population of mastlike cells. In addition, other progeny, such as a Thy-1-positive population, may be generated and may not survive.

6. IL-3-DEPENDENT LYMPHOMA CELL LINES

Although the above approach has suggested the possibility of a sequence of differentiation with the production of multiple phenotypes, the inability to clone and expand various transient phenotypes has made it difficult to precisely study the phenotypes and their continued responses to IL-3. This limitation has been partially overcome by the realization that a substantial percentage of the leukemias induced by Moloney leukemia virus (MoLV) are within an IL-3-regulated lineage of cells and can be established as stable long-term IL-3-dependent cell lines in culture. The phenotypes of these lymphoma cell lines are quite variable but represent the types present in cultures of *nu/nu* splenic lymphocytes differentiating in the presence of IL-3. The mechanisms involved in the induction of such lymphomas are not precisely known, although a number of observations have provided the basis for a hypothesis involving chronic immune stimulation. The details of many of the experiments on which this hypothesis was formulated have been described in previous reviews (Ihle and Lee, 1981; Ihle *et al.*, 1982a; Lee and Ihle, 1981a, 1981b). The essence of the hypothesis is that as a consequence of the establishment of an acute viremia, following inoculation of MoLV into newborn mice, there is the recruitment and expansion of helperlike T-cell populations which are immunologically specific for viral proteins and which can be readily detected *in vitro* in T-cell blastogenesis assays or by their ability to produce lymphokines such as IL-3 or IL-2 *in vitro* in response to specific viral proteins. *In vivo* the presence of helperlike T cells and the high levels of viral antigens are speculated to combine to continuously produce lymphokines, resulting in the dramatic expansion of lymphokine-responsive populations. In the presence of viremia but in the absence of helper T cells, this expansion does not occur nor does the viremia induce leukemia. The expanded lymphokine-responsive populations and particularly the IL-3-regulated populations are hypothesized to be necessary as a target cell population for either viral or somatic events associated with transformation. From an analysis of the lymphomas as discussed below, it has been suggested that transformation does not involve loss of factor dependency for proliferation but rather represents the inability to continue to terminally differentiate normally.

The properties of a few primary lymphomas induced by MoLV in BALB/c mice are given in Table III from which several IL-3-dependent cell lines were established. In all cases the primary tumors were splenic lymphosarcomas with no

TABLE III. Establishment of IL-3-Dependent Lymphoma Cell Lines

Animal No.	Thy-1	20αSDH	IL-3 binding	Growth in tissue culture	IL-3 dependence
118	++	46	0.65	—	
119	++	29	0.43	—	
120	++	62	0.95	—	
121	±	404	3.93	+DA-24	+
122	+	483	1.25	+DA-27	−
123	++	2736	1.30	+DA-28	+
124	−	142	0.92	+DA-29	+
125	−	66	0.41	—	
126	++	254	0.83	+DA-30	+
127	+	87	0.96	−DA-25	+
128	+	79	0.98	+DA-31	+
129	+	92	0.64	+DA-32	+

evidence of thymic involvement. Although the majority of the lymphomas were Thy-1 positive, there is a 5–10% frequency of Thy-1-negative lymphomas. A number of the lymphomas were found to have high levels of 20αSDH (> 100 pmoles/hr per 10^8 cells) which initially suggested that some of the tumors might be of an IL-3-regulated lineage. More importantly, however, a number of the lymphomas had cells which showed significantly elevated levels of specific binding of iodinated, homogeneous IL 3 (> 0.5%), suggesting the presence of IL-3-specific receptors on some of the lymphoma cells. When these various lymphomas were cultured in media containing IL-3, 8 of the 12 gave rise to continuous cell lines whereas in the absence of IL-3, none of the lymphomas could be established as continuous cell lines. Among these lines, in long-term culture one became independent of IL-3 for growth whereas seven remained factor dependent for growth. From such studies involving various strains of mice and various retroviruses, approximately 100 factor-dependent lines have been established and characterized for a variety of properties. All the factor-dependent cell lines examined have readily detectable receptors for IL-3 as characterized by standard radioreceptor binding assays (Palaszynski and Ihle, 1984). In contrast, the factor-independent lymphoma cell lines lack such receptors. Among the lines, therefore, there has been an absolute correlation between the presence of IL-3 receptors and a dependence on IL-3 for proliferation. In addition, all of the factor-dependent cell lines have readily detectable 20αSDH although the levels vary from 10^2 to 10^5 pmoles/hr per 10^8 cells. In contrast, the IL-3-independent lymphoma cell lines have values of 20αSDH which are < 10^2 pmoles/hr per 10^8 cells.

The cell surface phenotypes of the lymphomas as well as the presence or absence of IgE receptors and histamine vary but can be grouped into a limited number of phenotypes. The phenotypes are summarized in Fig. 2 where they are organized into a potential scheme of differentiation based on overlapping characteristics as well as experimental data from other approaches. The most common phenotypes observed in the lines are represented by the DA-3, DA-4, and P-cell

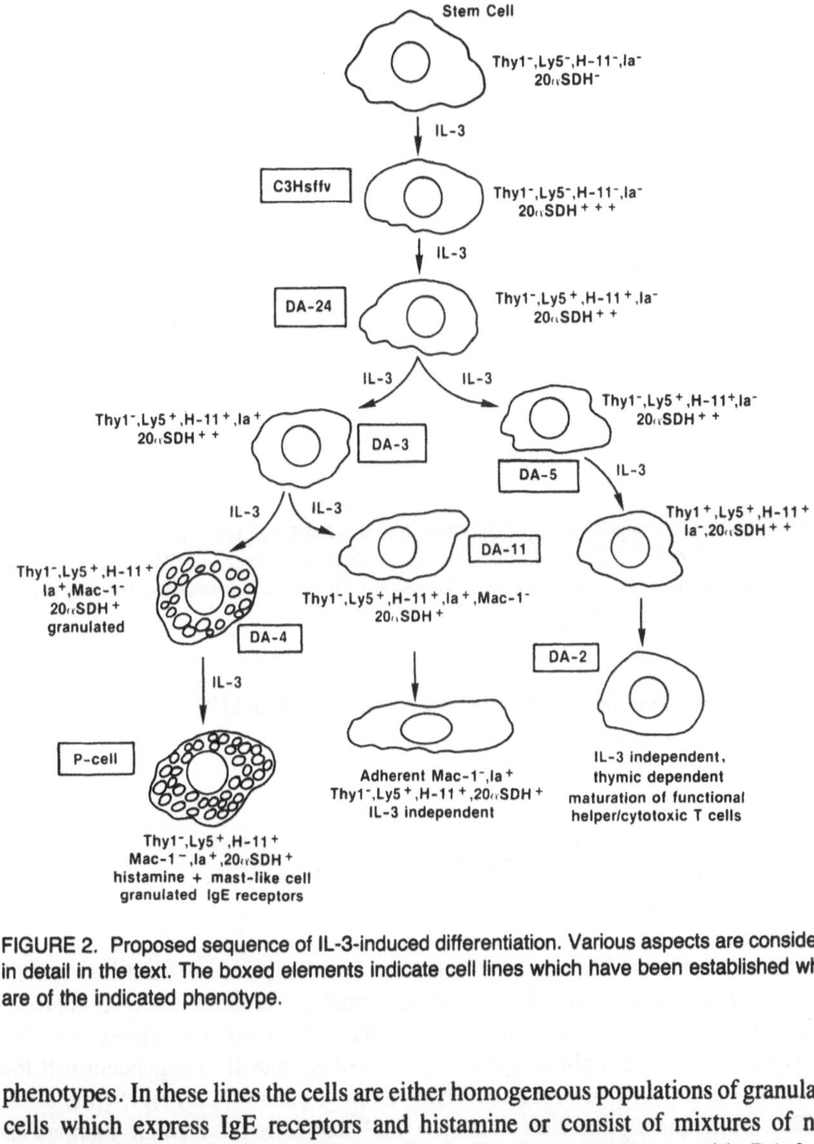

FIGURE 2. Proposed sequence of IL-3-induced differentiation. Various aspects are considered in detail in the text. The boxed elements indicate cell lines which have been established which are of the indicated phenotype.

phenotypes. In these lines the cells are either homogeneous populations of granulated cells which express IgE receptors and histamine or consist of mixtures of non-granulated cells and granulated mastlike cells. Occasionally as with DA-3 the lymphoma line continuously generates both granulated mastlike cells as well as an adherent nongranulated population. Most of these types of cell lines are Thy-1$^-$, Ly-5$^+$, H-11$^+$, and Ia$^+$ although the expression of Ia has been more variable than the other markers. A number of lymphoma cell lines (e.g. DA-24) are lymphoblastic and nongranulated in morphology and lack detectable histamine or IgE receptors. The phenotype represented by DA-5 is also relatively common in which case the cells are nongranulated and lymphoblastic in morphology. These lines are distinctly characterized, however, by the constant generation of Thy-1$^+$ cells which as shown

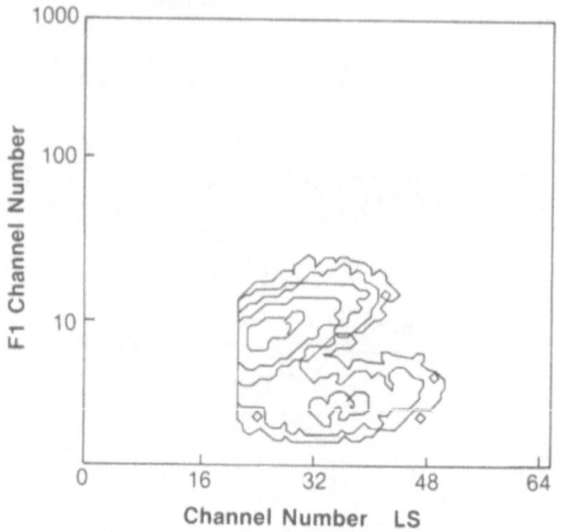

FIGURE 3. FACS analysis of Thy-1 expression in the DA-5 cell lines. Exponentially growing cultures of the DA-5 IL-3-dependent lymphoma cell lines were used in FACS analysis of Thy-1 expression. The results are plotted as light scatter (LS) versus fluorescence intensity (F) with the topographical lines indicating the number of cells.

in Fig. 3 can be identified as very distinct populations by FACS analysis. Curiously, none of the IL-3-dependent cell lines has been 100% positive for Thy-1 cells. In contrast, most of the IL-3-independent cell lines such as DA-2 are 100% Thy-1$^+$.

7. PROPOSED SEQUENCE OF IL-3-INDUCED DIFFERENTIATION

From the above data and others, the presumed sequence of IL-3-induced differentiation is as indicated in Fig. 2. In particular, we propose that IL-3 induces a stem cell to initially become a strongly 20αSDH-positive cell which initially is Thy-1$^-$, Ly-5$^-$, H-11$^-$. This phenotype is represented by one IL-3-dependent cell line which was derived from a long-term bone marrow culture (C3H sffv) (Ihle *et al.*, 1982b). Subsequently the cells become Ly-5$^+$ and H-11$^+$ which are markers expressed on all other Il-3-dependent cell lines. The major sequence of differentiation detected *in vitro* in cultures of *nu/nu* splenic lymphocytes results in a population of Thy-1$^-$, Ly-5$^+$, H-11$^+$, IgE receptor-bearing histamine-containing mastlike cells which have limited growth potential *in vitro*. This phenotype is identical to that described for P cells (Schrader *et al.*, 1981). In addition to mastlike cells, Thy-1$^+$ progeny are generated which fail to continue to proliferate in the cultures and may represent a progeny which is a precursor for further differentiation mediated by factors distinct from IL-3. In a comparable manner, the adherent cell population cannot be maintained and may require additional factors for continued differentiation.

8. ADDITIONAL BIOLOGICAL ACTIVITIES OF IL-3

The above scheme would predict that IL-3 should be similar to a number of recently described factors. To address this question, homogeneous IL-3 has been assayed and found to have the activities ascribed to a variety of factors. It should be noted that in all cases the dose–response curves were compared and shown to be identical to the curves obtained for 20αSDH induction (Ihle *et al.*, 1983). It should also be cautioned that in many cases IL-3 appears to be only one distinct protein capable of mediating a particular activity. The additional activities that can be mediated by homogeneous IL-3 include WEHI-3 growth factor activity or hemopoietic growth factor which is a factor derived from WEHI-3 cells which is required for the maintenance of a variety of factor-dependent cell lines established from long-term bone marrow cultures (Dexter *et al.* 1980; Greenberger *et al.*, 1979, 1980, 1981; Ihle *et al.*, 1982b). These cell lines have been characterized with regard to cell surface phenotypes, expression of 20αSDH, and presence of IL-3 receptors and are identical to a number of the MoLV-induced IL-3-dependent lymphoma cell lines (Ihle *et al.*, 1982b). Using either WEHI-3 conditioned media or media from mitogen-stimulated lymphocytes, it has been shown that IL-3 is either the major or exclusive factor required for the proliferation of a number of these cell lines *in vitro*.

The differentiation scheme also predicts that IL-3 may be similar to factors such as a mast cell growth factor described by several laboratories (Nagao *et al.*, 1981; Nabel *et al.*, 1981; Schrader *et al.*, 1981; Razin *et al.*, 1981; Yung *et al.*, 1981) and histamine-producing cell-stimulating factor activity (Dy *et al.*, 1981). As predicted, homogeneous IL-3 has been shown to have these activities (Ihle *et al.*, 1983). Moreover, from mitogen-stimulated lymphocytes, IL-3 appears to represent the major protein with this activity although the possibility exists that other proteins may mediate the activity but which constitute lesser levels of activity either by virtue of the physical mass of the protein or by virtue of the frequency of appropriate precursor cells. It should also be noted that terms such as mast cell growth factor may be misleading in implying strictly proliferation-inducing activity without reference to the more important parameter of differentiation.

IL-3 is also identical to a factor in conditioned media of mitogen-stimulated lymphocytes which is capable of inducing the expression of Thy-1 in cultures of bone marrow cells (Schrader *et al.*, 1982; Prystowsky *et al.*, 1983). Curiously in this case, the only activity detectable is associated with IL-3 and specifically IL-2 neither induces Thy-1 expression nor proliferation of bone marrow cells, suggesting that its activity is limited to mature thymic processed T cells. These observations are of interest for as noted above, the phenotype of medullary thymocytes is Thy-1^+, 20αSDH$^+$ which is the phenotype induced by IL-3 in cultures of Thy-1-depleted bone marrow cells. This observation coupled with the frequent occurrence of IL-3-dependent lymphoma cell lines which generate Thy-1^+ cells suggest the possibility that one progeny of IL-3-induced differentiation may be a prothymocyte which requires the thymic microenvironment for further maturation.

Homogeneous IL-3 also has been shown to induce bone marrow cells to form colonies in soft agar and therefore has a colony-stimulating activity. This activity, however, is clearly associated with a variety of biochemically and biologically distinct proteins and does not constitute a specific assay. In particular, conditioned medium from mitogen-stimulated lymphocytes contains a major colony-stimulating factor representing approximately 95% of the total activity which can be resolved from IL-3 on DEAE-cellulose columns. This activity has been designated CSF-2 to distinguish it from CSF-1 which is described in detail elsewhere in this volume. Thus, in terms of the total CSF activity, the IL-3-specific component represents < 5%. The CSF-2 activity is biologically distinguishable from IL-3 in that it does not induce 20αSDH, does not support the differentiation of mastlike cells, does not induce proliferation of IL-3-dependent cell lines, and does not induce Thy-1. Moreover, an antiserum against IL-3 which inhibits IL-3-induced colony formation does not inhibit CSF-2-induced colony formation.

9. CONCLUSIONS

The biological activities mediated by homogeneous IL-3 provide a relatively simple conceptual view of a lineage of cells regulated by a specific lymphokine. We would propose that a lineage exists which is uniquely promoted to differentiate in the presence of T-cell-derived IL-3. This IL-3 lineage is characterized by the expression of 20αSDH, the presence of receptors for IL-3, and an absolute requirement for the continual presence of IL-3 once differentiation is initiated. The progeny of this differentiation include a prothymocyte population, an adherent population which is not well characterized, and a population of mastlike cells. Thus, a stem cell population, regulated by other factors, generates a potential IL-3-responsive cell. In the presence of IL-3, this cell is *committed* to differentiate along the IL-3-lineage pathway. Additional factors in some cases may be required to terminally differentiate some of the progeny such as potential prothymocytes. In certain leukemias, transformation can occur within this lineage and phenotypically results in a block in the terminal differentiation of the cells with retention of factor dependence for growth. Although this hypothesis is somewhat different than contemporary views of hematopoiesis, it is consistent with the available data and represents our current working hypothesis to further explore the properties of IL-3.

REFERENCES

Dayhoff, M. O., 1976, *Atlas of Protein Sequence and Structure*, Vol. 5, Suppl. 2, p. 1, National Biomedical Research Foundation, Washington D. C.

Dexter, T. M., Garland, J., Scott, D., Scolnick, E., and Metcalf, D., 1980, Growth of factor-dependent hemopoietic precursor cell lines, *J. Exp. Med.* **152:**1036.

Dy, M., Lebel, B., Kamoun, P., and Hamburger, J., 1981, Histamine production during the anti-allograft response: Demonstration of a new lymphokine enhancing histamine synthesis, *J. Exp. Med.* **153:**293.

Greenberger, J. S., Gans, P. J., Davisson, P. B., and Moloney, W. C., 1979, *In vitro* induction of continuous acute promyelocytic cell lines in long-term bone marrow cultures by Friend or Abelson leukemia virus, *Blood* **53:**987.

Greenberger, J. S., Eckner, R. J., Ostertag, W., Colletta, G., Boschetti, S., Nagasawa, H., Weichselbaum, R. R., and Moloney, W. C., 1980, Release of spleen focus-forming virus (SFFV) in differentiation inducible granulocytic leukemia cell lines transformed *in vitro* by Friend leukemia virus, *Virology* **105**:425.

Greenberger, J. S., Humphries, S. K., Sakakeeny, M. A., Eckner, R. J., Ihle, J., Eaves, C., Cantor, H., Denberg, J., and Nabel, G., 1981, Demonstration of a permanent line of self-renewing multipotential hematopoietic stem cells *in vitro*, *Blood* **58** (Suppl. 1):97a.

Ihle, J. N., and Lee, J. C., 1981, Possible immunological mechanisms in C-type viral leukemogenesis in mice, *Curr. Top. Microbiol. Immunol.* **98**:85.

Ihle, J. N., Lee, J. C., and Rebar, L., 1981a, T cell recognition of Moloney leukemia virus proteins. III. T cell proliferative responses against gp70 are associated with the production of a lymphokine inducing 20 alpha hydroxysteroid dehydrogenase in splenic lymphocytes, *J. Immunol.* **127**:2565.

Ihle, J. N., Pepersack, L., and Rebar, L., 1981b, Regulation of T cell differentiation: *In vitro* induction of 20 alpha hydroxysteroid dehydrogenase in splenic lymphocytes is mediated by a unique lymphokine, *J. Immunol.* **126**:2184.

Ihle, J. N., Enjuanes, L., Lee, J. C., and Keller, J., 1982a, The immune response to C-type viruses and its potential role in leukemogenesis, *Curr. Top. Microbiol. Immunol.* **101**:31.

Ihle, J. N., Keller, J., Greenberger, J. S., Henderson, L., Yetter, R. A., and Morse, H. C., III, 1982b, Phenotype characteristics of cell lines requiring interleukin 3 for growth, *J. Immunol.* **129**:1377.

Ihle, J. N., Keller, J., Henderson, L., Klein, F., and Palaszynski, E. W., 1982c, Procedures for the purification of interleukin 3 to homogeneity, *J. Immunol.* **129**:2431.

Ihle, J. N., Keller, J., Oroszlan, S., Henderson, L., Copeland, T., Fitch, F., Prystowsky, M. B., Goldwasser, E., Schrader, J. W., Palaszynski, E., Dy, M., and Lebel, B., 1983, Biological properties of homogeneous interleukin 3. I. Demonstration of WEHI-3 growth factor activity, mast cell growth factor activity, P-cell stimulating factor activity, colony stimulating factor activity and histamine producing cell stimulating factor activity, *J. Immunol.* **131**:282.

Lee, J. C., and Ihle, J. N., 1981a, Increased responses to lymphokines are correlated with preleukemia in Moloney virus inoculated mice, *Proc. Natl. Acad. Sci, USA* **78**:7712.

Lee, J. C., and Ihle, J. N., 1981b, Chronic immune stimulation is required for moloney leukemia virus-induced lymphomas, *Nature* **209**:407.

Lee, J. C., Hapel, A. J., and Ihle, J. N., 1982, Constitutive production of a unique lymphokine (IL 3) by the WEHI-3 cell line, *J. Immunol.* **128**:2393.

Nabel, G., Galli, S. J., Dvorak, A. M., Dvorak, H. F., and Cantor, H., 1981, Inducer T lymphocytes synthesize a factor that stimulates proliferation of cloned mast cells, *Nature (London)* **291**:332.

Nagao, K., Yokoro, K., and Aaronson, S. A., 1981, Continuous lines of basophil/mast cells derived from normal mouse bone marrow, *Science* **212**:333.

Palaszynski, E. W., and Ihle, J. N., 1984, Evidence for specific receptors for interleukin 3 on lymphokine dependent cell lines established from long-term bone marrow cultures, *J. Immunol.* **132**:1872.

Pepersack, L., Lee, J. C., McEwan, R., and Ihle, J. N., 1980, Phenotypic heterogeneity of Moloney leukemia virus-induced T cell lymphomas, *J. Immunol.* **124**:279.

Prystowsky, M. B., Ely, J. M., Beller, D. I., Eisenberg, L., Goldman, J., Goldman, M., Goldwasser, E., Ihle, J., Quintans, J., Remold, H., Vogel, S., and Fitch, F. W., 1982, Alloreactive cloned T cell lines. VI. Multiple lymphokine activities secreted by cloned T lymphocytes, *J. Immunol.* **129**:2337.

Prystowsky, M. B., Ihle, J. N., Otten, G., Keller, J., Rich, I., Naujokas, M., Loken, M., Goldwasser, E., and Fitch, F. W., 1983, Two biologically distinct colony-stimulating factors are secreted by a T lymphocyte clone, in: *UCLA Symposia on Molecular and Cellular Biology, New Series*, Vol. 9 (D. W. Golde and P. A. Marks, eds.), Liss, New York.

Razin, E., Cordon-Cardo, C., and Good, R. A., 1981, Growth of a pure population of mouse mast cells *in vitro* with conditioned medium derived from concanavalin A-stimulated splenocytes, *Proc. Natl. Acad. Sci. USA* **78**:2559.

Schrader, J. W., Lewis, S. J., Clark-Lewis, I., and Culvenor, J. G., 1981, The persisting (P) cell: Histamine content, regulation by a T cell-derived factor, origin from a bone marrow precursor, and relationship to mast cells, *Proc. Natl. Acad. Sci. USA* **78**:323.

Schrader, J. W., Battye, F., and Scollay, R., 1982, Expression of Thy 1 antigen is not limited to T cells in cultures of mouse hemopoietic cells, *Proc. Natl. Acad. Sci. USA* **79**:4161.

Warner, N. L., Moore, M. A. S., and Metcalf, D., 1969, A transplantable myelomonocytic leukemia in BALB/c mice: Cytology, karyotype and muramidase content, *J. Natl. Cancer Inst.* **43:**963.

Weinstein, Y., 1977, 20αHydroxysteroid dehydrogenase: A T lymphocyte associated enzyme, *J. Immunol.* **119:**1223.

Weinstein, Y., 1981, Expression of 20αhydroxysteroid dehydrogenase in the mouse marrow cells: Strain differences, thymic effect on enzymatic activity, and possible localization in pre T lymphocytes, *Thymus* **2:**305.

Weinstein, Y., Linder, H. R., and Eckstein, B., 1977, Thymus metabolizes progesterone, a possible enzymatic marker for T lymphocytes, *Nature (London)* **266:**632.

Yung, Y. P., Eger, R., Tertian, G., and Moore, M. A. S., 1981, Long-term *in vitro* culture of murine mast cells. II. Purification of a mast cell growth factor and its dissociation from TCGF, *J. Immunol.* **127:**794.

Lymphotoxins, a Multicomponent Family of Effector Molecules

Purification of Human α-Lymphotoxin from a Cloned Continuous Lymphoblastoid Cell Line IR 3.4 to Homogeneity

GALE A. GRANGER, DIANE L. JOHNSON, J. MICHAEL PLUNKETT, IRENE K. MASUNAKA, SALLY L. ORR, and ROBERT S. YAMAMOTO

1. INTRODUCTION

Human lymphotoxins (LT) are a multicomponent family of related and distinct cell-lytic and growth inhibitory glycoproteins (Toth and Granger, 1979). They are inducible and can be released specifically when immune lymphoid cells are cocultured with the immunizing antigen or nonspecifically when nonimmune lymphoid cells are cocultured with lectins (Daynes and Granger, 1974; Granger and Williams, 1968; Granger *et al.*, 1969). An exception to the inducible mode of release of these effector molecules is that certain continuous human lymphoblastoid cell lines constitutively release low levels of LT *in vitro* (Fair *et al.*, 1979; Granger *et al.*, 1970; Khan *et al.*, 1982). There is now evidence that certain LT forms may have a role as lytic effectors in different classes of cell destructive reactions mediated by human lymphocytes *in vitro* (Weitzen *et al.*, 1983a,b; Yamamoto *et al.*, 1979, 1984b). However, the situation is complex for it is clear that different classes of effector

GALE A. GRANGER, DIANE L. JOHNSON, J. MICHAEL PLUNKETT, IRENE K. MASUNAKA, SALLY L. ORR, and ROBERT S. YAMAMOTO • Department of Molecular Biology and Biochemistry, University of California, Irvine, California 92717.

lymphocytes (T cells, NK cells) can release different LT forms (Harris *et al.*, 1981; Weitzen *et al.*, 1983a,b; Wright and Bonavida, 1982; Yamamoto *et al.*, 1984a). Moreover, in these reactions the different LT forms may assemble together to induce cell lysis as molecular complexes or alternatively may act as individual components on a target cell in a synergistic fashion. LT found in supernatants from stimulated normal human lymphocytes are heterogeneous and can be grouped into different molecular weight classes: [complex (> 200,000); α heavy (α_H, 120,000–140,000); α(70,000–90,000); β (30,000–50,000); and γ(12,000–20,000) (Granger *et al.*, 1978; Klostergaard *et al.*, 1981c)]. Members of the α and β classes can be further separated into subclasses based on differences in their charge by electrophoresis and ion-exchange chromatography (Granger *et al.*, 1978; Hiserodt *et al.*, 1976; Lee and Lucas, 1976). Recent findings indicate that certain of these molecular weight classes are an interrelated subunit system (Harris *et al.*, 1981; Yamamoto *et al.*, 1978). The α is a key form for it can assemble with immunoglobulin-like molecule(s) to form the larger molecular weight classes and either by proteolysis or by subunit disassociation act as a precursor for one set of β and γ forms (Harris *et al.*, 1981). However, there are multiple β-class forms which are distinct and released by different classes of effector cells (Harris *et al.*, 1981; Yamamoto *et al.*, 1984b).

Functional studies with LT-containing supernatants or partially purified molecular-weight classes derived from experimental animals and man reveal that they each can have different cell-lytic ability (Klostergaard *et al.*, 1981c; Lisafeld *et al.*, 1980; Yamamoto *et al.*, 1979). Because the murine L-929 fibroblast is lysed by LT forms from most animal species, it has been widely used as an indicator for many types of *in vitro* LT assays (Ross *et al.*, 1979). Initially, the L cell provided a sensitive assay for these molecules; however, the exclusive use of this target as an indicator also has had a negative effect for it is now clear that the L-929 cell is not sensitive to all human LT forms and is uniquely sensitive to others (Devlin *et al.*, 1981; Yamamoto *et al.*, 1979). This heterogeneity has made biochemical studies of these molecules more difficult than other lymphokines. Nevertheless, attempts have been made by various investigators to purify different LT forms from both human and animal sources.

The actual kinetics, levels, and species of LT molecule(s) released in a given culture vary according to the type of inducing stimulus, the class of lymphocyte responding, and the state of activation of the responding cell(s). For example, polyclonal stimulating agents can induce release of LT forms from multiple cell populations, whereas antigen only stimulates immune cell populations. Different types of lymphocytes can release both common and unique classes and subclasses of LT molecules; and preactivated cells can release higher levels of LT more rapidly upon secondary stimulation when compared to a primary stimulation (Daynes and Granger, 1974; Granger *et al.*, 1980).

What is now necessary is the identification, purification, and functional study of key LT forms from defined populations of lymphocytes. The α component is an important LT form for it can (1) be released by both T and B lymphocytes, (2) be assembled by T cells into larger multimeric forms with increased cell-lytic capabilities and associated with immunoglobulin-like molecules, and (3) can serve as a precursor for smaller β- and γ-LT forms (Devlin *et al.*, 1983; Yamamoto *et al.*,

1979). We have selected and cloned a human lymphoblastoid cell line (termed IR 3.4 cells) that releases high levels of α-LT *in vitro*. In addition, we have also developed methods to stimulate, produce, and purify significant quantities of α-LT from IR 3.4 cells (Yamamoto *et al.*, 1984a).

2. METHODOLOGY AND RESULTS

2.1. LT Assay

The α L-929 murine fibroblast, because of its sensitivity to human α-LT, was used as the target cell either to quantitate LT activity or to simply detect its presence or absence in a sample. The methods employed in these assay systems have been described (Kramer and Granger, 1972). The number of units of a particular LT preparation is identified from the reciprocal of the dilution causing 50% lysis of the L-cell targets (10^5 cells), and one unit of LT activity is defined as that amount of material that will destroy 50% of the target cells. This bioassay will detect α-LT activity in the low picomole range.

2.2. Selection of a Continuous Cell Line Releasing the α-LT Form

Previous studies identified α-LT-secreting continuous human lymphoid cell lines; however, the levels of LT released were very low and these cells could not be induced by stimulation with lectins to increase levels of LT secretion (Yamamoto *et al.*, 1984a). We have rescreened and identified a number of LT-secreting continuous human lymphoblastoid cell lines (Yamamoto *et al.*, 1984a). The selection of the cell line as a potential α-LT source was determined by two criteria: (1) the lytic product was identified as the α form by both biochemical and immunological methods, and (2) the cells release high levels of α-LT when induced with phorbol myristate acetate (PMA) under serum-free *in vitro* conditions. The continuous B-lymphoblastoid cell line GM3104A, while meeting both requirements, offered an additional advantage over the other cell lines—the HLA genotype had been established (HLA, A:3,3; B:35,35; C:4,4; D:1,1; DR:1,1) (Human Genetic Mutant Cell Repository, Camden, N.J.). A variant of the GM3104A cell line, termed IR 3.4, was developed by multiple rounds of cloning and selection for high-level α-LT production (Yamamoto *et al.*, 1984a).

2.3. Supernatant Production

The IR 3.4 cells were grown to a density of 10^6 cells/ml in RPMI 1640 supplemented with 10% heat-inactivated fetal calf serum (RPMI-10%) in 110 × 550-mm roller bottles containing 500 ml of culture medium. The cells from each roller bottle were washed three times by alternate sedimentation and resuspension with serum-free RPMI 1640; and finally resuspended at a cell density of 5 × 10^5 cells/ml in RPMI 1640 supplemented with 0.1% lactalbumin hydrolysate (LAH) and 20 ng PMA/ml. These production conditions were derived from careful studies of various parameters to achieve maximal α-LT release (Yamamoto *et al.*, 1984a). Next, 500 ml of cell suspension was added to each roller bottle and the bottles incubated at 37°C for 72–96 hr. After incubation, the supernatants were cleared of cells by centrifugation at 450*g* for 10 min. The cell pellets were then restimulated and supernatants were collected a second time under identical conditions. Under these culture conditions, the IR 3.4 cells routinely produced 1000–2000 U of LT activity/ml, whereas unstimulated IR 3.4 cells only produced 50–100 U/ml. Biochemical and immunological studies revealed 50–60% of this activity was due to α-LT; the remainder was due to an immunologically unrelated β-class LT form. It is important to note that this β-class LT form is not normally produced by these cells, providing evidence that the type of inducing stimulus can control the class of LT released by a lymphoid cell *in vitro* (Yamamoto *et al.*, 1984a). LAH is used as a serum substitute in these cultures because it: (1) appears to protect the α-LT molecule from proteolysis and results in twice as much α-LT from cultures containing LAH, and (2) it can be easily separated from the α-LT by molecular sieving (Yamamoto *et al.*, 1984a). During these studies, we found PMA would not induce the release of cell-lytic materials from various nonlymphoid cells (i.e., HeLa, L-929, WI-38), and that PMA did not render the target cell more sensitive to LT-lysis when treated with PMA (Yamamoto *et al.*, 1984a).

2.4. Purification of the α-LT Form from IR 3.4 Supernatants

Cell-free supernatants from PMA-activated IR 3.4 cells were first passed through a controlled-pore glass (CPG) column. Two liters of supernatant was passed over a 15-ml CPG column at a flow rate of 100 ml/hr. The CPG column routinely absorbed 85–90% of the lytic activity expressed in these supernatants. The amount of lytic activity desorbed from CPG columns with 50% ethylene glycol (in 10 mM Tris, 0.1 mM EDTA, pH 8.0 buffer) was 20% of that in the starting supernatant. The eluant contained from 9 to 14% of the total protein detected in the starting supernatant as determined by a Bio-Rad protein assay. We subsequently have found that the yield of bioactivity can be increased to 90% by eluting with 50% ethylene glycol in 1 M NaCl. However, all studies reported here were conducted with material eluted from CPG in glycol minus NaCl. The CPG elutant was next dialyzed against a low-salt Tris buffer and applied to and then eluted from a DEAE ion-exchange column. As can be seen in Fig. 1, 10–20% of the lytic activity routinely did not bind to the DEAE column. The remaining lytic activity eluted as a single major peak at 0.03 M NaCl while the majority of the protein eluted at 0.15 M NaCl. The two lytic peaks were pooled and assayed in the presence of an antiserum generated against a purified α-LT (Yamamoto *et al.*, 1978). The anti-α sera failed to neutralize the first lytic peak from the breakthrough of the DEAE column but completely neutralized the second lytic peak that eluted in the 0.03 M NaCl region of the DEAE column. Both of these lytic peaks were pooled separately and chromatographed over an Ultrogel AcA 44 molecular sieving column. The breakthrough peak off of DEAE eluted as a β-class LT and the second peak eluted as an α-class LT from the molecular sieving column. These studies revealed that most of the activity eluted from CPG with 50% ethylene glycol was due to the α molecule;

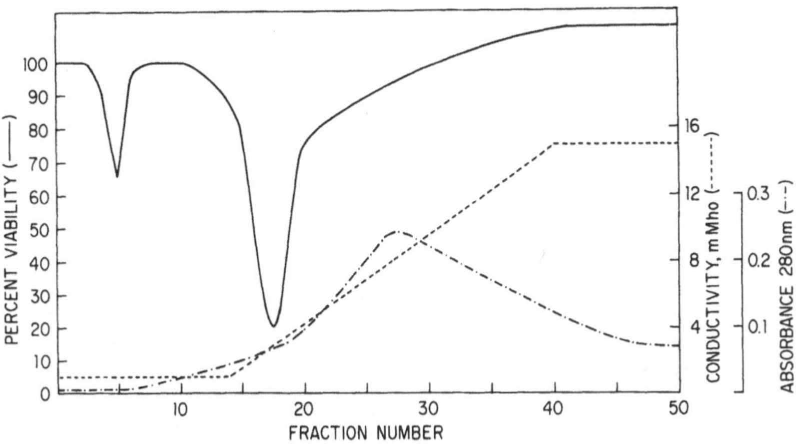

FIGURE 1. Elution profile of desorbed α LT fraction from CPG separated in a DEAE cellulose column. The column is equilibrated in 10 mM Tris-HCl, 0.1 mM EDTA (pH 8.0), and the protein eluted with a 0–0.3 M linear gradient of NaCl. Absorbence at 280 nm (-•• -•), conductivity at room temperature (- - - -), and lymphotoxin activity (——) are plotted versus the fraction number.

thus, the recovery of the α activity from CPG was from 40 to 80% of the amount in the unfractionated supernatant. The lytic activity from the major lytic peak was pooled and subjected to further biochemical separation on a 3-ml lentil-lectin affinity column. The pooled lytic activity from the DEAE column was passed through the lentil-lectin column at a flow rate of 12 ml/hr at 25°C. The lentil-lectin column removed from 90 to 98% of the lytic activity applied to the column. The lytic activity was then desorbed with 40 ml of 200 mM α-methyl-D-mannoside buffer. The amount of lytic activity desorbed from the lentil-lectin column was 40–45% of the activity applied and contained 3–11% of the total protein in the sample applied to this column. The lytic activity desorbed was next concentrated by ultrafiltration and aliquoted in 1-ml samples containing 2 × 10^5 U of lytic activity. A sample was radioiodinated according to the method of Klostergaard *et al.* (1981b) and then subjected to electrophoresis in 7% native PAGE tube gels according to the method of Davis (1964). The gels were then cut into 1-mm slices, counted in a Beckman Biogamma counter, and then LT activity was eluted from each slice by incubation in 0.3 ml of RPMI 1640 supplemented with 3% FCS for 18 hr at 4°C and tested for toxic activity on L cells as described previously (Klostergaard *et al.*, 1981a). The results of this study indicate that the lentil-lectin sample contained multiple components, but a good separation of lytic material (R_f 0.32) and nonlytic material (R_f 0.84) was achieved so final separation of the lentil-lectin material could be achieved on a high-capacity preparative native PAGE column run by the technique of Furlong *et al.* (1973). The bioactivity was run and eluted from the preparative PAGE as described previously (Klostergaard *et al.*, 1981a). This sample was then radiolabeled with ^{125}I and the labeled material was rerun on 5 and 7% native PAGE tube gels. The data shown in Fig. 2 and 3 illustrate that both radioactivity and bioactivity migrate as a single coincident peak in both 5 and 7% native

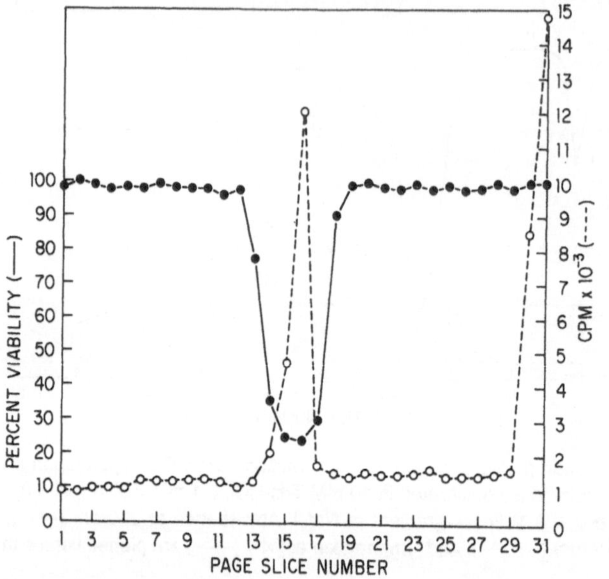

FIGURE 2. Migrations of bioactivity and radioactivity of α LT fraction from preparative gel electrophoresis on 5% acrylamide native PAGE tube gels. Radioactivity (○) and lytic activity (●) of each gel slice was determined.

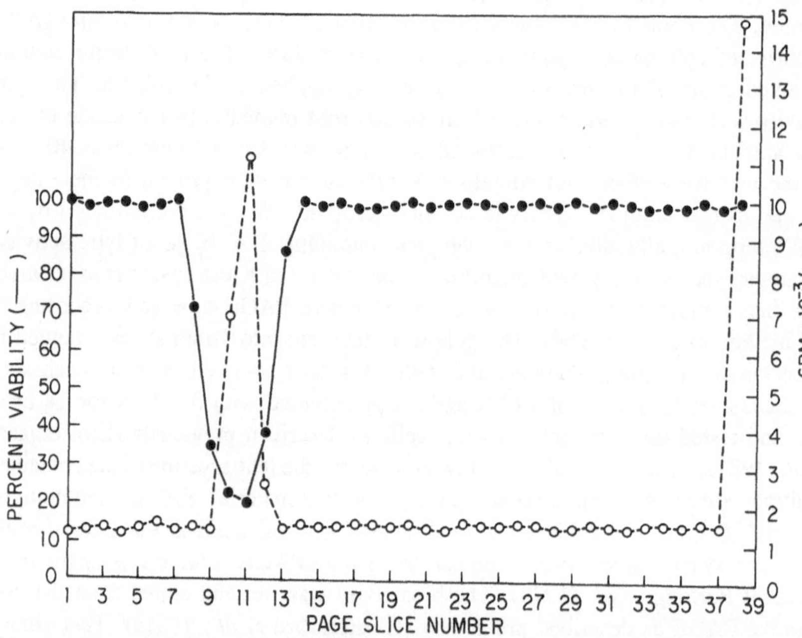

FIGURE 3. Migrations of bioactivity and radioactivity of α LT from preparative page electrophoresis on 7% acrylamide native PAGE tube gels. Radioactivity (○) and lytic activity (●) of each gel slice was determined.

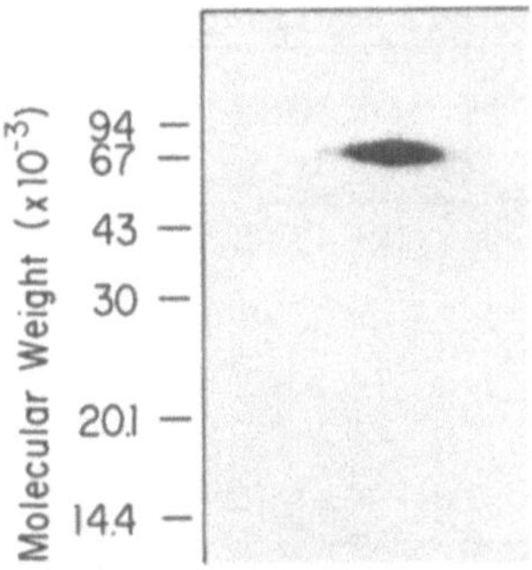

FIGURE 4. Two-dimensional (0.1% SDS reducing 15% acrylamide) slab gel of ^{125}I labeled IR 3.4 α LT fraction form by preparative PAGE. Molecular weight markers were phosphorylese B (94,000 MW), BSA (67,000 MW), ovalbumin (43,000 MW), carbonic amhydrase (30,000 MW), soybean trypsin inhibitor (20,000 MW), and lysozyme (14,400 MW).

PAGE gels, respectively. The sample was next analyzed on a two-dimensional SDS-reducing slab gel as described by O'Farrell (1975). Figure 4 shows the autoradiograph of one of these gels and it is evident that a single 68,000-dalton peptide is present in the sample. However, no bioactivity could be recovered from these gels because the SDS at the concentration used abolished all detectable lytic activity. Additional studies revealed the material obtained from preparative PAGE columns elutes from molecular sieving in Ultrogel AcA 44 columns as 70,000- to 90,000-dalton (α-class) LT activity. The bioactivity and radioactivity in these samples focus over a broad pH range (6.5–7.5) in column isoelectric focusing. This broad spread is also seen in the focusing dimension of the O'Farrell gels. This purification scheme has been adapted to large-scale preparation methods in our laboratory.

3. DISCUSSION

The use of PMA has made possible the production of high-activity supernatants from the continuous human B-lymphoblastoid cell line IR 3.4. These high-activity supernatants provide the means to obtain quantities of purified α-LT from a defined and cloned lymphoid cell type for structural and functional studies. We are now also studying a discrete β-LT form distinct from the α form from human cytotoxic T-cell clones (Yamamoto *et al.*, 1984b). The α-LT produced by PMA-activated IR 3.4 cells was purified to apparent homogeneity by a combination of several biochemical procedures. The data for complete purification runs for two supernatant lots from the IR 3.4 cells are shown in Table I. There is a 5500- to 6500-fold purification after the lentil-lectin step with approximately 3% recovery of the original LT activity of the whole supernatant. The preparative PAGE gel removed about 80% of the contaminating radiolabeled protein, resulting in an additional 5-fold purification. Thus, the final step resulted in a 25,000- to 30,000-fold purification

TABLE I. Biochemical Purification Scheme of the IR 3.4 α LT

Separation method	Supernatant lot	Percent recovery of LT units[a]	Protein	
			% Recovery[b]	Purification factor[c]
CPG	1	18	9	11.11
	2	22	14	7.14
DEAE	1	7.5	0.58	172.41
	2	7.6	0.15	666.67
Lentil Lectin	1	3.0	0.015	6666.67
	2	3.4	0.017	5882.35
Preparative PAGE	1	1.5	0.003	33333.34
	2	2.0	0.0034	29411.75

[a] Percent recovery of LT from the original starting IR 3.4 supernatant as calculated by the formula:

$$\frac{\text{Total LT units after biochemical step}}{\text{Total LT units from unseparated supernatant}} \times 100$$

[b] Percent recovery of protein from the original starting IR 3.4 supernatant as calculated by the formula:

$$\frac{\text{Total protein after biochemical step}}{\text{Total protein from unseparated supernatant}} \times 100$$

[c] Purification factor as calculated by the formula:

$$\frac{\text{Percent starting protein}}{\text{Percent recovery}}$$

of the α-LT. An exact calculation of specific activity of the α-LT in the final fractions is not possible because of the fluctuation in the bioassay itself. However, specific activity ranged from a low of 10^7 to a high of 8×10^7 U/mg protein. The α-LT from the IR 3.4 cell line appears to be a 68,000-dalton peptide which is in agreement with the α-LT molecule purified from lectin-stimulated normal human lymphocytes (Klostergaard *et al.*, 1981b) and α material immunoprecipitated from the human complex LT form by anti-α sera (Devlin *et al.*, 1983). It is interesting that Aggarwal *et al.* (1982) found a 20,000- to 25,000-dalton peptide(s) in SDS-reducing gels of their purified α-LT from the RPMI 1788 cell line. However, it is clear that these two LT forms are related for we have found that their antiserum prepared against the RPMI 1788 α-LT and our antiserum prepared against α-LT from tonsil and adenoid lymphocytes neutralize activity of both α forms. Studies are currently under way to examine the differences between the IR 3.4 α-LT and the RPMI 1788 α-LT peptides. This is an important issue for our model would have the α form create β- and γ-LT forms by protease clevage and their mode would indicate subunit disassociation.

A spectrum of studies are under way in our laboratory to examine functional capabilities of this purified α-LT form. Because of space and time limitations, a complete description of some of these data is not possible; however, we would like to point out several important new observations obtained from *in vitro* studies.

We have conducted preliminary studies examining the effects of purified α-LT on nondividing target cells of continuous and primary nature and have found three patterns of activity: (1) L-929 cells were highly sensitive; (2) HeLa, T-24 (human carcinoma of the bladder), and WI-38 VA-13 (virus-transformed human lung fibroblasts) were lysed with 700–800 U; and (3) non-virus-infected WI-38 cells, primary fibroblasts, and a human fibrosarcoma (HT-1080) were resistant to lysis by 1500 U of activity. While these are interesting results and suggest this molecule may be more effective on continuous vs. primary cells *in vitro,* a broader selection of target cells must be examined to clearly define the lytic effectiveness of this molecule on not only continuous tissue culture cells, but on primary short-term cultures of neoplastic cells isolated from freshly excised human tumors. In addition, we found α-LT is very effective in destroying various targets treated with sublethal levels of various anticancer drugs *in vitro*. This effect is pronounced and normally sublethal levels of the α molecule become cell-lytic when tumor cells are treated with very low levels of chemotherapeutic agents.

Another very interesting finding is the ability of purified α-LT to act in synergy or antagonistically with purified human interferons (IFN). Williams and Bellanti (1983) recently reported that 10–100 U of purified IFN from Nomalva cells and recombinant DNA sources dramatically increased the sensitivity of human WI-38 and HeLa to lysis by LT-containing unseparated supernatants from PHA-stimulated tonsil and adenoid cells. In extensive experiments, we have repeated these results with purified α-LT and human α-, β-, and γ-IFN. We have found that pretreatment of HeLa, HT-1080, and normal fibroblasts with 1–10 U of purified α-, β-, or γ-IFN for 8–24 hr makes these cells fully sensitive to nonlethal levels of IR 3.4 α-LT. The situation is complex for not all IFN are equally effective on each target, but vary on an individual basis with the type of target employed. Moreover, at higher levels of IFN (100–500 U), some degree of antagonism or resistance to lysis by these mediators was observed. It will be interesting to examine the collaborative effects of human IFN and α-LT on the outcome of virus infection on cells *in vitro*. These studies may provide information on a new mechanism(s) of host resistance to virus infection. All these data are currently being prepared for publication; however, they are supportive of the testing of α-LT as a possible anticancer agent perhaps in conjunction with one or more classes of human IFN or various chemotherapeutic agents.

4. SUMMARY

Studies with human LT have revealed a complex situation for LT forms are heterogeneous and different types of lymphocytes release different LT forms. Moreover, the types of LT released may vary with the type of inducing signal. To resolve this situation, we have begun studies of key LT forms released by defined populations of human lymphocytes *in vitro*. A human lymphoblastoid cell line was identified which releases high levels of α- and β-LT when stimulated with PMA *in vitro*. Methods to produce and purify quantities of the α-LT form from this cell line have been developed. The α molecule appears to be a 68,000-dalton peptide. Initial functional studies with the purified molecule indicate exciting new results which suggest in addition to a possible role in lymphocyte-mediated cytolytic reactions: (1) the α form may have antitumor effects *in vitro;* (2) the molecule has increased effects on cells treated with anticancer agents; and (3) the α form can both synergize and antagonize the effects of human α-, β-, and γ-IFN *in vitro*.

These findings have far-reaching effects on the possible role of α-LT acting alone as an anticancer agent in certain cases, or in combination with IFN or chemotherapy; and finally, IFN and LT synergy may have implications on host resistance to virus infections.

REFERENCES

Aggarwal, B., Moffat, B., and Harkins, R. N., 1982, Purification and characterization of lymphotoxin from human lymphoblastoid cell line 1788, Third International Lymphokine Workshop, Haverford, Pennsylvania.

Davis, B., 1964, Disc electrophoresis. II. Methods and application to human serum proteins, *Ann. N.Y. Acad. Sci.* **121**:404.

Daynes, R. A., and Granger, G. A., 1974, The regulation of lymphotoxin release from stimulated human lymphocyte cultures: The requirement for continual mitogen stimulation, *Cell. Immunol.* **12**:252.

Devlin, J. J., Yamamoto, R. S., and Granger, G. A., 1981, Stabilization and functional studies of high molecular weight murine lymphotoxins, *Cell. Immunol.* **61**:22.

Devlin, J. J., Klostergaard, J., and Granger, G. A., 1983, *J. Immunol.*, submitted.

Fair, D. S., Jeffes, E. W. B., and Granger, G. A., 1979, Release of LT molecules with restricted physical heterogeneity by a continuous human lymphoid cell line *in vitro*, *Mol. Immunol.* **16**: 186.

Furlong, C. E., Cirakoglu, C., Willis, R. C., and Santy, P. A., 1973, A simple preparative polyacrylamide disc gel electrophoresis apparatus: Purification of three branched-chain amino acid binding proteins from *E. coli*, *Anal. Biochem.* **15**:297.

Granger, G. A., and Williams, T. W., 1968, Lymphocyte cytotoxicity *in vitro:* Activation and release of a cytotoxic factor, *Nature (London)* **218**:1253.

Granger, G. A., Shacks, S. J., Williams, T. W., and Kolb, W. P., 1969, Lymphocyte *in vitro* cytotoxicity: Specific release of lymphotoxin-like materials from tuberculin sensitive lymphocyte cells, *Nature (London)* **221**:1155.

Granger, G. A., Moore, G. E., White, J. G., Matzinger, P., Sundsomo, J. S., Shupe, S., Kolb, W. P., Kramer, J., and Glade, P. R., 1970, Production of lymphotoxin and migration inhibitory factor by established human lymphocytic cell lines, *J. Immunol.* **104**:1476.

Granger, G. A., Yamamoto, R. S., Fair, D. S., and Hiserodt, J. C., 1978, The human LT system. I. Physical-chemical heterogeneity of LT molecules released by mitogen activated human lymphocytes *in vitro*, *Cell. Immunol.* **38**:388.

Granger, G. A., Yamamoto, R. S., Ware, C. F., and Ross, M. W., 1980, *In vitro* detection of human lymphotoxin, in: *Manual of Clinical Immunology* (N. R. Rose and H. Friedman, eds.), American Society for Microbiology, Washington, D. C.

Harris, P. C., Yamamoto, R. S., Crane, J., and Granger, G. A., 1981, The human LT system. X. The first cell-lytic molecule released by enriched lymphocytes is a 150,000 MW form in association with small non-lytic components, *J. Immunol.* **126**:2165.

Hiserodt, J. C., Fair, D. S., and Granger, G. A., 1976, Identification of multiple cytolytic components associated with the β-LT class of lymphotoxins released by mitogen activated human lymphocytes *in vitro*, *J. Immunol.* **117**:1503.

Khan, A., Weldon, D., Duvall, J., Pichyangkul, S., Hill, N. O., Muntz, D., Lanius, R., and Ground, M., 1982, A standardized automated computer assisted micro-assay for lymphotoxin, in: *Human Lymphokines: The Biological Immune Response Modifiers* (A. Khan and N. O. Hill, eds.), Academic Press, New York.

Klostergaard, J., Long, S., and Granger, G. A., 1981a, Purification of human α class lymphotoxin to electrophoretic homogeneity, *Mol. Immunol.* **18**:1049.

Klostergaard, J., Yamamoto, R. S., and Granger, G. A., 1981b, Lymphotoxins: Purification and biological parameters, in: *Lymphokines and Thymic Factors and Their Potential Utilization in Cancer Therapeutics* (A. L. Goldstein, ed.), Raven Press, New York.

Klostergaard, J., Yamamoto, R. S., Harris, P. C., and Granger, G. A., 1981c, The nature of a lymphocyte released cell toxin, in: *Handbook of Cancer Immunology* (H. Water, ed.), Garland STPM Press, New York.

Kramer, J. J., and Granger, G. A., 1972, An improved *in vitro* assay for lymphotoxin, *Cell. Immunol.* **3:**144–149.

Lee, S. C., and Lucas, Z. J., 1976, Regulatory factors produced by lymphocytes. I. The occurrence of multiple α-lymphotoxins associated with ribonuclease activity, *J. Immunol.* **117:**283.

Lisafeld, B. A., Minowada, J., Klein, E., and Holterman, O. A., 1980, Lymphotoxins: Selective cytotoxic effects, *Int. Archs. Allergy Appl. Immunol.* **62:**59.

O'Farrell, P. D., 1975, High resolution two-dimensional electrophoresis of proteins, *J. Biol. Chem.* **250:**4007.

Ross, M. W., Tiangco, G. J., Horn, P., Hiserodt, J. C., and Granger, G. A., 1979, The LT system in experimental animals. III. Physical-chemical characteristics and relationships of lymphotoxin (LT) molecules released *in vitro, J. Immunol.* **123:**325.

Toth, M. K., and Granger, G. A., 1979, The human LT system. VI. Identification of various saccharides on LT molecules and their contribution to cytotoxicity and charge heterogeneity, *Mol. Immunol.* **16:**671.

Weitzen, M. L., Innins, E., Yamamoto, R. S., and Granger, G. A., 1983a, Inhibition of human NK induced cell lysis and soluble cell-lytic molecules with anti-human LT antisera and various saccharides, *Cell. Immunol.* **77:**42.

Weitzen, M. L., Yamamoto, R. S., and Granger, G. A., 1983b, Identification of lymphocyte-derived lymphotoxins with binding and cell-lytic activity on NK sensitive cell lines *in vitro, Cell. Immunol.* **77:**30.

Williams, T. W., and Bellanti, J. A., 1983, *In vitro* synergism between interferons and human lymphotoxins: Enhancement of lymphotoxin-induced target cell killing, *J. Immunol.* **130:**518.

Williams, T. W., and Granger, G. A., 1969, Lymphocyte *in vitro* cytotoxicity: Mechanism of lymphotoxin induced cell destruction, *J. Immunol.* **102:**911.

Wright, S. C., and Bonavida, B., 1982, Selective lysis of NK-sensitive target cells by a soluble mediator released from murine spleen cells and human peripheral blood lymphocytes, *J. Immunol.* **126:**1516.

Yamamoto, R. S., Hiserodt, J. C., Lewis, J. E., Carmack, C. E., and Granger, G. A., 1978, The human LT system. II. Immunological relationships of LT molecules released by mitogen activated human lymphocytes *in vitro, Cell. Immunol.* **38:**403.

Yamamoto, R. S., Hiserodt, J. C., and Granger, G. A., 1979, The human LT system. V. A comparison of the relative lytic effectiveness of various MW human LT classes on ^{51}Cr labeled allogeneic target cells *in vitro, Cell. Immunol.* **45:**261.

Yamamoto, R. S., Johnson, D. L., Masunaka, I. K., and Granger, G. A., 1984a, Phorbal myristate acetate induction of lymphotoxins from continuous human B lymphoid cell lines *in vitro, J. Biol. Response Modifiers,* **3:**76.

Yamamoto, R. S., Ware, C. F., and Granger, G. A., 1984b, in preparation.

21

Chemical and Biological Properties of Human Lymphotoxin

BHARAT B. AGGARWAL, BARBARA MOFFAT, SANG HE LEE, and RICHARD N. HARKINS

1. INTRODUCTION

The cytotoxicity of lymphocytes toward allogeneic target cells *in vitro* was first shown by Govaerts (1960). This was later confirmed by Rosenau and Moon (1961). It was demonstrated that lymphocytes, when activated with antigen or mitogen, secrete a soluble cytotoxin which was named lymphotoxin (Granger and Kolb, 1968). Since then a number of established human lymphoid cell lines have also been reported to produce a similar cytotoxic molecule (Amino *et al.*, 1974; Granger *et al.*, 1970; Shacks *et al.*, 1973; Papermaster *et al.*, 1976). The activity of lymphotoxin has been tested against a number of cell lines both of animal and human origin. It has been demonstrated by several workers that lymphotoxin is both cytostatic and cytolytic to target cells (Evans and Heinbaugh, 1981; Rosenau, 1981; Sawada *et al.*, 1976). Most of the biological studies with lymphotoxin have been performed with relatively crude preparations. The isolation of lymphotoxin has been a rather difficult task due to its heterogeneous nature and also because of the small amounts produced by normal lymphocytes and various lymphoid cell lines. We have purified human lymphotoxin from several hundred liters of cell conditioned medium derived from the lymphoblastoid cell line RPMI 1788. Using purified material we have also examined the *in vitro* effects of lymphotoxin on human tumor and normal cells.

BHARAT B. AGGARWAL, BARBARA MOFFAT, and RICHARD N. HARKINS • Department of Protein Biochemistry, Genentech Inc., South San Francisco, California 94080. SANG HE LEE • Department of Pharmacological Sciences, Genentech Inc., South San Francisco, California 94080.

2. BIOASSAYS FOR LYMPHOTOXIN

Over the last 15 years, several qualitative and quantitative assays have been developed for detecting lymphotoxin activity. These assays can conveniently be separated into two distinct categories based on the metabolic effects of lymphotoxin on a target cell. The assays dependent on the metabolic effects of lymphotoxin have included the reduced uptake of [^{3}H]thymidine (Smith *et al.*, 1977), the decrease in protein synthesis as monitored by the uptake of ^{14}C-labeled amino acids (Granger and Williams, 1968), and the changes in cell permeability as indicated by the release of potassium (Walker and Lucas, 1972), release of cytoplasmic marker enzymes, and exclusion of vital stains (Smith *et al.*, 1977).

The majority of lymphotoxin assays are based on its cytolytic action on target cells. The lysis of cells by lymphotoxin can be enhanced by pretreatment with either actinomycin D (Walker and Lucas, 1972) or mitomycin C (Spofford *et al.*, 1974). Cell lysis is quantitated either by direct counting of unlysed cells (Kolb and Granger, 1968; Gately and Mayer, 1972) or by staining the residual cells with neutral red or crystal violet (Khan *et al.*, 1982a). The most sensitive target cell lines for this assay include mouse fibroblast L-929 cells (Kramer and Granger, 1972) and mouse lymphoma L-1210 cells (Smith *et al.*, 1977). The latter cell line is known to produce lymphotoxin activity and is therefore not as suitable as mouse L-929 fibroblast cells. We have developed an assay for lymphotoxin involving the lysis of mouse L-929 cells pretreated with actinomycin D. This assay is done in microtiter plates and it allows the screening of several samples during purification. The antibiotic (1 μg/ml)-treated cells were incubated with serially diluted lymphotoxin samples for 18 hr at 37°C and lysis was determined by staining unlysed cells with crystal violet, followed by monitoring the absorbance at 550 nm using a Microelisa Autoreader (Fig. 1). A 50% reduction in cell viability compared to control values is used as an operational standard unit of lymphotoxin.

3. SOURCES OF LYMPHOTOXIN

It has been shown that T and B lymphocytes (Shacks *et al.*, 1973), macrophages (Kramer and Granger, 1972), and natural killer cells (Wright and Bonavida, 1982) secrete cytotoxic factors. The relationship of these secreted factors derived from different cell types is not known. The cytotoxic activity released by lymphocytes was named lymphotoxin (Granger and Kolb, 1968). The most common sources of human lymphotoxin include peripheral blood lymphocytes and lymphocytes derived from tonsils, adenoids, and lymph nodes. These lymphocytes can be stimulated by a wide variety of substances to secrete lymphotoxin. Besides normal lymphocytes, several human lymphoblast cell lines have also been described which secrete factors having lymphotoxin activity (Granger *et al.*, 1970; Amino *et al.*, 1974; Papermaster *et al.*, 1976). Usually these cell lines exhibit B-cell surface markers. We have used the human lymphoblastoid cell line RPMI 1788 as a source of lymphotoxin. This

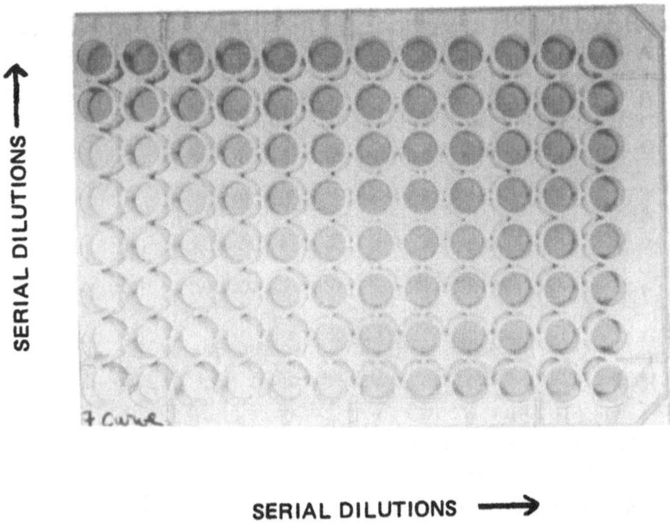

FIGURE 1. Effect of serially (twofold) diluted purified human lymphotoxin on actinomycin D-treated L-929 mouse fibroblast cells. Cells treated with control medium are shown in two top lanes. Both *x* and *y* axes represent serially diluted purified lymphotoxin sample.

cell line can be conveniently grown in both roller bottles and spinner flasks and is fairly stable for its production of lymphotoxin. We can grow these cells up to 65 hr in a serum-free medium and carry out the purification of lymphotoxin activity after the cell supernatants are harvested.

4. METHODS OF PURIFICATION OF LYMPHOTOXIN

Most of the methods employed for the purification of human lymphotoxin have included DEAE-cellulose chromatography, gel filtration, preparative gel electrophoresis, and isoelectric focusing (Russell *et al.*, 1972; Granger *et al.*, 1973; Peters *et al.*, 1973). Some workers have also used ammonium sulfate precipitation as a purification step for lymphotoxin (Amino *et al.*, 1974). The use of lectin affinity chromatography and of hydrophobic chromatography on alkyl agarose columns have also been reported (Klostergaard *et al.*, 1980). Due to the minute amount of protein encountered during purification, Klostergaard *et al.* (1980) performed iodination on partially purified lymphotoxin, which allowed them to purify lymphotoxin to electrophoretic homogeneity.

In our laboratory we have purified lymphotoxin from cell supernatants using controlled-pore glass chromatography, DEAE-cellulose and lentil-lectin Sepharose chromatography (Aggarwal *et al.*, 1982). After lectin chromatography, lymphotoxin could be further purified by preparative PAGE and by reverse-phase high-pressure liquid chromatography (HPLC) on a C_{18} column.

5. BIOCHEMICAL PROPERTIES OF HUMAN LYMPHOTOXIN

Kolb and Granger (1968) reported that lymphocytes from human adenoid tissues, when stimulated with phytohemagglutinin (PHA), secrete heat-sensitive and trypsin-resistant lymphotoxin with a molecular weight of 80,000–90,000 (see Table I) and an isoelectric point of 6.8–8.0 (Granger *et al.*, 1973). Peters *et al.*, (1973) showed that stimulated lymphocytes derived from human peripheral blood, tonsils, adenoids, and thoracic duct lymphs elaborate lymphotoxin of a molecular size 80,000–150,000, precipitable at 37–50% ammonium sulfate, and stable to heat at 56°C for 45 min but destroyed at 85°C in 15 min. The lymphotoxin derived from human lymphoid cell lines has been shown to be stable at 20°C for over a year, at 4°C for 3 weeks, and at 37°C for 1 week (Amino *et al.*, 1974). It was also shown to be stable at a pH range from 5 to 11. Walker *et al.* (1976) have reported that human blood lymphocytes secrete two different species of lymphtoxin which have molecular weights of 75,000 and 45,000, respectively. However, it was shown that antibodies against smaller forms would neutralize the activities of the larger species. In 1978, Granger's laboratory reported that stimulated human lymphocytes from tonsils and adenoid tissues secrete lymphotoxin molecules which differ in molecular size and charge (Granger *et al.*, 1978). They showed, based on molecular weight, four separate classes: complex (> 200,000), α (70,000–90,000), β (25,000–50,000), and γ (10,000–20,000), representing 5–20, 40–60, 20–40, and 0–10% of the total lymphotoxin activity, respectively. Based on the charge, the α class was further separated into α_1 (R_f 0.25), α_2 (R_f 0.37), and α_3 (R_f 0.50) subclasses. The β class was separated into β_1 (R_f 0.28) and β_2 (R_f 0.49). Whether these various lymphotoxin species are aggregates or breakdown products of a single form of lymphotoxin, or associated with other proteins, has not been fully resolved. However, it has been shown that the complex class of lymphotoxin (> 200,000) is a macromolecular assemblage of smaller molecular weight α-, β-, and γ-lymphotoxin classes (Yamamoto *et al.*, 1978). Klostergaard *et al.* (1980) have reported an α heavy (120,000–150,000) and an α light (70,000–90,000) subclass of lymphotoxin and claimed to have purified these forms to electrophoretic homogeneity (Klostergaard and Granger, 1981; Klostergaard *et al.*, 1981). This is the only report thus far that has been made on the purification of lymphotoxin to homogeneity; yet there were few data on the chemical characterization of the molecule.

Human lymphotoxin purified in our laboratory has been characterized by molecular sieving chromatography, SDS and native PAGE, isoelectric focusing, and reverse-phase HPLC. Furthermore, purified lymphotoxin was also examined by amino acid analysis and tryptic mapping. A preliminary amino acid sequence has also been determined. We have reported previously that human lymphotoxin from RPMI 1788 is a single molecular species with an apparent molecular weight of 64,000 on Sephadex G-100 (Aggarwal *et al.*, 1982). This molecule can dissociate into a monomeric form of molecular weight around 20,000 by SDS–PAGE. Furthermore, it has been found that antibodies against our RPMI 1788-derived lymphotoxin will neutralize most of the bioactivity of lymphotoxin derived from stimulated peripheral blood lymphocytes (Vilcek *et al.*, personal communication) or that derived from stimulated lymphocytes of tonsils and adenoids (Granger and

TABLE I. Properties of Human Lymphotoxin from Various Sources

Source	Molecular weight	Method of determination of MW	Isoelectric point	Effect of proteolytic enzymes on activity	Reference
Adenoid lymphocytes, PHA stimulation	80–90K	G-150		Trypsin resistant	Kolb and Granger (1968)
Lymphoid cell lines	80–150K	G-100	—	—	Granger *et al*. (1970)
Adenoid lymphocytes, PHA stimulation	60K	G-150	—	Trypsin resistant	Russell *et al*. (1972)
Blood lymphocytes, PHA stimulation	90–100K	G-150	6.8–8.0	—	Granger *et al*. (1973)
Tonsil lymphocytes, PHA stimulation	80–150K	G-150	—	—	Peters *et al*. (1973)
Lymphoid cell lines	70–150K	G-200	—	—	Amino *et al*. (1974)
Blood lymphocytes, PHA stimulation	α 75K β 45K	G-100	—	—	Walker and Lucas (1972)
Tonsil and adenoid lymphocytes, PHA stimulation	Complex > 200K α 70–90K β 25–50K γ 10–20K	Ultrogel AcA 44	—	—	Granger *et al*. (1978)
Lymphoblastoid cell line RPMI 1788	20–40K	SDS gel	4.0–4.1	Trypsin resistant	Papermaster *et al*. (1981)
	20K 64K	SDS gel G-100	5.8	Trypsin resistant Chymotrypsin resistant Thermolysin resistant Subtilisin resistant Staph. protease resistant	Aggarwal *et al*. (1982)

Yamamoto, personal communication). This suggests a fundamental immunological and structural similarity among many forms of lymphotoxin previously observed from these three sources.

The purified lymphotoxin derived from cell line 1788 was run on C_{18} reverse-phase HPLC. As shown in Fig. 2, a single major peak of protein was noted which eluted with about 50% acetonitrile and 0.1% trifluoroacetic acid. Purified lymphotoxin was also found homogeneous on SDS–PAGE. The relative mobility of lymphotoxin on native 7.5% polyacrylamide gels (Laemmli, 1970) was 0.33 and had an isoelectric point (pI) around 5.8 which is slightly higher than the theoretical pI of 4.4 obtained from the amino acid composition. The latter indicated that human lymphotoxin contains high amounts of proline, glutamic acid, and aspartic acid residues but low amounts of cysteine, methionine and tryptophan. Treatment of purified lymphotoxin with 5% trypsin cleaves the molecule into two major fragments of molecular weight 15,000 and 5000. Furthermore, it was also noted that lymphotoxin activity is trypsin resistant. Purified lymphotoxin was also examined for its heat stability (Table II). Lymphotoxin is stable for up to 30 min at 60°C but 90% of the activity is lost at 80°C in 5 min. These observations are consistent with those of Peters et al. (1973).

6. BIOLOGICAL ACTION OF LYMPHOTOXIN

It has been demonstrated by several workers that partially purified preparations of lymphotoxin derived from human lymphocytes cause lysis of target cells from

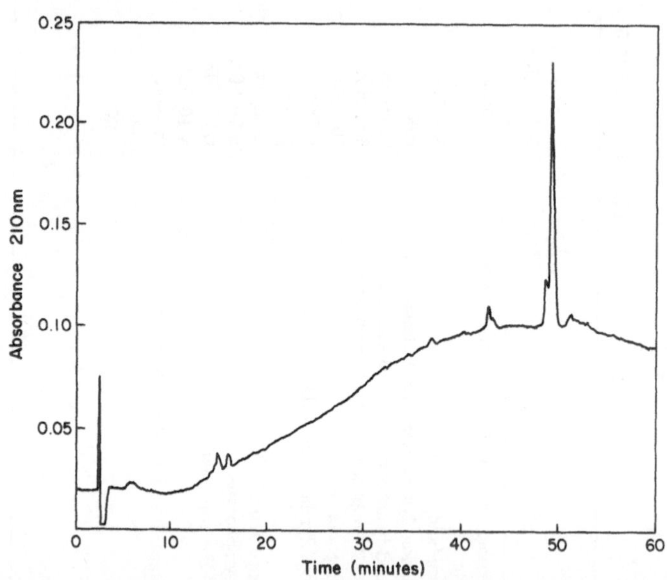

FIGURE 2. Reverse-phase high-pressure liquid chromatography of purified lymphotoxin from human lymphoblastoid cell line 1788.

TABLE II. Heat Stability of Human Lymphotoxin

Temperature (°C)	Time (min)	Bioactivity (units)
Control	—	233,100
60	5	259,700
	15	588,000
	30	334,100
	60	125,370
80	5	33,190
100	5	360

different species *in vitro* (Kramer and Granger, 1976). The relative sensitivity varies greatly among target cells (Kolb and Granger, 1970; Shacks *et al.*, 1973; Papermaster *et al.*, 1976). Walker and Lucas (1972) have found that the addition of actinomycin D greatly increases the sensitivity of the cytotoxic response of lymphotoxin. Rosenau and Tsoukas (1976) reported that simultaneous addition of lymphotoxin and actinomycin D produces a cytolytic response in a cell line that shows no effect to lymphotoxin alone. Furthermore, it was demonstrated in a stem cell assay that lymphotoxin acts as an antitumor agent against human lymphoma, ovarian carcinoma, and multiple myeloma (Khan *et al.*, 1982b). The tumor regression response of lymphotoxin is also enhanced by mitomycin C. The various antitumor and anticarcinogenic effects of lymphotoxin have been reviewed recently (Evans, 1982).

We have examined the growth inhibition properties of purified human lymphotoxin on human lung carcinoma A549 cells. As shown in Fig. 3, 60% growth inhibition was observed with 15,000 units. Recently, Williams and Bellanti (1983) have reported that human α-interferon could enhance the lymphotoxin-dependent lysis of HeLa cells. They used crude lymphotoxin for such studies. We have found that recombinant DNA-derived human γ-interferon acts synergistically with human lymphotoxin in inhibiting the growth of A549 cells (Fig. 3). Such synergism has also been observed when lymphotoxin and interferon were tested together on human lung fibroblast WI-38 cells transformed with SV-40 virus (Fig. 4). No effect was observed on normal lung fibroblast cells (WI-38). This suggests that lymphotoxin may be selective in its action to tumor cells and acts more effectively in combination with γ-interferon.

7. CONCLUSION

Lymphotoxin is an important lymphokine produced by lymphocytes and it has been purified to homogeneity and characterized. It is a protein with a molecular weight of 20,000 and the cytolytic activity is resistant to trypsin. It has also been shown that human lymphotoxin has a tendency to aggregate into higher molecular weight forms and cytolytic activity resides probably in a small protein fragment. Furthermore, it appears that lymphotoxin is selective in its cytostatic and cytolytic action toward tumor cells. However, the mechanism of such action is not well

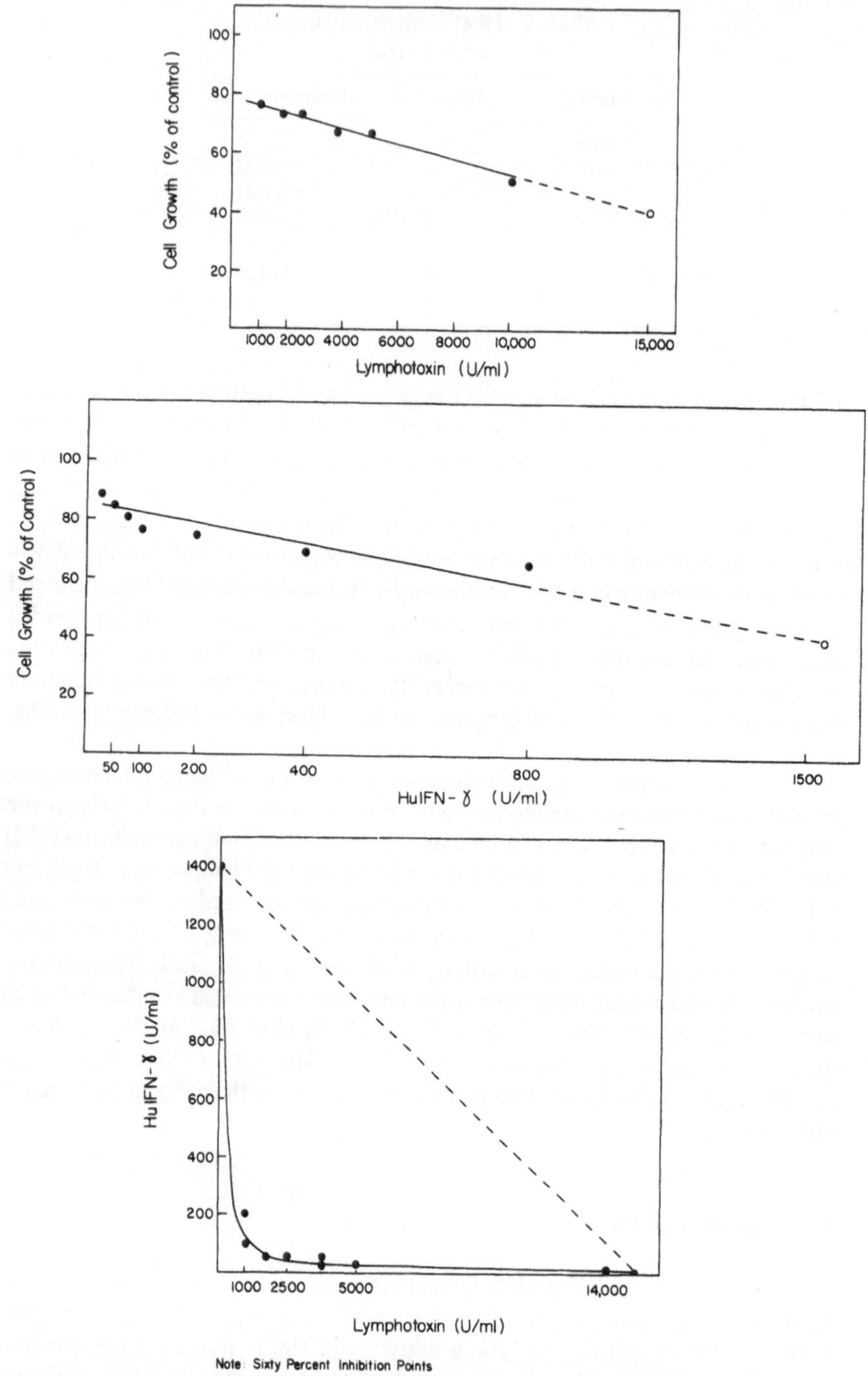

FIGURE 3. Effect of lymphotoxin (upper panel), recombinant human γ-interferon (middle panel), and both in combination (bottom panel) on the growth of human lung carcinoma cells A549.

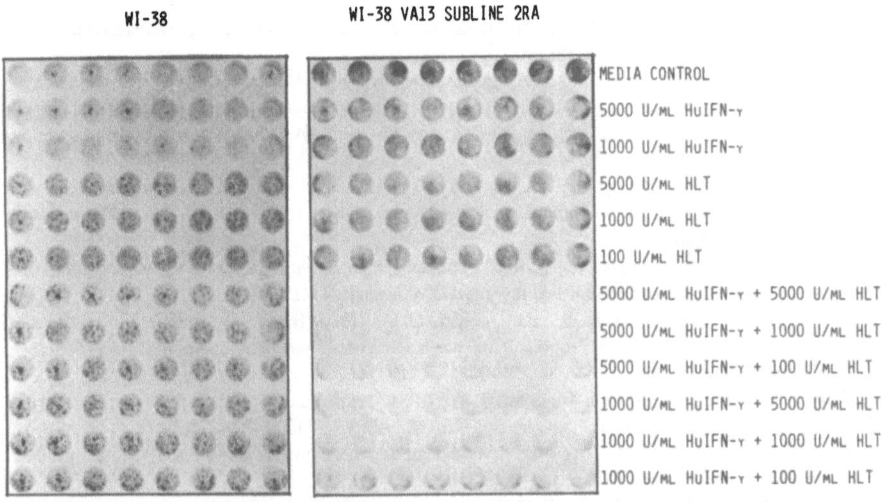

FIGURE 4. Effect of human lymphotoxin and human recombinant γ-interferon on normal and transformed human lung fibroblast cells.

understood. It is also not known what determines its specificity. Preliminary indications are that lymphotoxin could be a useful antitumor agent by itself or when used in combination with other interferons.

REFERENCES

Aggarwal, B. B., Moffat, B., and Harkins, R. N., 1982, Purification and characterization of lymphotoxin from human lymphoblastoid cell line 1788, in: *Interleukins, Lymphokines and Cytokines* (J.J. Oppenheim and S. Cohen, eds.), pp. 521–526, Academic Press, New York.

Amino, N., Linn, E. S., Pysher, T. J., Mier, R., Moore, G. E., and DeGroot, L. J., 1974, Human lymphotoxin obtained from established lymphoid lines: Purification characteristics and inhibition by antiimmunoglobulins, *J. Immunol.* **113**:1334–1345.

Evans, C. H., 1982, Lymphotoxin—An immunologic hormone with anticarcinogenic and antitumor activity, *Cancer Immunol. Immunother.* **12**:181–190.

Evans, C. H., and Heinbaugh, J. A., 1981, Lymphotoxin cytotoxicity, a combination of cytolytic and cytostatic cellular responses, *Immunopharmacology* **3**:347–359.

Gately, M. K., and Mayer, M. M., 1972, The effect of antibodies to complement components C2, C3 and C5 on the production and action of lymphotoxin, *J. Immunol.* **109**:728.

Govaerts, A., 1960, Cellular antibodies in kidney hemotransplantation, *J. Immunol.* **85**:516–522.

Granger, G. A., and Kolb, W. P., 1968, Lymphocyte *in vitro* cytotoxicity: Mechanism of immune and non-immune small lymphocyte mediated target L-cell destruction, *J. Immunol.* **101**:111–120.

Granger, G. A., and Williams, T. W., 1968, Lymphocyte cytotoxicity *in vitro:* Activation and release of a cytotoxic factor, *Nature (London)* **218**:1253–1254.

Granger, G. A., Moore, G. E., White, J. G., Matzinger, P., Sundsmo, J. S., Shupe, S., Kolb, W. P., Kramer, J., and Glade, P. R., 1970, Production of lymphotoxin and migration inhibiting factor by established human lymphocytic cell lines, *J. Immunol.* **104**:1476–1485.

Granger, G. A., Laserna, E. C., Kolb, W. P., and Chapman, F., 1973, Human lymphotoxin: Purification and some properties, *Proc. Natl. Acad. Sci. USA* **70**:27–30.

Granger, G. A., Yamamoto, R. S., Fair, D. S., and Hiserodt, J. C., 1978, The human LT system. I.

Physical-chemical heterogeneity of LT molecules released by mitogen activated human lymphocytes *in vitro*, *Cell. Immunol.* **38:**388–402.

Khan, A., Weldon, D., Duvall, J., Pichangkul, S., Hill, N. O., Mutx, D., Lanius, R., and Ground, M., 1982a, A standardized automated computer assisted micro-assay for lymphotoxin, in: *Human Lymphokines: The Biological Immune Response Modifiers* (A. Khan and N. O. Hill, eds.), pp. 23–32, Academic Press, New York.

Khan, A., Hill, N. O., Ridgway, H., and Webb, K., 1982b, Preclinical and phase I clinical trials with lymphotoxin, in: *Human Lymphokines* (A. Khan and N. O. Hill, eds.), pp. 621–632, Academic Press, New York.

Klostergaard, J., and Granger, G. A., 1981, Human lymphotoxin: Purification to electrophoretic homogeneity of the αh receptor bearing class, *Mol. Immunol.* **18:**455–458.

Klostergaard, J., Yamamoto, R. S., and Granger, G. A., 1980, Human and murine lymphotoxins as a multi-component system: Progress in the purification of the human αL component, *Mol. Immunol.* **17:**613–623.

Klostergaard, J., Long, S., and Granger, G. A., 1981, Purification of human alpha light class lymphotoxin to electrophoretic homogeneity, *Mol. Immunol.* **18:**1049–1054.

Kolb, W. P., and Granger, G. A., 1968, Lymphocyte *in vitro* cytotoxicity: Characterization of human lymphotoxin, *Proc. Natl. Acad. Sci. USA* **61:**1250–1255.

Kolb, W. P., and Granger, G. A., 1970, Lymphocyte *in vitro* cytotoxicity: Characterization of mouse lymphotoxin, *Cell. Immunol.* **1:**122–132.

Kramer, J. J., and Granger, G. A., 1972, The *in vitro* induction and release of cell toxins by immune C57 black 6 mouse peritoneal macrophages, *Cell. Immunol.* **3:**88–100.

Kramer, S. L., and Granger, G. A., 1976, The role of lymphotoxin in target cell destruction induced by mitogen activated human lymphocytes *in vitro*. II. The correlation of temperature and trypsin sensitive phases of lymphotoxin induced and lymphocyte mediated cytotoxicity, *J. Immunol.* **116:**562–567.

Laemmli, U. K., 1970, Cleavage of structural proteins during the assembly of the head of bacteriophage T-4, *Nature (London)* **227:**680–685.

Papermaster, B. W., Holterman, O. A., Klein, E., Parnrett, S., Dobkin, D., Laudico, R., and Djerass, I., 1976, Lymphokine properties of a lymphoid cultured cell supernatant fraction active in promoting tumor regression, *Clin. Immunol. Immunopathol* **5:**48–59.

Papermaster, B. W., Smith, M. E., and McEntire, J. E., 1981, Lymphotoxins: Soluble cytotoxic molecules secreted by lymphocytes, in: *The Lymphokines: Biochemistry and Biological Activity* (J. W. Hadden and W. E. Stewart II, eds.), pp. 149–180, Humana Press, Clifton, N.J.

Peters, J. B., Stratton, J. A., Stempel, K. E., Yu, D., and Cardin, C., 1973, Characteristics of a cytotoxin ("lymphotoxin") produced by stimulation of human lymphoid tissue, *J. Immunol.* **111:**770–782.

Rosenau, W., 1981, Lymphotoxin: Properties, role, and mode of action, *Int. J. Immunopharmacol.* **3:**1–8.

Rosenau, W., and Moon, H. D., 1961, Lysis of homologous cells by sensitized lymphocytes in tissue culture, *J. Natl. Cancer Inst.* **27:**471–483.

Rosenau, W., and Tsoukas, C. D., 1976, Lymphotoxin, A review and analysis, *Am. J. Pathol.* **84:**580–596.

Russell, S. W., Rosenau, W., Goldberg, M. L., and Kumitomi, G., 1972, Purification of human lymphotoxin, *J. Immunol.* **109:**784–790.

Sawada, J., Shioiri-Nakano, K., and Osawa, T., 1976, Cytotoxic activity of purified lymphotoxin against various cell lines, *Jpn. J. Exp. Med.* **46:**263–267.

Shacks, S. J., Chiller, J., and Granger, G. A., 1973, Studies on *in vitro* models of cellular immunity: The role of T and B cells in the secretion of lymphotoxin, *Cell. Immunol.* **7:**313–321.

Smith, M. E., Landico, R., and Papermaster, B. W., 1977, A rapid quantitative assay for lymphotoxin, *J. Immunol. Methods* **14:**243–251.

Spofford, B. T., Daynes, R. A., and Granger, G. A., 1974, Cell mediated Immunity *in vitro:* A highly sensitive assay for human lymphotoxin, *J. Immunol.* **112:**2111–2116.

Walker, S. M., and Lucas, Z. J., 1972, Cytotoxic activity of lymphocytes. I. Assay for cytotoxicity by rubidium exchange at isotopic equilibrium, *J. Immunol.* **109:**1223–1244.

Walker, S. M., Lee, S. C., and Lucas, Z. J., 1976, Cytotoxic activity of lymphocytes. IV. Heterogeneity of cytotoxins in supernatants of mitogen activated lymphocytes, *J. Immunol.* **116:**807–815.

Williams, T. W., and Bellanti, J. A., 1983, In vitro synergism between interferons and human lymphotoxin: Enhancement of lymphotoxin induced target cell killing, *J. Immunol.* **130:**518–520.

Wright, S. C., and Bonavida, B., 1982, Studies on the mechanism of natural killer (NK) cell mediated cytotoxicity (CMC). I. Release of cytotoxic factors specific for NK sensitive target cells (NKCF) during coculture of NK effector cells with NK target cells, *J. Immunol.* **129:**433–439.

Yamamoto, R. S., Hiserodt, J. C., Lewis, J. E., Carmack, C. E., and Granger, G. A., 1978, The human LT system. II. Immunological relationship of LT molecules released by mitogen activated human lymphocytes *in vitro*, *Cell. Immunol.* **38:**403–416.

T-Cell Growth Factor

Purification, Interaction with a Cellular Receptor, and *in Vitro* Synthesis

RICHARD J. ROBB and YUAN LIN

1. INTRODUCTION

T-cell growth factor (TCGF), also known as interleukin-2, is a small glycoprotein which provides a necessary signal for the proliferation of activated T cells (Morgan *et al.*, 1976; Gillis and Smith, 1977). The factor is released by T cells in response to antigen or lectin stimulation and requires the action of a second, macrophage-derived lymphokine, termed interleukin-1, for its production (Smith *et al.*, 1980). Although the bioactivity of TCGF is antigen-nonspecific and often crosses species lines, specificity of the immune response is maintained by the requirement that a responding T cell be specifically activated before it expresses receptors for the growth factor (Bonnard *et al.*, 1979; Larsson, 1981).

Although the role of TCGF in the T-cell immune response is known in general terms, many questions remain concerning its regulation and mechanism of action, as well as its involvement in non-T-cell responses. Such research has, in part, been hampered by the difficulty in purifying sufficient amounts of the factor to homogeneity. Moreover, the factor or defects in its production have been implicated in a number of pathological disorders (Altman *et al.*, 1981; Gootenberg *et al.*, 1981), making large quantities of purified factor valuable for clinical analysis. This chapter describes approaches to the large-scale production and purification of TCGF. In addition, recent experiments on the characterization of the cellular receptor and of the mRNA for the factor are discussed. These experiments provide a framework for examining TCGF activity and regulation in molecular terms.

RICHARD J. ROBB and YUAN LIN • Central Research and Development Department, E. I. du Pont de Nemours and Company, Glenolden Laboratory, Glenolden, Pennsylvania 19036.

TABLE I. Immunoaffinity Purification of JURKAT TCGF[a]

	Volume (ml)	Protein[b] (mg)	TCGF activity (U)	Recovery (%)	Specific activity (U/mg)	Fold purification
Cell supernatant	17,500	1600	2,200,000	(100)	1,380	(1)
Supernatant concentrate	1,200	—	2,160,000	98.2	—	—
Flowthrough and washes	1,400	1550	870,000	39.6	560	0.41
pH 2.5 eluate	14.0	4.1	1,260,000	57.3	307,000	220

[a] The cell supernatant was concentrated using an Amicon HIP5 hollow-fiber cartridge. The concentrated supernatant was chromatographed on a 4-ml column of Sepharose coupled with IgG_1 murine monoclonal antibody 46C8-A2 (8 mg antibody/ml).
[b] The quantity of protein was determined with the Lowry assay.

2. PURIFICATION AND BIOCHEMICAL CHARACTERIZATION OF TCGF

2.1. Immunoaffinity Purification

The human T-cell line JURKAT was chosen for production of TCGF. Sequential subcloning (by limiting dilution) of this cell line has allowed the selection of clones which are high producers of the growth factor and which retain good levels of its secretion over weeks and months in culture (Robb, 1982). Typically, such clones release 15–30 times more TCGF after stimulation with phytohemagglutinin (PHA) and phorbol myristate acetate (PMA) than a comparable number of normal peripheral blood lymphocytes.

TCGF in the supernatant of the stimulated JURKAT cells was enriched to > 98% purity by affinity column chromatography on Sepharose coupled with a murine monoclonal antibody (Table I) (Robb *et al.*, 1983). The antibody bound only about 60% of the TCGF made by these cells. This selectivity might be related to modification of the threonine residue in position 3 of the protein chain. Elution of the column at low pH resulted in recovery of 98% of the bound factor. Analysis by gel electrophoresis demonstrated that the final product migrated as a single spot on two-dimensional gels (isoelectric focusing/SDS–PAGE) and that the bioactivity comigrated with the protein (silver stain) on both SDS–PAGE and isoelectric focusing (Robb *et al.*, 1983).

2.2. Amino Acid Composition and Sequence Analysis

The amino acid composition of immunoaffinity-purified TCGF (Robb *et al.*, 1983) is given in Table II. The molecule is quite hydrophobic with an unusually high content of leucine (nearly 1 of 6 residues). The composition agrees well with that derived from the cDNA sequence published by Taniguchi *et al.* (1983). Most

TABLE II. Amino Acid Composition of JURKAT TCGF

Aspartic acid	12.0		
Threonine	11.8		
Serine	6.8		
Glutamic acid	18.2		
Proline	5.9		
Glycine	2.8		
Alanine	5.5		
Cysteine	2.7	Hydrophobic	42.3%
Valine	3.9	Acidic (amide)	23.1%
Methionine	2.8	Basic	13.8%
Isoleucine	8.2	Hydrophilic	20.8%
Leucine	22.7		
Tyrosine	2.9		
Phenylalanine	6.0		
Histidine	2.9		
Lysine	10.6		
Arginine	4.1		
Tryptophan	0.9		
	130.8		

of the minor discrepancies can be attributed to slight destruction of certain residues during hydrolysis (i.e., Thr, Ser, Met).

Sequence analysis (Robb *et al.*, 1983) of several independent preparations of the factor indicated a single N-terminal sequence (Fig. 1) which was later confirmed by cDNA data (Taniguchi *et al.*, 1983). Position 3 yielded an unidentifiable PTH-amino acid derivative. Compositional analysis of the N-terminal tryptic octapeptide together with sequence analysis of TCGF mRNA by the primer extension method indicated that this residue was a threonine. Posttranslational modification of this residue, which accounts for the aberrant characteristics of the PTH-derivative, appears to be involved in the recognition of the molecule by the monoclonal antibody used for purification (Table I). Further sequence data have been obtained using tryptic peptides fractionated by reverse-phase HPLC.

```
                  X                                         10
NH2 - ALA - PRO - THR - SER - SER - SER - THR - LYS - LYS - THR -

                                                            20
      GLN - LEU - GLN - LEU - GLU - HIS - LEU - LEU - LEU - ASP -

                                                            30
      LEU - GLN - MET - ILE - LEU - ASN - GLY - ILE - ASN - ASN -

      TYR - LYS - ASN - PRO - LYS - LEU -
```

FIGURE 1. Amino-terminal sequence of immunoaffinity-purified JURKAT TCGF. Position 3 has undergone posttranslational modification.

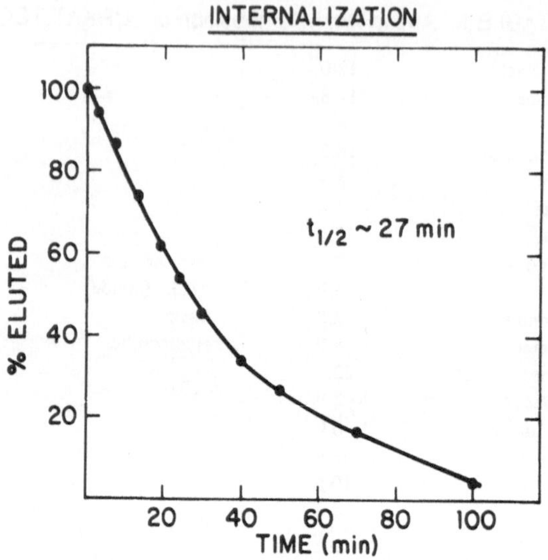

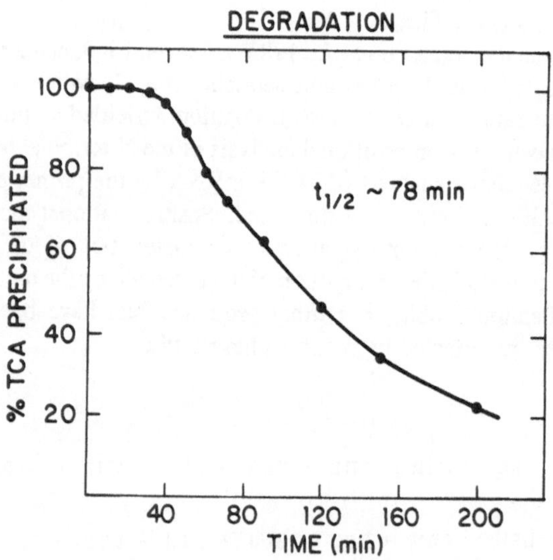

FIGURE 2. Time course of internalization and degradation of cell-bound, radiolabeled TCGF at 37°C. A murine, TCGF-dependent cell line was incubated with [^{3}H]-Leu, Lys TCGF at 0°C for 50 min followed by extensive washing. The cell suspension was then incubated at 37°C and aliquots were removed at the indicated times. Internalization was measured, in the presence of chloroquine to prevent degradation, by treating the cells for 30 sec at pH 4, a process which quantitatively removes intact, surface-bound factor. Degradation was estimated, in the absence of chloroquine, by treating the cell suspension at a 10% concentration of trichloroacetic acid and separating precipitable and nonprecipitable material.

3. ANALYSIS OF THE CELLULAR RECEPTOR FOR TCGF

3.1. Internalization and Degradation of TCGF

Binding studies on activated murine and human T cells using radiolabeled TCGF demonstrated the presence of a high-affinity receptor for the factor on TCGF-responsive cells (Robb *et al.*, 1981). As with other polypeptide factors (Olefsky and Kao, 1982), the cell-bound TCGF could be eluted intact at low pH if internalization of the receptor–factor complex was prevented (as at 4°C). Using this technique, it was possible to show that cell-bound factor was rapidly internalized at 37°C (Fig. 2). After internalization, the factor underwent degradation (Fig. 2) in a process which was sensitive to chloroquine, a lysomotropic agent.

3.2. Evidence for a Physiological Role

The high-affinity binding of labeled TCGF appeared to have physiological significance since the level of proliferation induced by a given concentration of factor correlated roughly with the proportion of cellular receptors occupied at that concentration (Robb *et al.*, 1981). As an additional demonstration of the role of these receptors, the concentration of an anti-TCGF monoclonal antibody (9B11-1E5) needed to block receptor binding and bioactivity was measured. As shown in Fig. 3, 9B11-1E5 was capable of blocking, in a dose-dependent fashion, both the proliferative response of a TCGF-dependent cell line and the direct binding of radiolabeled factor by the same cells. Control IgG at the same concentrations had no effect on either measurement. As the level of antibody necessary to block high-affinity binding correlated well with that required to block proliferation, the binding

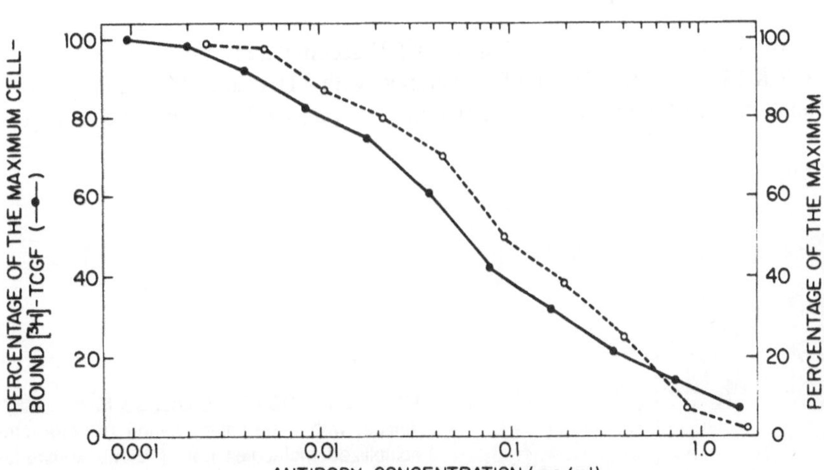

FIGURE 3. Inhibition by anti-TCGF monoclonal antibody 9B11-1E5 of the binding of [^{3}H]-Leu,Lys TCGF to a TCGF-dependent murine cell line (●) and of the proliferation ([^{3}H]thymidine) of the same cell line (○).

of the factor by these receptors appeared to be directly involved in the physiological response.

3.3. Comparison of Cellular Proteins Reactive with TCGF and Anti-Tac Antibody

Recently, Leonard *et al.* (1982) demonstrated that an antibody, termed anti-Tac, blocked the binding of labeled TCGF to human T cells. The authors suggested that the antibody recognized the TCGF receptor. As a direct demonstration that the Tac antigen is capable of binding TCGF, radiolabeled cellular proteins were fractionated on affinity supports coupled with either anti-Tac antibody or TCGF (Robb and Green, 1983). Both supports bound a single major component of 57,000–59,000 daltons from detergent-solubilized PHA-blast cells (Fig. 4).

In order to prove that the cellular components isolated in each instance were identical, sequential incubations were performed with the two affinity supports. Neither TCGF nor anti-Tac-coupled beads were able to bind appreciable radiolabeled molecules after two successive incubations with the alternative support (Table III). This result indicates that the Tac antigen, but not other surface molecules, contains a TCGF-binding site and that all Tac molecules appear to be capable of binding TCGF. Further study is necessary to determine how the binding to TCGF-coupled affinity supports relates to the high-affinity cellular binding detected at very low TCGF concentrations (Robb *et al.*, 1981).

4. CHARACTERIZATION OF THE mRNA FOR TCGF

4.1. Time Course of TCGF mRNA Induction

Although a high concentration of TCGF accumulated in the culture medium of JURKAT cells after 20 hr of stimulation with PHA and PMA, only a small amount of mRNA with TCGF activity could be recovered from these cells. Figure

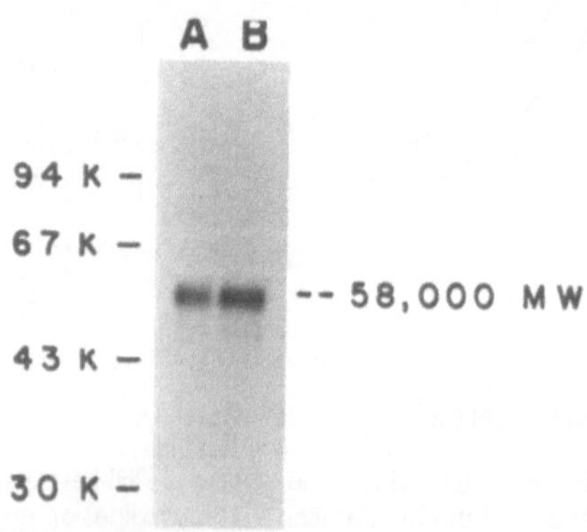

FIGURE 4. SDS–PAGE analysis (8.75% acrylamide, with 2-mercaptoethanol) of detergent-solubilized molecules from [^{35}S]methionine-labeled human PHA-blast cells which bound to (A) TCGF-coupled Affigel or (B) anti-Tac-coupled Sepharose.

TABLE III. Sequential Binding of [^{3}H]Glucosamine-Labeled Molecules from Human PHA-Blast Cells to TCGF-Coupled Affigel and Anti-Tac-Coupled Sepharose[a]

	Incubation No. 1		Incubation No. 2		Incubation No. 3	
Sample	TCGF–Affigel	Anti-Tac–Sepharose	TCGF–Affigel	Anti-Tac–Sepharose	TCGF–Affigel	Anti-Tac–Sepharose
1	13,870	—	3410	—	—	950
2	—	22,340	—	840	300	—

[a] Two identical samples of NP-40-solubilized molecules from [^{3}H]glucosamine-labeled blast cells were subjected to three sequential incubations with TCGF–Affigel (10 μl beads, 1 mg TCGF/ml) and/or anti-Tac–Sepharose (5 μl beads, 1 mg anti-Tac/ml). The results are given as dpm of ^{3}H radiolabel in the pellet. Control IgG-coupled Sepharose or Affigel bound less than 2% of the above values. All bound radiolabel migrated at 58,000 daltons on SDS–PAGE.

5 illustrates the TCGF mRNA activity from JURKAT cells after 5, 8, and 20 hr of stimulation. mRNA was isolated from the cytoplasmic fraction of the JURKAT cells by phenol–chloroform extraction and oligo-d(T) affinity column chromatography (Berger and Birkenmeier, 1979). The activity of the TCGF mRNA was determined by injecting the total mRNA into *Xenopus* oocytes (Gurdon *et al.*, 1971) which were able to synthesize and secrete biologically active TCGF (Lin *et al.*, 1981). After 8 hr of stimulation, high levels of TCGF mRNA activity were recovered from JURKAT cells.

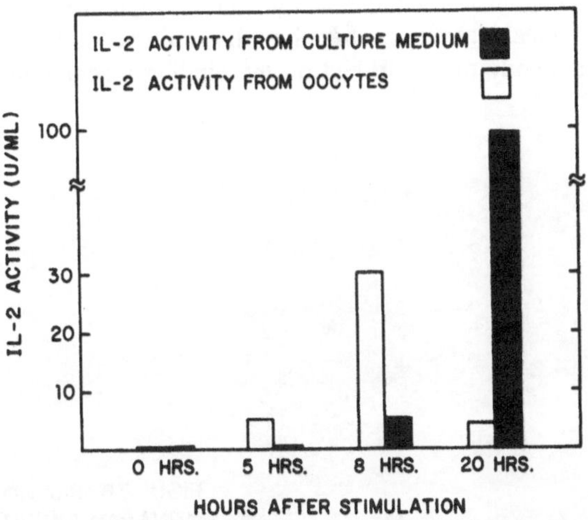

FIGURE 5. TCGF activity from oocytes injected with JURKAT mRNA after the JURKAT cells were stimulated with PHA and PMA for 0, 5, 8, and 20 hr. Ten oocytes were used for each time point with each oocyte receiving ~ 50 μl of 1 mg/ml mRNA. For comparison, TCGF activity from the JURKAT cell supernatants was assayed. TCGF bioactivity was determined by [^{3}H]thymidine incorporation in a cloned, TCGF-dependent murine cell line.

4.2. Translation of JURKAT mRNA in Oocyte and Wheat Germ Systems

When mRNA from JURKAT cells was used to direct protein synthesis in oocytes and wheat germ lysate in the presence of [^{35}S]methionine, a prominent band appeared at 16,000 daltons on SDS–PAGE when the mRNA was derived from induced cells, but not when it was derived from uninduced cells (Fig. 6). This band, which represented > 1% of the total proteins synthesized by the total mRNA, was shown to comigrate with TCGF biological activity after elution of proteins from the sliced gel (Fig. 7). Oocytes were shown to be able to process (i.e., glycosylate and remove the signal sequence) the newly synthesized proteins. The TCGF synthesized by oocytes was biologically active in stimulating the growth of T cells and the activity was neutralized by monoclonal antibody 9B11-1E5 (see Fig. 3). In the wheat germ system, however, if the 16,000-dalton protein is a form of TCGF, it does not have significant bioactivity.

5. CONCLUSIONS

1. TCGF was prepared in milligram quantities and purified to near homogeneity in a single step using an immunoaffinity column.

2. The molecule has a relatively high content of hydrophobic amino acids. Sequence analysis so far reveals no evidence of microheterogeneity or discrepancy with the cDNA data of Taniguchi *et al.* (1983).

3. After binding to a high-affinity cellular receptor, the factor was rapidly internalized and degraded. Such interaction with the high-affinity receptor appeared to be of physiological significance based on the similar dose-dependent effects that a monoclonal antibody to TCGF had on the binding reaction and cellular proliferation.

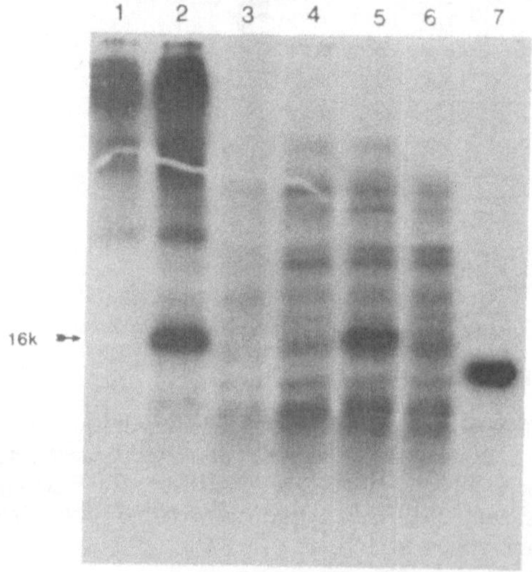

FIGURE 6. Proteins synthesized using mRNA from JURKAT cells. Lanes 1 and 2, from mRNA-injected oocytes: 1, no stimulation; 2, 8 hr of stimulation. Lanes 3–7, from wheat germ lysate: 3, no stimulation; 4, 5 hr of stimulation; 5, 8 hr of stimulation; 6, 20 hr of stimulation; 7, globin mRNA.

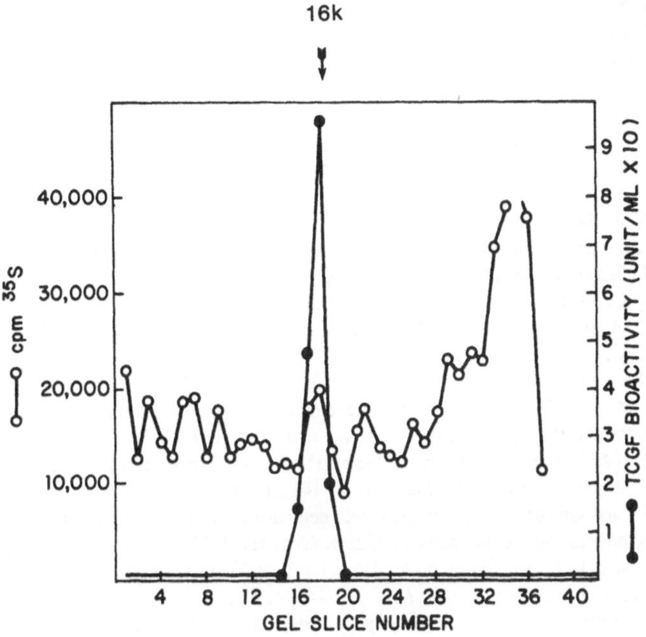

FIGURE 7. Radioactivity and TCGF bioactivity from gel slices. Oocytes were injected with mRNA from JURKAT cells after 8 hr of stimulation and incubated in Barth medium in the presence of [^{35}S]methionine. The secreted proteins were analyzed by SDS–PAGE (15% acrylamide). Gel slices (2 mm) were analyzed for radioactivity and TCGF bioactivity.

4. Direct comparison of the cellular proteins reacting with TCGF and anti-Tac-coupled affinity supports indicated that the Tac antigen is the cellular binding site for the growth factor.

5. A subclone of JURKAT cells, when induced by PHA and PMA for 8 hr, produced high levels of TCGF-specific mRNA which was used to direct the synthesis of TCGF in both oocyte and wheat germ systems. The TCGF produced by oocytes was biologically active while that produced by wheat germ appeared inactive.

REFERENCES

Altman, A., Theofilopoulos, A. N., Weiner, R., Katz, D. H., and Dixon, F. J., 1981, Analysis of T cell function in autoimmune murine strains: Defects in production of and responsiveness to interleukin 2, *J. Exp. Med.* **154:**791–808.

Berger, S. L., and Birkenmeier, C. S., 1979, Inhibition of intractable nucleases with ribonucleoside–vanadyl complexes: Isolation of mRNA from resting lymphocytes, *Biochemistry* **18:**5143–5149.

Bonnard, G. D., Yasaka, D., and Jacobson, D., 1979, Ligand-activated T-cell growth factor-induced proliferation: Absorption of T-cell growth factor by activated T-cells, *J. Immunol.* **123:**2704–2708.

Gillis, S., and Smith, K. A., 1977, Long term culture of tumor-specific cytotoxic T-cells, *Nature (London)* **268:**154–156.

Gootenberg, J. E., Ruscetti, F. W., Mier, J. W., Gazdar, A., and Gallo, R. C., 1981, Human cutaneous T cell lymphoma and leukemia cell lines produce and respond to T cell growth factor, *J. Exp. Med.* **154:**1403–1417.

Gurdon, J. B., Lane, C. D., Woodland, H. R., and Marbaix, G., 1971, Use of frog eggs and occytes for the study of mRNA and its translation in living cells, *Nature (London)* **233:**177–182.

Larsson, E.-L., 1981, Mechanism of T-cell activation. II. Antigen and lectin-dependent acquisition of responsiveness to TCGF is a nonmitogenic, active response of resting T-cells, *J. Immunol.* **126:**1323–1328.

Leonard, W. J., Depper, J. M., Uchiyama, T., Smith, K. A., Waldmann, T. A., and Greene, W. C., 1982, A monoclonal antibody that appears to recognize the receptor for human T-cell growth factor: Partial characterization of the receptor, *Nature (London)* **300:**267–269.

Lin, Y., Stadler, B. M., and Rabin, H., 1982, Synthesis of biologically active interleukin 2 by *Xenopus* oocytes in response to poly (A)-RNA from a gibbon T-cell line, *J. Biol. Chem.* **257:**1587–1590.

Morgan, D. A., Ruscetti, F. W., and Gallo, R., 1976, Selective in vitro growth of T-lymphocytes from normal human bone marrows, *Science* **193:**1007–1008.

Olefsky, J. M., and Kao, M., 1982, Surface binding and rates of internalization of ^{125}I-insulin on adipocytes and IM-9 lymphocytes, *J. Biol. Chem.* **257:**8667–8673.

Robb, R. J., 1982, Human T-cell growth factor: Purification, biochemical characterization, and interaction with a cellular receptor, *Immunobiol.* **161:**21–50.

Robb, R. J., and Greene, W. C., 1983, Direct demonstration of the identity of T-cell growth factor binding protein and the Tac antigen, *J. Exp. Med.* **158:**1332–1337.

Robb, R. J., Munck, A., and Smith, K. A., 1981, T-cell growth factor receptors: Quantitation, specificity and biological relevance, *J. Exp. Med.* **154:**1455–1474.

Robb, R. J., Kutny, R. M., and Chowdhry, V., 1983, Purification and partial sequence analysis of human T-cell growth factor, *Proc. Natl. Acad. Sci.* **80:**5990–5994.

Smith, K. A., Lachman, L. B., Oppenheim, J. J., and Favata, M. F., 1980, The functional relationship of the interleukins, *J. Exp. Med.* **151:**1551–1553.

Taniguchi, T., Matsui, H., Fujita, T., Takaoka, C., Kashima, N., Yashimoto, R., and Hamuro, J., 1983, Structure and expression of a cloned cDNA for human interleukin-2, *Nature (London)* **302:**305–310.

Demonstration and Characterization of Human Intracellular Interleukin-1

JOSE L. LEPE-ZUNIGA, J. SAMUEL ZIGLER, Jr., MARY LOUISE ZIMMERMAN, and IGAL GERY

1. INTRODUCTION

Since the initial report by Unanue and Kiely (1977), others have also shown that in addition to the interleukin-1 (IL-1) released into the medium (extracellular, EC), mononuclear phagocytes (MP) and related cell lines also contain high levels of intracellular (IC) IL-1 activity which can be measured following lysis of the cells (Mizel and Rosenstreich, 1979; Gery *et al.*, 1981a; Iribe *et al.*, 1982).

In mouse macrophages, the IC IL-1 levels can be markedly elevated by stimulation with certain agents such as lipopolysaccharide (LPS) (Gery *et al.*, 1981a). Little is known about the nature of this IC fraction and its relationship to the EC one. In the mouse, the IC IL-1 is not present in the quiescent MP, and in nonstimulated cultures it appears and disappears from the cells without being released into the medium, suggesting that the IC IL-1 may be different from the EC (Unanue and Kiely, 1977). However, the molecular weight profiles, obtained in that study, were remarkably similar for the two fractions with only a single peak of activity at 12K. Conversely, Mizel and Rosenstreich (1979) in the P388D1 cell line and Iribe *et al.* (1982) in guinea pig macrophages have found molecular weight heterogeneity of the IC fraction.

In human monocytes, preliminary results from this laboratory (Gery *et al.*,

JOSE L. LEPE-ZUNIGA, J. SAMUEL ZIGLER, Jr., MARY LOUISE ZIMMERMAN, and IGAL GERY • Laboratory of Vision Research, National Eye Institute, National Institutes of Health, Bethesda, Maryland 20205.

1981b) have also shown the existence of an IC IL-1 fraction in LPS-stimulated monocytes. As the identification and characterization of a potential IL-1 precursor in human cells could eventually lead to elucidation of a regulatory step in the production of this monokine, we have analyzed the production of the IC and EC IL-1 upon stimulation by several agents and compared their molecular weight profiles by gel exclusion chromatography. The results of these studies constitute the present report.

2. MATERIALS AND METHODS

2.1. Monocyte Cultures

Human mononuclear cells were obtained by differential centrifugation of peripheral blood from healthy adult volunteers on Isolymph gradients (Gallard–Schlesinger Chemical, Carle Place, N.Y.). After washing twice with Hanks' balanced salt solution (HBSS) and once with RPMI 1640 culture medium (GIBCO) containing 25 mM HEPES, penicillin (100 U/ml), streptomycin (100 μg/ml), and 2 mM glutamine (CRPMI), the cells were allowed to adhere to plastic wells (24 × 17 mm, Linbro, Hamden, Conn.) at a concentration of 5×10^6 cells/ml in CRPMI plus 10% autologous serum (HS). After 1 hr, the nonadherent cells were removed by washing the wells with HBSS. The remaining adherent cells (> 95% monocytes) were further incubated with 1-ml aliquots of CRPMI–5% HS either with or without any of the following agents: (1) quartz silica particles of less than 5 μm in diameter (50 μg/ml), (2) LPS W from *Salmonella typhimurium* (LPS, 20 μg/ml, Difco, Detroit, Mich.), (3) zymosan A from *Saccharomyces cerevisiae* (20 μg/ml, Sigma, St. Louis, Mo.), and (4) phorbol myristate acetate (PMA, 20 ng/ml, Sigma). In every case the supernatants were collected after 20 hr of culture, centrifuged at 900*g* for 10 min, and kept frozen at −20°C. The remaining monolayers were covered with 1 ml of fresh CRPMI–5% HS and frozen at −20°C. After thawing, the lysates were prepared by scraping the cells from the wells with a rubber policeman. The suspension was sonicated and then frozen again at −20°C until being tested for IL-1 activity.

2.2. Biochemical Analysis

For the purpose of biochemical analysis, LPS (20 μg/ml)-stimulated human monocyte cultures were used as a routine source of crude IL-1. Supernatants from 9 to 12 of such cultures were concentrated using Amicon PM-10 ultrafiltration membranes (Amicon Corp., Danvers, Mass.) and chromatographed at 4°C on a column (1.6 × 90 cm) of Sephadex G-75 superfine equilibrated with CRPMI. Fractions of 2 ml were collected and 0.1 ml of fetal calf serum (FCS) was added to each one. Fractions 21 through 60 were filter sterilized and their IL-1 activity determined.

2.3. IL-1 Assay

The levels of IL-1 activity in the samples were determined by their capacity to potentiate the mitogenic response of murine thymocytes to phytohemagglutinin (PHA), as described in detail elsewhere (Gery *et al.*, 1981a,b). The results are presented as units determined by comparing the activity in the tested samples with that of a standard preparation using a method similar to that of Mizel (1980).

3. RESULTS

3.1. Spontaneous Production of IL-1

Freshly separated human mononuclear cells do not contain measurable amounts of IL-1 activity. After 20 hr of culture with no added stimulus, a substantial amount of IL-1 is produced "spontaneously." Most of it can be detected in the monocyte lysates and only about 10% is detected in the supernatants (Fig. 1).

3.2. Effect of the Addition of Various Agents

The addition of any of the agents tested increased from 3 to 50 times the amount of IL-1 spontaneously produced by the same cultures. However, marked qualitative differences were observed among them. Silica increased both the IC and the EC IL-1 activity but was clearly more efficient at increasing the EC IL-1. PMA specifically increased the EC fraction, producing concomitantly an actual decrease in the IC pool when compared to nonstimulated cultures. LPS increased more the IC than the EC and zymosan markedly increased both fractions (Fig. 1).

3.3. Biochemical Characterization of EC and IC IL-1

3.3.1. Molecular Weight Profiles by Sephadex G-75 Chromatography

In agreement with several published reports (for review see Farrar and Koopman, 1979), the IL-1 activity in the supernatants of LPS-stimulated human monocyte

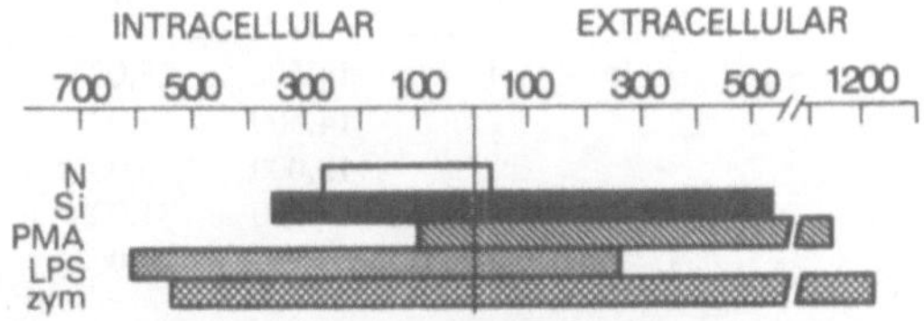

FIGURE 1. Levels of IL-1 activity in lysates ("intracellular") and supernatants ("extracellular") of monocyte cultures incubated for 20 hr with no additive (N), silica particles (Si, 50 μg/ml), PMA (20 ng/ml), LPS (20 μg/ml), or zymosan (zym, 20 μg/ml). The IL-1 activity is expressed as units/ml, with the unit values being determined as described in Materials and Methods (2.3).

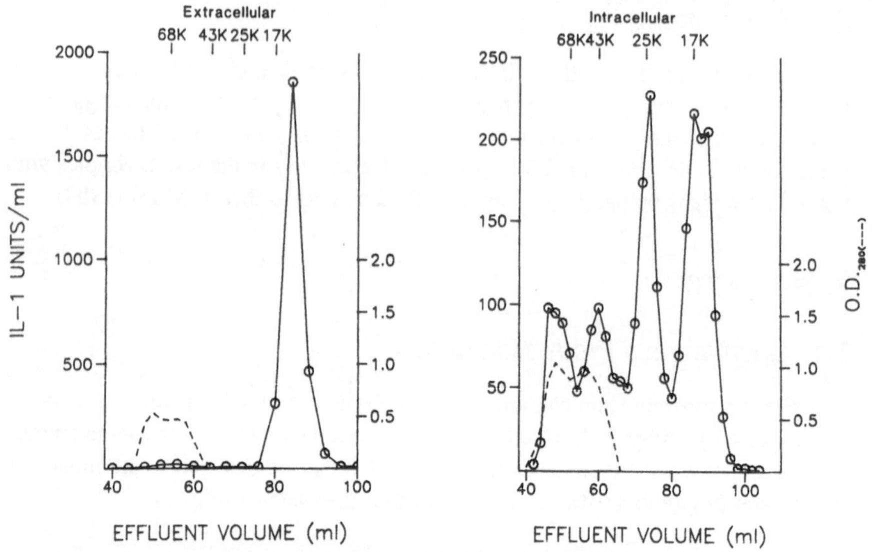

FIGURE 2. Chromatographic profiles of extracellular and intracellular IL-1 activities of cultured monocytes from a typical donor, incubated for 20 hr with LPS (20 μg/ml). Fractionation was carried out on a Sephadex G-75 superfine column and the molecular weight markers were: bovine serum albumin (68K), ovalbumin (43K), chymotrypsinogen (25K), and myoglobin (17K).

cultures was found mostly confined to a single peak in the 14–16K range (Fig. 2). In contrast, the IC IL-1 activity was consistently found distributed in four well-defined peaks (Fig. 2). The average molecular weights of these peaks in five different experiments were 13.2, 26.2, 45.7, and > 70K (exclusion limit of the gel) (Table I). The relative amount of IL-1 activity in each peak varied among different experiments.

TABLE I. Molecular Size of Different Peaks of Human Intracellular IL-1 Activity Obtained by Sephadex G-75 Chromatography in Five Different Experiments[a]

Expt. No.	Peak I	Peak II	Peak III	Peak IV
1	14,500	25,000	42,500	>70,000
2	14,500	23,000	47,000	>70,000
3	12,000	26,000	42,000	>70,000
4	15,500	31,000	50,500	>70,000
5	14,700	26,000	46,500	>70,000
Average	14,250	26,200	45,700	>70,000

[a] The cell lysates were obtained from LPS (20 μg/ml)-stimulated human monocytes.

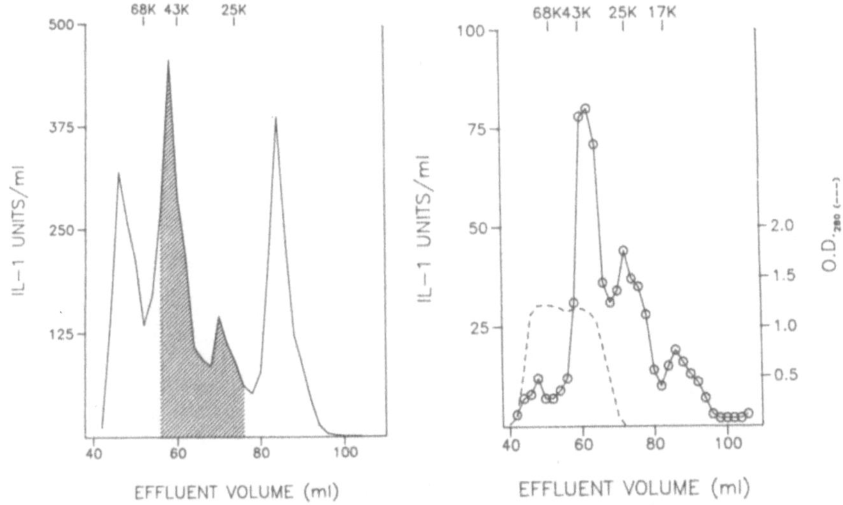

FIGURE 3. Rechromatography of the intermediate size peaks of intracellular IL-1. Fractions comprising the 26 and 45K peaks of activity from the Sephadex G-75 superfine chromatography (left panel, shaded area) were collected, incubated for 48 hr at 4°C, concentrated, and rechromatographed on the same column (right panel).

3.3.2. Rechromatography of the Intermediate Size Peaks

The intermediate size peaks (26 and 45K) were pooled and incubated for 48 hr at 4°C in an attempt to produce the 14K species. After concentration, the sample was rechromatographed on the same G-75 superfine column (Fig. 3). As can be seen, this simple procedure generated a small but significant amount of the 14K species.

3.3.3. Generation of the 14K Species by Treatment with a Detergent

As the amount of the smallest species generated by simple rechromatography was small, we tested whether the generation of the 14K peak could be improved by treating a pool of the three larger peaks with a deaggregating agent. Urea from 1 to 8 M was first used without success, for at concentrations above 2 M, this agent inactivated more than 80% of the human IL-1 (data not shown). This behavior differs from that of IL-1 obtained from the P388D1 cell line (Mizel, 1979; Mizel and Rosenstreich, 1979). We have therefore used a recently developed zwitterionic detergent, 3-[(3-cholamidopropyl)dimethylammonio]-1-propane-sulfonate · $2H_2O$ (CHAPS, Calbiochem, La Jolla, Calif.), which is gentle toward proteins and can be readily removed by dialysis (Hjelmeland, 1980; Bitonti *et al.*, 1982). At concentrations up to 8 mM, less than 30% of the IL-1 activity is lost by CHAPS treatment. For these experiments, a pool of the three larger peaks was concentrated 10 times, dialyzed against 8 mM CHAPS at 4°C for 24 hr, and rechromatographed

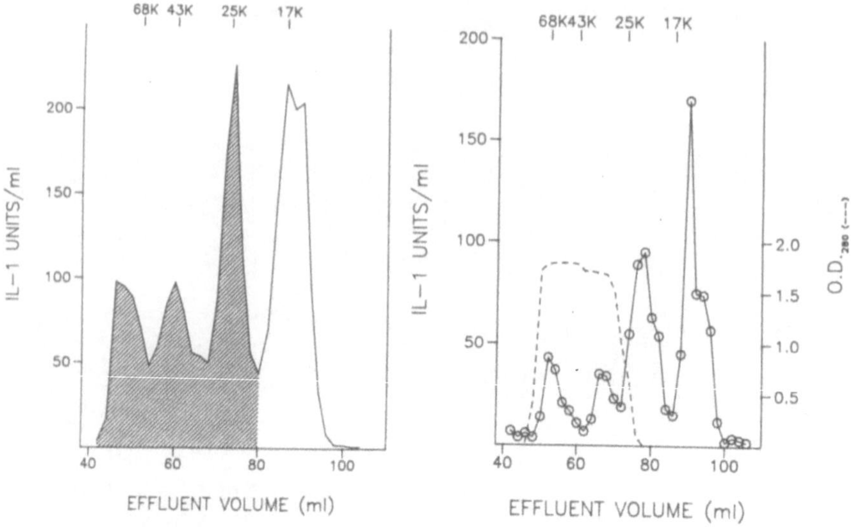

FIGURE 4. Generation of the small (14K) species IL-1 activity by treatment with CHAPS. Fractions from the three heavier molecular weight species of intracellular IL-1, obtained by chromatography on Sephadex G-75 superfine (left panel, shaded areas), were pooled and treated with CHAPS as described in Results (3.3.3). The chromatographic pattern of the treated preparation is shown in the right panel.

on Sephadex G-75 superfine equilibrated with the detergent at the same concentration. Two-milliliter fractions were collected and individually dialyzed against 50 volumes of PBS for 96 hr, changing the dialysis fluid twice daily. The fractions were then filter sterilized and tested for IL-1 activity. The results of one of these experiments are shown in Fig. 4. This procedure leads to the generation of large amounts of the 14K peak from the heavier species.

4. DISCUSSION

The present studies clearly demonstrate the existence of an IC pool of IL-1 in human monocytes. This pool is not normally present in the circulating cells and it appears "spontaneously" during the first 20 hr of culture. This spontaneous production is probably induced by the cell stimulation that usual culturing maneuvers cause to the monocytes. As in the mouse system, this type of IC IL-1 formation is not followed by the appearance of the factor in the medium, suggesting that secretion of IL-1 requires another stimulus or that the IC IL-1 is different from the EC IL-1 and is not released in normal culture conditions. In any case, the results strongly suggest the existence of two separate events: one that gives rise to the IC IL-1 and one that determines the appearance of EC IL-1.

To gain more information about the possible relationship of the IC and EC IL-1 fractions, they were analyzed by Sephadex chromatography. The profiles obtained were quite different. The EC IL-1 gave essentially a single peak of activity while the IC fraction contained four well-defined active species. This finding could be interpreted as evidence that the two fractions are essentially different. However, an alternative explanation is also possible. As the smallest IC species corresponded closely in size to the EC IL-1, these two species might be very similar; the additional IC species could then result from aggregation of the smallest species under the conditions present within the cell. This latter explanation was supported by the close correlation between the observed size of the higher molecular weight species and the expected size of "polymers" or multiples of the 14K species. It is further sustained by the spontaneous formation of the 14K species from the larger species in solution and by the clear generation of the 14K species following treatment of the heavier species with a deaggregating agent (CHAPS). Clearly the formation of 14K species in these experiments did not require cleavage of covalent bonds.

While these data are consistent with the idea that the two 14K species are similar, it remains to be explained why the IC species forms aggregates and the EC does not. This difference in behavior could result from conditions within the cell that promote such aggregation or from subtle structural differences between the two species themselves. *In vitro* we have been unable to induce aggregation of either species which may suggest that the IC environment is of primary importance in determining the formation of the larger molecular weight species. On the other hand, as it appears that the synthesis and release of IL-1 are directed by two separate mechanisms, it is conceivable that a posttranslational modification (e.g., glycosylation) precedes release. Such an event might alter the molecule in such a way that the tendency to aggregate is diminished. Further studies are necessary to determine whether such a structural difference is present.

REFERENCES

Bitonti, A. J., Moss, J., Hjelmeland, L., and Vaughan, M., 1982, Resolution and activity of adenylate cyclase components in a zwitterionic cholate derivative (3-((3-cholamidopropyl) dimethylammonio)-1-propanesulfonate), *Biochemistry* **21**:3650.

Farrar, J. J., and Koopman, W. J., 1979, Characterization of mitogenic factors and their effect on the antibody response in vitro, in: *Biology of Lymphokines* (S. Cohen, E. Pick, and J. J. Oppenheim, eds.), pp. 325–346, Academic Press, New York.

Gery, I., Davies, P., Derr, J., Krett, N., and Barranger, J. A., 1981a, Relationship between production and release of lymphocyte activating factor (interleukin-1) by murine macrophages, *Cell. Immunol* **64**:293.

Gery, I., Seminara, D., Derr, J., and Barranger, J. A., 1981b, Production and release of lymphocyte activating factor (interleukin-1) by human monocytes and their derived macrophages, in: *Mechanism of Lymphocyte Activation* (K. Resch and H. Kirchner, eds.), pp. 541–543, Elsevier/North-Holland, Amsterdam.

Hjelmeland, L. M., 1980, A non denaturing zwitterionic detergent for membrane biochemistry: Design and synthesis, *Proc. Natl. Acad. Sci. USA* **77**:6368.

Iribe, H., Koga, T., and Onoue, K., 1982, Production of T cell-activating monokine of guinea pig macrophages induced by MDP and partial characterization of the monokine, *J. Immunol.* **129**:1029.

Mizel, S. B., 1979, Physicochemical characterization of lymphocyte-activating factor (LAF), *J. Immunol.* **122**:2167.

Mizel, S. B., 1980, Studies on the purification and structure–function relationships of murine lymphocyte activating factor (interleukin-1), *Mol. Immunol.* **17**:571.

Mizel, S. B., and Rosenstreich, D. L., 1979, Regulation of lymphocyte activating factor (LAF) production and secretion in P388D1 cells: Identification of high molecular weight precursors of LAF, *J. Immunol.* **122**:2173.

Unanue, E. R., and Kiely, J.-M., 1977, Synthesis and secretion of a mitogenic protein by macrophages: Description of a superinduction phenomenon, *J. Immunol.* **119**:925.

24

Silicic Acid Batch Adsorption Procedure for the Partial Purification of Interleukin-2 from Cultures of Primary Lymphocytes and Long-Term T-Cell Lines

JOHN L. PAULY, CAROL J. TWIST, GERALDINE M. OVAK, CYNTHIA W. RUSSELL, and ANGELO CONSTANTINO

1. INTRODUCTION

Silicic acid (SA; approximately H_2SiO_3 or as hydrated silica, $SiO \cdot H_2O$) has been used by many investigators as a solid matrix for adsorption chromatography. This subject has been reviewed by Huaser (1955) and Unger (1979). Particularly noteworthy is the utilization of SA for the separation of natural lipid mixtures and highly polar complexes (Hirsch and Ahrens, 1958; Mangold and Malins, 1960). SA possesses a surface that can have widely varying properties because of the random distribution of different groups, including: (1) silanol groups (-Si-OH), (2) siloxan groups (-Si-O-Si), and (3) hydrated silanol groups (-Si-OH$\cdot H_2O$) (Smith, 1969). As with silica gel (SiO_2), the adsorptive activity of SA is due primarily to surface silanol groups, and these sites are hydrogen-binding, weakly acidic, and therefore polar (Freeman, 1982). A reaction of silica with chlorosilanes results in the formation of siloxane-type phase matrices that are widely used in reverse-phase high-performance liquid chromatography (rpHPLC) for protein isolation and analysis (Voelter, 1981).

JOHN L. PAULY, CAROL J. TWIST, GERALDINE M. OVAK, CYNTHIA W. RUSSELL, and ANGELO CONSTANTINO • Department of Molecular Immunology, Roswell Park Memorial Institute, Buffalo, New York 14263.

A number of different biological activities have been ascribed to the hormonelike lymphokine IL-2; however, its most salient feature is its ability to sustain the long-term proliferation of monoclones of immunocompetent human and murine T cells (Gillis, 1983; Ruscetti and Gallo, 1981). Results of attempts to purify IL-2 have been reported previously (Gillis, 1983; Mier and Gallo, 1982). In general, these procedures have been multiple-step sequences that included: (1) ammonium sulfate precipitation, (2) gel filtration, (3) anion-exchange chromatography, (4) isoelectric focusing, and (5) preparative gel electrophoresis. These purification schemes are tedious, and IL-2 recovery has been poor; the yield reported by one group (Mier and Gallo, 1982) following a purification of IL-2 to near homogeneity was less than 1%.

In a preliminary report (Pauly *et al.*, 1983), we have described the utility of SA for partially purifying IL-2 from supernatants of human lymphocytes activated with polyclonal stimulants. We report here the results of studies demonstrating the application of this preparative procedure for isolating IL-2 from supernatants of long-term T-cell lines derived from man, gibbon ape, and mouse.

2. MATERIALS AND METHODS

2.1. IL-2 Production

IL-2 from human peripheral blood lymphocytes (PBL) was produced using a procedure described previously (Pauly *et al.*, 1982a). In this two-phase culture method, PBL pooled from the blood of five to seven healthy donors were seeded ($\sim 1.25 \times 10^6$ cells/ml) in Phase I spinner cultures containing RPMI 1640 medium supplemented with fetal bovine serum (FBS, 5%) for 4 days prior to polyclonal activation. Following activation for 24 hr with phytohemagglutinin (PHA-M, 1% v/v), phorbol ester (PMA, 10 ng/ml), and nonirradiated B cells of Epstein–Barr virus-transformed human lymphoblastoid cells of established lines (PBL : B-cell ratio = 1 : 1). The cells were pelleted, resuspended in serum-free RPMI 1640, and cultivated as Phase II spinners for 18–24 hr. Thereafter, the serum-free supernatant was harvested, filtered (0.45 μm), and stored at −20°C.

IL-2 was also produced from different long-term T-cell lines derived from humans or animals with leukemia or lymphoma. Lines included in these studies, species of origin, investigators from whom they were obtained, and procedures used for producing IL-2 were as follows: (1) JURKAT-FHCRC, human, S. Gillis (Gillis and Watson, 1980); (2) MLA-144, gibbon ape, R. Neubauer (Rabin *et al.*, 1981); (3) EL-4, mouse, J. Farrar (Farrar *et al.*, 1980; and (4) LBRM-33-5A4, mouse, S. Gillis (Gillis *et al.*, 1980).

2.2. IL-2 Isolation

Details for the partial purification of IL-2 have been reported elsewhere (Pauly *et al.*, 1983). In brief, SA (Fisher Scientific Co., Pittsburgh, Penn.; 325 mesh) was washed twice before using with phosphate-buffered saline (PBS) to remove fines.

Almost (> 95%) complete adsorption of IL-2 has been achieved with SA concentrations as low as 3.75 mg/ml supernatant; however, a concentration of 15 mg SA/ml supernatant (~ 5% SA v/v) has been selected for routine use. The IL-2-containing supernatant was clarified by centrifugation (400*g*, 20 min, 20°C), added to the SA to form a slurry (usually 500–700 ml), and mixed in a Boston polyethylene bottle using a Teflon-coated magnetic stirring bar (30 min, 37°C). Thereafter, the SA was pelleted, washed twice with 0.9% NaCl, and desorbed by spin-mixing the SA particles with 5 volumes of 50% (v/v) ethylene glycol (EG) in phosphate-buffered (pH 7.2) high-salt (1.4 M NaCl) solution (designated EG/PBS; 30 min, 20°C). The EG/PBS eluates were collected and dialyzed (6.0Kd cutoff) against PBS (25 vol, five times) at 4°C to remove the EG. Following dialysis, some eluates were concentrated further by dialysis against polyethylene glycol, or dialyzed against ammonium bicarbonate and lyophilized. All samples were stored at −20°C until testing for protein content and biological activity.

2.3. IL-2 Bioassay

IL-2 activity was measured using the IL-2-dependent murine T-cell line (CTLL-2 (kindly provided by Dr. S. Gillis) and our modification (Pauly *et al.*, 1982b) of procedures described previously (Gillis *et al.*, 1978). A unit of IL-2 activity, as defined elsewhere (Popescu and Maneekarn, 1982), was calculated by the ratio of the end-point growth of the test supernatant at an optimal dose to that of a laboratory standard supernatant that had previously been assigned a value of 100 IL-2 U/ml (see Fig. 1).

2.4. SDS–PAGE

Sample aliquots incubated with 1% SDS were electrophoresed on 12% polyacrylamide gels (0.58% cross-linker) according to the method of Laemmli (1970). Molecular weight standards were phosphorylase B (92.5K), bovine serum albumin (66.2K), ovalbumin (45.0K), carbonic anhydrase (31.0K), soybean trypsin inhibitor (21.5K), and lysozyme (14.4K). Protein bands were detected by staining with Coomassie brillant blue R-250.

2.5. Protein Determination

Protein concentrations of the crude supernatants and eluates were determined by the fluorometric method of Böhlen *et al.* (1973) using bovine serum albumin as the standard.

3. RESULTS AND DISCUSSION

Presented in Fig. 1 are the results of a representative experiment defining IL-2 activity of a primary human lymphocyte culture supernatant before and after isolating IL-2 using the SA adsorption procedure. Maximal numbers of CTLL-2

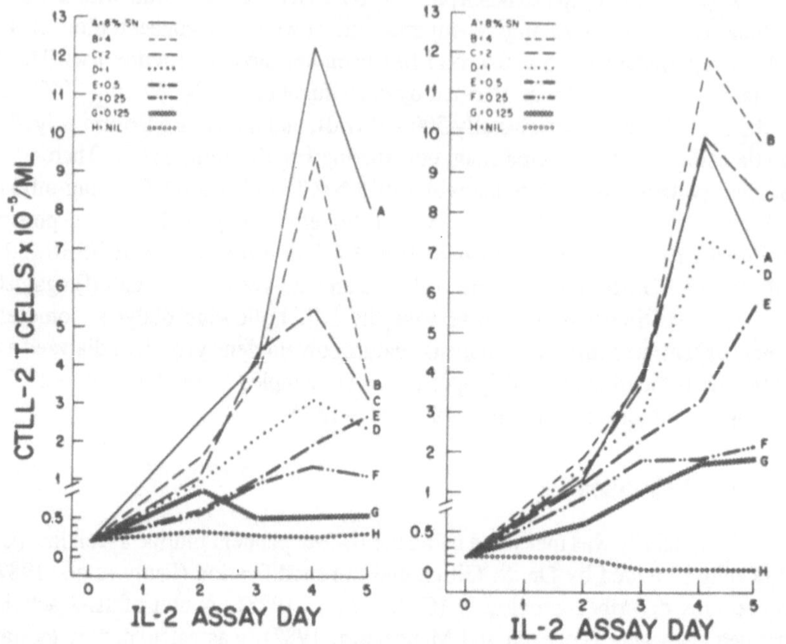

FIGURE 1. Comparison of the T-cell growth-promoting activity of a serum-free human primary lymphocyte supernatant before (left panel) and after (right panel) isolating IL-2 with the proposed SA adsorption procedure. Serial dilutions of the crude supernatant or processed eluate (dialyzed and diluted to the original volume) were added to 10-ml cultures containing IL-2-dependent CTLL-2 target T cells seeded at 0.2×10^5/ml to provide final sample concentrations ranging from 0.125 to 8% (v/v). Trypan blue dye-excluding viable cells were counted daily during the 5-day assay period using conventional hemocytometer procedures.

IL-2-dependent T cells (12×10^5/ml) were recorded on day 4, and this high level of growth was achieved using final concentrations of 8% crude supernatant or 4% PBS/EG eluate that had been dialyzed and diluted to its original volume. Cell death (< 5% viability) was recorded when IL-2-containing supernatant or eluate was absent from the culture medium (Group H, NIL). Accordingly, within the limits of this bioassay, almost all IL-2 activity had been recovered. Comparison of the two samples showed a protein and volume reduction of 13- and 234-fold, respectively, yielding an IL-2 purification of 3108-fold.

The ability of the proposed SA purification scheme to isolate concentrated IL-2 in a protein-poor eluate is illustrated in Fig. 2. Shown is an SDS–PAGE profile of Coomassie blue-stained proteins of samples obtained during the SA adsorption/desorption procedure. Noteworthy is that no major protein bands were detected in the dialyzed EG/PBS eluate containing the IL-2 concentrate. Also shown are profiles obtained when the protein-binding SA microparticles were placed into a particular channel and the proteins eluted from this solid matrix by the detergent SDS in the running buffer.

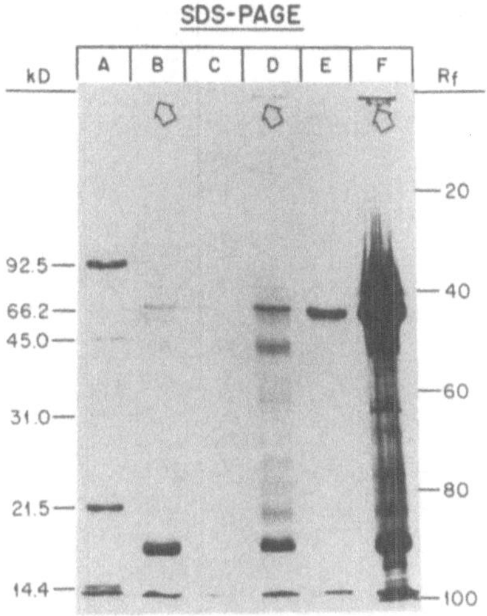

FIGURE 2. SDS–PAGE analysis of different samples obtained during the SA adsorption/desorption of IL-2 from supernatants of PHA-activated human primary lymphocyte cultures with 5% human serum. Arrows identify SA microparticles that were loaded onto a particular lane, and in which the proteins were eluted by the SDS detergent in the running buffer. Protein bands were visualized with Coomassie blue staining. Lane A, molecular weight standards (20 μg protein each); B, SA particles following adsorption of the IL-2-containing supernatant and desorption with EG/PBS eluent; C, EG/PBS eluate of sample B; D, washed SA particles before desorption with the EG/PBS eluant; E, unconcentrated supernatant from human PBL in primary culture; F, washed SA particles following adsorption of human serum.

To further define the bioactivity and stability of IL-2 when isolated by adsorption to SA, experiments were conducted to determine whether IL-2 bound to SA could be used for the prolonged cultivation of IL-2-dependent T cells. In this study, IL-2 from human primary lymphocyte cultures was adsorbed onto SA. The SA was then washed thoroughly, stored at 4°C, and rewashed before adding to the cultures. With this approach, CTLL-2 T cells have been propagated for more than 3 months with SA preparations that had been stored for over 4 months. Thus, IL-2 bound to SA for prolonged periods of time retains its biological activity and can be used for propagating IL-2-dependent T cells.

Presented in Table I are the results of an experiment demonstrating the utility of the SA adsorption procedure for isolating IL-2 from the supernatant of gibbon ape T cells of line MLA-144 and mouse T cells of line EL-4. The murine source of IL-2 (EL-4) required activation with PMA, whereas the primate source of IL-2 (MLA-144) produced IL-2 constitutively. With each line, the cells were seeded at 1×10^6/ml in either serum-free RPMI 1640 or the same medium enriched with 2% FBS that had been previously adsorbed with SA (FBS-Ad; 50 mg/ml, 2 hr, 37°C). Purification of IL-2 from serum-free supernatant of the MLA-144 and EL-4 cell lines was 1063- and 652-fold, respectively. When supplemented with 2% FBS-Ad, IL-2 purification values for the MLA-144 and EL-4 cell lines were 241- and 78.4-fold, respectively, indicating that serum represents a major source of SA-binding protein. We usually obtain values higher than those recorded in this experiment; nevertheless, these results illustrate the utility of the SA adsorption procedure as a preparative step in the purification of several mammalian forms of IL-2.

Relatively few proteins have been isolated using SA; a partial listing includes casein, gelatin, "immune" interferon (γ-IFN), and prostaglandin (PGE_2). In addition to IL-2 and γ-IFN other lymphokines are known to be hydrophobic, and the proposed

TABLE I. Partial Purification of IL-2 from Supernatants of Long-Term T-Cell Lines

Cell line (± FBS-Ad[a])	Purification step	Total volume (ml)	Total IL-2 activity ($U \times 10^{-4}$)	Total protein (mg)	IL-2 activity (U/mg)	Purification (-fold)	Activity recovered (%)
Gibbon MLA-144[b]							
+ FBS-Ad	Crude[d]	1425	5.0	6099	8.25	—	—
	Eluate[e]	37.5	4.3	21.8	1972	241	86
− FBS-Ad	Crude	1425	4.7	5842	8.07	—	—
	Eluate	40.5	2.4	2.75	8727	1063	51
Murine EL-4[c]							
+ FBS-Ad	Crude	1425	4.4	5757	7.65	—	—
	Eluate	39.5	1.3	2.28	586	78.4	29
− FBS-Ad	Crude	1425	3.4	5686	6.04	—	—
	Eluate	36.6	0.8	1.96	3929	652	21

[a] Medium RPMI 1640 without serum (−) or with (+) 2% fetal bovine serum (FBS) that had been adsorbed with SA (50 mg/ml, 2 hr, 37°C).
[b] Long-term lymphoblastoid T-cell line derived from a gibbon ape lymphosarcoma; produced IL-2 spontaneously; cultures seeded at 1×10^6 cells/ml.
[c] Long-term lymphoblastoid T-cell line derived from a mouse thymoma; stimulated with PMA (10 ng/ml); cultures seeded at 1×10^6 cells/ml.
[d] IL-2-containing crude supernatant was harvested after 48 hr.
[e] Supernatant was adsorbed with 1.5 mg SA/ml (2 hr, 37°C), and the IL-2-containing fraction was eluted with 50% ethylene glycol in high salt (1.4 M NaCl) PBS (pH 7.2), dialyzed extensively against PBS, and stored at 4°C.

batch adsorption procedure may be useful in isolating these immunoregulatory molecules.

We have recently learned of the studies by Henderson *et al.* (1983) who have developed an efficient method for isolating IL-2 from the MLA-144 cell line. In this two-step procedure, IL-2 was adsorbed onto controlled-pore glass beads, eluted with acetonitrile, and further purified to homogeneity by rpHPLC.

ACKNOWLEDGMENT. Supported in part by USPHS Grant CA-29635.

REFERENCES

Böhlen, P., Stein, S., Dairman, W., and Udenfriend, S., 1973, Fluorometric assay of proteins in the nanogram range, *Arch. Biochem. Biophys.* **155**:213–220.

Farrar, J. J., Fuller-Farrar, J., Simon, P. L., Hilfiker, M. L., Stadler, B. M., and Farrar, W. L., 1980, Thymoma production of T cell growth factor (interleukin-2), *J. Immunol.* **125**:2555–2558.

Freeman, D. H., 1982, Liquid chromatography in 1982, *Science* **218**:235–241.

Gillis, S., 1983, Interleukin-2: Biology and biochemistry, *J. Clin. Immunol.* **3**:1–13.

Gillis, S., and Watson, J., 1980, Biochemical and biologic characterization of lymphocyte regulatory molecules. V. Identification of an interleukin-2-producing human leukemia T cell line, *J. Exp. Med.* **152**:1709–1719.

Gillis, S., Ferm, M. M., Ou, W., and Smith, K. A., 1978, T cell growth factor: Parameters of production and a quantitative microassay for activity, *J. Immunol.* **120**:2027–2032.

Gillis, S., Scheid, M., and Watson, J., 1980, Biochemical and biologic characterization of lymphocyte regulatory molecules. III. The isolation and phenotypic characterization of interleukin-2 producing T cell lymphomas, *J. Immunol.* **125**:2570–2578.

Hauser, E. A., 1955, *Silicic Sciences*, Van Nostrand, Princeton, New Jersey.

Henderson, L. E., Hewtson, J. F., Hopkins, R. F., III, Sowder, R. C., Neubauer, R. H., and Rabin, H., 1983, A rapid, large scale purification procedure for gibbon interleukin-2, *J. Immunol.* **131**:810–815.

Hirsch, J., and Ahrens, E. H., Jr., 1958, The separation of complex lipid mixtures by the use of silicic acid chromatography, *J. Biol. Chem.* **233**:311–330.

Laemmli, U. K., 1970, Cleavage of structural proteins during assembly of the head of bacteriophage T4, *Nature (London)* **227**:680–685.

Mangold, H. K., and Malins, D. C., 1960, Fractionation of fats, oils and waxes on thin layers of silicic acid, *J. Am. Oil Chem. Soc.* **37**:383–385.

Mier, J. W., and Gallo, R. C., 1982, The purification and properties of human T cell growth factor, *J. Immunol.* **128**:1122–1127.

Pauly, J. L., Russell, S. W., Planinsek, J. A., and Minowada, J., 1982a, Studies of cultured human T lymphocytes. I. Production of the T cell growth-promoting lymphokine interleukin-2, *J. Immunol. Methods* **50**:173–186.

Pauly, J. L., Twist, C. J., Pirela, D. L., Reinertson, R. R., Callahan, J. J., Jr., and Russell, C. W., 1982b, Studies of cultured human T lymphocytes. III. A quantiative assay of the human T cell growth-promoting lymphokine interleukin-2, *IRCS Med. Sci. Biochem.* **10**:899–900.

Pauly, J. L., Twist, C. J., Pirela, D. L., and Russell, C. W., 1983, Studies of cultured human T lymphocytes. V. Isolation and partial purification of the human T cell growth-promoter interleukin-2 by adsorption onto silicic acid, *IRCS Med. Sci. Biochem.* **11**:184–185.

Popescu, M., and Maneekarn, N., 1982, Procedure for titration of interleukin-2 based on direct observation of factor-dependent growth of cells, *Immunol. Lett.* **5**:65–90.

Rabin, H., Hopkins, R. F., III, Ruscetti, F. W., Neubauer, R. H., Brown, R. L., and Kawakami, T. G., 1981, Spontaneous release of a factor with properties of T cell growth factor from a continuous line of primate tumor T cells, *J. Immunol.* **127**:1852–1856.

Ruscetti, F. W., and Gallo, R. C., 1981, Human T-lymphocyte growth factor: Regulation of growth and function of T lymphocytes, *Blood* **57**:379–394.

Smith, I., 1969, *Chromatographic and Electrophoretic Techniques*, 3rd ed., Vol. 1, pp. 457–493, Wiley, New York.

Unger, K. K., 1979, *Porous Silica*, Elsevier, Amsterdam.

Voelter, W., 1981, High performance liquid chromatography in peptide research, in: *High Performance Liquid Chromatography in Protein and Peptide Chemistry* (F. Lottspeich, A. Henschen, and K. P. Hupe, eds.), pp. 1–53, de Gruyter, Berlin.

TCGF (IL-2) Receptor (Tac Antigen) on ATL and Non-ATL Leukemic Cells

JUNJI YODOI, MICHIYUKI MAEDA, YUJI WANO, MITSURU TSUDO, KEISUKE TESHIGAWARA, and TAKASHI UCHIYAMA

1. MATERIALS AND METHODS

Cell Lines. Hut-102 and MT-1 cells were supplied by Dr. R. C. Gallo and Dr. I. Miyoshi, respectively. ATL-2, ATL-6, and ATL-7 cell lines were established from Japanese patients with typical ATL (Yodoi *et al.*, 1974; Uchiyama *et al.*, 1977). A non-ATL leukemic cell line (YT cells) was established from a 14-year-old boy with acute lymphoblastic lymphoma with thymoma (in preparation).

Analysis of Tac Antigen by SDS–PAGE. The cells were labeled with 100 μCi [^{35}S]methionine/5 × 10^6 cells for 6 hr. The cell lysate with extraction buffer containing 0.5% NP-40 and 1 mM PMSF was precipitated with anti-Tac antibody (Uchiyama *et al.*, 1981) and protein A–Sepharose 4B. The eluate of the beads was analyzed by SDS–PAGE with 7.5% gel, followed by autoradiography and fluorography.

2. RESULTS

2.1. Analysis of IL-2 Receptor (IL-2-R) Immunoprecipitated with Anti-Tac Antibody

Normal T cells activated with Con A, non-ATL YT cells, and HTLV$^+$ MT-1 cells were biosynthetically labeled with [^{35}S]methionine for 6 hr. The lysate was

JUNJI YODOI and KEISUKE TESHIGAWARA • Institute for Immunology, Faculty of Medicine, Kyoto University, Kyoto, Japan. MICHIYUKI MAEDA • Chest Disease Research Institute, Kyoto University, Kyoto, Japan. YUJI WANO, MITSURU TSUDO, and TAKASHI UCHIYAMA • First Department of Internal Medicine, Kyoto University, Kyoto, Japan.

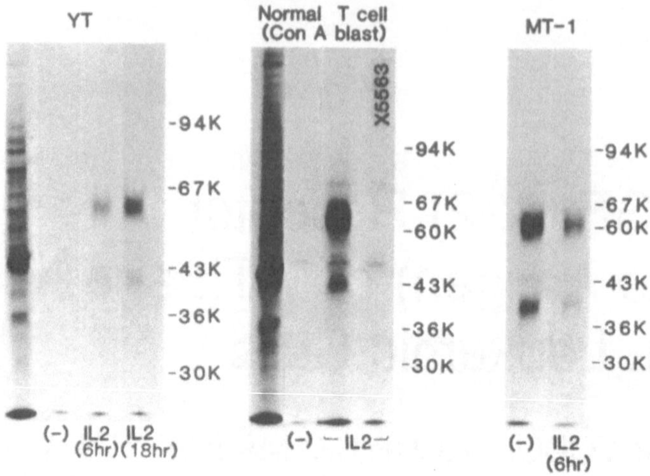

FIGURE 1. SDS–PAGE analysis of antigens reactive with anti-Tac antibody. YT cells, Con A-activated T cells, and MT-1 cells were labeled with [^{35}S]methionine in the presence or absence of crude TCGF. Con A-activated cells were prepared by culturing normal peripheral lymphocytes with 10 μgCon A/ml for 3 days. The lysate was precipitated with anti-Tac and protein A–Sepharose 4B and analyzed with 7.5% gel in the reducing condition.

immunoprecipitated with anti-Tac antibody plus protein A–Sepahrose, and was analyzed by SDS–PAGE (7.5% gel) in the reduced condition. As shown in Fig. 1, two broad bands with apparent molecular weight 60–65K (Tac 60–65) and around 40K (Tac 40) were precipitated from non-ATL YT cell line cells, Con A-activated T cells, and HTLV/ATLV$^+$ MT-1 cells. In Con A-activated T cells, crude IL-2 induced a marked increase of the radioactivity in both bands. Without incubation with crude IL-2, the incorporation of the radioactivity was not significant. Similar induction of the labeled band was also shown in YT cells which had regulatable IL-2-R. Without incubation with crude IL-2, there were no clear bands. After incubation with IL-2 for 6 hr, both Tac 60–65 and Tac 40 appeared. In the case with MT-1 cells, both components were precipitated without exposure to crude IL-2. Preincubation with crude IL-2 resulted in decreased synthesis of both components. Two similar components were also obtained from other ATL-derived T-cell lines (ATL-2, ATL-6) (not shown). The finding was similar in the nonreduced condition, indicating that these two components were not bound to each other by disulfide bond. In addition to these two components, an additional minor component of 115K was frequently obtained.

2.2. Lack of Down-Regulation of IL-2-R/Tac Antigen in ATL

Con A-activated T cells were cultured with or without 10 μg anti-Tac antibody/ml for 24 hr. After restaining with anti-Tac, the fluorescence was analyzed with FACS. As shown in Fig. 2A, the intensity of the fluorescence was markedly decreased when precultured with the antibody, suggesting that IL-2-R/Tac antigen

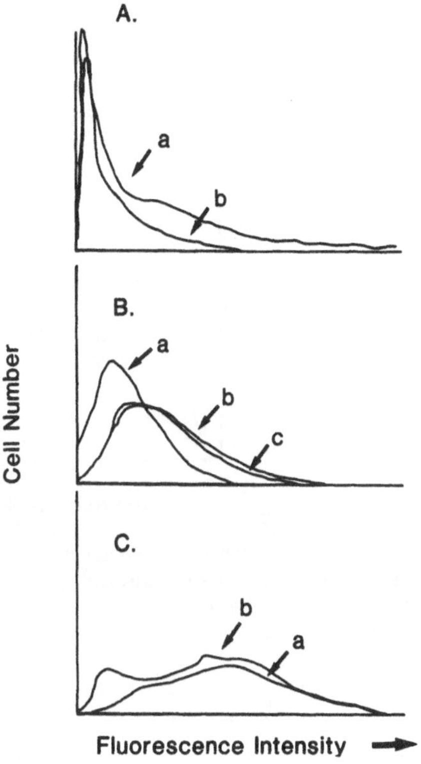

FIGURE 2. Lack of down-regulation of IL-2-R/Tac antigen expression on ATL cells and HTLV⁺ cell line. (A) Con A-activated cells were cultured for 48 hr in the presence (a) or absence (b) of anti-Tac antibody (10 μg/ml). (B) Fresh leukemic cells (a) from a patient with ATL were cultured with (b) or without (c) anti-Tac antibody for 48 hr. (C) MT-1 cells were cultured with (a) or without (b) anti-Tac antibody for 48 hr. The cells were stained with anti-Tac antibody and FITC–goat anti-mouse immunoglobulin antibody and analyzed with FACS.

on normal T cells was down-regulated by the antibody. Fresh leukemic cells from a typical ATL patient and MT-1 cells were precultured with or without anti-Tac. As shown in Figs. 2B and C, there was no significant reduction of the fluorescence after preculture with anti-Tac, indicating that IL-2-R/Tac antigen on these cells was not down-regulated by the antibody.

2.3. Up-Regulation of IL-2-R/Tac Antigen Expression by Crude TCGF on YT Cells

IL-2-R (Robb *et al.*, 1981) and Tac antigen on normal T cells were temporarily expressed after activation with mitogen or antigens. The expression of Tac antigen on the non-ATL leukemic cell line (YT cells) was enhanced by crude IL-2 preparations. The cells were precultured with medium, 0.1% PHA-P, or with crude IL-2 from normal PBL, and the expression of IL-2-R/Tac antigen was analyzed with FACS (Fig. 3). IL-2-R was weakly expressed on a significant proportion of YT cells without exposure to crude IL-2. The expression of IL-2-R on YT cells was markedly enhanced by 18 hr preculture with crude IL-2. As PHA-P did not enhance the expression of IL-2R/Tac antigen on YT cells, the up-regulation or enhancement seemed to be due to a soluble mediator(s) contained in the crude IL-2 preparation. As shown in Fig. 3, culture supernatant of the ATL-2 cell line contained IL-2-

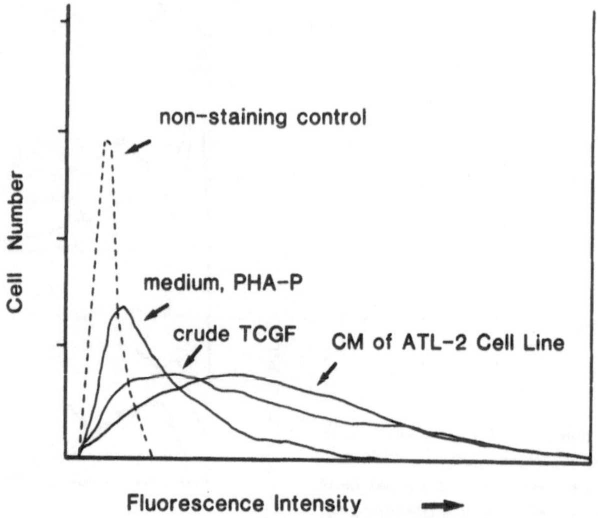

FIGURE 3. Up-regulation or enhancement of IL-2-R/Tac antigen expression on YT cells. YT cells were cultured in the culture medium, medium containing 0.1% PHA-P, crude TCGF obtained from PHA-stimulated human splenic cells, or conditioned medium (CM) of ATL-2 cells for 18 hr. The expression of IL-2-R was determined by FACS analysis with anti-Tac antibody.

R/Tac antigen inducing activity. In the cell line cells derived from ATL, IL-2-R were expressed constitutively, and there was no further enhancement of the receptor expression by exogenous IL-2 (not shown).

3. DISCUSSION

There is accumulating evidence indicating the close association between IL-2-R and Tac antigen (Leonard *et al.*, 1982). Our result showed that there is no overt structural difference of IL-2-R/Tac antigen between ATL-derived cell line cells and normal cells. However, the regulation of the receptor expression was abnormal in HTLV/ATLV$^+$ cell lines (Table I). IL-2-R/Tac antigen on these cells was "constitutively" expressed (Tsudo *et al.*, 1983). In non-ATL leukemic cell line YT cells, however, the expression of IL-2-R/Tac antigen was "regulatable" as was the case with normal T cells (Tsudo *et al.*, 1982). Constitutive expression of IL-2-R on ATL cells was confirmed by radioassay using anti-Tac antibody and [^{125}I]protein A.

The mechanism of "constitutive" IL-2-R expression is unknown. There may be a fine difference of IL-2-R which was not detectable by SDS–PAGE. However, it may be due to the regulation abnormality rather than the structural abnormality of the receptor.

Alternatively, ATL-derived cell lines may produce a factor having IL-2-R-inducing activity. The factor may not necessarily be IL-2, as the production of TCGF (IL-2) by ATL-derived cell lines proved not to be so significant as had been

TABLE I. Constitutive vs. Regulatable Expression of Tac Antigen (IL-2-R)

	HTLV (ATLV)	Tac antigen (IL-2-R)		Proliferation to TCGF	IL-2 production
		Up-regulation[a]	Down-regulation[b]		
Resting T	No	No	N.D.	No	No
Activated T	No	Yes	Yes	Yes	Yes
YT cells	No	Yes	Yes	Yes	No
MT-1	Yes	No	No	No	No
Hut-102	Yes	No	No	No	Yes/No
ATL-2	Yes	No	No	No	No

[a] Up-regulation by crude TCGF.
[b] Down-regulation by anti-Tac antibody.

expected (Gallo and Wong-Staal, 1982). Quite interestingly, IL-2-R/Tac antigen-inducing activity was present in ATL-2 culture supernatant, which did not contain a significant IL-2 activity. In any case, the "constitutive" IL-2-R expression may be involved in the abnormal growth of ATL-derived cell lines (Yodoi *et al.*, 1983).

ACKNOWLEDGMENTS. This work was supported by grants from the Ministry of Education, Science, and Culture and the Ministry of Health and Welfare.

REFERENCES

Gallo, R. C., and Wong-Staal, F., 1982, Retroviruses as etiologic agents of some animal and human leukemias and lymphomas and as tools for elucidating the molecular mechanism of leukemogenesis, *Blood* **60**:545–557.

Leonard, W. J., Depper, J. M., Uchiyama, T., Smith, K. A., Waldmann, T. A., and Greene, W. C., 1982, A monoclonal antibody that appears to recognize the receptor for human T-cell growth factor: Partial characterization of the receptor, *Nature (London)* **300**:267–269.

Robb, R. J., Munck, A., and Smith, K. A., 1981, T cell growth factor receptors: Quantitation, specificity and biological relevance, *J. Exp. Med.* **154**:1455–1474.

Tsudo, M., Uchiyama, T., Takatsuki, K., Uchino, H., and Yodoi, J., 1982, Modulation of Tac antigen on activated human T cells by anti-Tac monoclonal antibody, *J. Immunol.* **129**:592–595.

Tsudo, M., Uchiyama, T., Uchino, H., and Yodoi, J., 1983, Failure of regulation of Tac antigen/TCGF receptor on adult T cell leukemia cells by anti-Tac monoclonal antibody, *Blood* **61**:1014–1016.

Uchiyama, T., Yodoi, J., Sagawa, K., Takatsuki, K., and Uchino, H., 1977, Adult T cell leukemia: Clinical and hematologic features of 16 cases, *Blood* **50**:481–492.

Uchiyama, T., Broder, S., and Waldmann, T. A., 1981, A monoclonal antibody (anti-Tac) reactive with activated and functionally mature human T cells. I. Production of anti-Tac monoclonal antibody and distribution of Tac(+) cells, *J. Immunol.* **126**:1393–1397.

Yodoi, J., Takatsuki, K., and Masuda, T., 1974, Two cases of T-cell chronic leukemia in Japan, *N. Engl. J. Med.* **290**:572–573.

Yodoi, J., Uchiyama, T., and Maeda, M., 1983, T-cell growth factor receptor in adult T-cell leukemia, *Blood* **62**:509–510.

Part II Summary

Chemistry and Biological Activity of Lymphokines

MICHAEL A. CHIRIGOS and ROBERT C. GALLO

Unfortunately, there wasn't sufficient time to have some of the poster sessions presented orally because there were some excellent poster sessions and for session II concerning lymphokines there were actually 22 presentations, 9 oral and 13 posters. We have attempted to synthesize the information contained in the poster session with that of the oral presentations with a little bit more of an emphasis on the poster presentations.

Drs. Copeland and Sarngadharan discussed the biochemical and biological properties of TCGF (IL-2), describing their methods for purifying TCGF to homogeneity and sequencing it. They and their colleagues defined the necessity of activating normal T cells which then acquire receptors for a response to the TCGF. In contrast, neoplastic mature T cells respond directly to TCGF, a development that led to the isolation of a human type-C retrovirus known as HTLV. The specific TCGF mRNA in human peripheral blood lymphocytes was identified as an 11 to 12 S species and shown to program the synthesis of active TCGF in the *Xenopus* oocyte translation system. Several of the posters described isolation and purification procedures. John Pauly and his colleagues described a two-step IL-2 purification procedure employing an anti-IL-2 monoclonal antibody affinity column resulting in a highly purified IL-2 preparation which maintained greater than 95% of its biological activity with a greater than 99% reduction of protein content. Richard Robb and his colleagues also reported a TCGF (IL-2) purification procedure em-

MICHAEL A. CHIRIGOS • Immunopharmacology Section, Biological Therapeutics Branch, Biological Response Modifiers Program, Division of Cancer Treatment, NCI-Frederick Cancer Research Facility, Frederick, Maryland 21701. ROBERT C. GALLO • Laboratory of Tumor Cell Biology, Division of Cancer Treatment, National Cancer Institute, National Institutes of Health, Bethesda, Maryland 20205.

ploying an immune affinity column coupled with a murine monoclonal antibody. Results of binding experiments showed that their purified radiolabeled IL-2 was rapidly bound only by IL-2-responsive cells with internalization of the IL-2 occurring within 25–30 min followed by lysosome-dependent degradation which occurred within 75 min, which indicates the possible half-life of IL-2. J. Yodoi and his colleagues from Kyoto University described the TCGF receptor (Tac antigen) which appears on T leukemic cells but not on normal T cells. The TCGF receptor (IL-2-R/Tac antigen) induced in normal T cells by specific mitogens could be regulated by the use of the monoclonal antibody developed to the TCGF receptor site. This antibody, however, could not regulate the constitutively expressed receptor Tac antigen unique to the T -cell leukemias or cell lines containing the provirus of HTLV, indicating that there is a difference in the antigenic characteristics of both the normal T cells, converted normal T -cell line, and those that are mature T-cell leukemias. Y. Lin and co-workers, employing the *Xenopus* oocyte translation system, were able to enrich for and separate the IL-2 mRNA which was used for the synthesis of double-stranded cDNA. Through cloning procedures, they were able to isolate a clone which was shown to hybridize strongly with IL-2 mRNA, again indicating the increased specificity of these reactions with a more purified material. John Maples and his co-workers at the Naval Medical Research Institute examined highly active TCGF preparations for growth factors capable of supporting stem cell growth and differentiation. The impure TCGF-containing supernatants they were using markedly induced bone marrow cell growth, proliferation, and differentiation with the development of granulocytes, macrophages, erythroid and megakaryocyte colonies in both short- and long-term cultures.

Dr. Steven Rosenberg described the use of TCGF for therapeutic purposes. In four experimental tumor models he described how sensitized mouse lymphoid cells which were expanded in the presence of TCGF could be used in adoptive immunotherapy to cause a regression of mouse tumors. Of particular significance was his observation that the proliferative and not the cytolytic lymphoid cell clones are the most efficient in retarding tumor growth. These IL-2-activated lymphoid cells appear to be different from natural killer cells and cytotoxic T lymphocytes. He described a current project on the use of IL-2 for expanding and activating lymphoid cells collected by leukapheresis from tumor-bearing patients with the purpose of treating tumor-bearing patients with their autologous IL-2-activated cells. Since the IL-2-expanded proliferative lymphoid cells express a tumor-retarding or cytolytic effect but are not identified as NK cells, it would be interesting to consider what therapeutic response could be achieved by combining the adoptive immunotherapeutic approach described by Dr. Rosenberg with a BRM capable of augmenting NK cells and/or stimulating macrophage tumoricidal activity. Several BRMs have been identified that specifically augment NK cells and/or enhance macrophage tumoricidal activity.

The macrophage was the prominent cell in four presentations, three dealing with IL-1 and one with CSF. Steve Mizel described how macrophage-elicited IL-1 affects fibroblast proliferation, PG-2 and collagenase secretion from synovial cells, and causes liver cells to release acute-phase proteins and to induce B-cell maturation; but as important was the capacity of IL-1 to induce the secretion of IL-

2 from T cells. He described the purification of IL-1 by an immune-adsorbant column prepared with IL-1 antibodies elicited in the goat. Jose Lepe-Zuniga and colleagues characterized the IL-1 produced from LPS-stimulated human monocytes. The extracellular fluid contained a single peak of molecular weight 14,000 to 16,000. In contrast, the intracellular material contained four peaks of molecular weight 14,000, 26,000, 46,000, and > 70,000. Incubation of the three larger peaks generated *in vitro* the 14,000 molecular weight peak. The authors then suggest that the molecular weight heterogeneity of the intracellular IL-1 is due to the polymerization of the intracellular 14,000 to 16,000 molecular weight species. Lisa McKernan and colleagues described the secretion of a 15,000- to 30,000-dalton IL-1 from the RAW 264.7 monocyte/macrophage cell line. These two isolates differed in their capacity to induce TL-3, H-2K^k, and Thy-1.2 surface markers on immature thymocytes.

Dr. Stanley described the regulatory role of CSF-1. It appears to be most active on monoblasts, causing their differentiation to promonocytes, then to monocytes and macrophages. Depending on the amount of CSF present, the morphology of the macrophage was found to vary from that of a resting stage to that of a highly active vacuolated stage. He presented evidence that specific cell-surface receptors for CSF-1 occur only on mononuclear phagocytic cells and mediate the biological effects, as well as the degradation of the growth factor. Considering the critical role of CSF in regulating the survival, proliferation, and differentiation of cells of the mononuclear phagocytic lineage, it would be of interest to assess the role of BRMs on the regulation of CSF production. Several BRMs which have the capacity to stimulate macrophage activity also cause the production and release of CSF. Would autologous CSF induced in man or mouse by a BRM possess the same or different regulatory functions?

Jim Ihle reviewed the biochemical and biological properties of IL-3. Using several lines of investigation, he provided evidence that IL-3 induces T-cell differentiation. In one of the posters, his colleague Ed Palaszynski demonstrated through binding experiments with ^{125}I-labeled IL-3 that the majority of IL-3 receptor-positive cells were in the bone marrow, while the thymus, lymph node, and spleen cells had very few receptor-positive cells. This is an area that will be followed with the greatest interest.

We had two presentations on lymphotoxins. Dr. Gale Granger summarized some of the available information on the methods for separating so-called lymphotoxins into various groups. Granger and Aggarwal presented evidence that lymphotoxins possess tumor cell killing capacity, particularly when combined with γ-interferon.

All in all, the oral presentations and posters were informative, particularly in the development of methods for improved purification procedures. About one-third of the time dealt with specific purification procedures. Several investigators were of the opinion that it would be premature at this point to develop any systematic nomenclature for the various lymphokines. However, all agreed, particularly based on the clinical applications presentations, that lymphokines definitely have a strong potential for the treatment of various diseases such as cancer and autoimmune diseases.

III

Mechanism of Action and Immune Regulation

Lymphocyte Proliferation, Lymphokine Production, and Lymphocyte Receptors in Aging

A. L. DE WECK, F. KRISTENSEN, F. JONCOURT, F. BETTENS, G. D. BONNARD, and Y. WANG

Lymphocyte proliferation is one of the basic functions of the immune system, since it appears required as well for the acquisition of immunological memory as for clonal expansion of specific lymphocyte populations. In recent years, the combination of several techniques, such as analysis of the cell cycle by cytofluorometry (Darzinkiewicz *et al.*, 1976; Stadler *et al.*, 1980), measurement of [^{3}H]-TdR uptake, detection of membrane receptors by immunofluorescence or ligand binding (Munck and Vira, 1975), and quantitative assessment of various lymphokines produced by proliferating lymphoid cells, has permitted the establishment of an integrated picture of the various events associated with the proliferation of lymphocytes. Although this picture is still fragmentary, the analysis of lymphocyte functions which it makes possible has already been found relevant and informative in several clinical situations, where a dysregulation of lymphocyte functions is apparent. The purpose of this paper is to review briefly the current possibilities to analyze lymphocyte proliferation in clinical situations and their application to the aging process.

1. DYNAMIC ANALYSIS OF LYMPHOCYTE PROLIFERATION: AN INTEGRATED APPROACH

Techniques enabling lymphoid cells to be stained simultaneously for intracellular DNA and RNA allow the quantitation and study of lymphoid cells undergoing proliferation and their assessment throughout various phases of the cell cycle

A. L. DE WECK, F. KRISTENSEN, F. JONCOURT, F. BETTENS, G. D. BONNARD, and Y. WANG • Institute for Clinical Immunology, Inselspital, University of Bern, Bern, Switzerland.

by cytofluorometry (Darzinkiewicz *et al.*, 1976; Stadler *et al.*, 1980). Dynamic study during the first 48 hr of culture, i.e., while most of the cells have only completed their first proliferative cycle, reveals that several signals and concurrent events are required for the cell to process from a resting G0 phase through an early activation phase (G1a), a late activation phase (G1b), a DNA synthesis phase (S), a premitotic phase (G2), and mitosis (M) (Pardee *et al.*, 1978). It is by now well recognized that most substances formerly designated as "mitogens," such as lectins (e.g., PHA, Con A), are indeed only providing a first activating signal enabling the cell to initiate RNA synthesis and to proceed from the resting G0 phase to an early G1a phase (Stadler *et al.*, 1981; Kristensen and de Weck, 1982). In order to proceed further, the cell must receive a second signal provided by interleukin-2 (IL-2) alias T-cell growth factor (TCGF) (Smith *et al.*, 1980). The main but possibly not the only effect of this lymphokine seems to be to promote further RNA synthesis, manifested by the passage of the cells from the G1a into the G1b phase (Kristensen *et al.*, 1982a,b, 1983). While some lymphocyte populations seem to produce IL-2 constitutively and may even be activated by IL-2 produced endogenously without the participation of a membrane receptor available for exogenous IL-2 (Bettens *et al.*, 1983b), most activated lymphocytes appear to be dependent for their continued proliferation upon the interaction of some exogenous IL-2 with a specific IL-2 membrane receptor, formed during the early G1a phase (Smith *et al.*, 1980; Bettens *et al.*, 1983a,b). The formation of an IL-2 receptor during the activation of lymphocytes, which was originally followed indirectly from IL-2 absorption studies (Bonnard *et al.*, 1979) and by direct binding studies with labeled IL-2 (Robb *et al.*, 1981), has recently been kinetically followed in our laboratory by cytofluorographic studies (Bettens *et al.*, 1983a,b) with a monoclonal antibody (anti-Tac) recognizing membrane protein identical or closely associated with the IL-2 receptor (Uchiyama *et al.*, 1981; Leonard *et al.*, 1982; Miyawaka *et al.*, 1982; Niyawaki *et al.*, 1982; Bonnard *et al.*, 1983). Production of IL-2 is apparently the function of some subsets of activated T cells (Bonnard *et al.*, 1980; Smith, 1980, Chang *et al.*, 1982). As judged from kinetic analysis, IL-2 becomes detectable in the supernatant of activated lymphocyte cultures about 10–12 hr after the activating signal produced by a lectin (Bettens *et al.*, 1982). The production of IL-2 seems to be dependent in these activated lymphocytes upon a second signal provided by a monokine, IL-1, the product of activated monocytes or macrophages (Smith, 1980; Smith *et al.*, 1980). IL-1 becomes detectable in the supernatant of activated mononuclear cell populations already 2–4 hr after activation (Bettens *et al.*, 1982).

Following successful interaction of exogenous IL-2 with the IL-2 receptor, the cells proceed to a new burst of RNA synthesis (G1b phase) and will some 6 hr later start DNA synthesis. The passage of the G1b to the S phase does not seem to require some additional signal: all analyses performed so far on mouse, canine, and human lymphocytes indicate a close correlation between the number of cells in the G1b phase and [^{3}H]-TdR incorporation, and this independently of the cell source and of varying culture conditions, with or without addition of serum (Kristensen *et al.*, 1981, 1983). Accordingly, it seems that early events are really the most decisive for the fate of proliferating lymphocytes; once activated and furbished with the appropriate signals, they might pursue on their committed path throughout the cell cycle. It must be recognized, however, that further events and signals possibly required at later stages of lymphocyte proliferation, such as completion of the S phase, duration of the G2 phase, and initiation of the M phase, have not yet been analyzed for human peripheral blood lymphocytes proliferating *in vitro* as thoroughly as for lymphoid T-cell lines (Koponen *et al.*, 1982). Following mitosis, cells may either return to a resting G0 stage or directly enter a new proliferation cycle, provided the required activation and proliferation-inducing signals (e.g., IL-2) are continuously present. This is apparently the case for T cells continuously proliferating in the presence of exogenously provided IL-2, while deprival of IL-2 replaces the cells in a G0 phase, where they remain susceptible to renewed activation (Bettens *et al.*, 1983b). Our current dynamic conceptions about the cell cycle-associated events in human lymphocyte proliferation are illustrated in Fig. 1.

In order to better analyze such events, it is advantageous to be able to synchronize or at least to some extent to enrich cell populations at some defined phase of the cell cycle. This would be particularly valuable for analyzing the presence, density, and fate of various membrane receptors as well as the biochemical and molecular biological intracellular phenomena associated with cell proliferation. Recent studies on the effect of various manipulations on the cell cycle of human proliferating lymphocytes have permitted approaching this goal. As lymphoid cells taken from a living animal may be expected to have been, at least in part, activated *in vivo,* a variable proportion of peripheral blood lymphocytes are in fact already progressing throughout the cell cycle at time 0. This can be assessed as well by [^{3}H]-TdR incorporation at time 0 (de Weck, 1981; Joncourt *et al.*, 1982a; Kristensen and de Weck, 1982) as by cytofluorographic analysis. [^{3}H]-TdR uptake at time 0 reflects the activity of the cells having reached the G1b/S phases *in vivo* and has therefore also been used clinically as some kind of measure for "immunological activation" *in vivo;* its interpretation is, however, somewhat clouded by the observation that in several acute clinical conditions where this parameter ([^{3}H]-TdR uptake at T_0) has been followed, continuously proliferating myeloid cells from the bone marrow appear in the peripheral blood and provide high background values of proliferating cells. Cells activated *in vivo* but which have not yet proceeded beyond the G1a phase will proliferate *in vitro* only if they receive the appropriate

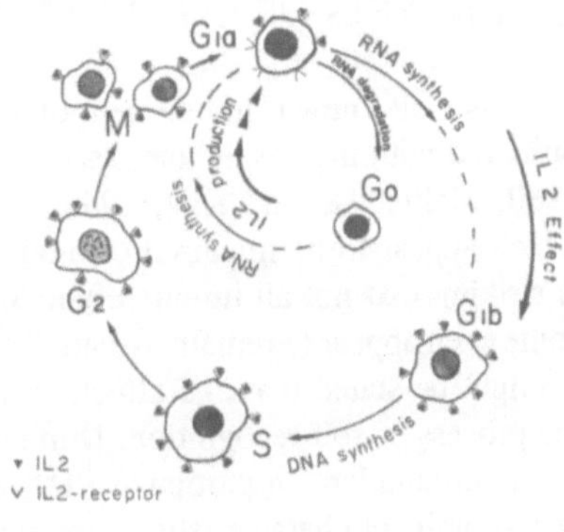

FIGURE 1. Lymphocyte cell cycle.

exogenous IL-2 signal. This explains why the addition of IL-2 at time 0 always appears to have some "mitogenic" effect: those cells having reached the G1a stage *in vitro* will proceed through the cell cycle *in vitro* (Kristensen and de Weck, 1982). If the same cell population, however, is allowed to rest in culture for 18 hr, the addition of exogenous IL-2 alone has no effect. At this time, the G1a cells produced *in vivo* are no longer responsive to IL-2 (Kristensen and de Weck, 1982; Kristensen *et al.*, 1982b), presumably because they have returned to the G0 stage (Bettens *et al.*, 1983b) although it cannot be excluded that some of them also die (Koponen *et al.*, 1982). Cells rested for 18 hr in culture contain therefore essentially G0 cells, they are susceptible to activation by lectins (Kristensen *et al.*, 1982b), and this procedure seems therefore a relatively simple way to synchronize the bulk of human peripheral blood lymphocytes for cell cycle studies, provided appropriate culture conditions for maintaining cell viability are used.

When lectins are used for providing the primary activating signal *in vitro*, several substances or procedures may be used for stopping the cells at some stage or providing cell populations enriched in some specific phase of the cell cycle. The addition of serotonin, for example, seems to block the cells quite early in the activation process provided by PHA, as judged from the low number of G1a cells produced and by the absence of (IL-2) receptors recognized by anti-Tac antibodies (Slauson *et al.*, 1983). In order to obtain cells in the G1a phase, several procedures may be used: culture of PHA-activated mouse thymocytes at low cell concentrations (the classical IL-1 and IL-2 mouse thymocyte assay), culture of human peripheral blood lymphocytes for 18–22 hr only (Wang *et al.*, 1983a), addition of dexamethasone (Bettens *et al.*, 1983a) or PGE_2 (Walker *et al., 1983), removal of IL-2 (Bettens et al.*, 1983b), and blocking of IL-2 receptors by anti-Tac antibodies (Bettens *et al.*, 1983b; Miyawaka *et al.*, 1982). If the cells are activated in such a way that IL-2 production and IL-2 receptors are not interfered with, addition of hydroxyurea, which prevents the passage of cells from the G1b into the S phase, will promote an accumulation of cells in the G1b phase (Stadler *et al.*, 1981; Bettens *et al.*, 1982). The use of these manipulations, the result of which are summarized in Table I, has already been shown useful in assessing the dynamic evolution of various lymphocyte receptors during the cell cycle (see below).

2. ANALYSIS OF LYMPHOCYTE PROLIFERATION AND AGING

As well known, impairment of various immune functions is a marked feature observed with increasing age, as well in humans as in experimental animals (Kay, 1980; Makinodan and Kay, 1980; Weksler, 1981). The main immune functions which appear to be impaired or modified with age are summarized in Table II. It is striking that not all immune functions appear to be affected to the same extent, some even appear to remain essentially intact until a very advanced age. Summarily, it might be stated that age affects in particular several T-cell functions and alters the process of self-recognition. Diminution of the proliferation of T cells following lectin stimulation, in particular PHA and Con A, has since long been recognized as a prominent characteristic of immunological aging (Makinodan and Kay, 1980;

TABLE I. Procedures Which Can Be Used to Enrich Human Peripheral Blood Lymphocytes in Early Phases of the Cell Cycle

	% of cells in G0	G1a	G1b
Enrichment in G0			
Fresh PBL incubated without stimulation for 18–30 hr	93	6.5	0.5
Fresh PBL + PHA + serotonin (10 M)	84	12	4
Enrichment in G1a			
PHA stimulation for 18 hr	62	34	4
PHA stimulation for > 18 hr + dexamethasone (10 M)	59	36	5
Enrichment in G1b			
PHA stimulation for 44 hr + IL-2 + hydroxyurea	49	26	25

Adler *et al.*, 1978). Although most studies on the impairment of T-cell proliferation with age have been performed with the [^{3}H]-TdR uptake technique, the integrated approach described above has enabled us to assess more precisely the mechanisms by which T lymphocytes of old mice and elderly humans fail to proliferate as efficiently as cells from young donors. In order to establish such a lymphocyte proliferation "profile" in a routine clinical test, the following parameters are studied following stimulation by PHA in autologous serum and in a standard AB serum:

1. Number of G1a and G1b cells (by cytofluorometry).
2. [^{3}H]-TdR incorporation.

TABLE II. Changes in Immune Functions with Age (Man)

Little change in IgG, IgM, IgE levels, slight rise in IgA
Decrease in primary antibody response
Little change in secondary antibody response
No change in number of circulating B lymphocytes
No change in immunoglobulin *in vitro* (PWM stimulation)
Little change in number of circulating T lymphocytes
Relative decrease in T-helper cells
Relative increase in T-suppressor cells
Decrease in delayed-type hypersensitivity
Decrease in T-lymphocyte proliferation
Decrease in induction of cytotoxic T cells
Decrease in production of some lymphokines
Decrease in some hormonal and lymphokine receptors
Decrease in serum thymic factors
Increase in autoreactive T lymphocytes
Increase in autoantibodies
Increase in circulating immune complexes

3. Effect on [^{3}H]-TdR incorporation of the addition of exogenous IL-2 (10 U/ml) or of indomethacin.
4. Production of IL-2 (in U/ml/24hr) under standardized culture conditions.
5. Number of G1 cells and [^{3}H]-TdR incorporation (T_0-4 hr) at time 0.

These investigations permit assessment of (1) the degree of cell activation *in vivo* prior to bloodletting; (2) the capacity of the cells to be activated by lectins; (3) the production of IL-2; (4) the capacity of response to IL-2 presumed to reflect indirectly the state of IL-2 receptors; (5) the internal "damping" effect of prostaglandin synthesis presumed to be due essentially to activated monocytes; and (6) the presence of autologous serum factors affecting one or more of the parameters mentioned above.

Such experiments have been carried out in old inbred and random-bred mice (Joncourt *et al.*, 1982a,b) and in three groups of humans (20–35 years, 50–65 years, and over 70 years old); the results are described in detail elsewhere (Joncourt *et al.*, 1983).

In addition to results obtained with peripheral blood lymphocytes in man, studies in mice have revealed several other age-associated changes, such as modifications in the composition of cellular subsets in various lymphoid organs and differences in lymphocyte compartmentalization (Joncourt *et al.*, 1981b). A summary of the results obtained to date is given in Table III. It is noteworthy that with this more sophisticated approach to the study of aging lymphocytes, age-associated dysfunctions already become detectable in middle age. Furthermore, there appear to be increasing individual variations with advancing age. Whether the maintenance of high T-lymphocyte proliferative capacity constitutes a selective advantage for the aging individual has not yet been demonstrated; preliminary prospective studies on a random-bred mice population (Joncourt *et al.*, 1981a) suggest that this may not be obligatorily the case. Many other factors, e.g., the dysregulation in self-recognition, may influence the overall performance of the immune apparatus in such a way that a slightly lowered T proliferative capacity instead represents an adaptive mechanism which favorably influences survival prospects. Marked to complete impairment of T-cell functions, on the other hand, as observed in anergic states and advanced cancer, has most certainly a detrimental effect on resistance to infections and thereby on survival.

TABLE III. Summary of Our Results on Immunological Changes of Human and Murine Lymphocytes with Age[a]

Shifts in T-cell subsets (T helper/suppressor ratio)
Changes in compartmentalization (T/B ratio in lymphoid organs)
No change in IL-1 production
Decrease in number of lectin-activatable T cells (G0–G1)
Decrease in IL-2 production
Decrease in sensitivity to IL-2 (IL-2 receptors)
Increase in number of *in vivo* activated cells (G1 at time 0)
No decrease in number of LPS-activatable B cells

[a] Joncourt *et al.* (1981a,b, 1982a,b, 1983, and unpublished results).

3. AGE-ASSOCIATED CHANGES IN LYMPHOCYTE RECEPTORS

It is well known among gerontologists that changes in cellular receptors often accompany the aging process and may be directly responsible for altered cellular functions. Age-associated changes in insulin and glucagon receptors in adipocytes (Sartin *et al.*, 1980), adrenergic reeptors in cardiac muscle cells (Lakatta, 1980), serotonin receptors in neuronal synapses (Roth, 1979), and various hormonal receptors in different tissues (Roth, 1979) are only some of the examples which may be quoted to illustrate the importance of cellular receptor modifications in aging. It could be stated summarily that "age is a disease of receptors"! Although this is manifestly an overstatement as some lymphocyte receptors do not change with age (Landman *et al.*, 1981; Abrass and Scarpace, 1981) or may even increase (Roth, 1979), it may well be that most of the physiological changes accompanying advancing age ultimately rest on modifications in expression, density and/or avidity of cellular receptors.

Studies on aging proliferating lymphocytes have clearly shown that the decreased proliferative capacity rests not only on diminished IL-2 production (Gillis *et al.*, 1981; Joncourt *et al.*, 1982a,b; Thoman and Weigle, 1981, 1982; Chang *et al.*, 1982) but also on diminished IL-2 receptivity (Joncourt *et al.*, 1982b; Gilman *et al.*, 1982), which is itself a consequence of a diminution in the density of IL-2 receptors available on the cell surface. It is not known at this stage whether changes in IL-2 receptor quality (e.g., molecular alterations in receptor structure leading to decreased affinity for IL-2) also participate in the overall decreased efficiency of IL-2–IL-2 receptor interactions.

These observations have led us (Wang *et al.*, 1983a; Slauson *et al.*, 1983) to investigate the fate of other receptors known to be present on lymphocytes, in function of the cell cycle, on the one hand, and of the age of the cell donor, on the other hand. Some of these other receptors seem to appear on the cell membrane only during the G1 phase following activation (e.g., insulin, transferrin)(Krug *et al.*, 1972; Helderman *et al.*, 1979), others are present on resting G0 cells but increase markedly during the G1 phase (e.g., corticosteroid receptors) (Crabtree *et al.*, 1980a,b), while others apparently change little during the cell cycle (OKT4 and OKT8 markers, HLA-A and B histocompatibility antigens). Lymphocytes carry a large number of membrane proteins functioning as receptors for a whole variety of ligands (Table IV); these cells could therefore be considered as an easily available cellular system to investigate the state of hormonal and other physiological receptors. Some of these receptors appear to be located not on the cell surface but in the cytoplasm (e.g., corticosteroid receptors) while the presence of specific membrane receptors for some ligands which exert functional influences on lymphocytes (e.g., histamine) is still controversial (Wang *et al.*, 1983b).

Our studies on the age-associated modifications of lymphocyte receptors have up to now been restricted to the insulin and corticosteroid receptors; these studies are described in detail elsewhere (Wang *et al.*, 1983a). The main result is shown in Table V. For both receptors, the increase observed during the early G1 phase appears similar in the young and middle-age groups; the middle-age group, however,

TABLE IV. Lymphocyte Receptors

Antigen-specific: Ig, T-cell receptor
Fc receptor
C′ components: C3a, C3b, C5a
Lectins: PHA, Con A, PWM
Histocompatibility antigens: DR, Ia
Lymphokines: IL-2, interferon
Hormones: glucocorticoids, insulin, growth hormone, β-adrenergic, thymosin, calcitonin, parathyroid hormone
Mediators: histamine, adenosine, acetylcholine, low-density lipoprotein

seems to lag behind in the late G1b phase, leading to an overall decreased number of receptors per activated cell at the end of the G1 phase. These results are consistent with previous publications indicating a decrease in sensitivity to hydrocortisone in aging lymphocyte populations. For IL-2 receptors, we do not yet have indications whether the observed decrease in number of IL-2 receptors already occurs during the early G1a phase or only becomes manifest during the later phase of RNA synthesis. It is tempting to speculate that the basic aging phenomenon occurs at the level of RNA synthesis and/or transcription, preventing the synthesis and expression of several membrane proteins at a similar rate. This, however, would require more extensive investigations. As shown by experiments with serotonin (Slauson *et al.*, 1983), the expression of IL-2 receptors and the production of IL-2 can be regulated independently. It would be interesting to assess whether age affects the RNA synthetic processes responsible for membrane receptor expression and lymphokine production to the same extent.

The changes in receptor expression and functions associated with age, at least in some other systems, do not seem to be entirely irreversible or to be impermeable to influence by exogenous factors. For example, dietary restrictions in middle age appear to affect favorably the rate of decay of glucagon and adrenergic receptors in adipocytes (Masoro *et al.*, 1980). Similar favorable effects of dietary restriction on the rate of decay of T-cell-dependent immune functions have been reported in old mice and rats (Walford *et al.*, 1973; Good *et al.*, 1980). *In vitro,* thymic hormones appear to modify the density of corticosteroid lymphocyte receptors. Accordingly, it is tempting to speculate that some therapeutic regimens acting

TABLE V. Changes in Binding of Insulin and Dexamethasone to Human Peripheral Blood Lymphocytes with Age, as a Function of the Cell Cycle

	G0		G1a		G1b	
Hormone	Young	Old	Young	Old	Young	Old
Insulin	0[a]	0	760	770	1,481	934
Dexamethasone	3085	4033	16,290	17,458	26,781	18,662

[a] Average numbers of molecules bound per cell in young (20–35 years old) or "old" (50–65 years old) human PBL. For experimental details, see Joncourt *et al.* (1983).

through effects related to receptor expression might reverse some of the functional immune defects observed with advancing age. As one of the main apparent causes of impaired T-lymphocyte proliferation is a decrease in the number of PHA-activatable mature T cells (possibly related to a decrease in the level of thymic hormones) and a deficiency in IL-2 production, it was logical to investigate whether concurrent administration of thymic hormones (in the form of a TF5-like calf thymic peptide extract) and of IL-2 to old mice would restore their impaired T-cell proliferative capacity. Preliminary experiments in our laboratory (Joncourt, unpublished results) suggest that such an approach is worth further consideration.

REFERENCES

Abrass, I. B., and Scarpace, P. J., 1981, Human lymphocyte betaadrenergic-receptors are unaltered with age, *J. Gerontol.* **35**:329.

Adler, W. H., Jones, K. H., and Brock, M. A., 1978, Aging and immune function, in: *The Biology of Aging* (J. A. Behnke, C. E. Finch, and G. B. Moment, eds.), p. 221, Plenum Press, New York.

Bettens, F., Kristensen, F., Walker, C., and de Weck, A. L., 1982, Human lymphocyte proliferation. II. Formation of activated (G1) cells, *Eur. J. Immunol.* **12**:948–952.

Bettens, F., Kristensen, F., Walker, C., Schwulera, U., Bonnard, G. D., and de Weck, A. L., 1983a, Lymphokine regulation of activated (G1) lymphocytes. II. Glucocorticoid and anti-Tac induced inhibition of human T lymphocyte proliferation, *J. Immunology,* **132**:261–265.

Bettens, F., Kristensen, F., Walker, C., Bonnard, G. D., and de Weck, A. L., 1983b, Lymphokine regulation of human lymphocyte proliferation: Formation of resting (G0) cells by removal of interleukin 2 in cultures of proliferating T lymphocytes, *Cell Immunol.* (in press).

Bonnard, G. D., Yasaka, K., and Jacobson, D., 1979, Ligand-activated T cell growth factor-induced proliferation: Absorption of T cell growth factor by activated T cells, *J. Immunol.* **123**:2704.

Bonnard, G. D., Yasaka, D., and Maca, R. N., 1980, Continued growth of functional T lymphocytes: Production of human T-cell growth factor, *Cell. Immunol.* **51**:390.

Bonnard, G. D., Grove, S., and Strong, D. M., 1983, A monoclonal antibody as a potential tool to isolate the human interleukin 2 (IL 2) receptor, in: *Roles of Lymphokines and Cytokines in Immunoregulation* (J. J. Oppenheim and S. Cohen, eds.), Academic Press, New York.

Chang, M. P., Makinodan, T., Peterson, W. J., and Strehler, B. L., 1982, Role of T cells and adherent cells in age-related decline in murine interleukin 2 production, *J. Immunol.* **129**:2426.

Crabtree, G., Munck, A., and Smith, K. A., 1980a, Glucocorticoids and lymphocytes. I. Increased glucocorticoid receptor level in antigen-stimulated lymphocytes, *J. Immunol.* **124**:2430.

Crabtree, G., Munck, A., and Smith, K. A., 1980b, Glucocorticoids and lymphocytes. II. Cell-cycle dependent changes in glucocorticoid receptor content, *J. Immunol.* **125**:13.

Darzinkiewicz, Z., Traganos, F., Sharpless, T., and Melamed, M. R., 1976, Lymphocyte stimulation: A rapid multiparameter analysis, *Proc. Natl. Acad. Sci. USA* **73**:2881.

de Weck, A. L., 1981, Lymphokines and other immunoactive soluble cellular products: Prospects for the future, in: *Lymphokines and Thymic Hormones: Their Potential Utilization in Cancer Therapeutics* (A. L. Goldstein and M. A. Chirigos, eds.), Raven Press, New York.

Gillis, S., Kozak, R., Durante, M., and Weksler, M. E., 1981, Immunologic studies of aging: Decreased production of and response to T-cell growth factor by lymphocytes from aged humans, *J. Clin. Invest.* **67**:937.

Gilman, S. C., Rosenberg, J. S., and Feldman, J. D., 1982, T lymphocytes of young and aged rats. II. Functional defects and the role of interleukin 2, *J. Immunol.* **126**:644.

Good, R. A., West, A. and Fernandes, G., 1980, Nutritional modulation of immune responses, *Fed. Proc.* **39**:3098.

Helderman, J. H., and Strom, T. B., 1979, Role of protein and RNA synthesis in the development of insulin binding sites on activated thymic-derived lymphocytes, *J. Biol. Chem.* **254**:2703.

Joncourt, F., Kristensen, F., and de Weck, A. L., 1981a, Ageing and immunity in outbred NMRI mice: Lack of correlation between age-related decline of the response to T cell mitogens, the antibody response to a T-dependent antigen and lifespan in outbred NMRI mice, *Clin. Exp. Immunol.* **44**:270.

Joncourt, F., Bettens, F., Kristensen, F., and de Weck, A. L., 1981b, Age-related changes of mitogen responsiveness in different lymphoid organs from outbred NMRI mice, *Immunobiology* **158**:39.

Joncourt, F., Kristensen, F., and de Weck, A. L., 1982a, Age-related changes in G0–G1 transition and proliferative capacity of mitogen-stimulated murine spleen cells, *Gerontology* **28**:281.

Joncourt, F., Wang, Y., Kristensen, F., and de Weck, A. L., 1982b, Aging and immunity: Decrease in interleukin 2 production and interleukin 2-dependent RNA synthesis in lectin-stimulated murine spleen cells, *Immunobiology* **163**:521.

Joncourt, F., Wang, Y., Kristensen, F., and de Weck, A. L., 1983, Age-related changes in the formation of glucocorticoid and insulin receptors during lectin-induced activation of human peripheral blood lymphocytes, *Immunoplogy*, in press.

Kay, M. M. B., 1980, Immunological aspects of aging, in: *Ageing*, Vol. 11 (M. B. Kay, J. E. Galpin, and T. Makinodan, eds.), Raven Press, New York.

Koponen, M., Grieder, A., and Loor, F., 1982, The effects of cyclosporins on the cell cycle of T lymphoid cell lines, *Exp. Cell. Res.* **140**:237.

Kristensen, F., and de Weck, A. L., 1982, Use of cytofluorometric techniques for assay of cell cycle regulating lymphokines, in: *Human Lymphokines: The Biological Immune Response Modifiers* (A. Khan and N. O. Hill, eds.), Academic Press, New York.

Kristensen, F., Joncourt, F., and de Weck, A. L., 1981, The influence of serum on lymphocyte cultures. II. Cell cycle specificity of serum action in spleen cells, *Scand. J. Immunol.* **14**:121.

Kristensen, F., Walker, C., Joncourt, F., Bettens, F., and de Weck, A. L., 1982a, Human lymphocyte proliferation. I. Correlation between activated and proliferating T lymphocytes, *Immunol. Lett.* **5**:59.

Kristensen, F., Walker, C., Bettens, F., Joncourt, F., and de Weck, A. L., 1982b, Assessment of IL 1 and IL 2 effects on cyclng and noncycling murine thymocytes, *Cell. Immunol.* **74**:140.

Kristensen, F., Bettens, F., Walker, C., Joncourt, F., and de Weck, A. L., 1983, Relationship between cell cycle events and interleukin 2 (IL 2) production, in: *Role of Lymphokines and Cytokines in Immunoregulation* (J. J. Oppenheim and S. Cohen, eds.), Academic Press, New York.

Krug, U., Krug, F., and Cuatrecasas, P., 1972, Emergence of insulin receptors on human lymphocytes during in vitro transformation, *Proc. Natl. Acad. Sci. USA* **69**:2604.

Lakatta, E. G., 1980, Age-related alterations in the cardiovascular response to adrenergic mediated stress, *Fed. Proc.* 39:3173.

Landman, R., Bittiger, H., and Bühler, F. R., 1981, High affinity beta-2-adrenergic receptors in mononuclear leucocytes: Similar density in young and old normal subjects, *Life Sci.* **29**:1761.

Leonard, W. J., Depper, J. M., Uchiyama, T., Smith, K. A., Waldmann, T. A., and Green, W. C., 1982, A monoclonal antibody that appears to recognize the receptor for human T cell growth factor: Partial characterization of the receptor, *Nature (London)* **300**:267.

Makinodan, T., and Kay, M. B., 1980, Age influence on the immune system, *Adv. Immunol.* **29**:287.

Masoro, E. J., Yu, B. P., Bertrand, H. A., and Lynd, F. T., 1980, Nutritional probe of the aging process, *Fed. Proc.* **39**:3178.

Miyawaka, T., Yachi, A., Uwadana, N., Ohzeki, S., Nagaoki, T., and Taniguchi, N., 1982, Functional significance of Tac antigen expressed on activated human T lymphocytes: Tac antigen interacts with T cell growth factor in cellular proliferation, *J. Immunol.* **129**:2474.

Munck, A., and Vira, C., 1975, Methods for assessing hormone–receptor kinetics with cells in suspension: Receptor bound and nonspecific bound hormone; cytoplasmic nuclear translocation, *Methods Enzymol.* **36**:255.

Pardee, A. B., Dubrow, R., Hamlin, H. L., and Kletzren, R. F., 1978, Animal cell cycle, *Annu. Rev. Biochem.* **47**:715.

Robb, R. J., Munck, A., and Smith, K. A., 1981, T cell growth factor receptors: Quantitation, specificity and biological relevance, *J. Exp. Med.* **154**:1455.

Roth, G. S., 1979, Hormone receptor changes during adulthood and senescence: Significance for aging research, *Fed. Proc.* **38**:1910.

Sartin, J., Chaudhuri, M., Obenrader, M., and Adelman, R. C., 1980, The role of hormones in changing adaptive mechanisms during aging, *Fed. Proc.* **39:**3163.

Slauson, D. O., Walker, C., Kristensen, F., Wang, Y., and de Weck, A. L., 1983, Mechanisms of serotonin-induced lymphocyte proliferation inhibition, *Cell. Immunol.* in press.

Smith, K. A., 1980, T cell growth factor, *Immunol. Rev.* **51:**337.

Smith, K. A., Lachman, L. B., Oppenheim, J. J., and Favata, M. F., 1980, The functional relationship of the interleukins, *J. Exp. Med.* **151:**1551.

Stadler, B. M., Kristensen, F., and de Weck, A. L., 1980, Thymocyte activation by cytokines: Direct assessment of G0–G1 transition by flow cytometry. *Cell. Immunol.* **55:**436.

Stadler, B. M., Dougherty, S., Farrar, J. J., and Oppenheim, J. J., 1981, Relationship of cell cycle to recovery of IL 2 activity from human mononuclear cells, human and mouse T cell lines, *J. Immunol.* **127:**1936.

Thoman, M. L., and Weigle, W. O., 1981, Lymphokines and aging: Interleukin 2 production and activity in aged animals, *J. Immunol.* **127:**2102.

Thoman, M. L., and Weigle, W. O., 1982, Cell-mediated immunity in aged mice: An underlying lesion in IL 2 synthesis, *J. Immunol.* **128:**2358.

Uchiyama, T., Broder, S., and Waldman, T. A., 1981, A monoclonal antibody (anti-Tac) reactive with activated and functionally mature human T cells. I. Production of anti-Tac monoclonal antibody and distribution of Tac (+) cells, *J. Immunol.* **126:**1393.

Walford, R. L., Liu, R. K., Gerbase-Delima, M., Mathies, M., and Smith, G. S., 1973, Longterm dietary restriction and immune function in mice: Response to sheep red blood cells and to mitogenic agents, *Mech. Ageing Dev* **2:**447.

Walker, C., Kristensen, F., Bettens, F., and de Weck, A. L., 1983, Lymphokine regulation of activated (G1) cells. I. Prostaglandin E2 induced inhibition of interleukin 2 production, *J. Immunol.* **130:**1770.

Wang, Y., Joncourt, F., Kristensen, F., and de Weck, A. L., 1983a, Cell cycle-related changes in number of T-lymphocyte receptors for glucocorticoids and insulin, *Int. J. Immunopharmacol.*

Wang, Y., Kristensen, F., Joncourt, F., Slauson, D. O., and de Weck, A. L., 1983b, Analysis of ^{3}H-histamine interaction with lymphocytes: Receptor or uptake?, *Clin. Exp. Immunol.* **54:**501.

Weksler, M. E., 1981, the senescence of the immune system, *Hosp. Prac.* October, p. 53.

Augmentation of Antibody Synthesis *in Vitro* by Thymosin Fraction 5

The Influence of Age

WILLIAM B. ERSHLER

1. INTRODUCTION

Previously, techniques available for the measurement of specific antibody synthesized *in vitro* have been complex and have usually required the addition of a nonspecific mitogen, such as pokeweed mitogen (PWM) (Fauci and Pratt, 1976; Thomson and Harris, 1977; Stevens and Saxon, 1978). We have recently described a one-step, non-mitogen-requiring assay for measuring antibody production *in vitro* (Ershler *et al.*, 1982; Moore *et al.*, 1984). One advantage of this assay is that with the addition to the culture medium of biological response modifiers such as thymosin or lymphokines, it allows the evaluation of these agents on this important immunological function. In this chapter, the effect of thymosin fraction 5 (TF5) on specific antitetanus and anti-influenza antibody production will be discussed. The *in vitro* effect of TF5 was found to vary with age, the most profound effect being observed in the advanced age groups (Ershler *et al.*, 1984b). The potential therapeutic implications of these observations will also be discussed.

2. THE MICROCULTURE ANTIBODY SYNTHESIS ENZYME-LINKED ASSAY (MASELA)

To measure specific antibody production *in vitro* we have developed a one-step microculture assay that is applicable to a variety of antigens. The details of this assay are published elsewhere (Ershler *et al.*, 1982; Moore *et al.*, 1984), but

WILLIAM B. ERSHLER • Department of Medicine, University of Vermont, Burlington, Vermont 05405.

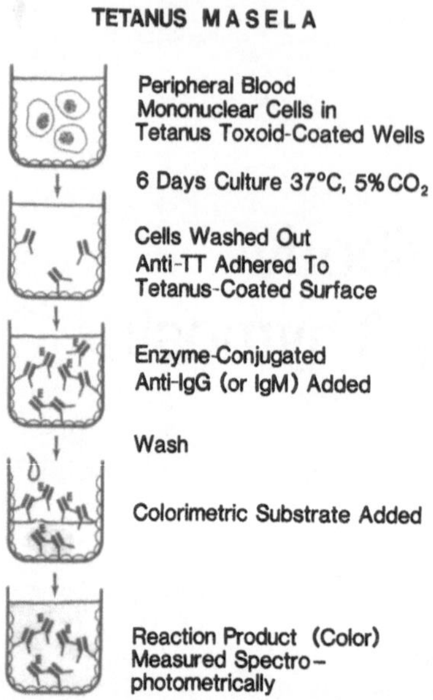

FIGURE 1. Schematic representation of the microculture antibody synthesis enzyme-linked assay for the measurement of *in vitro* production of antitetanus toxoid IgG or IgM. Peripheral blood mononuclear cells are cultured in tetanus toxoid-coated wells, washed out, and the antitetanus antibody measured by an enzyme-linked immunosorbent assay.

are schematically represented (for antitetanus antibody) in Fig. 1. Peripheral blood mononuclear cells are cultured in microtiter plates that are previously coated with antigen. After 7 days, the cultured cells are washed out and the antibody synthesized and secreted into the culture supernatant and adsorbed to the solid-phase antigen is measured by enzyme immunoassay. In the tetanus system, cells cultured prior

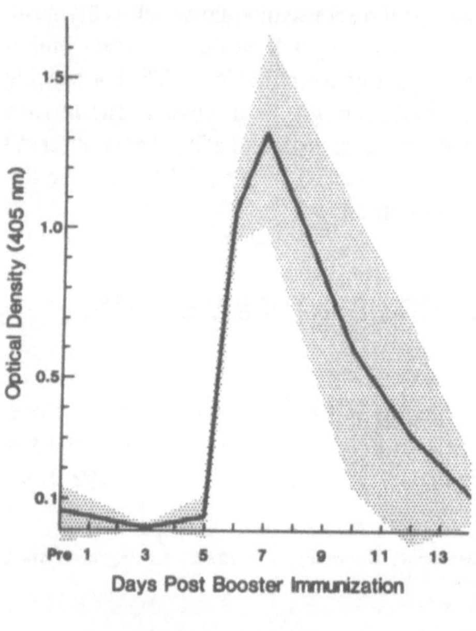

FIGURE 2. *In vitro* production of antitetanus IgG by peripheral blood mononuclear cells from 20 normal young volunteers cultured prior to 1 day and 3, 5, 6, 7, 9, 11, and 13 days after booster immunization. The antibody produced in the MASELA is measured spectrophotometrically (at 405 nm). The results depicted represent the mean (solid line) ± S.D. (shaded area) for the production at each of the study dates. Antibody production is negligible in cultures established up to 5 days after booster immunization, peaks at 7 days, and falls off abruptly thereafter. Cultures established 2 weeks after booster immunization produce antitetanus antibody at a level no different from those established in culture prior to immunization.

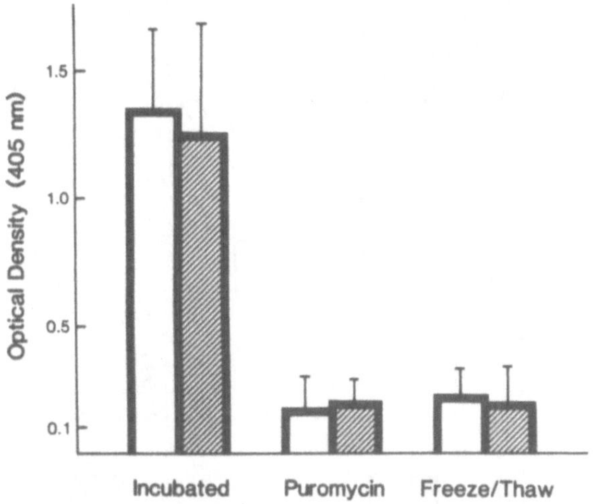

FIGURE 3. *In vitro* antitetanus antibody production in the MASELA by peripheral blood mononuclear cells from two normal volunteers (open bar and hatched bar) 7 days after immunization (incubated). In parallel cultures, antibody production was measured in wells that had added puromycin or were frozen and thawed. These experiments, and others, have indicated that intact, protein-synthesizing cells are required for antibody production in this assay.

to "booster" immunization produce negligible antitetanus antibody whereas production *in vitro* appears to be maximal in cultures established 7 days after the booster immunization (Fig. 2). Cells that were cultured after a brief freeze/thaw, or in the presence of puromycin, produced negligible antibody, indicating that intact cells, capable of protein synthesis, are required for antibody production *in vitro* (Fig. 3).

For the MASELA to be a useful tool in the assessment of biological response modifiers, it is important to know that cellular interactions are operant. Cell separation and remixture experiments (Moore *et al.*, 1984) have indicated that B cells or T cells cultured alone produce only small amounts of antibody, whereas when mixed, T cells augment the B-cell production of antibody in a dose–response manner (Fig. 4).

3. THYMOSIN AND *IN VITRO* ANTIBODY SYNTHESIS

Various thymic factors have been shown to influence T-cell function *in vitro* and *in vivo* (Pahwa *et al.*, 1979). Among these is the partially purified calf thymic preparation, TF5. This preparation has been demonstrated to enhance specific antibody sythesis when added (with PWM) to the culture medium with cells from recently immunized normal volunteers (Lowell *et al.*, 1980). Interestingly, nonspecific immunoglobulin was not enhanced by TF5. In the tetanus toxoid (TT)

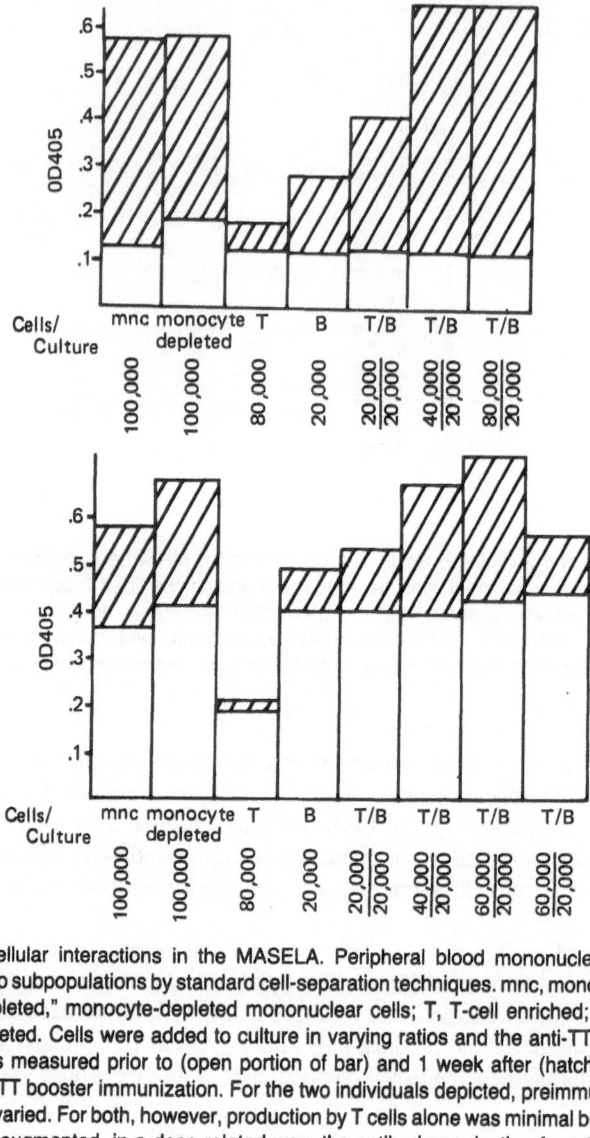

FIGURE 4. Cellular interactions in the MASELA. Peripheral blood mononuclear cells were fractionated into subpopulations by standard cell-separation techniques. mnc, mononuclear cells; "monocyte depleted," monocyte-depleted mononuclear cells; T, T-cell enriched; B, T-cell and monocyte depleted. Cells were added to culture in varying ratios and the anti-TT IgG antibody production was measured prior to (open portion of bar) and 1 week after (hatched plus open portion of bar) TT booster immunization. For the two individuals depicted, preimmunization anti-TT production varied. For both, however, production by T cells alone was minimal but the addition of this fraction augmented, in a dose-related way, the antibody production from the 7-day (but not the preimmunization) cultures of the B-cell fractions.

experiments in our laboratory, certain individuals demonstrated marked *in vitro* enhancement of specific antibody synthesis. The majority of young, healthy volunteers, however, had minimal if any change in *in vitro* antibody production with added TF5. This is in contrast to the results observed in samples from older volunteers. Nevertheless, when measuring specific anti-influenza antibody production, TF5 augmentation was observed in cultures established with cells from most young and old recently immunized donors (Fig. 5) (Ershler *et al.*, 1984a).

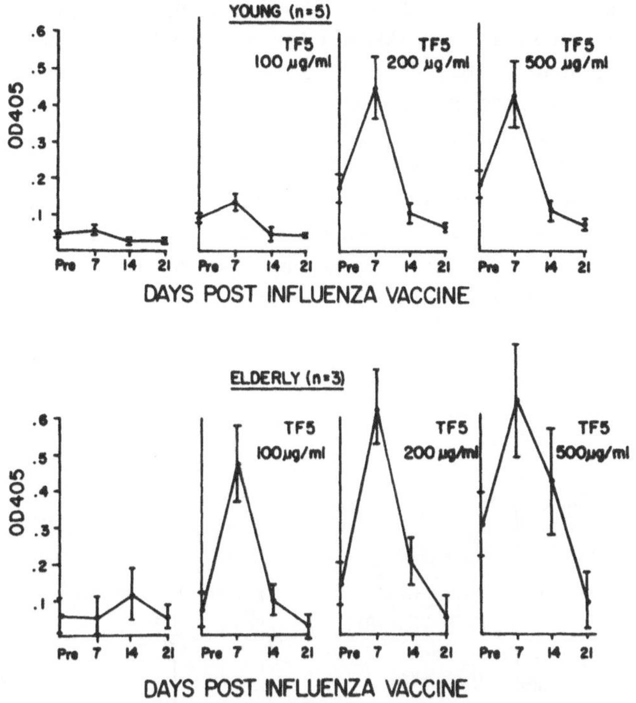

FIGURE 5. *In vitro* anti-influenza antibody production as determined by the MASELA technique in five young and three elderly volunteers sampled prior to and weekly × 3 after influenza vaccination. Without added TF5, no increase in *in vitro* antibody synthesis postimmunization is demonstrable under these conditions. There does, however, appear to be a dose-related enhancement of specific *in vitro* antibody synthesis which is maximal 1 week after immunization when TF5 is added. The TF5 effect is observed in both young and elderly volunteers.

4. AGING, SPECIFIC ANTIBODY SYNTHESIS, AND TF5

The immune deficiency associated with aging results primarily from thymic involution and loss of T-cell function (Makinodan *et al.*, 1980). Those specific antibody responses which require T-cell interaction have been shown to decrease with aging (Price and Makinodan, 1972; Weksler and Huuteroth, 1974), whereas response to T-cell-independent antigens is less altered with aging (Smith, 1976). TT is a complex protein antigen for which T-cell interactions are required for optimal antibody response (Willcox, 1975). To determine *in vitro* the reconstitutive capacity of TF5 for age-acquired immune deficiency, lymphocytes from elderly volunteers obtained prior to and after TT booster immunization were cultured in the MASELA with TF5 (50, 100, or 200 µg/ml) and the results compared to similar cultures established from young volunteers. Twenty-two young (< 30 years) and 12 elderly (> 65 years) volunteers were studied. After TT injection, plasma levels of anti-TT antibody, determined weekly by an enzyme-linked immunosorbent assay (ELISA), reached a significantly ($p < 0.01$) lower level in the older group (Fig. 6). *In vitro* production of anti-TT antibody (MASELA) was negligible prior to, and peaked 1 week after the immunization. When TF5 was added to the culture, there was a dose-related enhancement of anti-TT synthesis observed primarily in the elderly group. Weekly cultures were established in 10 from this group; in 7, striking enhancement with TF5 was observed. In 3 of the 12 younger volunteers similarly studied, there was some enhancement with TF5, but this was of a lesser magnitude ($p < 0.01$) (Fig. 7). Nonspecific immunoglobulin production *in vitro* was not enhanced by TF5, and thymosin α_1 or b_4 levels were not different in our two age groups (Ershler *et al.*, 1984b).

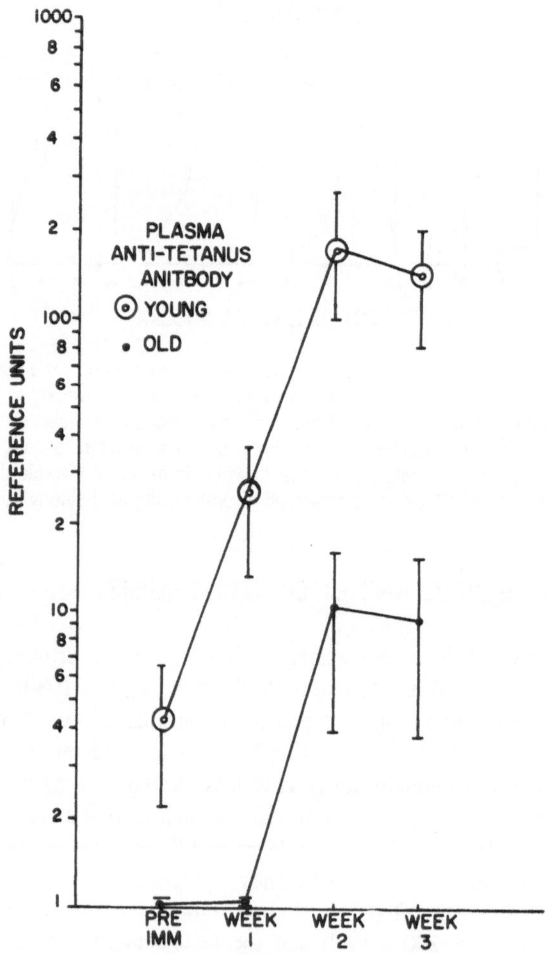

FIGURE 6. Plasma antitetanus antibody (IgG) levels (± S.E.M.). Plasma samples were diluted 1 : 1000 and anti-TT IgG was measured by ELISA. At each time point, young volunteers produced greater anti-TT IgG ($p < 0.01$) than the older volunteers.

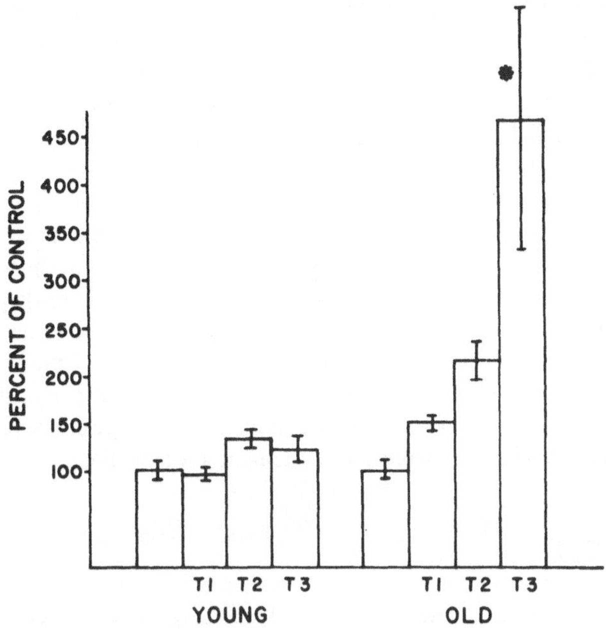

FIGURE 7. The effects of TF5 on *in vitro* antitetanus antibody synthesis. Peripheral blood lymphocytes obtained 1 week after booster immunization from young or old volunteers were cultured in tetanus-coated wells in the presence of TF5 (T1 = 50 μg/ml, T2 = 100 μg/ml, T3 = 200 μg/ml). Antibody synthesized in cultures without thymosin is assigned the control value of 100%. Change compared to the control is depicted as % change ± S.E.M. Antibody production increased slightly (but not significantly) in cultures from the young subjects, but increased dramatically in the older group. At the dose of 200 μgTF5/ml, the increased production was highly significant ($p < 0.01$).

5. CONCLUSIONS

The MASELA is a reliable and simple method for measuring specific antibody produced *in vitro*. It may prove to be a useful tool for assessing immune competence and for probing basic immune cellular interactions. Furthermore, it may become very useful for the *in vitro* evaluation of certain biological response modifiers or drugs. In this context it is apparent that TF5 may augment specific antibody synthesis. In our experiments, utilizing both TT and influenza antigen, *in vitro* antibody production was enhanced by TF5. These antigens are particularly relevant for the evaluation of TF5 because intact T-cell function in addition to the "effector" B-cell response is required for optimal production.

We have been concerned with the immune deficiency associated with aging. This acquired defect primarily involves the T-cell component of the immune response, and therefore specific antibody production in response to antigens which are known to require T-cell interaction (T-dependent) would be expected to be low in aged individuals. In this regard it is important to highlight the remarkable augmentation by TF5 of specific antibody production after the immunization of the elderly volunteers. In fact, this augmentation was greater for the elderly than for the young. We conclude, therefore, that with aging, the effector mechanism for antibody response is intact, but thymus-dependent-regulatory cells are deficient. These, in turn, may be reconstituted with the addition of the exogenous TF5.

These observations may be of some direct clinical importance. Immunization programs, such as for influenza, include the elderly, for this immune-deficient population is clearly at increased risk for infection. Immunization, however, is often not successful and vaccinated individuals have frequently proven not protected (Bentley, 1981; Ammann *et al.*, 1980). TF5 and other thymic hormone preparations have been utilized safely and with success in various immune-deficient states (Costanzi *et al.*, 1979; Goldstein *et al.*, 1982). Our studies would indicate that TF5 treatment concurrent with active immunization for immune-deficient elderly people would result in greater antibody production and, logically, more effective protection. Certainly such hypothesis is worthy of systematic investigation.

REFERENCES

Amman, A. J., Schiffman, G., and Austrian, R., 1980, The antibody response to pneumococcal capsular polysaccharides in aged individuals, *Proc. Soc. Exp. Biol. Med.* **164:**312.

Bentley, D. W., 1981, Pneumococcal vaccine in the institutionalized elderly: Review of past and present studies, *Rev. Infect. Dis.* **3**(Suppl.):61–70.

Costanzi, J., Daniels, J., Thurman, G., Goldstein, A., and Hokanson, J., 1979, Clinical trials with thymosin, *Ann. N.Y. Acad. Sci.* **322:**148–159.

Ershler, W. B., Moore, A. L., and Hacker, M. P., 1982, Specific *in vivo* and *in vitro* antibody response to tetanus toxoid immunization, *Clin. Exp. Immunol.* **49:**552–558.

Ershler, W. B., Moore, A. L., Hacker, M. P., Ninomiya, J., Naylor, P. B., and Goldstein, A. L., 1984b, Specific antibody synthesis *in vitro*. II. Age-associated thymosin enhancement of antitetanus antibody synthesis (submitted for publication).

Ershler, W. B., Moore, A. L., and Socinski, M. A., 1984a, Specific antibody synthesis *in vitro*. III. Correlation of *in vitro* antiboly response to influenza immunization in young and old subjects (submitted for publication).

Fauci, A. S., and Pratt, K. R., 1976, Activation of human B lymphocytes. I. Direct plaque-forming cell assay for the measurement of polyclonal activation and antigen stimulation of human B lymphocytes, *J. Exp. Med.* **144:**674–684.

Goldstein, A. L., Low, T. L. K., Thurman, G. B., Zatz, M., Hall, N. R., McClure, J. E., Hu, S.-K., and Schulof, R. S., 1982, Thymosins and other hormone-like factors of the thymus gland, in: *Immunological Approaches to Cancer Therapeutics* (E. Mihich, ed.), Wiley, New York.

Lowell, G. H., Smith, L. F., Klein, D., and Zollinger, W. D., 1980, Thymosin stimulates *in vitro* secretion of antibacterial antibodies by human peripheral blood lymphocytes, *Clin. Res.* **28:**A353.

Makinodan, T., and Kay, M. M. B., 1980, Age influence on the immune system, *Adv. Immunol.* **29:**287–330.

Moore, A. L., Ershler, W. B., and Hacker, M. P., 1984, Specific antibody synthesis in vitro. I. Technical considerations, *J. Immunol. Methods* (in press).

Pahwa, R., Ikehara, S., Pahwa, S. G., and Good, R. A., 1979, Thymic function in man, *Thymus* **1:**27–58.

Price, G. B., and Makinodan, T., 1972, Immunologic deficiencies in senescence. I. Characterization of intrinsic deficiencies, *J. Immunol.* **180:**403–412.

Smith, A. M., 1976, The effects of age on the immune response to type III pneumosaccharide (SIII) and bacterial lipopolysaccharide (LPS) in Balb/c, SJL/J and C3H mice, *J. Immunol.* **116:**469–474.

Stevens, R. H., and Saxon, A., 1978, Immunoregulation in humans: Control of antitetanus toxoid antibody production after booster immunization, *J. Clin. Invest.* **62:**1154–1160.

Thomson, P. D., and Harris, N. S., 1977, Detection of plaque-forming cells in the peripheral blood of actively immunized humans, *J. Immunol.* **118:**1480–1482.

Weksler, M. E., and Huuteroth, H., 1974, Impaired lymphocyte function in aged humans, *J. Clin. Invest.* **53:**99–104.

Willcox, H. N. A., 1975, Thymus dependence of the antibody response to tetanus toxoid in mice, *Clin. Exp. Immunol.* **22:**341–347.

28

The Production and Characterization of Monoclonal Antibodies to Lymphokines

KENDALL A. SMITH

1. INTRODUCTION

As the experimental procedures and results of the studies presented at this conference are described in detail in a recent publication (Smith *et al.*, 1983), this communication will provide a discussion of the implications of the data. The interested reader is referred to the primary publication for experimental details.

Antibodies reactive with potent biologically active molecules can be exceedingly useful reagents, as exemplified by the impact that antihormones had on endocrine research. The development of the radioimmunoassay for hormones provided the critical breakthrough necessary to proceed beyond ambiguous and lengthy bioassays: for the first time, hormonal effect could be examined with an exact knowledge of hormone concentration. The subsequent demonstrations, that the biological effects were mediated by very low levels of hormones and that the magnitude of the responses varied according to hormone concentration, led to a search for high-affinity, specific cellular receptors. Once it was realized that low hormone concentrations were active, reason dictated that such receptors had to exist to permit the hormone–cellular interaction. Thus, it is especially noteworthy that as endocrinology benefited from immunology, in the field of lymphokine research, endocrinology has provided the intellectual and technical foundation necessary for the development of monoclonal antibodies to these hormonelike moieties.

At the time the first antisera were raised against insulin, 30 years of physiological and pathophysiological research had led to the isolation and large-scale extraction of this important hormone. Consequently, the development of antisera reactive to the naturally occurring peptide was facilitated by the availability of

KENDALL A. SMITH • Department of Medicine, Dartmouth Medical School, Hanover, New Hampshire 03756.

sufficient quantities of the hormone so that immunization was possible. When considering the development of antibodies to lymphokines, the problems confronting the investigator appear more difficult, for unlike insulin most lymphokines exist as biological activities that have yet to be ascribed to single molecules. Moreover, adequate quantities of the activities or "factors" have not been prepared. Thus, the most immediate problem relates to the acquisition of sufficient quantities of the moiety so that immunization is feasible.

The term "factor" has been very appropriately used for most lymphokines, for their molecular nature and mechanism of interaction with target cells remain obscure. However, an endocrine approach to the lymphokine that promotes T-cell growth, interleukin-2, provided the understanding and insight necessary for a quantum leap forward, which facilitated the development of IL-2 monoclonal antibodies. Progress in the biochemical purification of IL-2, which was possible only because of the availability of the rapid and quantitative IL-2 bioassay (Gillis *et al.*, 1978), permitted the isolation of biosynthetically radiolabeled IL-2 and the construction of an IL-2 radioreceptor assay (Robb *et al.*, 1981; Smith, 1983). For the first time, a quantitative measurement of the number and affinity of IL-2 receptors became possible. In addition to the understanding this advance provided as to the mechanism whereby IL-2 interacts with target cells, these measurements allowed the expression of the biological activity of this "factor" in molecular terms. Detailed studies revealed that the binding and biological response curves for IL-2 were virtually identical. Thus, it was possible to establish that murine cytolytic IL-2-dependent T-cell clones bound and responded to IL-2 at concentrations ranging from 0.5 to 5×10^{-11} M. As well, 1 U/ml of IL-2 activity could be calculated as equal to 5×10^{-10} M. These numbers illuminated the fact that lymphocyte-conditioned medium containing 1 U/ml IL-2 activity (the concentration generally found in normal lymphocyte conditioned medium) contained only 8 μg/liter of IL-2 protein. Consequently, for the first time, it was realized that 50–100 liters of such conditioned medium would be required to accumulate only 1.0 mg of IL-2 protein.

2. PREPARATION OF THE ANTIGEN AND IMMUNIZATION

Fortunately, it was discovered simultaneously by several investigators (Farrar *et al.*, 1980; Gillis and Watson, 1980; Shimizu *et al.*, 1980) that neoplastic T-cell lines produced 10- to 100-fold higher quantities of IL-2 than could be obtained from normal lymphocytes. As well, under the proper conditions, IL-2 could be produced from these cell lines without a serum requirement. This finding was of considerable help in the concentration and purification of IL-2, for the inclusion of 10% fetal calf serum in the tissue culture medium results in a protein concentration of 7 g/liter. Thus, if IL-2 is produced under serum-containing conditions, one would expect to have 800 μg/liter (assuming 100 U/ml; 1 U = 8 ng) of IL-2 protein in the presence of 7 g/liter of serum protein. Accordingly, to obtain IL-2 free of contaminating proteins, an 8750-fold purification would be required. In practice, under serum-free conditions, we found that the protein concentration was generally 50–60 mg/liter. Thus, only a 75-fold purification was ultimately necessary. It is worthy of emphasis that it is not necessary to immunize with homogeneous antigen for the production of monoclonal antibodies, as would be the case for antigen-specific antisera. However, it is necessary to concentrate the antigen to a small volume if mice are to be immunized. The presence of 7 g of serum protein per liter of conditioned medium, even if 800 μg of antigen is present, effectively precludes concentration by any method other than ammonium sulfate precipitation. Unfortunately, salt fractionation results in significant losses (25–50%) of IL-2. Using serum-free conditioned medium, it is possible to concentrate several liters of fluid to a very small volume simply through filtration. Moreover, owing to the low protein concentration in the conditioned medium, it is possible to use only two separative steps [gel filtration and isoelectric focusing (IEF)] and obtain 50% recovery of starting material in a highly purified and concentrated form (Robb and Smith, 1981).

A standard protocol for immunization was found to be sufficient, provided microgram quantities (calculated on the basis of the number of units of IL-2 inoculated) of IL-2 were used. Numerous attempts where submicrogram amounts of IL-2 protein were inoculated failed to stimulate circulating anti-IL-2 activity, and fusions using the splenocytes of such mice were unproductive. This experience was predictable in view of the knowledge that the magnitude of the immune response is antigen-concentration dependent. However, these experiments were performed prior to the determination that 1 U of biological activity was equivalent to only 8 ng of IL-2 protein. After each immunization, mice were bled and the plasma tested for anti-IL-2 activity by three different assays: neutralization of IL-2 biological activity, precipitation of radiolabeled IL-2, and enzyme-linked immunoassay (ELISA). In this manner, it was possible to select those immunized mice prior to cellular fusion that contained antibody-producing cells.

3. SCREENING FOR ANTIBODY ACTIVITY

After multiple attempts to identify anti-IL-2-secreting hybridomas using the bioassay for the screening procedures, we turned to immunoassays. The major problem of the bioassay related to false-positive tests. Because, for the most part the hybridoma supernatants were taken from overgrown culture wells, it was difficult to discriminate between the presence of specific antibody and nonspecific suppression of cellular proliferation due to the depletion of nutrients from "spent" hybridoma supernatants. Moreover, two-stage assays involving precipitation of putative antibody–IL-2 complexes suffered from the same difficulty, in that the spent medium still had to be tested for suppression of cellular proliferation. In retrospect, had neutralization of biological activity been the only assay available, it is almost certain that all of the antibodies that were ultimately identified would have remained undetected, as relatively high concentrations of antibody were subsequently found necessary for neutralization. Moreover, the antibody subsequently found to be the most useful for affinity purification of IL-2 (DMS-3) lacks neutralizing activity and thus would have been missed entirely. The relatively high antibody concentrations required for neutralization of biological activity were anticipated considering the parameters of IL-2 receptor binding performed with intact cells and isolated plasma membranes. IL-2 receptors bind radiolabeled IL-2 with a remarkably high affinity (K_d = 5 to 20 × 10^{-12} M), whereas equilibrium dissociation constants for individual monoclonal antibodies have generally been reported to be considerably higher, in the range of 1 × 10^{-5} to 1 × 10^{-9} M (Ehrlich *et al.*, 1982; Peterfy *et al.*, 1983).

An ELISA using highly purified IL-2 to coat plastic microwells proved to be the most sensitive and rapid method to screen hybridoma supernatants. We found that it was posssible to use relatively small quantities of IL-2 for this purpose (30 ng/well), and therefore, only utilized the most highly purified IL-2 preparations. Again, this proved to be crucial to avoid ambiguities in the interpretation of positive assay results, especially if only partially purified material had been used for immunization. Although the material used for the ELISA was not subjected to analytical tests to ensure homogeneity, it contained a high concentration of IL-2 biological activity that was uniform with respect to charge when examined by IEF, and size when examined by SDS–PAGE. Although it is possible that non-IL-2 proteins with a pI and molecular size identical to IL-2 may have contaminated the material used for the screening ELISA, such putative contaminants were most probably present in minor amounts, as the ELISA screening assay identified the hybridomas that were subsequently found to secrete antibodies that react specifically to IL-2. It is evident that the use of only partially purified lymphokine for both the immunization and screening steps generates false-positive reactions that make interpretation of the immunoassay results almost impossible.

4. DEMONSTRATING ANTIBODY SPECIFICITY

Once putative antilymphokine-secreting hybridomas have been selected by a solid-phase antigen-binding immunoassay, it is necessary to prove the specificity of the antibody reactivity. We approached this problem through assays for the neutralization of IL-2 biological activity and assays for the immunoadsorption of biological activity. It is worthy of mention in this regard that two recent reports described neutralizing antibodies to IL-2 (Gillis and Henney, 1981; Stadler *et al.*, 1982). However, the data presented in both reports fall short of providing convincing evidence that the hybridoma products are in fact antibodies reactive with IL-2. As the biological activity of IL-2 is the promotion of T-cell proliferation, one must identify inhibition of proliferation by hybridoma products. As many additives to cultures can suppress cellular proliferation, either through the removal of essential nutrients or by suppressing general cellular metabolism, it is crucial to control for these possibilities.

We found that it was necessary to purify the monoclonal antibodies prior to testing them for neutralizing activity. In addition, several experimental approaches were required to provide conclusive data in support of the IL-2-specific reactivity of the antibodies. For example, it was demonstrated that the neutralizing antibodies also inhibited radiolabeled IL-2 T-cell binding and that the concentrations of antibodies required for both effects were identical. These findings suggested quite strongly that the mechanism of the neutralizing effect was unrelated to nonspecific suppression of cellular metabolism, as IL-2 binding does not require metabolic energy (binding occurs to glutaraldehyde-fixed cells and isolated plasma membranes) (Robb *et al.*, 1981; Smith, 1983). Moreover, the antibody neutralization was found to be IL-2-species specific: whereas human, murine, and gibbon ape IL-2 activities were all neutralized, rat IL-2 activity remained unaffected. This observation provided additional evidence indicating that IL-2 binding was responsible for the suppression of T-cell proliferation, rather than nonspecific suppression of cellular metabolism. If the inhibition of the T-cell proliferation were nonspecific, it should occur regardless of the source of IL-2. Further evidence against a nonspecific effect was the finding that the proliferation of IL-2-independent cells remained unaffected by antibody concentrations that completely suppressed IL-2-dependent T-cell proliferation. A final test for specificity, not included in the previous studies (Gillis and Henney, 1981; Stadler *et al.*, 1982), was provided by IL-2–antibody competition experiments. A maximal inhibitory concentration of antibody was chosen and increasing concentrations of IL-2 were added to the cultures. If the antibody-mediated neutralization of IL-2 activity resulted from binding to IL-2, then an excess of IL-2 would be expected to circumvent the antibody effect. This expectation was met exactly, for IL-2 overcame the antibody-mediated neutralization in a concentration-dependent fashion.

5. MONOCLONAL ANTIBODIES AS LYMPHOKINE-PURIFICATION REAGENTS

As the purification of lymphokines is one of the major obstacles to the assignment of a particular biological activity to a single molecule, rather than a "factor" or "factors," the use of monoclonal antibodies as solid-phase immunoadsorbants provides a particularly attractive approach. Our experience with the use of the DMS monoclonal antibodies in this way has been exceedingly satsifying. IL-2 present in crude conditioned medium can be concentrated from several liters to a few milliliters by using a high concentration of antibody complexed to a small volume of solid supporting material (e.g., 20 mg antibody coupled to 1 ml of gel). The theoretical molecular binding capacity of such an immunoadsorbant is 1.2×10^{17} molecules (assuming 15 mg antibody/ml gel). As the most active conditioned media contain IL-2 at a concentration of 5.0×10^{-8} M, 1 liter should contain 3×10^{16} IL-2 molecules. Thus, a 1-ml antibody affinity column can theoretically bind all of the IL-2 from 40 liters of conditioned medium. In practice, we find that the DMS-3 antibody or the DMS-3 antibody together with the DMS-1 antibody, yields recoveries ranging from 50 to 80% of the starting material after application of conditioned medium and extensive washing. The IL-2 activity that binds, and which can be subsequently eluted, retains biological activity and is homogeneous within the limits of detection by SDS–PAGE, reverse-phase liquid chromatography and amino acid sequence analysis (Copeland *et al.*, Chapter 17, this volume). Thus, as it is possible to obtain milligram quantities of human IL-2 in pure form, we can be cautiously optimistic that biological experiments performed with this material can be interpreted unambiguously.

6. CONCLUSIONS

Monoclonal antibodies reactive to lymphokines can be produced quite easily, provided the individual properties of each moiety are considered. The major initial obstacle relates to the accumulation of enough of the material in a concentrated and at least partially purified form. For IL-2 this was accomplished because a rapid unambiguous bioassay was available, so that the biological activity could be quantified during analytical and preparative biochemical separative procedures. Moreover, the availability of high-producer cells and the development of the radiolabeled IL-2 binding assay permitted, for the first time, the means to determine the molecular concentrations of the biological activity. With these characteristics firmly in mind, it was relatively simply to apply discriminative immunoassays that ultimately allowed for the identification of the hybridomas secreting anti-IL-2. Thus, the future of lymphokine research appears quite promising. With the concept that we are dealing with unique polypeptide hormones, and the availability of monoclonal antibodies, we should be able to proceed quite quickly to studies of the mechanisms of lymphokine action and their role in the regulation of the immune response.

REFERENCES

Ehrlich, P. H., Moyle, W. R., Moustafa, Z. A., and Canfield, R. E., 1982, Mixing two monoclonal antibodies yields enhanced affinity for antigen, *J. Immunol.* **128:**2709.

Farrar, J. J., Fuller-Farrar, J., Simon, P. L., Hilfiker, M. L., Stadler, B. M., and Farrar, W. L., 1980, Thymoma production of T-cell growth factor, *J. Immunol.* **125:**2555.

Gillis, S., and Henney, C. S., 1981, The biochemical and biological characterization of lymphocyte regulatory molecules. VI. Generation of a B cell hybridoma whose antibody product inhibits interleukin 2 activity, *J. Immunol.* **126:**1978.

Gillis, S., and Watson, J., 1980, Biochemical and biological characterization of lymphocyte regulatory molecules. V. Identification of interleukin 2-producing human leukemia T-cell line, *J. Exp. Med.* **152:**1709.

Gillis, S. M., Ferm, M., Ou, W., and Smith, K. A., 1978, T-cell growth factor: Parameters of production and a quantitative microassay for activity, *J. Immunol.* **120:**2027.

Peterfy, F., Kuusela, P., and Makela O., 1983, Affinity requirements for antibody assays mapped by monoclonal antibodies, *J. Immunol.* **130:**1809.

Robb, R. J., and Smith, K. A., 1981, Heterogeneity of human T-cell growth factor (TCGF) due to variable glycosylation, *Mol. Immunol.* **18:**1087.

Robb, R. J., Munck, A., and Smith, K. A., 1981, T-cell growth factor receptors: Quantitation, specificity, and biological relevance, *J. Exp. Med.* **154:**1455.

Shimizu, S., Konaka, Y., and Smith, R. T., 1980, Mitogen-initiated synthesis and secretion of T cell growth factor(s) by a T-lymphoma cell line, *J. Exp. Med.* **152:**1436.

Smith, K. A., 1983, T-cell growth factor, a lymphocytotrophic hormone, in: *Genetics of the Immune Response* (G. Moller, ed.), Plenum Press, New York.

Smith, K. A., Favata, M. F., and Oroszlan, S., 1983, Production and characterization of monoclonal antibodies to human interleukin 2: Strategy and tactics, *J. Immunol.* **131:**1808–1815.

Stadler, B. M., Berenstein, E. H., Siraganian, R. P., and Oppenheim, J. J., 1982, Monoclonal antibody against human interleukin 2 (IL-2). I. Purification of IL-2 for the production of monoclonal antibodies, *J. Immunol.* **128:**1620.

29

Modulatory Interactions between the Central Nervous System and the Immune System

A Role for Thymosin and Lymphokines

NICHOLAS R. HALL, JOSEPH P. McGILLIS, BRYAN L. SPANGELO, GEORGE V. VAHOUNY, and ALLAN L. GOLDSTEIN

1. INTRODUCTION

The immune system is comprised of varied cell types with diverse characteristics and functions, but which ultimately contribute to the maintenance of host defense. Within the central and peripheral immunological tissues are numerous mechanisms by which immunogenesis and active immunity are regulated. These include direct contact between cells as well as the elaboration of biologically active products. Most of these products have been characterized based on their biological function. Hence, T-cell growth factor, migration inhibitory factor, colony-stimulating factor, as well as a host of others, constitute a long list of growth-promoting or -regulating agents. Thymosin peptides, first isolated from extracted thymic tissue, also play an important immunoregulatory role during the course of T-cell differentiation. These as well as other functions of lymphokines and thymosins are discussed at greater length elsewhere in this volume. However, in addition to the lymphokines and thymic factors which were originally detected and defined during the course of

---

NICHOLAS R. HALL, JOSEPH P. McGILLIS, BRYAN L. SPANGELO, GEORGE V. VAHOUNY, and ALLAN L. GOLDSTEIN • Department of Biochemistry, The George Washington University School of Medicine and Health Sciences, Washington, D. C. 20037.

investigations of immune function, hormones that were classically defined within the context of a nonimmunological function can also influence host defense.

Glucocorticoids have been found to enhance or to inhibit lymphocyte function depending on their relative concentration (see Monjan, 1981, and Hall and Goldstein, 1984, for review of this topic). Reproductive hormones can also exert profound influences on the immune system, a phenomenon that is thought to underlie the fact that females and males differ in their immunological capacities (Nicol and Bilbey, 1960; Thompson *et al.*, 1969). Other hormones, such as insulin, growth hormone, prolactin, and circulating catecholamines, have immunomodulatory properties, but these are not as well defined as are the influences of reproductive and adrenal steroids. The effects of these various hormones are discussed at more deserving length elsewhere (Ahlqvist, 1976; Bourne *et al.*, 1974). They are briefly mentioned in this introduction to illustrate the fact that there are numerous immunomodulatory influences which are not produced by components of the classically defined immune system. Furthermore, these hormones are influenced by hypothalamic–hypophyseal pathways as well as by the sympathetic and parasympathetic branches of the autonomic nervous system (Renaud, 1984; Maclean and Reichlin, 1981).

An increasing body of evidence supports the hypothesis that hormonal influences on immunity are not unidirectional, but are part of a complex and reciprocal modulatory circuit in which the hormones can also be influenced by products of the activated immune system (Hall *et al.*, 1982; Besedovsky and Sorkin, 1977). This influence can occur at the level of the CNS, as well as at the receptor associated with the lymphoid target cell. As the complex interrelationships between immunological and other products become better defined, it is apparent that thymosins and lymphokines exert not only direct effects upon host defense, but may also have indirect effects via diverse neuroendocrine circuits. The evidence in support of this concept is as follows.

2. EFFECTS OF SYSTEMIC INJECTION OF THYMOSINS AND LYMPHOKINES ON THE GLUCOCORTICOID AXIS

During the course of the primary immune response, there occurs a significant rise in serum levels of corticosterone (Besedovsky and Sorkin, 1977). The elevation in steroid concentration coincides with the production of antibody and might be the consequence of thymosin and/or lymphokine release. Both families of immunomodulatory products have been shown to stimulate steroidogenesis *in vivo* (Besedovsky *et al.*, 1981; McGillis *et al.*, 1982).

Thymosin fraction 5, when injected intraperitoneally into adult male rats, results in a significant rise in circulating corticosterone that is both time and dose dependent. Table I reveals that although the handling procedure resulted in a presumed stress release of corticosterone, the rats that received thymosin fraction 5 displayed levels that were significantly greater than in vehicle-injected control rats. Furthermore, the levels remained elevated hours following the injection, returning to normal by 24 hr. The 2-hr values were comparable to those observed in rats that were injected with ACTH. Similar results have been observed following the systemic

TABLE I. Systemic Injection of Thymosin Fraction 5 (F5) Increases Serum Corticosterone in Wistar Rats

Treatment		Time after injection 0.5 hr	2 hr	24 hr
Noninjected	55.3 ± 15.3[a]			
Saline	—	146.5 ± 45.6	45.4 ± 11.2	62.9 ± 13.2
Thymosin F5, 2 μg/kg	—	161.6 ± 38.8	55.2 ± 9.5	40.0 ± 11.7
Thymosin F5 20 μg/kg	—	209.3 ± 44.6	136.1 ± 38.0*	58.2 ± 15.9
ACTH	—	—	115.0 ± 31.7	

[a] ng/ml, mean ± S.E.M.
*$p < 0.05$ vs. saline.

injection of thymosin fraction 5 into rabbits as well as monkeys (Sivas *et al.*, 1982; Healy *et al.*, Chapter 34, this volume). Lymphokines can also promote steroidogenesis. Supernatants of Con A-stimulated lymphocytes have been found to stimulate a significant rise in corticosterone in rats (Besedovsky *et al.*, 1981). Partially purified lymphokine preparations that are provided commercially have also been found to cause a rise in glucocorticoid levels in humans (Dumonde *et al.*, 1982).

It is possible that the corticogenic effects of both preparations are due to a shared component. Thymosin α_1, a 28-amino-acid polypeptide that is a constituent of thymosin fraction 5, is also present in the supernatants of Con A-stimulated lymphocytes and of certain cultured T-lymphoma cell lines (see Zatz *et al.*, Chapter 15, this volume). Furthermore, preliminary data suggest that thymosin α_1 is capable of stimulating steroidogenesis via a mechanism involving the CNS. These data will be reviewed in a subsequent section, in which the results of intracerebral injections of thymosin peptides will be discussed.

It is apparent from these observations that the immunomodulatory effects of glucocorticoids may not be simply the adverse consequence of generalized stress. Instead, they might constitute part of an immunoregulatory circuit by which homeostatic adjustments of immune function are brought about. The mechanism by which the hypothalamic–hypophyseal-adrenocortical axis is regulated appears to occur in part at the level of the CNS. This tentative conclusion is based on evidence that (1) there is no direct effect of thymosins or lymphokines on either adrenal fasciculata cells or pituitary tissue and (2) that the intracerebral injection of thymosin results in elevated corticosterone, but at concentrations that are ineffective when injected systemically.

3. EVIDENCE FOR A CNS SITE OF THYMOSIN-INDUCED CORTICOGENESIS

Isolated adrenal fasciculata cells normally respond to ACTH treatment by elaboration of cAMP as well as glucocorticoids. This model was used to evaluate the potential direct effects that various lymphokines and thymosin preparations might

TABLE II. Thymosin F5 and Thymosin α_1 Do Not Directly Stimulate Corticosterone Production by Isolated Adrenal Fasciculata Cells[a]

Test material	Corticosterone ($ng/2.5 \times 10^5$ cells)
ACTH (0.71 nM)	665 ± 15
Thymosin F5 (20 μg/ml)	11 ± 3
Thymosin α_1 (5×10^{-11} M)	23 ± 0
ACTH (0.28 nM)	323 ± 15
ACTH + F5	326 ± 10
ACTH + α_1	319 ± 40

[a] From Vahouny *et al.* (1983).

exert on these cells (Vahouny *et al.*, 1983). The results of this study are summarized in Tables II and III and reveal that none of the preparations tested had any effects on these adrenal–cortical cells. Neither cAMP nor corticosterone release was affected by the various incubation regimens. The possibility was considered that these substances might act to potentiate the stimulatory effects of ACTH. This question was addressed by adding the various preparations to a suboptimal dose of ACTH. Using the same assay system described above, there was no evidence of synergism as measured by levels of cAMP or by corticosterone release. Consequently, it is considered unlikely that the corticogenic effects of thymosins and lymphokines are due to direct stimulation of the adrenal cortex.

Another possible site of action are the ACTH-producing cells within the pituitary gland. However, preliminary studies suggest that this is not the case, at least with respect to thymosin fraction 5 (McGillis *et al.*, unpublished observations). To evaluate this possibility, minced pituitary tissue has been rapidly dissected from adult rats and superfused with thymosin fraction 5. Whereas corticotropin-releasing factor triggered a significant release of ACTH into the medium, thymosin fraction 5 was without effect. Concentrations that were tested ranged from 1 pg to 1 mg and as can be seen in Fig. 1, there was no alteration in the profile of ACTH release. The pituitary cells were still viable after thymosin treatment as superfusion with

TABLE III. Thymosin F5 and Thymosin α_1 Do Not Directly Stimulate cAMP Production by Adrenal Membranes[a]

Test material	cAMP (pmoles/mg/15 min)
ACTH (2.5×10^{-6} M)	10.7 ± 1.8
Thymosin F5 (1 μg)	1.1 ± 0.4
Thymosin F5 (10 ng)	0.8 ± 0.4
Thymosin α_1 (6.4×10^{-8} M)	1.5 ± 1.8
Thymosin α_1 (1.9×10^{-7} M)	1.6 ± 3.2

[a] From Vahouny *et al.* (1983).

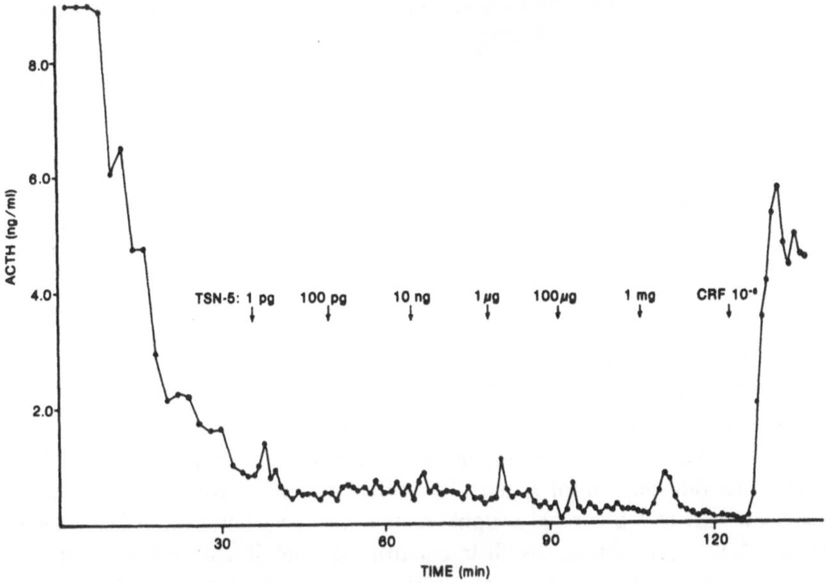

FIGURE 1. Profile of ACTH release from pituitary tissue following superfusion with thymosin fraction 5 (TSN 5) and corticotropin-releasing factor (CRF).

CRF was able to induce ACTH release after the thymosin had been evaluated. As a result of this study, thymosin fraction 5 does not appear to stimulate corticogenesis via direct stimulation of ACTH-producing cells. The possibility of synergism with CRF is currently being investigated.

An additional possibility is that thymosin fraction 5, as well as the other preparations, stimulate the adrenal axis at the level of the CNS. Ongoing studies indicate that this is the most likely site of action, for the injection of very small concentrations of thymosin into the brain can result in stimulation of the adrenal–cortical axis. This has been demonstrated using thymosin fraction 5 as well as its component peptide, thymosin α_1.

In the first investigation of intracerebrally injected thymosin, adult male rats were fitted with stainless steel guide tubes that were inserted into the calvarium overlying the anterior extent of the hypothalamus. Guide tubes were positioned stereotaxically while the rats were deeply anesthetized. All animals were allowed a 2-week recovery period before being randomly assigned to one of four experimental groups.

One group was injected with 10 μg of thymosin fraction 5 dissolved in 0.2 ml of phosphate-buffered saline. A peptide control group received the same concentration of kidney fraction 5 while a third group was not injected. A fourth group was subjected to the same surgery, but instead was injected with 10 μg of thymosin fraction 5 intraperitoneally. Injections were given for five consecutive days after which the animals were sacrificed. The brains were perfused with 10% formol-

TABLE IV. Effect of Intracerebroventricular (i.c.) Injection of Thymosin F5 on Endocrine Tissue Weights[a]

Tissue	Noninjected controls	+ Kidney F5 i.c. injection	+ Thymosin F5 i.c. injection	+ Thymosin F5 i.p. injection
Thyroid	26 ± 3	30 ± 2	33 ± 2	28 ± 6
Testes	370 ± 4	369 ± 17	375 ± 10	359 ± 10
Adrenal	70 ± 4	81 ± 9	114 ± 8*	74 ± 5

[a] mg wet weight ($\bar{X}$ ± S.E.M.).
*$p < 0.025$ vs. noninjected and i.p. control groups.

saline and serially sectioned for histological verification of the implant site. Wet tissue weights were determined at the time of sacrifice.

The results of this study are summarized in Table IV. The animals receiving intrahypothalamic injections of thymosin fraction 5 had significantly larger adrenal glands than did the control group. This effect was not attributable to injection-induced stress, as the adrenal weights of rats receiving intrahypothalamic kidney fraction 5 was not statistically different from the uninjected control group. The effect was not due to leakage of the thymosin out of the brain resulting in direct stimulation of the adrenal gland. This possibility was negated by the data from animals that received i.p. injections of thymosin. Their adrenal weights did not differ from the uninjected control values.

Histological evaluation of the brains from these animals revealed that most of the injections were in either the arcuate or the ventromedial nucleus. When animals were subgrouped on the basis of the precise hypothalamic nucleus injected, it was found that the heaviest adrenal weights occurred in rats that were injected with thymosin into the ventromedial nucleus (mean ± S.E.M. 127 ± 17) followed by the arcuate nucleus (111 ± 6). One animal, whose implant was in the medial-preoptic area, had a combined adrenal weight of 95 mg. All of these weights were greater than those from rats receiving vehicle injections in the same brain regions. It is perhaps noteworthy that during the time of peak antibody titer following primary sensitization with an immunogen, there is a threefold increase in the electrical firing rate of neurons in the ventromedial nucleus, implicating this hypothalamic nucleus in the modulation of immunity (Besedovsky and Sorkin, 1977).

TABLE V. Effect of Intracerebral Injection of Thymic Peptides on Serum Luteinizing Hormone and Corticosterone Concentration

Treatment group	Corticosterone (ng/nl ± S.E.M.)	Luteinizing hormone (ng/ml ± S.E.M.)
Noninjected control	118.1 ± 31.3	0.89 ± 0.45
+ Saline	92.2 ± 13.2	0.40 ± 0.07
+ Thymosin β_4 (1 μg)	82.6 ± 19.6	2.82 ± 0.75**
+ Thymosin α_1 (1 μg)	225.9 ± 57.5*	0.34 ± 0.75

*$p < 0.02$ vs. saline.
**$p < 0.01$ vs. saline.

TABLE VI. Intracerebral Injection of Thymosin F5 and Thymosin α_1 Inhibits Immunological Assays

	Control	Thymosin α_1
Experiment 1		
PFC[a]	269	152
Con A[b]	119	8
	Control	Thymosin F5
Experiment 2		
Hemagglutination titer	260	54
PHA	13	4
Con A	42	13
LPS	9	1

[a] Number of splenic PFC per 10^6 cells.
[b] Stimulation index.

Data generated using a mouse model indicate that the increase in adrenal weight is probably correlated with increased corticosteroid production (Hall *et al.*, 1984). Mice were injected into the lateral ventricle using precisely measured polyethylene cannulas that extended into the lateral ventricle. This route of intracerebral administration allows for transport of the injected material to multiple brain sites, including the hypothalamus. Although thymosin α_1 was effective in stimulating corticogenesis when injected in this manner (Table V), thymosin β_4 had no effect on the adrenal axis. This peptide was found to stimulate LH release, a phenomenon that has also been demonstrated using an *in vitro* model (Rebar *et al.*, 1981).

Of paramount importance is whether the thymosin-induced rise in corticosteroid levels is sufficient to induce changes in immunity. Data generated using the intracerebroventricular model suggest that it is. Mice received injections of either thymosin α_1, thymosin α_7, thymosin fraction 5, or vehicle. Both preparations that were effective in activating the adrenocortical axis, α_1 and fraction 5, were also found to be effective in modulating the immune capacity of the mice. As shown in Table VI, this influence was in the form of a diminished response which is consistent with the known inhibitory effects of certain glucocorticoids on immunity.

4. THYMOSIN α_1 CAN BE DETECTED IN BRAIN TISSUE

If thymosin peptides do indeed play an immunomodulatory role by acting on neuroendocrine circuits, then they should be present in those brain sites that influence the endocrine circuit in question. This was the case with respect to thymosin α_1 and the adrenal axis. Using a radioimmunoassay that was developed to measure thymosin α_1, the brains of several species have now been evaluated. These include mice, rats, and guinea pigs and in all species, thymosin α_1 cross-reactivity was found to be highest in the diencephalon and lowest in the cortex. The most extensive of these studies was carried out using rat brain and the results are summarized in

TABLE VII. Distribution of Thymosin α_1-like Immunoreactivity in Discrete Brain Regions of the Rat[a,b]

Brain region	Peptide concentration (pg/μg protein)
Hypothalamus and preoptic area	
Median eminence	6.33 ± 0.24
Arcuate n.	6.53 ± 0.54
Periventricular n.	3.33 ± 0.24
Paraventricular n.	3.18 ± 0.24
Supraoptic n.	3.52 ± 0.48
Ventromedial n.	3.68 ± 0.26
Dorsomedial n.	3.36 ± 0.23
Anterior hypothalamic n.	2.68 ± 0.37
Posterior hypothalamic n.	3.32 ± 0.08
Medial preoptic n.	2.4 ± 0.12
Thalamus	
Periventricular n.	4.36 ± 0.16
Septal area	
N. accumbens	3.59 ± 0.91
Lateral septum	2.30 ± 0.51
Anterior septum	3.55 ± 0.68
Midbrain	
Central gray	3.08 ± 0.21
Substantia nigra	3.10 ± 0.22
Cortex	
Cingulate cortex	1.37 ± 0.28
Striatum	1.01 ± 0.29

[a] Adapted from Palaszynski *et al.* (1983)
[b] Each value represents the mean ± S.E.M. of four determinations.

Table VII (Palaszynski *et al.*, 1983; Hall *et al.*, 1984). Highest concentrations were found in the median eminence and arcuate nucleus followed by the ventromedial nucleus. It is interesting that corticotropin-releasing factor activity can be modified by electrical stimulation of this region. Furthermore, this area corresponds to the sites where intrahypothalamic injections of thymosin fraction 5 were effective in stimulating increased adrenal weight.

5. DISCUSSION

The emotional state of an individual has a well-documented, but poorly characterized influence on health and disease. This includes anxiety-induced immunosuppression, as well as changes in immunocompetence, that are associated with psychotic behavior (Solomon *et al.*, 1974). Furthermore, animal studies have revealed that behavioral conditioning can influence not only the ability of mice to respond to antigenic challenge, but also their ability to tolerate autoimmune disorders (Ader and Cohen, 1974, 1982).

Pathways by which these effects might be conveyed to immunological tissues include the autonomic nervous system as well as neuroendocrine circuits. However, neural modulation of immunity is not simply a unidirectional influence. Just as target endocrine glands produce hormones capable of exerting a feedback influence on the hypothalamic–hypophyseal axis that regulates them, the immune system appears to provide signals that serve a similar function with regard to neural control of immunity. A hypothetical model that accounts for not only the thymosin results discussed in this review, but also the results of studies published by others, is illustrated in Fig. 2. It has as its premise that products of the activated immune system are capable of stimulating neuroendocrine circuits which have immunomodulatory influences. A possible reason for this influence is proposed to be the limitation of clonal expansion of lymphocytes (Besedovsky and Sorkin, 1977). Several types of evidence support this hypothesis.

During the course of the immune response, corticosterone levels become elevated reaching a peak at a time when antibody titers are at a maximum level (Besedovsky and Sorkin, 1977). T lymphocytes exerting effector functions have a reduced capacity to bind steroids after differentiation to this stage, whereas less mature cortical thymocytes are more sensitive to the inhibitory effects of steroids. Thus, elevated steroids at the time of peak antibody titers would have a minimal effect on the differentiated cells and a maximal effect on the less mature cells. It

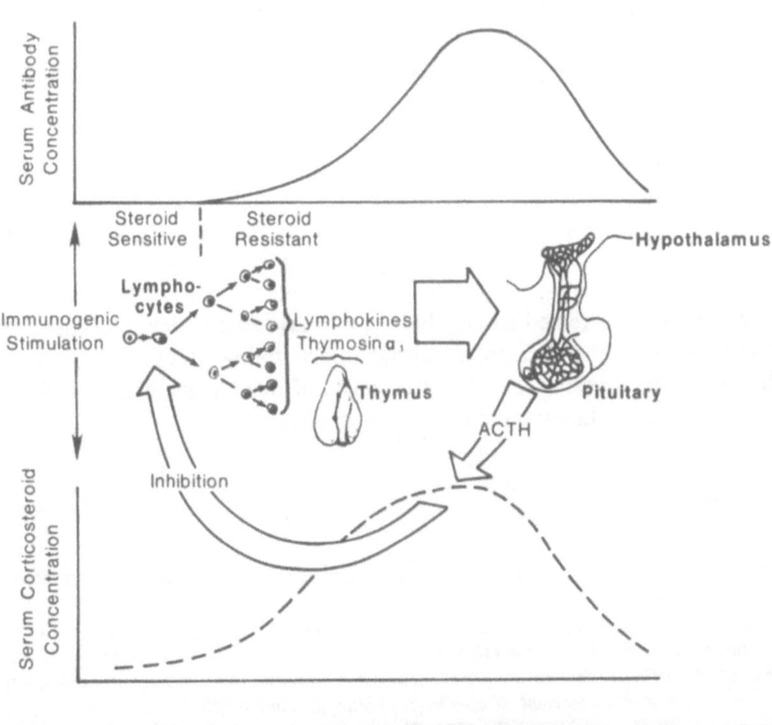

FIGURE 2. Hypothetical model of thymosin-induced immunoregulation involving the hypothalamic–hypophyseal–adrenocortical axis. (Adapted from Besedovsky and Sorkin, 1977.) See text for explanation.

has been proposed that the less mature cells, that might otherwise be responsive to various nonspecific lymphokines produced during the immune response, are prevented from doing so by the elevated corticosteroids. Consequently, the clonal expansion of cells with low or no affinity for the sensitizing immunogen would be restricted thereby preventing unnecessary differentiation. If this model is valid, then injecting a second immunogen at a time coinciding with the peak levels of steroid should result in a diminished antibody titer to this subsequent immunogen. This was demonstrated in a series of elegant experiments conducted by Besedovsky and Sorkin (1977). Not only did the predicted diminution of the response occur, but it could be prevented by prior adrenalectomy. Subsequent data reported by these investigators and from our laboratory, strongly suggest that the corticogenic factor is present in both lymphokine and thymosin preparations.

The mechanism by which the immune factor functions could involve one or more neurotransmitter systems. Serotonin and acetylcholine both stimulate corticotropin-releasing factor (Jones *et al.*, 1976). In addition, the same study revealed that norepinephrine is able to inhibit the stimulatory effect of serotonin and acetylcholine. Consequently, activation of the adrenal circuit could be subsequent to activation of serotonergic and/or cholinergic systems and/or to a reduction in norepinephrine activity. A disinhibition model involving norepinephrine is strongly suggested by recent evidence that a reduction in this neurotransmitter can occur following the administration of lymphokine-containing preparations (Besedovsky *et al.*, 1983).

In addition to the hypothalamic–hypophyseal–adrenal axis, there is evidence that other neuroendocrine circuits are part of a bidirectional network linking the central nervous and immune systems (Besedovsky and Sorkin, 1974; Rebar *et al.*, 1981; Hall *et al.*, 1982). The precise role that many of these potential pathways play in modulating immunity will have to await additional evidence. The adrenal influences on the immune system are the exception. The potential importance of such an immune–neuroendocrine circuit in health and disease is suggested by evidence that emotional states leading to increased anxiety or stress are generally associated with increased susceptibility to disease (see review by Hall, 1984). In addition to the well-documented adverse effects following overproduction of steroids, it would appear that during the course of mammalian evolution the system has become an additional immunoregulatory network.

REFERENCES

Ader, R., and Cohen, N., 1974, Behaviorally conditioned immunosuppression, *Psychosom. Med.* **37**:333–340.

Ader, R., and Cohen, N., 1982, Behaviorally conditioned immunosuppression and murine systemic lupus erythematosis, *Science* **215**:1534.

Ahlqvist, J., 1976, Endocrine influences on lymphatic organs, immune response, inflammation and immunity, *Acta Endocrinol. (Copenhagen)* **83**(Suppl. 206):1–136.

Besedovsky, H. O., and Sorkin, E., 1974, Thymus involvement in female sexual maturation, *Nature (London)* **249**:356–358.

Besedovsky, H. O., and Sorkin, E., 1977, Network of immune–neuroendocrine interactions, *Clin. Exp. Immunol.* **27**:1–12.

Besedovsky, H. O., del Rey, A., and Sorkin, E., 1981, Lymphokine containing supernatants from Con A-stimulated cells increase corticosterone blood levels, *J. Immunol.* **126:**385.

Besedovsky, H. O., del Rey, A., Sorkin, E., DaPrada, M., Burri, R., and Honegger, C., 1983, The immune response evokes changes in brain noradrenergic neurons, *Science* **221:**564–565.

Bourne, H. R., Lichtenstein, L. M., Melmon, K. L., Henney, C. S., Weinstein, Y., and Sheaver, G. M., 1974, Modulation of inflammation and immunity by cyclic AMP, *Science* **184:**19–28.

Dumonde, D. C., Pulley, M. S., Hamblin, A. S., Singh, A. K., Southcott, B. M., O'Connell, D., Paradinas, F. J., Robinson, M. R. G., Rigby, C. C., den Hollander, F., Schuurs, A., Verheul, H., and van Vliet, E., 1982, Short term and long-term administration of lymphoblastoid cell line lymphokine (LCL-LK) to patients with advanced cancers, in: *Lymphokines and Thymic Hormones: Their Potential Utilization in Cancer Therapeutics* (A. L. Goldstein and M. A. Chirigos, eds.), pp. 301–318, Raven Press, New York.

Hall, N. R., 1984, Behavioral and neuroendocrine interactions with immunogenesis, in: *Peptides, Hormones and Behavior* (C. Nemeroff and A. J. Dunn, eds.), pp. 913–938, Spectrum, New York.

Hall, N. R., and Goldstein, A. L., 1984, Endocrine regulation of host immunity: The role of steroids and thymosins, in: *Immunomodulating Agents: Properties and Mechanisms* (R. L. Fenichel and M. A. Chirigos, eds.), Dekker, New York (in press).

Hall, N. R., McGillis, J. P., Spangelo, B., Palaszynski, E., Moody, T., and Goldstein, A. L., 1982, Evidence for a neuroendocrine–thymus axis mediated by thymosin polypeptide, in: *Current Concepts in Human Immunology and Cancer Immunomodulation* (B. Serrou, C. Rosenfeld, J. C. Daniels, and J. P. Saunders, eds.), pp. 653–660, Elsevier/North-Holland, Amsterdam.

Hall, N. R., McGillis, J. P., and Goldstein, A. L., 1984, Activation of neuroendocrine pathways by thymosin peptides, in: *Stress, Immunity and Aging* (E. L. Cooper, ed.), Dekker, New York (in press).

Jones, M. T., Hillhouse, E. W., and Burden, J., 1976, Effect of various putative neurotransmitters on the secretion of corticotrophin-releasing hormone from the rat hypothalamus *in vitro*—A model of the neurotransmitters involved, *J. Endocrinol.* **69:**1–10.

McGillis, J. P., Feith, T., Kyeyune-Nyombi, E., Vahouny, G. V., Hall, N. R., and Goldstein, A. L., 1982, Evidence for an interaction between thymosin peptides and the pituitary adrenal axis, *Fed. Proc.* **41:**4918.

Maclean, D., and Reichlin, S., 1981, Neuroendocrinology and the immune process, in: *Psychoneuroimmunology* (R. Ader, ed.), pp. 475–520, Academic Press, New York.

Monjan, A., 1981, Stress and immunologic competence: Studies in animals, in: *Psychoneuroimmunology* (R. Ader, ed.), pp. 65–90, Academic Press, New York.

Nicol, T., and Bilbey, D. L. J., 1960, The effect of various steroids on the phagocytic activity of the reticuloendothelial system, in: *Reticuloendothelial Structure and Function* (J. H. Heller, ed.), pp. 301–320, Ronald Press, New York.

Palaszynski, E. W., Moody, T. W., O'Donohue, T. L., Goldstein, A. L., 1983, Thymosin α_1-like peptides: localization and biochemical characterization in the rat brain and pituitary gland, *Peptides,* **4:**463–467.

Rebar, R. W., Miyake, A., Low, T. L. K., and Goldstein, A. L., 1981, Thymosin stimulates secretion of luteinizing hormone-releasing factor, *Science* **214:**668.

Renaud, L. P., 1984, The neurophysiology of hypothalamic–pituitary regulation and of hypothalamic hormones in brain, in: *Peptides, Hormones and Behavior* (C. Nemeroff and A. J. Dunn, eds.), Spectrum, New York, pp. 173–215.

Sivas, A., Uysal, M., and Oz, H., 1982, The hyperglycemic effect of thymosin F5, a thymic hormone, *Horm. Metab. Res.* **14:**330.

Solomon, G. F., Amkraut, A. A., and Kasper, P., 1974, Immunity, emotions and stress, *Psychother. Psychosom.* **23:**209–217.

Thompson, J. S., Crawford, M. D., Reilly, R. W., and Severson, C. D., 1969, The effect of estrogenic hormones on immune responses in normal and irradiated mice, *J. Immunol.* **98:**331–335.

Vahouny, G. V., Kyeyune-Nyombi, E., McGillis, J.P., Tare, N. S., Huang, K.-Y., Tombes, R., Goldstein, A. L., and Hall, N. R., 1983, Thymosin peptides and lymphokines do not directly stimulate adrenal corticosteroid production *in vitro, J. Immunol.* **130:**791.

30

Effects of Thymic Peptides on Hypothalamic–Pituitary Function

ROBERT W. REBAR

1. INTRODUCTION

A role for the thymus gland within the immune system is now well recognized. Recent investigations have convincingly demonstrated that the thymus really functions as an endocrine organ, secreting peptides which influence lymphoid tissue structure and function (Goldstein *et al.*, 1981). However, studies from our laboratory also suggest that thymic peptides may play a broader role within the endocrine system and may function as modulators of the hypothalamic–pituitary axis. In this brief review I shall attempt to delineate what we have learned about the apparent relationship between the thymus gland and other endocrine organs. I shall focus particularly on the effects of the thymus on the reproductive system but conclude by speculating about other possible functions for thymic peptides and suggest that the neuroendocrine and immune systems are tightly coupled and function together in a coordinated fashion.

2. REPRODUCTIVE FUNCTION IN ATHYMIC MICE

2.1. Early Studies Suggestive of Thymic–Reproductive System Interaction

Calzolari (1898) first suggested a relationship between the thymus gland and reproductive function, noting hypertrophy of the thymus in castrated male rabbits. Subsequent studies have confirmed that thymic enlargement occurs after gonadec-

ROBERT W. REBAR • Department of Obstetrics and Gynecology, Northwestern Memorial Hospital, Northwestern University Medical School, Chicago, Illinois 60611.

tomy of either sex in almost every species of domestic and experimental animal, including man. Conversely, androgens and estrogens have been shown to induce atrophy of the thymus when administered to intact animals (Dougherty, 1952).

Nishizuka and Sakakura (1969) then demonstrated that neonatal thymectomy in the mouse within the first 72 hr of birth leads to ovarian dysgenesis which is first apparent at 2–3 months of age and which occurs independently of wasting disease. This effect of neonatal thymectomy upon the mouse ovary has since been confirmed by Besedovsky and Sorkin (1974) and Lintern-Moore and Norbaek-Sorensen (1976).

At about this time, Flanagan (1966) reported breeding a new strain of congenitally athymic nude (*nu/nu*) mice in which the females had decreased fertility. It has since been established that affected females also have delayed vaginal opening (Besedovsky and Sorkin, 1974) and accelerated follicular atresia such that the mice develop premature ovarian failure (Lintern-Moore and Pantelouris, 1975; Alten and Groscurth, 1975).

Histological examination of the ovaries of the athymic nude mice showed them to be quite different from the ovaries of normal mice by as early as 10 days of age, even though they appeared similar at birth. The number of small nongrowing follicles appears increased, and proportionately the number of developing follicles is decreased (Lintern-Moore and Pantelouris, 1976a). By 1 month of age, all healthy growing follicles are contracted in size even though the number and compositon of the follicular population appear normal (Lintern-Moore and Pantelouris, 1975). By 2 months of age, the number of follicles (and consequently of oocytes) in athymic mice is actually decreased compared to their normal littermates (Lintern-Moore and Pantelouris, 1975). Similar histological patterns with respect to follicular development and oocyte number have also been observed in neonatally thymectomized mice (Nishizuka and Sakakura, 1969, 1971; Lintern-Moore, 1977).

Lintern-Moore and Pantelouris (1976a,b) postulated that the ovarian changes observed in the athymic mice might be the result of insufficient gonadotropin stimulation. They demonstrated that the administration of pregnant mare serum gonadotropin (PMSG) to athymic mice eliminates the histological differences between the ovaries of athymic and normal mice observed at 10 days and 1 month of age. The histological similarity of the ovaries of the athymic animals to those of rodents challenged with antigonadotropin (Hardy *et al.*, 1974) was also apparent.

2.2. The Hormonal Basis of Reproductive Defects in Athymic Mice

Thus, prior to 1980, data suggested that gonadotropin deficiency during the prepubertal period might account for the ovarian abnormalities observed in athymic mice. What was required was the demonstration that gonadotropin secretion is diminished in such animals. Furthermore, if such a deficiency in gonadotropins existed in athymic animals, what was the mechanism by which the presence of the thymus prevented this deficiency?

We began our investigations in this area by determining concentrations of pituitary and circulating gonadotropins in congenitally athymic nude (*nu/nu*) mice and their endocrinologically normal heterozygous (*nu/*+) littermates. We noted decreased levels of both luteinizing hormone (LH) and follicle-stimulating hormone (FSH) in the pituitary gland and in the circulation prior to vaginal opening (i.e., puberty) and decreased circulating estrogen levels in adult athymic mice (Rebar *et al.*, 1981a; Fig. 1). Simultaneously, Michael *et al.* (1980) reported that circulating concentrations of LH and FSH are reduced in immature neonatally thymectomized mice. In addition, we observed reduced concentrations of immunoreactive luteinizing hormone-releasing factor (LRF) in the hypothalami of athymic mice relative to the heterozygotes at 10, 20, and 30 days of age (Rebar *et al.*, 1981b). We also tested the ability of hypothalamic extracts from 20-day-old athymic and heterozygous mice to release LH when injected intravenously into rats as an estimate of the biological activity of hypothalamic LRF. Hypothalamic extracts from athymic mice consistently released less immunoreactive rat LH than extracts from the heterozygotes, thus confirming the differences in immunoreactive LRF. Furthermore, the LH responses of the athymic mice to exogenous LRF administered *in vivo* were identical to those of the heterozygotes, and the maximal LH and FSH responses of dispersed pituitary cells *in vitro* to LRF from both groups of mice were also identical. These data support the conclusion that the pituitaries of the athymic animals can release gonadotropin appropriately and that the defect in the athymic mice is at or above the level of the hypothalamus.

Although the reduced levels of circulating estrogen observed in the adult thymic mice might be consequent to reduced gonadotropin secretion in immature mice, we

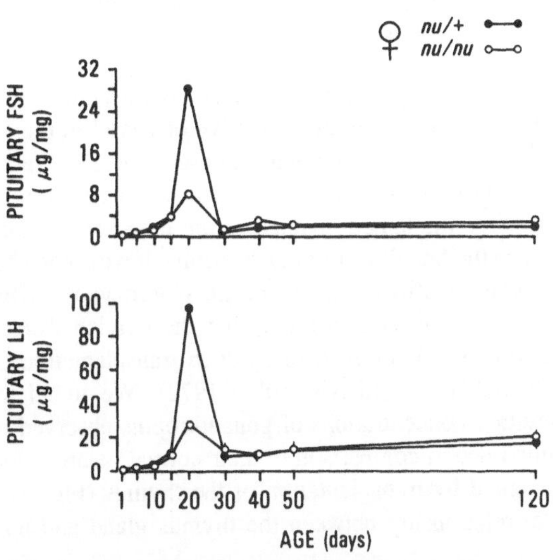

FIGURE 1. Pituitary concentrations of FSH and LH in female athymic nude mice (*nu/nu*) and their heterozygous littermates (*nu/+*) as determined by radioimmunoassay. The pituitary levels before 20 days of age represent the means of multiple determinations on pools from at least five mice. For mice 20 days of age and older, individual pituitary glands were assayed, and the means and standard errors of these latter determinations (generally too small to be visualized) are shown. Results are expressed in terms of μg rat FSH or LH equivalents per mg pituitary (wet weight). (Adapted from Rebar *et al.*, 1981a.)

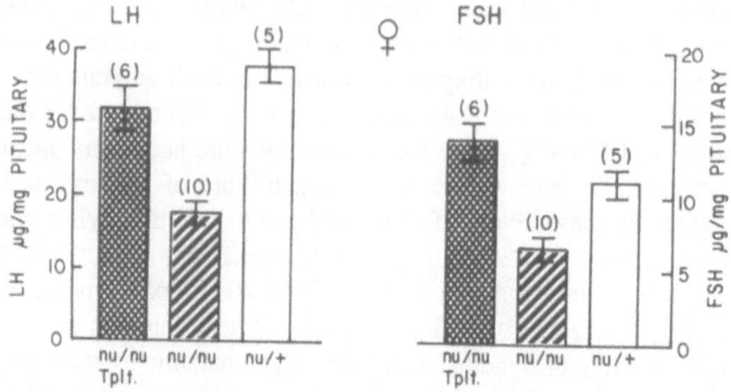

FIGURE 2. Pituitary concentrations of LH and FSH in female athymic nude mice undergoing transplantation (*nu/nu* Tplt.), compared with sham athymic (*nu/nu*) and sham normal heterozygous (*nu/+*) mice at 20 days of age. Results are expressed in terms of μg rat LH or FSH equivalents per mg pituitary (wet weight). The number of pituitary gonads assayed individually and included in each group is in parentheses. Means ± standard errors are shown by bars. (Reproduced with permission from Rebar, 1983, as modified from Rebar *et al.*, 1980.)

had not yet ruled out the possibility of a direct effect of the thymus on the gonad. Consequently, we examined the maximum binding of purified human chorionic gonadotropin (hCG; which is structurally similar to LH) to crude membrane fraction of ovarian tissue from athymic and normal heterozygous females of various ages (Rebar *et al.*, 1981b). Maximum binding of hCG was not decreased in the ovaries of the athymic mice relative to those of the heterozygotes. Thus, these preliminary data do not indicate any evidence of an inherent ovarian defect in the athymic mice.

Because the hormonal alterations that we observed in the athymic nude mice might be genetic in origin, we attempted to ascertain if the reduced gonadotropin concentrations found in the athymic mice are causally linked to the absence of the thymus. We therefore implanted thymuses from their heterozygous littermates into athymic animals on the first day of life to determine if we could abolish the hormonal differences between the athymic mice and the heterozygotes (Rebar *et al.*, 1980). Previous investigators had demonstrated that the ovarian dysgenesis observed in the athymic mice could be eliminated by such transplantation (Pierpaoli and Besedovsky, 1975; Sakakura and Nishizuka, 1972). We found that the significant reduction in pituitary concentrations of gonadotropins observed in 20-day-old congenitally athymic mice in comparison to their normal heterozygous littermates was completely prevented by transplantation of the thymus (Fig. 2). Thus, these data suggest a causal relationship between the thymus gland and normal reproductive function.

3. STUDIES WITH THYMIC PEPTIDES

The experiments with congenitally athymic nude mice led us to postulate that the thymus gland might exert its effects on reproduction through a humoral factor which influences the hypothalamic–pituitary unit. In collaboration with Drs. Allan

Goldstein and Theresa Low, who kindly provided us with thymic peptides, we demonstrated that thymosin fraction 5 and one of its synthetic peptide components, thymosin β_4, but not thymosin α_1, stimulated secretion of LRF from medial basal hypothalami from normal cycling female rats superfused *in vitro* (Rebar *et al.*, 1981c; Figs. 3, 4). We also observed release of LH from pituitary glands superfused together with hypothalami, but there was no release of LH in response to thymosin when pituitaries were superfused alone. In addition, preliminary studies have indicated that thymosin fraction 5 is ineffective in affecting growth hormone or prolactin release from pituitaries superfused with hypothalami (Rebar *et al.*, 1982).

Thymosin fraction 5 also stimulated increased secretion of LRF in a dose-dependent fashion from hypothalami from adult, castrate, and prepubertal male and female rats and mice incubated *in vitro* (Rebar *et al.*, 1982, 1983). Kidney fraction 5, prepared identically to the thymosin and with very little effect on tests of immunological function, and myoglobin, which has minor homologies to thymosin β_4, were ineffective in releasing significant LRF. In confirmation of our *in vitro* observations, Hall *et al.* (1982) recently noted that thymosin β_4, but not thymosin α_1, increased serum LH levels in adult female mice when administered by intraventricular injection.

To further determine if thymic peptides can directly effect release of LH or

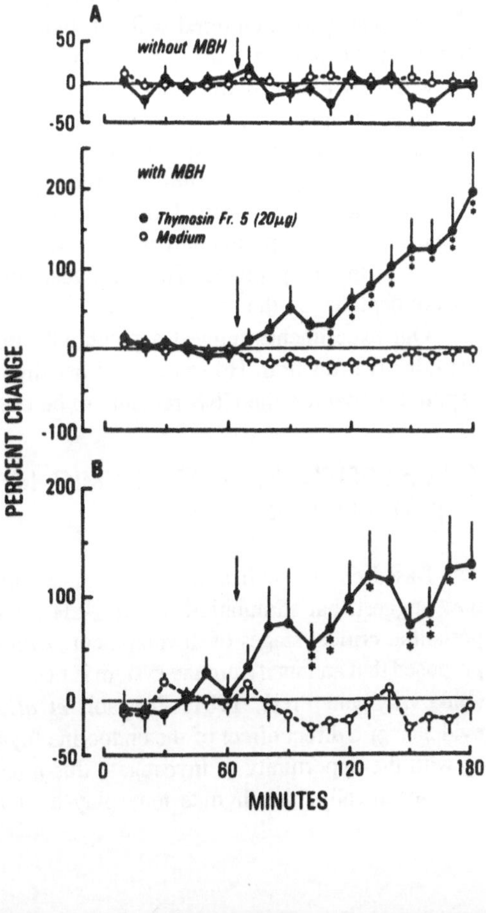

FIGURE 3. (A) Release of LH into the efflux from pituitary glands superfused alone (top panel) or in sequence with mediobasal hypothalami (MBH) (bottom panel) from random cycling female rats in response to thymosin fraction 5 (20 μg) or medium 199 injected at the arrows. For each experiment, the mean LH concentration in the fractions collected for 1 hr prior to injection was calculated. Each value during the experiment was then calculated as the percentage change from this baseline mean. The average percentage change (± S.E.) at each point for seven experiments is plotted. Significant changes from the control group are indicated by asterisks: *$p < 0.05$, **$p < 0.01$. The mean basal concentration (± S.E.) of LH was 323.4 ± 29.5 ng/ml. (B) Release of LRF from superfused MBH from random cycling female rats in response to thymosin fraction 5 or medium 199 expressed as the percentage change from mean basal levels. The mean basal concentration of LRF was 30.7 ± 6.6 pg/ml. (Modified from Rebar *et al.*, 1981c.)

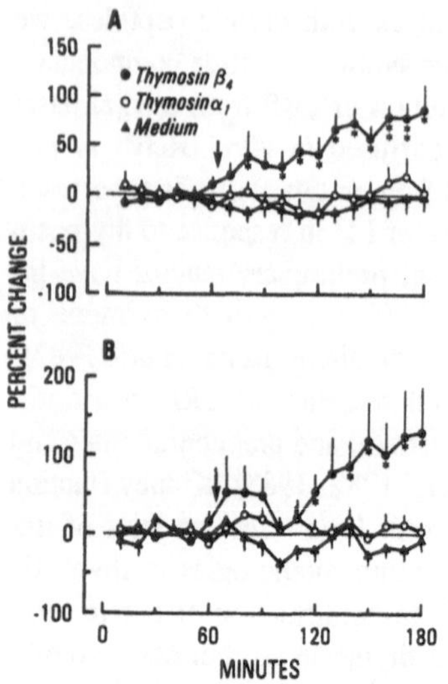

FIGURE 4. (A) Mean percentage changes (±S.E.) of LH from basal levels by pituitary glands superfused in sequence with MBH in response to synthetic thymosin α_1 (n = 8), β_4 (n = 6), and medium 199 alone (n = 7). The synthetic thymosin or medium 199 was injected at the arrow. The mean basal concentration of LH was 403.1 ± 45.6 ng/ml. (B) Mean percentage changes (±S.E.) of LRF from basal levels by superfused MBH in response to synthetic thymosin α_1 (n = 8), β_4 (n = 8), and medium 199 alone (n = 7). The mean basal concentration of LRF was 34.3 ± 3.7 pg/ml. The data in this figure were calculated and are presented as in Fig. 3. (Modified from Rebar *et al.*, 1981c.)

FSH from the pituitary, dispersed anterior pituitary cells from ovariectomized rats were cultured in Hepes buffer in the presence or absence of thymosin fraction 5, thymosin β_4, and thymosin α_1 (Rebar *et al.*, 1982, 1983). Some culture dishes were primed with thymosin fraction 5 (1 μg/ml) or thymosin β_4 or α_1 (2×10^{-12} M), and others were cultured without thymosin for 2 days. The cell cultures were then washed extensively and subsequently reincubated for 3 hr with or without LRF and with or without thymic peptides. Media were then assayed for LH and FSH. The thymic peptides had no effect on either basal or maximal LRF-stimulated release of LH or FSH. In similar experiments we have also examined the effects of thymic peptides on estrogen and androgen production by granulosa and theca cells, respectively, from rat ovarian follicles as well (Rebar *et al.*, 1982, 1983). No differences in steroid production *in vitro* were noted regardless of the presence or absence of thymic peptides. Thus, we also failed to document any direct effects of thymic peptides on the ovary.

Our experiments suggest a potentially important role for thymic peptides in reproductive function. However, whether thymosin β_4 itself or structurally similar peptides act within the CNS remains to be established.

4. A HYPOTHESIS FOR THYMIC–HYPOTHALAMIC INTERACTION

Based on our studies in athymic mice and with synthetic thymic peptides, we now suggest that stimulation of LRF–LH release by thymic peptides may be important at critical stages of development. Although a number of investigators have proposed that an intact immune system is necessary for normal reproductive function (Bukovsky and Presl, 1981; Pierpaoli *et al.*, 1977), our data provide the first evidence of a direct effect of the endocrine thymus on the hypothalamus and afford us with the opportunity to investigate this novel hypothesis.

Specifically, the thymus may play a role in regulating the rate of follicular (i.e., oocyte) degeneration, known as atresia, *in utero* in primates and early in neonatal life in rodents. As previously noted, Lintern-Moore and Pantelouris (1976a) demonstrated that follicular atresia is accelerated in athymic nude mice in the first 2 months of life. Maturation of the hypothalamic–pituitary–gonadal axis which occurs in the rodent in the first several days of life occurs in the human and other primates *in utero*. That thymic–endocrine interaction also occurs in the primate is suggested by the observation that young girls with congenital absence of the thymus gland have no oocytes in their ovaries on autopsy (Miller and Chatten, 1967). In collaboration with Drs. David Healy and Gary Hodgen of the NIH, we are now attempting to ascertain if fetal thymectomy of primates leads to reduced oocyte number of birth. Preliminary data have revealed that the ovaries of the thymectomized monkeys are significantly smaller than those of the controls. Additional evidence suggesting that the absence of the thymus in primates may lead to gonadotropin insufficiency and accelerated follicular atresia is provided by the earlier observation that fetal removal of the pituitary gland in rhesus monkeys leads to ovaries devoid of oocytes at birth (Gulyas *et al.*, 1977). Also providing circumstantial evidence is the observation that thymus glands are markedly increased in weight in anencephalic fetuses, which generally have small or absent pituitary glands (Benirschke, 1956). In collaboration with Drs. Allan Goldstein and Paul Naylor, we are now attempting to determine if the thymic peptide content of thymuses from anencephalic fetuses is increased as a result of decreased hormonal feedback from the hypothalamic–pituitary unit. In addition, thymic weight in the primate is maximal at the fetal age at which oocyte number is also greatest (Baker, 1963; Hendrickx, 1971). In the human, for example, the number of germ cells present at 20 weeks of fetal age has been estimated as approximately 7 million (Baker, 1963). Simultaneous with thymic regression, oocyte number decreases to about 2 million by the time of birth. Although very little is known about the mechanisms controlling follicular atresia, such observations suggest that the thymus gland may play a role in regulating the rate of atresia.

Even the suggestion that gonadotropins are necessary to prevent accelerated degenertion of primordial follicles is a radical one: Unsubstantiated dogma has maintained that LH and FSH are necessary only in later stages of follicular development. Recently, Ross and Vande Wiele (1981) have questioned this dogma and have suggested that gonadotropins may be necessary for maintenance of fetal oocytes. However, no data exist to support the dependence of follicles *in utero* on gonadotropins. Still, the evidence presented here is clearly consistent with the hypothesis that there is a role for thymic peptides (or structurally related peptides synthesized elsewhere) in stimulating LRF–gonadotropin secretion and thus in retarding the rate of follicular atresia *in utero*.

Furthermore, whether thymic peptides have other physiological roles in the regulation of LRF–LH secretion is not known. Any role in males *in utero* remains to be explored. In addition, is it possible that thymic peptides play a role in the increase in LH (and presumably LRF) secretion which occurs at puberty? In the human the size of the thymus gland increases a second time just prior to puberty (Kendall, 1981) coincident with activation of the hypothalamic–pituitary axis. In this regard, Michael *et al.* (1981) reported that thymosin fraction 5 accelerated the day of vaginal opening (i.e., puberty) when administered to immature mice.

Establishing the validity of this hypothesis of physiological thymic peptide-stimulated LRF–LH release will require extensive investigation. Thus far, we have not been able to document any increase in peripheral LH concentrations in response to intravenously administered thymosin fraction 5 in rodents (unpublished observations). We may not be administering the appropriate dose of thymosin in the appropriate manner. We have, however, recently reported that circulating levels of thymosin β_4 but not thymosin α_1 appear to differ in women with differing concentrations of circulating estrogen (Goldstein *et al.*, 1983).

Regardless of the validity of the hypothesis, the potential usefulness of synthetic derivatives of thymic peptides to stimulate or to inhibit the secretion of LRF and LH merits further study. One of the most common causes of anovulation in young women is hypothalamic in origin with apparent diminution of endogenous LRF secretion (Yen *et al.*, 1973; Lachelin and Yen, 1978). Is it possible that derivatives of thymosin β_4 might one day be utilized to induce ovulation in such women? Alternatively, is it possible that antagonists of thymic peptides might some day be used to inhibit LRF secretion and prevent ovulation? These possibilities warrant additional investigation.

5. EXTENSION OF THE HYPOTHESIS

Our studies with thymosin β_4 led other investigators to consider the possibility that other thymic peptides might also influence hypothalamic–pituitary function. McGillis *et al.* (1982) demonstrated that thymosin α_1 results in increased corticosterone secretion in mice. Hall *et al.* (1982) subsequently determined that intraventricular injection of thymosin α_1 results in increased corticosterone levels in rodents. At this symposium, Hall *et al.* (Chapter 29) report on their further studies with thymosin α_1 and Healy *et al.* (Chapter 34) report that intravenous injection of thymosin fraction 5 (10 mg/kg) into prepubertal female macaque monkeys consistently increased circulating levels of ACTH, β-endorphin, and cortisol. However, no changes in other pituitary hormones were noted. Thymosin α_1 and thymosin β_4 at doses of 75 μg/kg did not significantly elevate any pituitary hormone. It is possible that the dose of synthetic thymic peptide was inadequate or the mode of administration inappropriate.

In any event, such data suggest that a variety of thymic peptides may influence secretion of different hormones and neurosecretory products. In fact, one possibility is that at least some of the effects of the thymus and thymic peptides on the immune system represent the result of regulatory changes in hormonal secretion by other endocrine organs as well as alterations in lymphoid function. Three separate systems once thought to function quite independently of each other—the neurological, endocrine, and immune systems—may well interact in a finely coordinated fashion to regulate the internal milieu of all complex organisms. This may well be one reason why data are accumulating to suggest that peptides originally thought to be unique to one system or organ apparently can be synthesized in several sites. Although these concepts are purely speculative at this time, I believe that they are logical and worthy of further consideration. Even if these hypotheses ultimately are disproven, the experiments designed to test their validity should shed additional light on the interaction between thymic hormones and other endocrine organs.

ACKNOWLEDGMENTS. The studies from our laboratory summarized in this review were supported in part by NIH Grants HD-14362 and HD-12303. The author wishes to express his appreciation to all of his collaborators, without whom the investigations could not have been undertaken: Dr. Gregory F. Erickson, Dr. Allan L. Goldstein, Ms. Anna Latham, Dr. Teresa L. K. Low, Dr. Akira Miyake, Dr. I. Carlos Morandini, Mr. John Petze, Dr. M. Felipe Silva de Sa, and Dr. Gideon Strich.

REFERENCES

Alten, H.-E., and Groscurth, P., 1975, The postnatal development of the ovary in the "nude" mouse, *Anat. Embryol.* **145**:35–46.

Baker, T. G., 1963, A quantitative and cytological study of germ cells in human ovaries, *Proc. R. Soc. London Ser. B* **158**:417–433.

Benirschke, K., 1956, Adrenals in anencephaly and hydrocephaly, *Obstet. Gynecol.* **8**:412–425.

Besedovsky, H. O., and Sorkin, E., 1974, Thymus involvement in sexual maturation, *Nature (London)* **249**:356–358.

Bukovsky, A., and Presl, J., 1981, Control of ovarian function by the immune system: Commentary on the criticisms of Schenzle, *Med. Hypoth.* **7**:1–14.

Calzolari, A., 1898, Recherches experimentales sur un rapport probable entre la fonction du thymus et celle des testicules, *Arch. Ital. Biol. Torino* **30**:71–77.

Dougherty, J. F., 1952, Effect of hormones on lymphatic tissue, *Physiol. Rev.* **32**:379.

Flanagan, S. P., 1966, "Nude": A new hairless gene with pleiotrophic effects in the mouse, *Genet. Res.* **8**:295–309.

Goldstein, A. L., Low, T. L. K., Thurman, G. B., Zatz, M. M., Hall, N., Chen, J., Hu, S.-K., Naylor, P., and McClure, J. E., 1981, Current status of thymosin and other hormones of the thymus gland, *Recent Prog. Horm. Res.* **37**:369–415.

Goldstein, A. L., Naylor, P., and Rebar, R. W., 1983, Concentrations of thymic peptides in various clinical states, Abstracts of the 65th Annual Meeting of the Endocrine Society, Abstr. No. 471, p. 198.

Gulyas, B. J., Hodgen, G. D., Tullner, W. W., and Ross, G. T., 1977, Effects of fetal or maternal hypophysectomy on endocrine organs and body weight in infant rhesus monkeys *(Macaca mulatta)*: With particular emphasis on oogenesis, *Biol. Reprod.* **16**:216–227.

Hall, N. R., McGillis, J. P., Spangelo, B. L., and Goldstein, A. L., 1982, Evidence for an interaction between thymosin peptides and the pituitary–gonadal axis, *Fed. Proc. Abstr.* **41**:1267.

Hardy, B., Damon, D., Eshkol, A., and Lunenfeld, B., 1974, Ultrastructural changes in the ovaries of infant mice deprived of endogenous gonadotropins and after substitution with FSH, *J. Reprod. Fertil.* **36**:345–352.

Hendrickx, A. G., 1971, *Embryology of the Baboon*, pp. 174–180, University of Chicago Press, Chicago.

Kendall, M. D. (ed.), 1981, Introduction, in: *The Thymus Gland*, pp. 1–6, Academic Press, New York.

Lachelin, G. C. L., and Yen, S. S. C., 1978, Hypothalamic chronic anovulation, *Am. J. Obstet. Gynecol.* **130**:825–831.

Lintern-Moore, S., 1977, Effect of athymia on the initiation of follicular growth in the rat ovary, *Biol. Reprod.* **17**:155.

Lintern-Moore, S., and Norbaek-Sorensen, I., 1976, The effect of neonatal thymectomy upon follicle numbers in the postnatal mouse ovary, *Mech. Ageing Dev.* **5**:235.

Lintern-Moore, S., and Pantelouris, E. M., 1975, Ovarian development in athymic nude mice. I. The size and composition of the follicle population, *Mech. Ageing Dev.* **4**:385–390.

Lintern-Moore, S., and Pantelouris, E. M., 1976a, Ovarian development in athymic nude mice. III. The effect of PMSG and oestradiol upon the size and composition of the ovarian follicle population, *Mech. Ageing Dev.* **5**:33.

Lintern-Moore, S., and Pantelouris, E. M., 1976b, Ovarian development in athymic nude mice. V. The effects of PMSG upon the numbers and growth of follicles in the early juvenile ovary, *Mech. Ageing Dev.* **5**:259–265.

McGillis, J. P., Feith, T., Kyeyune-Nyombi, F., Vahouny, G. V., Hall, N. R., and Goldstein, A. L., 1982, Evidence for an interaction between thymosin peptides and the pituitary–adrenal axis, *Fed. Proc. Abstr.* **41**:111.

Michael, S. D., Taguchi, O., and Nishizuka, Y., 1980, Effect of neonatal thymectomy on ovarian development and plasma LH, FSH, GH, and PRL in the mouse, *Biol. Reprod.* **22**:343.

Michael, S. D., Allen, L. S., McClure, J. E., Goldstein, A. L., and Barkley, M. S., 1981, Interactions between estradiol and thymosin α_1 levels in the female mouse, Abstracts of the 63rd Annual Meeting of the Endocrine Society, Abstr. No. 308, p. 159.

Miller, M. E., and Chatten, J., 1967, Ovarian changes in ataxia telangiectasia, *Acta Paediatr. Scand.* **56**:559–561.

Nishizuka, Y., and Sakakura, T., 1969, Thymus and reproduction: Sex-linked dysgenesia of the gonad after neonatal thymectomy in mice, *Science* **166**:753–755.

Nishizuka, Y., and Sakakura, T., 1971, Ovarian dysgenesis induced by neonatal thymectomy in the mouse, *Endocrinology* **89**:886–893.

Pierpaoli, W., and Besedovsky, H. O., 1975, Role of the thymus in programming of neuroendocrine functions, *Clin. Exp. Immunol.* **20**:323–338.

Pierpaoli, W., Kopp, H. G., Müller, J., and Keller, M., 1977, Interdependence between neuroendocrine programming and the generation of immune recognition in ontogeny, *Cell. Immunol.* **29**:16–27.

Rebar, R. W., 1982, The thymus gland and reproduction: Do thymic peptides influence reproductive lifespan in females?, *J. Am. Geriatr. Soc.* **30**:603–606.

Rebar, R. W., Morandini, I. C., Benirschke, K., and Petze, J. E., 1980, Reduced gonadotropins in athymic mice: Prevention by thymic transplantation, *Endocrinology* **107**:2130–2132.

Rebar, R. W., Morandini, I. C., Erickson, G. F., and Petze, J. E., 1981a, The hormonal basis of reproductive defects in athymic mice. I. Diminished gonadotropin concentrations in prepubertal females, *Endocrinology* **108**:120–126.

Rebar, R. W., Morandini, I. C., Silva de Sa, M. F., Erickson, G. F., and Petze, J. E., 1981b, The importance of the thymus gland for normal reproductive function in mice, in: *Dynamics of Ovarian Function* (N. Schwartz and M. Hunzicker-Dunn, eds.), pp. 285–290, Raven Press, New York.

Rebar, R. W., Miyake, A., Low, T. L. K., and Goldstein, A. L., 1981c, Thymosin stimulates secretion of luteinizing hormone-releasing factor, *Science* **214**:669–671.

Rebar, R. W., Latham, A., and Petze, J., 1982, Thymic peptides stimulates secretion of luteinizing hormon-releasing factor (LRF), Abstracts of the 64th Annual Meeting of the Endocrine Society, Abstr. No. 11.

Rebar, R. W., Miyake, A., Erickson, G. F., Low, T. L. K., and Goldstein, A. L., 1983, The influence of the thymus gland on reproductive function: A hypothalamic site of action, in: *4th Biennial Workshop on the Ovary: Regulation of Ovarian Function*, (G.S. Greenwald and P.F. Terranova, eds.), pp. 465–469, Raven Press, New York.

Ross, G. J., and Van de Wiele, R. L., 1981, The ovaries, in: *Textbook of Endocrinology* (R.H. Williams, ed.), pp. 355–399, Saunders, Philadelphia.

Sakakura, T., and Nishizuka, Y., 1972, Thymic control mechanism in ovarian development: Reconstitution of ovarian dysgenesis in thymectomized mice by replacement with thymic and other lymphoid tissues, *Endocrinology* **90**:431–437.

Yen, S. S. C., Rebar, R. W., Van den Berg, G., and Judd, H., 1973, Hypothalamic amenorrhea and hypogonadotropinism: Response to synthetic LRF, *J. Clin. Endocrinol. Metab.* **36**:811–816.

31

Biochemistry of Lymphokine Action on Macrophages

Modulation of Macrophage Superoxide Production by Lymphokine

MAYA FREUND and EDGAR PICK

1. INTRODUCTION

Macrophages (MPs) respond to certain membrane stimulants by a sequence of biochemical reactions known as the oxidative burst (Johnston, 1978). The primary event in the sequence is the reduction of molecular oxygen to superoxide (O_2^-) catalyzed by a membrane-associated NADPH oxidase. Part of O_2^- is further converted to hydrogen peroxide (H_2O_2) by enzymatic or spontaneous dismutation.

MPs taken from animals infected with intracellular pathogens exhibit an enhanced, immunologically nonspecific, antimicrobial activity and are commonly referred to as "activated MPs." The biochemical mechanism responsible for the enhanced bactericidal capacity of activated MPs is incompletely understood. There is considerable experimental evidence indicating that MP activation is accompanied by an augmented propensity to produce oxygen radicals in response to membrane stimulation (reviewed in Johnston, 1981). A major, albeit not the only, pathway of MP activation is initiated by the specific antigen stimulation of T cells leading to the synthesis of a category of lymphokines (LKs) with MP-activating properties (Fowles *et al.*, 1973). An increasing amount of experimental data supports the idea that the enhanced oxidative metabolism of activated MPs is LK-induced. Thus, Nathan *et al.* (1979) found that mouse MPs incubated for 3 days with LK-containing

MAYA FREUND and EDGAR PICK • Laboratory of Immunopharmacology, Department of Human Microbiology, Sackler Faculty of Medicine, Tel-Aviv University, Tel-Aviv 69978, Israel.

spleen cell supernatants produced elevated amounts of H_2O_2 upon stimulation with phorbol myristate acetate (PMA). This property paralleled the ability of the cells to kill *Trypanosoma cruzi*. Enhanced production of O_2^- or H_2O_2 by LK-treated MPs upon membrane stimulation was also reported by other workers in mouse peritoneal MPs (Murray and Cohn, 1980), human monocytes (Nakagawara *et al.*, 1982; Seim, 1982), and in MP-like mouse (Murray, 1981) and human (Gately and Oppenheim, 1982) cell lines. The stimulants used included PMA, opsonized zymosan, and parasites such as *Toxoplasma* and *Leishmania*.

We have recently reported that guinea pig peritoneal MPs, elicited by mineral oil, respond by O_2^- production to an unusually wide array of stimulants (Pick and Keisari, 1981). The various stimulants were found to elicit an oxidative burst by activating distinct biochemical pathways (Bromberg and Pick, 1983). On the other hand, the effect of LK on the oxidative metabolism of guinea pig MPs has been investigated only to a limited extent. Two reports deal with enhanced reduction of nitroblue tetrazolium (NBT) by guinea pig MPs incubated with LK-containing culture supernatants (Krueger *et al.*, 1976; Alföldy and Lemmel, 1979) and another report describes an increase in oxygen consumption by LK-treated guinea pig MPs (Block *et al.*, 1980).

We have, therefore, undertaken a systematic study of the modulatory effect of LK on O_2^- production by chemically elicited guinea pig MPs responding to a battery of nine chemically unrelated stimulants. We demonstrate that, in the guinea pig, LK has a bidirectional modulatory effect on the oxidative metabolism of MPs, meaning that, depending on the stimulant, LK can both enhance and inhibit O_2^- production.

2. MATERIALS AND METHODS

Animals. Male Hartley guinea pigs weighing 300–400 g were used for obtaining peritoneal MPs and as lymphocyte donors for LK production.

Production of LK-Containing Culture Supernatants. Guinea pigs were immunized with 1 ml complete Freund's adjuvant (CFA), containing 0.4 mg/ml of killed *Mycobacterium tuberculosis* $H_{37}Ra$ (Difco) divided among the four footpads and the nuchal region. Four weeks later the draining lymph nodes were removed and made into a cell suspension in Eagle's minimum essential medium (MEM) supplemented with L-glutamine (2 mM), nonessential amino acids, and sodium pyruvate (1 mM) and containing 100 U/ml penicillin, 50 μg/ml gentamicin, and 25 U/ml mycostatin. Lymph node cells from 10 animals were pooled and purified on glass bead columns as described (Manheimer and Pick, 1973). LK production was induced by pulse exposure of lymph node lymphocytes, at a concentration of 10^7 trypan blue-excluding cells/ml, to 10 μg/ml Con A (Miles-Yeda) for 2 hr at 37°C, followed by culture in mitogen-free medium for 24 hr (Pick and Kotkes, 1977). A batch of the same lymphocyte pool which was not subject to Con A pulse was cultured for 24 hr at 37°C and served as control. Cell-free supernatants were prepared by centrifugation, and material from up to five Con A-stimulated and control cultures was pooled and dialyzed extensively for 24 hr at 4°C against fresh

Dulbecco's modified Eagle's medium containing 4.5 g/liter D-glucose (DMEM, Gibco) supplemented with L-glutamine (2 mM), sodium pyruvate (1 mM), penicillin (100 U/ml), gentamicin (50 μg/ml), and mycostatin (25 U/ml). The dialyzed culture supernatants, referred to as Con A LK and control LK, were divided in aliquots and kept frozen at −20°C until use. As an indicator of LK activity, each supernatant was tested for its ability to inhibit MP migration by the capillary tube assay and only material exhibiting percent inhibition of migration values higher than 30% was used to make up pools. Con A LK was also assayed for residual Con A as described (Pick and Kotkes, 1977). Free Con A never exceeded 0.5 μg/ml.

MP Culture. Peritoneal exudate cells (PEC) were obtained from guinea pigs injected intraperitoneally, 4 days earlier, with 10–15 ml mineral oil. The PEC were harvested, washed three times with Earle's balanced salt solution (BSS), and suspended in DMEM containing 15% heat-inactivated fetal calf serum (FCS, Gibco) to a concentration of 2.5×10^6 cells/liter. The cell suspension was added to 96-well flat-bottom tissue culture plates (Nunclon, Nunc, or Microtest II, Falcon), 100 μl/well, and the MPs allowed to adhere for 1 hr at 37°C in 95% air–5% CO_2. Following this, the cell monolayers in the wells were washed three times with 100-μl volumes of warm BSS per well to remove nonadherent cells. The plates were now either used for testing O_2^- production in response to stimulants (time 0) or for culturing the MPs with LK. For the latter purpose, 100-μl amounts of an appropriate dilution of Con A LK, Control LK, or DMEM, all supplemented with 15% FCS, were added per well and the plates incubated at 37°C in 95% air–5% CO_2 for 1, 2, or 3 days. When MPs were cultured for longer than 1 day, fresh LK or medium was added every 24 hr, after aspirating the LK left in the wells from the previous 24-hr period.

Assay for O_2^- Production. O_2^- production by freshly explanted MPs or by MPs cultured in LK medium was assayed by the reduction of ferricytochrome *c* directly in the 96-well plates with the cells *in situ*, using the automated microassay described by us (Pick and Mizel, 1981). The MP monolayers were rinsed three times with warm phenol red-free BSS, covered with 100 μl/well of a 160 μM solution of ferricytochrome *c* (type III, Sigma) in BSS, containing the various elicitors of O_2^- production, and incubated for 90 min at 37°C. Normally, one complete 96-well plate of MPs was incubated with either DMEM, control LK, or Con A LK and each one of the nine stimulants was applied to all eight wells in a vertical row. At the completion of the incubation period, the amount of reduced cytochrome *c* was measured with the aid of an automatic enzyme immunoassay reader (Titertek-Multiskan, Flow Laboratories) fitted with a 550-nm interference filter. Mean absorbance values from eight identical wells were calculated and results expressed as nmoles O_2 produced per mg MP protein per 90 min, based on an extinction coefficient of reduced minus oxidized cytochrome *c* of $21 \times 10^3\ M^{-1}\ cm^{-1}$. MP protein concentrations were determined on MP monolayers treated identically to those used for assaying O_2^- production. For each treatment, cells in 24–32 replicate wells were covered with 50 μl/well of 1 N NaOH and incubated overnight at 37°C in a humidified incubator. The following day, the protein hydrolysate from identical wells was pooled and the protein content determined by the method of Lowry *et al.* (1951), using bovine serum albumin as standard.

Stimulants of O_2^- Production. PMA was obtained from Consolidated Midland Corporation. Zymosan, wheat germ agglutinin (WGA), *N*-formyl-L-methionyl-L-leucyl-L-phenylalanine (fMet-Leu-Phe), phospholipase C (type I, from *Clostridium welchii*, 8.1 U/mg), and superoxide dismutase (SOD, type I, 3000 U/mg protein) were purchased from Sigma. A23187 was obtained from Calbiochem–Behring Corp. NaF was from E. Merck and Na nitroprusside, from J. T. Baker. Zymosan was opsonized with fresh guinea pig serum. Concentrated solutions of stimulants were prepared and stored as previously reported (Pick and Keisari, 1981). Stimulants were added to the cytochrome *c* solution in a volume not exceeding 10 μl/ml. All elicitors of an oxidative burst were employed at a concentration found to induce maximal O_2^- production in freshly explanted oil-elicited peritoneal MPs.

3. RESULTS

3.1. Effect of Con A LK on Stimulated O_2^- Production—Influence of LK Concentration

The majority of earlier reports on LK modulation of the oxidative metabolism of MPs demonstrate that prolonged incubation (3 days) of MPs with LK is required for the enhancing effect to become evident. As preliminary experiments performed with guinea pig MPs suggested the same pattern of response, we investigated the effect of pretreating MPs for 3 days with various dilutions of Con A LK and control LK on their ability to produce O_2^- in response to a battery of nine stimulants. These experiments were performed under strictly standardized conditions and the availability of the automatic microassay permitted the assay of each LK dilution–stimulant combination on eight replicate wells in each experiment. The fact that results were expressed as nmoles O_2^- per mg MP protein provided a correction for loss of cells during the 3-day culture period.

Data from 7 to 15 experiments performed with each stimulant are summarized in Table I. It is apparent that oil-elicited guinea pig MPs maintained in culture for 3 days exhibited vigorous O_2^- production in response to all nine stimulants, shown in the past to be capable of activating freshly explanted cells (Pick and Keisari, 1981). Their order of potency was similar to that found with fresh cells; PMA and opsonized zymosan were the most active, Na nitroprusside was the least effective. Incubation of MPs in Con A LK affected O_2^- production in two distinct and opposite directions, depending on the nature of the stimulant. Thus, O_2^- production in response to PMA, opsonized zymosan, Con A, the Ca^{2+} ionophore A23187, and NaF was partially inhibited by Con A LK. This effect was more pronounced at higher concentrations of LK (1/4 and 1/8 dilutions) but with some stimulants, it was evident even at a 1/16 dilution (PMA, opsonized zymosan, and Con A). O_2^- production in response to WGA, fMet-Leu-Phe, phospholipase C, and Na nitroprusside was enhanced by pretreatment of the MPs with Con A LK. The enhancing effect was present at all concentrations of Con A LK, down to the highest dilution tested (1/16). The most pronounced enhancing effect of a 1/16 dilution of Con A

TABLE I. Modulatory Effect of Lymphokine on Stimulated O_2^- Production by Cultured Macrophages—Effect of Lymphokine Concentration[a]

Stimulant	Concentration	Produced by cells preincubated with:		Dilution of lymphokine 1/4	1/8	1/16
PMA	20 nM	Medium	867 ± 110			
		Control LK		900 ± 107	905 ± 110	890 ± 178
		Con A LK		598 ± 86	487 ± 74	634 ± 88
Opsonized zymosan	0.5 mg/ml	Medium	1104 ± 136			
		Control LK		878 ± 131	782 ± 124	745 ± 109
		Con A LK		706 ± 74	552 ± 78	657 ± 74
Con A	50 μg/ml	Medium	688 ± 90			
		Control LK		859 ± 114	813 ± 91	883 ± 180
		Con A LK		483 ± 54	423 ± 64	600 ± 97
WGA	50 μg/ml	Medium	269 ± 35			
		Control LK		359 ± 60	279 ± 35	289 ± 48
		Con A LK		403 ± 57	331 ± 52	426 ± 46
A23187	10 μM	Medium	148 ± 32			
		Control LK		149 ± 50	123 ± 30	152 ± 44
		Con A LK		101 ± 38	73 ± 28	146 ± 33
fMet-Leu-Phe	1 μM	Medium	216 ± 30			
		Control Lk		432 ± 77	341 ± 40	253 ± 40
		Con A LK		396 ± 56	352 ± 51	435 ± 59
NaF	10 mM	Medium	390 ± 43			
		Control LK		373 ± 87	381 ± 105	438 ± 88
		Con A LK		276 ± 48	230 ± 44	395 ± 65
Phospholipase C	0.08 U/ml	Medium	283 ± 44			
		Control LK		363 ± 66	292 ± 33	260 ± 45
		Con A LK		382 ± 65	272 ± 46	344 ± 39
Na nitroprusside	1 mM	Medium	88 ± 22			
		Control LK		235 ± 49	156 ± 34	75 ± 19
		Con A LK		372 ± 41	316 ± 45	310 ± 52

[a] Macrophages were incubated with medium, control LK, or Con A LK at the indicated dilutions for 72 hr (with daily replacement with fresh material) and exposed to stimulants for 90 min. Values are expressed as nmoles O_2^-/mg macrophage protein per 90 min. Data represent means from 7 to 15 experiments ± S.E.M. In each experiment, eight replicate wells were used for each lymphokine–stimulant combination. Basal (unstimulated) O_2^- production values were deducted from the stimulated O_2^- production values.

LK was on the response to Na nitroprusside (about 3.5-fold increase in O_2^- production). The response to fMet-Leu-Phe was enhanced 2-fold and that to WGA, 1.6-fold. The least pronounced enhancing effect was on the phospholipase C-elicited O_2^- production (1.2-fold increase).

The effect of control LK was clearly distinguishable from that of Con A LK. Control LK did not significantly inhibit O_2^- production in response to PMA, Con A, A23187, and NaF at any of the concentrations tested. O_2^- production elicited by these agents was inhibited by Con A LK. A certain degree of inhibition was seen with zymosan-stimulated cells but this was clearly less pronounced than that caused by the same dilution of Con A LK. Interestingly, the Con A-elicited O_2^- production was enhanced by pretreatment of MPs with control LK at concentrations at which Con A LK was inhibitory. On the other hand, control LK, at higher concentrations (1/4 and 1/8 dilutions), had a certain enhancing effect on O_2^- production elicited by stimulants the response to which was enhanced by Con A LK (WGA, fMet-Leu-Phe, phospholipase C, and Na nitroprusside) but this effect was almost absent at a dilution of 1/16, at which Con A LK was clearly enhancing. In conclusion, the specificity of enhancement by LK of the oxidative metabolism of MPs was most evident when control and LK-containing supernatants were compared at high dilutions. Again, the most pronounced difference between Con A LK- and control LK-treated MPs was found with Na nitroprusside-stimulated cells; at a dilution of 1/16, Con A LK-treated MPs produced almost 4 times more O_2^- than control LK-treated MPs.

3.2. Kinetics of Modulation of O_2^- Production by LK

Treatment of MPs with either Con A LK or control LK had no effect on the basal (unstimulated) level of O_2^- production at any time during a 3-day culture period (Table II).

O_2^- production in response to stimulants by MPs cultured in DMEM varied with time in culture (Table III). This variation was not uniform and depended on the nature of the stimulant. The responsiveness to all stimulants increased during

TABLE II. Effect of Lymphokine on Basal O_2^- Production by Cultured Macrophages[a]

Macrophages preincubated with:	Days in culture			
	0	1	2	3
Medium	17.5 ± 5	55 ± 23	35 ± 24	50 ± 15
Control LK		26 ± 15	66 ± 33	40 ± 14
Con A LK		36 ± 12	47 ± 24	51 ± 20

[a] Macrophages were preincubated with medium or a 1/16 dilution of control LK or Con A LK for 1, 2, or 3 days (with daily replacement with fresh material) and basal O_2^- production measured during incubation for 90 min in the absence of stimulants. Valves are expressed as nmoles O_2^-/mg macrophage protein per 90 min.

TABLE III. Modulatory Effect of Lymphokine on Stimulated O_2^- Production by Cultured Macrophages—Kinetics of Modulation[a]

Stimulant	Concentration	Produced by cells preincubated with:	Days in culture: 0	1	2	3
PMA	20 nM	Medium	325 ± 64	451 ± 46	570 ± 127	474 ± 167
		Control LK		443 ± 15	433 ± 62	498 ± 98
		Con A LK		420 ± 18	360 ± 35	419 ± 94
Opsonized zymosan	0.5 mg/ml	Medium	315 ± 101	403 ± 125	496 ± 141	500 ± 206
		Control LK		421 ± 84	429 ± 103	416 ± 129
		Con A LK		452 ± 69	431 ± 88	413 ± 132
Con A	50 μg/ml	Medium	132 ± 52	308 ± 68	426 ± 65	265 ± 84
		Control LK		311 ± 22	359 ± 83	411 ± 59
		Con A LK		297 ± 18	295 ± 26	373 ± 92
WGA	50 μg/ml	Medium	181 ± 22	392 ± 71	312 ± 24	130 ± 51
		Control LK		388 ± 32	294 ± 49	180 ± 33
		Con A LK		434 ± 12	347 ± 38	300 ± 76
A23187	10 μM	Medium	85 ± 10	188 ± 61	194 ± 120	125 ± 68
		Control LK		186 ± 57	224 ± 141	175 ± 89
		Con A LK		197 ± 66	225 ± 132	165 ± 45
fMET-Leu-Phe	1 μM	Medium	77 ± 16	316 ± 89	214 ± 22	149 ± 58
		Control LK		309 ± 88	227 ± 29	211 ± 68
		Con A LK		327 ± 76	234 ± 26	356 ± 82
NaF	10 mM	Medium	184 ± 23	342 ± 8	259 ± 51	290 ± 113
		Control LK		353 ± 16	247 ± 32	293 ± 45
		Con A LK		323 ± 5	235 ± 34	298 ± 58
Phospholipase C	0.08 U/ml	Medium	96 ± 11	274 ± 47	224 ± 43	158 ± 46
		Control LK		260 ± 29	206 ± 15	190 ± 21
		Con A LK		262 ± 32	248 ± 39	258 ± 37
Na nitroprusside	1 mM	Medium	82 ± 9	111 ± 27	116 ± 42	87 ± 17
		Control LK		99 ± 14	116 ± 23	102 ± 38
		Con A LK		109 ± 21	194 ± 31	286 ± 74

[a] Macrophages were incubated with medium or a 1/16 dilution of control LK or Con A LK for 1, 2, or 3 days (with daily replacement with fresh material) and exposed to stimulants for 90 min. Values are expressed as nmoles O_2^-/mg macrophage protein per 90 min. Data represent means of three to six experiments ± S.E.M. In each experiment, eight replicate wells were used for each lymphokine–stimulant combination. Basal (unstimulated) O_2^- production values were deducted from the stimulated O_2^- production values.

the first day in culture. Maximal O_2^- production was evident on day 1 for WGA, fMet-Leu-Phe, NaF, and phospholipase C; on days 1 and 2 for A23187 and Na nitroprusside; on day 2 for PMA and Con A; and on days 2 and 3 for zymosan.

We next examined the kinetics of the LK-induced modulation of the oxidative metabolism of MPs by incubating MPs with Con A LK and control LK diluted 1/16 for 1 to 3 days and assaying stimulated O_2^- elicited by nine agents at daily intervals.

As apparent in Table III, the LK-induced enhancement of O_2^- production elicited by WGA, fMet-Leu-Phe, phospholipase C, and Na nitroprusside was most pronounced after 3 days of culture. For three out of the four stimulants, only a marginal enhancing effect was found after 1 and 2 days of culture; however, the response to Na nitroprusside was almost doubled after 2 days of culture with Con A LK. We found that a promoting action of LK on O_2^- production was most common with stimulants, the response to which peaked after 1 day of culture in medium and decreased in the course of the following 2 days. The moderate enhancing effect on O_2^- production to the above-mentioned four stimulants exhibited by control LK diluted 1/16 was only apparent after 3 days of culture.

The LK-mediated inhibition of O_2^- production, described in the previous section, was little pronounced with material diluted 1/16 and was limited to the PMA,-zymosan-, and Con A-elicited response. Maximal inhibition was seen after 2 days (PMA- and Con A-elicited response) or 2 and 3 days (zymosan-elicited response) of incubation in Con A LK and coincided with the time when the response of the cells incubated in culture medium, to the respective stimulant, was at its peak. A certain degree of inhibition of O_2^- production elicited by PMA, zymosan, and Con A was also caused by control LK diluted 1/16, and this exhibited similar kinetics. In the case of two stimulants, A23187 and NaF, for which an inhibitory effect could be demonstrated with MPs incubated for 72 hr with Con A LK diluted 1/4 and 1/8 (see Table I), no inhibitory effect was detected with a 1/16 dilution of LK at any time during culture.

4. DISCUSSION

The results presented in this paper allow us to derive the following conclusions:

1. Mitogen-stimulated guinea pig lymphocyte culture supernatants known to contain MP activating and MP migration inhibitory activities have no effect on basal (unstimulated) O_2^- production by inflammatory (chemically elicited) peritoneal MPs.

2. Such supernatants are, however, capable of modulating the oxidative burst of MPs in response to a variety of stimulants. This modulation is expressed in two diametrically opposite directions. O_2^- generation to one category of stimulants (WGA, fMet-Leu-Phe, phospholipase C, and Na nitroprusside) is enhanced 1.2- to 3.5-fold while the oxidative burst elicited by PMA, opsonized zymosan, A23187, and NaF is inhibited to various degrees.

3. The enhancing effect is detectable down to a 1/16 dilution of LK-containing

culture supernatant, which was the highest dilution tested. The inhibitory effect is mostly present at higher concentrations of LK.

4. Maximal enhancement of stimulated O_2^- production is found after 3 days of exposure to the LK.

5. Both enhancement and inhibition of O_2^- production are independent of the effect of LK on cell survival as evident in the absolute change (increase or decrease) in O_2^- production per cell protein. On the other hand, an enhancing effect of LK appears to be associated with stimulants the response to which decays progressively during culture.

Our finding that LK treatment by itself does not stimulate O_2^- production by MPs is in agreement with most earlier work, although direct stimulation of NBT reduction in LK-treated MPs was occasionally reported (Krueger *et al.*, 1976; Alföldy and Lemmel, 1979). Recent results from our laboratory, however, suggest that the lack of effect of LK on basal O_2^- generation cannot be automatically extended to H_2O_2 production. Thus, we found that LK treatment of guinea pig peritoneal MPs for 48–72 hr induces peroxide production without the need for further stimulation and in the absence of concomitant O_2^- formation (Pick and Freund, 1984). The mechanism of this unexpected result is under investigation.

Our results confirm in the guinea pig similar reports of enhanced stimulated O_2^- and/or H_2O_2 release by LK-treated mouse and human MPs and are in good agreement with numerous reports of an enhanced oxidative metabolism of MPs activated *in vivo,* presumably by an LK-mediated mechanism. However, while PMA and opsonized zymosan were the two triggers most commonly employed to demonstrate the enhancement of oxidative metabolism by LK in mouse and man, we, in the guinea pig, found an inhibitory effect of LK and O_2^- production elicited by these stimulants. The reason for this discrepancy remains to be determined. The inhibition of oxygen radical production by LK-containing material has, to the best of our knowledge, not been described before. Boraschi *et al.* (1982) recently reported a dose-dependent inhibitory effect of β-interferon on the zymosan-stimulated O_2^- production by resident mouse peritoneal MPs.

Our study differs from earlier investigations of a similar design performed on mouse and human MPs by the range of stimulants employed. We demonstrate that LK treatment has a different effect on the responsiveness of MPs to various stimulants. We are incapable, at present, of offering an explanation for these differences but they are in keeping with our other findings indicating that various elicitors of an oxidative burst in guinea pig MPs activate distinct intracellular biochemical pathways (Bromberg and Pick, 1983).

One of the main drawbacks of this and similar earlier studies is the use of whole, unfractionated culture supernatants containing an unknown number of LKs with various and possibly opposite effects on MP functions. It is thus likely that the enhancing and inhibitory effects of Con A LK on O_2^- production result from the action of distinct molecular entities. One possibility, suggested by the results of Boraschi *et al.* (1982), is that the inhibitory effect is exerted by interferon probably present in the supernatants derived from Con A-stimulated lymphocyte cultures. This issue can only be resolved by subjecting the supernatants to fractionation and tionation and testing various components for their effect on O_2^- production. Fractionation studies are also required in order to correlate effects on the oxidative burst with accepted criteria of MP activating activity such as the enhancement of bactericidal and tumoricidal capacities of MPs.

Bacterial endotoxin was also shown to exhibit an enhancing effect on O_2^- production by MPs in the mouse and man (Pabst and Johnston, 1980; Pabst *et al.*, 1982). Although no tests were performed for determining the level of endotoxin contamination of our culture supernatants, similar amounts of endotoxin would be expected to be present in control LK and Con A LK derived from the same culture. This makes it very unlikely that LK-induced enhancement of O_2^- production is due to endotoxin.

An important question raised by our findings is the mechanism of the enhancing and inhibitory effects of LK on the oxidative response of MPs to membrane stimulation. A technical limitation of our experiments was the fact that only endpoint readings of O_2^- production were performed over a 90-min time period. This type of measurement does not permit the detection of differences in the rate of activation (lag time) of the O_2^- producing enzyme and in the maximum rate of O_2^- production by the enzyme. It was recently reported that mouse MPs activated *in vivo* by the injection of *Corynebacterium parvum* and stimulated by PMA produced O_2^- at a maximum rate 13-fold higher than nonactivated cells, suggesting an increase in NADPH oxidase in activated MPs (Bryant *et al.*, 1982). Preliminary data indicating a lowering of the K_m of NADPH oxidase, assayed on subcellular fractions derived from endotoxin-activated mouse MPs, were also reported (Johnston, 1981).

The enhancing effect of Con A LK on O_2^- generation found by us could be due to: (1) an increase in membrane receptors for stimulants; (2) an increase in O_2^- producing enzyme content or a change in enzyme characteristics; (3) an increase in NADPH content or availability; (4) a more efficient transduction mechanism from the cell surface to the enzyme; (5) an increased rate of O_2^- transport to the exterior of the cells; and (6) a decreased activity of the O_2^- degrading enzyme, SOD. It is unlikely that we are dealing with an LK-mediated increase in membrane receptors because the most marked enhancement was seen with Na nitroprusside, a stimulant for which no receptor-mediated mechanism was demonstrated. Also, the response to Con A and opsonized zymosan, two stimulants acting via membrane receptors, was inhibited rather than enhanced. An increase in NADPH oxidase is also unlikely as this would be difficult to accommodate with the dependence of the enhancement by LK on the nature of the stimulant. For this same reason, changes in O_2^- transport or cellular levels of SOD are also unsatisfactory explanations for the enhancement. MPs treated with LK for 3 days were found to have a more active hexose monophosphate shunt (Nathan *et al.*, 1971) which might lead to an elevated cellular level of NADPH. Such a mechanism would, however, be expected to enhance O_2^- production in response to all stimulants, which is not the case.

In light of the evidence suggesting that various stimulants utilize different transduction mechanisms for the activation of NADPH oxidase, probably involving membrane phospholipids (Bromberg and Pick, 1983), it appears likely that the site of LK action is at this level. Studies are under way to investigate this possibility.

ACKNOWLEDGMENTS. This research was supported by a grant from the Arpad Plesch Research Foundation. We thank Mrs. Patricia Bar-On for superb editorial and secretarial assistance.

REFERENCES

Alföldy, P., and Lemmel, E. M., 1979, Reduction of nitroblue tetrazolium for functional evaluation of activated macrophages in the cell-mediated immune reaction, *Clin. Immunol. Immunopathol.* **12**:263–270.

Block, B., Bernheim, H. A., Wenc, K., and Jaksche, H., 1980, Purified human MIF: Effect on oxidative metabolism and surface architecture of effector cells, in: *Biochemical Characterization of Lymphokines* (A. L. De Weck, F. Kristensen, and M. Landy, eds.), pp. 79–83, Academic Press, New York.

Boraschi, D., Ghezzi, P., Salmona, M., and Tagliabue, A., 1982, IFN-β-induced reduction of superoxide anion generation by macrophages, *Immunology* **45**:621–628.

Bromberg, Y., and Pick, E., 1983, Unsaturated fatty acids as second messengers of superoxide generation by macrophages, *Cell. Immunol.* **79**:240–252.

Bryant, S. M., Lynch, R. E., and Hill, H. R., 1982, Kinetic analysis of superoxide anion production by activated and resident murine peritoneal macrophages, *Cell. Immunol.* **69**:46–58.

Fowles, R. E., Fajardo, I. M., Leibowitch, J. L., and David, J. R., 1973, The enhancement of macrophage bacteriostasis by products of activated lymphocytes, *J. Exp. Med.* **138**:952–964.

Gately, C. L., and Oppenheim, J. J., 1982, Activation of macrophages by lymphokines: A factor which induces hydrogen peroxide production by macrophage-like cell lines, *Fed. Proc.* **41**:768.

Johnston, R. B., 1978, Oxygen metabolism and the microbicidal activity of macrophages, *Fed. Proc.* **37**:2759–2764.

Johnston, R. B., 1981, Enhancement of phagocytosis-associated oxidative metabolism as a manifestation of macrophage activation, *Lymphokines* **3**:33–56.

Krueger, G. G., Ogden, B. E., and Weston, W. L., 1976, *In vitro* quantitation of cell mediated immunity in guinea pigs by macrophage reduction of nitroblue tetrazolium, *Clin. Exp. Immunol.* **23**:517–524.

Lowry, O. H., Rosebrough, N., Farr, A. L., and Randall, R. J., 1951, Protein measurement with the Folin phenol reagent, *J. Biol. Chem.* **193**:265–275.

Manheimer, S., and Pick, E., 1973, The mechanism of action of soluble lymphocytic mediators. I. A pulse exposure test for the measurement of macrophage migration inhibitory factor, *Immunology* **24**:1027–1034.

Murray, H. W., 1981, Interaction of *Leishmania* with a macrophage cell line: Correlation between intracellular killing and the generation of oxygen intermediates, *J. Exp. Med.* **153**:1690–1695.

Murray, H. W., and Cohn, Z. A., 1980, Macrophage oxygen-dependent antimicrobial activity. III. Enhanced oxidative metabolism as an expression of macrophage activation, *J. Exp. Med.* **152**:1596–1609.

Nakagawara, A., de Santis, N., Nogueira, N., and Nathan, C. F., 1982, Lymphokines enhance the capacity of human monocytes to secrete reactive oxygen intermediates, *J. Clin. Invest.* **70**:1042–1048.

Nathan, C. F., Karnovsky, M. L., and David, J. R., 1971, Alterations of macrophage functions by mediators from lymphocytes, *J. Exp. Med.* **133**:1356–1376.

Nathan, C. F., Nogueira, N., Juangbhanich, C., Ellis, J., and Cohn, Z., 1979, Activation of macrophages in vivo and in vitro: Correlation between hydrogen peroxide release and killing of *Trypanosoma cruzi*, *J. Exp. Med.* **149**:1056–1068.

Pabst, M. J., and Johnston, R. B., 1980, Increased production of superoxide anion by macrophages exposed in vitro to muramyl dipeptide or lipopolysaccharide, *J. Exp. Med.* **151**:101–114.

Pabst, M. J., Hedegaard, H. B., and Johnston, R. B., 1982, Cultured human monocytes require exposure to bacterial products to maintain an optimal oxygen radical response, *J. Immunol.* **128**:123–128.

Pick, E., and Freund, M., 1984, Biochemical mechanisms in macrophage activation by lymphokines:

Intracellular peroxide production by lymphokine-treated macrophages, in: *Progress in Immunology, V* (T. Tada, ed.), Academic Press, New York, in press.

Pick, E., and Keisari, Y., 1981, Superoxide anion and hydrogen peroxide production by chemically elicited peritoneal macrophages—Induction by multiple nonphagocytic stimuli, *Cell. Immunol.* **59**:301–318.

Pick, E., and Kotkes, P., 1977, A simple method for the production of migration inhibitory factor by concanavalin A-stimulated lymphocytes, *J. Immunol. Methods* **14**:141–146.

Pick, E., and Mizel, D., 1981, Rapid microassays for the measurement of superoxide and hydrogen peroxide production by macrophages in culture using an automatic enzyme immunoassay reader, *J. Immunol. Methods* **46**:211–226.

Seim, S., 1982, Production of reactive oxygen species and chemiluminescence by human monocytes during differentiation and lymphokine activation in vitro, *Acta Pathol. Microbiol. Immunol. Scand. Sect. C* **90**:179–185.

32

Modulation of Macrophage Phenotypes by Migration Inhibitory Factors

CLEMENS SORG, ELMAR MICHELS, URSULA MALORNY, and CHRISTINE NEUMANN

I. INTRODUCTION

Macrophage migration inhibitory factor (MIF) was originally defined by its inhibitory effect on random migration of macrophages (Bloom and Bennett, 1966; David, 1966). In the years after its discovery a number of other biological activities have been associated with MIF activity such as a number of macrophage-activating activities (David and Remold, 1976) and interferons (Neta and Salvin, 1982).

In the past years we have identified and characterized several molecular weight species of MIF in the guinea pig (Sorg and Bloom, 1973), man (unpublished data), and mouse (Sorg, 1980). The molecular weights of mouse MIF were determined to be 56, 42, 28, and 14K. Using these purified materials it was determined that MIF was not identical with chemotactic factors for mononuclear as well as polymorphonuclear cells, with macrophage-activating factors inducing tumor cytotoxicity or tumor cytostasis, with macrophage growth (mitogen) factor, inducers of plasminogen activator (PA), interferons, interferon inducers, and T-cell-replacing factor (Sorg, 1982).

In our previous studies concerned with the functional heterogeneity of macrophages it was found that only certain types of macrophages are cabable of responding to lymphokines by production of interferons (Neumann and Sorg, 1977,

---

CLEMENS SORG, ELMAR MICHELS, URSULA MALORNY, and CHRISTINE NEUMANN • Department of Experimental Dermatology, Universitäts-Hautklinik, D-4400 Münster, Federal Republic of Germany.

1978) or of PA (Klimetzek and Sorg, 1976, 1977). Furthermore, it was found that only certain macrophage phenotypes were able to migrate and to respond to MIFs (Neumann and Sorg, 1980).

With regard to the dissociation of MIFs from a series of other biological activities and with regard to the phenotype-associated response of macrophages to MIFs we addressed ourselves to the following questions:

1. What are the characteristics of the MIF response-responding macrophage phenotypes?
2. What are the functional changes induced by MIF on macrophages in addition to migration inhibition?

2. RESULTS

2.1. Characteristics of the MIF-Responsive Macrophage Phenotype

Murine bone marrow cells were cultivated in the presence of L-cell-conditioned medium. As described (Neumann and Sorg, 1980), the cells proliferate and differentiate within several days to mature macrophages. A series of functions was recorded daily such as proliferation as determined by [^{3}H]-TdR incorporation, PA production (Overwien *et al.*, 1980), expression of intracellular transglutaminase (TGase) (Schroff *et al.*, 1981), and responsiveness to MIF. As seen in Fig. 1, the

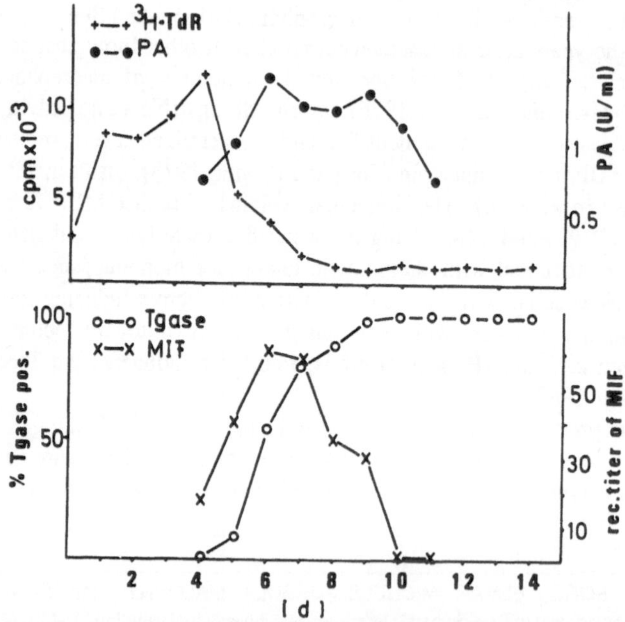

FIGURE 1. Kinetics of bone marrow cultures.

TABLE I. Separation of Bone Marrow-Derived Macrophages (Day 6) by Hypotonic Percoll Gradient Centrifugation

		Unseparated cells	Band 1	Band 2	Band 3
Density (g/ml)			1.055	1.06	1.065
S phase (%)		19.3	18.5	17.3	24.2
G2/M (%)		7.0	7.0	8.5	9.1
CSF stimulation	4 hr		5.3	8.9	5.5
([^{3}H]-TdR	18 hr		10.1	13.0	12.0
uptake,	30 hr		11.0	17.0	13.0
cpm × 10^{-3})	48 hr		15.0	18.0	13.0
TGase (%)		17.9	4.5	55.7	16.7
PA release after 4 hr (U/ml)		+	−	+	+
Random migration (arbitrary units)		20	18	30	20
MIF response (reciprocal titer of 30% inhibition)		27	0	12	25

four functions are each expressed at different stages of the maturation process. In particular, sensitivity to MIF is only expressed between day 5 and 9 of culture. The cells were harvested at day 6, i.e., the day of optimal MIF sensitivity, and were separated on a hypotonic Percoll gradient (Feige *et al.*, 1982). As seen in Table I, three bands of cells at the densities 1.055, 1.06, and 1.065 g/ml were obtained. Cells of all three bands did not differ with regard to their content of S or G2/M phases as determined by flow cytofluorometry, and to their response to L-cell-conditioned medium [which contains colony-stimulating factor (CSF)] as determined by thymidine uptake. However, TGase seems to be expressed most by cells in band two. Furthermore, spontaneous production of PA after 4 hr of culture was only detectable in cells of band two and three. Random migration was highest in band two; however, the response to MIF was most sensitive in cells of band three.

For further characterization, cells of the different bands were serologically typed using monoclonal antibodies which had been generated against different phenotypes of murine macrophages (Malorny and Sorg, 1982). Cytopreparations were stained using the indirect immunoperoxidase technique with 2-aminoethylcarbazole as chromogenic substrate and Mayer's Hemalaun as counterstain. The data were expressed as percent positively stained cells. As shown in Table II, the unseparated cells reacted from 18 to 54% with various monoclonal antibodies. In general, an accumulation of positive cells seems to take place which is most striking for BM 11, where 82 or 73% positive cells accumulated in band two or three, respectively.

Another series of experiments concerned whether a MIF-insensitive phenotype may be converted to a MIF-sensitive one. Bone marrow cells at day 16 of culture were characterized by their moderate random migration and their total unrespon-

TABLE II. Antigen Expression on Bone Marrow-Derived Macrophages Fractionated by Hypotonic Percoll Gradient Centrifugation

Monoclonal antibody	Unseparated cells	% positively stained Band 1 (Density 1.055 g/ml)	Band 2 (1.06)	Band 3 (1.065)
BM 1	34	30	45	41
BM 6	37	24	69	53
BM 11	40	35	82	73
NP 2	20	22	11	44
NP 6	18	11	29	25
NP 7	54	36	36	50

siveness to MIF (Fig. 2). Furthermore, at this stage of the cultures, proliferating cells are no longer present. By the addition of L-cell-conditioned medium (CSF), the cells are pushed again into the cell cycle and within 2 days a vigorous proliferative response of the cells takes place. After the proliferative response was subsiding, the random migration improved and the response to MIF was fully restored.

Serological typing of the cells in the course of culture revealed that CSF had a modulatory influence on a number of phenotype-associated surface markers, particularly on BM 11 which was down-regulated upon the addition of CSF and increased dramatically with the return of MIF sensitivity. Thus, it appears that the BM 11 antigen is closely associated with the MIF-sensitive state. Whether this antigen is related to the MIF receptor remains subject to future investigations.

2.2. Modulation of Macrophage Phenotypes by MIFs

In order to investigate the short-and long-term effects of MIFs on macrophage functions and phenotypic differentiation, bone marrow-derived macrophages on day

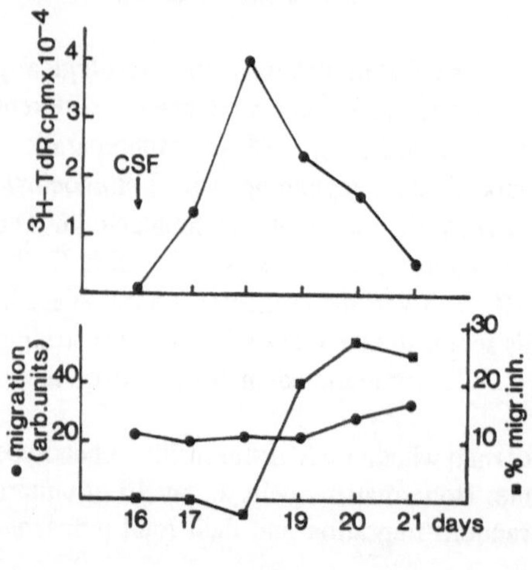

FIGURE 2. Restimulation of BM macrophages with L-cell-conditioned medium.

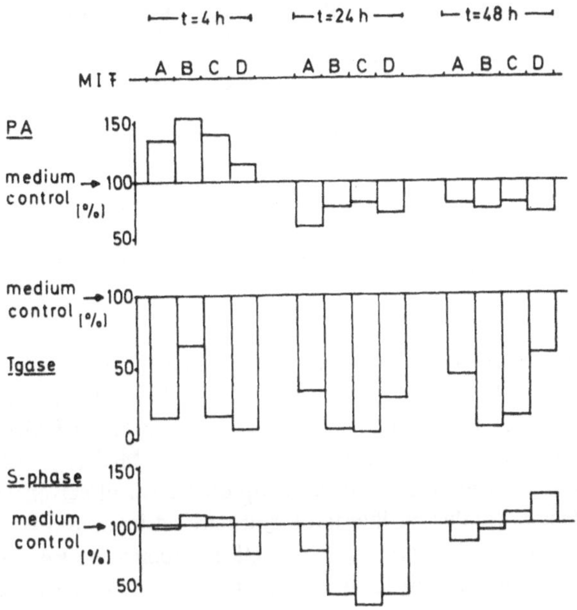

FIGURE 3. Effects of MIFs on BM macrophages (day 6).

6 of culture were incubated with MIF (A–D). PA production, TGase expression, and proliferation were monitored as it had been shown earlier that these functions could be attributed to certain stages of the cell cycle (Sorg and Neumann, 1981). As shown in Fig. 3, PA production was enhanced during the first 4 hr of culture and subsequently suppressed for the next 24–48 hr TGase expression was immediately suppressed and remained at a very low level for the remaining period. In order to determine the effect of MIFs on proliferation, bone marrow cells were cultured in the presence of CSF. As also seen in Fig. 3, during the first 4 hr no suppressive effect was seen. However, after 24 hr the number of S phases was drastically reduced but was back to normal after 48 hr of culture.

In order to obtain more information on the nature of the MIF signal, it was compared to a number of well-defined natural or synthetic signals, whose effect on macrophages has been well studied (Neumann and Sorg, 1981, 1983). As seen in Table III, known mitogens such as Con A, PMA, and Pronase are also inducers of PA, yet they suppress TGase expression and do not induce interferon. Interferon is only induced by the known inducers LPS and poly I : C which in turn do not affect proliferation, PA production, or TGase expression. The effect of dexamethasone on proliferation and PA production has been described before (Neumann and Sorg, 1983). On the other hand, interferon is not induced and TGase expression is not modulated by dexamethasone. Retinoic acid, on the other hand, is suppressive of proliferation, PA production, and TGase expression and does not induce interferon. It thus comes closest in its effects to MIF. Retinoic acid, on the other hand, does not inhibit macrophage migration, thus suggesting a similar effect on internal cell regulation, yet different mechanisms of membrane interactions.

TABLE III. Effects of Various Signals on Bone Marrow-Derived Murine Macrophages (Day 7)[a]

	Con A	PMA	Pronase	Retinoic acid	Dexamethasone	LPS	Poly I:C
Proliferation	+	+	+	−	−	0	0
PA	+	+	+	−	−	0	0
Interferon	0	0	0	0	0	+	+
TGase	−	−	−	−	0	0	0

[a] +, increase; −, decrease; 0, no effect.

As MIF had been associated with interferon activity for a number of reasons (Neta and Salvin, 1982) and as MIF also showed an antiproliferative effect on macrophages, we undertook a comparative study using purified MIFs and highly purified interferon (α,β) kindly provided by Dr. E. De Maeyer, Paris. As seen in Table IV, interferon and MIF share an antiproliferative effect on macrophages. However, MIF is not inducing interferon and purified interferon is not inhibiting macrophage migration. Furthermore, only MIF is suppressing PA production and TGase expression. In order to further define the phenotypic changes induced by MIFs, bone marrow-derived macrophages were typed with a panel of monoclonal antibodies before and after incubation with MIFs. As can be seen from Table V, the antigens MH 7, 9, and 14 are not modulated under MIF influence. However, antigens MH 1 and 8 are down-regulated, whereas MH 11 (later renamed BM 11) was enhanced.

3. CONCLUSIONS

Our data further substantiate our hypothesis put forward earlier on macrophage heterogeneity and their interaction with lymphokines (Sorg and Neumann, 1981). Accordingly, the events in a bone marrow liquid culture system may be described as follows. Macrophages differentiate from precursors in the presence of CSF. After a phase of intensive proliferation of precursors and differentiation into macrophages, the young macrophages driven by an excess of CSF keep on cycling, thus accumulating in "late" GI which is a phase characterized by PA production, inducibility

TABLE IV. Effects of Purified IFN and MIFs on Macrophages[a]

	IFN	MIF
Proliferation	−	−
PA	0	−
TGase	0	−
Migration inhibition	neg.	
Interferon activity		neg.

[a] −, decrease; 0, no change.

TABLE V. Modulation by MIF of Phenotype-Associated Surface Antigens on Bone Marrow-Derived Macrophages as Detected with Monoclonal Antibodies

Monoclonal antibody	Control	% positively stained MIF A	B	C	D
MH 1	34	26	26	8	6
MH 5	0	0	0	0	0
MH 6	22	30	28	21	15
MH 7	100	100	100	100	100
MH 8	75	41	37	49	25
MH 9	100	100	100	100	100
MH 11	42	57	57	50	64
MH 14	100	100	100	100	100

of interferon, and the responsiveness to MIF and certain chemotactic factors. When CSF is consumed, proliferation is fading away, the cells differentiate gradually in G1, losing a series of constitutive functions and pass on to a G0-like state which is characterized by production of fibrinolysis inhibitors (Klimetzek and Sorg, 1979). This process is reversible insofar as the addition of CSF to mature resting macrophages induces proliferation and reexpression of functions associated with "late" G1. Similar events may take place in peritoneal exudate cells, i.e., peripheral blood monocytes migrate into the peritoneal cavity, differentiate to macrophages, and, depending on the time and the stimulus, will finally arrive at different phases of cell cycle, e.g., in "late" G1 (thioglycollate-induced macrophages) or "early" G1 (proteose peptone-elicited macrophages). Due to the lack of mitogenic signals, peritoneal exudate cells no longer proliferate.

Based on the experimental evidence, we are able to map tentatively the various phenotype-associated functions within the cell cycle. PA expression is a constitutive function of "late" G1. Hence, mitogenic signals (Con A, PMA, Pronase, CSF) always induce PA secretion. Agents that suppress proliferation also reduce PA secretion. As found in the bone marrow-derived macrophages this process is reversible. TGase expression, on the other hand, is not associated with proliferation as it is decreasing upon a mitogenic signal. This enzyme therefore is expressed during early to mid-G1. As resident peritoneal cells and proteose peptone-induced macrophages are negative or low in TGase expression and mineral oil- or thioglycollate-induced and noncycling bone marrow-derived macrophages express highest amounts of TGase, it is concluded that TGase is expressed in G1 and decreases when proceeding from G1 to S. Empirically, mineral oil-induced macrophages are responding best to MIF. Thioglycollate-induced macrophages also migrate and respond to MIF, whereas resident and proteose peptone-induced macrophages are not or only poorly migrating and do not respond to MIF. The MIF-sensitive bone marrow-derived macrophage is characterized by its position postmitosis, its low content of TGase, and its secretion of moderate amounts of PA, suggesting its position in mid-G1.

Interaction of macrophages with MIF not only suppresses proliferation but also seems to shift macrophages from postmitosis or late to mid-G1. This conclusion is also corroborated by our data using monoclonal antibodies against phenotype-associated surface markers.

In summary, our data suggest that MIF is a differentiation signal, turning "young" macrophages into more mature macrophages. Whether MIF can act on macrophages in G0 or early G1 and cause a shift to mid-G1, i.e., to "younger" macrophages, remains to be determined. In earlier experiments by Geczy *et al.* (1976), it was shown that local or systemic administration of a specific antibody against MIF could prevent the elicitation of a delayed-type hypersensitive skin response in a sensitized guinea pig. The question now is, why is the MIF-induced macrophage phenotype so central to the development of a cellular immune reaction?

4. SUMMARY

Macrophages are heterogeneous with respect to a number of constitutive and inducible functions. In order to study the underlying biological principle, a bone marrow liquid culture system was adopted in which bone marrow cells proliferate and differentiate into macrophages. It was found that maturing macrophages express various constitutive or inducible functions in an ordered sequence. The kinetics of their appearance and disappearance are dependent on the proliferative activity of macrophages. Macrophages in late G1 of the cell cycle express constitutive functions like PA production and are inducible by bacterial LPS, poly I : C, and lymphokines to release interferons. The response to lymphokines like MIF and chemotactic factors is also transiently expressed during maturation. Using purified MIF, its influence on proliferation, differentiation, and activation of macrophages was investigated. The changes induced were monitored following the expression of marker enzymes and of phenotype-associated cell surface antigens using monoclonal antibodies. The results showed that functional changes induced by MIF on macrophages are limited and are not related to certain macrophage-activating activities (MAF). As determined by flow cytofluorometry, TGase expression and proliferation are consistently down-regulated by MIFs. This together with the shift and the expression of surface antigens indicates that MIFs provide a differentiation signal for a "young" macrophage to become more mature.

REFERENCES

Bloom, B. R., and Bennett, B., 1966, Mechanism of a reaction in vitro associated with delayed hypersensitivity, *Science* **153**:80.

David, J. R., 1966, Delayed hypersensitivity in vitro: Its mediation by cell-free substances formed by lymphoid cell interaction, *Proc. Natl. Acad. Sci. USA* **56**:62.

David, J. R., and Remold, H. G., 1976, Macrophage activation by lymphocyte mediators and studies on the interaction of macrophage inhibitory factor (MIF) with its target cell, in: *Immunobiology of the Macrophage* (D. S. Nelson, ed.), pp. 401–426, Academic Press, New York.

Feige, U., Overwien, B., and Sorg, C., 1982, Purification of human blood monocytes by hypotonic density gradient centrifugation in Percoll, *J. Immunol. Methods.* **54**:309.

Geczy, C. L., Geczy, A. F., and De Weck, A. L., 1976, Antibodies to guinea pig lymphokines. II. Suppression of delayed hypersensitivity reactions by second generation goat antibody against guinea pig lymphokines, *J. Immunol.* **117**:66.

Klimetzek, V., and Sorg, C., 1976, Production of plasminogen activator by murine macrophages after exposure to lymphokines, *Cell. Immunol.* **27**:350.

Klimetzek, V., and Sorg, C., 1977, Lymphokine induced production of plasminogen activator by macrophages, *Eur. J. Immunol.* **7**:185.

Klimetzek, V., and Sorg, C., 1979, The production of fibrinolysis inhibitors as a parameter of the activation state in murine macrophages, *Eur. J. Immunol.* **9**:613.

Malorny, U., and Sorg, C., 1982, Monoclonal antibodies specific for different phenotypes of murine macrophages, *Immunobiology* **162**:389.

Neta, R., and Salvin, S. B., 1982, Lymphokines and interferon: Similarities and differences, *Lymphokines* **7**:137.

Neumann, C. and Sorg, C., 1977, Immune interferon. I. Production by lymphokine activated murine macrophages, *Eur. J. Immunol.* **8**:719.

Neumann, C., and Sorg, C., 1978, Immune interferon. II. Different cellular sites for the production of murine macrophage migration inhibitory factor and interferon, *Eur. J. Immunol.* **8**:582.

Neumann, C., and Sorg, C., 1980, Sequential expression of functions during macrophage differentiation in murine bone marrow liquid cultures, *Eur. J. Immunol.* **10**:834.

Neumann, C., and Sorg, C., 1981, Independent induction of plasminogen activator and interferon in murine macrophages, *J. Reticuloendothel. Soc.* **30**:79.

Neumann, C., and Sorg, C., 1983, Regulation of plasminogen activator secretion, interferon induction and proliferation in murine macrophages, *Eur. J. Immunol.* **13**:143.

Overwien, B., Neumann, C., and Sorg, C., 1980, Detection of plasminogen activator in macrophage culture supernatants by a photometric assay, *Hoppe-Seyler's Z. Physiol. Chem.* **361**:1251.

Schroff, G., Neumann, C., and Sorg, C., 1981, Transglutaminase as a marker for subsets of murine macrophages, *Eur. J. Immunol.* **11**:637.

Sorg, C., 1980, Characterization of murine macrophage migration inhibitory activities (MIF) released by concanavalin A stimulated thymus or spleen cells, *Mol. Immunol.* **17**:565.

Sorg, C., 1982, Modulation of macrophage functions by lymphokines, *Immunobiology* **161**:352.

Sorg, C., and Bloom, B. R., 1973, Products of activated lymphocytes. I. The use of radiolabelling techniques in the characterization and partial purification of the migration inhibitory factor of the guinea pig, *J. Exp. Med.* **137**:148.

Sorg, C., and Neumann, C., 1981, A developmental concept for the heterogeneity of macrophages in response to lymphokines and other signals, *Lymphokines* **3**:85.

The Anticancer Action of Lymphotoxin

CHARLES H. EVANS and JANET H. RANSOM

Study of the mechanisms of natural immunity at the target cell level possessing the potential to prevent and intervene to modulate the development of carcinogenesis has identified lymphotoxin to be a major component (Evans *et al.*, 1983). Lymphotoxin anticancer activities include: (1) cancer prevention measured by irreversible inhibition of chemical carcinogen or radiation transformation of syngeneic Syrian hamster cells *in vitro* (Evans and DiPaolo, 1981) or in a combined *in vivo–in vitro* transplacental carcinogenesis assay (Ransom *et al.*, 1982b, 1983a); (2) cytostasis assessed by reversible growth inhibition of allogeneic or syngeneic tumor cells (Evans and Heinbaugh, 1981; Evans *et al.*, 1983); (3) cytolysis indicated by radionuclide release from allogeneic or syngeneic tumor cells (Evans and Heinbaugh, 1981); and (4) cytoreductive stimulating activity demonstrated by enhanced sensitivity of lymphotoxin-treated tumor cells to natural killer cell cytolysis (Ransom and Evans, 1982, 1983).

The lymphotoxin employed in defining these anticancer actions has been obtained from the serum-free RPMI 1640 culture medium of 24-hr antigen or phytohemagglutinin-stimulated normal freshly isolated splenic, peritoneal, or peripheral blood guinea pig, human, and Syrian hamster lymphocytes. The lymphokine preparations were concentrated by diafiltration over an Amicon YM-10 membrane and equilibrated against 0.01 M sodium phosphate-buffered saline, pH 7.4, containing 0.1% 4000-dalton polyethylene glycol (Ransom *et al.*, 1982a). Diafiltered lymphotoxin preparations were further fractionated by column isoelectric focusing (Ransom *et al.*, 1982a) and HPLC size-exclusion chromatography (Fuhrer and Evans, 1983). Lymphotoxin activity prior to and during fractionation procedures was followed for comparative purposes by the ability of preparations/fractions to lyse the

CHARLES H. EVANS and JANET H. RANSOM • Tumor Biology Section, Laboratory of Biology, National Cancer Institute, National Institutes of Health, Bethesda, Maryland 20205.

exquisitely sensitive α L929 murine tumor cell (Evans and Heinbaugh, 1981). One unit was defined as that quantity of lymphotoxin causing release of 50% of the [^{3}H]-TdR from 10^4 α L929 cells in a 1-ml culture volume after 72 hr culture of α L929 cells with lymphotoxin (Evans and Heinbaugh, 1981).

Lymphotoxin exhibits biological and biochemical heterogeneity resulting from differential target cell responses to the actions of lymphotoxin (Fig. 1), the presence of several lymphotoxin biochemical forms possessing inequality of anticancer action (Ransom and Evans, 1983), and the heterogeneity among individuals and their lymphocytes producing lymphotoxin.

The anticancer actions of Syrian hamster lymphotoxin reside within glycoprotein(s) possessing pI of 4.6–5.2, average molecular weights of 45,000, and which are free of detectable interferon (Evans, 1982), MMIF, and T-lymphocyte mitogenic activity (Ransom and Evans, 1983). Syrian hamster lymphotoxin is noncytotoxic for normal cells *in vivo* (Ransom *et al.*, 1983a) and *in vitro* (Evans and DiPaolo, 1981) with 8000 U injected intravenously in a hamster inhibiting carcinogenesis > 97% *in vivo* (Ransom *et al.*, 1983a). The anticarcinogenic action of lymphotoxin is specifically directed to the target cell, with 6 U/ml culture medium preventing 50% of the morphological transformation in either γ-, X-, or UV-irradiated or chemical carcinogen-treated cells *in vitro* (Evans *et al.*, 1983).

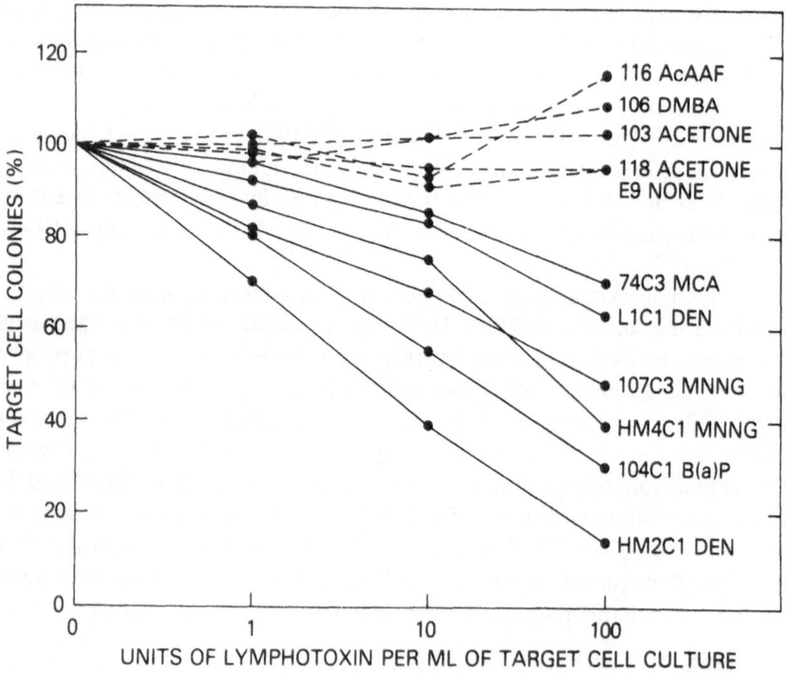

FIGURE 1. Susceptibility of tumor-producing (●—●) and nontumorigenic (●---●) guinea pig cell lines (Evans and DiPaolo, 1975) to the growth inhibitory activity of guinea pig lymphotoxin. Aliquots of lymphotoxin were incubated with 100 target cells in 4 ml RPMI 1640 medium–10% FBS/60-mm dish and after 7 days the number of colonies formed in the presence of lymphotoxin was compared to the number in medium alone expressed as a percentage.

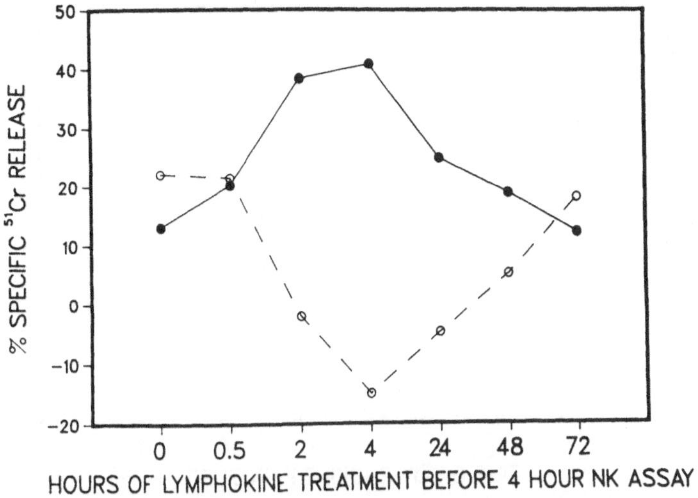

FIGURE 2. ^{51}Cr-labeled K562 human erythroleukemia cells were treated with 40 U lymphotoxin (●—●) or 500 U α-interferon/ml (O---O) for 0, 0.5, 2, 4, 24, 48, or 72 hr, mixed with natural killer lymphocytes at a ratio of 25 : 1 natural killer lymphocytes : K562 cell, and natural killer cell cytolysis assessed by the degree of specific radionuclide release 4 hr later (Ransom and Evans, 1982).

Inhibition of the growth of transformed and fully tumorigenic cells requires 20-fold more lymphotoxin than is necessary to prevent carcinogenesis (Evans *et al.*, 1983). Lymphotoxin inhibition of tumor growth, moreover, is reversible except at very high lymphotoxin concentrations where cytolytic lymphotoxin activity is observed (Evans and Heinbaugh, 1981; Evans *et al.*, 1983).

Exposure of tumorigenic cells to lymphotoxin followed by interaction with natural killer cells can result in lymphotoxin-induced lysis resulting from lymphotoxin sensitization of target cells to natural killer cell lysis (Ransom and Evans, 1982, 1983). Sensitization occurs within 30 min precedes interferon-induced target cell resistance to natural killer cell lysis, and like lymphotoxin induction of resistance to carcinogenesis (DiPaolo *et al.*, 1984) is transient (Fig. 2). Lymphotoxin induction of an anticarcinogenic state is accompanied by increased synthesis of cell membrane glycoproteins (Fuhrer and Evans, 1983), whereas lymphotoxin inhibition of tumor cell growth is accompanied by a transient decrease in the synthesis of high-molecular-weight membrane glycoproteins (Fuhrer and Evans, 1983) and a transient increase in membrane microviscosity (Fig. 3).

Isoelectric focusing of Syrian hamster or human lymphotoxin reveals that there are two classes of lymphotoxin separable according to molecular charge. Syrian hamster anticarcinogenic and tumor growth inhibitory activities each have two pIs, one near 4.6 and the second near 5.0 with the lymphotoxin cytolytic ativity residing within the second, pI 5.0, lymphotoxin class (Ransom and Evans, 1983). Human peripheral blood lymphocyte lymphotoxin activities can be similarly separated with the acidic class I lymphotoxin having a pI of about 5 and the more basic class II lymphotoxin possessing a pI of 7.1 and also containing the cytolytic activity.

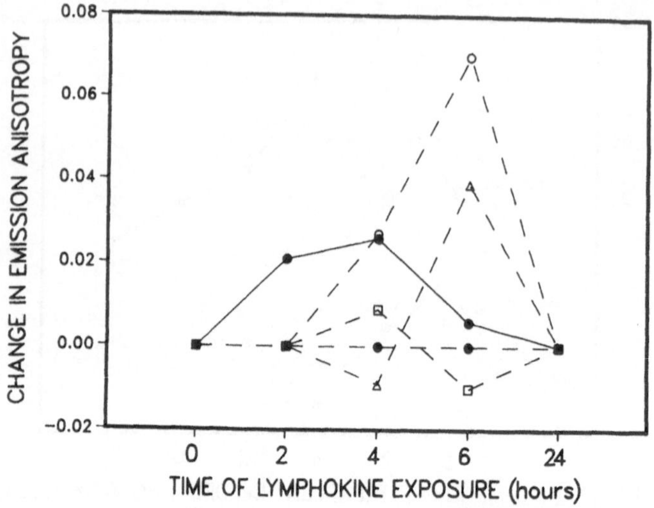

FIGURE 3. K562 cells were treated for 2, 4, 6, or 24 hr with 20 U lymphotoxin/ml (●—●), 50 (●---●) or 200 (○---○) U α-interferon/ml, or a combination of 20 U lymphotoxin and 40 U α-interferon/ml (□---□), or 45 U lymphotoxin and 100 U α-interferon/ml (△---△). The cells were then labeled with 3 μM 1,6-diphenyl-1,3,5-hexatriene for 15 min at 37°C. The emission anisotropy of the 1,3,5-hexatriene-labeled cells was measured in a microflow cytometer using UV excitation with the 351- to 363-nm argon laser lines and a combination of LP400-nm and polarizing emission filters. Values for emission anisotropy were calculated by the LSI/11 computer within the FACS IV flow cytometer. Emission anisotropy is expressed as the change in emission anisotropy of the treated compared to the nontreated cells. An increase in emission anisotropy is indicative of increased membrane microviscosity or decreased membrane fluidity.

Four observations suggest that the cytolytic activity present in the class II lymphotoxin may be yet another form of lymphotoxin. The first is that the ratio of the cytolytic/cytostatic activities within different lymphotoxin preparations is inconstant (Table I). The second is that the cytolytic and cytostatic activities are differentially susceptible to protease and neuraminidase attack. For example, in lymphotoxin preparation HuLT-11 (Table I), cytolytic activity is reduced 96 and

TABLE I. Relative Amounts of Cytolytic and Cytostatic Lymphotoxin Activities in Human Lymphotoxin Preparations from Different Individuals

Preparation	α L929 cytolytic (U/ml)	K562 cytostatic (U/ml)	Ratio
HuLT-8C	10,000	8	1250
HuLT-11C	971	42	23
HuLT-12C	5,000	4	1250
HuLT-13C	1,500	93	16
HuLT-17C	5,415	20	271

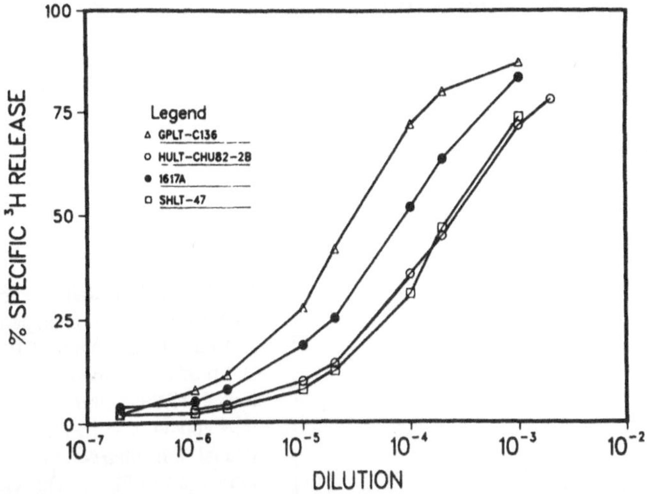

FIGURE 4. Cytolytic activity of guinea pig (△), Syrian hamster (□), human peripheral blood-derived (○), and homogeneous RPMI 1788 cell line-produced human lymphotoxin (●) for murine α L929 cells (Evans and Heinbaugh, 1981) were adjusted to yield approximately 50% lysis at a dilution of 10^{-4} and their ability to release ^3H from [^3H]-TdR α L929 cells compared in a 72-hr assay (Evans and Heinbaugh, 1981). The RPMI 1788 lymphoblastoid lymphotoxin was the generous gift of Dr. Bharat Aggarwal, Genentech Inc., South San Francisco, California.

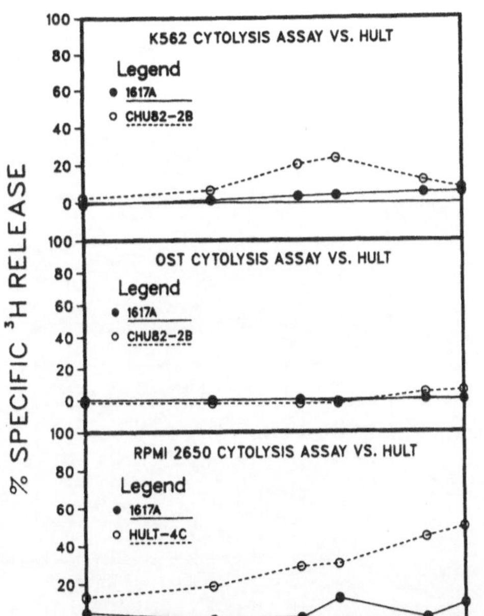

FIGURE 5. Cytolytic activity of human peripheral blood (○) and homogeneous RPMI 1788 cell line-produced human lymphotoxin (●) for human K562 erythroleukemia, OST osteosarcoma, and RPMI 2650 nasopharyngeal carcinoma cells. Assays were performed as in Fig. 4.

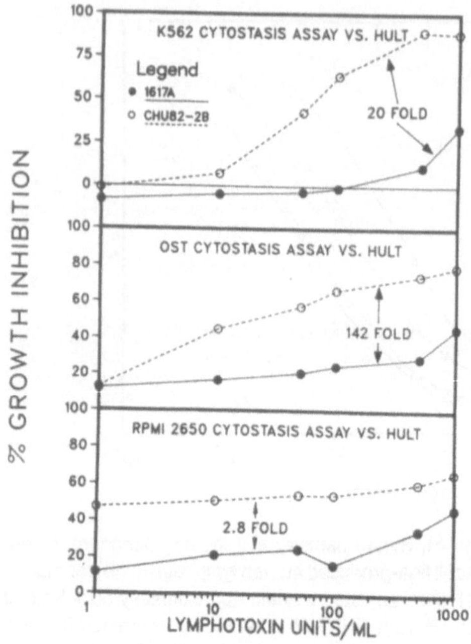

FIGURE 6. Cytostatic (tumor growth inhibitory) activity of human peripheral blood (○) and homogeneous RPMI 1788 cell line-produced human lymphotoxin (●) for human K562 erythroleukemia, OST osteosarcoma, and RPMI 2650 nasopharyngeal carcinoma cells. Assays were performed as in Fig. 1. The values for the differences in the activities of the lymphotoxins are the ratio of the quantity of human peripheral blood to homogeneous RPMI 1788 cell line lymphotoxin producing a 50% effect as calculated by linear regression analysis.

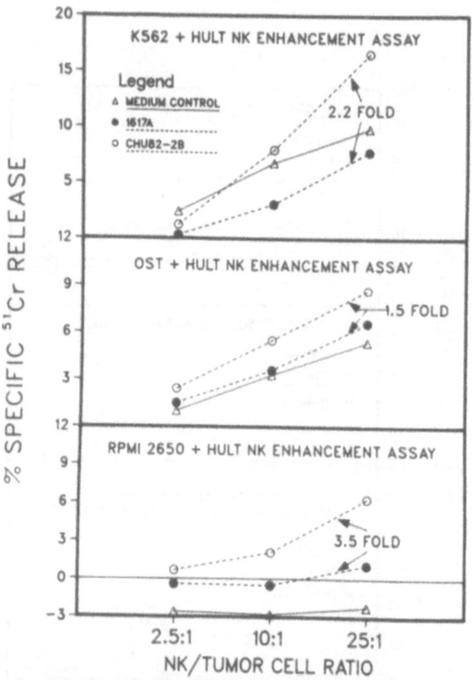

FIGURE 7. Lymphotoxin amplification of tumor cell sensitivity to natural killer cell destruction. ^{51}Cr-labeled human K562 erythroleukemia, OST osteosarcoma, or RPMI 2650 nasopharyngeal carcinoma cells were treated with human peripheral blood (○) or homogeneous RPMI 1788 cell line-produced human lymphotoxin (●) for 30 min and the radionuclide release measured 4 hr after addition of natural killer cells at the ratios shown (Ransom and Evans, 1982). The values for the differences in the activities of the lymphotoxins are the ratio of the quantity of human peripheral blood to homogeneous RPMI 1788 lymphoblastoid lymphotoxin producing a 50% effect as calculated by linear regression analysis.

49% after incubation for 1 hr at 37°C with 32 U trypsin and 0.2 U neuraminidase, respectively; however, cytostatic activity was reduced 32% by trypsin and not at all by neuraminidase. Third, generation of Syrian hamster lymphotoxin in the presence of tunicamycin, an inhibitor of protein *O*-linked glycosylation, results in a shift in the pI of the cytolytic activity to approximately 5.8 without alteration of the other lymphotoxin anticarcinogenic and tumor cell growth inhibitory activities with pIs of 4.6 and 5.0. The fourth observation indicating that lymphotoxin cytolytic activity is a third lymphotoxin form or class is the activity of the purified human lymphotoxin (Aggarwal *et al.*, 1982) obtained from the RPMI 1788 lymphoblastoid cell line (Figs. 4–7). The 1788 lymphoblastoid lymphotoxin has the same cytolytic activity for α L929 cells as does guinea pig peritoneal, Syrian hamster peritoneal, and human peripheral blood lymphocyte lymphotoxin (Fig. 4). The 1788 lymphoblastoid lymphotoxin, however, possesses little cytolytic (Fig. 5), cytostatic (Fig.

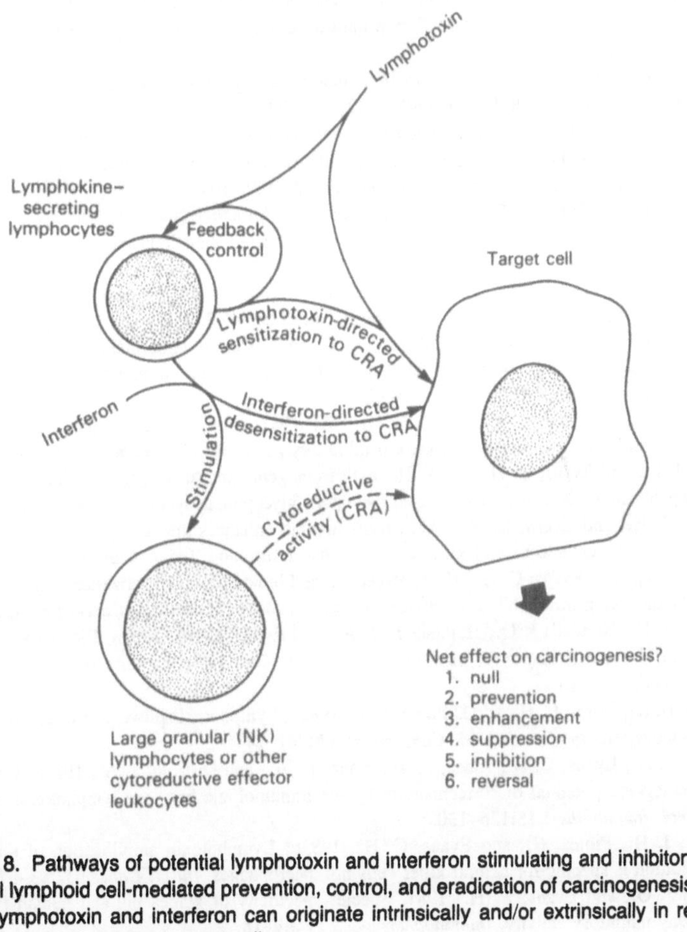

FIGURE 8. Pathways of potential lymphotoxin and interferon stimulating and inhibitory actions in natural lymphoid cell-mediated prevention, control, and eradication of carcinogenesis (Evans, 1983). Lymphotoxin and interferon can originate intrinsically and/or extrinsically in relation to the reaction of effector and target cells.

(Fig. 6), and cytoreductive sensitizing activity for natural killer cells (Fig. 7) compared to lymphotoxin obtained from freshly isolated peripheral blood lymphocytes when evaluated against human carcinoma, leukemia, and sarcoma cells.

The direct and "indirect" actions of lymphotoxin upon the target cell (Fig. 8) can result in prevention or inhibition of carcinogenesis and tumor growth. Due to the existence of two and probably three lymphotoxin activity classes, it is, therefore, important to select the appropriate assay and target cells when evaluating the molecular basis of the anticancer actions of lymphotoxin.

REFERENCES

Aggarwal, B., Moffat, B., and Harkins, R., 1983, Purification and characterization of lymphotoxin from the human lymphoblastoid cell line 1788, in: *Interleukins, Lymphokines, and Cytokines* (J.J. Oppenheim, S. Cohen, and M. Landy, eds.), pp. 521–526, Academic Press, New York.

DiPaolo, J. A., Evans, C. H., DeMarinis, A. J., and Doniger, J., 1984, Inhibition of radiation initiated and promoted transformation of Syrian hamster embryo cells by lymphotoxin. *Cancer Res.* **44:** 1465–1471.

Evans, C. H., 1982, Lymphotoxin—An immunologic hormone with anticarcinogenic and antitumor activity, *Cancer Immunol. Immunother.* **12:**181–190.

Evans, C. H., 1983, Lymphokines, homeostasis, and carcinogenesis, *J. Natl. Cancer Inst.* **71:**253–257.

Evans, C. H., and DiPaolo, J. A., 1975, Neoplastic transformation of guinea pig fetal cells in culture induced by chemical carcinogens, *Cancer Res.* **35:**1035–1044.

Evans, C. H., and DiPaolo, J. A., 1981, Lymphotoxin: An anticarcinogenic lymphokine as measured by inhibition of chemical carcinogen or ultraviolet irradiation induced transformation of Syrian hamster cells, *Int. J. Cancer* **27:**45–49.

Evans, C. H., and Heinbaugh, J. A., 1981, Lymphotoxin cytotoxicity, a combination of cytolytic and cytostatic cellular responses, *Immunopharmacology* **3:**347–359.

Evans, C. H., Cooney, A. M., and DiPaolo, J. A., 1975, Colony inhibition mediated by nonimmune leukocytes *in vitro* as indices of tumorigenicity of guinea pig cultures transformed by chemical carcinogens, *Cancer Res.* **35:**1045–1052.

Evans, C. H., Heinbaugh, J. A., and DiPaolo, J. A., 1983, Comparative effectiveness of lymphotoxin anti-carcinogenic and tumor cell growth inhibitory activities, *Cell. Immunol.* **76:**295–303.

Fuhrer, J. P., and Evans, C. H., 1983, The anticarcinogenic and tumor growth inhibitory activities of lymphotoxin are associated with altered membrane glycoprotein synthesis, *Cancer Lett.* **19:**283–292.

Ransom, J. H., and Evans, C. H., 1982, Lymphotoxin enhances the susceptibility of neoplastic and preneoplastic cells to natural killer cell mediated destruction, *Int. J. Cancer* **29:**451–458.

Ransom, J. H., and Evans, C. H., 1983, Molecular and biological characterization of anti-carcinogenic and tumor cell growth-inhibitory activities of Syrian hamster lymphotoxin, *Cancer Res.* **43:**5222–5227.

Ransom, J. H., Rundell, J. O., Heinbaugh, J. A., and Evans, C. H., 1982a, Biological and physicochemical characterization of keyhole limpet hemocyanin-induced guinea pig lymphotoxin, *Cell. Immunol.* **67:**1–13.

Ransom, H. H., Evans, C. H., and DiPaolo, J. A., 1982b, Lymphotoxin prevention of diethylnitrosamine carcinogenesis *in vivo, J. Natl. Cancer Inst.* **69:**741–744.

Ransom, J. A., Evans, C. H., Jones, A. E., Zoon, R. A., and DiPaolo, J. A., 1983a, Control of the carcinogenic potential of ^{99m}technetium by the immunologic hormone lymphotoxin, *Cancer Immunol. Immunother.* **15:**126–130.

Ransom, J. H., Pintos, C., and Evans, C. H., 1983b, Lymphotoxin amplification of tumor growth inhibition is specific for natural killer cells not macrophages, *Int. J. Cancer* **32:**93–97.

Rundell, J. O., and Evans, C. H., 1981, Species specificity of guinea pig and human lymphotoxin colony inhibitory activity, *Immunopharmacology* **3:**9–18.

Pituitary Responses to Acute Administration of Thymosin and to Thymectomy in Prepubertal Primates

DAVID L. HEALY, NICHOLAS R. HALL, HEINRICH M. SCHULTE, GEORGE P. CHROUSOS, ALLAN L. GOLDSTEIN, D. LYNN LORIAUX, and GARY D. HODGEN

I. INTRODUCTION

There is firm clinical and laboratory evidence that adrenal glucocorticoids induce thymic involution, reduce mitotic activity in thymus-dependent (T) lymphocytes, and inhibit phagocytic activity of human leukocytes (Ishidate and Metcalf, 1963; Monjan, 1981). By contrast, low concentrations of glucocorticoids enhance thymocyte differentiation and stimulate antibody formation *in vitro* (Ambrose, 1964; Ritter, 1977). How this interaction between the hypophyseal–adrenal axis and the immune system is controlled is unclear.

Recent studies indicate that thymosins, a family of peptides isolated from calf thymus which stimulate maturation of T lymphocytes (Goldstein *et al.*, 1981), can also produce various endocrinological responses. Thymosin fraction 5 (TF5) and a component peptide, thymosin B_4, have been reported to release gonadotropin-

DAVID L. HEALY and GARY D. HODGEN • Pregnancy Research Branch, NICHD, National Institutes of Health, Bethesda, Maryland 20205. HEINRICH M. SCHULTE, GEORGE P. CHROUSOS, and D. LYNN LORIAUX • Developmental Endocrinology Branch, NICHD, National Institutes of Health, Bethesda, Maryland 20205. NICHOLAS R. HALL and ALLAN L. GOLDSTEIN • Department of Biochemistry, The George Washington University School of Medicine and Health Sciences, Washington, D.C. 20037.

releasing hormone (GnRH) from the rat hypothalamus (Rebar *et al.*, 1981). TF5 has also been shown to stimulate adrenal cortical secretion in rats and rabbits (Deschaux *et al.*, 1979; Sivas *et al.*, 1982).

In man and nonhuman primates, the thymus is functionally a fetal and juvenile organ which atrophies after puberty and whose weight, expressed as a function of body weight, inexorably declines with advancing age (Healy *et al.*, 1983a). Does the thymus of prepubertal primates influence developmental endocrinology? To begin to address this question, we undertook this study with the following aims: (1) to determine if acute administration of TF5 alters anterior pituitary hormone secretion and (2) to evaluate the pituitary response to prepubertal primate thymectomy.

2. MATERIALS AND METHODS

2.1. Thymosin Preparation

TF5 was prepared from calf thymus, as described by Hooper *et al.* (1975). The thymic tissue was homogenized and centrifuged and the supernatant filtered through glass wool. The filtrate was then processed through 80°C acetone precipitation and ammonium sulfate precipitation. The 25–50% ammonium sulfate precipitate was subjected to ultrafiltration in an Amicon DC-2 hollow-fiber system and desalted on a Sephadex G-25 column to yield fraction 5. Thymosins α_1 and B_4, two component peptides of TF5, were synthesized, dissolved in phosphate-buffered saline (pH 9.0), and sterilized by filtration (2-μm pore size; Millex-GS filter unit, Millipore Corp., Bedford, Mass.) immediately prior to use. All peptides were endotoxin-free.

2.2. Primate Preparation: The Mobile Tether Assembly

Premenarchial cynomolgus monkeys (*Macaca fascicularis*), median age 22 months (range 11–27 months) (n = 18), were used in these studies. Each animal was fitted with a vest and mobile tether assembly which permitted chronic femoral vein cannulation for blood collection (Fig. 1). This allowed plasma to be harvested from monkeys which were unanesthetized, freely moving, and undisturbed, with personnel in an adjacent room (Williams *et al.*, 1981).

2.3. Radioimmunoassays (RIA)

2.3.1. Corticotropin (ACTH)

ACTH and β-endorphin were extracted from plasma using octodecasilyl-silica cartridges (Waters Assoc., Milford, Mass.). The peptides were eluted from the cartridges with 60 ml (60 : 40, v : v) acetonitrile : 0.1% trifluoroacetic acid (Pierce, Rockford, Ill.). Samples were lyophilized and reconstituted in assay buffer (62 mM Na_2HPO_4, 13 mM Na_2 EDTA, pH 7.4, containing 0.02% NaN_3 and 0.1% Triton X-100). The RIAs were performed at 4°C in assay buffer containing 250 kIU/ml

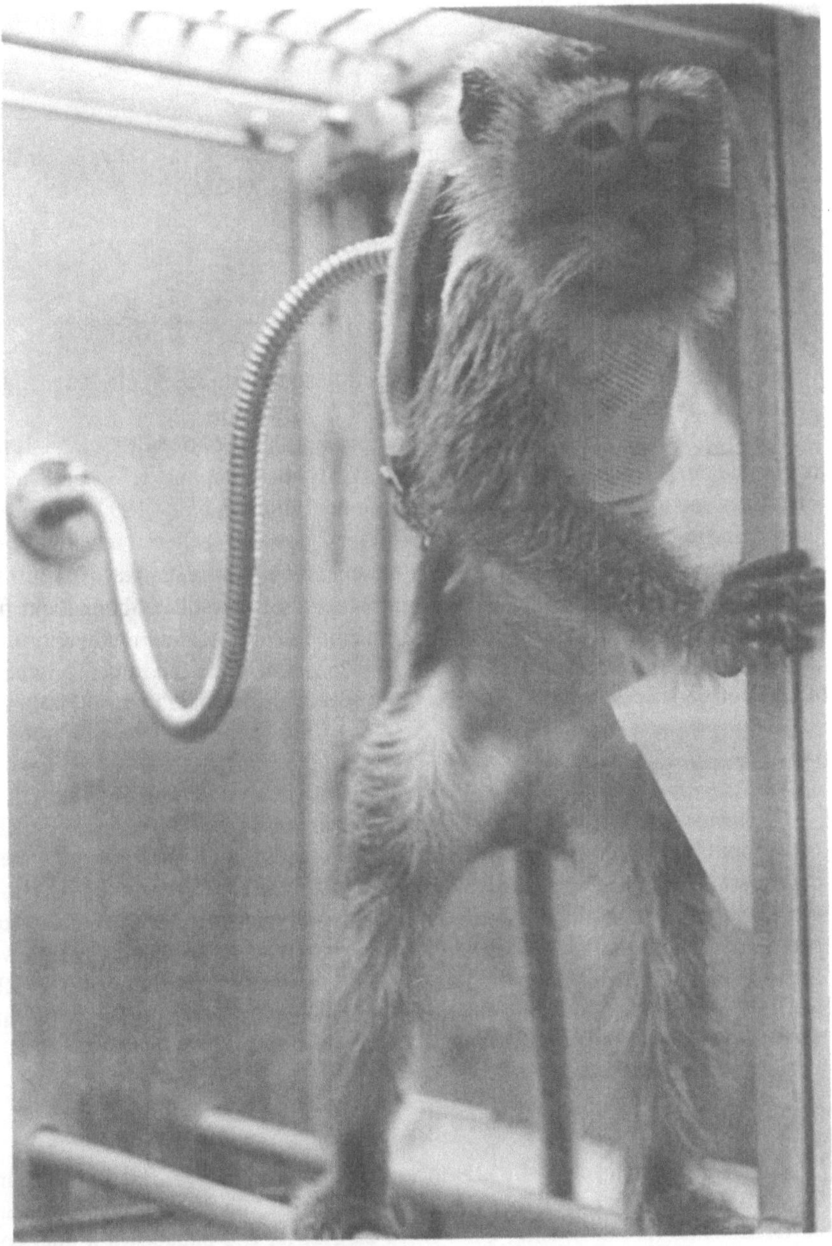

FIGURE 1. The primate vest and mobile tether assembly. The cannula passes subcutaneously from the femoral canal to the interscapular region.

of Trasylol. ACTH was measured using a specific antibody against ACTH (IgG-ACTH-1) (IgG Corporation, Nashville, Tenn.). The detection limit of this RIA was 5 pg/ml. The intra- and interassay coefficients of variation were 4.3 and 12.1%, respectively.

2.3.2. β-Endorphin (β-END)

β-Endorphin was measured using an antibody raised in rabbits against β-endorphin (antibody HS-7 at a final dilution of 1 : 60,000). The sensitivity of this RIA was 10 pg/ml. Intra- and interassay coefficients of variation were 5.5 and 16.0%, respectively. The antibody had 100% cross-reactivity with β-lipotropin.

2.3.3. Corticotropin-Releasing Factor (CRF)

CRF was also measured by RIA in unextracted plasma using the same buffer system and ^{125}I-labeled synthetic ovine CRF$^{1-41}$ as tracer.

The antiserum had a binding affinity for [^{125}I]-CRF of 5×10^{11} M^{-1}. It had negligible cross-reactivity ($< 0.001\%$) for ACTH, β-endorphin, TSH, and GnRH. The assay mixtures contained antiserum in a final dilution of 1 : 180,000 in assay buffer. Incubations were carried out at 4°C for 2 days. To enhance sensitivity, the antibody was first allowed to react with the unknown plasma samples (0.1 ml) for 24 hr and was then incubated with the tracer for 24 hr. Antibody-bound and free CRF were separated by a second antibody method using goat anti-rabbit antiserum at a 1 : 20 dilution. The detection limit was 25 pg/ml. Intra- and interassay coefficients of variation were 5.0 and 13.0%, respectively (Schulte *et al.*, 1983).

2.3.4. Other Hormones

Follicle-stimulating hormone (FSH), luteinizing hormone (LH), prolactin (PRL), thyrotropin (TSH), growth hormone (GH), and cortisol were measured as previously described (Williams *et al.*, 1981; Chrousos *et al.*, 1981). Thymosin α_1 RIA used a synthetic analog of thymosin α_1, *N*-Ac-(Tyr^1)-thymosin α_1, with tyrosine substituted for the N-terminal serine to allow iodination. Assay methodology followed that previously reported (McClure *et al.*, 1981). The detection limit of this RIA was 0.5 ng/ml. The intra- and interassay coefficients of variation were 10.0 and 15.0%, respectively.

2.4. Chromatographic Studies

TF5 was examined for the presence of immunoreactive ACTH, β-endorphin, and CRF. Peptides were extracted from a 1 mg/ml solution of TF5 in phosphate-buffered saline (pH 8.6) using ODS-silica cartridges, and aliquots in a final concentration of 250 μg/ml to 2.5 pg/ml were tested for displacement of ^{125}I-labeled peptide in the ACTH, β-endorphin, or CRF RIA. In a second study, TF5 (0.5 ml of a 10 mg/ml solution) was chromatographed on a Sephadex G-50 fine column (55×0.9 cm) and eluted with buffer (62 mM Na_2HPO_4, 13 mM Na_2 EDTA, pH

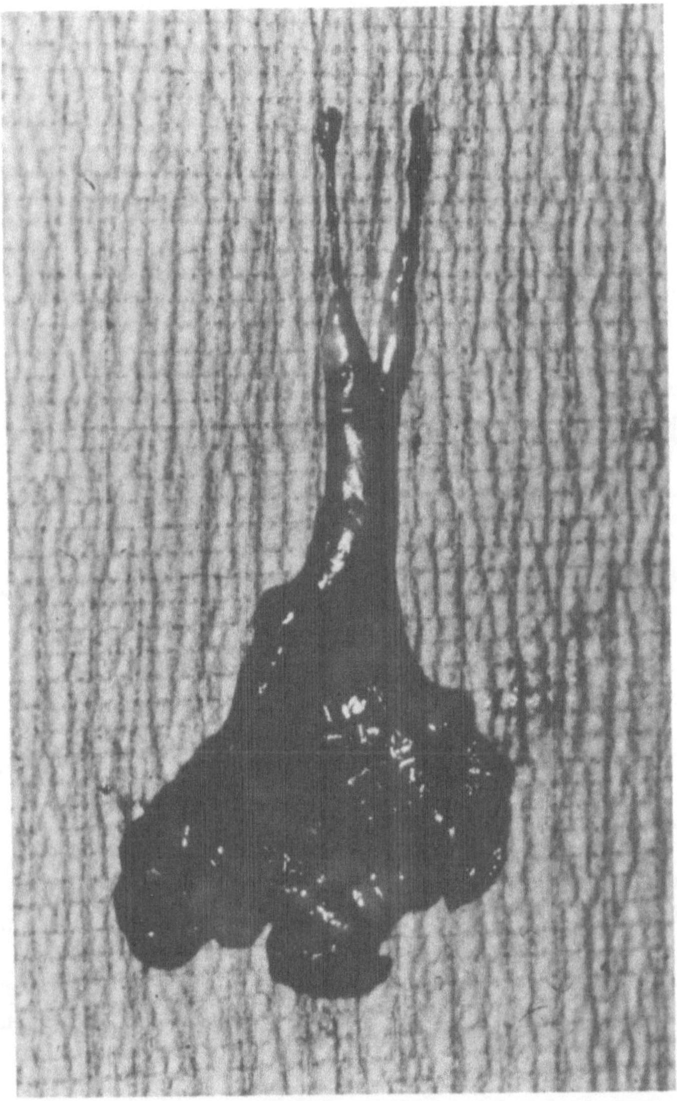

FIGURE 2. Gross anatomy of the prepubertal macaque thymus. Note that the upper half of the gland lies in the neck reflecting the embryological anlage close to the thyroid and parathyroid glands.

7.4, and 0.04% NaN_3) at a flow rate of 6 ml/hr. Fractions (0.5 ml) were collected and compared with the elution position of $ACTH^{1-39}$ as a standard. In a third study, monkey thymus (n = 4), sheep thymus (n = 3), and sheep pituitary were homogenized, extracted, lyophilized, and analyzed for their immunoreactive CRF, β-endorphin, and ACTH content.

2.5. Thymectomy Technique

Thymectomy proceeded through a cervico-thoracotomy via blunt gauze dissection. Surgical details are provided in an earlier report (Healy *et al.*, 1983a). Extensive neck dissection is vital in these primates as a considerable volume of thymic tissue exists near the thyroid and trachea (Fig. 2). All interstitial tissues between the thymus and adjacent organs were excised and multiple, representative 5-μm histological sections examined for thymic remnants. In addition, three animals were sacrificed 3–12 weeks after surgery and serial section of the anterior mediastinum and lower neck performed to examine for thymic residua. No thymic remnants were identified.

3. RESULTS

3.1. Pituitary Hormone Changes after Thymosin Administration

In the first experiment, TF5, at doses of 10.0 and 1.0 mg/kg, or normal saline was injected intravenously via the cannula at 0700 hr. TF5 produced significant increases in plasma ACTH, β-endorphin, and cortisol concentrations which peaked 30, 60, and 90 min, respectively, following injection (Fig. 3). Basal ACTH, β-endorphin, and cortisol values were 24.06 ± 3.91 pg/ml, 37.3 ± 7.65 pg/ml, and 35.3 ± 3.16 μg/dl (n = 16), respectively, in these juvenile monkeys. By contrast, no change in plasma PRL, GH, TSH, FSH, or LH concentrations were observed after TF5 administration.

Additional monkeys (n = 3) received intravenous administration of 75 μg/kg thymosin α_1 or β_4. α_1 has been reported in mice to elevate corticosterone levels (Hall *et al.*, 1983). Plasma α_1 concentrations rose from a basal value of 1.47 ± 0.80 ng/ml (mean ± S.E.) to a peak value of 14.7 ± 2.1 ng/ml 30 min after exogenous α_1 injection. Similar peak plasma concentration of α_1 followed TF5 administration. However, no significant increase in any pituitary hormone resulted from injection of synthetic α_1 or β_4 alone.

3.2. ACTH, β-Endorphin, and CRF Content of Thymosin and the Thymus

TF5 did not displace [^{125}I]-ACTH, β-endorphin, or CRF in their respective RIAs: this result was observed when TF5 was prepared by either ODS-silica or Sephadex chromatography extraction. Thymic tissue contained no measurable CRF or β-endorphin immunoreactivity (<5 pg/100 mg wet wt). Monkey thymus, but not that of sheep, contained small amounts of immunoreactive ACTH (29.7 ± 14.9

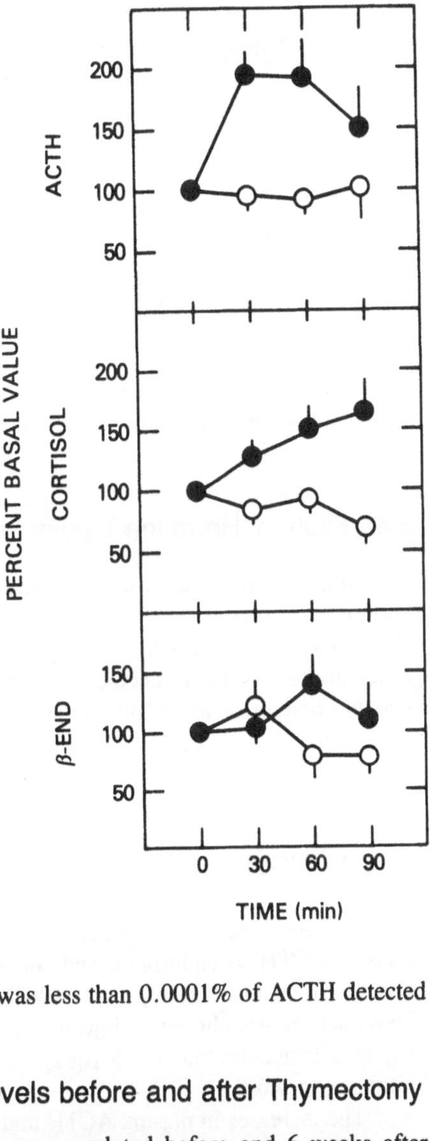

FIGURE 3. Time course of the plasma ACTH, cortisol, and β-endorphin concentrations following i.v. administration of TF5, 10 mg/kg (●) or placebo (○). Significant elevations ($p < 0.05$) of ACTH occurred at 30 and 60 min and of cortisol at 90 min after TF5 injection. Hormonal increases are expressed as percent basal value. The vertical bar indicates ±S.E.

pg/100 mg wet wt; mean ± S.E.); this was less than 0.0001% of ACTH detected in the ovine anterior pituitary gland.

3.3. Cortisol and Thymosin α_1 Levels before and after Thymectomy

In this experiment, six primates were cannulated before and 6 weeks after thymectomy to compare plasma cortisol and thymosin α_1 concentrations. Table I

TABLE I. Plasma Cortisol and Thymosin α_1 before and after Thymectomy in Premenarchial Primates ($n = 6$)

Study	Plasma cortisol (μg/dl) (mean ± S.E.)	Plasma thymosin α_1 (ng/ml) (mean ± S.E.)
Prethymectomy	29.0 ± 3.5	3.49 ± 0.38
Postthymectomy[a]	19.5 ± 2.5*	3.70 ± 0.34

[a] Primates examined 6 weeks after surgery.
* $p < 0.025$.

TABLE II. Plasma ACTH, β-Endorphin, and Cortisol Concentration in Athymic and Age-Matched Control Monkeys ($n = 8$)[a]

	Plasma hormone concentrations (mean ± S.E.)		
Primate group	ACTH (pg/ml)	β-Endorphin (pg/ml)	Cortisol (μg/dl)
Athymic	39.6 ± 5.7*	85.8 ± 11.3**	33.1 ± 4.0
Control	90.7 ± 26.6	147.1 ± 24.0	46.5 ± 7.6

[a] Primates examined 6–10 weeks after surgery.
* $p < 0.05$; ** $p < 0.025$.

indicates that plasma cortisol values fell significantly following removal of the thymus. By contrast, there was no change in plasma α_1 concentrations pre- and postthymectomy.

3.4. Pituitary Hormone Values in Athymic and Control Primates

In this second study, eight additional athymic monkeys were cannulated and hormonal profiles compared with age-matched, intact, control animals. Table II shows that athymic primates had significantly lower plasma ACTH and β-endorphin concentrations. Also in accord with the complementary data displayed in Table I was the finding of lower plasma cortisol levels in the athymic animals, although the decrease did not reach statistical significance. Concentrations of all other anterior pituitary hormones were equivalent between the athymic and the control populations.

4. DISCUSSION

TF5 administration to premenarchial primates does elevate plasma concentrations of ACTH, β-endorphin, and cortisol. These responses are not only time but also dose dependent (Healy *et al.*, 1983b). Concentrations of other anterior pituitary hormones are not altered by injection of these peptides. Moreover, thymosin α_1 or β_4, at a concentration of 75 μg/kg, causes no change in plasma levels of any pituitary hormone.

The increases in plasma ACTH and β-endorphin appear to be endogenous and not derived from the injected TF5. Indeed, no ACTH or β-endorphin immunoreactivity was detected in extracted TF5 material. The elevations in plasma levels of ACTH, β-endorphin, and cortisol after TF5 administration also did not appear to result from nonspecific stress. No change in primate behavior occurred, heart rates and total leukocyte counts were unaltered after TF5, and no hormonal changes followed injection of normal saline or BSA.

The structure of the putative peptide(s) in TF5 responsible for these changes has not been determined. Thymosin α_1 does not appear to stimulate glucocorticoid secretion in primates, unlike its effect in mice (Hall *et al.*, 1983). Thymosin β_4 had no effect on pituitary hormone secretion: in particular, plasma FSH and LH levels remained unchanged. This absence of a rise in gonadotropins might reflect

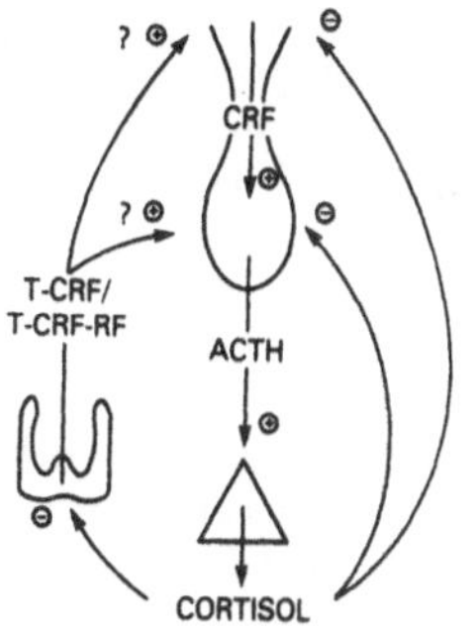

FIGURE 4. Schematic representation of the putative immunoregulatory circuit between the thymus and the pituitary–adrenal axis. A thymic CRF (T-CRF) or CRF-releasing factor contributes to the total CRF input controlling ACTH release. The circuit is completed by cortisol negative feedback upon the thymus.

species variation from earlier *in vitro* murine studies (Rebar *et al.*, 1981) or that premenarchial primates, with low or absent endogenous GnRH, FSH, and LH release, are suboptimal models to examine *in vivo* gonadotropin responses to thymosins.

Our interpretation from all data in this study is that the juvenile primate thymus contains a peptide(s) which contributes to the total CRF input impinging on the pituitary corticotroph to secrete ACTH. Precisely how this occurs is uncertain. Figure 4 describes one hypothesis. Thymic CRF, or possibly a thymic -CRF releasing factor circulates to reach the pituitary gland and/or the hypothalamus via the median eminence. This putative thymic CRF then serves as a subsidiary source to hypothalamic CRF for ACTH secretion. In this schema, regulation of thymic CRF release is controlled by cortisol thymolytic negative feedback action. Given the ontogeny of the thymus, one would predict this immune–endocrinological regulatory circuit should become less important to homeostasis as primates age.

5. SUMMARY

TF5 elevates plasma ACTH, β-endorphin, and cortisol concentrations in a time-dependent fashion. By contrast, TF5 does not alter plasma levels of other anterior pituitary hormones, while thymosin α_1 or β_4 at 75 μg/kg has no effect on pituitary secretion. Juvenile primate thymectomy decreases plasma ACTH, β-endorphin, and cortisol concentrations. Our data suggest that the prepubertal primate thymus secretes a CRF which contributes to a physiological immunoregulatory circuit between the developing immunological and pituitary adrenal axes.

REFERENCES

Ambrose, C. T., 1964, The requirement for hydrocortisone in antibody-forming tissue cultivated in serum-free medium, *J. Exp. Med.* **119:**1027.

Chrousos, G. P., Poplack, D., Kostolich, M., Wiede, C., Olitt, A., Brown, T., and Bercu, B., 1981, Hypothalamic–adenohypophyseal functions in male rhesus monkeys, *J. Med. Primatol.* **10:**61.

Deschaux, P., Massengo, B., and Fontages, R., 1979, Endocrine interaction of the thymus with the hypophysis, adrenals and testes: Effects of two thymic extracts, *Thymus* **1**:95.

Goldstein, A. L., Low, T. L. K., Thurman, G. B., Zatz, M. M., Hall, N., Chen, J., Hu, S., Naylor, P. B., and McClure, J. E., 1981, Current status of thymosin and other hormones of the thymus gland, *Recent Prog. Horm. Res.* **37**:369.

Hall, N. R., McGillis, J. P., and Goldstein, A. L., 1983, Activation of neuroendocrine pathways by thymosin peptides, in: *Stress, Immunity and Aging* (E. L. Cooper, ed.), Dekker, New York.

Healy, D. L., Bacher, J., and Hodgen, G. D., 1983a, A method of thymectomy in macaques, *J. Med. Primatol.* **12**:89.

Healy, D. L., Hodgen, G. D., Schulte, H. M., Chrousos, G. P., Loriaux, D. L., Hall, N. R., and Goldstein, A. L., 1983b, The thymus–adrenal connection: Thymosin has corticotropin releasing activity in primates, *Science* **222**:1353.

Hooper, J. A., McDaniel, M., Thurman, G. B., Cohen, G. H., Schulof, R. S., and Goldstein, A. L., 1975, Purification and properties of bovine thymosin, *Ann. N.Y. Acad. Sci.* **249**:125.

Ishidate, M., and Metcalf, D., 1963, The pattern of lymphopoiesis in the mouse thymus after cortisone administration or adrenalectomy, *Aust. J. Exp. Biol. Med. Sci.* **41**:637.

McClure, J. E., Lameris, N., Wara, D. W., and Goldstein, A. L., 1981, Immunochemical studies on thymosin: Radioimmunoassay of thymosin α_1, *J. Immunol.* **128**:368.

Monjan, A. A., 1981, Stress and immunologic competence: Studies in animals, in: *Psychoneuroimmunology* (R. Ader, ed.), pp. 65–90, Academic Press, New York.

Rebar, R. W., Miyake, A., Low, T. L. K., and Goldstein, A. L., 1981, Thymosin stimulates secretion of luteinizing hormone-releasing factor, *Science* **214**:668.

Ritter, M. A., 1977, Embryonic mouse thymocyte development: Enhancing effect of corticosterone at physiological levels, *Immunology* **33**:241.

Schulte, H. M., Chrousos, G. P., Booth, J. D., Oldfield, E. H., Gold, P. W., Cutler, G. B., and Loriaux, D. L., 1984, Corticotropin releasing factor: Pharmacokinetics in man, *J. Clin. Endocrinol. Metab.* **58**: 192.

Sivas, A., Uysal, M., and Oz, H., 1982, The hyperglycemic effect of thymosin F5, a thymic hormone, *Horm. Metab. Res.* **14**:330.

Williams, R.F., Barber, D. L., Cowan, B. D., Lynch, A., Marut, E. L., and Hodgen, G. D., 1981, Hyperprolactinemia in monkeys: Induction by an estrogen–progesterone synergy, *Steroids* **38**:2842.

Expression of T-Cell Markers on Chicken Bone Marrow Precursor Cells Incubated with an Avian Thymic Hormone

KRISHNA K. MURTHY, FRANCES G. BEACH, and WILLIAM L. RAGLAND

1. INTRODUCTION

Thymic extracts from a number of mammals (mice, rats, calves, pigs, and humans) have been shown to contain a variety of hormones or factors which play an important role in the differentiation and maturation of T lymphocytes. Some of the well-characterized hormones or factors include thymosin (Goldstein, 1976), thymulin (Bach *et al.*, 1978), thymopoietin (Goldstein, 1975), and thymic humoral factor (Kook *et al.*, 1975). These and several additional factors have also been demonstrated to be present in the serum of some of the mammals. Most of these hormones are active in bioassays both *in vitro* and *in vivo*. Thymosin fraction 5 induces the expression of Thy-1 and Lyt antigens on murine precursor cells and functional maturation of T cells (Ahmed *et al.*, 1978). It also has been shown to induce HTLA surface marker on human bone marrow cells (Touraine *et al.*, 1975) and to enhance the number of E-rosette-forming T cells (Incefy *et al.*, 1975). Similar biological activity has been reported for thymulin (Bach *et al.*, 1978). Mouse bone marrow cells or spleen cells express TL and Thy-1 antigens when incubated *in vitro* with thymopoietin (Basch and Goldstein, 1974).

A thymus-specific antigen, designated as T_1 and found present in soluble extracts of chicken thymus, has been described (Pace *et al.*, 1978). Further puri-

KRISHNA K. MURTHY, FRANCES G. BEACH, and WILLIAM L. RAGLAND • Poultry Disease Research Center, Department of Avian Medicine, College of Veterinary Medicine, University of Georgia, Athens, Georgia 30605.

fication and biochemical characterization of T_1 (Barger *et al.*, 1977) indicated that it was acid and heat stable with a molecular weight ranging from 3000 to 10,000. Amino acid analysis revealed that T_1 was rich in acidic amino acids and had a blocked N-terminus. Recent studies by Murthy *et al.* (unpublished) have extended these observations by demonstrating the presence of T_1 within the reticuloepithelial-like cells of the chicken thymus. The similarities in physiochemical characteristics and *in situ* localization of T_1 and other mammalian thymic hormones prompted us to examine the possibility that T_1 may be an avian thymic hormone (ATH) involved with T -cell differentiation.

2. MATERIALS AND METHODS

2.1. Chickens

Specific-pathogen-free chickens were obtained from a commercial source and were maintained in isolation units. One- to three-week-old chickens were used for collection of bone marrow samples.

2.2. T_1 Antigen

Purified and lyophilized endotoxin-free T_1 was reconstituted in RPMI 1640 medium (GIBCO, Long Island, N.Y.) at a concentration of 1 mg/ml, sterilized by passing through a 0.45-μm filter, and stored at −20°C for *in vitro* assays.

2.3. Antisera and Conjugate

Antisera to T and B lymphocytes were prepared by immunizing rabbits with thymic or bursal lymphocytes from three-week-old chickens. Rabbits were injected intravenously with 2×10^8 cells at 2-week intervals for a total of three times. One week after the last antigenic stimulation, serum samples were collected and heat inactivated at 56°C for 30 min. Antisera were absorbed with chicken erythrocytes and either T cells or B cells to eliminate cross-reactivity. Specificity of antisera was ascertained by indirect fluorescent antibody (IFA) tests.

Fluorescein-conjugated goat anti-rabbit globulin (GARG) was obtained from Schwarz/Mann (Orangeburg, N.Y.) and used for IFA tests as described (Murthy *et al.*, 1983).

2.4. Bone Marrow Cells

Long bones (femur and tibia), aseptically collected from several chickens, were used as the source of bone marrow. The diaphyses were cut and the marrow was flushed out with cold phosphate-buffered saline (0.1 M PBS, pH 7.3). Pooled samples were pushed gently through a 60-μm mesh screen (Tetko, Elmsford, N.Y.) to obtain single-cell suspensions of bone marrow cells. Enrichment of precursor cells was achieved by using a bovine serum albumin (BSA) discontinuous gradient or by differential agglutination with peanut agglutinin (PNA).

2.5. Discontinuous BSA Gradients

Discontinuous BSA gradients were prepared as described (Incefy *et al.*, 1975), using BSA obtained from Sigma (St. Louis, Mo.). The bone marrow cell suspensions (1–4 × 10^8 cells/ml), in PBS containing 5% BSA, were layered on top of the gradient and centrifuged at 750*g* for 30 min at 4°C. Individual fractions were collected, washed with RPMI 1640 medium, and cultured *in vitro*.

2.6. Differential Agglutination with PNA

Bone marrow cells previously subjected to centrifugation (2–3 times) on Histopaque (Sigma) were used for agglutination with PNA (Sigma). Equal volumes of bone marrow cell suspensions (1–4 × 10^8 cells/ml) and PNA solution (1 mg/ml) were mixed and incubated at room temperature for 20 min. After incubation, the agglutinated cells (PNA^+) were separated from nonagglutinated cells (PNA^-) at 1*g* sedimentation. Cell suspensions were layered on top of a 50% bovine fetal serum gradient made up in PBS and the tubes kept at room temperature for 20 min. The PNA^-cells remained at the top of the gradient and the PNA^+ cells settled to the bottom. The two populations were collected separately and washed 2–3 times in PBS containing 0.2 M D-galactose (Sigma) and twice in RPMI 1640 for *in vitro* culture.

2.7. Cell Culture and Incubation

Enriched precursor cells obtained by using BSA gradients or by agglutination with PNA were cultured at a concentration of 5 × 10^6 cells/ml in 35-mm tissue culture plates (Costar, Cambridge, Mass.) to test the biological activity of T_1. Culture medium was RPMI 1640 supplemented with 5% bovine fetal serum, 2 mM L-glutamine, 1 × 10^{-5} M 2-mercaptoethanol, and 50 mg gentamicin per liter of the medium. Replicate cultures were set up with various concentration of T_1 ranging from 5 to 100 μg/ml. Control cultures were incubated with bursal extract (BE) or medium alone.

To test the specificity of T_1, additional experiments were conducted with other factors. BE was a crude supernatant obtained by centrifugation at 45,000*g* of homogenized bursal tissue. Chicken neurotensin (CNT) and bovine neurotensin (BNT) and thymic origin were generous gifts from Dr. R. Galloway of the University of Amherst Medical Center. Ubiquitin (UB) was kindly provided by Dr. G. Goldstein of Ortho Pharmaceuticals, Raritan, New Jersey. These factors were added to the precursor cells in culture at various concentrations. All cultures were incubated overnight at 40°C in 5% CO_2 atmosphere.

2.8. IFA Test

To determine the expression of T-cell markers on precursor cells incubated with various factors, IFA tests were conducted as described in an earlier study (Murthy *et al.*, 1984).

3. RESULTS

3.1. Discontinuous BSA Gradients

Fractionation of bone marrow cells on BSA gradients resulted in six separate fractions, as shown in Fig. 1. Cells from each fraction were incubated with 50 μg/ml of T_1 to detect fractions enriched for precursor cells. Replicate experiments indicated that fractions I, II, and III contained a majority of precursor cells (data not shown). Therefore, for subsequent studies, cells from the first three fractions were pooled. The optimum concentration of T_1 that induced the expression of T-cell markers was determined by dose–response studies and the average from three experiments is represented in Fig. 2. At a concentration of 25 μg/ml, T_1 induced the expression of T-cell markers on approximately 26% of the cells (range 15–30%). Incubation with BE did not appreciably increase the percentage of cells with T-cell markers when compared to cells cultured with medium alone. In contrast, BE or UB markedly increased the percentage of B cells but T_1 did not induce differentiation of precursor cells to B cells (data not shown).

3.2. Differential Agglutination with PNA

Separation of bone marrow cells into PNA^- and PNA^+ cell types after a short incubation period with PNA is in agreement with other studies (Reisner *et al.*, 1978). T_1 induced the expression of T-cell markers on both PNA^- and PNA^+ cells (Table I). However, a wide variation in the percentage of positive cells was evident between the two cell types. Although a two- to fourfold increase in the percentage of positive cells as compared to the untreated controls was observed with PNA^- cells, a dose effect was not discernible. In contrast, a dose response was observed

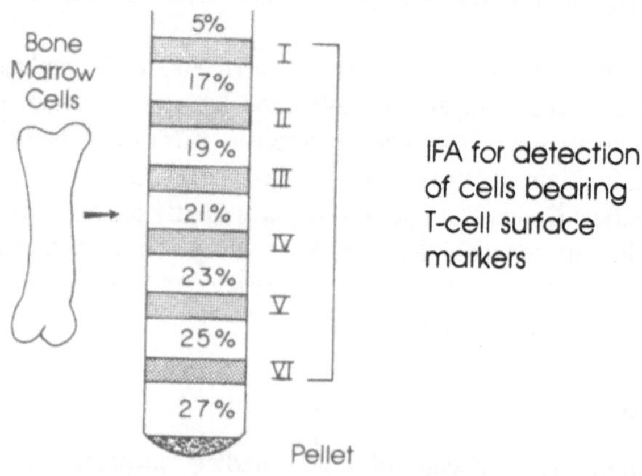

FIGURE 1. BSA gradient separation of marrow cells.

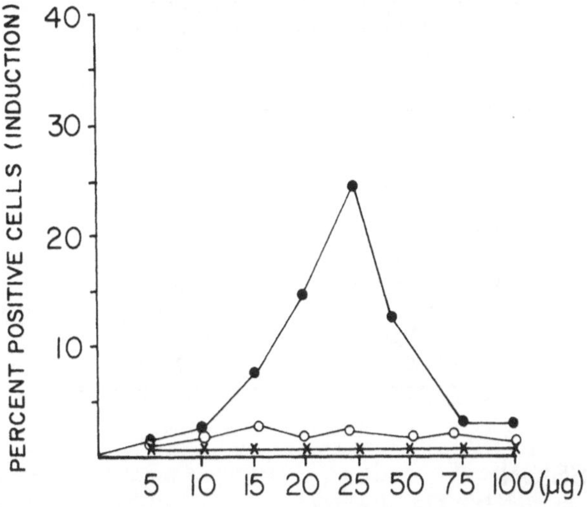

FIGURE 2. Effect of various soluble factors on the induction of T-cell markers. T_1 (●), BE (○), and RPMI 1640 medium (X).

with PNA^+ cells. A marked increase in the percentage of positive cells was evident with increasing concentrations of T_1, up to the 50-μg level. At 100 μg/ml, T_1 was inhibitory to PNA^+ cells and had no effect on PNA^-cells.

3.3. Incubation with Other Factors

Lastly, the data in Table II clearly demonstrate the specificity of induction of T-cell markers by T_1. Incubation of precursor cells with various soluble factors of thymic origin (BNT, CNT) or nonthymic origin (UB, BE) resulted in little or no induction of T-cell markers. On the other hand, T_1 induced T-cell markers on a markedly high percent (27%) of the precursor cells.

TABLE I. Induction of T-Cell Markers on Bone Marrow Cells Separated by PNA

Incubated with T_1 (μg)	Cells expressing T-cell markers (%)	
	PNA^-	PNA^+
0	2.8	1.3
10	9.6	8.6
25	4.3	17.3
50	8.4	26.2
100	11.5	10.5

TABLE II. *In Vitro* Attempts to Induce T-Cell Markers with Other Factors

Inducing factor		Cells expressing T-cell markers (%)
T_1	30 μg	26.8
UB	30 μg	5.5
	50 μg	4.0
BNT	10 μg	1.5
	40 μg	1.0
CNT	500 nM	0
	1000 nM	6.5
BE	30 μg	1.2
RPMI 1640		2.0

4. DISCUSSION

Earlier studies from this laboratory demonstrated the presence of a soluble antigen, designated as T_1, in extracts of chicken thymus (Pace *et al.*, 1978). Physiochemical characterization of T_1 suggested that it shared several properties attributed to mammalian thymic hormones (Barger *et al.*, 1977). Recently, we have shown that T_1 is localized within the reticuloepithelial-like cells as well as in small mononuclear cells of the thymus (Murthy *et al.*, unpublished). Based on these observations, we hypothesized that T_1 could be the avian counterpart of the mammalian thymic hormone(s). The next logical step was to examine its biological activity.

An *in vitro* assay was set up using enriched precursor cells obtained either by centrifugation on discontinuous BSA gradients or by differential agglutination with PNA. Both techniques appeared to be efficient in enriching precursor cell population. Precursors of T cells were concentrated in the lighter density regions of BSA gradients and in the PNA^+ cell fractions. In contrast, the heavier density regions and the PNA^- cell fractions had only a small percent of inducible cells.

Incubation of enriched precursor cells with T_1 induced a marked increase in the number of cells expressing T-cell markers (27%) when compared to background levels of 1–2% (Fig. 1 and Table I). Expression of T-cell markers on a high percentage of precursor cells after exposure to T_1 is especially interesting in view of earlier studies which indicated that avian bone marrow cells do not label with anti-T-cell sera (Hudson and Roitt, 1973). Various other factors of thymic (CNT, BNT) or nonthymic (UB, BE) origin induced either low expression (4–6%) or no expression of T-cell markers, thus clearly establishing the specificity of T_1 in inducing differentiation of precursors of T cells.

Our results are consonant with a number of observations made using thymic extracts or factors from mammalian species. A soluble factor obtained from murine thymus was shown to induce *in vitro* the expression of TL and Thy-1 antigens on precursor cells found in bone marrow and spleen of adult mice (Komuro and Boyse, 1973). Similarly, a variety of extracts or purified derivatives of calf thymus and thymulin have been shown to induce *in vitro* differentiation and maturation of precursor cells to T lymphocytes (Ahmed *et al.*, 1978; Bach *et al.*, 1978; Basch and Goldstein, 1974; Komuro *et al.*, 1975; Miller *et al.*, 1973). Further, soluble factors obtained from either calf or human thymus induce the differentiation of human bone marrow precursor cells as determined by the expression of HTLA antigen (Touraine *et al.*, 1975) and acquisition of E-rosette receptors (Incefy *et al.*, 1975). Results of studies pertaining to the location, physiochemical properties, and apparent functional properties of T_1 provide compelling evidence that it is an avian thymic hormone.

5. SUMMARY

Bone marrow cells from one- to three-week-old chickens were separated into six fractions on discontinuous BSA gradients (5–27%). The top three fractions of enriched precursor cells were pooled for *in vitro* assays. Enriched precursor cells were also obtained by differential agglutination of bone marrow cells with PNA. Precursor cells obtained by either method were cultured in RPMI 1640 (5×10^6/ml) containing various concentrations of an avian thymic antigen (T_1) free of endotoxin. Control cultures were incubated with BE or ubiquitin UB. Cultures were examined 18 hr later by indirect immunofluorescence for the expression of T- and B-cell surface markers. The T_1 antigen, at concentrations of 20–50 μg/ml culture medium, induced the expression of T-cell markers on 15–30% of the enriched precursor cell population. No more than 6% differentiation to T cells was observed in cultures incubated with BE or UB. The percentage of T cells in culture medium alone was 2%. Incubation with UB and BE markedly increased the percentage of B cells but T_1 did not induce differentiation of precursor cells to B cells. These results indicate that T_1 is an avian thymic hormone.

REFERENCES

Ahmed, A., Smith, A. H., Wong, D. M., Thurman, G. B., Goldstein, A. L., and Sell, K. W., 1978, In vitro induction of Lyt surface markers on precursor cells incubated with thymosin polypeptides, *Cancer Treat. Rep.* **62**:1739–1747.

Bach, J.-F., Bach, M.-A., Charreire, J., Dardenne, M., and Pléau, J. M., 1978, The mode of action of thymic hormones, *Ann. N.Y. Acad. Sci.* **332**:23–32.

Barger, B. O., Pace, J. L., Inman, F. P., and Ragland, W. L., 1977, The purification and physicochemical characterization of a chicken thymus specific antigen, *Fed. Proc.* **36**:1237.

Basch, R. S., and Goldstein, G., 1974, Induction of T-cell differentiation in vitro by thymin, a purified polypeptide hormone of the thymus, *Proc. Natl. Acad. Sci. USA* **71**:1474–1478.

Goldstein, A. L., 1976, The history of the development of thymosin: Chemistry, biology and clinical applications, *Trans. Am. Clin. Climatol. Soc.* **88**:79–94.

Goldstein, G., 1975, The isolation of thymopoietin (thymin), *Ann. N.Y. Acad. Sci.* **249**:177–185.

Hudson, L., and Roitt, I. M., 1973, Immunofluorescent detection of surface antigens specific for B and T lymphocytes in the chicken, *Eur. J. Immunol.* **3**:63–67.

Incefy, G. S., L'esperance, P., and Good, R. A., 1975, In vitro differentiation of human marrow cells into T lymphocytes by thymic extracts using the rosette technique, *Clin. Exp. Immunol.* **19**:475–483.

Komuro, K., and Boyse, E. A., 1973, Induction of T lymphocytes from precursor cells in vitro by a product of the thymus, *J. Exp. Med.* **138**:479–482.

Komuro, K., Goldstein, G., and Boyse, E. A., 1975, Thymus-repopulating capacity of cells that can be induced to differentiate to T cells in vitro, *J. Immunol.* **115**:195–198.

Kook, A. I., Yakir, Y., and Trainin, N., 1975, Isolation and partial chemical characterization of THF, a thymus hormone involved in immune maturation of lymphoid cells, *Cell. Immunol.* **19**:151–157.

Miller, H. C., Schmiege, S. K., and Rule, A., 1973, Production of functional T cells after treatment of bone marrow with thymic factor, *J. Immunol.* **111**:1005–1009.

Murthy, K. K., Odend'hal, S., and Ragland, W. L., 1984, Demonstration of T lymphocytes in the bursa of Fabricius of the chicken following cyclophosphamide treatment, *Dev. Comp. Immunol.* **8**: 213–218.

Pace, J. L., Barger, B. O., Dawe, D. L., and Ragland, W. L., 1978, Specific antigens of chicken thymus, *Eur. J. Immunol.* **8**:671–678.

Reisner, Y., Itzicovitch, L., Meshorer, A., and Sharon, N., 1978, Hemopoietic stem cell transplantation using mouse bone marrow and spleen cells fractionated by lectins, *Proc. Natl. Acad. Sci. USA* **75**:2933–2936.

Touraine, J. L., Touraine, F., Incefy, G. S., and Good, R. A., 1975, Effect of thymic factors on the differentiation of human marrow cells into T-lymphocytes in vitro in normals and patients with immunodeficiencies, *Ann. N.Y. Acad. Sci.* **249**:335–342.

Transferrin Receptor Induction in Mitogen-Stimulated Human T Lymphocytes Is Required for DNA Synthesis and Cell Division and Is Regulated by Interleukin-2 (TCGF)

LEONARD M. NECKERS and JEFFREY COSSMAN

1. INTRODUCTION

Since the discovery of T-cell growth factor (TCGF, IL-2) by Gallo and co-workers nearly 8 years ago (Morgan *et al.*, 1976), much interest has been focused on its ability to support the long-term growth of activated T lymphocytes (Ruscetti *et al.*, 1977; Mier and Gallo, 1980). At the same time, others have shown that lymphocyte proliferation is dependent on transferrin, a major serum glycoprotein (Dillner-Centerlind *et al.*, 1979). Because both transferrin and TCGF exert their effects through interaction with cell surface receptors (Robb *et al.*, 1981; Neckers and Cossman, 1983), it is of interest that resting lymphocytes do not express receptors for either substance (Smith *et al.*, 1979; Tormey *et al.*, 1972; Larrick and Cresswell, 1979). Making use of recently available monoclonal antibodies which recognize each receptor (Goding and Burns, 1981; Sutherland *et al.*, 1981; Uchiyama *et al.*, 1981a,b; Trowbridge and Lopez, 1982; Leonard *et al.*, 1982), we have investigated the sequence of events in which lymphocytes become responsive to both TCGF and transferrin. In the course of these experiments we have determined that the

LEONARD M. NECKERS and JEFFREY COSSMAN • Laboratory of Pathology, National Cancer Institute, National Institutes of Health, Bethesda, Maryland 20205.

mitogenic potential of TCGF is dependent on transferrin receptor appearance on T cells and that TCGF receptor stability is also dependent on maintenance of transferrin receptor expression.

2. MATERIALS AND METHODS

2.1. Cell Preparation

Human peripheral blood mononuclear cells (PBM) were obtained by Ficoll–Hypaque separation of heparinized venous blood drawn from healthy volunteers. For macrophage/monocyte depletion, cells were resuspended at 5×10^6/ml in RPMI 1640 (GIBCO, Grand Island, N.Y.) containing 10% heat-inactivated fetal calf serum and 1% penicillin–streptomycin solution. Twenty milliliters of this suspension was placed in Corning 75-cm^2 culture flasks and incubated for 1 hr at 37°C in 5% CO_2. At the end of 1 hr, nonadherent cells were removed and reincubated in a fresh flask as above. This process was repeated three to five times until surface marker analysis [using monoclonal antimonocyte antibody (63XD3, BRL, Gaithersburg, Md.)] revealed < 1% monocytes. Initial analysis routinely found 7–10% monocytes in our PBM preparations. T cells comprised 70–80% of PBM [detection by LYT-3, a monoclonal anti-T-cell reagent (New England Nuclear, Boston, Mass.)] and 80–90% of peripheral blood lymphocytes (PBL; following monocyte depletion).

2.2. Culture Methods

Nonadherent cells were plated out in 24-well microtiter plates (Costar) at 1×10^6/ml (2 ml/well). Phytohemagglutinin (PHA; Sigma Chemical Co., St. Louis, Mo.) was used at a final concentration of 2 μg/ml. Tetradecanoyl-13-phorbol acetate (TPA) was used at a final concentration of 50 ng/ml (81 nM). Monoclonal antibody to the IL-2 receptor (anti-Tac, a gift from Dr. Thomas Waldmann, NIH) was used at a final dilution of ascites fluid of 1 : 500. Monoclonal antibody to the transferrin receptor (42/6, a gift from Dr. Ian Trowbridge, Scripps Clinic, San Diego, Calif.) was used at a final concentration of 2 μg purified antibody/ml. 42/6 is a mouse IgA antibody which blocks the transferrin-binding site and only minimally interferes with the binding of OKT9 (Trowbridge and Lopez, 1982). OKT9 (Ortho Pharmaceutical Corp., Raritan, N.J.) is directed against the transferrin receptor (Goding and Burns, 1981; Sutherland *et al.*, 1981), but does not block transferrin binding (Trowbridge and Lopez, 1982). In this study, OKT9 was used to detect receptors by immunofluorescence (see below). Anti-Leul, a pan T-cell reagent, was purchased from Becton–Dickinson (Sunnyvale, Calif.) and used at a final concentration of 2 μg/ml. All reagents were added to cultures as described in the text. Antibodies containing azide were dialyzed overnight at 4°C against sterile PBS containing 2% albumin and were filtered through 0.2-μm filters before use. IL-2-dependent human T cells were cultured at 1×10^5/ml in RPMI 1640 containing 10% heat-inactivated human AB serum, 1% penicillin–streptomycin solution, and 10% partially purified IL-2 concentrate (Grimm and Rosenberg, 1982). This amount of IL-2 has previously been shown to sustain the growth of these cells for 1 week (Grimm and Rosenberg, 1982).

2.3. Immunofluorescence

IL-2 receptors and transferrin receptors were detected by indirect immunofluorescence with FACS analysis. Cells were prepared by incubating 2–4 × 10^5 cells with an appropriate dilution of primary antibody or mouse ascites control for 30 min on ice. After two washes, fluorescein-labeled goat anti-mouse IgG antibody (Kirkegaard and Perry, Gaithersburg, Md.) was added and the cells incubated an additional 30 min on ice in the dark. After three final washes, the cells were resuspended in medium and analyzed on a FACS (FACS II, Becton–Dickinson) using the 488-nm line of an argon ion laser at 500-mW power. Data analysis was based on the collection of 10,000 to 50,000 cells per sample. Dual staining of cells for transferrin receptor, using OKT9, and for DNA content, using propidium iodide (Calbiochem, La Jolla, Calif.), was performed as described in detail by Braylan *et al.* (1982). Two-color fluorescence studies were performed using an avidin–Texas Red conjugate (Molecular Probes, Inc., Junction City, Oreg.) together with biotinylated goat anti-mouse immunoglobulin (TAGO, Inc., Burlingame, Calif.) and OKT9 to tag transferrin receptors, while directly fluoresceinated anti-Tac (a gift from Dr. Thomas Waldmann, NCI) was used to tag IL-2 receptors. FACS analysis was performed using the 568-nm line of a krypton laser at 170 mW and the 488-nm line of an argon laser at 500 mW. Data analysis was carried out on 50,000 cells.

2.4. DNA Synthesis

At appropriate times, 1 μCi of [*methyl*-^{3}H]thymidine (Amersham, 5 Ci/mmole) was added to 1 × 10^6 cells in 1 ml culture medium. After 4 hr at 37°C, tube contents were filtered, washed, and analyzed by liquid scintillation spectrometry. Results are expressed as a stimulation index = cpm stimulated culture/cpm unstimulated culture.

3. RESULTS

3.1. Requirements for the *in Vitro* Expression of Receptors for Transferrin and IL-2 and DNA Synthesis in PHA-Stimulated Lymphocytes

The mononuclear cell population contained < 1% monocytes (63XD3-positive cells) following adherent cell depletion. These PBL showed negligible thymidine incorporation following a 4-day exposure to PHA when compared to nondepleted controls containing at least 7% monocytes (Table I). Concurrently, only low levels of transferrin receptors and IL-2 receptors were expressed on PBL if monocytes were depleted. By contrast, in the PHA-stimulated, monocyte-containing population, 76% of the cells expressed transferrin receptors and 84% expressed IL-2

TABLE I. Requirements for Expression of Transferrin Receptors, IL-2 Receptors, and DNA Synthesis in PHA-Treated Lymphocytes[a,b]

Additives	Receptor-positive cells (%) IL-2 (Tac⁺)	Receptor-positive cells (%) Transferrin (OKT9⁺)	DNA synthesis (stimulation index)
Monocytes + serum	84	76	185
Monocytes	74	56	25
Serum	0	0	13
None	0	0	3

[a] Modified from Neckers and Cossman (1983).
[b] Receptor-positive cells and thymidine incorporation were measured 4 days post PHA treatment (2 μg/ml). PBL were depleted of monocytes. Monocytes (7%) and/or serum (10%) were added back to these PBL. The stimulation index was calculated based on cells freshly prepared from peripheral blood with no further incubation (80–90 cpm).

receptors. DNA synthesis was maximally stimulated in this group (Table I). However, when serum was removed from the cultures, DNA synthesis was markedly reduced, even though the percentage of $OKT9^+$ and TAC^+ cells remained essentially unchanged (Table I). Thus, PHA-induced transferrin receptor and IL-2 receptor expression and maintenance required that monocytes but not serum be present, while PHA-stimulated DNA synthesis required the presence of both monocytes and serum.

3.2. Substitution of Phorbol Ester for Monocytes

As the phorbol ester TPA can adequately substitute for monocytes in mitogen-driven lymphocyte activation (Koretsky *et al.*, 1982), TPA (8.1×10^{-8} M) was selected as a substitute for monocytes in the remainder of our studies. The ability of TPA to promote mitogenesis, while not being mitogenic itself, is shown in Table II. It was also determined that monocyte-depleted cells, exposed to PHA for only 6 hr (PHA then being washed out) followed by exposure to TPA, could still maximally synthesize DNA (Table II). Thus, TPA did not have to be present together

TABLE II. TPA Replaces Monocytes in PHA-Induced Lymphocyte DNA Synthesis[a,b]

Additives	DNA synthesis (stimulation index)
PHA* + monocytes (10%)	33
PHA* + TPA§	23
PHA (0–6 hr)† + TPA (added at 6 hr)¶	24
PHA (0–6 hr)†	8
TPA§	5

[a] Modified from Neckers and Cossman (1983).
[b] PHA (2 μg/ml) was present either throughout the experiment (*) or for the first 6 hr only (†). TPA (50 ng/ml) was either added with PHA (§) or after PHA had been removed (¶). DNA synthesis was measured 72 hr after the beginning of the experiment. The stimulation index was calculated based on cells depleted of monocytes and kept in culture for 4 days with no other additions (500–600 cpm).

with PHA to replace monocytes in this system. However, one cannot rule out the continued presence of membrane-bound PHA affecting the cellular response to TPA.

3.3. Kinetics of IL-2 Receptor and Transferrin Receptor Induction in PHA-Stimulated Lymphocytes

Monocyte-depleted PBL were treated with PHA, followed 6 hr later by the addition of TPA (8.1×10^{-8} M). In one group, PHA was removed after 6 hr and TPA was added to one-half of the cells. The expression of IL-2 receptors and transferrin receptors was determined at several times after PHA addition (Fig. 1). Cells became IL-2 receptor positive in the presence of PHA alone, but receptor appearance was unstable and did not persist for 72 hr post PHA. In the presence of PHA alone, cells never became transferrin receptor positive. When TPA was added to the cultures 6 hr after PHA, both IL-2 receptors and transferrin receptors were induced in the majority of the cells. Removal of PHA 6 hr after its addition had no effect on receptor expression as long as TPA was added. IL-2 receptors were expressed prior to transferrin receptors. Transferrin receptors did not become maximally expressed until 72 hr post PHA. Two-color fluorescence studies demonstrated that, 72 hr post PHA, the majority of cells possessed both transferrin receptors and IL-2 receptors. A large proportion were positive only for IL-2 receptors, but less than 1% of the cells were transferrin receptor positive and IL-2

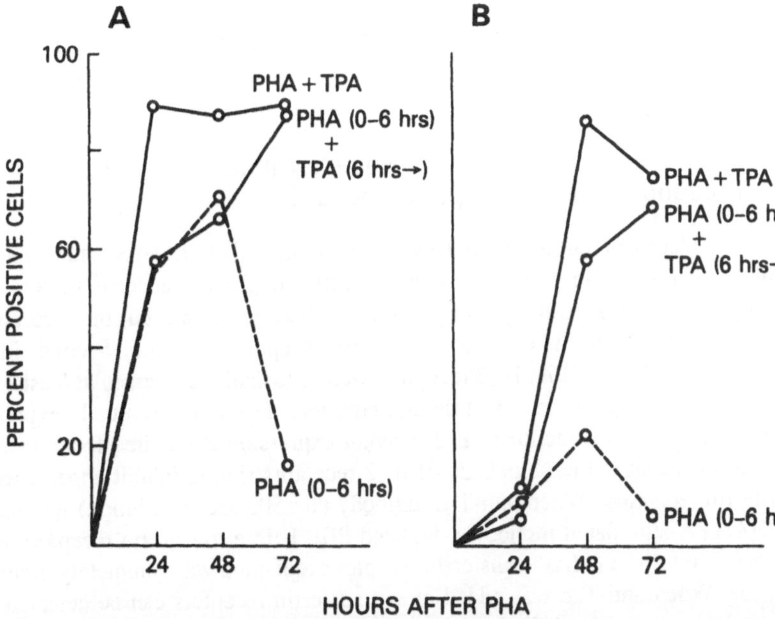

FIGURE 1. PHA (2 μg/ml) was added to monocyte-depleted PBL. Six hours later it was washed out of some cultures and not others. TPA (50 ng/ml) was added at this time. Cells expressing IL-2 receptors (Tac⁺, panel A) and transferrin receptors (OKT9⁺, panel B) were examined at various time points after PHA addition. (From Neckers and Cossman, 1983, with permission.)

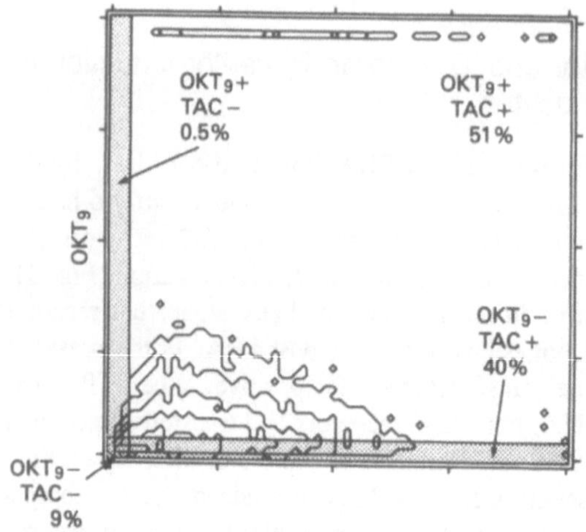

FIGURE 2. Monocyte-depleted PBL were treated with PHA (2 μg/ml) and TPA (50 ng/ml) for 3 days. The cells were then simultaneously stained for transferrin receptors using OKT9 and for IL-2 receptors using anti-Tac and subjected to FACS analysis. The data (collected for 50,000 cells) are expressed as a contour plot with OKT9 fluorescence intensity increasing on the *y* axis and anti-Tac fluorescence intensity increasing on the *x* axis. The contours represent cell numbers. Greater than 50% of the cells possess both transferrin and IL-2 receptors. Forty percent of the cells possess only IL-2 receptors, but less than 1% of the cells possess only transferrin receptors. (From Neckers and Cossman, 1983, with permission.)

receptor negative (Fig. 2). When cells were treated with TPA alone, less than 25% were IL-2 receptor and transferrin receptor positive 48 hr later (data not shown).

When transferrin receptor expression was examined as a function of the cell cycle, receptor-positive cells were found in all phases, including G0/G1. Forty-eight hours post PHA, only 25% of G0/G1 cells were transferrin receptor positive while cells in S and G2 + M were nearly all receptor positive (84 and 88%, respectively; data not shown).

3.4. Dependence of Transferrin Receptor Induction on the Presence of IL-2 Receptors and IL-2

Two pieces of circumstantial evidence suggest that IL-2 receptors may be involved in transferrin receptor induction. First, in the presence of PHA alone, monocyte-depleted PBL temporarily expressed IL-2 receptors, but these receptors soon disappeared. In this case, no transferrin receptors appeared. Second, in the presence of PHA and TPA, IL-2 receptors were maximally expressed at least 1 day before transferrin receptors, even though both receptors were eventually expressed on the same cells. If functional IL-2 receptor expression is required for transferrin receptor expression, then blockade of IL-2 receptors should inhibit appearance of transferrin receptors. When anti-Tac antibody (1 : 500 ascites dilution) was added to PHA/TPA-stimulated monocyte-depleted PBL before transferrin receptors were expressed (6 hr post PHA), transferrin receptor acquisition was completely inhibited (Fig. 3). When anti-Tac was added after transferrin receptors can be detected (48 hr post PHA), the antibody had no effect on continued transferrin receptor expression (Fig. 3). Brief exposure of PBL to mitogen has been reported to result in IL-2 receptor induction. To test the requirement for IL-2 in transferrin receptor induction,

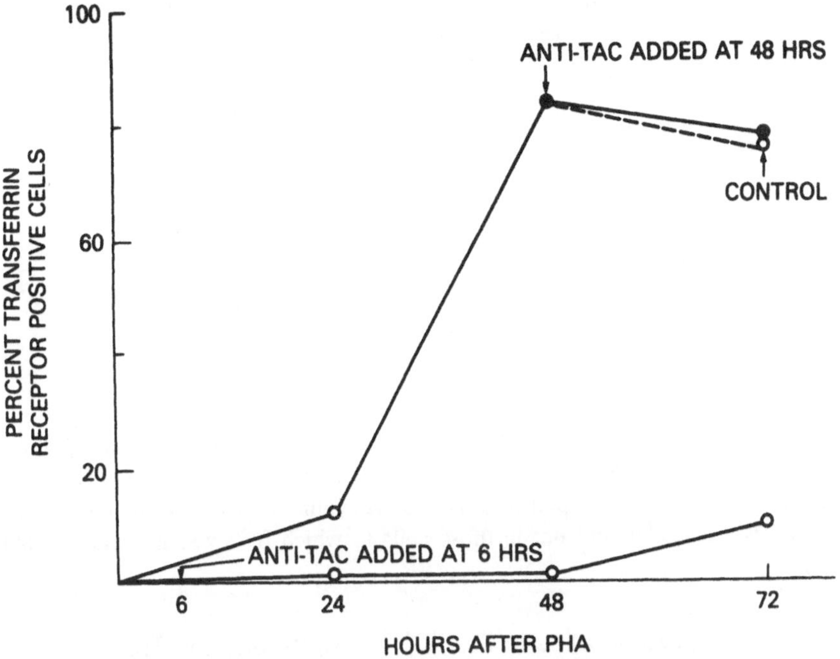

FIGURE 3. Anti-Tac antibody (anti-IL-2 receptor) was added to cultures at either 6 or 48 hr post PHA. Transferrin receptor-positive cells were measured at various time points after PHA addition in antibody-treated and control cultures. Expression of transferrin receptors was reduced only if anti-Tac antibody was added to cultures prior to the expression of these receptors. (From Neckers and Cossman, 1983, with permission.)

we added PHA to monocyte-depleted PBL for 3 hr, then washed the lymphocytes twice and resuspended them in fresh medium in the presence or absence of lectin-free (IL-2 (a gift of Drs. S. A. Rosenberg and E. A. Grimm). Three days later, the cells were analyzed for the presence of transferrin receptors, IL-2 receptors, and DNA synthesis (Table III). After a brief pulse of PHA, a significant portion

TABLE III. IL-2 Is Required to Induce Transferrin Receptors on PHA-Treated Lymphocytes[a,b]

	Receptor-positive cells (%)		
Additives	IL-2 receptor (Tac$^+$)	Transferrin receptor (OKT9$^+$)	DNA synthesis (stimulation index)
PHA + IL-2	43	35	52
PHA	12	7	4

[a] Modified from Neckers and Cossman (1983).
[b] PHA (2 μg/ml) was added to the cultures at the initiation of the experiment and removed 3 hr later. IL-2 was added to half the cells at this point. The cultures were maintained for 3 days at which time receptor-positive cells and thymidine incorporation were determined. The stimulation index was calculated based on cells depleted of monocytes and kept in culture for 3 days with no further treatment (300–400 cpm).

TABLE IV. Effect of Time of Addition of Anti-Tac on PHA-Treated Lymphocytes[a,b]

Time of addition of anti-Tac (post PHA)	Transferrin receptor-positive cells (%)	DNA synthesis (stimulation index)
None	75	23
+6 hr	10	3.5
+48 hr	76	18

[a] Modified from Neckers and Cossman (1983).
[b] Anti-Tac (1 : 500 dilution of ascites) was added at the times indicated. Transferrin receptor-positive cells and DNA synthesis were determined 72 hr after PHA was added to the cultures. The stimulation index was based on cells depleted of monocytes and kept in culture for 3 days with no further treatment (400–600 cpm).

of the cells cultured in IL-2-containing medium for 3 days were IL-2 receptor and transferrin receptor positive, while those cells cultured in the absence of IL-2 were essentially transferrin receptor negative. DNA synthesis was markedly enhanced in the IL-2-treated cells but not in those cells to which IL-2 was not added (Table III).

3.5. Effect of Antibodies to the IL-2 Receptor and the Transferrin Receptor on DNA Synthesis and Cell Growth.

Although functionally active IL-2 receptors must be present before transferrin receptors will appear, this in itself does not demonstrate that transferrin receptors are required for mitogenesis. To test this directly, we measured thymidine incorporation (DNA synthesis) in cells where anti-Tac antibodies were added either pre- or post-transferrin receptor expression (6 and 48 hr post PHA, respectively) (Table IV). DNA synthesis was significantly reduced (15% of control) only when anti-Tac was added before transferrin receptors were expressed. When anti-Tac was added 48 hr post PHA, DNA synthesis was reduced by only 20%. Thus, the ability

TABLE V. Effect of Anti-Transferrin Receptor Antibody (42/6) on PHA-Treated Lymphocytes[a,b]

Antibody added	Receptor-positive cells (%)		DNA synthesis (stimulation index)
	IL-2 receptor (Tac^+)	Transferrin receptor ($OKT9^+$)	
None	86	63	32
Anti-Leu1	86	76	28
Anti-transferrin receptor (42/6)	55	6	8

[a] Modified from Neckers and Cossman (1983).
[b] Antibodies (2 μg/ml) were added 15 hr post PHA. Receptor-positive cells were determined 48 hr post PHA. DNA synthesis was measured 72 hr post PHA. The stimulation index was based on cells depleted of monocytes and kept in culture for 3 days with no further additions (300–400 cpm).

TABLE VI. Anti-Transferrin Receptor Antibody (42/6) Inhibits Growth of IL-2-Dependent T Cells[a,b]

Additives	Growth rate ($\times 10^5$/ml) Expt 1	Expt 2
None	1.9	1.6
Mouse ascites	1.7	1.7
Anti-transferrin receptor (42/6)	0.6	0.3

[a] Modified from Neckers and Cossman (1983).
[b] 42/6 or nonreactive mouse ascites (2 μg/ml) were added to 1×10^5 cells in 1 ml. Cell counts were taken 72 hr later and growth rates determined as the increase in cell number per milliliter above 1×10^5.

of anti-Tac antibody to inhibit DNA synthesis was closely correlated to its ability to block transferrin receptor induction.

When anti-transferrin receptor antibody (42/6) was added to these cultures, it consistently reduced DNA synthesis by approximately 75% (Table V). These antibodies also appeared to modulate transferrin receptors off the surface of cells, for the receptors could no longer be detected by OKT9. IL-2 receptor expression was not significantly reduced under these conditions. As a control, a pan T-cell antibody, anti-Leul, when added to cultures, had no effect on either receptor expression or DNA synthesis. Thus, 42/6 antibody caused the disappearance of transferrin receptors from the cell surface and greatly inhibited DNA synthesis, while leaving IL-2 receptors essentially unaffected.

To test the growth-regulating potential of the transferrin receptor in the presence of excess IL-2, we added anti-transferrin receptor antibody (42/6) to cultures of IL-2-dependent T cells seeded at 1×10^5 cells/ml in the presence of excess partially purified IL-2. Cells were counted 3 days later and antitransferrin receptor antibody was found to have nearly completely inhibited their growth (by 68 and 81% in two experiments) (Table VI). Thus, even in the presence of excess IL-2, blockade of transferrin receptors inhibits the growth of IL-2-dependent T cells.

3.6. Continued IL-2 Receptor Expression Requires the Expression of Transferrin Receptors

In the experiments described in Fig. 1 and Table V, when transferrin receptors did not appear or were blocked by antibody, IL-2 receptor expression seemed to rapidly decline. In the last experiment, we tested whether maintenance of IL-2 receptor expression actually required the continued presence of transferrin receptors.

Lymphocytes were depleted of monocytes and cultured with PHA and IL-2 either in the presence or absence of the anti-transferrin receptor antibody 42/6. TAC^+ (IL-2 receptor positive) and $OKT9^+$ (transferrin receptor positive) cells were determined on days 1 and 2 (Fig. 4). Prevention of transferrin receptor expression resulted in a rapid decline of the number of TAC^+ cells on day 2 even though the

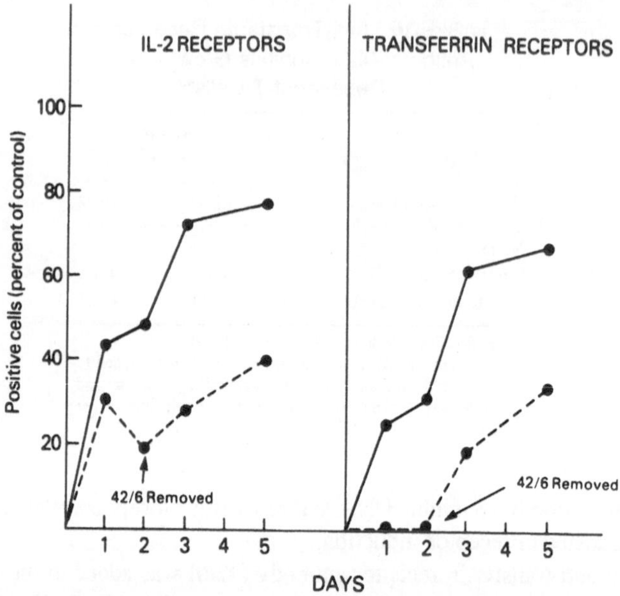

FIGURE 4. IL-2 receptor expression in the presence and absence of 42/6. 42/6 (5 μg/ml) was removed on day 2 of the experiment. Receptor-positive cells were determined by FACS analysis. Control cultures, ●—●; 42/6-treated cultures, ●---●.

number on day 1 was nearly the same as control. On day 2, 42/6 was washed out and the cultures continued for three more days. As OKT9 staining returned, the number of TAC^+ cells increased above the day 2 value. Thus, blockade of the transferrin receptor prevented continued expression of the IL-2 receptor, even though initial expression was still permitted. Removal of the transferrin receptor blockade allowed for reexpression of the IL-2 receptor at the same time that transferrin receptors were being expressed. Given the fact that IL-2 receptors appear on cells before transferrin receptors, these data would suggest that IL-2 receptor maintenance is somehow dependent on continued transferrin receptor expression.

4. DISCUSSION

Our results demonstrate that transferrin receptor appearance on T cells is dependent on prior expression by the same cells of receptors for IL-2. Sequential expression of the two receptors is essential for DNA synthesis. The presence of monocytes or a monocyte-replacing factor (TPA) (Koretsky *et al.*, 1982) is absolutely required for induction of both of these receptors. The interaction of IL-2 with its receptor leads to transferrin receptor expression. Furthermore, as shown in Fig. 4, continued expression of IL-2 receptors depends on continued transferrin receptor expression; i.e., when transferrin receptors are blocked by antibody, IL-2 receptor expression is lost. Maximal DNA synthesis occurs only in the presence of serum (once receptors have been expressed). Because transferrin can replace serum in mitogen-induced lymphocyte proliferation (Dillner-Centerlind *et al.*, 1979), the requirement for serum in our system is presumably a requirement for transferrin (a major serum component). The low level of DNA synthesis observed in the absence of transferrin (but in the presence of monocytes) may be dependent on transferrin produced and secreted by the cultured lymphocytes themselves (Imrie and Mueller, 1968).

Recently, it has been proposed that IL-2 provides a signal necessary for activated lymphocytes to enter S phase (Maizel *et al.*, 1981). Our data would suggest that induction of transferrin receptors is this signal. If transferrin receptors are blocked, DNA synthesis and cell growth are inhibited even in the presence of excess IL-2. The present study indicates that the growth-promoting effects of IL-2 are mediated via the induction of cell surface receptors for the serum glycoprotein transferrin. As transferrin is required by most, if not all, cells for proliferation (Barnes and Sato, 1980), it is intriguing to speculate that the role of cell-specific growth factors, such as TCGF, is to induce expression of a receptor for a universal growth factor, transferrin. Such a scheme would allow for specifically regulated proliferation of unique cells types through a common mechanism. Support for this concept comes from the ability of antibodies to the transferrin receptor to block the growth of cell lines derived from a number of different tissues (Neckers and Cossman, 1982; Neckers, unpublished observations; Trowbridge and Lopez, 1982). We have preliminary data that B-cell growth factor functions in this way (Neckers *et al.*, unpublished observations) and kidney proliferation and differentiation have recently been reported to require transferrin as well (with an initial step being prior sensitization of the tissue to transferrin by interaction with closely associated tissues, perhaps involving induction of transferrin receptors on the kidney cells) (Ekblom *et al.*, 1983). Thus, transferrin receptor expression in specific tissues may be regulated by tissue-specific growth factors (such as TCGF).

ACKNOWLEDGMENTS. We thank Dr. T. A. Waldmann for supply of anti-Tac, Dr. I. Trowbridge for supply of 42/6 antibody, and Dr. E. A. Grimm and Dr. S. A. Rosenberg for supply of IL-2 and IL-2-dependent T cells.

REFERENCES

Barnes, D., and Sato, G., 1980, Serum-free cell culture: A unifying approach, *Cell* **22**:649–655.

Braylan, R. C., Benson, N. A., Nourse, V., and Kruth, H. S., 1982, Correlated analysis of cellular DNA, membrane antigens and light scatter of human lymphoid cells, *Cytometry* **2**:337–343.

Dillner-Centerlind, M.-L., Hammarstrom, S., and Perlmann, P., 1979, Transferrin with or without Fe can replace serum for in vitro growth of mitogen-stimulated T lymphocytes, *Eur. J. Immunol.* **9**:942–948.

Ekblom, P., Thesleff, I., Saxen, L., Miettinen, A., and Timpl, R., 1983, Transferrin as a fetal growth factor: Acquisition of responsiveness related to embryonic induction, *Proc. Natl. Acad. Sci. USA* **80**:2651–2655.

Goding, J. W., and Burns, G. F., 1981, Monoclonal antibody OKT-9 recognizes the receptor for transferrin on human acute lymphocytic leukemia cells, *J. Immunol.* **127**:1256–1258.

Grimm, E. A., and Rosenberg, S. A., 1982, Production and properties of human IL-2, in: *Isolation,*

Characterization and Utilization of T Lymphocyte Clones (G. Fathman and F. Fitch, eds.), pp. 57–82, Academic Press, New York.

Imrie, R. C., and Mueller, G. C., 1968, Release of a lymphocyte growth promoter in leucocyte cultures, *Nature (London)* **219**:1277–1279.

Koretzky, G. A., Daniele, R. P., and Nowell, P. C., 1982, A phorbol ester (TPA) can replace macrophages in human lymphocyte cultures stimulated with a mitogen but not with an antigen, *J. Immunol.* **128**:1776–1780.

Larrick, J. W., and Cresswell, P., 1979, Modulation of cell surface iron transferrin receptors by cellular density and state of activation, *J. Supramol. Struct.* **11**:579–586.

Leonard, W. J., Depper, J. M., Uchiyama, T., Smith, K. A., Waldmann, T. A., and Greene, W. C., 1982, A monoclonal antibody that appears to recognize the receptor for human T-cell growth factor; partial characterization of the receptor, *Nature (London)* **300**:267–269.

Maizel, A., Mehta, S. R., Hauft, S., Franzini, D., Lachman, L. B., and Ford, R. J., 1981, Human T lymphocyte/monocyte interaction in response to lectin: Kinetics of entry into the S-phase, *J. Immunol.* **127**:1058–1064.

Mier, J. W., and Gallo, R. C., 1980, Purification and some characteristics of human T-cell growth factor from phytohemagglutinin-stimulated lymphocyte-conditioned media, *Proc. Natl. Acad. Sci. USA* **77**:6134–6138.

Morgan, A. D., Ruscetti, F. W., and Gallo, R. C., 1976, Selective in vitro growth of T lymphocytes from normal human bone marrow, *Science* **193**:1007–1008.

Neckers, L. M., and Cossman, J., 1982, Transferrin receptor expression and growth of leukemia cell lines, *Fed. Proc.* **41**:740a.

Neckers, L. M., and Cossman, J., 1983, Transferrin receptor induction in mitogen-stimulated human T lymphocytes is required for DNA synthesis and cell division and is regulated by interleukin-2 (TCGF), *Proc. Natl. Acad. Sci. USA* **80**:3494–3498.

Robb, R. J., Munck, A., and Smith, K. A., 1981, T-cell growth factor receptors, *J. Exp. Med.* **154**:1455–1474.

Ruscetti, F. W., Morgan, A. D., and Gallo, R. C., 1977, Functional and morphologic characterization of human T cells continuously grown in vitro, *J. Immunol.* **119**:131–138.

Smith, K. A., Gillis, S., and Baker, P. E., 1979, The role of soluble factors in the regulation of T cell immune reactivity, in: *The Molecular Basis of Immune Cell Function* (J. G. Kaplan, ed.), pp. 223–237, Elsevier/North-Holland, Amsterdam.

Sutherland, R., Delia, D., Schneider, C., Newman, R., Kemshead, J., and Greaves, M., 1981, Ubiquitous cell-surface glycoprotein on tumor cells is proliferation-associated receptor for transferrin, *Proc. Natl. Acad. Sci. USA* **78**:4515–4519.

Tormey, D. C., Imrie, R. C., and Mueller, G. C., 1972, Identification of transferrin as a lymphocyte growth promoter in human serum, *Exp. Cell Res.* **74**:163–169.

Trowbridge, I. S., and Lopez, F., 1982, Monoclonal antibody to transferrin receptor blocks transferrin binding and inhibits human tumor cell growth in vitro, *Proc. Natl. Acad. Sci. USA* **79**:1175–1179.

Uchiyama, T., Broder, S., and Waldmann, T. A., 1981a, A monoclonal antibody (anti-Tac) reactive with activated and functionally mature human T cells. I. Production of anti-Tac monoclonal antibody and distribution of Tac (+) cells, *J. Immunol.* **126**:1393–1397.

Uchiyama, T., Nelson, D. L., Fleisher, T. A., and Waldmann, T. A., 1981b, A monoclonal antibody (anti-Tac) reactive with activated and functionally mature human T cells. II. Expression of Tac antigen on activated cytotoxic killer T cells, suppressor cells, and on one of two types of helper T cells, *J. Immunol.* **126**:1398–1403.

Part III—Chairman's Essay

The Relationship of Terminology to Conceptualization

JOOST J. OPPENHEIM

Since all of you have heard or can read the papers in this volume, I do not consider it productive to actually summarize the session on "The Mechanisms of Action and Immune Regulation of Lymphokines and Thymic Hormones." Instead, I would like to discuss how a number of observations that were made during the meeting illustrate the crucial role that terminology can play in the process of conceptualization. The nomenclature with which we communicate our ideas can either stifle our ability to perceive new concepts or help to rapidly promulgate the acceptance of a new observation.

This issue was actually brought to mind by Dr. Robert Gallo's provocative comments in his opening presentation on T-cell growth factor (TCGF). Dr. Gallo, for a number of reasons, vociferously objected to the fact that TCGF has been renamed "interleukin-2" (IL-2) by most immunologists. He pointed out that the major biological activity of TCGF/IL-2 is to induce T-cell proliferation. Since we have little or no knowledge as to the mechanism of action of this factor, it is presumptuous to name it on any basis other than its major biological activity. The idea that TCGF/IL2 acts on a variety of leukocytes has not actually been borne out by the experimental data, and even the concept that it is induced by interleukin-1 (IL-1), and therefore acts in sequence with that monocyte-derived factor, is questioned by some investigators.

The proposal to rename TCGF "IL-2" was made by a group of immunologists at the Second International Lymphokine Workshop in Ermatingen, Switzerland, in

JOOST J. OPPENHEIM • Laboratory of Molecular Immunoregulation, Biological Response Modifiers Program, Division of Cancer Treatment, NCI-Frederick Cancer Research Facility, Frederick, Maryland 21701.

1979, in an effort to reduce the existing terminological redundancies. A number of factors which were named by immunologists on the basis of their various biological activities, were observed to share the biological activities and biochemical characteristics of TCGF. Based on these observations, it was proposed that an umbrella term such as "IL-2" could encompass the older descriptive terms (Aarden *et al.*, 1979). "IL-2" was proposed to include a factor that was termed "thymocyte-stimulating factor" (TSF), a helper factor (HF) that augmented B-cell antibody production, a killer helper factor (KHF) that promoted cytotoxic lymphocyte activities, as well as TCGF, all of which cochromatographed and were active in all the bioassays.

On the other hand, Dr. Gallo is quite right in pointing out that some of these activities are biologically indistinguishable (e.g., TSF and TCGF) and that the other biological effects of TCGF/IL-2 have been shown to be indirect (e.g., HF and KHF) and are in reality based on the clonal expansion of lymphocytes induced by TCGF. In addition, the HF and KHF activities have been shown to be mediated by factors other than TCGF. Therefore, some of the reasons for renaming TCGF "IL-2" have not been supported by experimental findings. In retrospect, in the absence of more complete knowledge regarding the functional and biochemical characteristics of TCGF, it was probably premature to rename it. In addition, I can appreciate that the scientists who discovered TCGF as well as those who first described lymphocyte-activating factor (LAF), which was renamed "IL-1" at the same meeting, may have serious reservations regarding the changes in nomenclature.

There are a number of compelling reasons, however, for not rejecting the revised terminology at this time. The scientific reasons are as follows. (1) A fortuitous but relevant series of observations made by many laboratories since the Second Lymphokine Workshop clearly documents that one of the many activities that can be ascribed to IL-1 is that it induces lymphocytes to produce IL-2; thus, these factors appear to exhibit a sequential relationship. (2) Renaming LAF "IL-1" appears to be more readily justifiable because IL-1 clearly appears to have many more biological activities on diverse target cells than does IL-2. As Dr. Steven Mizel pointed out in his presentation, IL-1 appears to induce fever and stimulates fibroblasts, hepatocytes, and B lymphocytes, as well as T lymphocytes, and therefore, strictly speaking, functions as more than an "interleukin" (Oppenheim and Gery, 1982). However, T lymphocytes are also not the only target of IL-2, since IL-2 also stimulates the growth and functional activities of large granular lymphocytes (LGL) with natural killer (NK) activity (Timonen *et al.*, 1982). IL-2 is also produced by a subset of NK cells as well as T lymphocytes (Kasahara *et al.*, 1983). Therefore, IL-2 is not limited to classic T lymphocytes in its activities and production.

There are also several nonscientific reasons for the continued use of the interleukin nomenclature. Obviously, one of the major purposes of terminology is to foster communication, and since all four terms, i.e., TCGF/IL-2 and LAF/IL-1, are widely recognized, they fulfill that need. These terms should not be abolished by decree but should be allowed to disappear by attrition in true Darwinian fashion when they are no longer useful and are replaced by more appropriate terms based on better data. For the same reason, namely respect for usage, even though IL-2 induced the production of lymphocyte-derived γ-interferon, to advocate that the interferon terminology should be replaced with the term "IL -3" would perturb many investigators, especially Dr. Jim Ihle, who, as is well known, has christened his factor "IL-3." Despite the fact that many investigators contend that IL-3 which is actually a colony stimulating factor, may not be the most appropriate name for Dr. Ihle's factor since this now represents a well-known acronym, for the sake of good communication, the use of "IL-3" can be continued. Ironically, the controversies that have been generated by the interleukin terminology have had the beneficial, albeit irrational, effect of focusing considerable scientific interest on these cytokines. We therefore have unexpectedly benefited from the controversy, since it has promoted increased experimental study of the properties and relevance of these factors.

The issue of terminology is also critical to a number of the presentations at this meeting. The session was highlighted by several reports that may necessitate some drastic changes in our present thinking and a conceptual redefinition of several of the cytokine activities that were discussed. For example, the findings of N. Hall and R. Rebar that the so-called thymic peptides that Allan Goldstein has extracted from thymus glands act directly on the CNS to regulate hypothalamic–pituitary functions certainly require a redefinition of the role of these thymic factors. On the basis of their thymic source, it is logical to propose that the thymic peptides may provide a link between the immune and neuroendocrine systems. However, the fact that these peptides are recovered from the thymus may be entirely fortuitous. We have to beware of being rigidified by the "thymic" term, and we may have to accept that the immunological role of the peptides may actually be minor but that these peptides may be analogs of important CNS signals. Regardless of what future data will reveal, we must also accept the reality that the immune function is not isolated but is intricately interdigitated with other physiological functions.

The issue of terminology also is of vital concern to the observation of Dr. Charles Evans that a factor(s) that is identical or closely related to lymphotoxin (LT), rather than merely being cytotoxic for L929 target cells, may actually have a much more general antitumor role. Dr. Evans' data revealed that a factor(s) that chromatographs close to or along with LT is able to inhibit chemically or radiation-induced neoplastic transformation of cells and act synergistically with NK cells to lyse tumor cells. These observations certainly demand a redefinition of our previously rather limited concept of LT, which could be achieved by rechristening the factor(s). However, Dr. Evans has wisely decided that more data are needed regarding the biochemical properties of these anticarcinogenic activities to more precisely define their molecular relationship to LT before any new acronyms are generated. Once again, his paper serves to illustrate how terminology can inhibit conceptual changes: because LT was defined as having very limited functions, most investigators question its relevance. Fortunately, Dr. Evans' ability to redefine the role of LT even without any change in terminology should rekindle interest in this lymphokine.

It is fitting that this essay conclude with a comment regarding Dr. Clemens Sorg's studies of the "granddaddy" of the lymphokines, macrophage migration inhibitory factor (MIF). Dr. Sorg's data also suggest that the primary role of classical MIF may not be comprised by its name. For the past 18 years, MIF has been defined as an inhibitor of mobility. Now we finally hear the refreshing news from

Dr. Sorg that MIF may signal macrophages to differentiate. Perhaps MIF should be renamed "macrophage differentiation factor," but what is really needed is a drastic renovation of our fossilized concept of this classic lymphokine rather than a change in its name.

In conclusion, although nomenclature functions to promote communication, it can actually impede acceptance of change especially in the case of descriptive terms. At times a change in nomenclature is needed to draw the attention of fellow scientists to a new scientific finding or concept. Changes in terminology or appropriate choice of a name may galvanize interest in new ideas and observations. However, arbitrary changes in terminology are not acceptable, since they may confuse rather than simplify issues. As illustrated in this essay, there are certain prerequisites for changes in nomenclature, such as evidence of biochemical, rather than merely biological uniqueness; data regarding the relationship of the entity to other biological activities; and a consensus among scientists actively engaged in the relevant studies. Until a factor has been purified and its activities identified, it is appropriate to name it descriptively on the basis of its biological activities. On the other hand, once a number of biological activities can be ascribed to a single biochemical entity, the concerned investigators should consider devising an appropriate term for the entity within the existing system of nomenclature, rather than continuing to use a multiplicity of descriptive terms.

ACKNOWLEDGMENTS. We are grateful for the constructive comments and discussion of Drs. William Farrar, Ronald Herberman, Frank Ruscetti, Luigi Varesio, Edward Kimball, and Mike Chirigos.

REFERENCES

Aarden, L. A., *et al.*, 1979, Revised nomenclature for antigen-nonspecific T cell proliferation and helper factors, *J. Immunol.* **123**:2928–2929.

Kasahara, T., Djeu, J. Y., Dougherty, S. A., and Oppenheim, J. J., 1983, Capacity of large granular lymphocytes (LGL) to produce multiple lymphokines: Interleukin 2, interferon and colony stimulating factor, *J. Immunol.* **131**:2379–2385.

Oppenheim, J. J., and Gery, I., 1982, Interleukin 1 is more than an interleukin, *Immunol. Today* **3**:113–119.

Timonen, T., Ortaldo, J. R., Stadler, B. M., Bonnard, G. D., Sharrow, S., and Herberman, R. B., 1982, Cultures of purified human natural killer cells: Growth in the presence of IL-2, *Cell. Immunol.* **72**:198.

IV

Pharmacology and *in Vivo* Action

Phase I Trial of Intravenous Lymphokine MAF

BEN W. PAPERMASTER, RALPH D. REYNOLDS, JOHN E. McENTIRE, PAMELA A. DUNN†, JEANNE WALTER, MARY K. DOYLE, and WILLIAM EATON

1. INTRODUCTION

Preparation of large enough quantities of purified, chemically characterized biological response modifiers (BRMs) has required years of preparative and analytic studies leading to isolation and characterization of biologically active, pure peptides. Prior to the presently limited use of such peptides, phase I studies were carried out to determine biological effects of more complex preparations in patients, as in the case of human leukocyte interferon (Strander *et al.*, 1973) and thymosin fraction 5 (Costanzi *et al.*, 1977). Complex lymphokine fractions have been used for inducing local skin inflammatory reactions (Papermaster *et al.*, 1976) and to treat dermal tumor lesions by direct intralesional injection (Paradinas *et al.*, 1982) and also have been administered intravenously by Dumonde *et al.* (1981) to patients with disseminated cancer at low to moderate doses [1–2 ml delivering a total of 70 mg protein (Organon preparation)] and roughly equivalent to our unitage of between 200 and 500 units (Paradinas *et al.*, 1982).

We report here a phase I study of a MAF lymphokine containing supernatant preparation from the lymphoblastoid cell line RPMI 1788 (Papermaster *et al.*, 1976). The objectives of the study were to (1) determine any toxicity of administered lymphokine preparations and, in particular, side effects on major organ systems

BEN W. PAPERMASTER, RALPH D. REYNOLDS, JOHN E. McENTIRE, PAMELA A. DUNN†, JEANNE WALTER, MARY K. DOYLE, and WILLIAM EATON • Cancer Research Center and Ellis Fischel State Cancer Center, Columbia, Missouri 65205. Pamela A. Dunn died February 9, 1983.

(i.e., hematological, renal, hepatic, and CNS); (2) evaluate local cutaneous inflammatory reactions at the injection site; and (3) define a maximum tolerable dose necessary to enhance cell-mediated immunity reactions by subcutaneous and intravenous administration. Overall, the preparations were well tolerated, inducing positive changes in peripheral blood neutrophil, lymphocyte, and monocyte responses, and inducing reversal of previously negative microbial antigen recall skin test test reactions.

The major criticism of these studies has been the availability of a consistent method for preparation of purified lymphokines. Preparation of purified, chemically characterized BRMs in large enough quantities for biological and clinical studies has only been achieved, practically, for thymosin α_1 and α- and β-interferon to date. These achievements required years of preparative and analytical studies leading to the isolation and characterization of biologically active pure peptides from complex starting preparations of cell culture supernatants. Recently, pure peptides of these BRMs have been produced by direct chemical synthesis or recombinant DNA technology in *E. coli* (Pestka *et al.*, 1981). An α-interferon preparation, Wellferon, has been purified to 95% homogeneity from supernatants of the NAMALWA lymphoblastoid cell line (Sarna *et al.*, 1983). Thus, there is much information that can be gained from a consistent cell line source in preliminary characterization and bioassay development for complex supernatant preparations which can be purified to increasing levels of specific activity for any given lymphokine.

MAF have been summarized recently. Characterization has been hampered by even greater assay problems than with MIF. Assays for MAF are more complex and variable in readout than those for MIF. Less agreement exists on their acceptance and standardization. However, nearly all investigators in the field agree that one assay necessary to test MAF is the macrophage-mediated cytotoxic assay, various types of which have been discussed by Nathan *et al.* (1971), David (1975), Piessens *et al.* (1975), and Fidler *et al.* (1976).

Recently, Kniep *et al.* (1981) have claimed to have separated MAF from MIF on the basis of isoelectric focusing in urea and Ultragel chromatography. McDaniel (1980) has described multiple molecular species of human MAF in 6 M guanidine. Pace *et al.* (1981) have pointed out the role of endotoxin and prostaglandins as factors in macrophage activation and expression. Both Sorg (1982) and Ruco and Meltzer (1978) have stressed the qualitatively different stages of differentiation necessary for macrophage-mediated tumor cell killings.

The problem involves characterizing polypeptide chains of complex lymphokine molecules which, like toxins and hormones, may have different subunits for attachment and induction of response. In addition, it appears that the target cells must be at an appropriate stage of differentiation to carry receptors or they must be encapsulated in liposomes in order to bypass receptors in the absence of any requirement for antigenic stimulation. Preparations from cell line 1788 have been used to study skin inflammatory reactions following direct intradermal injection in humans and guinea pigs (Rios *et al.*, 1979), local tumor regression in cancer patients with dermal metastases, tumor-bearing mice, and parenterally in patients with disseminated cancers (Dumonde *et al.*, 1981).

2. MATERIALS AND METHODS

MAF was obtained as the partially purified supernatant from the culture of RPMI 1788 human lymphoblastoid cells grown in RPMI 1640 medium containing 2% human serum at a density of 1.5×10^6 cell/ml. Techniques of culture, purification, standardization, and characterization have been described (McEntire *et al.*, 1981). MAF has been described (Dunn *et al.*, 1982) and induces phagocytosis, macrophage-mediated cytotoxicity in mouse and human target cells, and lysosomal hydrolase activity in macrophages. Each vial of MAF was stoppered and tested for sterility and endotoxin. The resultant biological preparation is known to contain lymphotoxin but is essentially free of interferon. Each dose of MAF was reconstituted first in 10 ml of normal saline and then injected into 250 ml normal saline for administration as a 4-hr infusion.

Each patient entered on the study met the requirements of having histologically proven metastatic cancer, absence of reasonable acceptable alternative therapy, and a detailed informed consent process. The affixation of the signature of the patient on the informed consent document was permitted only after it was clear that the patient understood all aspects of alternative treatments, potential anticipated toxicities, dose schedules, the research nature of the study, and the lack of prior treatment studies on which to base information with regard to toxicity and/or response. Both the treatment protocol and the consent form had approval of the Institutional Review Board and the Joint Research Committee of Ellis Fischel State Cancer Center and Cancer Research Center. The protocol had received prior approval by the Biological Response Modifiers Division of the National Cancer Institute and was carried out under IND No. BB-1584 issued by the FDA. Each treatment was given with continuous cardiac monitoring in the Intensive Care Unit (ICU) as a precautionary measure for the treatment of sudden life-threatening toxicities.

Pretreatment laboratory analysis included white blood cell count, differential count, hemoglobin, hematocrit, platelet count, reticulocyte count, serum protein electrophoresis, serum immunoglobulin determination, peripheral B cells by immunoglobulin staining, T cells by E-rosette formation, BUN, creatinine, glucose, uric acid, total protein, albumin, calcium, LDH with isoenzymes, CPK with iso-

TABLE I. Schedule of Doses Given in Preliminary MAF Trial

	Dose of MAF administered (ml)					
	Day 1	Day 3	Day 5	Day 8	Day 10	Day 12
Patient No. 1	(see text)		10	—	10	—
Patient No. 2	10	15	20	40	17	17
Patient No. 3	10	15	20	40	65	100
Patient No. 4	10	15	20	40	70	100
Patient No. 5	10	15	20	40	70	100

TABLE II. Initial Laboratory Data on Five Patients Treated with Intravenous MAF in This Study

Laboratory test	Baseline data				
	Patient No. 1	Patient No. 2	Patient No. 3	Patient No. 4	Patient No. 5
Complete blood count					
WBC (× 1000 mm³)	8.0	5.7	4.5	4.3	8.2
Neutrophils (%)	81	71	78	86	72
Lymphocytes (%)	6	6	16	6	14
Monocytes (%)	10	16	6	2	5
Eosinophils (%)	3	7	0	10	9
Basophils (%)	0	0	0	0	0
Hemoglobin (g/dl)	9.7	12.1	12.5	13.7	11.7
Hematocrit (vol%)	28	36	38	42	34
Reticulocytes (%)	1.6	0.8	1.2	0.8	0.8
Platelets (× 1000 mm³)	259	267	188	96	297
Chemistries					
BUN (mg/dl)	14	21	10	16	24
Creatinine (mg/dl)	0.8	0.8	1.2	1.3	1.2
Glucose (mg/dl)	102	94	85	74	88
Uric acid (mg/dl)	6.8	6.8	2.9	6.7	6.4
Total protein (g/dl)	7.1	6.8	6.1	7.2	7.7
Albumin (g/dl)	4.1	3.8	2.4	4.8	3.8
Calcium (mg/dl)	9.4	8.9	9.9	9.0	9.5
Total bilirubin (mg/dl)	0.3	0.5	0.9	1.0	0.3
Alkaline phosphatase (i.u.)	196	105	70	98	123
SGPT (i.u.)	26	18	33	26	19
GGTP (i.u.)	327	34	16	168	20

enzymes, total bilirubin, alkaline phosphatase, SGOT, SGTP, fasting AM cortisol, serum iron, serum magnesium, serum zinc, serum copper, serum sodium, serum chloride, serum potassium, serum bicarbonate, C-reactive protein, TSH, T_3 uptake, T_4, T_4 index, EKG, MAF, MAF skin test, and delayed hypersensitivity skin tests with mumps, PPD, dermatophytin, and histoplasmin. Skin tests were read as er-

TABLE III. Hematological Parameters

	Pretreatment	Posttreatment	% change
Mean WBC	6.140	7.240	+17.9
Mean absolute neutrophils (×1000 mm³)	4.728	5.080	+7.4
Mean absolute lymphocytes (×1000 mm³)	590	981	+68.0
Mean absolute mononuclears (×1000 mm³)	1.085	1.448	+33.5
Mean hemoglobin (g/dl)	11.9	10.7	−10.1
Mean hematocrit (vol %)	36	33	−8.3
Mean platelets (×1000 mm³)	221	227	+2.7
Mean reticulocytes (%)	1.1	1.3	+18.2

TABLE IV. Serum Chemistry Results

Test	Pretreatment	Posttreatment	% change
Creatinine	1.06	1.04	−1.9
BUN	17	17	0
Fasting glucose	89	93	−4.5
Uric acid	5.9	5.1	−13.6
Calcium	9.1	8.4	−7.7
Total bilirubin	0.6	0.5	−16.7
Alkaline phosphatase	118	97	−16.9
SGOT	24	19	−20.8
γ-GTP	93	94	+1.1
Iron	97	232	+139.2
Magnesium	1.7	2.3	+35.3
Zinc	76	72	−7.2
Copper	185	191	+3.2
Total LDH	164	165	+0.6
% LDH-1	18	18.5	+2.8
% LDH-2	28.5	32.5	+14.0
% LDH-3	21.5	21	−2.3
% LDH-4	16	14	−12.5
% LDH-5	16	14	−12.5
Total CPK	—	91	—
% MM (CPK-3)	—	100	—

TABLE V. Immunological Studies

	Pretreatment	Posttreatment	% change
Serum protein electrophoresis			
Total protein	6.7	6.9	+3.0
Albumin	2.4	2.2	−8.3
α-1	0.4	0.6	+50.0
α-2	1.1	1.4	+27.3
β-1	0.8	0.9	+12.5
β-2	0.5	0.6	+20.0
γ	1.5	1.5	0
Immunoelectrophoresis			
IgG	851	983	+15.5
IgA	225	259	+1.8
IgM	97	117	+20.6
Absolute B cells	140	188	+34.3
Absolute T cells	279	245	−12.2
Absolute null cells	391	693	+77.2
% positive lymphokine skin test	80	80	0
Mean positive delayed hypersensitivity skin tests	1.2	1.5	+25.0

TABLE VI. Hormonal Observations

	Pretreatment	Posttreatment	% change
TSH	4.0	6.0	+50
T_3 uptake	46	47	+2.2
T_4	7.8	8.7	+11.5
T_4 index	1.3	1.4	+7.7
ACTH	—	19	—
Cortisol	22	19	−13.6
PTH	—	—	—

ythema and induration at 12, 24, 48, and 72 hr. ACTH and PTH were measured in three patients. The complete blood count, serum chemistries, serum iron, serum cortisol, TSH, PTH, serum zinc, serum copper, serum magnesium, C-reactive protein, isoenzymes of alkaline phosphatase, LDH, serum protein electrophoresis and serum immunoglobulins were measured at 24 hr in selected patients. All studies were repeated at the completion of the six treatments, given on an alternate-day basis over a 12-day period. Because of the volume of blood needed to perform each test, some variations in the requirements were permitted to help minimize the degree of anemia. This was done because of the presence of underlying organic heart disease in four of the five patients. The results of these tests are given in Tables I–VI.

3. CASE HISTORIES

Patient No. 1. An otherwise healthy 34-year-old white female was diagnosed as having breast cancer in June 1981, at which time seven of eight axillary nodes were involved with tumor. The estrogen receptor assay on the primary tumor was negative. She was treated with cyclophosphamide, methotrexate, fluorouracil, vincristine, and prednisone for 3 months but developed recurrent skin nodules; therefore, she was then treated with doxorubicin, mitomycin C, and cyclophosphamide without response. In June 1982, she received 4080 rads of radiotherapy to the chest wall. In October 1982, she was treated with γ-interferon without response. In December 1982, she was found to have metastatic disease to multiple skin sites, lung, bone, and liver. She was treated with dihydroxyanthracenedione and vincoleukoblastine, but the disease progressed. A trial of tamoxifen was given and, at her insistence, she was then treated with MAF. Initially, 10 ml was given intradermally, in divided doses, in rosette fashion about the skin lesions. This treatment was associated with greater than 50% shrinkage of two of five skin nodules, but with the remaining three nodules demonstrating an increase in tumor size. The same dose of MAF was given on five additional occasions over a 12-day span, with increasing increments of the dose being given intramuscularly. She developed both increased existing tumor growth and new metastatic lesions. She was then given

10 ml intravenous lymphokine on January 20 and 25, 1983. Pre- and posttreatment studies were obtained and are included in the tables of this paper. The intravenous MAF was given as shown in Table I. She experienced one shaking chill, which lasted several minutes, with the second infusion as well as a temperature elevation to 99°F lasting about 3 hr. No other adverse reactions were noted. She was then given 50 ml MAF intradermally and intramuscularly over a 9-day span giving a total administered MAF dose of 130 ml over a period of 37 days. While the two responding skin lesions remained small, the disease otherwise progressed and she died of pulmonary tumor involvement that was confirmed by autopsy.

Patient No. 2. An 83-year-old white female had wide excision of the right upper arm for Clark's Level IV malignant melanoma in August 1980. Between January 1981 and March 1982, she had seven local excisions for recurrent melanoma. Between July 31, 1981, and August 31, 1981, she also received 4200 rads of radiotherapy to the right shoulder area. She declined chemotherapy on repeated occasions. In July 1982, she was found to have bone and cutaneous metastases. The multiple skin lesions gradually increased in size and in January 1983, she was also found to have pulmonary metastases. She had a past history of hypertension, Parkinson's disease, old anterioseptal myocardial infarction, left ventricular hypertrophy with strain, first degree AV block, intra-atrial conduction delay, intraventricular conduction delay, and ST-T wave changes. She expressed a desire for alternative therapy and these were discussed with her. At the time, the left shoulder cutaneous lesion measured 4.0 × 3.5 cm in perpendicular transverse diameters and the right shoulder cutaneous lesion measured 5.0 × 5.0 cm, with each lesion 2.0 cm thick. She elected to receive intravenous MAF therapy and this was begun on February 9, 1983 (Table I). The plan was to give alternate-day treatments for six doses, in increasing doses of 10, 15, 20, 40, 70, and 100 ml MAF. On day 10, after 17 ml of a planned 70-ml dose was given, she experienced severe shaking chills, abdominal bloating, and temperature elevation to 102.6°F, lasting 2 hr. Two days later, this dose was repeated using an 8-hr infusion, and while associated with a transient temperature elevation of 101°F and mild abdominal bloating, there were no shaking chills and the treatment was well tolerated. No other toxicity was noted. She was discharged and followed in the outpatient department. When seen on February 22, 1983, there was no palpable tumor mass on the right shoulder and the left shoulder mass measured 3.0 × 2.0 cm. No change could be detected by chest X-ray of a left pulmonary nodule. She continues in partial tumor regression (PR) status 2 months following treatment, having received a total of 119 ml MAF.

Patient No. 3. A 57-year-old white female was diagnosed as having stage III bilateral adenocarcinoma of the ovaries in February 1975. She was treated with 4500 rads to the whole abdomen in March 1975; alkeran between June 1975 and October 1978; hexamethylmelamine between November 1978 and November 1979; cis-platinum plus doxorubicin plus cyclophosphamide between February 1982 and May 1982; fluorouracil plus alkeran between June 1982 and August 1982; and no therapy in spite of progressive disease between that time and February 1983. She developed persistent chronic diarrhea following radiotherapy and the electrocardiogram showed occasional premature atrial contractions. Physical examination revealed an easily palpable lower abdominal mass that measured 11.0 × 8.0 cm in perpendicular transverse diameters.

She was treated with gradually increasing doses of MAF on alternate days (Table I) over a 12-day period. On day 10, after receiving 65 ml of a planned 70 ml of MAF, she developed shaking chills and a fever to 100.6°F, which lasted less than 1 hr. On day 12, she was given the planned increment of 100 ml MAF, as an 8-hr intravenous infusion, without fever or chills. No other side effects were noted. She was discharged and followed in clinic and has stable disease 2 months following therapy.

Patient No. 4. A 61-year-old white male had an anterior resection for Duke's B-2 adenocarcinoma of the colon. On March 10, 1981, he was found to have liver metastases at the time of exploratory surgery. He was treated with radiation therapy between May 1, 1981, and June 16, 1981, and with FUDR plus mitomycin chemotherapy from June 20, 1982, to November 16, 1982. He had a history of inferior myocardial infarction, prolonged QT interval and premature ventricular contractions for which he took quinidine, hypertension, and Parkinson's disease for which he took L-Dopa. The CEA gradually increased from 18 in May 1982 to 198 in February 1983. The SGOT and GGTP became more abnormal and the radioisotope liver scan showed enlarging masses. Additional chemotherapy was not indicated because of persistent thrombocytopenia and bone marrow hypoplasia secondary to prior therapy with mitomycin. MAF treatment was given as six 4-hr infusions on days 1, 3, 5, 8, 10, and 12 in increasing increments of 10, 15, 20, 40, 70, and 100 ml, respectively (Table I). No side effects were noted. He was discharged and remains stable 2 months after treatment.

Patient No. 5. An 80-year-old white male had a resection of a malignant fibrous histiocytoma from the left deltoid region in September 1976. He was treated with 5000 rads of radiation therapy to the local area. He was then noted to have a lung nodule and was treated with vincristine, actinomycin, and cyclophosphamide with temporary regression of the lesion. He was next treated with DTIC and doxorubicin with regression of the lung lesion for 13 months, lasting until January 1981. He had a thoracotomy with wedge resection of a metastatic lesion in March 1981. Multiple pulmonary nodules were noted in March 1982 and gradually increased in both size and number until February 1983. He had a history of chronic obstructive pulmonary disease, premature ventricular contractions, wandering atrial pacemaker, left axis deviation, nonspecific ST-T changes, intraventricular conduction defect, and hypertension. The creatinine was 1.2 mg/dl, BUN 24 mg/dl, and serum alkaline phosphatase elevated to 123 units (normal 36–90). The liver–spleen and bone scans were normal. The metastatic bone series was also normal. He was treated with MAF using the same schedule as patient No. 4. No toxicities were observed and he has remained stable 2 months after treatment.

4. DISCUSSION

We were pleased with the relative lack of toxicity in this group of high-risk patients who had received large amounts of prior treatment. Patient No. 1 had no toxicity from the intravenous MAF. At the time of the initial dose of intradermal injection, she developed malaise and a low-grade fever of 99°F, lasting about 3 days. These symptoms did not recur in spite of repeated MAF injections by various systemic routes and were, therefore, most likely related to a concurrent viral illness. Patient No. 2 developed the most severe symptoms. On the fifth treatment day, she developed severe shaking chills followed by a temperature elevation of 102.6°F and abdominal bloating after 17 ml of MAF had been given. Treatment was stopped and all symptoms subsided within 2 hr. The dose was repeated as an 8-hr infusion 2 days later, and the resulting symptoms were much less severe. No shaking chill developed, although she did develop a fever of 101°F and had mild abdominal bloating. These symptoms again subsided within 2 hr. Additional treatment was not given as she had completed the planned course of six MAF infusions in the prescribed period of time. Patient No. 3 developed chills and fever to 100.6°F on day 10 of the 12-day scheduled infusion period, with symptoms lasting about 1 hr. The final dose was escalated and given as an 8-hr infusion without the appearance of these or other signs of toxicity. Patients No. 4 and No. 5 had no toxicity.

Unlike phase I studies of new chemotherapy agents, the effectiveness of BRMs may not be related to the maximum tolerable dose. As biological effectiveness may be detectable by laboratory parameters before a clinical response in tumor reduction is observed, laboratory studies that reflect these potential changes need to be carefully correlated with each dose level and schedule used. We chose to use MAF as our unit of measurement both because of its acceptance as a unit of biological measurement and because of our expertise in this area. While we would have preferred to use a different therapeutic approach, the cost of production and the cost of the performance of laboratory analysis prevented this from being done. As we analyzed the situation, there were two issues that needed to be addressed. The first was whether the treatment would be tolerated, and the second was whether a biological effect could be detected. The answer to both of these questions seems to be yes, and we can now address other issues in the use of this product in the treatment of cancer. As cases with a high tumor burden are not likely to be ideal candidates for treatment with BRMs, emphasis in other MAF studies will concentrate on the documentation of biological effectiveness as measured by the various tests of the immune and endocrine systems. The response in the patient with far advanced malignant melanoma (patient No. 2) is gratifying, but responses of this type should not be the only emphasis placed on future studies.

The present study suggests that MAF, as measured in this manner, may be only one of several measurements that need to be performed as potential indicators of effective treatment. A wide variety of laboratory tests have been developed at various centers throughout the world. In addition, several commonly used laboratory studies appear to reflect important biological changes resulting from treatment. Although the number of patients studied in our series is too small to draw conclusions that are statistically significant, the observations strongly suggest that there are important immunological and endocrinological effects that need additional investigations. The fall in the serum iron and the fasting plasma cortisol 24 hr after the initiation of treatment have been observed by Dumonde *et al.* (1981). In our studies, we made additional measurements of these parameters and found that the values return to normal either as treatment continues or after cessation of treatment. The cause for the fall in serum iron may represent an increase in plasma clearance, but this appears to be a transient phenomenon for the serum iron appears to return to normal within 2 weeks, and subsequent measurements suggest that there is an increase in iron binding by transferrin. The fall in serum cortisol may be related to a CNS effect for random measurements of ACTH have shown values that are lower than those normally found in healthy persons. None of the patients in this study had any evidence of adrenal or pituitary malfunction, although these glands were not specifically evaluated in this study.

The fall in serum calcium and serum alkaline phosphatase suggest an effect on parathyroid hormone. The two measurements of this hormone showed normal values. Future attention to this relationship should also focus on the potential action at the osteoclast and osteoblast level, although the kidney and liver may theoretically be involved if these observations prove to be valid.

The changes in thyroid function are subtle but suggest that there is either an effect on the hypothalamus or a direct stimulation of TSH. The increase in thyroxine (T_4) giving an elevation in the T_4 index without increasing the T_3 uptake remains to be explained. Other observations or changes in blood chemistry measurements are also interesting, but the discussion is beyond the scope of this paper and should be analyzed in future studies.

The immunological studies are of special interest as they more closely reflect the known relationship of the BRMs on the growth and development of cancer. While there was an increase in the total white blood cell count, the major portion of this increase was found to be due to the increase in circulating lymphocytes and monocytes. The fall in hemoglobin was attributed to the extensive laboratory analysis that was performed, and as none of the patients became symptomatic in spite of known underlying heart disease in most instances, we elected to follow this parameter without clouding the issue by administering either packed red cells or whole blood transfusions. The increase in lymphocytes was found to be initially due to an increase in both B lymphocytes and null cells and a slight fall in T lymphocytes as measured by sheep cell rosette formation. The determination of various subsets of the lymphocyte populations was not done in this study but represents a worthy area for future investigation. Lymphocyte measurements made 1 month after therapy with MAF suggest that the absolute numbers of T cells and null cells return to normal while the number of B cells continue to increase. Additional follow-up is indicated to observe this pattern so that the duration of future studies can be planned. The same is true of the observations made with regard to the serum protein electrophoresis and the serum immunoglobulins.

We chose doses of MAF that were felt likely to produce an upper level of acceptable toxicity, but such was not the case. Our fears of the potential toxic effects of intravenous MAF proved to be unfounded. Furthermore in three patients (Nos. 1, 2, and 3), the adverse symptoms of fever and shaking chills were significantly reduced by simply prolonging the rate of infusion. In one instance, an increase in dose (patient No. 3) was successfully given without a recurrence of these symptoms. We had also been wary that adverse cardiac problems might be encountered. Therefore, we were pleased when none developed in spite of the presence of underlying heart disease in the majority of our cases.

The administration of intravenous MAF has, therefore, been shown to be relatively safe, although similar precautions should be taken in future studies. The demonstration of *in vivo* changes that may reflect beneficial antitumor effects is encouraging. The additional preliminary observations that the MAF effect may involve other important body symptoms open up a new area for potential study. The remarkable response in our patient with advanced malignant melanoma gave us an unexpected boost in the hope that the pursuit of further studies of this product will result in additional beneficial tumor responses.

In future studies, we plan to pursue both concepts that are usually accepted as the basis for phase I and phase II studies of BRMs. While it is desirable that the tolerable toxic dose be determined, it is clear from this study that such levels may not be the final factor in obtaining the desired biological and, therefore, antitumor effect.

ACKNOWLEDGMENTS. The authors' research reported herein was supported by NCI Grants RO1-CA-21540 and CA-29145 and private funds available to the Cancer Research Center, the Order of the Eastern Star, Fraternal Order of Eagles, and A. P. Green Foundation.

REFERENCES

Costanzi, J. J., Gagliano, R., Delaney, F., Harris, N., Thurman, G., Sakai, H., Goldstein, A., Loukas, D., Cohen, G., and Thomson, P. D., 1977, The effect of thymosin on patients with disseminated malignancies, *Cancer* **40:**14–19.

David, J. R., 1975, Macrophage activation by lymphocyte mediators, *Fed. Proc.* **34:**1730–1735.

Dumonde, D. C., Paradinas, F., Pulley, M., Southcott, B., O'Connell, D., Robinson M., and den Hollander, F., 1981, A histological study of intradermal and intralesional injection of human lymphoid cell line lymphokine (LCL-LK) in patients with advanced cancer, *J. Clin. Hematol. Oncol.* **11:**130.

Dunn, P. A., Eaton, W. R., Lopatin, E., Tyrer, H., McEntire, J. E., Papermaster, B. W., Miller, A., and Vosika, G., 1982, Macrophage activation by lymphokine factors (MAF) as measured by phagocytosis, metabolic products, and tumor cell cytotoxicity: Correlation of multiparameter functional assays, in: *Human Lymphokines: The Biological Immune Response Modifiers* (A. Khan and N. O. Hill, eds.), pp. 67–82, Academic Press, New York.

Fidler, I. J., Darnell, J. H., and Budman, M. B., 1976, In vitro activation of mouse macrophage by rat lymphocyte mediators, *J. Immunol.* **117:**666–673.

Goldstein, A. L., Low, T. L. K., Thurman, G. B., Zatz, M. M., McClure, J. E., Hall, N., and Hu, S. K., 1981, Recent developments in the chemistry, biology and clinical applications of thymosin, in: *Cellular Responses to Molecular Modulators* (L. W. Mozes, J. Schultz, W. A. Scott, and R. Werner, eds.), pp 237–247, Academic Press, New York.

Kniep, E. M., Domzis, W., Lohmann-Matthes, M. L., and Kickhofen, B., 1981, Partial purification and chemical characterization of macrophage cytotoxicity factor (MCF, MAF) and its separation from migration inhibitory factor (MIF), *J. Immunol.* **127:**417–422.

McDaniel, M. C., 1980, Human macrophage activation factors. I. Multiple molecular species of MAF produced by a human lymphoid cell line, *Inflammation* **4:**125–135.

McEntire, J. E., Dunn, P. A., Gehrke, C. W., and Papermaster, B. W., 1981, Isolation and purification of an acid-soluble polypeptide with lymphokine properties from a human lymphoblastoid cell line, in: *Lymphokines and Thymic Hormones: Their Potential Utilization in Cancer Therapeutics* (A. L. Goldstein and M. A. Chirigos, eds.) pp. 109–119, Raven Press, New York.

Nathan, C. F., Karnovsky, M. L., and David, J. R., 1971, Alteration of macrophage functions by mediators from lymphocytes, *J. Exp. Med.* **133:**1356–1376.

Pace, J. L., Taffet, S. M., and Russell, S. W., 1981, Endotoxin in eliciting agents affects activation of mouse macrophages for tumor killing, *J. Reticuloendothel. Soc.* **30:**15–21.

Papermaster, B. W., Holterman, O. A., Klein, E., Parmett, S., Dobkin, D., Laudico, R., and Djerassi, I., 1976, Preliminary observations on tumor regressions induced by local administration of a lymphoid cell culture supernatant fraction in patients with cutaneous metastatic lesions, *Clin. Immunol. Immunopathol.* **5:**48–59.

Paradinas, F. J., Southcott, B. M., O'Connell, D., den Hollander, F., Schuurs, A. H., Pulley, M. S., and Dumonde, D. C., 1982, Changes induced by local injection of human lymphoid cell lymphokine into dermal metastases of breast carcinoma: A light and electron microscopial study, *J. Pathol.* **138:**309–323.

Pestka, S., Maeda, S., Hobbs, D. S., Levy, W. P., McCandliss, R., Stein, S., Moschera, J., and Staehelin T., 1981, The human interferon, in: *Cellular Responses to Molecular Modulators* (L. W. Mozes, J. Schultz, W. A. Scott, and R. Werner, eds.), pp. 455–489, Academic Press, New York.

Piessens, W. F., Churchill, W. H., Jr., and David J. R., 1975, Macrophages activated in vitro with lymphocyte mediators kill neoplastic but not normal cell, *J. Immunol.* **114:**293–299.

Rios, A., Hersh, E. M., Gutterman, J. U., Mavligit, G. M., Schimek, H., McEntire, J. E., and Papermaster, B. W., 1979, The use of a leukocyte cell line culture supernatant for skin reaction testing in malignant melanoma. *Cancer* **44:**1615–1621.

Ruco, L. P., and Meltzer, M. S., 1978, Macrophage activation for tumor cytotoxicity: Development of macrophage cytotoxic activity requires completion of a sequence of short-lived intermediary reactions, *J. Immunol.* **121:**2035–2041.

Sarna, G., Figlin, R., and McCarthy, S., 1983, Phase I study of Wellferon (human lymphoblastoid α-interferon) as cancer therapy: Clinical results, *J. Biol. Resp. Modif. 2:*187–195.

Sorg, C., 1982, Heterogeneity of macrophages in response to lymphokines and other signals, *Mol. Immunol.* **19:**1275–1278.

Strander, H. K., Cantell, K., Carlstrom, G., and Jakobson, P. A., 1973, Clinical and laboratory investigations on man: Systemic administration of the potent interferon to man, *J. Natl. Cancer Inst.* **51:**733–740.

38

Interaction of Thymic Hormones with Prostaglandins

ENRICO GARACI, CARTESIO FAVALLI, and CRISTINA RINALDI-GARACI

1. INTRODUCTION

Prostaglandins have been implicated as possible mediators of the biological activity of thymic factors. Our recent studies reported that thymic hormones [i.e., thymosin fraction 5(TF5) and thymosin α_1] restore lowered thymic functions, such as serum thymic-like activity (STA) and azathioprine (AZ) sensitivity of spontaneous rosette-forming cells (sRFC), when administered to adult thymectomized (ATx) mice (Garaci *et al.*, 1981; Rinaldi-Garaci *et al.*, 1982, 1983a). This effect was completely abolished by indomethacin administration, which is a potent inhibitor of prostaglandin biosynthesis. These data suggested that the action of thymic hormones would require prostaglandin biosynthesis.

The present report illustrates our studies on the ability of a synthetic analog of PGE_2 [16,16-dimethyl-PGE_2-methyl ester (di-M-PGE_2)], acting via PGE_2 biosynthesis, to mimic the effect of thymic hormones (i.e., induction of STA and AZ-sensitive sRFC) when administered to ATx mice.

Furthermore, we report the results of a study of the mechanism of action of thymic factors on lymphocytes, showing that TF5 and thymosin α_1 were able to induce an early release of PGE_2 by immature lymphocytes, such as spleen cells collected from ATx mice and thymocytes.

ENRICO GARACI and CARTESIO FAVALLI • Department of Experimental Medicine, 2° Rome University, 00173 Rome, Italy. CRISTINA RINALDI-GARACI • Institute of General Pathology, 1° Rome University, 00173 Rome, Italy.

2. MATERIALS AND METHODS

Animals. Male, 4-week-old C57BL/6 Cr strain mice were obtained from Charles River, Calco, Milan, Italy.

Thymectomy. Thymus removal was performed by suction technique after anesthesia with ether. The animals were employed 20 days after the operation. The absence of thymic remnants was always verified macroscopically when the mice were killed.

Drugs. di-M-PGE_2 (Upjohn Co., Kalamazoo, Mich.) and indomethacin (Sigma, St. Louis, Mo) were dissolved in absolute ethanol (10 mg/ml) and kept at $-20°C$. For injection the drugs were diluted in sterile 0.9% NaCl solution to the desired concentration and injected i.p. in a total volume of 10 ml/kg. Control mice were injected with 1% ethanol solution alone. Stock solutions of TF5 and thymosin α_1 (Hoffmann–LaRoche, Nutley, N.J.) were prepared in sterile 0.9% NaCl solution and in 1.4% $NaHCO_3$ solution, respectively.

AZ Rosette Inhibition Assay. This assay was based on different sensitivities to AZ of spleen cells of ATx and intact mice with regard to their ability to form spontaneous rosette. The assay was performed by a modification of Bach's technique, as described elsewhere (Garaci *et al.*, 1979). Briefly, 1×10^6 splenocytes were incubated with various concentrations of AZ (ranging from 1 to 50 μg/ml) in a total volume of 0.5 ml. After 60 min incubation at 37°C, 4×10^6 normal sheep erythrocytes were added; cell suspensions were centrifuged at 500*g* and resuspended gently for 10 min at 4°C. Rosettes were counted in a hemocytometer. Each determination was performed in duplicate. A control sample, without AZ, was carried out for each experiment. The minimal inhibitory concentration of AZ is defined as the amount of the drug causing 50% inhibition of rosette formation, compared with that detectable in control samples.

Evaluation of the Level of STA. Determination of STA was performed according to the AZ rosette inhibition assay. The principle of the determination consisted in the induction of AZ sensitivity on AZ-negative T-cell precursors after incubation with serum containing the STA. Spleen cells from ATx mice were incubated with serial dilutions of test sera with a low concentration of AZ (10 μg/ml), normally inhibiting the 50% of sRFC of intact mice. Before incubation, the sera were ultrafiltrated on Amicon UM-10 in order to exclude molecules over 10,000 daltons.

Preparation of Cells for Determination of PGE_2 Release. Lymphocytes, collected from the spleen or from the thymus, were separated on sodium metrizoate/Ficoll solution (Lymphoprep), washed twice in RPMI 1640, and adjusted to a final concentration of 1×10^6/ml. They were incubated with different concentrations of the drugs at 37°C in a 5% CO_2–95% air humidified incubator. Cells were harvested at 15 min, centrifuged, and the supernatants collected for radioimmunoassay (RIA)

RIA for PGE_2 Determination. PGE_2 concentration in supernatant was measured by a specific RIA as described by Jaffe *et al.* (1971) utilizing a specific anti-PGE_2 antibody. All prostaglandin determinations were made corrected for appropriate blank. The assay was quite specific: none of the compounds added to the media, including di-M-PGE_2, cross-reacted with the anti-PGE_2 antibody used.

TABLE I. The Effects of di-M-PGE$_2$ Administration on AZ Sensitivity and on STA in ATx Mice

	Time (hr)	AZ concentration (μg/ml)[a]	STA (log$_2$)[b]
Me$_2$-PGE$_2$	1	1 ± 1.3	5 ± 0.5
	6	1 ± 0.5	5 ± 0.6
	24	7 ± 4	4 ± 0
	48	20 ± 3	2 ± 0.5
di-M-PGE$_2$ + indomethacin	1	18 ± 4	2 ± 0.8
	24	50 ± 0	2 ± 0
Control diluent	1	50 ± 0	2 ± 0

[a] Values represent the arithmetic mean (± S.D.) of three determinations, the spleen of each of three mice being pooled. Data presented are from a representative experiment.
[b] Values represent the arithmetic mean (± S.D.) of the log$_2$ of three STA determinations (i.e., serum dilution), the sera of each of three mice being pooled.

3. RESULTS

3.1. The *in Vivo* Effects of di-M-PGE$_2$ on Thymic Functions

The effect of the administration of di-M-PGE$_2$ on STA and AZ-sensitivity of sRFC in ATx mice is reported in Table I. ATx mice were treated with either di-M-PGE$_2$ (0.5 mg/kg), indomethacin (5 mg/kg) followed by di-M-PGE$_2$ 30 min later, or control diluent. At different time intervals mice were bled for determination of STA and their spleens collected for determination of AZ-sensitive sRFC.

A single administration of di-M-PGE$_2$ was able to restore both STA and AZ-sensitive sRFC in ATx mice. This effect appeared at 1 hr, reached a maximum at 6 hr, and returned to normal values at 48 hr. Treatment with indomethacin abolished the effect of di-M-PGE$_2$.

In order to confirm that the mechanism of action of the analog was mediated by PGE$_2$ biosynthesis, we examined the release of PGE$_2$ in the medium of splenocytes from ATx mice after incubation with di-M-PGE$_2$. Two concentrations of di-M-PGE$_2$ (1 and 5 μg/ml) were tested (Table II). Fifteen minutes of incubation was

TABLE II. Release of PGE$_2$ by Splenocytes of ATx Mice after 15 min Incubation with di-M-PGE$_2$

	PGE$_2$ release (ng/ml)[a]
di-M-PGE$_2$ 1 μg/ml	16 + 2.4
di-M-PGE$_2$ 5 μg/ml	157 + 17.3
Control diluent	6 + 3.4

[a] Values are arithmetic mean (± S.D.) of three determinations, each made in duplicate.

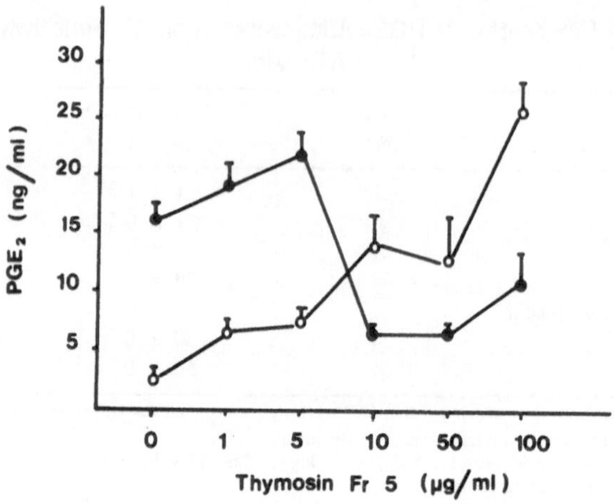

FIGURE 1. Release of PGE_2 by lymphocytes collected from ATx and intact mice after 15 min incubation with different concentrations of TF5. ○, ATx lymphocytes; ●, intact lymphocytes.

enough to increase the release of PGE_2 in the medium. The highest release was achieved by incubation with 5 μg/ml of the analog.

3.2. The Effects of Thymic Hormones on PGE_2 Release by Lymphocytes

The release of PGE_2 by splenic purified lymphocytes after 15 min incubation with different concentrations of TF5 is illustrated in Fig. 1. Lymphocytes were collected from spleens of ATx or intact mice. TF5 induced a dose-dependent release

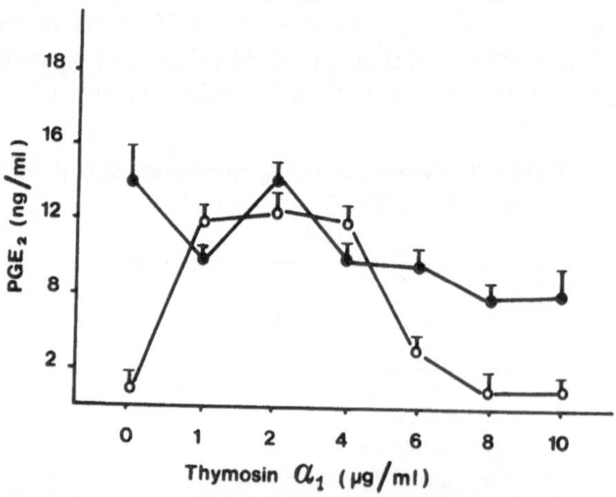

FIGURE 2. Release of PGE_2 by lymphocytes collected from ATx and intact mice after 15 min incubation with different concentrations of thymosin α_1. ○, ATx lymphocytes; ●, intact lymphocytes. Data are expressed as arithmetic mean ± S.D.

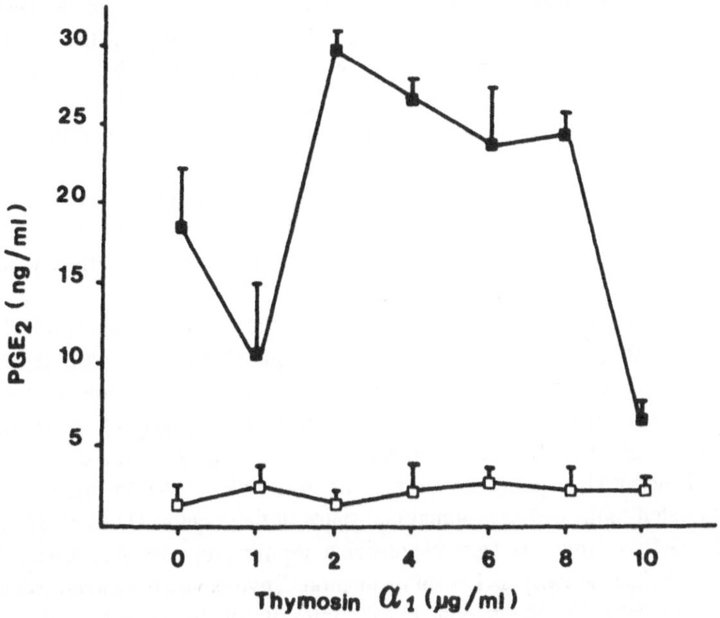

FIGURE 3. Release of PGE_2 by thymocytes after 15 min and 3 hr incubation with different concentrations of thymosin α_1. □, 15 min; ■, 3 hr. Data are expressed as arithmetic mean ± S.D.

of PGE_2, which was maximal with 100 μg/ml. Conversely, the release of PGE_2 by splenic lymphocytes from intact mice was significantly reduced with concentrations higher than 10 μg/ml.

Figure 2 illustrates the amounts of PGE_2 released by splenic lymphocytes obtained from ATx and intact mice, after 15 min incubation with various concentrations of thymosin α_1. Concentrations of α_1 ranging from 1 to 4 μ/ml were active in inducing release of PGE_2. Higher concentrations were inactive. On the contrary, the same concentrations of α_1 reduced significantly the release of PGE_2 by lymphocytes from intact mice.

Figure 3 shows the release of PGE_2 by thymocytes after 15 min and 3 hr incubation with different concentrations of α_1. Fifteen minutes of incubation did not modify PGE_2 release compared to untreated thymocytes. After 3 hr, α_1 induced a progressive increase of PGE_2. A high concentration of α_1 (10 μg/ml) inhibited instead the PGE_2 release.

4. DISCUSSION

The object of this report was to study the interaction between thymic factors and PGE_2. The results of our experiments show that a synthetic analog of PGE_2 could restore the lowered thymic functions, such as STA and AZ-sensitivity of sRFC, when administered to ATx mice. It is well known that indomethacin inhibits the synthesis of prostaglandins (Robinson *et al.*, 1974). Therefore, antagonism between di-M-PGE_2 and indomethacin (Table I) appears to support the hypothesis that the analog acts via prostaglandin biosynthesis as it has been shown *in vivo* and *in vitro* (Favalli *et al.*, 1980; Santoro *et al.*, 1979) and in Table II.

TF5 and α_1 were able to stimulate an early and consistent release of PGE_2 by splenocytes from ATx mice and by thymocytes.

The specific release of PGE_2 was different in splenocytes of intact mice as compared to that detectable in splenocytes of ATx mice. Splenocytes from intact mice did not show increase of PGE_2 released. A slight inhibitory effect was also observed with high concentrations of α_1.

The two thymic preparations had a differential action in stimulating PGE_2 release by splenocytes from ATx mice. TF5 stimulated PGE_2 in a dose-dependent manner, whereas α_1 induced PGE_2 release only within a narrow range of concentration (i.e., 1–4 μg/ml). Thymocytes required a longer period of incubation to respond to α_1, and the effect was dependent on critical concentrations.

The differential response of normal and ATx splenocytes to thymic factors could be related to the different functional states of these cells. Thus, lymphocytes from spleen of ATx mice are very responsive to thymic preparations, releasing high amounts of PGE_2 at as early as 15 min incubation. Thymocytes have a more delayed response, requiring 3 hr incubation, but the intensity of the response is the same. Conversely, splenocytes from intact mice, being mature cells, are not activated by thymic factors to produce PGE_2. Furthermore, high amounts of thymic factors appear to inhibit PGE_2 production.

In our previous studies we had demonstrated that TF5 and α_1 were able to restore thymic functions both *in vivo* and *in vitro* (Rinaldi-Garaci *et al.* 1983a,b). These effects were abolished by Indomethacin administration, suggesting that the action of thymic hormones requires prostaglandins biosynthesis.

It is well known that PGE modulates the early stages of lymphocyte maturation, promoting lymphocyte differentiation (Webb *et al.*, 1979), and stimulates lymphokine production (Gordon *et al.*, 1976). Furthermore, studies *in vivo* showed that di-M-PGE_2 restores the humoral and the cellular immune responses in immunosuppressed mice, such as tumor-bearing hosts (Favalli *et al.*, 1980) and mice treated with antineoplastic agents (Leport *et al.*, 1982). Similarly, thymosin was effective in inducing the formation of helper and cytotoxic T cells (Low *et al.*, 1979; Ahmed *et al.*, 1979) and the production of lymphokines (Svedersky *et al.*, 1982).

On the basis of the data presented in this report, we suggest that the biological effects of thymic factors on their target cells could be related to PGE_2 biosynthesis and production.

5. SUMMARY

Some thymosin peptides (TF5 and α_1) were able to stimulate a specific release of PGE_2 from immature lymphocytes, derived from spleen of ATx mice or from thymus of intact mice.

A long-acting synthetic analog of PGE_2 (di-M-PGE_2) was able to increase some of the lowered thymic functions in ATx mice, mimicking the action of thymic hormones. In fact, this analog restores the lowered STA and the AZ sensitivity of spontaneous spleen RFC in ATx mice.

Such results strongly support the hypothesis that PGE_2 could act as mediators of thymosin function.

REFERENCES

Ahmed, A., Wong, D., Thurman, G. B., Low, T. L. K., Goldstein, A. L., Sharkis, S., and Goldschneider, I., 1979, T lymphocyte maturation: Cell surface marker and immune functions induced by lymphocyte cell-free products and thymosin polypeptides, *Ann. N.Y. Acad. Sci.* **332**:81.

Favalli, C., Garaci, E., Etheredge, E., Santoro, M. G., and Jaffe, B. M., 1980, Influence of PGE on the immune response in melanoma-bearing mice, *J. Immunol.* **125**:897.

Garaci, E., Del Gobbo, V., Santucci, L., Rossi, G. B., and Rinaldi-Garaci, C., 1979, Changes of serum thymic like factor levels in Friend leukemia virus-infected mice, *Leuk. Res.* **3**:67.

Garaci, E., Rinaldi-Garaci, C., Del Gobbo, V., Favalli, C., Santoro, M. G., and Jaffe, B. M., 1981, A synthetic analog of PGE_2 is able to induce *in vivo* theta antigen on spleen cells of adult thymectomized mice, *Cell. Immunol.* **62**:8.

Gordon, D. M., Bray, M. A., and Morley, J., 1976, Control of lymphokine secretion by prostaglandin, *Nature (London)* **262**:401.

Jaffe, B. M., Smith, J. W., Newton, W. T., and Parker, C. W., 1971, Radioimmunoassay for prostaglandins, *Science* **171**:494.

Leport, P., Favalli, C., Santoro, M. G., Rinaldi, C., and Jaffe, B. M., 1982, PGE restores the immune response in chemotherapy-treated, tumor-bearing mice, *Life Sci.* **30**:1219.

Low, T. L. K., Thurman, G. B., McAdoo, M., McClure, J., Rossio, J. L., Naylor, P. H., and Goldstein, A. L., 1979, The chemistry and biology of thymosin. I. Isolation, characterization and biological activities of thymosin α_1 and polypeptide β_4 from calf thymus, *J. Biol. Chem.* **254**:981.

Rinaldi-Garaci, C., Del Gobbo, V., Favalli, C., Garaci, E., Bistoni, F., and Jaffe, B. M., 1982, Induction of serum thymic like activity in adult thymectomized mice by a synthetic analog of PGE_2, *Cell. Immunol.* **72**:97.

Rinaldi-Garaci, C., Garaci, E., Del Gobbo, V., Favalli, C., Jezzi, T., and Goldstein, A. L., 1983a, Modulation of endogenous prostaglandin by thymosin α_1 in lymphocytes, *Cell. Immunol.* **80**:57.

Rinaldi-Garaci, C., Favalli, C., Del Gobbo, V., Garaci, E., and Jaffe, B. M., 1983b, Is thymosin action mediated by prostaglandin release?, *Science* **220**:1163.

Robinson, H. J., Phares, H. F., and Graessle, O. E., 1974, Prostaglandin synthetase inhibitors and infection, in: *Prostaglandin Synthetase Inhibitors* (H. J. Robinson and J. R. Vane, eds.), Raven Press, New York.

Santoro, M. G., Benedetto, A., and Jaffe, B. M., 1979, Effect of endogenous and exogenous prostaglandin E on Friend erythroleukemia cell growth and differentiation, *Br. J. Cancer* **39**:259.

Svedersky, L. P., Hui, A., May, L., McKay, P., and Stebbing, N., 1982, Induction and augmentation of mitogen-induced immune interferon production in human peripheral blood lymphocytes by N-desacetylthymosin α_1, *Eur. J. Immunol.* **12**:244.

Webb, D. R., Rogers, T. J., and Nowowiejski, 1979, Endogenous Prostaglandin synthesis and the control of lymphocytes function, *Ann. N.Y. Acad. Sci.* **332**:262.

A Small-Molecular-Weight Growth Inhibitory Factor (SGIF) in Stationary-Phase Supernatants of Tissue Cultures

A Preliminary Report

ISAAC DJERASSI

Tissue cultures stop growing and in time self-destroy when the cell concentration in the medium reaches a certain maximum. This phenomenon has been known for as long as tissue cultures have been used. Periodic addition of medium is a prerequisite for maintaining the cultured cells in adequate growing phase. The common explanation for this phenomenon is that contact inhibition occurs in overgrown cultures. Depletion of nutrients in the culture medium by the growing and metabolizing cells is an alternative explanation. The contact inhibition mechanism was self-suggested by the observation that cells growing in monolayers come in very intimate contact with each other in overgrown cultures, which do not grow further. This explanation of growth arrest was not adequate, in our opinion, to account for the arrest of growth of cells in suspension cultures, such as the human lymphoid cells. In such cultures, the total mass of cells is insignificant compared to the volume of medium in which they float. The depletion of nutrients, on the other hand, is an equally inadequate explanation, for we, and others, have found that cells can grow in media diluted with saline or other solutions for various purposes. We considered, therefore, the hypothesis that the lack of cell growth in stationary-phase cultures is due to the accumulation of materials produced by the cells themselves, which are capable of arresting cell growth when reaching a critical concentration. In fact, the effect of addition of tissue culture medium to keep a cell culture growing may be only the dilution effect on the growth inhibitory factor.

ISAAC DJERASSI • Mercy Catholic Medical Center, Philadelphia, Pennsylvania 19143.

To explore this hypothesis we studied the growth-supporting ability of stationary-phase supernatants. The parent cells were harvested in their growth phase and resuspended in stationary-phase supernatant, but at concentrations optimal for sustained growth. Using direct cell counting we found that stationary-phase supernatants of a normal human lymphoid cell line (No. 802, obtained through the courtesy of Dr. Robert Gallo) failed to support the growth of the parent cells, even when the latter were in optimal numbers. Growth inhibition after 48 or 72 hr was complete whenever stationary-phase supernatant was used instead of fresh medium (RPMI 40). Dilution of the stationary-phase supernatants with fresh medium (RPMI 40) showed increased inhibition as the time of stationary-phase growth was increased from 2 to 4 and 6 days. For example, 10% fresh medium in the supernatant from a culture kept stationary for 48 hr only was sufficient to promote growth of the cells to practically normal rate. On the other hand, the supernatants from stationary-phase cultures kept for 4 and 6 days were highly inhibitory even when present in 10% concentrations only. Supernatants from a 6-day stationary cell culture, in a ratio of 1 : 9 with fresh medium, inhibited the parent cell growth by 94% after 48 and 72 hr. It must be noted that the control cultures, diluted in the same ratio with saline, grew normally, thus redemonstrating that depletion of nutrients is not a major cause of arrest of growth in cell cultures.

The growth inhibitory activity of the total supernatant from stationary-phase cell culture was then investigated for association with specific molecular size fractions. The total supernatant was fractionated by filtration with an Amicon filtration apparatus, using filters with decreasing pore size, namely with 100,000-, 50,000-, 5000-, 2000-, and 1000-dalton pores. The addition of these fractions to cultures with optimal cell concentrations and fresh medium showed that the inhibitory activity of the supernatants could be recovered in the lower molecular size fractions. The material with components smaller than 5000 daltons was highly inhibitory even when added in concentrations of 1 part to 9 parts of fresh medium. A 90% inhibition was observed with these low concentrations of material obtained from 6-day stationary-phase supernatants. Activity was also demonstrated in materials smaller than 2000 daltons and even, in a number of experiments, with materials smaller than 1000 daltons.

Similar experiments were carried out with three other cell lines, namely a human acute leukemia cell line (CCRF-SB), a mouse acute leukemia (L1210) and a human lung cancer cell line (A-549). The human and the mouse leukemia cell lines were grown in suspension as the human lymphoid line (No. 802), while the human lung cancer line was grown in monolayer cultures. Stationary-phase supernatants from all three cultures inhibited the parent cells as did the human lymphoid line. The inhibition of growth was shown in the absence or presence of human serum (1%) in the assay cultures for all three human lines. The mouse leukemia (L1210) supernatant was inhibitory to the parent cells only in the absence of serum. The inhibitory activity in the supernatants of these three other cell lines was also present in the small-molecular-weight fractions as described above. Cross experiments using cells from one line and fractions from stationary-phase supernatants from another line showed reactivity of all human cells among themselves with their own and with the fractions from the other lines, suggesting a lack of tissue specificity for these inhibitory activities. The supernatant fractions from the mouse L1210 leukemia cells on the other hand, failed to inhibit the human lymphoid cells (No. 802) but had some activity against the human acute leukemia cells. Species specificity of this inhibitory activity, therefore, cannot be ruled out on the basis of the studies carried out so far.

The growth inhibitory activity in the stationary-phase supernatants of tissue cultures appears to be associated in our experiments with the small-molecular-weight fractions of the supernatant. For this reason we propose to name this activity a "small growth inhibitory factor" or SGIF. Previously known inhibitors of DNA synthesis produced by cells in tissue cultures have usually been of substantial molecular size, in the range of 30,000 to 80,000 daltons (Attalah *et al.*, 1975; Smith *et al.*, 1970; Vesole *et al.*, 1979). Even the growth inhibitory factor (Oncostatin) reported at this symposium by Todaro is in the range of 10,000 daltons or more. It is possible, however, that the conditions of our experiments, and specifically the prolonged stationary-phase incubation of the cell cultures, have favored a degradation, or splitting off, of active, 1000- to 2000-dalton fragments from the large-molecular-weight inhibitors. In the process of their separation from the large molecule, the active fragments may be losing the specificity they otherwise may possess when associated with the large-molecular-weight material.

Small-molecular-weight inhibitors of cell growth from cellular origin are of substantial interest. They may indeed add better understanding of the mechanisms involved in regulating normal or malignant cell growth. Obviously, they may also present important opportunities in the therapeutic field of disease processes associated with abnormal growth.

It should be remembered, however, that the association of this phenomenon with nonspecific artifacts in the test system cannot be fully excluded at this time. For example, endotoxins or even cAMP could be suspected to account for similar effects. In preliminary studies we have already determined that low concentrations of endotoxin (*E. coli;* Difco) do not equal the effects of the fractions we described above. cAMP was not measurable in the supernatants from the growing or stationary-phase cultures (assays performed at Smith, Kline and French Laboratories, Philadelphia). Whatever changes or artifacts could or may have been introduced into the system, during or after the fractionation of the original supernatant, the fact remains that unprocessed stationary-phase supernatants, especially from prolonged stationary-phase cultures, definitely do not support cell growth and in fact inhibit it.

REFERENCES

Attalah, A. M., Sunshine, G. H., Hunt, C. V., and Houck, J. C., 1975, The specific endogenous mitotic inhibitor of lymphocytes (chalone), *Exp. Cell Res.* **93**:283–292.

Smith, R. T., Bauscher, J. A. C., and Adler, W. H., 1970, Studies of an inhibitor of DNA synthesis and a nonspecific mitogen elaborated by human lymphoblasts, *Am. J. Pathol.* **60**:495–504.

Vesole, D. H., Goust, J. M., Fett, J. W., Arnaud, P., and Fudenberg, H. H., 1979, An inhibitor of DNA synthesis produced by established lymphoid cell lines, *Clin. Immunol. Immunopathol.* **14**:489–501.

Efficacy of Thymosin α_1 in Animal Models

HIDEO ISHITSUKA, YUKIO UMEDA, EMIKO TEZUKA, YUMIKO OHTA, and YASUO YAGI

1. INTRODUCTION

The thymus gland produces many factors which are known to play a major role in the differentiation and maturation of T cells (Trainin *et al.*, 1983). Considering the importance of T cells in immunoregulatory systems, thymic factors are expected to be useful as pharmaceutical agents to counteract immunodeficiency states. Thymic factors have been given in many clinical instances and shown to improve immunodeficiency diseases, particularly those with congenital T-cell defects (Wara *et al.*, 1975). However, their efficacy in patients with the secondary immunodeficiencies has not been fully investigated.

Such secondary immunodeficiencies leading to serious problems such as opportunistic infections and recurrence of tumors are often seen among cancer patients, mostly due to immunosuppressive side effects accompanying the therapeutic treatments of the neoplasms. Although the immunocompetence alone may not be the sole factor for rejection of tumor cells, the thymic factors may be useful in cancer treatment through prevention and improvement of immunological adverse effects of cancer therapies. Therefore, we have explored animal model systems, which are relevant to such clinical situations, using mice immunosuppressed with cytostatics or X-irradiation.

Treatment of mice with cytostatics or X-irradiation causes various deleterious effects. The activity of thymic factors to counteract such deleterious effects can be shown by assessing (1) various immunological parameters, (2) susceptibility to opportunistic pathogens, and (3) metastasis and progression of tumors. In these

HIDEO ISHITSUKA, YUKIO UMEDA, EMIKO TEZUKA, YUMIKO OHTA, and YASUO YAGI • Nippon Roche Research Center, Kamakura City, Kanagawa 247, Japan.

systems, synthetic thymosin α_1 (Goldstein *et al.*, 1977), one of the thymic factors, was proven to be effective (Ishitsuka *et al.*, 1983; Ohta *et al.*, 1983; Umeda *et al.*, 1983), and suggested to be useful in cancer adjuvant therapies.

2. MATERIAL AND METHODS

Animals. Female mice of ddY (6 weeks old) and BDF_1 (11 or 12 weeks old), and male mice (6 weeks old) of DBA/2, C57BL/6, and CDF_1 strains were purchased from Shizuoka Agricultural Cooperative Association for Laboratory Animals, Hamamatsu, Japan.

Microorganisms. *Candida albicans* ATCC 10231, *Listeria monocytogenes* EGD, *Pseudomonas aeruginosa* 5E81-1, and *Serratia marcescens* 5A412-1 were used in the present study. The culture condition of these pathogens and mode of infection of mice were described elsewhere (Ishitsuka *et al.*, 1983).

Tumor Cells. L1210 leukemic cells resistant to 5-fluorouracil (5-FU), P815 mastocytoma, RL ♂1 leukemic cells, and B16 melanoma were used. L1210 was maintained by continuous passage in DBA/2 mice and other tumor cells were maintained by passage in tissue culture.

Delayed-Type Hypersensitivity (DTH). For active immunization, female BDF_1 mice (11 or 12 weeks old) were sensitized by injection of 10^8 chicken red blood cells (CRBC) in the left footpad and 4 days later challenged with 10^8 CRBC in the right footpad. The footpad swelling was measured with a dial thickness gauge 24 hr after the challenge. The DTH response was expressed by the difference in the thickness of footpad between the two hind feet. Seven mice were used for each group.

CFU-c Assay. The assay was carried out in the soft agar system described by Bradley and Metcalf (1966). Briefly, bone marrow cells pooled from five mice in a group were suspended in a lukewarm agar medium. The medium used was Eagle's MEM (Grand Island Biological Co.) containing a final concentration of 0.3% agar (Difco), 0.0075% DEAE-dextran (Sigma, St. Louis, Mo.), 20% horse serum (Pel Freeze), and 10% of conditioned medium from L929 cells. The conditioned medium was used as the source of colony-stimulating factor. A 1-ml portion (10^5 cells) was placed into a 35-mm Falcon petri dish in quadruplicate and, after gelling, incubated at 37°C in a humidified atmosphere of 5% CO_2 in air. Colonies (more than 50 cells) were counted 7 days later under a dissecting microscope.

Treatment with Immunosuppressive Agents and Thymosin α_1. Immunosuppressed mice used for microbial infection or tumor inoculation were obtained by treatment with cytostatic agents or X-ray irradiation as follows. Mice were treated with cyclophosphamide (CY; 100 mg/kg, i.p.) every other day for three times, or with 5-FU (25 or 30 mg/kg, i.p.) daily for 7 to 10 days before inoculation of microorganisms or tumor cells. Alternatively, mice received 600 rads of X-radiation once at 5 days before inoculation of tumor cells. Thymosin α_1 (40 μg/kg, i.p.) was given daily for 7 to 10 days concurrently with the cytostatic agents or after X-irradiation.

For DTH reaction, mice were treated once with 5-FU (100 mg/kg, i.p.) 4 days before immunization. The mice were then treated three times with saline, thymosin α_1, or fraction 5 at 1, 4, and 7 days after the 5-FU treatment.

Preparation of Radiolabeled Cells and Isotopes. [^{125}I]-UdR- or ^{51}Cr-labeled tumor cells were prepared by a method described elsewhere (Fidler, 1970). Lymph node and bone marrow cells for lymphocyte homing were labeled by ^{51}Cr as described elsewhere (Kasai *et al.*, 1980).

Reagents. 5-FU (Roche) was dissolved in physiological saline. CRBC were obtained from Nippon Biotest, Tokyo. Thymosin α_1 synthesized chemically and fraction 5 were obtained from Drs. J. Meienhofer and A. M. Ramel, respectively, Hoffmann–La Roche Inc., Nutley, New Jersey. Mouse monoclonal anti-Thy-1.2 and Lyt-5.1 sera were purchased from Flow Laboratory and New England Nuclear, respectively. Rabbit anti-asialo G_{M1} serum was obtained from Wako Pure Chemical Industries, Tokyo.

Statistical Analyses. Statistical significance was analyzed using Student's *t* Test and the Mann–Whitney U test (rank sum test). Differences were considered to be significant when probability values $p < 0.05$ were obtained.

3. RESULTS

3.1. Restoration of Immunological Damages Caused by 5-FU

Treatment of mice with 5-FU causes various immunological damage (Ohta *et al.*, 1980). Activity of thymosin α_1 to counteract these deleterious effects was examined by measurement of restoration of DTH response, which is reduced by the 5-FU treatment. In this experiment, a group of mice ($n = 7$) were treated with 5-FU (100 mg/kg, i.p.) at day 0 and subsequently with thymosin α_1 or fraction 5 at days 1, 4, and 7. Other groups of mice were treated with either saline (control) or 5-FU alone. All mice were immunized with CRBC (10^8 cells) at day 4 and challenged with the same dose of the antigen at day 8 for DTH measurement at day 9. As shown in Table I, both thymosin α_1 and fraction 5 restored the DTH response. The potency of thymosin α_1 was about 100 to 1000 times that of fraction 5 on a weight basis.

On the other hand, when normal mice without 5-FU administration were treated with thymosin α_1 according to the above schedule, the effect of thymosin α_1 on the DTH response was not detected (data not shown).

3.2. Mechanism for Restoring the DTH Response by Thymosin α_1

In order to investigate the effector cells affected by thymosin α_1 in the DTH response, lymph node cells from donor immune mice, which had been pretreated with saline, 5-FU, or 5-FU and thymosin α_1, were transferred with CRBC to the normal recipient mice for their footpad reaction. As shown in Table II Expt A, adoptive transfer of lymph node cells from mice treated with 5-FU and thymosin

TABLE I. Dose Responses of Thymosin Fraction 5 (TF5) and Thymosin α_1

Mice treated with	Dose (μg/kg)	DTH response[a] Mean ± S.E. (0.1 mm)	DTH response[a] %
Expt A			
Saline (control)	—	15.6 ± 0.9**	100
5-FU	—	8.0 ± 1.0	51
5-FU plus TF5	2,500	11.7 ± 1.7	75
	5,000	12.1 ± 1.0*	77
	10,000	8.4 ± 1.3	53
5-FU plus thymosin α_1	50	12.6 ± 0.8**	80
	500	11.2 ± 0.7*	71
Expt B			
Saline (control)	—	11.6 ± 0.6***	100
5-FU	—	3.6 ± 0.5	31
5-FU plus thymosin α_1	0.05	6.3 ± 1.8	54
	0.5	8.0 ± 1.2**	69
	5	10.1 ± 1.3***	87
	50	7.0 ± 1.4*	60

[a] p value: statistical difference from the corresponding group treated with 5-FU.
*$p < 0.05$; **$p < 0.01$; ***$p < 0.001$.

TABLE II. Effect of Thymosin α_1 on the Induction Phase and the Expression Phase of the DTH

Group	Treatment with: Donor 5-FU	Donor α_1	Recipient 5-FU	Recipient α_1	DTH response[a] [mean ± S.E. (0.1 mm)] Expt 1	Expt 2
Expt A[b]						
1	−	−	−	−	3.7 ± 0.4***	3.8 ± 0.4*
2	+	−	−	−	2.2 ± 0.3	2.7 ± 0.3
3	+	+	−	−	4.1 ± 0.3***	3.7 ± 0.3**
4	−	−	(no transfer)[d]		(8.0 ± 1.0)	(7.1 ± 0.7)
Expt B[c]						
11	−	−	−	−	3.7 ± 0.4*	3.8 ± 0.4**
12	−	−	+	−	3.0 ± 0.2	2.9 ± 0.3
13	−	−	+	+	3.9 ± 0.2**	3.9 ± 0.3**
14	−	−	(no transfer)[d]		(8.0 ± 1.0)	(7.1 ± 0.7)

[a] p value: statistical difference from group 2 or 12 (5-FU alone).
*$p < 0.1$; **$p < 0.05$; ***$p < 0.01$.

[b] Donor mice were treated with 5-FU (100 mg/kg, i.p.) at day 0 and thymosin α_1 (5 μg/kg, i.p.) at days 1, 3, and 6, and immunized with CRBC (10^8 cells, i.p.) at day 3. Lymph node cells (nonadherent, 2×10^6) obtained at day 7 were transferred to normal recipient mice with CRBC (10^8) into the right footpad. The DTH response was measured 24 hr after the transfer.

[c] Lymph node cells (nonadherent, 2×10^6) obtained from normal donor mice immunized with CRBC 4 days before were transferred into the right footpad with CRBC (10^8) to the recipient mice at day 0, which had been treated with 5-FU at day −7 and thymosin α_1 (5 μg/kg, i.p.) at day −6, −4, and −1. The DTH response was measured 24 hr thereafter.

[d] Mice were immunized directly with CRBC but not by the adoptive transfer of lymph node cells.

α_1 restored the DTH response in the recipient mice. These results indicate that T cells damaged by 5-FU can be restored to normal by an additional treatment with thymosin α_1.

On the other hand, when lymph node cells from mice immunized with CRBC were transferred with the antigen to the 5-FU-treated recipient mice, an additional treatment of the recipient mice with thymosin α_1 enhanced the DTH response (Table II, Expt B). Thymosin α_1 might restore the function of macrophages damaged by 5-FU.

In order to ascertain the possibility that thymosin α_1 affects macrophages or their progenitor in the DTH response, thymosin α_1 was examined for its effect on CFU-c (colony-forming units in culture) in bone marrow cells, which are known to be progenitor cells of macrophages and granulocytes. Groups of mice injected with 5-FU (100 mg/kg) were subsequently treated intraperitoneally with either saline (control) or thymosin α_1, 4 hr, 1 day, and 2 days after the 5-FU injection. Then bone marrow cells were prepared 3 days after the 5-FU injection for the CFU-c test. As shown in Table III, both the number of bone marrow cells and the proportion of CFU-c in bone marrow were markedly reduced by 5-FU, while the subsequent treatment with thymosin α_1 corrected dose-dependently the damages caused by 5-FU.

3.3. Prevention of Microbial Infections in Immunosuppressed Mice

Treatment of mice with 5-FU at a relatively high dose (25 mg/kg per day for 7 to 10 days, i.p.) increased susceptibility to opportunistic pathogens at low doses, at which normal mice were highly resistant. Differences between LD_{50}s in normal and the 5-FU-treated mice of *L. monocytogenes*, *C. albicans*, *P. aeruginosa*, and *S. marcescens* were 40, 20, 2000, and 2200 times, respectively. In all these models for opportunistic infections, thymosin α_1 given concurrently with 5-FU prevented mice from lethal infections (Table IV).

The protective activity of thymosin α_1 was then investigated together with carrageenan and antithymocyte serum (ATS), depressors of macrophages and T cells, respectively. Treatment with carrageenan or ATS abrogated the activity of thymosin α_1 against *Candida* infection, but not against *Pseudomonas* infection (Table V). These results indicated that thymosin α_1 exerts activity against *Candida* infection through T cells and/or macrophages either directly or indirectly. On the other hand, the activity shown in *Pseudomonas* infection may be through its effector other than T cells and macrophages, probably neutrophils or their progenitors.

3.4. Prevention of Progressive Tumor Growth in Immunosuppressed Mice

Impairment of the host defense systems against tumor growth by treatment with cytostatics is well known (Gorelik *et al.*, 1980; Hanna and Burton, 1981). Intravenous inoculation of B16 melanoma caused a high incidence of pulmonary tumor nodules in mice pretreated with CY, 5-FU, or X-irradiation, while inoculation

TABLE III. Restoration by Thymosin α_1 of CFU-c Suppressed in the 5-FU-Treated Mice

Mice treated with	Dose of thymosin α_1 (μg/kg)	BM cells		CFU-c (mean ± S.E.)[a]			
		Per leg × 10^5	(%)	Per 10^5 cells	(%)	Per leg	(%)
Saline (control)	—	81.1	(100)	114.0 ± 4.7*	(100)	9245 ± 381*	(100)
5-FU	—	29.5	(36)	23.5 ± 4.1	(21)	693 ± 121	(8)
5-FU plus thymosin α_1	50.0	55.3	(66)	64.8 ± 4.4*	(57)	3454 ± 235*	(37)
	5.0	29.6	(36)	54.3 ± 4.8*	(56)	1903 ± 142*	(21)
	0.5	31.3	(39)	16.0 ± 1.6	(14)	501 ± 50	(5)

[a] p value: statistical difference from the corresponding 5-FU-treated group. *$p < 0.001$.

TABLE IV. Effect of Pretreatment with Thymosin α_1 on Lethal Infection with Microorganisms in 5-FU-Treated Mice

Pretreatment[a]			Survivors/tested at day		
Thymosin (i.p.)		5-FU (25 mg/kg, i.p.)	4	8	p value (U test)[b]
Expt 1					
L. monocytogenes[c]					
Control (saline)		−	10/10	10/10	
		+	9/10	0/10	
Thymosin α_1	4 μg/kg	+	10/10	7/10	0.001
	40	+	10/10	6/10	0.001
Expt 2					
C. albicans					
Control (saline)		−	7/7	7/7	
		+	2/7	2/7	
Thymosin α_1	4 μg/kg	+	5/7	5/7	0.02
	40	+	7/7	7/7	0.005
Expt 3					
P. aeruginosa					
Control (saline)		−	10/10	10/10	
		+	0/10	0/10	
Thymosin α_1	4 μg/kg	+	4/10	4/10	0.001
	40	+	8/10	8/10	0.001
Expt 4					
S. marcescens					
Control (saline)		−	10/10	10/10	
		+	1/10	1/10	
Thymosin α_1	4 μg/kg	+	6/10	6/10	0.016
	40	+	7/10	7/10	0.013

[a] Pretreated daily for 8 days (Expt 1) or 10 days (Expts 2, 3, and 4).
[b] Compared with the control mice pretreated with 5-FU, and then infected.
[c] Infected with *L. monocytogenes* (1×10^3, i.v.), *C. albicans* (2×10^5, i.v.), *P. aeruginosa* (5×10^4, i.v.), or *S. marcescens* (5×10^5, i.v.) at day 0.

of the tumor cells in normal mice caused little or no metastasis (Table VI). In these systems, thymosin α_1 given concurrently with the cytostatics or after X-irradiation suppressed the incidence of the pulmonary metastasis.

A similar activity was shown when the 5-FU-treated mice were inoculated with leukemic cells. When mice pretreated with 5-FU at relatively high doses and then inoculated with L1210, the mice died much more rapidly than normal mice which received the same number of L1210 cells (control) (Table VII). In this system, thymosin α_1 given concurrently with 5-FU prevented the rapid death, although the mice eventually died with leukemia similarly to those in the control group.

In order to have some insight as to the type of cells involved in the above phenomena, adoptive transfer of spleen cells from donor mice pretreated with 5-FU and/or thymosin α_1 to the 5-FU-treated recipient mice was performed (Table VII). When the spleen cells from donor mice treated with thymosin α_1 plus 5-FU were transferred into the recipient mice and L1210 leukemic cells then inoculated into the recipients, the rapid death was prevented, whereas the spleen cells from

TABLE V. Effect of ATS or Carrageenan (CA) on the Protective Activity of Thymosin α_1 against *Candida* and *Pseudomonas*

Pretreatment[a]	Survival/tested at day 15[b] *C. albicans*	*P. aeruginosa*
Expt 1		
Saline	7/7	8/8
5-FU	1/7	1/8
5-FU + ATS	1/7	1/8
5-FU + thymosin α_1	6/7*	7/8*
5-FU + thymosin α_1 + ATS	3/7	8/8
Expt 2		
Saline	8/8	7/7
5-FU	2/7	0.7
CA	1/8	5/7
5-FU + thymosin α_1	8/8*	5/7*
5-FU + thymosin α_1 + CA	3/8	5/7*

[a] C57BL/6 (Expt 1) or ddy (Expt 2) were pretreated with 5-FU (25 mg/kg per day, i.p.) or thymosin α_1 (40 μg/kg per day, i.p.) daily for 10 days from days −10 to −1. ATS (0.2 ml, i.v.) was injected at days −11, −8, −5, −3, and −1. CA (200 mg/kg, i.p.) was injected at day −1.
[b] Infected with *C. albicans* (2.5×10^5) or *P. aeruginosa* (5×10^4 for Expt 1, 2×10^4 for Expt 2) by the i.v. route at day 0.
[c] *p* value (U test), compared with 5-FU or 5-FU plus ATS or CA, then infected. $*p < 0.05$.

TABLE VI. Protective Activity of Thymosin α_1 against Pulmonary Metastasis of B16 Melanoma

Treatment (i.p.)[a]	No. of pulmonary tumors (mean ± S.D.)[b]	% inhibition
Expt 1		
Saline	0	
Cyclophosphamide (CY; 100 mg/kg, i.p. × 3)	142 ± 83	
CY + thymosin α_1 (40 μg/kg, i.p., × 10)	61 ± 33*	57
Expt 2		
Saline	6 ± 14	
5-FU (25 mg/kg, i.p., × 7)	18 ± 12	
5-FU + thymosin α_1 (40 μg/kg, i.p., × 7)	2 ± 3*	89
Expt 3		
Saline	0	
X-ray (600R, × 1)	22 ± 12	
X-ray + thymosin α_1 (40 μg/kg, i.p., × 10)	8 ± 4*	74

[a] C57BL/6 mice were treated with CY three times at day −5, −3, and −1, with 5-FU 10 times daily from day −5, or with X-ray at day −5. B16 melanoma cells (Expt 1, 10^5; Expt 2, 5×10^4; Expt 3, 10^5) were inoculated by the i.v. route at day 0.
[b] The number of pulmonary nodules was counted at day 14 (Expt 1), day 30 (Expt 2), and day 11 (Expt 3).

TABLE VII. Prevention by Adoptive Transfer of Spleen Cells of Rapid Death Caused by Tumor Inoculation in Immunosuppressed Mice

	Pretreatment[a]		Recipient mice[b]		Survival (days, mean ± S.D.)[c] (No. of mice tested)	
	Donor mice	Donor spleen cells	Pretreatment	Tumor inoculation	Expt 1	Expt 2
1.	—	—	5-FU	–	No death (6)	No death (7)
2.	—	—	Saline	–	12.3 ± 0.7* (5)	13.0 ± 0.5** (7)
3.	—	—	5-FU	+	6.7 ± 4.7 (6)	4.1 ± 3.4 (7)
4.	—	—	5-FU + thymosin α_1	+	12.0 ± 0.5 (6)	11.7 ± 3.6** (7)
5.	Saline	—	5-FU	+	10.0 ± 4.0 (6)	
6.	5-FU	—	5-FU	+	5.8 ± 4.8 (6)	4.3 ± 4.4 (7)
7.	5-FU + thymosin α_1	—	5-FU	+	12.5 ± 0.5***(6)	12.0 ± 4.1***(7)
8.	Saline	Anti-Thy-1.2 + C′	5-FU	+	12.5 ± 0.5***(7)	
9.	5-FU + thymosin α_1	Anti-Thy-1.2 + C′	5-FU	+	13.0 ± 0.5***(7)	10.5 ± 4.4 (4)
10.	5-FU + thymosin α_1	Anti-Lyt-5.1 + C′	5-FU	+	9.4 ± 4.4 (7)	
11.	5-FU + thymosin α_1	Anti-asialo G_{M1} + C′	5-FU	+		4.5 ± 4.0 (6)

[a] 5-FU (25 mg/kg, i.p.) and/or thymosin α_1 (40 μg/kg, i.p.) were given daily for 7 days. The mice were sacrificed 24 hr after the last treatment, and spleen cells were prepared.
[b] Recipient mice were treated daily for 7 days (day −7 to −1) with 5-FU (25 mg/kg, i.p.) and/or thymosin α_1 (40 μg/kg, i.p.), injected with the donor spleen cells at day −1, and then inoculated subcutaneously with L1210 leukemic cells (10^4) at day 0.
[c] $*p < 0.05$, $**p < 0.01$ as compared with group 3; $***p < 0.05$ as compared with group 6.

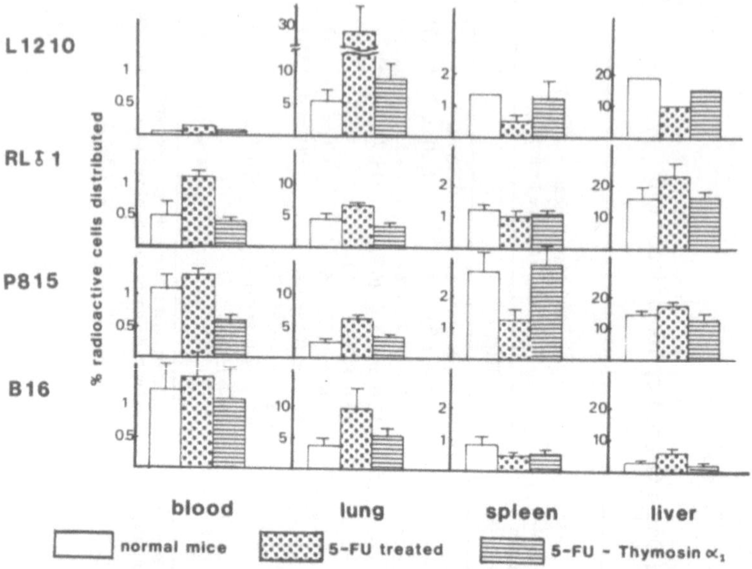

FIGURE 1. *Tissue distribution of radiolabeled tumor cells.* Mice treated daily for 7 days with or without 5-FU (25 mg/kg, i.p.) and thymosin α_1 (40 μg/kg, i.p.) were inoculated intravenously with [^{125}I]-UdR-labeled L1210 leukemic cells (10^6 cells/5.7×10^5 cpm), ^{51}Cr RL ♂1 (10^6 cells/138,060 cpm), ^{51}Cr P815 mastocytoma cells (10^6 cells/197,900 cpm), or B16 melanoma cells (6×10^5 cells/3.5×10^4 cpm). The first three tumor cells and B16 melanoma cells were inoculated into DBA/2 and C57BL/6 mice, respectively. The radioactivity in each tissue was measured at 5 hr after the inoculation of the tumor except at 24 hr for L1210.

the donor mice treated with 5-FU alone did not show such preventive activity. Subsequently, the donor spleen cells from the thymosin α_1-treated mice were incubated with various antisera and complement, and the adoptive transfer then performed. The donor cells treated with anti-asialo G_{M1} and Lyt-5 lost their protective activity, but those deprived of T or B cells by incubation with anti-Thy-1 or anti -immunoglobulin, respectively, maintained the activity.

The possibility that the effector cells in the spleen cells were NK cells was further demonstrated in a separate experiment (Umeda *et al.*, 1983), where the donor mice treated with 5-FU and thymosin α_1 were injected with antisera and then inoculated with L1210 cells. Injection with anti-asialo G_{M1} but not ATS abrogated the preventive activity of the spleen cells.

3.5. Prevention of Abnormal Tissue Distribution of Tumor Cells and Homing of Lymphocytes

The mechanisms leading to the rapid death and causing a high incidence of the pulmonary metastasis were investigated from a different aspect, the tissue distribution of tumor cells inoculated in the 5-FU-treated mice. ^{51}Cr- or ^{125}I-labeled

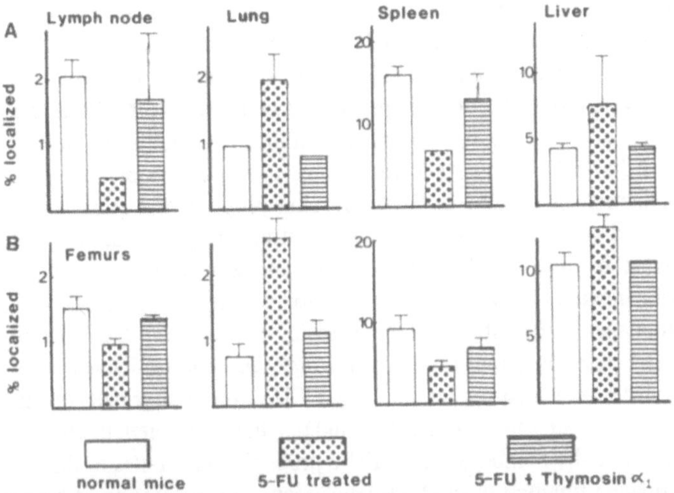

FIGURE 2. *Homing of ^{51}Cr-labeled lymphoid cells.* DBA/2 mice treated with 5-FU and thymosin α_1 as described in Fig. 1 were inoculated with ^{51}Cr inguinal lymph node cells (10^6 cells/1.97×10^5 cpm, i.v.) or (B) ^{51}Cr bone marrow (femurs) (5×10^6 cells/1.46×10^5 cpm, i.v.). The radioactivity of each tissue was measured at 24 hr after the inoculation and expressed as a percent of total radioactivity inoculated.

tumor cells were inoculated intravenously into normal mice or the 5-FU-treated mice, and the tissue distribution of the radioactivity then compared (Fig. 1). In the 5-FU-treated mice, the level of the radioactivity was higher in blood and lungs but lower in spleen than that in normal mice, although there was some difference in the distribution pattern depending on the type of tumor cells. On the other hand, when thymosin α_1 was given concurrently with 5-FU, the distribution pattern was similar to that in normal mice.

These phenomena were also observed in the homing of ^{51}Cr-labeled lymph node or bone marrow cells inoculated into normal mice by the i.v. route. A fairly large number of the cells homed to the tissues of their origin and distributed to spleen (Fig. 2). On the other hand, in the mice treated with 5-FU, the cells were trapped more in the lungs, substantially reducing the number of the homing cells. Additional treatment with thymosin α_1 prevented the change of the tissue distribution pattern. Apparently, the treatment with 5-FU disturbed trapping systems against lymphocytes as well as tumor cells, while thymosin α_1 prevented such an effect.

4. DISCUSSION

The present study has dealt with the effect of thymosin α_1 in mice immunosuppressed by cytostatics or X-irradiation. In the immunosuppressed mice, thymosin α_1 restored the DTH response which was reduced by a cytostatic, 5-FU, and prevented microbial infections and suppressed tumor growth. Thymosin α_1 has been known to affect mainly the function and differentiation of T cells (Ahmed *et al.*, 1978; Goldschneider *et al.*, 1981). However, some data presented in this report indicate that thymosin α_1 also affects cells other than T cells such as macrophages, neutrophils, NK cells, or their progenitor cells either directly or indirectly.

The studies utilizing the adoptive transfer technique in the DTH reaction indicated that 5-FU causes severe damage to the cell population involved, both in the induction phase and in the expression phase of the delayed footpad reaction, and that thymosin α_1 restores the cells affected by this damage. This means that thymosin α_1 restores the function (or number) of not only the mature T cells transferring the delayed footpad reaction to the recipient mice, but also of the macrophages responsible in the expression phase of the DTH reaction.

The possibility that thymosin restores the depressed function of macrophages was further indicated by the facts that CFU-c, progenitor cells of macrophages and granulocytes, are sensitive to 5-FU, and that the subsequent treatment with thymosin α_1 increases the number of the CFU-c probably contributes to the restoration of the macrophage function through replenishment and maturation of hematopoietic cells necessary for the DTH response, which had been suppressed by the 5-FU treatment.

It is known that macrophages are exclusively important for elimination of *Candida*, and the treatment of mice with a depressor of macrophages, carrageenan, increases the susceptibility to the pathogen (Dobius, 1964). Because macrophages are activated by sensitized T cells or their products, lymphokines, it is quite likely that thymosin α_1 prevented *Candida* infection in the immunosuppressed mice, by T-cell-mediated macrophage killing as indicated by the fact that the activity of thymosin α_1 was abrogated by ATS as well as carrageenan.

On the other hand, the mechanism of the protecting activity of thymosin α_1 against *Pseudomonas* infection in immunosuppressed mice is not yet clear. For elimination of this pathogen, neutrophils, but not macrophages or T cells, are known to be quite important (Tatsukawa *et al.*, 1979), and the activity of thymosin α_1 against this pathogen was not affected by ATS or carrageenan. The present results indicate that the activity was mediated through cells other than T cells or macrophages. Probably neutrophils or their progenitors such as pluripotent stem cells in bone marrow might be the targets of thymosin α_1.

NK cells participate in the suppression of tumor metastasis (Talmadge *et al.*, 1980). Their damage is known to cause a high incidence of pulmonary metastasis by inoculation of B16 melanoma in immunosuppressed mice (Hanna and Burton, 1981; Riccardi *et al.*, 1981). As thymosin α_1 prevented the damage to the NK cells, its activity was at least in part through maintenance of NK cell activity. The NK cells must also be important against leukemic cells. This was shown by the experiment of the adoptive transfer of spleen cells, where inoculation of NK-enriched spleen cells from the 5-FU and thymosin α_1-treated mice to the 5-FU-treated recipient mice prevented the rapid death caused by L1210 leukemic cells.

Another possible mechanism causing the rapid death and increasing metastasis is the damage to the barrier system for spread of tumor cells. In the 5-FU-treated mice, the trapping system against the tumor cells in organs such as spleen and liver might be disturbed, resulting in poor clearance of the tumor cells from blood and an accumulation into the lungs. In such mice, the tumor cells could migrate into various sites. The migration of tumor cells to some specific site and growth of the tumor there may lead to the rapid death. Abnormal homing of lymph node and bone marrow cells in the 5-FU-treated mice also suggests that cells recognizing surface markers of lymphocytes as well as tumor cells are damaged, or that the number of such cells are reduced in the immunosuppressed mice. Thymosin α_1 may maintain the capacity and the number of such cells.

Thymosin α_1 seemed to affect various cell populations, although it is not known whether the effects were mediated through T cells or not. As thymosin α_1 appeared to affect various progenitor cells in bone marrow, its pleiotropic effect is likely. Another explanation for the multiple activity of thymosin α_1 is that thymosin α_1 affects hormonal balance. It is reported that the circadian rhythm of endogenous thymosin α_1 correlated to circadian fluctuation in the immune system, but inversely correlated to that of glucocorticoid affect the immune system (McGills *et al.*, 1983), exogenous thymosin α_1 may exert its multiple activity through hormonal regulation.

Thymosin has been shown to have beneficial effects on fungal infections (Bistoni *et al.*, 1982), viral infections (Hang *et al.*, 1981), animal tumors (Chirigos, 1978), and human neoplasms (Lipson *et al.*, 1979). The present studies give some insights on the mechanism of the thymosin activities. Thymosin α_1 restored immunocompetence reduced by cytostatics or X-irradiation. Consequently, thymosin α_1 protected mice from progression of tumor and microbial infections caused by treatment with the immunosuppressive agents. This means that thymosin α_1 may reduce undesirable toxic effects of cytostatics or X-irradiation, and suggests that in combination with thymosin α_1 an extensive chemotherapy and/or radiotherapy for a longer duration are possible.

REFERENCES

Ahmed, A., Smith, A. H., Wond, K. M., Thurman, G. B., Goldstein, A. L., and Sell, K. W., 1978, *In vitro* induction of *Lyt* surface markers on precursor cells incubated with thymosin polypeptides, *Cancer Treat. Rep.* **62:**1739.

Bistoni, F., Marconi, P., Farati, L., Bonmassar, E., and Garaci, E., 1982, Increase of mouse resistance to *Candida albicans* infection by thymosin α_1, *Infect. Immun.* **36:**609.

Bradley, T. R., and Metcalf, D., 1966, The growth of mouse bone marrow cells *in vitro, Aust. J. Exp. Biol. Med. Sci.* **44:**287.

Chirigos, M. A., 1978, *In vitro* and *in vivo* studies with thymosin, in: *Immune Modulation and Control of Neoplasia by Adjuvant Therapy* (M. A. Chirigos, ed.), p. 305, Raven Press, New York.

Dobias, B., 1964, Specific and nonspecific immunity in *Candia* infections, *Acta. Med. Scand.* **176:**79.

Fidler, I. J., 1970, Metastasis: Quantitative analysis of distribution and fate of tumor emboli labeled with ^{125}I-5-iodo-2′-deoxyuridine, *J. Natl. Cancer Inst.* **45:**775.

Goldschneider, I., Ahmed, A., Bollum, F. J., and Goldstein, A. L., 1981, Induction of terminal deoxynucleotidyl transferase and Lyt antigens with thymosin: Identification of multiple subsets of prothymocytes in mouse bone marrow and spleen, *Proc. Natl. Acad. Sci. USA* **78:**2469.

Goldstein, A. L., Low, T. L. K., McAdoo, M., McClure, J., Thurman, G., Rossio, J., Lai, C., Chang, D., Wang, D., Wang, S., Marvey, C., Ramel, A. H., and Meienhofer, J., 1977, Thymosin α_1: Isolation and sequence analysis of an immunologically active thymic polypeptide, *Proc. Natl. Acad. Sci. USA* **74:**725.

Gorelik, E., Segal, S., and Feldman, M., 1980, Control of lung metastasis progression in mice: Role of growth kinetics of 3LL Lewis lung carcinoma and host immune reactivity, *J. Natl. Cancer Inst.* **65**:1257.

Hang, K-Y., Kind, P. D., Jagoda, E. M., and Goldstein, A. L., 1981, Thymosin treatment modulates production of interferon, *J. Interferon Res.* **1**:411.

Hanna, N., and Burton, R. C., 1981, Definitive evidence that natural killer cells inhibit experimental tumor metastasis *in vivo*, *J. Immunol.* **127**:1754.

Ishitsuka, H., Umeda, Y., Namakura, J., and Yagi, Y., 1983, Protective activity of thymosin against opportunistic infections in animal models, *Cancer Immunol. Immunother.* **14**:145.

Kasai, M., Iwamori, M., Nagai, Y., Okumura, K., and Tada, T., 1980, A glycolipid on the surface of mouse natural killer cells, *Eur. J. Immunol.* **10**:175.

Lipson, S. D., Chretien , P. B., Makuch, R., Kenady, D. E., and Cohen, M. H., 1979, Thymosin immunotherapy in patients with small cell carcinoma of the lung: Correlation of *in vitro* studies with clinical course, *Cancer* **43**:863.

McGillis, J. P., Hall, N. R., and Goldstein, A. L., 1983, Circadian rhythm of thymosin α_1, *Fed. Proc.* **42**:576.

Ohta, Y., Sueki, K., Kitta, K., Takemoto, K., Ishitsuka, H., and Yagi, Y., 1980, Comparative studies on the immunosuppressive effect among 5′-deoxy-5-fluorouridine, ftorafur and 5-fluorouracil, *Gann* **71**:190.

Ohta, Y., Sueki, K., Yoneyama, Y., Tezuka, E., and Yagi, Y., 1983, Immunomodulating activity of thymosin fraction 5 and thymosin α_1 in immunosuppressed mice, *Cancer Immunol. Immunother.* **15**:108.

Riccardi, C., Barlozzari, T., Santoni, A., Herberman, R., and Cesarini, C., 1981, Transfer of cyclophosphamide-treated mice of natural killer (NK) cells and *in vivo* natural reactivity against tumor, *J. Immunol.* **126**:1284.

Talmadge, J. E., Meyers, K. M., Prieur, D. J., and Starkey, J. R., 1980, Role of NK cells in tumor growth and metastasis in beige mice, *Nature (London)* **284**:622.

Tatsukawa, K., Mitsuyama, M., Takeya, K., and Nomoto, K., 1979, Differing contribution of polymorphonuclear cells and macrophages to protection of mice against *Listeria monocytogenes* and *Pseudomonas aeruginosa*, *J. Gen. Microbiol.* **115**:161.

Trainin, N., Pecht, M., and Handzel, T., 1983, Thymic hormones: Inducers and regulators of the T-cell system, *Immunol. Today* **4**:16.

Umeda, Y., Sakamoto, A., Nakamura, J., Ishitsuka, H., and Yagi, Y., 1983, Thymosin α_1 restores NK-cell activity and prevents tumor progression in mice immunosuppressed by cytostatics or X-rays, *Cancer Immunol. Immunother.* **14**:145.

Wara, D. W., Goldstein, A. L., Doyle, W., and Ammann, A. J., 1975, Thymosin activity in patients with cellular immunodeficiency, *N. Engl. J. Med.* **292**:70.

The *in Vivo* Effect of Thymosin on Cell-Mediated Immunity

S. B. SALVIN and RUTH NETA

1. INTRODUCTION

Inbred strains of mice, similar to individual humans, vary in their resistance to infections. Variations also exist in the capacity of inbred strains of sensitized mice to release lymphokines into the circulation (Neta *et al.*, 1981). As little evidence is present on the role of individual *in vivo*-induced lymphokines in resistance to infections, correlation of these variations in the different murine strains may lead to the development of a better experimental model in which to study their association. A possible presence in some strains of a defect in both resistance and the *in vivo* release of lymphokines, together with the means of corrective treatment of either or both parameters, will present a model in which such studies may be conducted.

In this work, (1) the resistance or susceptibility of inbred strains of mice to infection with *Candida albicans*, (2) the release of two lymphokines, γ-interferon (IFN-γ) and migration inhibitory factor (MIF), by these strains, and (3) their capacity to elicit delayed hypersensitivity were compared. The low-responder, susceptible strains were treated in turn by a thymic hormone, thymosin fraction 5. The differences in the responses of the several inbred murine strains to this treatment are described.

S. B. SALVIN • Department of Microbiology, University of Pittsburgh, School of Medicine, Pittsburgh, Pennsylvania 15261. RUTH NETA • Armed Forces Radiobiology Research Institute, Bethesda, Maryland 20814. Research was conducted according to the principles enunciated in the "Guide for the Care and Use of Laboratory Animals" prepared by the Institute of Laboratory Animal Resources, National Research Council.

2. MATERIALS AND METHODS

Inbred strains of mice were used in the various *in vivo* experiments and were purchased from The Jackson Laboratory, Bar Harbor, Maine.

The procedures for determination (1) of degree of resistance to infection to *C. albicans*, (2) of titers of MIF and IFN-γ released *in vivo* into the circulation, and (3) of development of delayed footpad reactions are the same as those previously described (Neta and Salvin, 1983; Salvin and Neta, 1983; Salvin and Nishio, 1969).

Thymosin fraction 5 was obtained through the courtesy of Dr. Allan L. Goldstein, Department of Biochemistry, The George Washington University School of Medicine and Health Sciences, Washington, D.C., and of Dr. Bernard L. Horecker, Roche Institute of Molecular Biology, Nutley, New Jersey.

3. RESULTS

Resistance to i.v. Challenge with C. albicans. With regard to resistance to an i.v. challenge of 4×10^4 cells of *C. albicans*, inbred strains may be divided into two categories: those that are resistant, and those that are susceptible. For example, such strains as C57BL/10SNJ, C57BL/6J, and C57BL/KsJ are highly resistant, whereas such strains as AKR/J, CBA/CaJ, C3H/HeJ, and DBA/1J are susceptible. The resistance, as measured at intervals after challenge by the number of units of *C. albicans* in the kidneys, became most apparent in the resistant strains from about 10 to 21 days after infection. However, where the resistant strains showed a decline or elimination of yeast in the kidneys by about 10 days after infection, the susceptible strains had a gradual increase in the number of fungus cells in the kidney, so that by 10–14 days, most of the animals were sickly or dying of infection. Thus, a genetic factor was present which resulted in some inbred strains being resistant, and others being susceptible to intravenous challenge with *C. albicans* (Table I).

Capacity to Release MIF and IFN-γ in Vivo into the Circulation. Both resistant and susceptible strains were studied for their capacity to release MIF and IFN-γ *in vivo* into the circulation. The mice were sensitized i.v. with 700 μg merthiolate-killed, lyophilized cells of *C. albicans* in Drakeol–Tween 80 emulsion, challenged 3 weeks later with a *C. albicans* extract, and bled 4 hr later.

The resistant strains, such as C57BL/KsJ and C57BL/6J, released high titers of MIF and IFN-γ into the circulation, while the susceptible strains, such as CBA/CaJ and AKR/J, did not release detectable quantities of either MIF or IFN-γ into the circulation. Thus, a correlation existed between the capacity of a strain to resist infection with *C. albicans* and its capacity to release MIF and IFN-γ *in vivo* into the circulation in response to the specific antigen (Table I).

Capacity to Elicit Delayed Hypersensitivity. The two types of murine strains were also compared with regard to their capacities to elicit delayed footpad hypersensitivity. The mice were sensitized with 700 μg lyophilized cells of *C. albicans* and 2–3 weeks later challenged in the footpads with a cell extract of the yeast. The resistant, high-responder strains showed a capacity to elicit marked delayed hypersensitivity responses, while the susceptible, low-responder (for MIF and IFN-γ) strains did not elicit any delayed reactions to the specific antigen.

Effect of Thymosin Fraction 5 on Cell-Mediated Responses of Genetically Susceptible Murine Strains. Because these three immunological responses are related to T-lymphocyte activity, experiments were initiated wherein the effect of *in vivo* injection of thymosin fraction 5 was assayed on these three parameters. When susceptible strains, such as C3H/HeJ, were treated daily i.p. with 5 μg thymosin fraction 5 in 0.5 ml saline, beginning on the day of infection, a marked increase in resistance was observed. Similarly, when C3H/HeJ mice, sensitized with 700 μg dead lyophilized *C. albicans* cells in Drakeol–Tween 80 emulsion and challenged with *C. albicans* extract (150 μg nitrogen), were treated i.p. with 5 μg thymosin fraction 5 for 3 days before sensitization and for 3 days before challenge, the titers of MIF in the sera rose from a nondetectable level to a level where dilutions of the sera of 1 : 32–1 : 64 induced migration inhibition of greater than 20%. This is the approximate titer which develops in normal, untreated responder strains, such as C57BL/10SNJ and C57BL/6J.

Although treatment of susceptible strains with thymosin fraction 5 enhanced (1) resistance to infection with *C. albicans* and (2) the capacity to release MIF *in vivo* in response to antigens from *C. albicans,* treatment with thymosin fraction 5 did not enhance the capacity of susceptible strains to elicit delayed footpad reactions to specific antigen.

Effect of Thymosin Fraction 5 on Cell-Mediated Responses of Alloxan-Diabetic Mice. As *C. albicans* is an opportunistic pathogen, its infectivity was tested in an immunocompromised host. Thus, 200–225 mg alloxan was injected i.p. into a strain, such as C57BL/10SNJ or C57BL/KsJ, which is normally resistant before it becomes hyperglycemic. Where normal mice have a serum glucose level of 100–200 mg/dl, alloxan-diabetic mice typically have a level of 400–600 mg/dl.

Such alloxan-diabetic mice, which in the normal nonhyperglycemic state are resistant to an intravenous challenge of 4×10^4 *C. albicans,* as indicated by the virtual absence of the yeast in the kidneys at 10–14 days after challenge (see above), become highly susceptible and may have 10,000 or more *C. albicans* cells per

TABLE I. Relationship of Resistance to Infection with *Candida albicans* and *in Vivo* Release of MIF and IFN-γ in Inbred Murine Strains

Murine strain	No. of *C. albicans* per kidney on day 14 after infection	Titers of lymphokines in circulation	
		MIF	IFN-γ
C57BL/6J	0	> 64	> 10,000
C57BL/KsJ	0	> 64	> 10,000
CBA/CaJ	4000	< 4	< 100
AKR/J	6000	< 4	< 100

kidney. Daily i.p. injections of 5 μg thymosin fraction 5 beginning on the day of infection restored normal resistance to the alloxan-diabetic mouse.

Similarly, the alloxan-diabetic mice of the C57BL/10SNJ or C57BL/KsJ strains lose their capacity to release MIF *in vivo* into the circulation, a capacity which is restored when the mice are treated i.p. for 3 days both before sensitization and before challenge with 5 μg thymosin fraction 5.

Sensitized alloxan-diabetic mice also did not elicit delayed footpad reactions on challenge with specific antigen, *C. albicans* extract, in mice sensitized with *C. albicans* cells. Delayed responses did develop, however, when such mice were treated daily i.p. with 5 μg thymosin fraction 5. Thus, thymosin fraction 5 had a pronounced effect on restoring or enhancing three parameters of cell-mediated resistance, i.e., cell-mediated resistance to infection with *C. albicans* and *in vivo* release of MIF and IFN-γ. Thymosin was effective in enhancing delayed hypersensitivity only in those immunocompromised hosts which normally were highly responsive. Apparently, a specific precursor cell must be present before thymosin could enhance an immunological response.

4. DISCUSSION

Different inbred strains of mice vary in their susceptibility to infections with such microorganisms as *C. albicans*. Those strains that are inherently resistant, such as C57BL/10SNJ and C57BL/KsJ, are capable of releasing high titers of MIF and IFN-γ *in vivo* into the circulation upon appropriate sensitization and challenge with specific antigen. Those strains that are inherently susceptible, such as C3H/HeJ and DBA/1J, are not capable of releasing high titers of MIF and IFN-γ into the circulation upon appropriate sensitization and challenge with specific antigen. Genetically resistant strains after sensitization also have the capacity to elicit strong delayed footpad reactions on challenge with specific antigen.

Administration of a thymic hormone, thymosin fraction 5, enhanced the cellular immune capabilities of the susceptible, nonresponder mice, with regard to resistance to infection and *in vivo* release of lymphokines. This effect was apparent not only in those mice that were normally unresponsive because of a genetically determined deficiency, but also in those mice that were immunocompromised because of the induction of a diabetic state. In the case of delayed hypersensitivity, thymosin was effective in enhancing the response only in those strains, such as C57BL/10SNJ and C57BL/KsJ, which normally had the capacity to develop strong delayed footpad reactions, but which had their immunological responses compromised by treatment with alloxan and the development of hyperglycemia. The difference in the response of the murine strains to thymosin in the expression of resistance to infection and *in vivo* release of lymphokines vs. elicitation of delayed hypersensitivity suggests that the development of delayed hypersensitivity may depend on a cellular mechanism different from that responsible for resistance to infection and *in vivo* release of MIF and IFN-γ.

The foregoing procedures serve as an excellent laboratory model for studying the effect of thymosin on such parameters of cellular immunity as resistance to infection, *in vivo* release of MIF and IFN-γ, and elicitation of delayed hypersensitivity. Of special significance is the demonstration that thymosin fraction 5 does enhance resistance to infection with such microorganisms as *C. albicans* in certain immunodeficient mice.

ACKNOWLEDGMENT. The authors' research reported herein was supported by USPHS Grants AI-16064 and AM-27727.

REFERENCES

Neta, R., and Salvin, S. B., 1983, Resistance and susceptibility to infection in inbred murine strains. II. Variations in the effect of treatment with thymosin, *Cell. Immunol.* **75**:173–180.

Neta, R., Salvin, S. B., and Sabaawi, M., 1981, Mechanisms in the *in vivo* release of lymphokines. I. Comparative kinetics in the release of six lymphokines in inbred strains of mice, *Cell. Immunol.* **64**:203–219.

Salvin, S. B., and Neta, R., 1983, Resistance and susceptibility to infection in inbred murine strains. I. Variations in the response to thymic hormones in mice infected with *Candida albicans*, *Cell. Immunol.* **75**:160–172.

Salvin, S. B., and Nishio, J., 1969, *in vitro* cell reactions in delayed hypersensitivity, *J. Immunol.* **103**:138–141.

Effect of Thymosin α_1 on Immunoregulatory T Lymphocytes

GINO DORIA and DANIELA FRASCA

1. INTRODUCTION

The extensive work performed by A. L. Goldstein and co-workers has led to the isolation, chemical characterization, and biological and clinical applications of thymosins present in "fraction 5," a partially purified extract from bovine thymus (Hooper *et al.*, 1975). Fraction 5 consists of 40–50 peptides whose molecular weights range from 1000 to 15,000, as demonstrated by analytical polyacrylamide gel electrophoresis and isoelectric focusing (Goldstein *et al.*, 1977). Some of these peptides exhibit a wide range of biological activities in animal models (Goldstein, 1978). It has been established that there is no homology between the biologically active thymosin peptides and thymopoietin (Goldstein, 1975) or facteur thymique sérique (Bach and Dardenne, 1972). Among the several peptides present in fraction 5, α_1 was the first to be purified, sequenced (Goldstein *et al.*, 1977), and synthesized (Goldstein, 1978). It is a highly acidic molecule, consisting of 28 amino acid residues. Several tests performed both *in vivo* and *in vitro* have shown that α_1 is 10–1000 times more active than fraction 5 in promoting T-cell differentiation. It has been found that *in vivo* administration of α_1 enhances the lymphoid cell responses to mitogens as well as lymphotoxin production (Schulof and Goldstein, 1983). Results of α_1 treatment in conjunction with intensive conventional chemotherapy show that growth of plasmacytoma MOPC-315 in Balb/c mice is reduced or prevented. Moreover, the surviving animals reject retransplanted tumor cells (Goldstein *et al.*, 1983). Immunosuppressed mice infected with *Candida* or *Cryptococcus* prolong their survival after *in vivo* injection of α_1. This thymic factor also enhances the production of interferon in mice infected with Newcastle disease virus (Zatz *et al.*, 1982). *In vitro* studies indicate that thymosin α_1 induces appearance of Thy-1.2$^+$ cells; Lyt-1.2.3$^+$ phenotype is also induced by α_1 at concentrations of 1 μg/

GINO DORIA and DANIELA FRASCA • ENEA-EURATOM Immunogenetics Group, Laboratory of Pathology, C.R.E. Casaccia, Rome, Italy.

ml (Ahmed *et al.*, 1979). Moreover, α_1 increases the percentage of TdT^+ cells in the bone marrow and spleen when used at high concentrations, while low concentrations of α_1 suppress TdT activity in murine thymocytes (Wetzel *et al.*, 1980). In the bone marrow of normal mice, 20% of α_1-induced Lyt-1.2.3$^+$ cells are TdT^- while in the spleen 50% of the induced cells are TdT^-. In nude mice, the percentages of TdT^-, α_1-induced Lyt-1.2.3$^+$ cells are 80% in the bone marrow and 75% in the spleen (Zatz *et al.*, 1982). It has also been reported that short-term incubation of thymocytes with α_1 results in an increase in the percentage of cortisone-resistant cells (Osheroff, 1981). The *in vitro* incubation of lymphocytes with thymosin α_1 enhances the production of macrophage inhibitory factor (MIF) (Thurman *et al.*, 1977), T-cell-dependent IgG, IgM, and IgA secondary antibody responses (Schulof and Goldstein, 1983), as well as T-cell-dependent specific antibody production and helper T-cell activities (Ahmed *et al.*, 1979). The *in vitro* treatment of lymphoid cells with α_1 also raises the levels of intracellular cGMP but not cAMP (Goldstein, 1978) and increases E-rosette formation (Low *et al.*, 1979). Also in humans *in vitro* incubation of peripheral blood lymphocytes (PBL) with α_1 increases the percentage of E-rosette-forming cells and autologous rosette-forming cells in patients with diseases such as primary immunodeficiency, cancer, viral infection, and autoimmune pathology. If PBL are derived from healthy individuals, α_1 does not affect the percentage of lymphocytes expressing E-rosette receptors. However, mitogen responses or mixed lymphocyte reactions of PBL from healthy subjects can be either enhanced or suppressed by α_1 treatment. Moreover, in patients immunized with tetanus toxoid or group C meningococcal polysaccharide, PBL incubated with α_1 exhibit significant enhancement of specific antibody responses. α_1 has also been shown to decrease the abnormally elevated $T\mu/T\gamma$ ratios in PBL from cancer patients (Schulof and Goldstein, 1983).

The present report describes the effect of injecting immunodeficient old mice with synthetic α_1 on helper T-cell activity. It has been previously established that T-cell-mediated antibody response declines with advancing age (Doria *et al.*, 1980). Spleen cells from old animals were shown to mount decreased antibody responses when stimulated *in vitro* with T-dependent immunogens, even when enriched T-cell populations were used (Kay, 1978).

In the present study, helper activity of spleen cells from mice of different ages, uninjected or injected with α_1 before horse red blood cell (HRBC) priming, has been determined *in vitro* by adding limiting numbers of the primed cells to cultures containing normal spleen cells from young mice and the conjugate 2,4,6-trinitrophenyl (TNP)–HRBC. Results demonstrate that helper T-cell activity, which is markedly reduced in senescence, can be efficiently repaired by injection of synthetic α_1.

2. MATERIALS AND METHODS

2.1. Animals

Male (C57BL/10 × DBA/2)F_1 mice have been bred, maintained in our animal facilities, and used at different ages.

2.2. Antigens

HRBC and sheep red blood cells (SRBC) in Alsevier's solution were obtained from Sclavo (Siena, Italy). 2,4,6-Trinitrobenzene sulfonic acid (TNBS) was purchased from Eastman Organic Chemicals (Rochester, N.Y.) and further purified by recrystallization from 1 NHCl solution. TNP–HRBC was prepared by heavy conjugation of TNBS with HRBC (Kettman and Dutton, 1970) and used *in vitro* as T-dependent immunogen. TNP–SRBC was prepared by light reaction of TNBS with SRBC (Rittenberg and Pratt, 1969) and used as test antigen in the Cunningham and Szenberg (1968) technique to detect anti-TNP plaque-forming cells (PFC).

2.3. Cell Culture

Spleen cell suspensions were prepared from normal and carrier-primed mice and cultured in microtissue culture plates (Falcon Plastics No. 3040, Oxnard, Calif.), according to Mishell and Dutton (1967). Cells in culture were suspended in medium RPMI 1640 (GIBCO, Grand Island, N.Y.), supplemented with 10% fetal calf serum, 1 mM sodium pyruvate, 0.1 mM nonessential amino acid mixture, 2 mM L-glutamine, 20 mM Hepes buffer solution, 100 IU/ml penicillin, and 100 μg/ml streptomycin sulfate. Fetal calf serum was obtained from Rehatuin (Phoenix, Ariz.); Hepes buffer from Eurobio (France); all the other reagents from Microbiological Associates (Bethesda, Md.). 2-Mercaptoethanol was not added to cultures.

2.4. Induction and *in Vitro* Titration of Helper Cell Activity

To induce helper activity, mice of a given age, untreated or treated with thymosin α_1, were carrier-primed by injecting i.v. 2×10^5 HRBC in 0.2 ml phosphate-buffered saline (PBS) 4 days before sacrifice.

Helper activity of carrier-primed spleen cells in the *in vitro* anti-TNP antibody response was titrated by a modification of the method originally described by Kettman and Dutton (1971). Normal spleen cells immunized *in vitro* with TNP–HRBC give rise to negligible anti-TNP PFC responses. If graded numbers of carrier-primed spleen cells are added to a constant number of normal spleen cells, the anti-TNP antibody response increases with the number of carrier-primed cells added in a range in which helper activity is limiting. The titration procedure was the following: each culture well received 0.1 ml medium alone or containing immunogen (2×10^5 TNP–HRBC) as well as 1×10^6 nucleated spleen cells from a pool of 8–10 uninjected, normal mice. One group of culture wells received no further addition of cells and was used as control. Other groups of culture wells received graded numbers (5, 10, 15, 20, 25, 30 $\times$ 10^4) of nucleated spleen cells pooled from four to six carrier-primed mice. In some experiments, spleen cells from carrier-primed mice were T-cell enriched and/or nylon wool-fractionated before testing their helper activity. The anti-TNP antibody response was evaluated at the peak response (day 4 or 5 of culture) and processed as follows. The antibody response of 1×10^6 nucleated spleen cells from uninjected, normal mice, generally less than 10% of the response given by 1×10^6 carried-primed nucleated spleen cells, was subtracted from the antibody response exhibited by 1×10^6 normal spleen cells supplemented with 5–30 × 10^4 carrier-primed spleen cells. The net anti-TNP antibody response was found to increase with the number of carrier-primed cells added to the culture well, so that the log number of direct anti-TNP PFC is a linear function of the log number of carrier-primed spleen cells added. When helper activity of whole spleen cells was titrated, linear regressions were calculated by the least-squares method, which yields the expected log PFC number (Y) for any log number (X) of carrier-primed spleen cells added to culture and the regression standard deviation. The anti-TNP antibody response in cultures containing the maximal number of helper cells was evaluated as follows. The expected number of PFC induced by 30 × 10^4 carrier-primed spleen cells is the antilog of the corresponding Y value. The antilog of the regression standard deviation value represents a factor by which the expected number of PFC should be multiplied or divided to obtain the variation due to one standard error.

Contributions of B cells and macrophages to the helper cell activity of HRBC-primed cells, as titrated by this method, have been ruled out by the results of previous control experiments (Doria *et al.*, 1980).

2.5. T-Cell Enrichment of Carrier-Primed Spleen Cells

Pooled spleen cells from carrier-primed mice, uninjected or injected with thymosin α_1, were suspended in medium supplemented with Hepes buffer, fetal calf serum (10%), and antibiotics, and then washed. The supernatants were discarded and the pellets resuspended in 2 ml trichloroacetic acid for 2 min at 37°C, in order to lyse RBC. Cells were washed again and adjusted at a concentration of 10 × 10^6 cells/ml in medium. Cell suspensions were enriched in T lymphocytes (immunoglobulin-negative cells) by positive selection on plastic plates, coated overnight with 5 ml rabbit anti-mouse immunoglobulin antibody (100 μg/ml). Each plate was repeatedly washed with PBS before receiving about 100 × 10^6 carrier-primed spleen cells, and then incubated at 4°C for 1 hr. Nonadherent immunoglobulin-negative cells were recovered by gentle swirling.

2.6. Separation of Nylon Wool-Adherent and Nonadherent T Cells

T-cell-enriched suspensions were further fractionated by filtration through a nylon wool column into nylon-adherent and nylon-passed T cells. According to Julius *et al.* (1973), the columns were made with 0.7 g of nylon wool in a 10-ml plastic syringe, sterilized, and equilibrated with medium supplemented with 5% heat-inactivated fetal calf serum, antibiotics, and Hepes buffer, at 37°C. About 2–3 ml of T-cell-enriched suspensions (~100 × 10^6 nucleated cells) were applied to each column and incubated at 37°C for 45 min. Nylon-passed T cells were collected from the column by washing with warm (37°C) medium. Subsequently, the column was incubated at 4°C for 30 min and then the nylon wool was vigorously pressed with chilled medium to recover the adherent T-cell population.

2.7. Thymosin α_1 Treatment

Mice were injected i.p. with 0.2 ml saline containing 1 or 10 μg thymosin α_1, over a period of five consecutive days before carrier-priming. Alternatively, treatment with α_1 was performed by injecting mice i.p. with 0.2 ml saline containing 1 μg α_1, 3 days before carrier-priming. α_1 was a generous gift of Dr. A. L. Goldstein.

3. RESULTS AND DISCUSSION

Table I illustrates the results of four different experiments in which 3-, 15-, 18-, and 24-month-old mice were left uninjected or injected with 1 or 10 μg α_1 in 0.2 ml saline, over a period of five consecutive days before carrier-priming. The age control reference group (carrier-primed 3-month-old mice) has also been reported for each experiment. Each PFC number represents the expected anti-TNP response to the TNP–HRBC as induced in culture by the helper activity of 30×10^4 whole spleen cells from mice uninjected or injected with α_1 and carrier-primed at a given age. The number in parentheses is a factor by which the PFC number should be multiplied or divided to obtain the variation due to one standard error. Table I

TABLE I. Enhancement of Helper Cell Activity in Mice Injected with Synthetic Thymosin α_1 at Different Ages[a]

Experiment	Age (months)	Thymosin α_1 (μg)	Anti-TNP PFC/culture
1	3	none	158 (1.28)
	3	1	169 (1.13)
2	3	none	234 (1.08)
	15	none	60 (1.09)
	15	1	217 (1.09)
3	3	none	not done
	18	none	113 (1.08)
	18	1	179 (1.21)
	18	10	312 (1.08)
4	3	none	203 (1.10)
	24	none	28 (1.26)
	24	1	91 (1.07)
	24	10	92 (1.37)

[a] Mice of different ages were left uninjected or injected i.p. with 1 or 10 μg α_1 in 0.2 ml saline for five consecutive days before carrier-priming. The helper activity of HRBC-primed spleen cells from mice injected with α_1 or from uninjected age controls was titrated by adding graded numbers ($5–30 \times 10^4$) of carrier-primed spleen cells from mice of a given age to cultures containing 2×10^5 TNP–HRBC and 1×10^6 normal spleen cells from 3-month-old mice as described under Materials and Methods. Each PFC number in the last column represents the expected anti-TNP response to TNP–HRBC as induced in culture by the helper activity of 30×10^4 HRBC-primed spleen cells. The number in parentheses is a factor by which the PFC number should be multiplied or divided to obtain the variation due to one standard error.

indicates an age-related impairment in helper cell activity which is reduced by 75% at 15 months and by 87% at 24 months of age when referred to age control reference groups. The injection of 1–10 μg α_1 into 15-, 18-, or 24-month-old mice has been effective in recovering helper cell activity, while the injection of 1 μg α_1 at the age of 3 months had no sizable effects. The induced recovery of helper cell activity was complete in 15- and 18-month-old mice, while incomplete in 24-month-old mice regardless of the α_1 injected dose. The helper cell activity of these mice, however, was enhanced three- to fourfold as compared to α_1-uninjected mice of the same age.

Figure 1 illustrates the helper activity of carrier-primed, T-cell-enriched spleen cells from mice of 3 (young) and 24 (old) months of age. Old mice have been injected with 1 μg α_1 3 days before carrier-priming or left uninjected. Graded numbers (2.5, 5, 7.5, 10, 12.5 × 10^4) of T-cell-enriched spleen cells from carrier-primed young or old mice, untreated or treated with α_1, have been added to culture wells containing 1 × 10^6 normal spleen cells from young mice and 2 × 10^5 TNP–HRBC. As shown in the two separate experiments of Fig. 1, the age-related impairment of helper T-cell activity is repaired to some extent by a single injection of 1 μg α_1. This finding is in line with the results of Table I suggesting that full recovery of helper T-cell activity may not be achieved in 24-month-old mice after α_1 injection.

T-cell-enriched spleen cell populations from carrier-primed young or old mice have been dissected out by nylon wool filtration in passed (Th1) and adherent (Th2) T cells to investigate whether aging affects helper activity of either one or both cell types. Normal spleen cells (1 × 10^6) from young mice and 2 × 10^5 TNP–HRBC have been cultured alone or with 5 × 10^4 Th1 cells and/or 7.5 × 10^4 Th2 cells. Results in Table II demonstrate impairment of helper activity of both Th1 and Th2 cells from old mice when tested either separately or upon their recombination. The aging effect was also evident when one cell type from old mice was recombined with the other cell type from young mice.

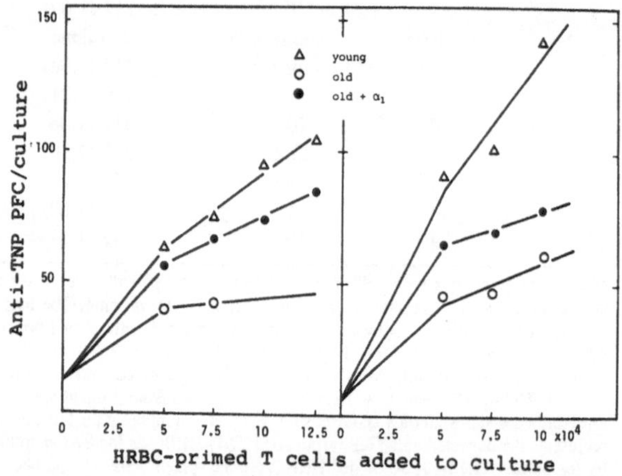

FIGURE 1. Effect of thymosin α_1 injection on helper activity of T-cell-enriched spleen cell populations. Old mice were left uninjected or injected with 1 μg α_1 in 0.2 ml saline 3 days before carrier-priming.

TABLE II. Effect of Aging on Helper Activity of Nylon Wool-Separated T Cells[a]

	Adherent T cells		
Passed T cells	None	Young	Old
None	29	143	83
Young	207	565	380
Old	155	210	180

[a] Splenic T cells from HRBC-primed young (3 month) or old (24 month) mice were separated by nylon wool filtration. Passed (5×10^4) and/or adherent (7.5×10^4) cells were added to cultures containing normal spleen cells (1×10^6) from young mice and 2×10^5 TNP-HRBC.

Helper activity of Th1 and Th2 cells from old mice injected with 1 μg α_1 3 days before carrier-priming was assessed as described above. Results in Table III confirm the age-related impairment of Th1 and Th2 cells and indicate that upon α_1 treatment, helper activity of both cell types recovers to a large extent when Th1 and Th2 cells are tested separately. However, following recombination, enhancement of helper activity is evident only when Th1 but not Th2 cells are derived from α_1-injected old mice, suggesting that α_1 also induces nylon-adherent suppressor T cells that counteract the helper activity of Th1 cells.

In conclusion, the present results confirm and extend previous findings (Frasca *et al.*, 1982) demonstrating that injection of immunodeficient aged mice with α_1 can restore helper T-cell activity. Furthermore, α_1 appears to act on the precursors of both Th1 and Th2 cells which synergize in the generation of helper activity (Tada *et al.*, 1978) a process that is impaired by aging (Doria *et al.*, 1980). In addition, our results indicate that α_1 also induces nylon-adherent suppressor T cells which negatively influence the helper activity of Th1 cells. However, the prevailing effect

TABLE III. Effect of Aging and α_1 Injection on Helper Activity of Nylon Wool-Separated T Cells[a]

	Adherent T cells			
Passed T cells	None	Young	Old	Old + α_1
None	54	231	111	276
Young	487	948	432	426
Old	105	216	171	108
Old + α_1	209	402	306	216

[a] Spleen cells from HRBC-primed young (3 month) or old (24 month) mice were separated by nylon wool filtration. Passed (5×10^4) and/or adherent (7.5×10^4) cells were added to cultures containing normal spleen cells (1×10^6) from young mice and 2×10^5 TNP–HRBC. Old mice were injected i.p. with 1 μg α_1 3 days before HRBC priming.

of α_1, at the doses used, appears to be enhancement of helper cell activity. Current studies in our laboratory investigate whether α_1 at different doses modulates the balance between helper and suppressor cell activity.

ACKNOWLEDGMENT. The authors' research reported herein was supported by an ENEA-EURATOM contract. This is publication No. 2131 from the EURATOM Biology Division.

REFERENCES

Ahmed, A., Wong, D. M., Thurman, G. B., Low, T. L. K., Goldstein, A. L., Sharkis, S. J., and Goldschneider, I., 1979, T-lymphocyte maturation: Cell surface markers and immune functions induced by T-lymphocyte cell-free products and thymosin polypeptides, *Ann. N.Y. Acad. Sci.* **332:**81.

Bach, J.-F., and Dardenne, M., 1972, Thymus dependency of rosette forming cells: Evidence for a circulating thymic hormone, *Transplant. Proc.* **4:**345.

Cunningham, A. J., and Szenberg, A., 1968, Further improvements in plaque technique for detecting single antibody-forming cells, *Immunology* **14:**559.

Doria, G., D'Agostaro, G., and Garavini, M., 1980, Age-dependent changes of B-cell reactivity and T cell–T cell interaction in the in vitro antibody response, *Cell. Immunol.* **53:**195.

Frasca, D., Garavini, M., and Doria, G., 1982, Recovery of T-cell functions in aged mice injected with synthetic thymosin α_1, *Cell. Immunol.* **72:**384.

Goldstein, A. L., 1978, Thymosin: Basic properties and clinical potential in the treatment of patients with immunodeficiency diseases and cancer, *Antibiot. Chemother.* **24:**47.

Goldstein, A. L., Low, T. L. K., McAdoo, M., McClure, J., Thurman, G. B., Rossio, J., Lai, C.-Y., Chang, D., Wang, S.-S., Harvey, C., Ramel, A. H., and Meienhofer, J., 1977, Thymosin α_1: Isolation and sequence analysis of an immunologically active thymic polypeptide, *Proc. Natl. Acad. Sci. USA* **72:**725.

Goldstein, A. L., Low, T. L. K., Zatz, M. M., Hall, N. R., and Naylor, T. H., 1983, Thymosins, *Clin. Immunol. Allergy* **3:**119.

Goldstein, G., 1975, The isolation of thymopoietin (thymin), *Ann. N.Y. Acad. Sci.* **249:**177.

Hooper, J. A., McDaniel, M. C., Thurman, G. B., Cohen, G. H., Schulof, R. S., and Goldstein, A. L., 1975, The purification and properties of bovine thymosin, *Ann. N. Y. Acad. Sci.* **249:**125.

Julius, M. H., Shimpson, E., and Herzenberg, L. A., 1973, A rapid method for the isolation of functional thymus-derived lymphocytes, *Eur. J. Immunol.* **3:**645.

Kay, M. M. B., 1978, Effect of age on T cell differentiation, *Fed. Proc.* **37:**1241.

Kettman, J., and Dutton, R. W., 1970, An in vitro primary immune response to 2,4,6-trinitrophenyl substituted erythrocytes: Response against carrier and hapten, *J. Immunol.* **104:**1558.

Kettman, J., and Dutton, R. W., 1971, Radioresistance of the enhancing effect of cells from carrier-immunized mice in an in vitro primary immune response, *Proc. Natl. Acad. Sci. USA* **68:**699.

Low, T. L. K., Thurman, G. B., Chincarini, C., McClure, J. E., Marshall, G. D., Hu, S.-K., and Goldstein, A. L., 1979, Current status of thymosin research: Evidence for the existence of a family of thymic factors that control T-cell maturation, *Ann. N.Y. Acad. Sci.* **332:**33.

Mishell, R. J., and Dutton, R. W., 1967, Immunization of dissociated spleen cells from cultures of normal mice, *J. Exp. Med.* **126:**423.

Osheroff, P. L., 1981, The effect of thymosin on glucocorticoid receptors in lymphoid cells, *Cell. Immunol.* **60:**376.

Rittenberg, M. B., and Pratt, K. L., 1969, Antitrinitrophenyl (TNP) plaque assay: Primary response of Balb/c mice to soluble and particulate immunogen, *Proc. Soc. Exp. Biol. Med.* **132:**575.

Schulof, R. S., and Goldstein, A. L., 1983, Clinical applications of thymosin and other thymic hormones, in: *Recent Advances in Clinical Immunology*, Vol. 3 (R. A. Thompson and N. R. Rose, eds.), pp. 243–286, Churchill Livingstone, New York.

Tada, T., Takemori, T., Okumura, K., Nonaka, M., and Tokuhisa, T., 1978, Two distinct types of helper T cells involved in the secondary antibody response: Independent and synergistic effects of Ia^- and Ia^+ helper T cells, *J. Exp. Med.* **147**:446.

Thurman, G. B., Rossio, J. L., and Goldstein, A. L., 1977, Thymosin induced enhancement of MIF production by peripheral blood lymphocytes of thymectomized guinea pigs, in: *Regulatory Mechanisms in Lymphocyte Activation* (D. O. Lucas, ed.), pp. 629–631, Academic Press, New York.

Wetzel, R., Heyneker, H. L., Goeddel, D. V., Jhurani, P., Shapiro, J., Crea, R., Low, T. L. K., McClure, J. E., Thurman, G. B., and Goldstein, A. L., 1980, Production of biologically active N^{α}-desacetyl thymosin α_1 through expression of a chemically synthesized gene, *Biochemistry* **19**:6096.

Zatz, M. M., Low, T. L. K., and Goldstein, A. L., 1982, Role of thymosin and other thymic hormones in T-cell differentiation, in: *Biological Responses in Cancer*, Vol. 1 (E. Mihich, ed.), pp. 219–247, Plenum Press, New York.

43

Treatment of the Diabetic, Autoimmune *db/db* Mouse with Thymosin

ANITA D. NOVITT, ROSEMARY P. FIORE, and HELEN R. STRAUSSER

1. INTRODUCTION

The subjects of the present study are the genetically diabetic mouse C57BL/KsJ-*db/db* (or *db/db*) and its nondiabetic heterologous strain C57BL/KsJ-*db/m* (*db/m*). These animals have a life span of approximately 6 to 8 months and exhibit the glycosuria (at least 2 mg/100 ml as adults), polyuria, polydipsia, and obesity (maximum weight of 45 g) expected in diabetes mellitus. Hummel *et al.* (1966) first described the *db/db* mouse from an endocrinological viewpoint. Pathological changes are limited in most cases to degranulation and degeneration of pancreatic β cells due to the short life span of the mouse (Heiniger and Dorey, 1981). The infiltration of lymphocytes around the pancreatic islet cells has been reported with juvenile-onset insulin-dependent diabetes (LeCompte, 1958; Gepts, 1965; Egeberg *et al.*, 1976; MacCuish and Irvine, 1975) and occasionally with the adult-onset type (LeCompte and Legg, 1972). It has been shown that this disease may be transferred using spleen cells of the animals with streptozotocin-induced diabetes (Buschard and Rygaard, 1977) or by injecting human diabetic peripheral blood lymphocytes into athymic nude mice (Buschard *et al.*, 1978). Handwerger *et al.* (1980) suggested that this ability to induce diabetes in nondiabetic subjects is due to the transfer of lymphocytes reactive to antigens on β cells of the islets of Langerhans. Cell-mediated immunity has also been shown to be directed at pancreatic antigens (Nerup

ANITA D. NOVITT, ROSEMARY P. FIORE, and HELEN R. STRAUSSER • Department of Zoology and Physiology, Rutgers University, Newark, New Jersey 07102.

et al., 1971, 1973a,b, 1974), insulin (Berson and Yalow, 1959a,b, 1965), proinsulin (Kumar and Miller, 1973), glucagon, and somatostatin (Bottazzo and Lendrum, 1976).

The connection between the thymus gland and diabetes mellitus has led us to utilize thymosin fraction 5 and two of its component peptides in an attempt to alleviate the symptoms of this disease and restore immune function in the *db/db* mouse.

2. MATERIALS AND METHODS

2.1. Animals

Genetically diabetic *db/db* mice and nondiabetic *db/m* animals were obtained from The Jackson Laboratory, Bar Harbor, Maine, at 5 weeks of age. They were maintained on a 12-hr light/dark cycle and given food (Purina Rat Chow) and water *ad libitum* for 6 weeks prior to experimentation.

2.2. Thymosin Administration

Thymosin fraction 5 and thymosin peptides α_1 and α_7 were donated by Dr. A. L. Goldstein, Department of Biochemistry, The George Washington University School of Medicine and Health Sciences, Washington, D.C., and Hoffman–La Roche Inc., Nutley, New Jersey. Animals received a total of 10 i.p. injections over a 20-day period. Fraction 5 was administered at 500 μg/injection while α_1 animals received 10 μg and the α_7 group received 1 μg/injection.

2.3. Urine Glucose Testing

Urine glucose levels were determined at time of injection using Diastix (Miles Laboratories, Elkhart, Ind.). All animals in experimental and control groups were fasted for 2 hr and then allowed to feed for 2 hr. Food was removed and glucose levels were tested 30 to 60 min postfeeding. The weight of food consumed during the 2-hr feeding period was determined to see if a correlation exists between urine glucose concentration and food intake or thymosin administration.

2.4. Indirect Cytotoxicity Assay

Animals were sacrificed by cervical dislocation. Single-cell suspensions from the spleens (Mishell and Shiigi, 1980) were prepared using Cedarlane Cytotoxicity Medium (Accurate Chemical, Westbury, N.Y.). The spleen suspension was layered onto Lymphocyte M separation medium (Cedarlane Laboratories, Ontario, Canada) and centrifuged at room temperature, 1200 rpm for 15 min. The lymphocyte layer was pipetted off and washed three times.

Cell concentration was adjusted to 6×10^6 cells/ml. Viability was determined by staining with ethidium bromide/acridine orange solution (Bottazzo and Lendrum, 1976).

The cytotoxicity assay was performed using arsanilate-conjugated anti-Lyt-1^+ and anti-Lyt-2^+ monoclonal antibodies (1 : 750) and rabbit antiarsanilate serum (1 : 500; Becton–Dickinson, Sunnyvale, Calif.). Optimal concentrations for the use of monoclonal antibodies and antiarsanilate were predetermined by optimal killing analysis.

2.5. Plaque-Forming Cell (PFC) Assay

Animals were sacrificed by cervical dislocation. Spleen single-cell suspensions were prepared (Mishell and Shiigi, 1980) in Eagle's minimum essential medium (GIBCO, Grand Island, N.Y.). The cells were washed twice by centrifuging for 10 min, 1200 rpm at 4°C. Cell concentration was adjusted to 4×10^6 cells/ml. The PFC assay was performed according to the Cunningham method as described by Lefkovits and Cosenza (1979).

3. RESULTS

3.1. Urine Glucose Testing

Urine glucose levels in all *db/db* animals were 2 mg/100 ml or higher prior to thymosin administratiron (Table I). Control *db/m* animals were negative for urine glucose at all times. After 10 injections of fraction 5, five of the eight *db/db* mice showed a drop in urine glucose levels to 1 or 0.5 mg/100 ml. The diabetic animals receiving α_1 or α_7 had a glucose value of 1 mg/100 ml.

There was an overall trend toward increasing food intake within the groups of diabetic animals receiving thymosin (Fig. 1). This was especially true of the α_7 mice which exhibited a 32% increase in food consumption. Fraction 5 and α_1 recipients increased food intake by 21 and 10%, respectively.

TABLE I. Urine Glucose Levels

		No. of animals with urine glucose ≥ 2 mg/100 ml	
	Treatment	Before treatment	After 10 injections
db/db	Fraction 5 (500 μg/injection)	8 of 8	3 of 8
	α_1 (10 μg/injection)	2 of 2	0 of 2
	α_7 (1 μg/injection)	3 of 3	0 of 3
	Saline	8 of 8	7 of 8
db/m	Fraction 5 (500 μg/injection)	0 of 8	0 of 8
	No treatment	0 of 7	0 of 7

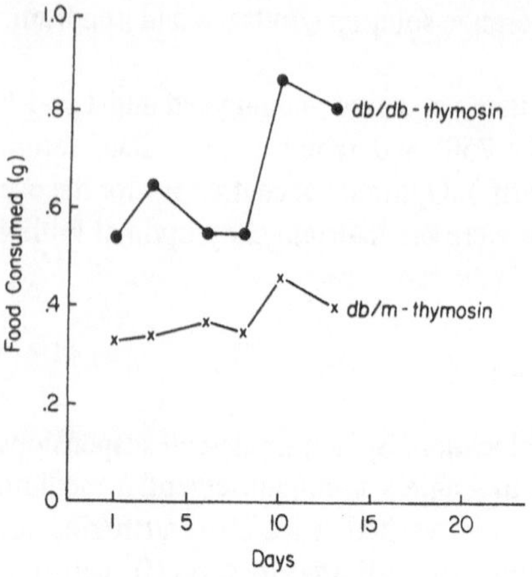

FIGURE 1. Food consumption in thymosin-treated mice.

3.2. Indirect Cytotoxicity Assay

Among *db/db* mice, fraction 5 caused a significant change in the T-helper subset (Fig. 2A): cytotoxicity increased from 11.6% to 16.3% following treatment. Suppressor cell increase was even more marked, rising from 4.3% to 8.9% (Fig.

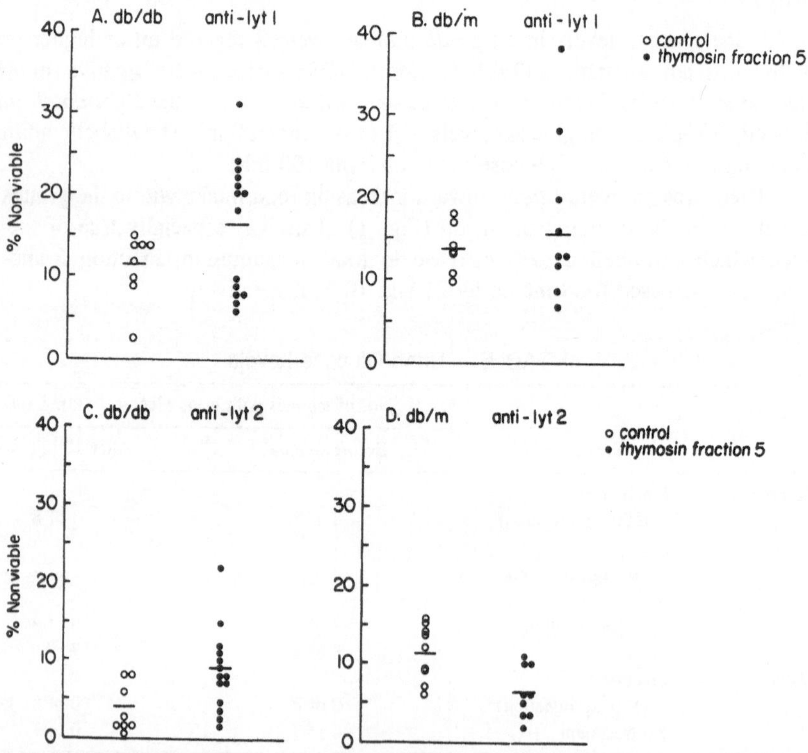

FIGURE 2. Cytotoxicity in treated and control animals.

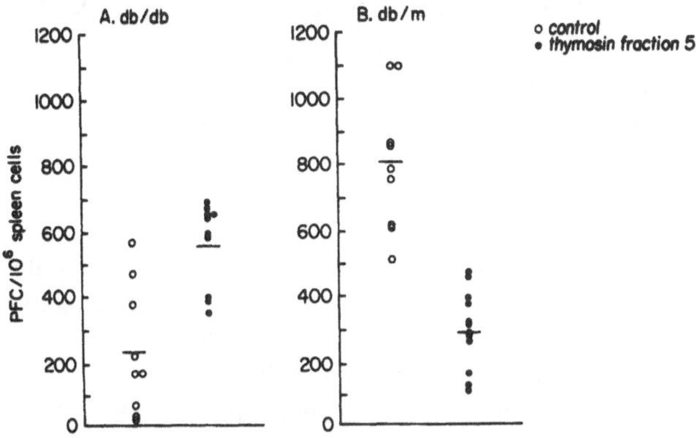

FIGURE 3. Plaque-forming cells in experimental and control groups.

2C). Helper cells among *db/m* mice rose from 14% to 17% with treatment (Fig. 2B), while cells bearing the Lyt-2$^+$ receptor dropped from 10.2% to 6.6% (Fig. 2D). The groups of mice given α_1 and α_7 were too few in number to provide significant results but, similar to the fraction 5 group, they both produced an increase in T-helper and -suppressor subsets among the *db/db* mice, with no effect on the *db/m* helper cells and a slight increase among *db/m* suppressor cells. Autoclaved thymosin had no effect on percentages of Lyt-1$^+$ and Lyt-2$^+$ cells.

3.3. PFC Assay

The ability of *db/db* mice to form hemolytic plaques to sheep red blood cells was markedly depressed as compared to *db/m* controls (Fig. 3). Plaques per 10^6 cells increased from an average of 240 to 550 in the thymosin-treated *db/db* spleens, while in the treated *db/m* animals, PFC dropped from an average of 800 to 290. Results with both α_1 and α_7, while not conclusive, showed a decrease in plaque-forming capacity.

4. DISCUSSION

The results obtained in this investigation indicate that diabetes in the *db/db* mouse is an autoimmune-like disease with symptoms similar to those of human adult-onset diabetes mellitus. As in classic autoimmune diseases, such as lupus erythematosus and rheumatoid arthritis, the autoimmune phenomenon in the *db/db* is characterized by depressed numbers of T-suppressor cells (cells bearing Lyt-2$^+$ receptors). Fernandes *et al.* (1978) have demonstrated that these mice have a diminished ability to reject skin grafts and to generate cytotoxic cells as well as decreased thymic weights and DNA synthesis in the thymus. In addition, thymic pathology and lymphocyte dysfunction were shown by Dardenne *et al.* (1983), evidenced by an accelerated age-dependent decline in histological structure and a reduced number of cells with serum thymic factor (FTS) receptors in the thymus.

Abnormalities in the immune function of lupus mice (Koike *et al.*, 1979; Krakauer *et al.*, 1980) and spontaneously hypertensive rats (Takeichi *et al.*, 1980) have been attributed to a natural thymocytotoxic antibody directed preferentially against T-suppressor cells. If a similar autoantibody is generated in the *db/db* mouse, an agent useful in treating diabetes should possess not only a hypoglycemic action, but should be immunologically therapeutic. Increasing suppressor cells or function could, theoretically, block the action of those cells or antibodies which inhibit insulin release. Similarly, increasing T-helper cells could increase the anti-idiotype antibody and thereby inhibit the anti-β cell or anti-insulin antibody. Both T suppressors and helpers could also function to alleviate immune blocking of insulin receptors, if receptor blocking is involved in this disease.

Data presented here indicate that thymosin increases both T suppressors and helpers in the *db/db* mouse. Although the evidence obtained on α_1 and α_7 was inconclusive due to the small number of animals treated, it did appear that all thymosin fractions increased T-helper and -suppressor populations. Concurrently, glycosuria was diminished and food intake was increased. The extremely small amounts of the α_1 and α_7 fractions used as compared to thymosin fraction 5 should be noted. It would appear that dosages necessary to produce results decline with isolation of active peptides of thymosin.

As the *db/db* mouse has been known as the prototype of insulin-resistant diabetes, due to pathological thymic function, it would appear possible that treatment with a thymic extract such as thymosin could increase the number of T cells or B-"suppressor" or anti-idiotypic clones and thereby ameliorate the disease.

5. SUMMARY

The genetically diabetic C57BL/KsJ-*db/db* mouse illustrates the autoimmune phenomenon present in insulin-resistant diabetes mellitus. In this study, fraction 5, a hormone of the bovine thymus gland and a stimulator of T-cell subpopulations, was used *in vivo* to determine its effects on the splenic immune system and the diabetic syndrome. Preliminary experiments were also performed using two thymosin peptides, α_1 and α_7. Ten injections of these substances given i.p. over a 20-day period reduced urine glucose levels in 10 of 13 of the diabetic animals while food intake was increased an average of 33%. Indirect cytotoxicity testing showed an increase in T-helper (Lyt-1$^+$) and T-suppressor (Lyt-2$^+$) cell numbers in diabetic mice, but not in the nondiabetic controls. In addition, PFC increased in the diabetic mice. Thymosin extracts appear to be inducers of T-helper and -suppressor cells and function as hypoglycemic agents in these diabetic autoimmune mice.

ACKNOWLEDGMENT. The authors' research reported herein was supported in part by the Guggenheim Foundation.

REFERENCES

Berson, S. A., and Yalow, R. S., 1959a, Recent studies on insulin-binding antibodies, *Ann. N.Y. Acad. Sci.* **82**:338–344.

Berson, S. A., and Yalow, R. S., 1959b, Species specificity of human anti-beef, pork insulin serum, *J. Clin. Invest.* **38**:2017–2025.

Berson, S. A., and Yalow, R. S., 1965, Some current controversies in diabetes research, *Diabetes* **14**:549–572.

Bottazzo, G. F., and Lendrum, R., 1976, Separate autoantibodies to human pancreatic glucagon and somatostatin cells, *Lancet* **2**:873–876.

Buschard, K., and Rygaard, J., 1977, Passive tranafer of streptozotocin-induced diabetes mellitus with spleen cells, *Acta Pathol. Microbiol. Scand. Sect. C* **85**:469–472.

Buschard, K., Madsbad, S., and Rygaard, J., 1978, Passive transfer of diabetes mellitus from man to mouse, *Lancet* **1**:908–910.

Dardenne, M., Savino, W., Gastinel, L., Nabarra, B., and Bach, J.-F., 1983, Thymic dysfunction in the mutant diabetic (db/db) mouse, *J. Immunol.* **130**:1195.

Egeberg, J., Junker, K., Kromann, H., and Nerup, J., 1976, Autoimmune insulitis: Pathological findings in experimental animal models and juvenile diabetes mellitus, *Acta Endocrinol. Suppl.* **205**:129–137.

Fernandes, G., Handwerger, B. S., Yunis, E. J., and Brown, D. M., 1978, Immune response in the mutant diabetic C57BL/Ks-db$^+$ mouse, *J. Clin. Invest.* **61**:243.

Gepts, W., 1965, Pathologic anatomy of the pancreas in juvenile diabetes mellitus, *Diabetes* **14**:619–633.

Handwerger, B. S., Fernandes, G., and Brown, D. M., 1980, Immune and autoimmune aspects of diabetes mellitus, *Hum. Pathol.* **11**:338–352.

Heiniger, H., and Dorey, H. H. (eds.), 1981, *Handbook on Genetically Standardized Jax Mice*, pp. 5.11–5.13, The Jackson Laboratory, Bar Harbor, Maine.

Hummel, K. P., Dickie, M. M., and Coleman, D. L., 1966, Diabetes, a new mutation in the mouse, *Science* **153**:1127.

Koike, T., Kobayashi, S., Yoshiki, T., Itoh, T., and Shirai, T., 1979, Differential sensitivity of functional subsets of T cells to the cytotoxicity of natural T-lymphocytic autoantibody of systemic lupus erythematosis, *Arthritis Rheum.* **22**:123–129.

Krakauer, R. S., Clough, J. D., Alexander, T., Sundeen, J., and Sauder, D. N., 1980, Suppressor cell defect in SLE: Relationship to DNA binding, *J. Clin. Exp. Immunol.* **40**:72–76.

Kumar, D., and Miller, L. U., 1973, Proinsulin-specific antibodies in human serum, *Diabetes* **22**:361–366.

LeCompte, P. M., 1958, "Insulitis" in early juvenile diabetes, *Arch. Pathol.* **66**:450–457.

LeCompte, P. M., and Legg, M. A., 1972, Insulitis (lymphocytic infiltration of pancreatic islets) in late-onset diabetes, *Diabetes* **21**:762–769.

Lefkovits, I., and Cosenza, H., 1979, *Immunological Methods*, p. 277, Academic Press, New York.

MacCuish, A. C., and Irvine, W. J., 1975, Autoimmunological aspects of diabetes mellitus, *Clin. Endocrinol. Metab.* **4**:435–471.

Mishell, B. B., and Shiigi, S. M. (eds.), 1980, *Selected Methods in Cellular Immunology*, pp. 1–21, Freeman, San Francisco.

Nerup, J., Andersen, O., Bendixin, G., Egeberg, J., and Poulson, J. E., 1971, Antipancreatic cellular hypersensitivity in diabetes mellitus, *Diabetes* **20**:424–427.

Nerup, J., Andersen, O., Bendixin, G., Egeberg, J., and Poulson, J. E., 1973a, Antipancreatic cellular hypersensitivity in diabetes mellitus: Antigenic activity of fetal calf pancreas and correlation with clinical type of diabetes, *Acta Allergol.* **28**:223–230.

Nerup, J., Andersen, O., Bendixin, G., Egeberg, J., Vilien, M., and Westrup, M., 1973b, Anti-pancreatic, cellular hypersensitiivty in diabetes mellitus: Experimental induction of anti-pancreatic, cellular hypersensitivity and associated morphological β-cell changes in the rat, *Acta Allergol.* **28**:231–249.

Nerup, J., Andersen, O., Bendixin, G., Egeberg, J., Gunnarson, R., Kromann, G., and Poulson, J., 1974, Cell mediated immunity in diabetes mellitus, *Proc. R. Soc. Med.* **67**:506–513.

Takeichi, N., Suzuki, K., Okayasu, T., and Kobayashi, H., 1980, Immunological suppression in spontaneously hypertensive rats, *Clin. Exp. Immunol.* **40**:120–126.

Role of Thymosin Fraction 5 in Host Defenses to Leukemia in Nutritionally Stressed Mice

RONALD ROSS WATSON, THOMAS M. PETRO, and GEORGE L. MANDERINO

1. INTRODUCTION

Lymphoid cells are continually generated throughout life and are susceptible to changes in nutritional intake. The thymus gland seems to be very sensitive to nutritional stresses which has important consequences for immunoregulation, thymic hormone production, disease resistance, and health (Petro and Watson, 1982). The relationship between nutritional stresses, thymus functions, and immune defenses has recently been reviewed in detail elsewhere (Manderino and Watson, 1984) and is summarized briefly here. Clearly, thymus growth and T-lymphocyte functions can be enhanced or suppressed by high (Watson, 1984) or low (Manderino and Watson, 1983) intakes of various nutrients. For example, high vitamin A or retinoid intakes enhance thymus growth and cellular immune functions (Watson, 1984). The effects and/or role of thymic hormones on these immunological changes is unclear. Therefore, studies of thymosin fraction 5 (TF5) and its effects on resis-

RONALD ROSS WATSON • Department of Family and Community Medicine, University of Arizona Medical School, Tucson, Arizona 85724. THOMAS M. PETRO • Department of Food and Nutrition, Purdue University, West Lafayette, Indiana 47907. *Present address:* Microbial Biochemistry Branch, Food and Drug Administration, Cincinnati, Ohio 45266. GEORGE L. MANDERINO • Abbott Laboratories, North Chicago, Illinois 60064.

tance to leukemia growth in malnourished mice were undertaken and are detailed herein.

2. MALNUTRITION AND THYMUS FUNCTION

Smythe *et al.* (1971) examined thymuses from children who died with kwashiorkor and with marasmus. The abnormalities included acute or chronic thymic involution characterized by depletion of thymocytes with loss of distinction between the cortex and medullary areas, narrow lobules, and replacement of Hassall's corpuscles with fibrous tissue. Despite the extreme atrophy of the thymus induced by malnutrition, regeneration to normal thymic weight and cellularity has been achieved by feeding malnourished mice a standard high-protein diet (Suskind, 1977). This suggested that thymic atrophy induced by malnutrition does not lead to premature thymic involution with its implication of permanence. However, renutrition of children showed short-term increases in cellular immune functions with subsequent decline, even though the children remained well nourished (McMurray *et al.*, 1981).

Lymphocyte subpopulations in severely malnourished individuals have demonstrated a reduction of T lymphocytes identified by their ability to bind spontaneously to sheep red blood cells in a rosette configuration (McMurray *et al.*, 1981). Many possible mechanisms may contribute to the reduction in T-lymphocyte numbers commonly observed in protein-energy malnutrition. Thymic atrophy and depletion of thymus-dependent areas of peripheral lymphoid tissues may be the most significant factors. The pathogenesis of these histological changes has not been elucidated. Nutritional deficiencies may, however, directly affect lymphocyte differentiation by inhibiting the metabolism of rapidly dividing thymocytes or by suppressing the production of thymic epithelial cell products essential for normal lymphocyte differentiation.

It has been reported that the reduction of T lymphocytes in malnourished patients correlated with an increase of null cells (Chandra, 1979). Chandra (1979) has reported a marked increase in levels of terminal deoxynucleotidyl transferase (TdT) in lymphocyte extracts obtained from severely malnourished children. The increase in TdT in these malnourished children correlated with an increase in the incidence of circulating null cells and a decrease in T lymphocytes. These findings suggest that reduced levels of T lymphocytes with a concomitant increase in null cell levels in malnourished children may be partly responsible for immunodeficiencies observed in this group. A recent study of the levels of TdT in the circulation of certain hospitalized, malnourished adults in the United States has confirmed these findings (Duncan *et al.*, 1983).

Jackson and Zaman (1980) have shown that thymopoietin mixed with peripheral blood lymphocytes of malnourished children increases the number of rosetting cells *in vitro*. In addition, TF5 has been shown to induce rosette formation in peripheral blood lymphocytes of children with kwashiorkor (Olusi *et al.*, 1980). In animal studies, thymic factor activity was significantly reduced in rats fed calorie-restricted diets (Heresi and Chandra, 1980). Diminished thymic hormone activity has also been observed in the sera of malnourished children.

3. THYMOSIN RESTORATION OF RESISTANCE TO LEUKEMIA CELL GROWTH IN PROTEIN-MALNOURISHED MICE

We have recently performed experiments based on the assumption that nutritional stresses cause altered T-cell function via reduction of thymic hormone production. Young Balb/c female mice (3 weeks of age) were fed any of three isocaloric diets containing varying amounts of protein (0, 4, and 20% casein) which were models of severe protein malnutrition, moderate protein malnutrition, and normal nutrition.

We obtained female Balb/c mice 3 weeks of age from Harlan–Sprague–Dawley (Indianapolis, Ind.). After consuming the respective diets for 6 weeks, each mouse was given i.p. injections of either 100 μg of TF5 (Hoffman–LaRoche, Nutley, N.J.) or phosphate-buffered saline (PBS) every other day for 2 weeks. Immediately after the last thymosin or PBS treatment, the mice were injected i.p. with 10^7 L1210 cells. Mice were sacrificed prior to, 7, and 14 days postinoculation with the L1210 cells. The number of peritoneal exudate cells (PEC), spleen cell cytotoxicity against L1210 cells, and mitogenesis of spleen cells induced by PHA were assayed. Each datum consisted of the mean values from at least five mice.

TF5 consistently showed an immunopotentiating effect in the severely protein-deficient, immunosuppressed mice (Chien *et al.*, 1983) as measured by mitogen-induced lymphocyte transformation. Thus, severe protein malnourishment is immunosuppressive and TF5 partially restores the suppressed responses.

Moderate or mild protein malnutrition (4% protein) does not suppress these responses; rather, it may enhance them on a per cell basis. Lymphocyte mitogenesis and leukemia growth in the moderately protein-malnourished mice injected with TF5 were also altered. Both PHA- and LPS-induced lymphocyte transformations in the moderately protein-malnourished mice injected with TF5 were elevated as

TABLE I. Thymosin Treatment and Resistance of Protein-Malnourished Mice against L1210 Leukemia Cell Growth[a]

Dietary protein level (%)	Treatment type (i.p. injection)	Time after last treatment (days)	PEC ($\times 10^6$)[b] at following times after injection of L1210 cells		
			Day 0	Day 7	Day 14
20	PBS	1	7.2 ± 1.1[c]	398.3 ± 105.4	14.5 ± 2.7
20	Thymosin	1	11.3 ± 1.6[d]	377.3 ± 68.7	15.0 ± 6.0
4	PBS	1	1.2 ± 0.8[d]	166.2 ± 41.2[d]	19.2 ± 9.5
4	Thymosin	1	0.2 ± 0.1[d]	4.27 ± 24.4[d]	15.0 ± 2.6
20	PBS	7		562.0 ± 43.3	18.9 ± 3.4
20	Thymosin	7		353.4 ± 64.5[d]	19.2 ± 6.1
4	PBS	7		450.7 ± 113.5	11.3 ± 1.6[d]
4	Thymosin	7		300.9 ± 73.9[d]	11.7 ± 1.9[d]

[a] Modified from Petro and Watson (1982).
[b] Challenged with 1×10^7 L1210 mouse leukemia cells.
[c] Mean ± S.E.
[d] Significantly different from PBS-treated mice fed the control diet: $p < 0.05$.

compared with mice fed normally. In contrast, TF5 did not increase the immune responsiveness in the mice fed normally, but had a suppressive effect (Petro and Watson, 1982). Moderately protein-malnourished mice injected with L1210 cells exhibited reduced growth of the cells (Table I). This reduction in the moderately malnourished mice. The nutritionally normal mice injected with TF-5 showed no similar enhancement (Petro and Watson, 1982).

4. HIGH- AND LOW-FAT INTAKES: IMMUNE FUNCTIONS AND RESISTANCE TO LEUKEMIA AFTER TF5 INJECTION

A similar experiment has been performed with different dietary regimens, including a dietary excess. Young Balb/c female mice were fed three different fat diets [low (0% corn oil), normal (5% corn oil), or high (20% corn oil) fat] beginning at the age of 4 weeks. Mice from each dietary group were injected i.p. with either TF5 or PBS as described above.

There was no significant difference between the TF5-treated and the PBS-treated mice within each dietary group as to the number of splenic lymphocytes isolated. The thymosin-treated groups for each time point showed no significant difference when controls were compared to thymosin-treated mice fed the fat-free or high-fat diet. After dietary stress and TF5 injection, PHA mitogenesis showed no significant effect of TF5 injection except in mice fed the 0% fat diet, where there was a statistically significant decline. Otherwise, the 20% fat and 0% fat diets did not change the PHA mitogenesis compared to mice fed the control (5% fat) diet.

There was a significant increase 7 and 14 days postinjection in cell-mediated cytotoxicity (CMC) in all groups of mice. At day 0 the mice fed the fat-free diet

TABLE II. Effect of Thymosin Treatment on Resistance against L1210 Leukemia Cell Growth of Mice Fed High- and Low-Fat Diets

Dietary fat level	Treatment type	PEC ($\times 10^6$)[a] at following times after injection of L1210 cells: Day 0	Day 7	Day 14
20	PBS	—	374.2 ± 88.1[c]	12.5 ± 2.0
20	Thymosin	—	15.7 ± 4.5[c,d]	8.3 ± 1.1[c]
5	PBS	7.2 ± 1.1[b]	267.5 ± 36.8	16.0 ± 2.2
5	Thymosin	11.3 ± 1.6[c]	194.2 ± 110.3	9.2 ± 0.7[c]
0	PBS	—	173.3 ± 57.5[c]	9.0 ± 1.0[c]
0	Thymosin	—	95.2 ± 36.5[c,d]	11.0 ± 1.0[c,d]

[a] Challenged with 1×10^7 L1210 mouse leukemia cells at day 0 by i.p. injection.
[b] Mean ± S.E. Each data point represents at least five mice. Mice aged 3 weeks were fed diets for 6 weeks, then treated wtih PBS or TF5 (0.7 mg) i.p. every other day for 2 weeks, prior to injection of L1210 cells.
[c] Significantly different from PBS-treated mice fed control diet (5% fat): $p < 0.05$.
[d] Significantly different from PBS-treated mice fed same diet (either 0 or 20% fat): $p < 0.05$.

had significantly lower CMC activity than control mice. The TF5-treated mice, fed the 5% fat diet, had a significant reduction in CMC activity only at day 14.

Seven days after injection of leukemia cells the number of leukemia cells isolated from the peritoneal cavity all increased significantly (Table II). However, the increase was significantly less in the mice injected with TF5 and fed the high-fat and fat-free diets ($p < 0.05$). A similar reduction was indicated in the mice fed the 5% fat diet and TF5-treated. In the PBS-injected mice, the high-fat diet resulted in increased growth of leukemia cells while the fat-free diet reduced it compared to controls. By day 14, all mice had few leukemia cells. The number isolated were significantly smaller in all groups of TF5-treated mice than in PBS-treated control mice fed the control (5% fat) diet. Similar results were observed (Table I) in mice stressed with a 4% protein diet (Petro and Watson, 1982).

5. CONCLUSIONS

There is an enhancing effect of a high-fat diet on growth of leukemia cells. This has been observed by others who have found increased tumor growth in animals fed high-fat diets. A possible explanation is an alteration in some key immune function. Our data do not significantly explain the suppressed resistance to leukemia growth as we did not find large decreases in immunological functions we measured. There are several other important host defenses which could have been affected including phagocytosis which were not studied. On the other hand, mice fed the 0% fat diet and injected with PBS had a somewhat enhanced resistance to leukemia cell growth. This change also did not correlate with any enhancement of the immune parameters we measured: spleen cell number, PHA mitogenesis, and CMC activity. However, we have recently shown that a low-fat diet enhances serum corticosteroid levels (Chien *et al.*, 1983). High corticosteroid levels are suppressive for leukemia growth *in vitro* and *in vivo*. Elevated levels observed in mice fed similar diets are a partial explanation for enhanced resistance to leukemia cell growth in nutritionally stressed mice. TF5 produced additional resistance to leukemia cell growth while reducing serum corticosteroid levels moderately (Chien *et al.*, 1983).

TF5 injection had a significant enhancing effect on resistance to leukemia cell growth at day 7 and 14 postinjection of the leukemia cells in mice which were nutritionally stressed with high- or low-fat diets (Table II), but not the mice fed the control (5% fat) diet. Thus, these data confirm our previous observations. Our moderately protein-deficient diet produced nutritional stress and TF5 injection was applied to overcome alterations to thymic function and disease resistance. We found that TF5 greatly stimulated resistance to *Listeria monocytogenes* and to L1210 cell growth in the moderately protein-stressed mice but not in nutritionally normal animals (Petro *et al.*, 1982; Petro and Watson, 1982). Although a partial explanation could have been enhanced lymphocyte mitogenesis to PHA, the greatly reduced number of hyperactive lymphocytes makes this changed immune function probably only a minor component of enhanced resistance. It is clear that TF5 injection in nutritionally stressed animals enhanced resistance but the key immunological change promoting this has not been identified. The importance of TF5 to enhanced resistance and some immune functions needs to be defined in humans, particularly cancer patients who are often nutritionally stressed. In addition, the effects of various immunostimulating therapies, such as high retinoid intakes, on cancer resistance and thymus function need to be understood (Watson, 1984).

ACKNOWLEDGMENTS. Research which stimulated this review was supported in part by the National Livestock and Meat Board, and the Phi Beta Psi Sorority.

REFERENCES

Chandra, R. K., 1979, T and B lymphocyte subpopulations and leukocyte terminal deoxynucleotidyl transferase in energy-protein undernutrition, *Acta Paediatr. Scand* **68:**841.

Chien, G., Watson, R. R., and Chung, C., 1983, Thymosin treatment: serum corticosterone and lymphocyte mitogenesis in moderately and severely protein-malnourished mice, *J. Nutr.* **113:**483–494.

Duncan, J. L., Moldawer, L. L., Manderino, G. L., Bistrian, B. R., and Blackburn, G. L., 1984, Terminal transferase in adult malnutrition, *Nutr. Res.* (in press).

Heresi, G., and Chandra, R. K., 1980, Effects of severe calorie restriction on thymic factor activity and lymphocyte stimulation responses in rats, *J. Nutr.* **110:**1888.

Jackson, T. M., and Zaman, S. N., 1980, The in vitro effect of the thymic factor thymopoietin on a subpopulation of lymphocytes from severely malnourished children, *Clin. Exp. Immunol.* **39:**717.

McMurray, D. N., Watson, R. R., and Reyes, M. A., 1981, Effect of renutrition in humoral and cell-mediated immunity in severely malnourished children, *Am. J. Clin. Nutr.* **34:**2117–2126.

Manderino, G. L., and Watson, R. R. (eds.), 1984, Role of the thymus gland and thymosin in nutritionally mediated immunosuppression, in: *Malnutrition, Disease Resistance and Immune Function*, pp. 285–298, Dekker, New York.

Olusi, S. O., Thurman, G. B., and Goldstein, A. L., 1980, Effect of thymosin on T-lymphocyte rosette formation in children with kwashiorkor, *Clin. Immunol. Immunopathol.* **15:**687–691.

Petro, T. M., and Watson, R. R., 1982, Resistance to L-1210 mouse leukemia cells in moderately protein malnourished BALB/c, *Cancer Res.* **42:**2139–2145.

Petro, T. M., Chien, G., and Watson, R. R., 1982, Alteration of cell-mediated immunity to *Listeria monocytogenes* in protein-malnourished mice treated with thymosin fraction V, *Infect. Immun.* **37:**601–608.

Smythe, P.M., Brereton-Stile, G. G., Grace, H. J., Mafoyane, A., Schorland, M., Cocvadia, H. M., Loening, W. E. K., Purent, M. A., and Vos, G. H., 1971, Thymolymphatic deficiency and depression of cell-mediated immunity in protein-calorie malnutrition, *Lancet* **2:939.**

Suskind, R. M., 1977, *Malnutrition and the Immune Response*, pp. 135–139, Raven Press, New York.

Watson, R. R. (ed.), 1984, Regulation of immunological resistance to cancer by beta carotene and retinoids, in: *Malnutrition, Disease Resistance and Immune Function*, pp. 345–355, Dekker, New York.

Tumor Growth, Interleukins, and Immune Complexes

THANJAVUR RAVIKUMAR, GLENN STEELE, Jr., MARY RODRICK, JEANNE MARRAZO, DONALD ROSS, and STEPHEN LAHEY

1. INTRODUCTION

Using an antigen-nonspecific assay for circulating immune complexes (PEG-CIC) that detects immune complexes in regions of both antigen and antibody excess, we have shown in our recent studies that changes in serial PEG-CIC levels correlate with changes in tumor volume in a variety of human and animal tumors (Jerry *et al.*, 1976; Rodrick *et al.*, 1983). Furthermore, the elevations in CIC levels were shown to precede decreases in T-cell mitogenic responses to PHA. The precise mechanism of immune modulation in tumor-bearing hosts still remains unclear although regulatory interactions between host immunocytes are undoubtedly involved. As interleukins (IL) play a central role in the mediation and amplification of immune response, and as our earlier data pointed to a temporal relationship between changes in PEG-CIC and tumor growth, we postulated that immune complexes by perturbing intercellular interactions might decrease IL generation.

The current investigation was, therefore, undertaken to serially monitor PEG-CIC levels and the capacity to generate IL-1 and IL-2 by the tumor-bearing hosts in a syngeneic rat colon cancer model at successive stages of tumor progression. Subsequently, experiments were performed to delineate the mechanism of inhibition of IL generation. The animals bearing progressively growing tumors exhibited a decrease in the ability to elaborate IL-1 and IL-2, and this correlated with serial increase in CIC levels. Preliminary evidence is also presented that suggests a causal relationship between CIC and inhibition of IL-1 generation.

THANJAVUR RAVIKUMAR, GLENN STEELE, Jr., MARY RODRICK, JEANNE MARRAZO, DONALD ROSS, and STEPHEN LAHEY • Department of Surgery, Brigham & Women's Hospital and Dana-Farber Cancer Institute, Harvard Medical School, Boston, Massachusetts 02115.

2. MATERIALS AND METHODS

Male Wistar/Furth (W/Fu) rats were used at 8–10 weeks of age. Rats were fed normal pellet diets and given water *ad libitum*. 2×10^6 live DMH-W163 cells (DMH-W163 is a 1,2-dimethylhydrazine-induced colon adenocarcinoma which is being maintained by continuous *in vivo* passage) suspended in sterile phosphate-buffered saline (PBS) were injected s.c. into the right leg of experimental rats. Normal age-matched control rats were given a similar volume of PBS alone. Tumor growth was measured twice weekly with Vernier calipers and the volume determined in cubic centimeters. Experimental and control rats were bled before isografting and weekly thereafter. Serum was frozen and stored at −20°C until tested. CIC in serial serum samples were quantitated using the 3.75% polyethylene glycol 6000 assay (PEG-CIC), and the results were expressed as $\Delta OD_{450} \times 10^3$ (Rodrick *et al.*, 1983).

Peripheral blood mononuclear cells (PBMC) were isolated by Ficoll–Hypaque separation of heparinized blood. For IL-1 generation experiments, adherent mononuclear cells (AMC; 1×10^6 cells/well) were cultured with lipopolysaccharide (LPS; 5 μg/well) for 24 hr. IL-2 was generated by coculturing PBMC (final concentration 1×10^6 cells/ml) with Con A (final concentration 2.5 μg/ml) for 48 hr. To minimize interassay variability, the IL-1 and IL-2 thus generated were frozen at −70°C until the assay was performed at the conclusion of the experiment. IL-1 assay was based on augmentation of PHA-stimulated C3H/HeJ mouse thymocyte proliferation as measured by [^{3}H]-TdR incorporation. IL-2 was tested by the microassay based on [^{3}H]-TdR incorporation by the IL-2-dependent cell line CTLL-2.

In the subsequent experiment, peripheral AMC, pooled from six normal rats, were preincubated with serial fourfold dilutions of either normal rat sera (NRS) or sera from rats bearing 1.5-to-2-cm DMH-W163 tumors (tumor-bearer sera; TBS). IL-1 generation and assay were carried out as described above.

3. RESULTS

Tumor Growth in Isografted Animals. The injection of 2×10^6 live DMH-W163 cells resulted in a progressive tumor growth after a latent period of 1 week. Regional metastases were detected during the fourth week, and all tumor-bearing rats died within a week after regional spread of tumor.

Serial PEG-CIC Levels during Tumor Growth. While there was no difference in the PEG-CIC levels between isografted and control rats before and 1 week after isografting, a moderate but not significant rise in CIC levels occurred in tumor-bearers at week 2 (Table I). At week 3, when the mean tumor volume was 7.5 cm^3, there was a significant rise in the CIC levels in tumor-bearing animals (tumor bearers: 431 ± 115; controls: 277 ± 35; $p < 0.05$). Large tumor burden and regional spread of tumor resulted in a relative decrease of CIC; the levels were, however, still higher than the control levels.

TABLE I. CIC Levels and Interleukin Generation during Tumor Growth

Weeks after isograft	CIC levels (ΔOD_{450}) × 10^3 (mean ± S.E.M.)		Mean tumor volume (cm^3)	IL generation: % of control in tumor-bearers	
	Control	Tumor-bearers		IL-1[a]	IL-2[b]
Baseline	246 ± 32	193 ± 23	—	100%	100%
1	259 ± 18	258 ± 14	—	51%	29%
2	244 ± 48	358 ± 53[c]	0.4	56%	32%
3	277 ± 35	431 ± 115[d]	7.5	16.6%	23%
4	277 ± 35	342 ± 128	32	0.9%	10%

[a] The baseline levels of IL-1 were similar for control rats (n = 5) and tumor-bearers (n = 15) and were arbitrarily assigned an activity of 10 units/ml. The weekly IL-1 levels were compared against this baseline level of IL-1.
[b] Fifty percent of the maximum [^{3}H]-TdR incorporation in the baseline samples was obtained by probit analysis and was given an arbitrary value of 1 unit/ml. The value of the weekly sample divided by that of baseline value gives the relative IL-2 concentration.
[c] $p < 0.1$.
[d] $p < 0.05$.

IL and Tumor Growth. The units of IL-1 activity were determined as described by Mizel *et al.* (1978). IL-2 responses were analyzed according to the method described by Gilles *et al.* (1978). The baseline level of IL was used as the standard against which serial IL levels were compared. The control rats showed a decline during the successive weeks due to immunosuppressive effects of anesthesia, repetitive bleeding, etc. To account for these variables, the levels of IL-1 and IL-2 are expressed as the percent in tumor-bearers when compared with control rats each

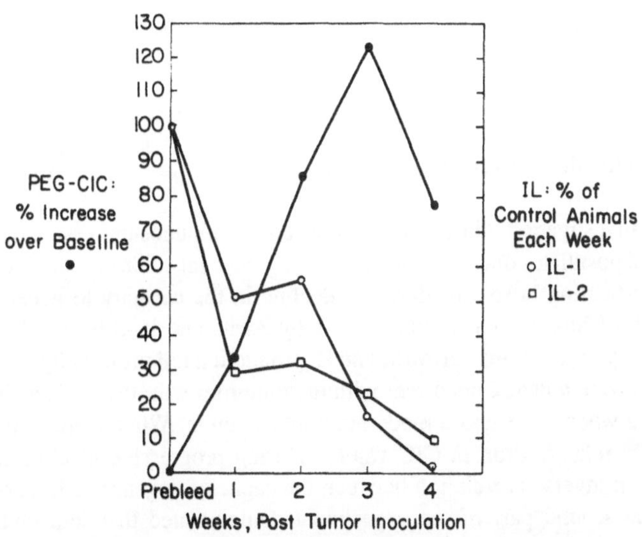

FIGURE 1. Correlation of PEG-CIC and IL production.

TABLE II. Comparison of Inhibition of IL-1 Generation of Normal AMC by Normal and Tumor-Bearer Rat Sera

Serum dilution used for incubation with AMC	Units of IL-1[a] activity		% IL-1 generation by incubation with TBS, compared to NRS
	NRS	TBS	
1 : 4	3.7	0.9	24.3
1 : 16	2.7	1.0	37
1 : 64	3.0	< 0.1	< 3

[a] Background counts subtracted. Maximum cpm for the assay (2245) taken as 100%. Data plotted on probit plot and EC_{50} calculated for each assay and expressed as IL-1 units.

week (Table I). The ability of AMC to elaborate IL-1 and PBMC to elaborate IL-2 decreased progressively in tumor-bearing animals. The fall was greatest when the tumors were large and metastatic (IL-1: 0.9%, IL-2: 10% of control animals in tumor-bearing rats; $p < 0.01$, analysis of variance).

Correlation of PEG-CIC and IL Production (Fig. 1). During local tumor growth, there was a good negative correlation ($r = -0.768$) between serial PEG-CIC levels and IL-1 and IL-2 generation by AMC and PBMC, respectively.

The Comparative Role of NRS and TBS in the Inhibition of IL-1 Generation (Table II). Normal AMC from peripheral blood of six W/Fu rats were plated in the same cell concentration as the previous experiments but were incubated with sera from either normal or tumor-bearing rats. TBS were obtained from rats harboring 1.5- to 2-cm tumors, for the peak CIC level correlated to moderate tumor volumes. Serial fourfold dilutions of normal serum did not show a significant difference in IL-1 generation, but the AMC, preincubated with TBS at different dilutions, generated only 37% to less than 3% of IL-1 produced by those treated with normal rat sera.

4. DISCUSSION AND SUMMARY

This investigation addresses the question of immunosuppression during tumor growth and postulates one mechanism in a syngeneic rat colon cancer isograft model. The tumor-bearing hosts manifested a decline in the capacity to generate IL with the growth of tumor. Generation of IL-1 by AMC and IL-2 by PBMC decreased progressively during tumor growth, and IL production fell profoundly when regional metastases were noted. Concurrent with the tumor growth, the CIC levels increased and peaked when there was a moderate tumor volume. When metastases occurred, there was a relative drop in CIC values. During progressive local tumor growth, there was an inverse correlation between the capacity to generate IL and PEG-CIC levels. The second part of the experiment demonstrated that sera obtained from tumor-bearing rats at the time of peak CIC levels could inhibit significantly the generation of IL-1 by normal host AMC.

The development of various *in vitro* assays capable of detecting and accurately quantitating monokines and lymphokines in murine, rat, and human systems has resulted in the recent description of a number of these factors necessary for immunocyte regulation. Of these, IL-1 and IL-2 have occupied a central role in providing the signals needed for activation of the cellular components of the immune system and in amplification of the resultant response. Though Hoffman and Pollack (personal communication) have shown that IL-1 release by peripheral blood monocytes in response to LPS was decreased in many cancer patients, the results were conflicting when correlated to the stage of the disease. As the baseline variability could be an important factor, longitudinal studies are in order. Our current investigation addresses this question and shows that not only IL-1 but also IL-2 generating capacity of PBMC drop during tumor growth. The study also shows that there was a profound decrease in the capacity of the host to elaborate IL-1 and IL-2 when there were regional metastases.

The increase in PEG-CIC levels correlated to a decrease in IL response, corresponding to a postulated *in vivo* role for CIC in decreasing host antitumor immune response. In addition to the correlative evidence for the role of CIC in cancer, immune complexes have been shown to interfere with antigen recognition and inactivation of various effector cells (Sedlacek, 1980). In T-cell-dependent antibody production, CIC have been shown to inhibit T- and B-cell cooperation; macrophages have been shown to be involved in this process (Taylor and Basten, 1976). Although the mechanism of immune complex effect on IL production has not been defined in this paper, we speculate that immune complexes may interfere with the ability of macrophages to elaborate IL-1, by adhering to Fc receptors. They could interfere with antigen processing or could inhibit the induction of IL-2 receptors. Mizel *et al.* (1978) and Farr and Unanue (1977) have shown that the stimulatory effects of activated T cells on IL production are dependent on physical contact or close proximity of the interacting cells. Immune complexes perhaps interfere with the physical contact in cell–cell interactions.

Preliminary data are presented from the first of a series of experiments performed in an attempt to delineate the mechanism of inhibition of IL generation. Generation of IL by AMC from peripheral blood of normal rats was carried out after preincubation with sera of either normal or tumor-bearing rats. The TBS were obtained at the time of peak CIC response. The results show a rather significant inhibition of IL-1 generation by TBS when compared with NRS. These preliminary results show that decreased IL-1 generation during tumor growth could be caused by CIC; reduction in IL-1 generation may then lead to diminished IL-2 production, thereby dampening the amplification of immune response.

Acknowledgment. The authors wish to acknowledge the expert manuscript preparation by Ms. Jeannette Clowes.

REFERENCES

Farr, A. G., and Unanue, E. R., 1977, T cell–macrophage interaction leading to increased production of lymphocyte stimulatory molecules, *Fed. Proc.* **36**:1194.

Gilles, S., Ferm, M. M., Ou, W., and Smith, K. A., 1978, T cell growth factor: Parameters of production and a quantitative microassay for activity, *J. Immunol.* **120:**2027–2032.

Jerry, L. M., Lewis, M. G., and Cano, P., 1976, Anergy, anti-antibodies and immune complex disease—A syndrome of disordered immune regulation in human cancer, in: *Immunocancerology in Solid Tumors* (M. Martin and L. Dionne, eds.), pp. 63–80, Stratton Intercontinental Medical Book Corporation, New York.

Mizel, S. B., Oppenheim, J. J., and Rosenstreich, D. L., 1978, Characterisation of lymphocyte activating factor (LAF) produced by the macrophage cell line $P388D_1$, *J. Immunol.* **120:**1497–1503.

Rodrick, M. L., Steele, G., Jr., Ross, D. S., Lahey, S. J., Deasy, J. M., Rayner, A. A., Harte, P. J., Wilson, R. E., Munroe, A. E., and King, V. P., 1983, Serial circulating immune complex levels and mitogen responses during progressive tumor growth in Wistar/Furth rats, *J. Natl. Cancer Inst.*, **70:**1113–1118.

Sedlacek, H. H., 1980, Pathophysiological aspects of immune complex disease. Part I. Interaction with plasma enzyme systems, cell membranes, and the immune response, *Klin. Wochenschr.* **48:**543–550.

Taylor, R. B., and Basten, A., 1976, Suppressor cells in humoral immunity and tolerance, *Br. Med. Bull.* **32:**152–157.

Interleukin-3 Production and Action in Tumor-Bearing Hosts

CAROL J. BURGER and KLAUS D. ELGERT

1. INTRODUCTION

Examination of the differentiation pathway leading to clonal expansion of cytotoxic T lymphocytes (CTL) has been approached by many methods (Cantor and Boyse, 1975; Gillis *et al.*, 1979; Reinherz *et al.*, 1980). Detection of soluble factors that provide proliferative signals has afforded a new tool for dissecting this pathway. The proposed sequential cascade model, resulting in CTL maturation, begins with the interaction of macrophage (Mϕ)-derived interleukin-1 (IL-1) with antigen- or mitogen-primed helper T (T_h) cells (Farrar *et al.*, 1980). T_h cells then produce both IL-2 (Hancock *et al.*, 1981) and IL-3 (Ihle *et al.*, 1981a). IL-2 stimulates CTL precursor proliferation and immune interferon production which leads to the production of specific, mature CTL (Farrar *et al.*, 1981).

Description and characterization of IL-3 (Ihle *et al.*, 1981a, 1982) allowed evaluation of an earlier step of the CTL cascade. Several unique characteristics of IL-3 have been established: (1) IL-3 induces 20-α-steroid dehydrogenase (20αSDH) expression on athymic *nu/nu* mouse splenocytes (Ihle *et al.*, 1981b); (2) it promotes splenocyte but not thymocyte proliferation (Ihle *et al.*, 1981a); (3) it is a blastogenic factor associated with antigen-induced proliferation (Enjuanes *et al.*, 1981); and (4) it supports the growth of Thy-1^+ lymphocytes, adherent Thy-1^-,Ia^+ cells, and mastlike cells that are Thy-1^-,Ia^+.

In previous studies (Elgert and Farrar, 1978), we reported the dualistic inhibition of splenocyte proliferation in normal and tumor-bearing hosts (TBH) by suppressor T (T_s) cells and Mϕ. With tumor growth there is a concomitant decrease in splenocyte responsiveness to mitogens (Elgert and Farrar, 1978; Farrar and Elgert,

CAROL J. BURGER and KLAUS D. ELGERT • Department of Biology, Microbiology Section, Virginia Polytechnic Institute and State University, Blacksburg, Virginia 24061.

1978a) or alloantigens (Elgert and Connolly, 1978), and IL-2 activity subsides (Burger *et al.*, 1984). This reduction in immune proliferation may result in fewer CTL. A recent report indicates that an absence of IL-2 results in a slowdown or cessation of cytotoxic activity against tumor cells (Mills and Paetkau, 1980). Because IL-3 is reputed to act at an earlier differentiation step (Ihle *et al.*, 1982), its activity during the growth of fibrosarcoma cells was evaluated. Results demonstrated that the decrease in IL-3 activity with increasing tumor burden did not appear to be due to the presence of a mildly nylon wool-adherent T_s cell. Absorption studies suggested that fresh cells have more accessible IL-3 receptors than do Con A-induced blast cells. Used with Con A or phytohemagglutinin (PHA), IL-3 augmented normal splenocyte responses but significantly suppressed the TBH response. Inoculation of IL-3 into normal or TBH also resulted in suppression of TBH splenocyte blastogenesis.

2. RESULTS AND DISCUSSION

2.1. Kinetics of IL-3 Activity during Tumor Growth

Spleen cells were removed from Balb/c Dub mice at specific intervals after fibrosarcoma cell inoculation (Elgert and Farrar, 1978), Con A-stimulated for 24 hr, and the resulting conditioned medium assessed for IL-3 activity. IL-3 activity decreased over the 28-day assay period when measured by the 20αSDH assay (Ihle *et al.*, 1981 (Fig. 1) or by growth of FDC-P1 cells which have an absolute requirement for IL-3 (cells were a kind gift from Dr. James Ihle, Frederick Cancer Research Center). Although at day 0 approximately 600 units of IL-3 was measured, by day 28 only 62 units was detected.

2.2. Splenocyte Production of IL-3 after Nylon Wool Fractionation

As our past research shows that tumor-induced T_s cells (which are mildly nylon wool-adherent) inhibit cell proliferation (Elgert and Farrar, 1978), we examined whether nylon wool fractionation of splenocytes would alter the suppression of IL-3 synthesis. Both normal and TBH nonadherent populations (Fig. 2) had increased IL-3 production [perhaps due to T_h cell enrichment (Elgert and Farrar, 1978)]. However, as TBH nonadherent cells did not produce normal levels of IL-3 when T_s cells were removed, IL-3-producing cells may be: (1) found in low numbers in TBH; (2) regulated by a cell (or factor from) other than an adherent T_s cell; or (3) fully activated but high numbers of responder cells removed IL-3 from the cellular milieu as it was formed (Ihle *et al.*, 1981a).

2.3. Absorption Studies

In an attempt to discover the receptor population for IL-3, absorption studies were done in the manner of Bonnard *et al.* (1979). Fresh cells (both normal and

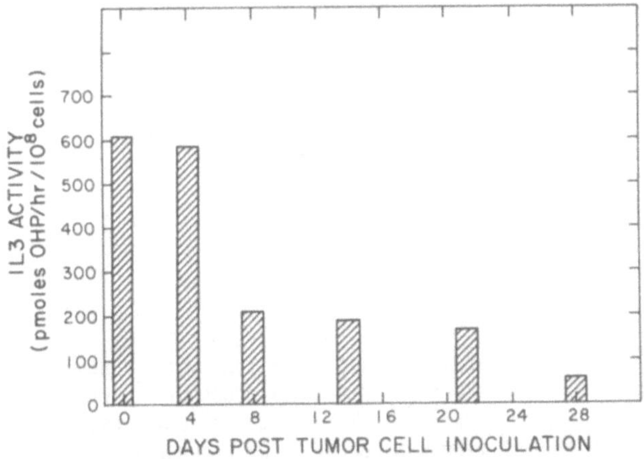

FIGURE 1. IL-3 activity decreased with increased tumor growth. Fibrosarcoma cells (10^6 cells/0.1 ml) were transplanted into left hind leg of normal Balb/c recipients at day 0. At times indicated, spleens were harvested and 10^7 cells/ml cultured for 24 hr with 2 μg/ml Con A to stimulate IL-3 production. Supernatants were tested for IL-3 activity using the enzyme assay for 20αSDH. Control mice sham-injected with 10^6 normal host spleen cells at day 0 had consistent IL-3 activity of 450–550 pmoles hydroxyprogesterone (OHP)/hr per 10^7 cells throughout the experiment.

TBH) (Fig. 3) appeared to have more accessible IL-3 receptors than Con A-induced blast cells, for the fresh cells absorbed significantly more IL-3 than did blast cells. The specificity of these receptors awaits testing using purified radiolabeled IL-3 or monoclonal anti-IL-3 antibody. If these were definite IL-3 receptors, they were evidently not needed once the cell underwent blast transformation and clonal proliferation.

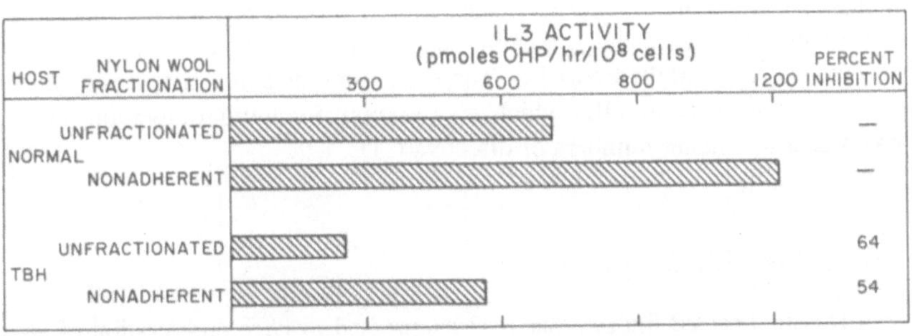

FIGURE 2. IL-3 activity following nylon wool separation of normal and TBH splenocytes. Cells were incubated on nylon wool columns for 45 min at 37°C. After elution with RPMI medium, 10^7 cells/ml were cultured with Con A. After 24 hr, supernatants were harvested and tested for IL-3 activity. Percentage of inhibition of IL-3 activity by TBH splenocytes was compared to its normal host treatment counterpart.

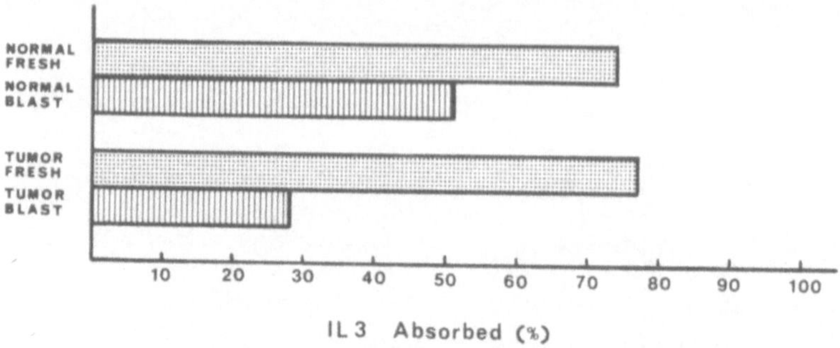

FIGURE 3. Comparison of absorptive capabilities of fresh and activated splenocytes. Spent medium from cultures of WEHI-3 cells (constitutive producers of IL-3) was pretested for IL-3 activity using the IL-3-dependent cell line FDC-P1. 0.3 ml IL-3-containing supernatant was incubated with 10^7 fresh or Con A-induced blast cells for 1 hr at 37°C. As a control, a tube with IL-3 supernatant, but no added cells, was incubated in the same way. No loss of IL-3 activity was noted after a 1-hr incubation. Serial dilutions of the resulting absorbed supernatants were tested on FDC-P1 cells and the units of IL-3 activity calculated using probit analysis with WEHI-3 cell-derived supernatant as a control with 100 U IL-3 activity in 1.0 ml. There was a significant difference in the amount of IL-3 absorbed by fresh cells as compared to that absorbed by Con A-induced blast cells.

2.4. Modulation of *in Vitro* Splenocyte Responsiveness by Purified IL-3

The blastogenic effects of IL-3 on normal splenocytes seen by Enjuanes *et al.*, (1981) were substantiated by our findings (Table I). This effect was not apparent in TBH splenocyte cultures as purified IL-3 alone did not enhance cellular proliferation and, furthermore, was significantly suppressive when cocultured with the mitogens Con A and PHA. Active suppression was suggested by the fact that this effect could be titrated away. Viability checks indicated no cytotoxicity to account for the low [^{3}H]-TdR incorporation. These data suggested the existence of a feedback mechanism at work in TBH which may account for low proliferation levels and could lead to smaller numbers of effective CTL.

2.5. *In Vivo* Effects of IL-3

The unexpected finding that IL-3 suppressed *in vitro* proliferation of already hyporeactive TBH splenocytes led us to examine the consequences of IL-3 administration *in vivo*. The responsiveness of normal or TBH splenocytes alone or with mitogen was tested 24 hr after a tail vein injection of purified IL-3 (Table II). The same pattern of suppression seen *in vitro* emerged when IL-3 was administered *in vivo*. IL-3 inoculation into normal hosts caused a 22% drop in baseline (treated) splenocyte proliferation as compared to splenocytes from medium-injected mice.

TABLE I. *In Vitro* Proliferation of Normal and TBH Splenocytes in the Presence of IL-3

		Mitogen added (cpm)[a]		
Host	IL-3 (U/ml)	None	PHA	Con A
Normal	—	3178	18,829	207,455
	500	7841	45,784	243,581
	250		43,511	238,055
	50		19,793	242,196
	0.05		18,316	281,068
TBH	—	2353	12,943	57,574
	500	2071	879	5,665
	250		1,627	27,208
	50		7,499	51,017
	0.5		10,534	45,492

[a] Test wells received 0.05 ml of an optimal concentration of either PHA (10 μl/ml) or Con A (16 μg/ml). 0.1 ml of purified IL-3 was added to 2 × 10^5 cells/well to a final volume of 0.2 ml. After 66 hr incubation (37°C, 5% CO_2), the cells were pulsed with 1 μCi/well [^{3}H]-TdR (Amersham, Arlington Heights, Ill.) for 6 hr, harvested, and counted by liquid scintillation spectrometry.

There was little difference between medium- and IL-3-treated splenocyte proliferation in response to Con A (a 3% decrease). However, when PHA was added, the blastogenic response of treated splenocytes increased 37%. Spleen cells from treated TBH had decreased proliferation in all cases. The baseline response decreased 18% compared to control animals, quite similar to the normal host baseline difference between treated and untreated counterparts. Addition of Con A to the treated mouse splenocytes caused a 48% decrease in proliferation, whereas PHA addition caused a 29% decrease. The similarity between *in vitro* and *in vivo* IL-3-induced suppression of blastogenic responses in TBH pointed to the existence of a tumor-promoted feedback mechanism which may prevent the formation of mature CTL.

TABLE II. Splenocyte Responsiveness following *In Vivo* IL-3 Administration

		Mitogen added (cpm)		
Host	Treatment[a]	None	PHA	Con A
Normal	Medium control	12,964	44,234	243,546
	IL-3	10,089*	60,642*	237,116
TBH	Medium control	5,566	12,687	162,861
	IL-3	4,548	8,998*	84,888*

[a] Mice received a tail vein injection of 10 U purified IL-3 or RPMI medium 24 hr before mitogen assay.
* Significant difference from sham-inoculated counterpart at $p \leq 0.01$.

3. CONCLUSION

These data indicate that splenocytes from TBH are incapable of producing normal levels of IL-3 *in vitro*. Further, the deficient blastogenic response in TBH cannot be reconstituted using addition of exogenous IL-3. The biological relevance of the suppressive effect exerted by IL-3 *in vitro* was substantiated by similar *in vivo* findings. The apparent lack of IL-3 receptors on proliferating cells lends credence to the scheme suggesting IL-3 stimulates progenitor or stem cells rather than mature, differentiated cells.

Combining the role of IL-3 in T-cell differentiation, tumor growth-mediated diminution of IL-3 activity, and the contribution of CTL to tumor cell killing, the importance of IL-3 becomes obvious. Therefore, delineation of the IL-3 mode of action in tumor-burdened animals may allow a fuller understanding of the way tumor cells subvert the cell-mediated immune response and escape destruction.

ACKNOWLEDGMENTS. This work was supported by NIH Grant CA-25943 and a Whitehall Foundation Grant.

REFERENCES

Bonnard, G. D., Yasaka, K., and Jacobson, D., 1979, Ligand-activated T cell growth factor-induced proliferation: Absorption of T cell growth factor by activated T cells, *J. Immunol.* **123**:2704.

Burger, C. J., Elgert, K. D., and Farrar, W. L., 1984, Interleukin 2 (IL-2) activity during tumor growth: IL-2 production kinetics, absorption of and responses to exogenous IL-2, *Cell. Immunol.* **84** (in press).

Cantor, H., and Boyse, E. A., 1975, Functional subclasses of T lymphocytes bearing different Ly antigens. I. The generation of functionally distinct T-cell subclasses is a differentiative process independent of antigen, *J. Exp. Med.* **141**:1376

Elgert, K. D., and Connolly, K. M., 1978, Macrophage regulation of the T cell allogeneic response during tumor growth, *Cell. Immunol.* **35**:1.

Elgert, K. D., and Farrar, W. L., 1978, Suppressor cell activity in tumor-bearing mice. I. Dualistic inhibition by suppressor T lymphocytes and macrophages, *J. Immunol.* **120**:1345.

Enjuanes, L., Lee, J. C., and Ihle, J. N., 1981, T cell recognition of Moloney sarcoma virus proteins during tumor regression. I. Lack of a requirement for macrophages and the role of blastogenic factors in T cell proliferation, *J. Immunol.* **126**:1478.

Farrar, W. L., and Elgert, K. D., 1978a, Inhibition of mitogen and immune blastogenesis by two distinct populations of suppressor cells present in the spleen of fibrosarcoma-bearing mice: Adoptive transfer of suppression, *Int. J. Cancer* **22**:142.

Farrar, W. L., and Elgert, K. D., 1978b, *In vitro* immune blastogenesis during contact sensitivity in tumor-bearing mice. II. Mechanisms of inhibition, *Cell. Immunol.* **40**:365.

Farrar, W. L., Mizel, S. B., and Farrar, J. J., 1980, Participation of lymphocyte activating factor (interleukin 1) in the induction of cytotoxic T cell responses, *J. Immunol.* **124**:1371.

Farrar, W. L., Johnson, H. M., and Farrar, J. J., 1981, Regulation of the production of immune interferon and cytotoxic T lymphocytes by interleukin 2, *J. Immunol.* **126**:1120.

Gillis, S., Union, N. A., Baker, P. E., and Smith, K. A., 1979, The *in vitro* generation and sustained culture of nude mouse cytolytic T-lymphocytes, *J. Exp. Med.* **149**:1460.

Hancock, E. J., Kilburn, D. G., and Levy, J. B., 1981, Helper cells active in the generation of cytotoxicity to a syngeneic tumor, *J. Immunol.* **127**:1394.

Ihle, J. N., Lee, J. C., and Rebar, L., 1981a, T cell recognition of Moloney leukemia virus proteins. III. T cell proliferative responses against gp70 are associated with the production of a lymphokine inducing 20-alpha-hydroxysteriod dehydrogenase on splenic lymphocytes, *J. Immunol.* **127**:2565.

Ihle, J. N., Peppersack, L., and Rebar, L., 1981b, Regulation of T cell differentiation: *In vitro* induction of 20α-hydroxysteriod dehydrogenase in splenic lymphocytes from athymic mice by a unique lymphokine, *J. Immunol.* **126**:2184.

Ihle, J. N., Hapel, A., Greenberg, J., Lee, J. C., and Rein, A., 1982, Possible roles of interleukin 3 in the regulation of lymphocyte differentiation, in: *The Potential Role of T Cells in Cancer Therapy* (A. Fefer and A. Goldstein, eds.), pp. 93–112, Raven Press, New York.

Mills, G. B., and Paetkau, V., 1980, Generation of cytotoxic lymphocytes to syngeneic tumor using co-stimulator (interleukin 2), *J. Immunol.* **125**:1897.

Reinherz, E. L., Moretta, L., Roper, M., Breard, J. M., Mingari, M. C., Cooper, M. D., and Schlossman, S. F., 1980, Human T lymphocyte subpopulations defined by Fc receptors and monoclonal antibodies: A comparison, *J. Exp. Med.* **151**:969.

47

Lymphokine-Mediated Enhancement of Antibody Formation by Leukemic Cells

R. CHRISTOPHER BUTLER, JERI M. FRIER, and HERMAN FRIEDMAN

1. INTRODUCTION

Infection of susceptible strains of mice with Friend leukemia virus (FLV) results in the development of lymphatic leukemia and focus formation within the spleen. This leukemogenesis is associated with a progressive impairment of immunological responses, including those mediated by both T and B lymphocytes (Notkins *et al.*, 1970). When cultured *in vitro* the splenocytes from these leukemic mice demonstrate a markedly inhibited ability to respond to antigenic stimulation by sheep erythrocyte (SRBC) antigens (Specter *et al.*, 1976). It has been generally accepted that this impairment is due to the interaction of the virus with antibody-producing B cells or their precursors.

Recent studies, however, have provided evidence that B cells from FLV-infected hosts retain the ability to produce specific antibodies under appropriate conditions. For example, even though the development of an antibody response to SRBC is suppressed, the "background" antibody response of unsensitized FLV-infected splenocytes remains at normal levels (Bendinelli and Friedman, 1980). Therefore, the B cells do not lose the capacity to produce specific antibody, but for some reason they do not respond to antigenic stimulation. Second, the treatment

R. CHRISTOPHER BUTLER and JERI M. FRIER • Departments of Medical Microbiology and Immunology, Arlington Hospital, Arlington, Virginia 22205. HERMAN FRIEDMAN • Departments of Medical Microbiology and Immunology, University of South Florida College of Medicine, Tampa, Florida 33612.

of FLV-infected splenocytes with immunostimulants such as lipopolysaccharides (LPS) or muramyl dipeptide (MDP) can enhance the development of an antibody response to SRBC (Butler and Friedman, 1980). This indicates that with proper stimulation the FLV-infected splenocytes can mount an immunological response. A third observation was that the addition of normal macrophages to FLV-infected cultures could partially restore the response capacity, thus indicating that the mechanism for this unresponsiveness might be related to these accessory cells (Specter *et al.*, 1976.

As the mechanism for the suppression of antibody responses by FLV is now apparently not due solely to infection of B cells by the virus but may involve macrophages, we sought to determine the mechanism by which FLV might exert its suppressive effects via macrophages. In previous studies we have demonstrated that the development of antibody responses by normal splenocytes is greatly enhanced by the production of antibody response helper factors, including interleukin 1 (IL-1), by macrophages (Butler *et al.*, 1979). These factors are released "spontaneously" by cultured macrophages but their production is greatly enhanced by treatment with LPS. This study was designed to determine the role of these antibody response helper factors in the mechanism of suppression of the antibody response by FLV.

2. MATERIALS AND METHODS

Experimental Animals. Inbred male Balb/c mice, 6 to 8 weeks of age, were obtained from Cumberland View Farms, Clinton, Tennessee. Mice were infected by injection of a 100 ID_{50} dose of FLV contained in 0.1 ml of a 1% clarified homogenate of infected splenocytes.

LPS. Serratia marcescens LPS was prepared by the trichloroacetic acid extraction procedure as previously described (Nowotny *et al.*, 1966). We wish to thank Dr. Nowotny for the generous donation of this preparation.

Antigen. SRBC in Alsever's solution were obtained from Baltimore Biological Laboratories, Baltimore, Maryland. The erythrocytes were washed several times in media and resuspended to a 0.1% concentration.

In Vitro Immunization. A suspension of 8×10^6 viable splenocytes in 2.0 ml of complete tissue culture medium was cultured in multiwell dishes as described previously (Kamo *et al.*, 1976). For *in vitro* immunization, 0.1 ml of the 0.1% suspension of SRBC was added to each culture. All cultures were incubated for 5 days at 37°C in a humidified atmosphere containing 10% CO_2.

Assay for Antibody-Forming Cells. The numbers of direct hemolytic plaque-forming cells (PFC) to SRBC were determined for 8 to 24 cultures using the micromethod (Cunningham and Szenburg, 1968).

Post-LPS-Serum. Normal or FLV-infected mice were injected i.p. with 20 μg LPS and exsanguinated 2 hr later by aseptic cardiac puncture. Serum was separated from cells and kept on ice until use.

In Vitro Factor Production. Suspensions of 10^7 splenocytes/ml from normal or FLV-infectd mice were incubated in RPMI 1640 plus 10% fetal bovine serum and antibiotics at 37°C under CO_2. Experimental cultures received 10 μg/ml of LPS at the time of culture initiation. Supernatants were collected after 5 days and either stored on ice or frozen at −70°C until tested.

3. RESULTS

Antibody response helper factor (IL-1) is normally released from macrophages in response to stimulation with LPS or its nontoxic PS derivative (Butler *et al.*, 1979). Table I demonstrates the production of these lymphokines both *in vivo* (post-LPS serum) and *in vitro* (culture supernatants) by normal mice or splenocytes in response to LPS. When mice were infected with FLV at different time intervals, they gradually lost the ability to produce these helper factors *in vivo* as the leukemia progressed. The effects of normal unstimulated serum did not change, thus indicating that the observed decrease in stimulatory activity of the serum was not due to the buildup of any immunosuppressive factors. The stimulatory effects of the post-LPS sera were not due to the presence of residual LPS which has been demonstrated to be several orders of magnitude below a stimulatory concentration (Butler *et al.*, 1979).

The sequential suppression of factor production was further illustrated by obtaining factor production *in vitro* by splenocytes from leukemic mice. Similar to the results obtained *in vivo*, the *in vitro* cultures of FLV-infected cells showed a depressed ability to produce helper factor activity both "spontaneously" and in response to LPS stimulation. The increasing spleen weights of infected mice are

TABLE I. FLV Suppresses Helper Factor Production *In Vivo* and *In Vitro*

Source of cells or serum[a]	LPS treatment to induce factor	Source of factor		% control PFC	p[d]
		Serum[b]	Supernatant[c]		
None	−	−	−	100	
Normal mouse	−	+	−	92	
	−	−	+	139	
	+	+	−	189	0.005
	+	−	+	302	0.005
7-day FLV-infected mouse	−	+	−	83	
	−	−	+	118	
	+	+	−	147	0.05
	+	−	+	188	0.05
21-day FLV-infected mouse	−	+	−	97	
	−	−	+	98	
	+	+	−	106	
	+	−	+	118	

[a] Mice were preinfected with 100 ID_{50} of FLV 7 or 21 days prior to sacrifice to obtain splenocytes or serum.
[b] Normal or FLV-infected mice received an i.p. injection of 20 μg of LPS 2 hr prior to collection of serum. PFC cultures were treated with 0.5% serum on day 0.
[c] Culture supernatants were obtained from cultures of 10^7 splenocytes/ml incubated 5 days with or without 10 μg/ml LPS. Cultures of normal mouse splenocytes were treated with 5% supernatant at the time of primary *in vitro* sensitization.
[d] p values relative to the normal untreated control.

TABLE II. Helper Factors Restore Immunocompetence to Leukemic Cells

Helper factor source[a]	Responder cells[b]	% control PFC[c]	p^d
None	Normal	100	
	Leukemic	34	0.001
Untreated splenocytes	Normal	97	
	Leukemic	47	0.001
LPS-treated splenocytes	Normal	198	0.01
	Leukemic	138	0.05

[a] Cultures of 10^7 normal Balb/c splenocytes/1.0 ml were incubated for 5 days with or without 10 μg/ml of *S. marcescens* LPS.
[b] Cultures of 8×10^6 normal or 21-day FLV-infected splenocytes (spleen weight = 2.5 g) were treated with 5% factor supernatant at the time of *in vitro* primary sensitization with SRBC.
[c] After 5 days the cultures were collected and assayed for direct PFC. Control cultures received no treatment.
[d] *p* values relative to the normal responder cell control were determined by Student's *t* test.

an indication both of the dramatic proliferative response resulting from FLV infection and transformation of cells and of the state of progression of the disease.

As FLV infection caused splenocyte cultures to lose the capacity to produce antibody response helper factor activity, we hypothesized that the restoration of this helper factor activity to FLV splenocytes might restore the antibody response capacity. Table II demonstrates the suppressive effect of FLV infection on the ability of leukemic cells to develop an antibody response. The addition of physiological levels of exogenous helper factors to FLV leukemic splenocyte cultures restored the antibody response to normal or even slightly elevated levels.

4. DISCUSSION

The immunosuppressive effects of FLV on antibody responses to SRBC antigens have been demonstrated *in vivo* (Kately *et al.*, 1974; Bendinelli and Friedman, 1980) and *in vitro* (Butler and Friedman, 1979, 1980; Butler *et al.*, 1980; Specter *et al.*, 1976). The results shown here demonstrated that the ability of murine splenocytes to mount an antibody response to SRBC *in vitro* decreases steadily with time following infection with FLV. The mechanism for this loss of response capacity has not been clearly defined. However, these studies have established a strong relationship between the loss of ability to mount a primary antibody response and the loss of ability to produce antibody response helper factors (IL-1) either "spontaneously" or in response to stimulation with LPS. This decrease in helper factor activity production was demonstrated to occur both *in vivo* (post-LPS serum) and *in vitro* (splenocyte culture supernatants).

Our previous studies have demonstrated the role of these helper factors in antibody responses by normal splenocytes (Butler *et al.*, 1979, 1981; Friedman and Butler, 1980). The factor is produced by macrophages and its target is a cell of bone marrow lineage. It can enhance SRBC antibody responses in the absence of T cells. A dose–response relationship exists between helper factor concentration and the magnitude of the antibody response (Butler *et al.*, 1979). Therefore, it is feasible that the decrease in helper factor production by FLV-infected cells could be causally related to the depressed antibody response. If this were the case, then the addition of exogenous helper factor should restore the antibody response of FLV cells to normal levels. The results in Table II show that this reversal does occur and that B lymphocytes from FLV leukemic spleens retain the capacity to produce a normal antibody response under the condition of adequate levels of helper factor.

Previous studies have suggested that the suppression of antibody responses in FLV-infected mice is due to a suppression of macrophages (Specter *et al.*, 1976). They demonstrated that the addition of normal macrophages to FLV splenocyte cultures could restore the antibody response capacity. Our results clarify the mechanism for a macrophage-induced suppression by indicating that it is the inability of cultures to produce sufficient quantities of a specific macrophage product—an antibody response helper factor—which inhibits the development of antibody responses by FLV-infected splenocytes.

This depression of antibody responses could be due to either a numerical decrease or a selective suppression of macrophage populations capable of producing antibody helper factor in response to antigen or immunoadjuvant stimulation. It is conceivable that these macrophages are refractory due to either the effects of direct infection, a depression of stimulatory factors from other cells, or prolonged stimulation by FLV viral antigens which might block receptors. These possibilities are currently being evaluated in our laboratories.

ACKNOWLEDGMENTS. This investigation was supported by NCI Grant 7R23CA27415-02 and by awards from the Elaine R. Shepard Foundation and K. S. Associates.

REFERENCES

Bendinelli, M., and Friedman, H., 1980, B and T lymphocyte activation by murine leukemia virus infection, *Adv. Exp. Biol. Med.* **121B**:91–97.

Butler, R. C., and Friedman, H., 1979, Leukemia virus induced immunosuppression: Reversal by subcellular factors, *Ann. N.Y. Acad. Sci.* **332**:446–450.

Butler, R. C., and Friedman, H., 1980, Restoration of leukemia cell immune responses by bacterial products, in: *Current Chemotherapy and Infectious Disease* (J. D. Nelson and C. Grassi, eds.), pp. 1719–1720, American Society for Microbiology, Washington, D.C.

Butler, R. C., Nowotny, A., and Friedman, H., 1979, Macrophage factors that enhance the antibody response, *Ann. N.Y. Acad. Sci.* **332**:564–578.

Butler, R. C., Friedman, H., and Nowotny, A., 1980, Restoration of depressed antibody responses of leukemic splenocytes treated with LPS-induced factors, *Adv. Exp. Biol. Med.* **121A**:315–322.

Butler, R. C., Nowotny, A., and Friedman, H., 1981, Induction of immunomodulatory factors by LPS and nontoxic derivatives, in: *Immunomodulation by Bacteria and Their Products* (H. Friedman, T. W. Klein, and A. Szentivanyi, eds.), pp. 181–198, Plenum Press, New York. .

Cunningham, A. J., and Szenburg, A., 1968, Further improvements in the plaque technique for detecting single antibody-forming cells, *Immunology* **14**:599–601.

Friedman, H., and Butler, R. C., 1980, Immunomodulatory effects of endotoxin-induced factors, in: *Microbiology 1980* (D. Schlessinger, ed.), pp. 44–48, American Society for Microbiology, Washington, D.C.

Kamo, I., Pan, S. H., and Friedman, H., 1976, A simplified procedure for *in vitro* immunization of dispersed spleen cell cultures, *J. Immunol. Methods* **11**:55–62

Kately, J. R., Kamo, I., Kaplan, G., and Friedman, H., 1974, Suppressive effect of leukemia virus-infected lymphoid cells on *in vitro* immunization of normal splenocytes, *J. Natl. Cancer Inst.* **53**:1371–1378.

Notkins, A. L., Megenhagen, S. E., and Howard, R. J., 1970, Effect of virus infections on the function of the immune system, *Annu. Rev. Microbiol.* **24**:525–538.

Nowotny, A., Cundy, K. R., Neale, N. L., Nowotny, A. M., Radvany, P., Thomas, S. P., and Tripodi, D. J., 1966, Relation of structure to function in bacterial O-antigens. IV. Fractionation of the components, *Ann. N.Y. Acad. Sci.* **133**:586–603.

Specter, S., Patel, N., and Friedman, H., 1976, Restoration of leukemia virus-suppressed immunocytes *in vitro* by peritoneal exudate cells, *Proc. Soc. Exp. Biol. Med.* **151**:163–167.

Part IV Summary

Pharmacology and *in Vivo* Actions

JOHN W. HADDEN and ENRICO GARACI

I will not summarize the session since the material is available for your review in these proceedings. We thought instead to offer general comments on the progress of this field as exemplified by the presentations of this session and specifically in terms of its immunopharmacology and its pharmacology. My cochairman, Dr. Garaci, will emphasize some additional points at the end of this section.

I had an opportunity to make a similar synthesis at the Montpellier meeting three years ago (see International Symposium on New Trends in Human Immunology and Cancer Immunotherapy, 1980), and at that time I made the suggestion that thymic hormones seemed to be best applied within the context of immunoprophylaxis, that is, in individuals who have defined immunodeficiency in an effort to forestall the complications of that immunodeficiency rather than in treatment of the complications once they have arisen. This suggestion was in contrast to what was then an almost exclusive focus on thymic hormones within the context of primary immunodeficiency diseases and cancer. I was most delighted to have in the symposium and in the abstract session, seven papers which utilized this kind of approach. The examples that were presented included murine studies, in which mice with diabetes, with low-fat or high-fat diet, with protein malnutrition, with treatment by cytostatics or by X-rays, in mice that were genetic low responders, and in mice with infections were studied (Table I). In these circumstances, the condition was associated in most of the studies with a demonstrated immunodeficiency and the primary effort of immunotherapy was to reverse that immunodeficiency. Indeed, thymosin fraction 5 and α_1, as the two major substances under

JOHN W. HADDEN • Department of Immunopharmacology, University of South Florida College of Medicine, Tampa, Florida 33612. ENRICO GARACI • Department of Experimental Medicine, 2° Rome University, 00173 Rome, Italy.

TABLE I. Mouse Immunoprophylactic Study

Author	Immunodepressive influence	Nature of immunosuppression	Corrective therapy
Ohta	5-FU	↓ DTH	Thymosin fraction 5
Salvin	Genetic low-responder	↓ DTH, lymphokines	Thymosin fraction 5
	Alloxan diabetes	↓ resistance to *Candida*	Thymosin fraction 5
Strausser	*db/db* diabetic mice	↓ PFC, ↓ Lyl and 2 cells	Thymosin fraction 5
Watson	Low- and high-fat diet Low protein	↑ resistance to tumor cell challenge	Thymosin fraction 5 further ↑ resistance
Doria	Aged mice	↓ helper T cell ↓ anti-TNP PFC response	Thymosin α_1
Ohta	5-FU	↓ DTH, NK, and CSF	Thymosin α_1
Ishitsuka	5-FU or X-ray	↓ resistance to pathogen and tumor challenge	Thymosin α_1

study, seem to be rather effective in doing that. In one study, this effect to restore immune response was then subsequently translated into increased resistance to pathogen challenge and was also associated with enhanced tumor resistance. These then seem to me the most natural circumstances for studying thymic hormones in their application to human disease and I think the future will bring more of this kind of focus. I think you can imagine, with the examples of the immunodeficiencies studied in mice ranging from diabetes to aging, a very significant potential in terms of a great variety of human disorders is implied.

There was a paucity, I think, of comparative studies in which more than one thymic hormone was compared; often thymosin α_1 was compared to fraction 5 and showed equivalent magnitude of activity at a much lower dose. The use of the various thymic hormone preparations in the same study will in the future be informative in showing their comparative effects. Insofar as there are differences in their actions, it will lend support concerning the potential for their combined application in the future. I think for reasons that were brought out in this morning's session, the use of crude preparations of 35 peptides, like fraction 5, may be effective clinically but there is a potential hazard with its use in terms of possible inhibitory substances present in the mixture. If one were to theorize on the issue, an approach to primary immunodeficiency might very well be a combination of the existing purified hormones, e.g., thymosin α_1, thymopoietin, and FTS.

Now, the only real pharmacology we had presented in our session had to do with my cochairman's presentation relating prostaglandin (PG) metabolism to the possible action of thymic hormones, particularly facteur thymique sérique. I think that that work lends itself to a restricted definition within a context of facteur thymique sérique and the Bach assay. Under these circumstances prostaglandin metabolism seems to relate to thymic hormone action. This does not seem to be the case in other circumstances, i.e., the action of thymopoietin and thymosin fraction 5 in the Komuro–Boyse assay is not inhibited by blockade of PG metabolism with indomethacin.

I would like to emphasize some of the aspects of cyclic nucleotide pharmacology in the mediation of thymic hormone action for two reasons: because Nathan Trainin's recent review in *Immunology Today* carries some misinterpretations of the history of the study of cyclic nucleotide pharmacology in relation to thymic hormone action, and because I think that the study of the action of the purified hormones represents a very exciting area of endocrinology. It offers an extraordinary opportunity to go past changes in cyclic nucleotide levels to probe the enzymatic mechanisms involved and the relationship between the cyclic nucleotides and the processes of lymphocyte proliferation and differentiation. We and Naylor and Goldstein have demonstrated that thymopoietin and thymosin fraction 5, respectively, modulate cyclic GMP levels in mature T cells but *not* in the immature T-cell population. Neither they, nor us, have been able to show increases of cyclic AMP in the precursor population and what this implies is that there are two different mechanisms by which these substances modulate. One in mature cells relates to cyclic GMP, and some other mechanism of unknown nature in immature cells. Several of the preparations have been shown to modulate cyclic AMP levels, one of which I understand, has been removed from the scene, that is the serum thymic factor of Astaldi, which turned out to be adenosine. The thymic humoral factor of Trainin is really the most extensively studied with respect to its capacity to raise cyclic AMP levels in lymphocytes and I accept, tentatively, this as a possible action of a thymic hormone. The legitimacy of his thymic hormone, however, will remain in doubt until the chemical structure is known and further study performed demonstrating a thymic origin. At present we can tentatively accept that there are different thymic hormones having different mechanisms relating to these cyclic nucleotide pathways. In general, insofar as it has been studied, the correlations hold that the increase in cyclic GMP, when seen, was related to proproliferative or prosecretory functions in mature cells and that the increase in cyclic AMP, when seen, was related to differentiation. In any case, as thymic hormones are purified and characterized there need to be studies on their mechanisms of action which involve more depth than mere changes in cellular cyclic nucleotide levels.

There was a lack of conventional pharmacology presented in our session. It was apparent in the various studies that thymosin fraction 5 or α_1 was being used over quite a range of concentration and almost randomly with respect to the frequency with which it was administered. There seemed to be no effort in the material that I saw to determine the time period a thymic hormone has an effect on various responses. As is the case with other immunomodulators, I think you will find that there will be a dissociation of the pharmacokinetic curves of the presence of the material in the serum and the pharmacodynamics, i.e., the substantive biological or immunological responses which take place. Were we to know this information, we would know better how to apply these substances and how to keep them, other than just on a dose basis, in a positive mode versus a suppressive mode of action. There was evidence of differences among mouse strains in their responses to thymic hormones, low responders being augmented by thymic hormones and normal responders being unaffected or suppressed. This seems to be a general kind of observation shared also by the chemically defined immunostimulating compounds and obviously deserves more elucidation as it relates to thymic hormones.

In this session were also studies on cytokines, interleukins, and lymphokines. We had introduced into our session oncostatin, which I don't believe qualifies as a lymphokine but it was certainly interesting to hear about as a potential antiproliferative mechanism for regulating cancer cells. The notion of using an oncostatin therapeutically is, of course, immensely enticing. The notion of modifying a tumor growth factor to make an inactive analog which would bind to the cells and prevent the action of an endogenously produced tumor growth factor seems like another interesting anticancer approach. Apparently forgotten in this session were the chalones. It seems to me a shame not to have these represented in the sense that since they have been hypothesized, in general, to act through cyclic AMP, they may very well have an identity with the material that Trainin is studying. The chalones may well be very useful in terms of therapeutic approaches not just to inhibit lymphoid malignancies but, for example, if these chalones were used on a short-term protective basis to keep the immunologic system out of cycle while cytotoxic chemotherapy was being administered, one might be able to obviate immune suppression produced by our currently employed anticancer therapies.

We heard a good deal about interleukins. What was interesting to me was that more and more evidence is accumulating to indicate that interleukin production or action is disturbed in various diseases. We had presented evidence that IL-1, IL-2, and IL-3 are abnormal in various disease states. The application of interleukins individually or collectively, still seems to be a rational therapeutic approach although years behind the efforts with the thymic hormones and with the interferons. The notion that these substances could be used not just as collections but individually to regulate immune response seems to have a tremendous power, and I see this as an area that is going to grow rapidly over the next few years. We do, of course, have our problems with purity and nomenclature but I think these will be resolved over time. I think the combined studies of Dumonde and Papermaster would indicate to us that these materials are safe to administer to animals and to humans and that they offer some significant prospect in their own right.

The issue of using combinations of thymic hormones and interleukins was raised by us based on our own studies with TCGF and thymic hormones. Since this work is not presented in a separate manuscript I will summarize this. I presented collaborative work performed with Drs. S. S. Chen and R. A. Good (*Proc. Natl. Acad. Sci. USA* **80:**5980–5984, 1983), in which we examined the effects of thymic hormones to induce intrathymic maturation using the peanut agglutinin (PNA)-positive immature thymocytes as a model. We sought to determine if thymic hormones would induce changes in their surface markers (TL, H2, Qa2,3, and gp70) corresponding to the known maturation scheme, i.e., induce them to a surface marker status comparable to the PNA-negative mature thymocyte. Thymopoietin, thymosin α_1, and FTS (thymulin) had little or no effect and thymosin fraction 5 had a small effect. Interestingly, TCGF (IL-2) was very active to induce maturation. These results indicate that since intrathymic evolution may be more importantly regulated by IL-2 than thymic hormones, reconstitution efforts may require IL-2 therapy in addition to thymic hormones. Allan Goldstein tells me he has some unpublished experimental support for the notion in that a combination of interleukin and thymosin may be synergistic in their anticancer therapeutic effects. These kinds of observations of synergism have been made with other drugs and biologicals; isoprinosine and interferon have a potentiative interaction, so also muramyl dipeptide and lymphokines, and endotoxin and lymphokines. These examples offer an interesting repertoire of strategies but it involves getting together in order to do these things. Strategies for approaching immunotherapeutically what are, I think, very complicated diseases like cancer will be necessarily complex. I can see thymic hormones or lymphokines being used individually in relationship to infection or other secondary immunodeficiencies but I think with respect to the cancer problem we are going to have to use every weapon in our repertoire and that will involve combinations of agents.

The future? A future is made by imaginative individuals; the presence of something that is interesting, as we have with the thymic hormones and the interleukins; their availability in adequate quantities as purified substances, as will come from contributions of genetic engineering; and finally with money. It is obvious that in contrast to the not so lucrative state of the BRM program, private industry is infusing large amounts of capital in this sector. This is obviously very important to the life of the endeavor; it will also be very important to the science of the endeavor. While some of the research to date has been lacking in scientific rigor, when you start having to do studies that relate to licensing by various federal agencies for administration of these substances in humans, clearly the studies will improve in their precision and quality and I think we will all enjoy the results. With it will come, of course, the hard pharmacology that is required to prepare biological substances and drugs for clinical administration. I see, of course, a bright future in this regard and how rapidly it occurs we will all have to wait and see.

J. W. H.

An area of primary interest that emerges from our session is the treatment with thymic factors of infectious diseases in the disabled host. This is of major importance because of a newly emerging clinical pathology, the pathology of immunocompromised patients. The immunodeficient host may die because of opportunistic infections with microorganisms like *Serratia* or *Proteus,* which develop antibiotic resistance very rapidly. About 40 years ago, in the beginning of the antibiotic era, Domagk, the German chemotherapist, said that "bacteria cannot be eliminated from the body without host collaboration," a fact that we now realize in its complete and full actuality; therefore, a number of experimental models are required to investigate the right antibacterial therapy in the immunodeficient host. The data presented by H. Ishitsuka *et al.* illustrate an interesting model which could be reasonably applied to the clinic. The authors induced immunosuppression with cyclophosphamide followed by treatment with thymosin. The protective effect obtained with thymosin shows that this treatment could represent a good model of immunopharmacological prophylaxis. However, if a protective effect of thymic factors against infectious agents has been clearly evidenced, the mechanism of this anti-infective activity remains to be elucidated. It also remains to be determined if the protection against microorganisms was established by T-cell reconstitution. In this respect, what is the role of lymphokines? Are macrophages and/or other cells different from T cells influenced directly or indirectly by thymic hormones? These are the kinds of questions and approaches that have to be studied in the immediate future and which are relevant for understanding the basic mechanism of thymic hormone activity. An important consideration which may have practical importance is that in our experiments of anti-*Candida* protection, thymosin fraction 5 or TP-1 were more active than thymosin α_1. This fact doesn't mean that thymic extracts are preferable to the use of purified peptides, but should be a stimulus to study the other still undefined peptides in thymic hormone preparations to identify perhaps other potent molecules. In the clinic the use of purified molecules would obviously be safer. In this respect, given the positive effects of thymus products in treating infectious diseases, it would be important to test the purified well-defined peptides in several models of experimental infections to establish efficacy. These studies should include:

1. Testing of combinations of different thymic peptides.
2. Testing of the effects of thymic peptides in combination with other lymphokines and immunostimulating agents.
3. Testing of the ability of thymic peptides to reduce the toxicity of chemotherapeutic and antibiotic drugs.

The last point, which I think is really important to this discussion, is the need to expand the very elegant experiments of Dr. G. Doria who studied the role of thymic hormones on T-helper cell function in the reconstitution of aged mice. I think that thymic factors may find a very successful application in the reconstitution of the immune systems of the elderly.

E. G.

V

Clinical Applications

A Phase I Trial of Immune Interferon

A Preliminary Report

ROBERT K. OLDHAM, STEPHEN A. SHERWIN, PAUL G. ABRAMS, ANNETTE MALUISH, CEDRIC W. LONG, THELMA WATSON, and KENNETH A. FOON

1. INTRODUCTION

A variety of interferons have now been tested in clinical trials. Most of these trials have utilized leukocyte (α) interferon preparations. Early trials were conducted with partially purified material derived from the supernatants of virus-stimulated leukocytes that were of low purity and inconsistent pharmaceutical quality. More recently, trials have been conducted with a lymphoblastoid cell line interferon (α) of high purity and good reproducibility (Knost *et al.*, 1983). Several recent studies have utilized recombinant α-interferon derived by cloning a gene for α-interferon in an *E. coli* expression system (Sherwin *et al.*, 1982, 1983). Despite the major differences between these α-interferon preparations, many of the toxicities, immunological modulating effects, and therapeutic effects have been similar. Fever, chills, headache, fatigue, and anorexia have been rather constant side effects of these interferon preparations. At higher doses, mild hematological depression and transient hepatic enzyme abnormalities have been seen. Occasional cardiac effects including arrhythmias and ischemic effects have been observed in the context of these trials. Some central nervous toxicity including confusion, decreased ability

ROBERT K. OLDHAM, STEPHEN A. SHERWIN, PAUL G. ABRAMS, ANNETTE MALUISH, CEDRIC W. LONG, THELMA WATSON, and KENNETH A. FOON • Biological Response Modifiers Program, Division of Cancer Treatment, NCI-Frederick Cancer Research Facility, Frederick, Maryland 21701.

to concentrate, and rarely seizures at very high doses have been seen in these studies. It is unclear whether all these effects are due to the direct action of the interferon preparation since the induction of fever, tachycardia, and fatigue may have secondary effects (Oldham, 1983a). The phase I trials for the α-interferons are virtually complete. While further information can be derived from studies of dose and schedule with respect to phase I toxicology studies, most investigators are now concentrating on phase II studies to determine the therapeutic effects of the α-interferons using the lymphoblastoid or the genetically engineered α -preparations. These phase II trials are well under way and have demonstrated antitumor activity of these forms of interferon in lymphoma, renal carcinoma, melanoma, and a few other types of cancer (Oldham, 1983b).

Limited trials have been done with β-interferon extracted from the supernatant of virus-stimulated fibroblasts (McPherson and Tan, 1980). Trials are just getting under way with genetically engineered fibroblast (β) interferon. These early trials indicate that many of the side effects are similar to those observed with α-interferon, but certain *in vitro* studies seem to indicate that the β-interferons may have stronger immunomodulatory or antiproliferative effects when compared to the α-interferons. Less consistent absorption from intramuscular administration has been seen, necessitating intravenous administration of β preparations.

The third form of interferon has been termed immune or γ-interferon. In contrast to α-interferon for which as many as 12 genes have been identified, there appears to be only one molecular species of γ-interferon. *In vitro* studies conducted thus far suggest that γ-interferon may have greater antiproliferative effects than α- or β-interferon, although these observations are complicated by difficulties comparing specific activities of the preparations. Nonetheless, γ-interferon, in contrast to the other interferons, appears to have unique immunomodulatory effects *in vitro* and perhaps should be viewed as a lymphokine of the interleukin type.

Our trials at the National Cancer Institute have focused on phase I and then phase II studies in cancer patients. In the phase I trial, described in this paper, we have used a γ-interferon preparation prepared from the supernatant of cells derived from peripheral blood leukocytes chemically stimulated to produce γ-interferon. This preparation was then utilized in an escalating-dose phase I trial to determine its biological effects including its toxicity, immunomodulating capability, and pharmacokinetics.

2. MATERIALS AND METHODS

2.1. Interferon Preparation

This extracted form of γ-interferon was prepared by Meloy Laboratories under National Cancer Institute contract. Human γ-interferon was produced from buffy coats obtained from healthy volunteers. Leukocytes were diluted in Dulbecco's minimum essential medium to 5×10^6 cells/ml. Cultures were then induced with two mitogenic heterocyclic compounds: mezerein and A23187, a lipophilic calcium ionophore. Induction was carried out at 37°C, in open air for 2–4 hr in 3% calf serum. Subsequently, the human γ-interferon was purified from the crude material

TABLE I. Properties of Human γ-Interferon[a]

Lot	Specific activity[b] (U/mg)	Bacterial sterility	Endotoxin/LAL (ng/ml)	Rabbit pyrogenicity	Con A (ng/ml)	General safety	Antiviral activity (VSV challenge virus) (U/ml)[b]
I-3	2.6×10^7	Sterile	3.1	Positive	< 1	Pass	2.0×10^6
I-5	3.4×10^7	Sterile	1.6	Negative	466	Pass	4.1×10^6
I-8	2.6×10^7	Sterile	0.8	Negative	293	Pass	2.1×10^6
I-11	4.9×10^7	Sterile	0.8	Negative	1381	Pass	2.9×10^6
I-12	3.0×10^7	Sterile	0.8	Negative	967	Pass	2.0×10^6
I-14	2.4×10^7	Sterile	0.05	Negative	963	Pass	2.0×10^6

[a] Preparation had antiviral activity on WISH cells but none on bovine cells (indicating γ- rather than α-interferon activity) and was negative for hepatitis antigen. Low levels of neomycin (< 10 ng/ml) and bovine serum albumin (< 200 ng/ml) were present in the final preparation.

[b] In reference to NIH α standard.

by sequential chromatography first by controlled-pore glass-adsorption chromatography, followed by Con A affinity chromatography, heparin–Sepharose affinity chromatography, and gel filtration on Sephadex G-100 (superfine). This procedure resulted in about a 10,000- to 20,000-fold purification and a recovery of 40%, with a specific activity of 10^7 U/mg protein by assay of vesicular stomatitis virus (VSV) on human WISH cells. The NIH α-interferon standard was used to calculate the units as a γ-interferon standard was not available at the time this material was produced. This γ-interferon was subjected to the usual tests for safety and was relatively nonpyrogenic and contained only low levels of endotoxin (Table I).

2.2. Patient Population

Patients with a variety of disseminated malignancies, refractory to standard curative therapy with a Karnofsky performance status greater than 60% and normal baseline hematological, hepatic, and renal function tests, were eligible for this study. After informed consent, the trial was carried out by the Biological Response Modifiers Program in our Frederick Memorial Hospital clinical unit under an approved Investigational New Drug (IND) application filed with the Office of Biologics, Food and Drug Administration.

2.3. Study Design

Patients were treated twice weekly with escalating doses of the γ-interferon preparation according to the following schedule: 0.2, 0.5, 1, 2, 5, 10, 15, 20, 30, 40, 50, and 60 × 10^6 U/m^2. Each patient was taken up through this dose escalation schedule as tolerated. The initial seven patients were treated by the intramuscular route to determine the absorption of this γ-interferon. The remaining 25 patients were given the material by intravenous administration once it was clear that the intramuscular administration did not result in significant serum levels. Patients were continued on study, as long as unacceptable toxicity did not develop, until a maximum tolerated dose (MTD) was reached or until the tumor progressed requiring other therapeutic approaches.

2.4. Monitoring

Patients were monitored for clinical toxicity, immunological effects, and antitumor effects during their dose escalations. Immunological functions were followed including natural killer cell activity, monocyte growth inhibitory activity, surface T-cell antigens, and lymphocyte proliferative responses to mitogens and alloantigens. In addition, percentages of leukocyte populations and lymphocyte subpopulations were monitored using monoclonal antibodies by flow cytometry. Serum interferon activity was monitored and careful pharmacokinetics were done in each patient afer each interferon dose. Antiviral assay activity was determined with the VSV/WISH assay used in the calculation of units for this preparation. Blood samples were drawn and interferon levels determined at 30 min, 1, 2, 3, 4, 5, 6, and 24 hr after administration.

NK cell cytotoxicity was measured as previously described (Ortaldo *et al.*, 1977) and the results interpreted after three baseline determinations were done to determine the level of NK activity of the individual patient. Methods have been developed to determine the "normal range" for each patient based on the patient's pretreatment activity and based on information from a bank of normal donors tested in this assay (Maluish *et al.*, 1983).

Monocyte growth inhibition was measured using a modification of a previously described method (Jerrells *et al.*, 1979) and with similar considerations with respect to pretreatment levels and the determination of the normal range.

Lymphoproliferative responses were carried out in response to Con A and allogeneic cells. The NK assay was done on freshly isolated peripheral blood cells and the monocyte and lymphoproliferative assays were carried out on cryopreserved lymphocytes. The lymphoproliferative response was calculated according to the relative proliferation index (RPI) as previously described (Dean *et al.*, 1977) using this index to compare the activity of the patient with appropriate normal controls.

3. RESULTS

Thus far, 32 patients have been entered on this trial. The first seven patients received interferon by intramuscular injection in escalating doses up to 2×10^7 U/m^2. No further dose escalations were possible with this route of administration due to excessive injection volumes. After seven patients were treated intramuscularly without evidence of clinical effects or detectable antiviral activity, the next 25 patients were treated using intravenous administration. The majority of the patients were given a 2-hr infusion but a few patients were treated by rapid infusion (5 min) and a few by a prolonged infusion (24 hr).

3.1. Toxicity

The γ-interferon preparation used in this study was available in lots of approximately 10^9 units. Each lot was tested for safety, pyrogenicity, and endotoxin contamination, and as shown in Table I, there was some lot-to-lot variation. The intramuscular study was done exclusively with Lot 3, and other than low-grade fever which could be attributed to pyrogenic material in this lot, no other reproducible clinical toxicities were observed. We subsequently determined on intravenous administration that Lot 3 was the most highly pyrogenic of all the lots tested.

Table II describes the toxicities associated with 2-hr intravenous administration of γ-interferon from Lots 8, 11, 12, and 14, all of which were nonpyrogenic in preclinical studies. The most prominent toxicities were fever, chills, fatigue, and anorexia and were similar to those seen with other interferon preparations (Sherwin *et al.*, 1983). However, the onset of fever was somewhat sooner and the duration was more prolonged than seen with the α preparations. In addition, hypotension, a toxicity not previously seen to any degree with other interferon preparations, was the dose-limiting toxicity for this γ-interferon preparation when given in single doses. Two of three patients treated at 60×10^6 U/m^2 developed significant hy-

TABLE II. Toxicity for 2-hr Intravenous Infusion (Lots 8, 11, 12, 14)

Dose × 10^6 U/m^2	Fever > 103	Fever < 103	Chills	Fatigue	Anorexia	Hypotension
2	0/5	4/5	3/5	2/5	0/5	0/5
5	1/5	4/5	4/5	5/5	3/5	0/5
10	1/5	4/5	4/5	4/5	3/5	0/5
15	1/4	3/4	3/4	4/4	2/4	0/4
20	2/9[a]	7/9	8/9	6/9	5/9	0/9
30	2/5	3/5	5/5	5/5	3/5	0/5
40	3/9	6/9	9/9	7/9	3/9	0/9
50	1/6	5/6	3/6	4/6	3/6	1/6
60	0/3	3/3	2/3	3/3	3/3	2/3

[a] Some patients were escalated only through 2, 5, 10, 15, and 20 × 10^6 U/m^2 dose levels and others began at 20 × 10^6 U/m^2 to be escalated up to 60 × 10^6 U/m^2. Some patients progressed on treatment and could not undergo the whole dose escalation. Thus, the denominator reflects the number of patients actually treated at each dose level.

potension with greater than a 30% fall in their systolic and diastolic blood pressure lasting at least 30 min. Because some hypotension effect was also seen at 50 × 10^6 U/m^2, the MTD was defined as 50 × 10^6 U/m^2.

3.2. Monitoring

The antiviral activity and pharmacokinetics for one patient treated at two dose levels are shown in Fig. 1. Significant antiviral activity was seen at 2 × 10^6 U/m^2

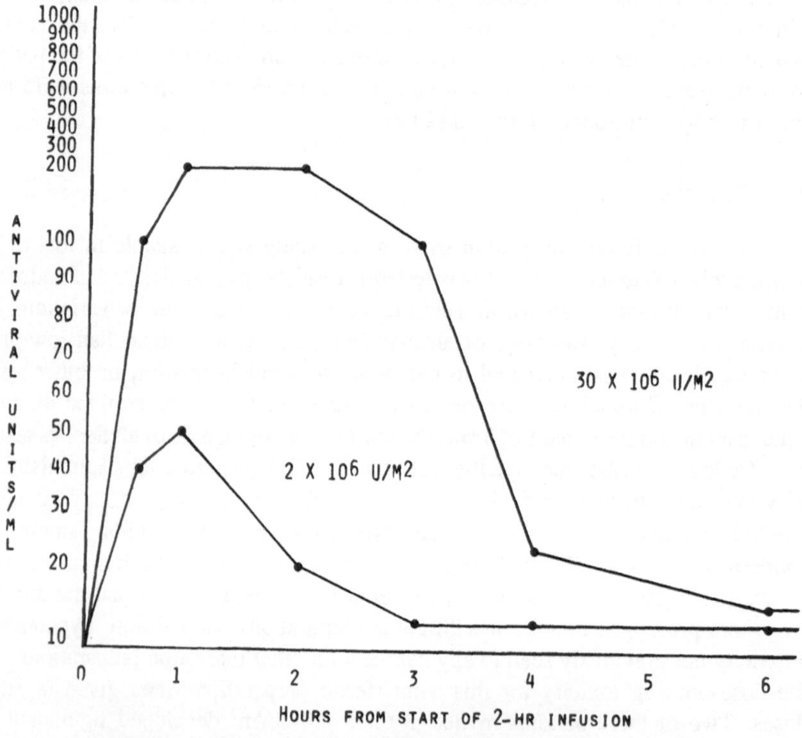

FIGURE 1. Serum antiviral activity (VSV/WISH assay).

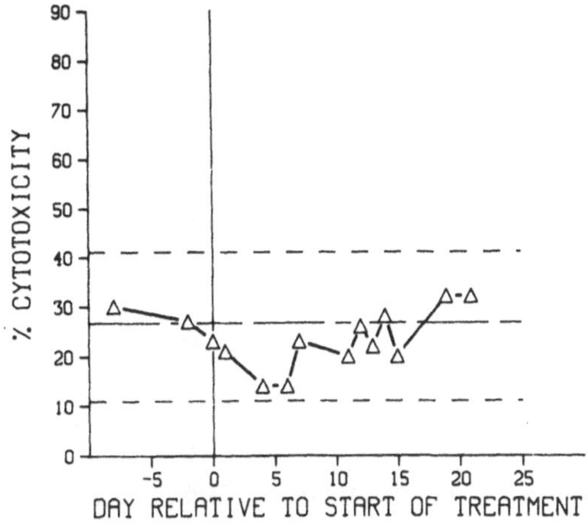

FIGURE 2. NK cytotoxicity.

with peak levels up to 200 to 300 U at the end of the 2-hr infusion with 30 × 10^6 U/m^2. In general, these levels were lower on a per unit basis and were less sustained than with the α-interferon preparations. While this preparation cannot be compared with α-interferons on a unit per unit basis, the dose-limiting side effects (MTD) of this γ-interferon were obvious at 60 × 10^6 U/m^2. At the MTD, the serum levels of this interferon were much lower than those for the MTD of the α preparations.

NK cytotoxicity was measured in all patients, and results for a typical patient are shown in Fig. 2. The dashed lines depict the normal range for this patient and while some drop within this range was noted, no significant change in the NK activity was seen for this patient or the others studied in this trial. Most of these patients had reasonably normal levels of NK cells and NK cytotoxicity when entered on this study.

With regard to the monocyte growth inhibition assay, most of the patients had

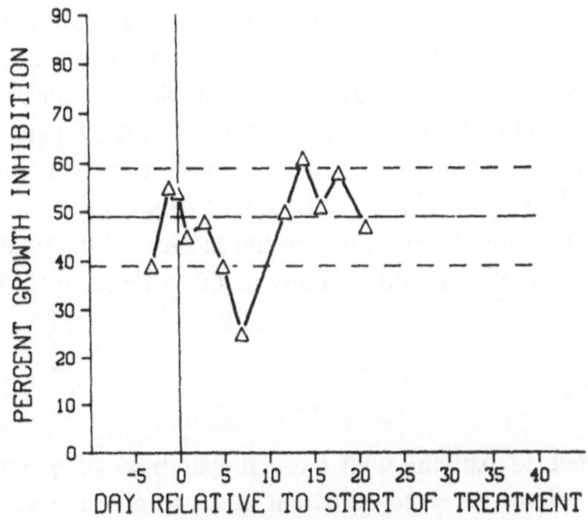

FIGURE 3. Monocyte growth inhibition.

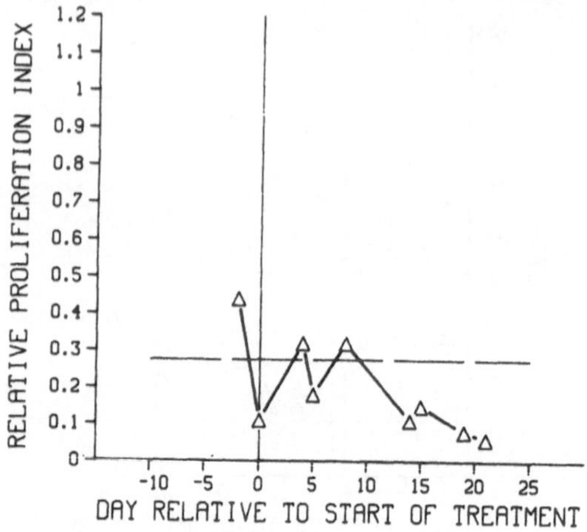

FIGURE 4. Lymphoproliferative response to Con A.

normal initial levels as compared to normal controls. Although an occasional value did go outside the normal range, the vast majority of follow-up determinations did not reveal any significant effect clearly related to interferon administration (Fig. 3).

In contrast, the lymphoproliferative responses of these patients were generally low with RPI of 0.2 to 0.5 when compared to a normal index of 1.0. In this context, some of the patients did show a decrease in their lymphoproliferative response to Con A during this trial. An example of this kind of suppression of lymphoproliferation is seen in Fig. 4.

Lymphocyte surface markers using monoclonal antibodies OKT3, T4, T8, T10, B1, and MO2 by cytofluorograph analysis did not reveal any significant changes during or after the administration of this interferon preparation.

3.3. Antitumor Activity

Of the seven patients evaluable with intramuscular administration, three remained stable during the treatment and four progressed while on study. Twenty-five patients were treated intravenously; 19 were stable and four progressed during the study. However, only 11 patients received doses in the ranges associated with detectable antiviral activity in the serum and those were generally the patients receiving 2×10^6 U/m^2 and above. Subsequent follow-up of these patients for several months has revealed further progression in most of the treated patients and no evidence of an antitumor response in any of the patients treated in this study.

4. DISCUSSION

We have treated 32 patients with these preparations of γ-interferon. Unfortunately, only 11 patients received sufficient doses to give measurable serum antiviral titers. Fever, chills, fatigue, and anorexia, common side effects with interferon preparations, were seen with these γ-interferon preparations. The MTD was defined by hypotension, a side effect not often seen with previous α-interferon preparations. It should be clearly understood that this preparation could easily contain other clinically active biological molecules given the methods used in the preparation of extracted γ-interferon. Thus, it cannot be stated with certainty that the dose-limiting toxicity of this material was due to the γ-interferon molecule or due to other biological substances that copurified with it.

The biological response modifying activities of this preparation were not dramatic. No significant changes were seen in NK activity or in monocyte growth inhibitory activity. Some suppression of Con A-induced lymphoproliferation was seen and this may reflect some antiproliferative activity of this preparation. Lymphocyte and monocyte surface markers were unaltered during this trial.

As this was a phase I trial, a complete assessment of therapeutic activity for this preparation is not possible. Only a few patients were treated and even fewer had detectable serum antiviral activity during therapy. Moreover, patients were not given repetitive courses of treatment at the same dose, and the lack of therapeutic activity can only be a preliminary observation for this γ-interferon preparation.

Given the lot-to-lot variability, in terms of pyrogenicity, endotoxin contamination, and biological activity, it would be difficult to pursue this extracted form of gamma interferon in phase II studies. As with the α-interferons, it is preferable to use a preparation of high pharmaceutical quality and one with reproducible biological activity and reproducible purity. For such studies, it will be preferable to investigate the genetically engineered γ-interferons which are now available and in phase I studies. Large-scale phase II studies will be necessary to determine whether γ-interferon has greater anticancer activity than the α or β forms (Oldham, 1983b).

ACKNOWLEDGMENT. This project has been funded in part with federal funds from the Department of Health and Human Services under Contract NO1-CO-23910 with Program Resources, Inc. The contents of this publication do not necessarily reflect the views or policies of the Department of Health and Human Services, nor does mention of trade names, commercial products, or organizations imply endorsement by the U.S. Government.

REFERENCES

Dean, J. H., Connors, R., Herberman, R. B., Silva, J., McCoy, J. L., and Oldham, R. K., 1977, The relative proliferation index as a more sensitive parameter for evaluating lymphoproliferative response of cancer patients to mitogens and alloantigens, *Int. J. Cancer* **20**:359–370.

Jerrells, T. R., Dean, J. H., Richardson, G., Cannon, G. B., and Herberman, R. B., 1979, Increased monocyte-mediated cytostasis of lymphoid cell lines in breast and lung cancer patients, *Int. J. Cancer* **23**:768–76.

Knost, J. A., Sherwin, S. A., Abrams, P. G., Ochs, J. J., Foon, K. A., Williams, R., Tuttle, R., and Oldham, R. K., 1983, The treatment of cancer patients with human lymphoblastoid interferon: A comparison of two routes of administration, *Cancer Immunol. Immunother.* **15**:144–148.

McPherson, T. A., and Tan, Y. H., 1980, Phase I pharmacotoxicology study of human fibroblast interferon in human cancer, *J. Natl. Cancer Inst.* **65**:75–79.

Maluish, A. E., Ortaldo, J. R., Conlon, J. C., Sherwin, S. A., Leavitt, R., Strong, D. M., Weirnik,

P., Oldham, R. K., and Herberman, R. B., 1983, Depression of natural killer cytotoxicity following *in vivo* administration of recombinant leukocyte interferon, *J. Immunol.* **131**:503–507.

Oldham, R. K., 1983a, Toxic effects of interferon, *Science* **219**:902.

Oldham, R. K., 1983b, Biologicals and biological response modifiers: The fourth modality of cancer treatment, *Cancer Treat. Rep.* **68**:221–231.

Ortaldo, J. R., Bonnard, G. D., and Herberman, R. B., 1977, Cytotoxic reactivity of human lymphocytes cultured, *in vitro, J. Immunol.* **119**:1351–1357.

Sherwin, S. A., Knost, J. A., Fein, S., Abrams, P. G., Foon, K. A., Ochs, J. J., Schoenberger, C., Maluish, A. E., and Oldham, R. K., 1982, A multiple dose phase I trial of recombinant leukocyte A interferon in cancer patients, *J. Am. Med. Assoc.* **248**:2461–2466.

Sherwin, S. A., Mayer, D., Ochs, J. J., Abrams, P. G., Knost, J. A., Foon, K. A., Fein, S., and Oldham, R. K., 1983, Recombinant leukocyte A interferon in advanced breast cancer: Results of a phase II efficacy trial, *Ann. Intern. Med.* **98**:598–602.

The Response of Tumor-Bearing Patients to the Injection of Lymphoid Cell Line Lymphokine

D. C. DUMONDE, MELANIE S. PULLEY, ANNE S. HAMBLIN, BARBARA M. SOUTHCOTT, and F. C. DEN HOLLANDER

1. RATIONALE AND DESIGN OF STUDY

1.1. Background

We began this work in 1976 at a time when lymphokines were becoming accepted as mediators of cellular immune responses and when consideration was being given to the immunodepression associated with progressive cancer and with its treatment. At that time, work on leukocyte interferon, dialyzable leukocyte transfer factor, and thymic extracts had established a precedent for the therapeutic investigation of biological materials in patients with otherwise irreversible neoplastic disease; and it seemed that lymphokines could also have therapeutic potential in human cancer (Hamblin *et al.*, 1978). We were examining lymphokines generated by the cultured B-lymphoblastoid cell line RPMI 1788 as potential standards in the leukocyte migration test (see Hamblin *et al.*, 1982); and when Papermaster *et al.* (1976) reported on the safe intralesional (i.l.) injection of RPMI 1788 lymphokine into human cutaneous metastases, we decided to use this source of lymphokine for the present study. With ethical permission and informed consent, we set out to extend

D. C. DUMONDE, MELANIE S. PULLEY, and ANNE S. HAMBLIN • Department of Immunology, St. Thomas' Hospital, London SE1 7EH, England. BARBARA M. SOUTHCOTT • Department of Radiotherapy, Charing Cross Hospital, London W6, England. F. C. DEN HOLLANDER • Organon International BV, Oss, The Netherlands.

knowledge of the histopathological responses to the i.d. and i.l. injection of RPMI 1788 lymphokine ("LCL-LK") in patients with advanced cancer and to investigate the clinical, hematological, biochemical, and immunological responses to single and repeated i.v. injections of LCL-LK.

1.2. Design of Study

Intradermal reactions to graded doses of LCL-LK and to corresponding amounts of a control bland protein (culture medium) were studied clinically (in 28 patients) and histologically (in 10 patients) between 30 min and 72 hr and were compared with the features of the tuberculin skin reaction. Single and multiple (3–6) i.l. injections of graded amounts of LCL-LK were given into a total of 10 accessible skin metastases in three patients with recurrent breast carcinoma and the histology compared with that of saline-injected nodules; likewise, six patients with prostatic carcinoma received up to five intraprostatic injections of LCL-LK and prostatic biopsies were subsequently examined.

Responses to single and repeated i.v. injection of LCL-LK were studied in an initial series of 20 patients (mostly with advanced breast carcinoma) in three groups: (1) in the short term (1–3 weeks, 10 patients), (2) in the medium term (2–4 months, 6 patients), and (3) in the long term (9–25 months, 4 patients). A variety of injection schedules were used ranging from weekly maintenance (as an outpatient) to daily escalation (while hospitalized).

1.3. Lymphokine Preparations and Patient Management

LCL-LK were prepared at Organon Laboratories, Holland, by large-scale culture of the RPMI 1788 cell line in the absence of added plasma. Culture supernatants were treated with β-propiolactone, ultrafiltered (5–100 or 10–300Kd),desalted and lyophilized, redissolved in pyrogenfree water, and filtered through a 0.22-μm membrane into vials and relyophilized. Control protein preparations were similarly processed and dispensed. Quality control included tests for sterility, endotoxin, hepatitis B surface antigen, pyrogenicity, and acute toxicity (up to 90–120 mg/kg in mice and guinea pigs). These LCL-LK preparations contained a number of lymphokine activities including LIF, MIF, MAF (*Listeria* clearance), lymphotoxin, and skin inflammatory activities. Their endotoxin content was low (*20 ng/mg by the Limulus* assay) and there was no or little interferon (IFN) activity (0–1000 U/mg). In a comparative study (Pulley *et al.*, 1983), the control protein preparation had twice the endotoxin content of the corresponding active LCL-LK. All active preparations were pyrogenic in rabbits and more recently we have found similar preparations to have IL-1 activity (on C3H/HeJ mouse thymocytes) but not IL-2 activity (on human PHA-blasts). The LCL-LK preparations therefore had multiple lymphokine activities, no or negligible IFN activity, and very low quantities of endotoxin.

Patients were cared for individually; all had advanced cancer refractory to other treatment modalities (Southcott *et al.*, 1982). Symptomatic treatment was maintained where indicated and all patients were hospitalized for an initial study period of responses to single injections of LCL-LK. Patients were given escalating i.v. doses of LCL-LK up to a maximum symptomatic dose (usually limited by pyrexia and rigors; see below) and were maintained on this either daily or thrice weekly (in the short term) or weekly (in the longer term). Four patients studied over 9–25 months were hospitalized at intervals for 3-week courses of LCL-LK given daily or thrice weekly. Six inpatients had extensive biochemical and endocrinological as well as hematological monitoring (Pulley *et al.*, 1982, 1983).

2. RESPONSE TO INTRADERMAL INJECTION

Dose-related skin reactions of erythema appeared within the first hour and became progressively indurated to reach a clinical maximum between 12 and 18 hr, sustained for up to 72 hr. Histologically, these were characterized by early polymorph infiltration followed by a progressive mononuclear cell accumulation around small blood vessels with a characteristic endothelial cell hypertrophy closely resembling the established tuberculin skin reaction (Dumonde *et al.*, 1982). The LCL-LK reaction seemed to superimpose an early phase of polymorph exudation upon an inflammation otherwise characteristic of delayed hypersensitivity. Different batches of LCL-LK differed in their ability to induce a sustained (12–18 hr) skin reaction and in individual patients their relative potency could be compared readily from dose–response relationships by determining the amount (e.g., 1–20 μg) required to elicit a sustained erythema of 20- or 30-mm diameter (ED_{20} or ED_{30}). Patients usually gave prominent skin reactions to LCL-LK whether or not they were anergic to recall antigens (e.g., tuberculin); however, LCL-LK reactions became weaker as the clinical condition deteriorated. At present we are studying the clinical and therapeutic factors that may determine the extent, duration, and quality of the skin test to LCL-LK.

3. RESPONSE TO INTRALESIONAL INJECTION

Injection of LCL-LK into nodular cutaneous metastases of breast carcinoma led to regression of the nodules, sometimes with ulceration. Excision biopsy showed tumor cell necrosis and mononuclear cell infiltration; in some nodules biopsied after several LCL-LK injections the scirrhous stroma was still evident, bereft of recognizable tumor (Paradinas *et al.*, 1982). In one patient (with hepatic secondaries) whose clinical condition was deteriorating rapidly, islands of intact tumor cells appeared to be regrowing in skin nodules removed 7 days after the last of six local LCL-LK injections. Repeated (alternate-day) intraprostatic injection of substantial amounts of LCL-LK in six patients with metastatic prostate carcinoma led to local mononuclear cell infiltration and some tumor cell necrosis. Despite short-term pyrexia and leukocytosis (see below) this was well tolerated: two patients experienced a dramatic relief of metastatic bone pain; a third showed a prominent flare of a previous tuberculin reaction; and two patients showed a fall in prostatic acid phosphatase levels. We conclude that the study of i.l. responses to LCL-LK is a useful and integral part of their characterization in the therapeutic investigation of tumor-bearing patients.

4. ACUTE RESPONSES TO INTRAVENOUS INJECTION

4.1. Clinical and Hematological Responses

A typical clinical response to a symptomatic i.v. dose of LCL-LK consisted of mild rigors between 30 and 60 min after injection, sometimes with nausea, headache, and occasional vomiting; and the development of characteristic pyrexia within 30–90 min, peaking at 2–4 hr. When any one preparation was used in one individual, the pyrexia was dose-related but different individuals had different thresholds of symptomatic response. Pyrexia and subjective responses had usually disappeared by 8 hr and could be terminated by giving oral aspirin or i.v. hydrocortisone. Tachycardia and a rise in pulse pressure were related to the rigors and pyrexia. Hematological responses typically consisted of a transient lymphopenia and polymorph leukocytosis, most evident at 3–5 hr; sometimes a small (20%) drop in platelet count at that time; but no other effect on the red cell compartment or on the clotting system. In the more prominent reactions the polymorph count showed a "left shift"; but the white count had usually returned to preinjection values by 8 hr (Dumonde *et al.*, 1981; Pulley *et al.*, 1982). In patients with experience of cytotoxic drug therapy, such injections were particularly well tolerated.

4.2. Biochemical, Endocrine, and Immunological Effects

A detailed study of six patients (Pulley *et al.*, 1982) revealed an acute-phase reaction with increases in serum cortisol, growth hormone, and ACTH peaking at 2–4 hr postinjection; a decrease in plasma zinc and iron, most prominent at 6–12 hr postinjection; and a rise in acute-phase proteins (CRP; α_1-AG) peaking later at 24–48 hr. A control study was done to show that neither these responses nor the associated pyrexial or hematological responses could be attributed to the (low) amount of endotoxin in the LCL-LK preparation (Pulley *et al.*, 1983). There were no changes in T_3, T_4, thyroid-stimulating hormone, prolactin, or glucose levels; this indicated differences from a simple "stress" response (to pyrexia). In fact, on abolishing a pyrexial response in one patient with paracetamol, there was still an increase in cortisol; and pyrexia and leukocytosis still occured in another patient on steroids whose cortisol response was suppressed. These findings suggested the operation of more than one response pathway to LCL-LK.

In studying other parameters, no acute effects have been observed on autoantibody or antibacterial antibody levels, on complement components (including C_3 activation), or on liver or renal function. To date we have yet to undertake systematic studies of white cell function; but so far we have not confirmed Papermaster's report of an increased phagocytic activity of PMN after a single systemic injection of RPMI 1788 lymphokine (Rodes *et al.*, 1982).

5. LONGER-TERM SYSTEMIC ADMINISTRATION

5.1. Effects of Repeated Injection

Repeated daily i.v. injections with a symptomatic dose of LCL-LK induced in some patients a tolerance of pyrexia with a lessening of the leukocytosis and acute-phase protein responses. However, when such injections were more widely spaced (even on alternate days), pyrexial tolerance was less often encountered. With an escalating dose-schedule of injections every 2 or 3 days, a maximum symptomatic dose was generally attainable and was dependent on the extent to which rigors, nausea, and pyrexia were accomodated. In four of the patients receiving five alternate-day intraprostatic injections of LCL-LK, pyrexial responses up to 40–41°C were regularly seen with a tendency to a gradual increase in WBC count during the course of LCL-LK. We are now studying the extent to which different components of the acutephase response may be dissociable by the phenomenon of tolerance, and the extent to which this may be dependent on the dose and administration schedule of LCL-LK.

5.2. Patient Compliance: Lack of Long-Term Toxicity

With individual compassionate care, patient compliance was high; and remarkably, patients receiving i.v. courses of LCL-LK actually welcomed each successive injection. Not only was there a complete lack of long-term toxicity (over a 2-year period) but also the majority of patients volunteered feelings of well-being. In individual patients there have been sufficient indications of temporary clinical benefit to encourage us to continue with this approach (Dumonde *et al.*, 1981; Southcott *et al.*, 1982) but in view of the small numbers of patients currently maintained for more than 1 year on regular LCL-LK, we reserve judgement concerning the effect on the host–tumor relationship until more individuals have been studied (in protocols which also incorporate the monitoring of white-cell function). In this context, Gouveia *et al.* (1982) have reported a rise in NK activity during a course of daily injections of RPMI 1788 lymphokine; and in this symposium the reports of Papermaster *et al.* (Chapter 37) and Joshua and Djerassi (Chapter 50) also give reinforcement to continued studies of RPMI 1788 lymphokine.

6. IMMUNOPHARMACOLOGICAL IMPLICATIONS

6.1. Intradermal, Intralesional, and Systemic Effects

Our evidence indicates that characterization of i.d. and i.l. responses to preparations of LCL-LK is a useful component in their investigation. The work supports the view that lymphokines are mediators of local inflammatory reactions of delayed hypersensitivity and tumor cell necrosis in man and raises the interesting question of whether IL-2 is necessarily involved in either phenomenon. It could be argued that the presence of IL-1 or an IL-1 inducer ("MAF") in the LCL-LK preparations could well lead to the production of "effective" IL-2 (Grimm and Rosenberg, 1984) by lymphocytes recruited into a local environment where additional mediator effects on the microvasculature were also operating. At present it is not known what moieties present in LCL-LK preparations initiate local inflammation and we speculate that local modulation of leukocyte–endothelial interactions may turn out to be one important pathway of lymphokine action (Dumonde *et al.*, 1983).

The pyrexial, hematological, and biochemical responses to i.v. LCL-LK support the view that lymphokines also mediate systemic reactions associated with the state of delayed hypersensitivity (Rich, 1951). It is tempting to ascribe the pyrexial, cation, and acute-phase protein effects to the presence of IL-1 (or an IL-1 inducer lymphokine) in the LCL-LK preparations (Kluger and Cannon, 1983). It is less clear how a direct action of interleukins or lymphokines could promote a temporary granulocytosis or lymphopenia. The timing and appearance of immature granulocytes suggest an effect on recruitment from bone marrow rather than mobilization (e.g., by rigor) from the marginated pool; while the finding of raised cortisol levels suggests that the lymphopenia is cortisol-mediated. Again, further analysis of the acute-phase response is required to study selective modification by tolerance-inducing schedules of LCL-LK and by steroidal and nonsteroidal drugs.

The striking elevation of ACTH, cortisol, and growth hormone levels supports evidence that lymphokines may provide a homeostatic link between the immunological and the neuroendocrine systems (Besedovsky and Sorkin, 1977; Besedovsky *et al.*, 1981) by signaling the state of lymphoid cell activation to the hypothalamus. The question naturally arises of whether this phenomenon is ascribable simply to IL-1 per se or to some other polypeptide molecule present in preparations of lymphokines that have been examined to date. Recent studies from Goldstein's group indicate that some of the thymic peptides may have related biological effects (see Hall and Goldstein, 1983) and it will be of interest to determine if the LCL-LK preparations that we have been studying exert relevant endocrinological effects upon more direct application to the hypothalamus (see also Hall *et al.*, Chapter 29, this volume). In any event, there would seem to be two distinct mechanisms of systemic response to LCL-LK; for the pyrexial and cortisol responses were dissociable (Pulley *et al.*, 1982). On this basis, the pyrexia could be prostaglandin-mediated while selective activation of hormone-releasing factors could be mediated via altered neuropeptide/catecholamine levels in the hypothalamus (see Besedovsky *et al.*, 1983).

6.2. Relationship between Local and Systemic Responses

These findings strongly implicate lymphokines in the activation and homeostasis of host-defense systems which may be "at risk" in the tumor-bearing patient, either due to the tumor itself or to treatment modalities such as irradiation, chemotherapy, and endocrinological manipulation. Both local and systemic effects of LCL-LK are highly complex and are unlikely to be the result of a single response mechanism; however, neither can be attributed to the presence of endotoxins, of interferons, or possibly of IL-2 in the LCL-LK preparations. If IL-1 (or IL-1 induction) is involved in these responses, it might act both intradermally and on the hypothalamus by stimulation of prostaglandin production, one result being a local or general rise in temperature. It therefore seems relevant to determine whether the ability of different batches (or fractions) of LCL-LK to cause local erythematous inflammation runs parallel with their ability to induce the systemic responses of pyrexia, metal cation depression, and acute-phase protein synthesis (Dinarello and Wolff, 1982). It has recently been suggested that fever, as a host-defense mechanism in cancer (Coley-Nauts, 1978), may act by facilitating T-cell responses to both IL-2 and IL-1 itself (Duff and Durum, 1983). On this basis the study of local inflammatory responses to LCL-LK may have a broader significance than appreciated hitherto.

It remains a matter for further investigation whether the ability of LCL-LK preparations to induce local tumor regression and tumor cell necrosis is related to their apparent effects on the compartmentation of lymphocytes and other leukocytes. At present it seems that these local and systemic effects may be brought about by different mechanisms and it is therefore possible that they are initiated by different entities in the LCL-LK preparations. Taken separately and together, the local and systemic reactions to LCL-LK can be viewed in terms of the sequential activation of host-defense and homeostatic mechanisms; and as providing a striking illustration of the concept of biological response modification.

7. CLINICAL IMPLICATIONS

7.1. "Tactics": The Choice of Lymphokines and Host Response

In considering lymphokine preparations as therapeutic agents in cancer, chronic infections, and even other diseases, a first tactical problem is whether or when to use purified or unpurified preparations. Table I (Dumonde and Hamblin, 1983) conveys the view that expression of the cellular immune response is probably effected by different combinations of lymphokines acting in physiological concert with different combinations of lymphoid and accessory cell populations producing and responding to lymphokines and to lymphokine-induced "secondary" mediators. This view is also supported by evidence of lymphokine synergy (Dinarello *et al.*, 1982).

In most forms of human cancer it is likely that no single lymphokine function is deficient; and with the logistic problems presented by quantitative lymphokine bioassay and lymphoid cell compartmentation, it seems tactically unrealistic to hope to exhibit lymphokines in an analogous way to endocrine replacement therapy. Moreover, the host response to cancer (and its treatment) involves more than just the immune system; and different patients will be expected to be in different states of pathophysiological risk. It seems attractive to administer highly purified lymphokines locally, systemically, or even extracorporeally (Grimm and Rosenberg, 1983) if only to find out more about them; yet at present there is no scientific or clinical basis for putting individual lymphokines in any ethical order of preference or priority (see Table I). Accordingly, our present approach is simply to study and compare host response to a limited range of eukaryotic lymphokine preparations of broad spectrum whose prominent activities can be manipulated (e.g., presence or absence of IFN; presence or absence of IL-2; presence or absence of IL-1) and also specified by means of acceptable methods of measurement. By this means we hope to provide information complementary to that arising from the investigation of highly purified or recombinant lymphokine preparations.

The second major tactical problem is to decide what host responses to monitor. It has already been implied that objective tumor shrinkage may be an unrealistic endpoint (Carter, 1980; Oldham, 1981) and that in cancer the goal of biological response modifier therapy should be directed at improving the ability of the host to "live with" both the tumor and the oncologist. On the other hand, it is impractical to utilize every quality-controlled test of inflammatory capability, white-cell function, other hematology and immunology, clinical chemistry and endocrine status, with the regularity and frequency that one might use in matching groups of syngeneic laboratory rodents. The choice of what to monitor (other than clinical toxicity) will depend on the category of patients investigated, the quality of the lymphokine preparations themselves, and the value judgement of the investigator: it is the latter that will provide the diversity of information necessary for future assessment of this field.

TABLE I. A Simplified Approach to the Significance of Lymphokines in the Pathophysiology of the Immune System[a]

Physiological event	Suggested participation of lymphokine mediators
Allergic inflammation	
Delayed hypersensitivity, granuloma formation, allograft rejection, graft-vs.-host reactions, etc.	"Inflammatory" (complex), chemotactic, cytotoxic, migration-inhibitory, macrophage-aggregating, fibroblast-activating, endothelial cell-activating, etc.
Protective cellular immunity	
Restriction of tumor growth and parasite multiplication; destruction of parasitized host cells; activation of host macrophages; etc.	Cytotoxic, macrophage-activating, chemotactic, migration-inhibitory, interferon-like, colony-stimulating, etc.
Immunoregulation	
Control of lymphocyte compartmentation; cellular cooperation during immune induction; physiological suppression of immune responses; nonspecific Ig production; maintenance of diversity despite phenotypic restriction; etc.	Lymph node-activating (complex), lymphocyte-mitogenic (complex), macrophage-activating, lymphocyte and macrophage differentiation factors, endothelial-activating factors, suppressor factors, helper factors, colony-stimulating factors, etc.

[a] Modified from Dumonde and Hamblin (1983).

7.2. Strategy: The Choice of Study Design

It is our working hypothesis that one or more "lymphokine strategies" will be found which will improve the host's ability to tolerate his (neoplastic) disease, to tolerate existing treatment modalities, and to recover more quickly from their depressant effect on immunological and other host-defense mechanisms. Although our LCL-LK preparations contain prominent lymphotoxin activity and induce tumor cell necrosis on local injection, it is unlikely that therapeutic benefit following systemic injection is due to a direct cytotoxic action of the lymphokine on tumor cells sequestered in a distant extravascular compartment. Accordingly, the classical approach of chemotherapy involving simply the ascertainment of MTD needs modification in phase I/II studies (Carter, 1980; Oldham, 1981). What is needed is a double-blind evaluation of the lymphokine against a placebo, the criterion being impact on a group of clinical features and laboratory assays accepted as favorable indications of host integrity: this is the strategy we have chosen to adopt in longer-term studies of patients with the common forms of solid tumors, in addition to undertaking "open" investigations of the acute-phase response to preparations of different but known quality.

On the basis of our present results a limited number of related objectives emerge. A first is to determine whether recurrent acute-phase responses stimulate white-cell function more effectively than partial responses, modified by tolerance schedules or by symptomatic drugs, in which greater amounts of LCL-LK may be accepted. A second objective is to determine whether the form and extent of a patient's response to LCL-LK injection are related to the state of prior immunodepression or to the ability of that LCL-LK preparation to stimulate the patient's white-cell function *in vitro*. A third objective is to evaluate whether recurrent or sequential longer-term administration of lymphokines to the tumor-bearing patient results in a greater ability to tolerate or recover from the depressive effects upon white-cell function of extensive radiotherapy or chemotherapy; and to determine whether there are certain criteria of the acute response of such patients which would appear to be predictive of this favorable outcome. The formulation of these objectives illustrates how strategic thinking influences the tactics of study design.

8. CONCLUSIONS

Lymphokines are nonantibody proteins, generated by lymphocyte activation, that act as intercellular mediators of the immunological response. Our studies indicate that lymphokines activate both intrinsic and extrinsic (neuroendocrine) pathways by which the immunological response may be regulated; that they exert additional acute-phase biochemical responses also relevant to host defense; and that they induce significant effects upon lymphocyte and leukocyte compartmentation. We conclude that RPMI 1788 lymphokine (LCL-LK) is safe for long-term systemic administration to tumor-bearing patients and that further study of its effects will yield criteria on which to base strategies and protocols for the general evaluation of lymphokines in cancer research therapy.

ACKNOWLEDGMENTS. We acknowledge this collaboration of: Dr. F. J. Paradinas, Charing Cross Hospital, London, for histopathological studies; Professor A. Fleck, Dr. G. Carter, and Dr. B. Muller, Department of Chemical Pathology, Charing Cross Hospital, London, and Professor Lesley Rees, Department of Endocrinology, St. Batholomew's Hospital, London, for hormone, acute-phase protein, Zn, and Fe assays; Dr. Alan Morris, Department of Biological Sciences, University of Warwick, for interferon assays; Dr. A. H. Gordon, National Institute for Biological Standards and Control, Hampstead, London, for IL-1 assays; Mr. M. R. G. Robinson, Pontefract General Infirmary, Yorkshire, England, in the investigation of patients with prostatic carcinoma; and Drs. Elenie van Vliet and A. H. W. M. Schuurs of Organon BV, Holland, for much encouragement and discussion.

REFERENCES

Besedovsky, H., and Sorkin, E., 1977, Network of immune–neuroendocrine interactions, *Clin. Exp. Immunol.* **27**:1–12.

Besedovsky, H. O., del Ray, A., and Sorkin, E., 1981, Lymphokine-containing supernatants from Con A stimulated cells increase corticosterone blood levels, *J. Immunol.* **126**:385–387.

Besedovsky, H. O., del Ray, A., and Sorkin, E., 1983, Neuroendocrine immunoregulation, in: *Immunoregulation* (N. Fabris, E. Garaci, J. Hadden, and N. A. Mitchison, eds.), pp. 315–339, Plenum Press, New York.

Carter, S. K., 1980, Biological response-modifying agents: What is an appropriate phase I–II strategy?, *Cancer Immunol. Immunother.* **8**:207–210.

Coley-Nauts, H., 1978, Bacterial vaccine therapy of cancer, *Dev. Biol. Stand.* **38**:487–494.

Dinarello, C. A., and Wolff, S. M., 1982, The molecular basis of fever in humans, *Am. J. Med.* **72**:799–819.

Dinarello, C. A., Dempsey, R. A., Mier, J. W., Rosenwasser, C. J., and Parkinson, D. R., 1982, Fever, interleukin-1 and host defence against malignancy, in: *Lymphokine Research*, Vol. 1 (L. B. Lachman and S. Gillis, eds.), pp 59–68, Mary Ann Liebert Inc., New York.

Duff, G. W., and Durum, S. K., 1983, The pyrogenic and mitogenic actions of interleukin-1 are related, *Nature (London)* **304**:449–451.

Dumonde, D. C., and Hamblin, A. S., 1983, Lymphokines, in: *Immunology in Medicine* (E. J. Holborow and W. G. Reeves, eds.), 2nd ed., pp. 121–150, Academic Press, New York.

Dumonde, D. C., Pulley, M. S., Hamblin, A. S., Singh, A. K., Southcott, B. M., O'Connell, D., Paradinas, F. J., Robinson, M. R. G., Rigby, C. C., den Hollander, F., Schuurs, A., Verheul, H., and van Vliet, E., 1981, Short-term and long-term administration of lymphoblastoid cell line lymphokine (LCL-LK) to patients with advanced cancer, in: *Lymphokines and Thymic Hormones: Their Potential Utilization in Cancer Therapeutics* (A. L. Goldstein and M. A. Chirigos, eds.), pp. 301–318, Raven Press, New York.

Dumonde, D. C., Pulley, M. S., Paradinas, F. J., Southcott, B. M., O'Connell, D., Robinson, M. R. G., den Hollander, F., and Schuurs, A. H., 1982, Histological features of skin reactions to human lymphoid cell line lymphokine in patients with advanced cancer, *J. Pathol.* **138**:289–308.

Dumonde, D. C., Hamblin, A. S., Kasp-Grochowska, E., Pulley, M. S., and Wolstencroft, R. A., 1983, Lymphokines and the lymphoendothelial system: An illustration of immunoregulatory integration, in: *Immunoregulation* (N. Fabris, E. Garaci, J. Hadden, and N. A. Mitchison, eds.), pp. 177–199, Plenum Press, New York.

Gouveia, J., Ribaud, P., Goutner, A., and Mathé, G., 1982, Phase I study of systemic lymphokine (LCL-LK) administration in advanced cancer patients, in: *Human Lymphokines: The Biological Immune Response Modifiers* (A. Khan and N. O. Hill, eds.), pp. 633–639, Academic Press, New York.

Grimm, E. A., and Rosenberg, S. A., 1984, The human lymphokine-activated killer cell phenomenon, *Lymphokines* **9** (in press).

Hall, N. R., and Goldstein, A. L., 1983, The thymus–brain connection: Interactions between thymosin and the neuroendocrine system, in: *Lymphokine Research,* Vol. 2 (L. B. Lachman and S. Gillis, eds.), pp. 1–6, Mary Ann Liebert Inc., New York.

Hamblin, A. S., Wolstencroft, R. A., Dumonde, D. C., den Hollander, F. C., Schuurs, A. H. W., Backhouse, B. M., O'Connell, D. and Paradinas, F., 1978, The potential of lymphokines in the treatment of cancer, *Dev. Biol. Stand.* **39**:335–345.

Hamblin, A. S., Zawisza, B., Shipton, U., Dumonde, D. C., den Hollander, F. C., and Verheul, H., 1982, The use of human lymphoid cell line lymphokine preparations (LCL-LK) as working standards in the bioassay of human leucocyte migration inhibition factor (LIF), *J. Immunol. Methods* **59**:317–329.

Kluger, M. J., and Cannon, J. G., 1983, Endogeneous pyrogen, in: *Advances in Immunopharmacology,* Vol. 2 (J. W. Hadden, L. Chedid, P. Dukor, F. Spreafico, and D. Willoughby, eds.), pp. 579–584, Pergamon Press, Elmsford, N.Y.

Oldham, R. K., 1981, Biological response modifier therapy—An overview, *Cancer Bull.* **33**:244–250.

Papermaster, B. W., Holtermann, O. A., Klein, E., Djerassi, I., Rosner, D., Dao, T., and Costanzi, J. J., 1976, Preliminary observations on tumour regressions induced by local administration of a lymphoid cell culture supernatant fraction in patients with cutaneous metastatic lesions, *Clin. Immunol. Immunopathol.* **5**:31–43.

Paradinas, F. J., Southcott, B. M., O'Connell, D., den Hollander, F., Schuurs, A. H., Pulley, M. S., and Dumonde, D. C., 1982, Changes induced by local injection of human lymphoid cell line lymphokine into dermal metastases of breast carcinoma: A light and electron microscopical study, *J. Pathol.* **138**:309–323.

Pulley, M. S., Dumonde, D. C., Carter, G., Muller, B., Fleck, A., Southcott, B. M., and den Hollander, F. C., 1982, Hormonal, haematological and acute phase protein responses of advanced cancer patients to the intravenous injection of lymphoid-cell lymphokine (LCL-LK), in: *Human Lymphokines: The Biological Immune Response Modifiers* (A. Khan and N. O. Hill, eds.), pp. 651–662, Academic Press, New York.

Pulley, M. S., Dumonde, D. C., Carter, G., Muller, B., Southcott, B. M., and den Hollander, F. C., 1983, Lymphokines as mediators of extrinsic immunoregulatory circuits, in: *Interleukins, Lymphokines and Cytokines* (J. J. Oppenheim and S. Cohen, eds.), pp. 723–728, Academic Press, New York.

Rich, A. R., 1951, *The Pathogenesis of Tuberculosis,* 2nd ed., Thomas, Springfield, Ill.

Rodes, N. D., McEntire, J. E., Dunn, P. A., Gay, C., Decker, M., Kardinal, C. G., Gilliland, C. D., Oxenhandler, R. W., and Papermaster, B. W., 1982, Phase I clinical studies with lymphoblastoid lymphokine preparations from the RPMI 1788 cell line, in: *Human Lymphokines: The Biological Immune Response Modifiers* (A. Khan and N. O. Hill, eds.), pp. 699–718, Academic Press, New York.

Southcott, B. M., Dumonde, D. C., Pulley, M. S., van Vliet, E., den Hollander, F. C., and Schuurs, A., 1982, Systemic lymphokine treatment of selected cancer patients, in: *Human Lymphokines: The Biological Immune Response Modifiers* (A. Khan and N. O. Hill, eds.), pp. 641–649, Academic Press, New York.

Clinical Studies of Lymphoid Cell Lymphokines in 102 Patients with Malignant Solid Tumors

HENRY JOSHUA and ISAAC DJERASSI

1. INTRODUCTION

Immune responses, whether humoral or cellular, are subtly regulated and balanced by various subsets of lymphocytes. The latter, when stimulated, elaborate and release a variety of nonantibody molecules, named lymphokines (LK) by Dumonde, which act as mediators and regulators of various immune reactions. LK are especially important in so -called cell-mediated immunity, which is responsible for the rejection and destruction of foreign tissue. Malignant tumor cells are sensed by lymphocytes as "nonself" and should therefore be subject to immune attack. For poorly understood reasons this immune response to cancer cells is insufficient in most cases to either prevent the growth of the tumor or to help destroy it when already established. The possibility that the reinforcement of some aspects of the immune response by the administration of exogenously produced LK may help destroy *in vivo* tumor tissue has been considered by many.

One readily available source of LK is the culture medium of human lymphoblastoid lines. The supernatants of such cultures are rich in various LK which can be concentrated and partially purified. The clinical study of LK from lymphoid cell cultures was initiated some time ago in this clinic with the intralesional injections of the materials in a variety of superficial metastatic cancers (Papermaster *et al.*, 1974). The procedure was followed by reduction of the size or the elimination of the lesion injected with LK.

The study presented here was aimed to determine the safety of systemic administration of LK from lymphoid cell cultures to patients with solid tumors, while

HENRY JOSHUA and ISAAC DJERASSI • Mercy Catholic Medical Center, Philadelphia, Pennsylvania 19143.

studying some immunological parameters of their cell-mediated immunity, along with observation of the effect on directly visible or palpable tumor masses. The observations made in the course of this study are summarized in this report.

2. METHOD AND MATERIALS

LK were prepared from the stationary-phase supernatants of two human lymphoid cell lines, namely No. 1788 and No. 802. The first line has been in existence for some time and has been used widely by others. The second is a cell line established by Gallo's group from mixed normal human lymphocytes in their early study on T-cell growth factor (Gallo *et al.*, 1976). Both lines were grown in suspension medium using RPMI 40 medium with 10% calf serum initially and with 1% human serum more recently. The two preparations of LK were found to have equivalent properties *in vitro* and *in vivo*. Following the growth phase in suspension cultures, the lymphocytes were concentrated by centrifugation, washed with medium, and resuspended in fresh medium at a concentration of 1000 cells/mm^3, without calf serum, or more recently, with 1% human serum. The cells were maintained for 48 hr in this stationary-phase culture. The supernatant was then collected and fractionated for various molecular size components using an Amicon filtration apparatus. The fraction with molecular size of 5000 to 50,000 daltons was used as the LK preparation in these studies. The material was sterilized using filtration through Millipore filters. Aliquots were frozen and stored at 30°C until used.

The patient population consisted of adults with a variety of disseminated malignant solid tumors, previously untreated or between pulses of chemotherapy. LK were administered by s.c. injection (1 or 2 ml) or by i.v. infusion of 1 ml in 250 ml 5% dextrose in water in 4 hr. At least 50 ml of blood was collected in heparin prior to and 3 hr following the administration of the LK for the study of its effects on the properties of the patients' lymphocytes. The leukocytes were collected by the inverted syringe technique (Klein *et al.*, 1958) in the presence of 1% hydroxyethylstarch (Roy *et al.*, 1971). The buffy coats were submitted to Ficoll–Hypaque gradient centrifugation (Boyum, 1968) and the mononuclear layer was aspirated with a long needle. The lymphocytes were then studied for their ability to form E-rosettes with sheep red cells (Wybran and Fudenberg, 1973) and for their effects in the GVH reaction in immunosuppressed rats (Shohat and Joshua, 1976). Both early E-rosetting and total E-rosettes after 24 hr incubation were determined. All patients were observed closely for local reaction at the site of injection as well as for systemic changes. Cutaneous or subcutaneous tumor masses were measured and outlined with indelible ink and followed for tumor size changes with daily inspection.

3. RESULTS

A total of 112 adult patients were studied. These included patients with breast cancer (36), lung cancer (32), prostate cancer (7), colon cancer (8), pancreas cancer

(4), rhabdomyosarcoma (3), fibrosarcoma (3), ovarian cancer (6), non-Hodgkin's lymphoma (3), melanoma (2), bladder cancer (1), hypernephroma (1), osteogenic sarcoma (2), Hodgkin's disease in remission (1), myeloma (1), and cancer of the larynx (2). The immunological parameters described above were studied in 63 of these patients. The effects on subcutaneous tumor masses were observable in 18 patients. The remaining patients were observed only for reactions to LK administration.

Local redness and slight edema at the site of the s.c. injection were observed in 83 of 92 patients who received LK in this fashion. Remarkably, the site of a previous and healed tuberculin test reactivated in one patient. The site of an old BCG injection, which had already completely healed, reactivated, becoming infiltrated, swollen, and red, suggesting increased reactivity of the patient's lymphocytes to residual antigens. Mild fever of short duration was observed in 16 of the 92 patients receiving s.c. LK.

Six patients were given i.v. LK diluted in 250 ml of 5% dextrose in water. Five of these patients had severe pyrogenic reactions with vigorous chills and shaking and fever of 103 to 105°F. The shaking responded promptly to the administration of 50 mg Demerol (Djerassi, 1975). The body temperature returned to normal usually with 2 to 3 hr.

The E-rosetting lymphocytes in 63 patients, both early and total, were substantially fewer than in normal controls. The patients had a average of 273 ± 193 early rosetting cells compared to 370 ± 129 in a group of normal controls. The total number of E-rosetting cells, determined after 24 hr incubation with sheep red cells, was 421 ± 259 in the patients as compared to 1329 ±243 in the normal controls. Although neither the early nor the total E-rosetting cells in the patients reached the levels in normal controls, there was a substantial increase of both types of cells following the administration of LK, reaching 414 ± 305 and 614 ± 301, respectively. These increases were statistically significant with a t value of 0.001 for the total and 0.01 for the early rosettes.

The GVH activity of lymphocytes from cancer patients, before and after LK, was determined by injecting them i.d. in immunosuppressed rats. A substantial decrease of the reactivity of patients' lymphocytes was observed prior to the administration of LK. While the area of rat skin rejection by the lymphocytes from normal controls had an average diameter of 5.4 ± 1.1 mm, the patients with cancer produced only 1.9 ± 1.04 mm lesions. The administration of LK caused a marked increase of the lesions produced by patients' lymphocytes, namely to an average diameter of 4.9 ± 1.4 mm. The difference between the GVH effectiveness of lymphocytes from patients versus from normal controls, as well as the patients' lymphocytes before and after LK were highly significant (t value 0.0001). The increased E-rosetting and the GVH reactivity usually returned to the pretreatment level, or slightly above it, 24 hr later, but occurred again on second, third, and fourth daily administration of LK. A tendency of decreased response, however, was observed, especially with respect to the E-rosetting cells. Administration of LK after intervals of 1 week or more was associated with renewed response of the E-rosetting.

Directly observable and palpable tumor masses were present in 18 of the patients. No response at all of the tumor mass was noticed in two patients. Two other patients had minimal response, defined as less than 10 to 20% reduction in size, as determined by the product of the two largest opposing diameters. The other 14 patients showed readily observable decrease of tumor size ranging from 20% to 50+%. The tumor size change was usually observable in 24 to 48 hr. It lasted usually for the duration of administration of LK (4 to 5 days). The tumors returned to pretreatment size within days following LK administration. Typically, pain and discomfort associated with the subcutaneous masses, or with bone lesions, were promptly relieved. On no occasion was there a demonstrable reduction of bone or other internal lesions.

4. DISCUSSION

This extensive clinical trial of systemic administration of LK to patients with solid tumors clearly demonstrates the safety of this material in the doses indicated. The s.c. administration is extremely well tolerated and the side effects are negligible. The i.v. infusion is clearly pyrogenic and could be difficult to tolerate. No attempts were made to determine whether the pyrogenic reactions were due to presence of IL-1 or to common pyrogens. Although the doses were relatively small, the changes of the immune parameters showed conclusively that these doses were sufficient to modify favorably the immune status of the recipient. The increase of the E-rosetting cells (T lymphocytes) within 3 hr following administration of LK was consistent and substantial. The increased ability of patients' mononuclear cells (lymphocytes and monocytes) to attack the skin of an immunosuppressed rat, upon i.d. injection, was indeed dramatic. Remarkably, this increase of activity occurred very promptly after LK administration. The significance of this observation is further increased by the fact that this whole phenomenon was produced *in vivo,* in the intact patient. Lymphocytes collected before and after LK were tested by s.c. injection into the rat with minimal, if any, extracorporeal manipulation. The GVH assay clearly reflected an increased ability of the mononuclear cells of the patient to recognize better the foreign nature of the rat skin and either attack it directly or produce changes which the pretreatment patient cells could not induce.

The mechanism responsible for these changes of the immune parameters of the patients following LK is not clear. The increase of the E-rosetting cells occurs too fast to be accounted for by increased production of T cells by the recipients. Release of T cells from tissues into circulation is a more likely explanation. The possibility, however, that a prompt development of T-cell receptors in noncommitted lymphocytes, under the influence of the LK, cannot be excluded. Similarly, the increased reactivity of the patients' mononuclear cells to rat tissue is not readily understood. The presence of MIF in the LK preparation could account for the damage of the rat skin by activating the monocytes or natural killer cells. The time delay, however, suggests that some mechanism of recognition of the foreign nature of rat skin may be involved. If the improved GVH effect of the injected human mononuclears is due to an improved recognition of rat antigens by the lymphocytes, the presence of a previously undescribed LK capable of increasing the recognition ability of the lymphocytes must be assumed. Another possibility explaining the improved GVH is the increased number of T cells (E-rosetting lymphocytes) following the injection of LK. Again a previously unknown LK could explain either the release of lymphocytes or the differentiation of noncommitted cells to E - rosetting lymphocytes. Another mechanism to explain all of the changes observed in the patients' immune parameters is the release of other substances (such as prostaglandins) by the patient under the influence of LK, without a direct effect on the patients' lymphocytes.

The clinical effect of the small doses of systemic LK on superficial tumor masses is of substantial interest. Remarkable is the speed of producing these effects, most tumors having reduced in size within 24 hr. It remains unexplained why the only changes in tumor size occurred with subcutaneous masses, while all internal lesions remained unchanged. The relief of pain due to bone metastases in patients with breast and prostate cancers was also dramatic and, to a limited extent, of clinical value. The pain improvement also occurred very promptly, often within 24 hr. Objective changes of the bone lesions were not, of course, demonstrable. It is likely that the pain relief was due to a minimal decrease of tumor size, insufficient for accurate detection by measurement but sufficient to relieve pressure on sensory nerves. As histological studies of the course of events in the tumors following the administration of LK were not carried out, we have no evidence that the tumor size reduction or the pain relief were at all associated with actual tumor cell destruction by the relevant cytolytic cells in the tumor (macrophages, cytotoxic lymphocytes, or natural killer cells). It is entirely possible that the reduction of subcutaneous masses was due to decreased swelling or hydration of the mass with resulting moderate reduction of size. The inability to enhance this effect by continued administration of LK beyond the initial change is suggestive of such anti-inflammatory or tumor-dehydrating process. Similarly, the prompt recurrence of growth to pretreatment size, often within 3–4 days, suggests that the tumor reduction may not have been due to actual tumor destruction. The failure of continuous administration of LK to reduce tumor size beyond the very early response is supportive of this possibility.

The observations described above demonstrate and establish that LK produced by human lymphoid cells in tissue culture can be used safely for the study of immune mechanisms in patients. The specific crude preparation of mixed LK we used, had very little, if any, clinical therapeutic impact. The limited clinical studies of this nature reported by others (Dumonde *et al.*, Chapter 49, this volume; Papermaster *et al.*, Chapter 37, this volume) seem to support these conclusions. Emphasis must now be placed on the separation, identification, and clinical study of more purified LK with attempts to associate specific immune responses to individual molecular species.

REFERENCES

Boyum, A., 1968, Separation of leucocytes from blood and bone marrow, *Scand. J. Clin. Lab. Invest.* **21**(Suppl. 97):1–29.

Djerassi, I., 1975, Transfusions of filtered granulocytes [Editorial], *N. Engl. J. Med.* **292:**803.

Gallo, R., Morgan, D. A., and Ruscetti, F. W., 1976, Selective in vitro growth of T lymphocytes from normal human bone marrows, *Science* **193:**1007–1008.

Klein, E., Eridani, S., Djerassi, I., and Resnick, R., 1958, A simple method for the separation of leukocytes from whole blood, *Am. J. Clin. Pathol.* **29:**550–552.

Papermaster, B. W., Djerassi, I., Holterman, O., Rosner, D., Klein, E., and Dao, T., 1974, Regression produced in breast cancer lesions by a lymphokine fraction from a human lymphoid cell line, *Res. Commun. Chem. Pathol. Pharmacol.* **8:**2.

Roy, A. J., Franklin, A., Simmons, W. B., and Djerassi, I., 1971, A method for separation of granulocytes from normal human blood using hydroxyethyl starch, *Prep. Biochem.* **1:**197–203.

Shohat, B., and Joshua, H., 1976, Assessment of the functional activity of human T lymphocytes in malignant disease by the local graft-versus-host reaction in rats and the T-rosette forming cell test, *Clin. Exp. Immunol.* **24:**534.

Wybran, J., and Fudenberg, H. H., 1973, Thymus derived rosette forming cells in various human disease states: Cancer, lymphoma, bacterial and viral infections and other diseases, *J. Clin. Invest.* **52:**1026.

Regulation of T-Cell Proliferation by Interleukin-2 in Male Homosexuals with Acquired Immune Deficiency Syndrome

MICHAEL A. PALLADINO, KARL WELTE, NICULAE CIOBANU, CORA STERNBERG, ROLAND MERTELSMANN, and HERBERT F. OETTGEN

1. INTRODUCTION

In 1981, a sudden increase in the incidence of Kaposi's sarcoma (KS), *Pneumocystis carinii* pneumonia, and other opportunistic infections was recognized among highly promiscuous homosexual men (Centers for Disease Control, 1981). More recently, these diseases are being observed among heterosexual men who are intravenous drug users, the sexual partners of these heterosexuals, and Haitian refugees (Centers for Disease Control, 1982). The disease has been named the acquired immune deficiency syndrome (AIDS).

The immunosuppression which is found in the population at the highest risk to develop AIDS appears to be the main reason for the development of the opportunistic infections (Masur *et al.*, 1981). The nature of the cellular defect is not known; however, a decrease in the ratio of OKT4 versus OKT8, i.e., T_H/T_S T-cell subsets, has been one of the characteristics (Gottlieb *et al.*, 1981).

In this study we have examined male homosexuals with AIDS/KS to determine whether the defect in cellular immunity correlates to their ability to produce IL-2 and whether this defect can be corrected by exogenous IL-2.

MICHAEL A. PALLADINO, KARL WELTE, NICULAE CIOBANU, CORA STERNBERG, ROLAND MERTELSMANN, and HERBERT F. OETTGEN • The Memorial Sloan-Kettering Cancer Center, New York, New York 10021. *Present address of M.A.P.*: Genentech Inc., South San Francisco, California 94080.

CTLL-AII

↓ Remove IL-2 and starve for 6 hours at 37°C.

Remove from flask with 0.05% EDTA

↓ Wash ×2

Dilute to 5×10^4/ml and add 100 μl to serially diluted samples in 96-well-flat bottom plates

↓ Incubate for 18–20 hours at 37°C.

Add 1 μc of ^{3}H-thymidine

↓ Incubate for 4–6 hours at 37°C.

Add 20 μl of 2% EDTA

↓ Incubate for 15 minutes at 37°C.

Remove cells with multicell harvester

↓ Count in Liquid-Scintillation counter

Determine Units of IL-2/ml by Probit Analysis

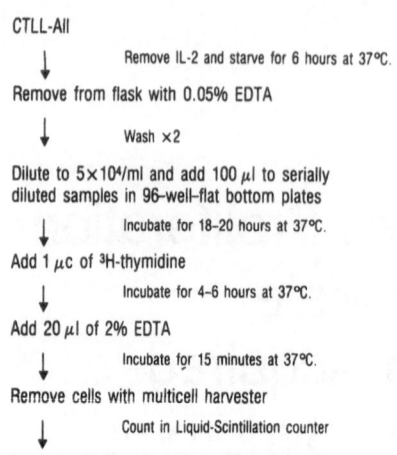

FIGURE 1. IL-2 proliferation assay.

2. MATERIALS AND METHODS

2.1. Separation and Mitogen Stimulation of Peripheral Blood Mononuclear Cells

Mononuclear cells were isolated by Ficoll–Hypaque density centrifugation from heparinized patient and normal donor peripheral blood lymphocytes. Mononuclear cells (1×10^6/ml) were tested for proliferation and IL-2 production using phytohemagglutinin (PHA-P, 1/500 final dilution, Miles Laboratories, Elkhart, Ind.). Proliferation on day 3 was measured by [^{3}H]-TdR incorporation in the absence or presence of highly purified exogenous IL-2 (10 U/ml).

2.2. IL-2 Purification and Assay for IL-2 Activity

Highly purified IL-2 was prepared as previously described. This IL-2 preparation is 37,000-fold enriched from lymphocyte-conditioned medium, shows a specific activity of 10^6 U/mg protein, and consists of two active bands on a silver-stained SDS-polyacrylamide gel (Welte *et al.*, 1982).

The IL-2 assay was previously described (Palladino *et al.*, 1983) and is outlined in Fig. 1. The IL-2 concentration in each sample was calculated by probit analysis using a standard containing 420 U rat IL-2/ml.

Our IL-2 units are defined as follows:

Mononuclear source		IL-2 U/ml at 48 hr
CD rat spleen cells	10^6/ml + 5 μg Con A	75–150
Balb/c spleen cells	10^6/ml + 5 μg Con A	30–50
C57BL/6 spleen cells	10^6/ml + 5 μg Con A	2–5
Human peripheral blood mononuclear cells	10^6/ml + 1/500 PHA	5–15

TABLE I. Production of IL-2 in Peripheral Blood Mononuclear Cells of Patients with AIDS/KS or Other Cancers

	IL-2 U/ml after induction with:[a]			
Source	None	1/500 PHA	10^{-6} M Indocin[b]	1/500 PHA + 10^{-6} M Indocin
Normal	0.10	13.10	0.10	15.50
AIDS/KS	0.10	0.25	0.31	0.33
Lymphoma	0.15	0.10	0.10	7.49
Melanoma	0.10	4.10	0.10	7.98

[a] 10^6 Ficoll–Hypaque (Pharmacia Fine Chemicals, Piscataway, N.J.)-separated peripheral blood monocytes were stimulated for 24 hr in Eagle's minimum essential medium supplemented with 10% heat-inactivated fetal bovine serum, 1% nonessential amino acids, 2 mM L-glutamine, 100 U penicillin/ml, 100 μg streptomycin/ml (complete MEM), and 5 × 10^{-5} M 2-mercaptoethanol.

[b] Indocin (indomethacin, Sigma Chemical Co., St. Louis, Mo.) was added at the start of culture.

3. RESULTS

3.1. Endogenous IL-2 Production

Peripheral blood monocytes of patients with AIDS/KS, melanoma, or lymphoma produced significantly lower amounts of IL-2 than normal controls after stimulation with PHA-P (Table I) (0.25, 0.10, and 4.1 U/ml, respectively, as compared to 13 U/ml for the normal controls). Addition of 10^{-6} M indomethacin resulted in restoration of IL-2 production in the patients with melanoma or lymphoma. However, at present, no treatment has been able to restore IL-2 production in AIDS patients. Similar results have now been obtained with eight additional AIDS/KS patients and patients with sarcoma, breast, renal, lung, or larynx tumors (Palladino, unpublished observations).

3.2. Effect of Purified IL-2 on Mitogen-Induced Proliferation

The addition of IL-2 did not have a significant effect on the proliferation of mononuclear cells from AIDS/KS patients or controls (Table II). In the AIDS/KS

TABLE II. IL-2-Induced Proliferation of Peripheral Blood Mononuclear Cells from Patients with AIDS/KS

	[^{3}H]-TdR incorporation (cpm) in mononuclear cells stimulated with:[a]			
Source	Cells alone	10 U IL-2	1/500 PHA	1/500 PHA + 10 U IL-2
Normal	268	1506	20,569	25,723
AIDS/KS	118	639	451	11,174

[a] Proliferation was measured on day 3 by the addition of 1 μCi [^{3}H]-TdR (6.7 Ci/mmole, New England Nuclear, Boston, Mass.) for 4–6 hr prior to cell harvest.

group, addition of purified IL-2 increased the level of [^{3}H]-TdR incorporation to within 50% of the normal range of PHA-stimulated normal donor cultures.

4. DISCUSSION

In the present study we have described a defect in IL-2 production in peripheral blood monocytes of patients with AIDS/KS, melanoma, or lymphoma. The defect in the AIDS/KS patients appears to differ from the immunological depression seen in patients with the other cancers which could be corrected by indomethacin. The defect in AIDS/KS, however, does not appear to be at the level of the IL-2 receptor as exogenous IL-2 was able to significantly increase the proliferative response of the peripheral blood monocytes in the presence of PHA.

The depressed IL-2 production in AIDS could suggest that production of IL-2 by OKT8$^+$ T cells is less than that of OKT4$^+$ cells. In general, cells of the helper (OKT4, Lyt-1) T-cell subset produce significantly greater amounts of IL-2 than do cells of the cytotoxic or suppressor (OKT8, Lyt-2) phenotype. More likely though, the difference in IL-2 production may be due to the greatly reduced number of OKT4$^+$ cells in the circulation (Gottlieb *et al.*, 1981). In addition, the IL-2 production defect may be due to the altered production by both T-cell subsets possibly induced by a viral infection. Earlier studies have demonstrated that the epidemiology of hepatitis B virus, while not suspected of being the etiological agent in AIDS/KS, closely parallels that of AIDS/KS. Cytomegalovirus (CMV) and Epstein–Barr virus (EBV) are considered as highly suspect etiological agents for this disease. High levels of circulating CMV and EBV antibodies are also seen in these high-risk groups (Giraldo *et al.*, 1980). However, the primary agent which is most likely a biological one has not as yet been defined.

Treatment for AIDS has initially been directed to treatment of the immune deficiency and opportunistic infections. In this regard, our studies indicate that *in vivo* administration IL-2 may be beneficial. The availability of purified IL-2 through biotechnology and recombinant DNA technologies will be eagerly awaited.

REFERENCES

Centers for Disease Control, 1981, Kaposi's sarcoma and *Pneumocystis* pneumonia among homosexual men—New York City and California, *U.S. Morbidity and Mortality Weekly Report* **30**:305–308.

Centers for Disease Control, 1982, Epidemiologic aspects of the current outbreak of Kaposi's sarcoma and opportunistic infections, *N. Engl. J. Med.* **306**:248–252.

Giraldo, G., Beth, E., and Huang, E., 1980, Kaposi's sarcoma and its relationship to cytomegalovirus. III. CMV, DNA and CMV early antigens in Kaposi's sarcoma, *Int. J. Cancer* **26**:23–29.

Gottlieb, M., Schroff, R., Schanker, H., Weisman, J., Fan, P., Wolf, R., and Saxon, A., 1981, Pneumocystis carinii pneumonia and mucosal candidiasis in previously healthy homosexual men, *N. Engl. J. Med.* **305**:1425–1431.

Masur, H., Michelis, M., Greene, J., Onorato, I., Stouwe, R., Holzman, R., Wormser, G., Brettman, L., Lange, M., Murray, H., and Cunningham-Rundles, S., 1981, An outbreak of community-acquired Pneumocystis carinii pneumonia, *N. Engl. J. Med.* **305**:1431–1438.

Palladino, M. A., Ranges, E. E., Scheid, M. P., and Oettgen, H. F., 1983, Suppression of T-cell cytotoxicity by nude spleen cells: Reversal of monosaccharides and interleukin-2, *J. Immunol.* **130**:2200–2202.

Welte, K., Wang, C. Y., Mertlesmann, R., Venuta, S., Feldman, S. P., and Moore, M. A. S., 1982, Purification of human interleukin-2 to apparent homogeneity and its molecular heterogeneity, *J. Exp. Med.* **156**:454–464.

Incorporation of Porcine Thymic Peptides into the Treatment of Fulminant Viral Hepatitis

S. SU, Z. Y. CUI, S. Y. CHANG, J. Y. SHI, W. P. DAO, C. X. ZHENG, S. L. LIU, C. S. HSU, L. X. CUI, G. C. YANG, and K. TAO

1. INTRODUCTION

Fulminant viral hepatitis is a serious disorder resulting from massive hepatic necrosis or sudden and severe impairment of liver function in a patient who has had no evidence of previous liver disease. Lately, exchange blood transfusion, charcoal hemoperfusion and hemodialysis with polyacrylonitrile have recently been introduced into the treatment of the disease, but its mortality remains high (>70%) (Gimson, 1982). Recent investigation of the pathogenesis in China reveals that most patients with this disease are complicated with severe deficiency of cellular immunity in addition to defect in general host defense. With these new findings in mind, we have conducted a trial using a low-molecular-weight thymic hormone preparation (thymic peptides) in our treatment protocol. To our regret, we have been unable to find in the English language medical literature any report of incorporation of thymic peptide into the treatment of patients with fulminant viral hepatitis.

In this paper, we shall review briefly our preliminary experience on intravenous administration of thymic peptide preparation in the treatment of fulminant viral hepatitis, and give a detailed account of the clinical and histopathological changes in the liver during the course of the treatment.

S. SU, Z. Y. CUI, S. Y. CHANG, J. Y. SHI, and W. P. DAO • The First Hospital for Infectious Diseases, Beijing, China C. X. ZHENG • Beijing University, Beijing, China. S. L. LIU, C. S. HSU, L. X. CUI, G. C. YANG, and K. TAO • Chinese Academy of Medical Sciences, Beijing, China.

2. PATIENTS AND METHODS

2.1. Patients

During the period between January 1981 and November 1982, 15 patients were observed in the First Hospital for Infectious Diseases, Beijing. All of them suffered initially from fulminant viral hepatitis with grade II coma and eventually progressed to grade IV or V hepatic encephalopathy. The diagnosis was based on the following criteria: (1) clinical diagnosis of acute viral hepatitis with no previous history of liver disease; (2) after onset, rapid progression of the disease within a few days to further than grade II hepatic coma; (3) clinical and biochemical evidence of severe impairment in liver function; (4) duration of the disease not longer than 3 weeks; and (5) histopathological examination revealing varying degrees of liver necrosis, including massive necrosis, submassive necrosis, confluent focal necrosis, or multilobular necrosis. The clinical features of the patients on admission are shown in Tables I and II.

2.2. Liver Specimens

Liver specimens for histopathological examination were obtained by liver biopsy from three surviving and two fatal cases at the initial stage of hepatic coma and from four recovered cases during the early stage after recovery from unconsciousness. Liver specimens of the remaining six fatal cases were obtained promptly after death by liver puncture. The hepatic morphological findings of 15 patients with fulminant viral hepatitis are given in Tables III and IV.

2.3. Treatment

The treatment protocol included (1) supportive measures, consisting of i.v. drip of 1500 ml of 15% glucose once a day, to which commonly used insulin (1 U to every 6 g of glucose) was added; infusion of 500 ml dextrose daily and i.v. drip of potassium chloride 4–6 g/day (for cases with normal renal function) and i.v. drip of albumin 20 g/day for as long as 1–2 weeks; (2) corticosteroid therapy: prednisolone 60 mg/day given by i.v. injection; (3) symptomatic management; (4) glucagon 1 mg plus insulin 10 U dissolved in 10% glucose 500 ml i.v. drip once a day; and (5) use of porcine thymic peptide: 10–20 mg of lyophilized porcine thymic peptide preparation provided by Drs. Zheng and Liu was dissolved in 100–200 ml of 10% glucose and given by i.v. drip once a day.

3. RESULTS

Seven of the fifteen patients survived after treatment. The survivors generally recovered from hepatic coma after being treated with the combined therapy within 5 to 7 days and their clinical and biochemical evidence of hepatitis returned to

TABLE I. Clinical Data of Nonsurviving Cases

Case	Sex	Age	Grade of coma	Liver size	Bil. (mg/100 ml)	GPT (U)	TTT (U)	A/G (g/100 ml)	Prothrombin activity (%)
1	F	28	II→IV	Reduced	16.0	564	16	3.22/3.48	40
2	F	28	II→V	Reduced	34.9	564	13	4.16/3.34	20
3	M	33	II→IV	Reduced	7.0	752	9	4.5/2.07	—
4	F	27	IV	Reduced	6.4	854	6	3.61/1.89	<20
5	F	24	III→IV	Reduced	8.65	545	3	3.3/1.74	<20
6	M	63	III→V	Reduced	6.75	580	2	4.8/2.4	—
7	M	34	II→V	Reduced	30.6	960	6	2.50/2.30	<20
8	M	64	II→III	Reduced	9.6	735	8	2.32/2.33	<20

TABLE II. Clinical Data of Survivors

Case	Sex	Age	Grade of coma	Liver size	Bil. (mg/100 ml)	GPT (U)	TTT (U)	A/G (g/100 ml)	Prothrombin activity (%)
1	M	25	II→V	Reduced	8.8	576	2	3.06/2.02	48
2	F	25	IV	Reduced	16.95	520	5	2.6/3.17	40
3	M	18	II→IV	Reduced	28.3	754	3	3.8/3.0	40
4	F	27	II→III	Reduced	17.19	124	3	1.96/1.89	75
5	M	61	IV	Reduced	10.95	528	11	2.78/2.95	—
6	F	40	II→V	Reduced	17.96	903	5	3.17/2.4	—

TABLE III. Histopathological Findings in Nonsurviving Cases

Case	Time of liver puncture	Histopathological findings
1	Promptly after death	Massive hepatic necrosis
2	Promptly after death	Massive hepatic necrosis
3	Promptly after death	Massive hepatic necrosis
4	In deep coma	Massive hepatic necrosis
5	Promptly after death	Massive hepatic necrosis
6	In deep coma	Submassive hepatic necrosis
7	Promptly after death	Massive hepatic necrosis
8	Promptly after death	Massive hepatic necrosis

normal after 3 to 6 months of treatment. Repeated liver biopsy performed at an early stage after recovery from unconsciousness in four recovered cases showed that the degree of liver necrosis markedly decreased, but portal and periportal inflammation and liver cell swelling and degeneration remained quite marked. The biopsy specimen obtained at recovery stage showed no liver necrosis or portal inflammation. However, hepatocyte degeneration persisted. No fibrosis could be found on repeated biopsy of a specimen obtained in the period of convalescence. Two patients received follow-up liver biopsy at 12 months after the fulminant illness and by that time they had completely recovered clinically and their liver function tests were normal. Their follow-up histopathological examination showed that the hepatic morphological findings had returned to nearly normal.

4. CASE REPORTS

4.1. Case 1

A 34-year-old male was admitted to the hospital in September 1982 with a clinical diagnosis of fulminant viral hepatitis. The patient developed fulminant liver failure with grade II hepatic coma and deep jaundice. Physical examination showed marked reduction in liver size and this was confirmed by ultrasound examination. The biochemical liver function test performed on admission showed a high level

TABLE IV. The Results of Liver Biopsies in Survivors

Case	Time of liver biopsy	Histopathological findings
1	In deep coma	Confluent multilobular focal necrosis
2	One day after recovery from hepatic coma	Multilobular focal necrosis
3	8 days after recovery from hepatic coma	Focal liver necrosis
4	10 days after recovery from hepatic coma	Focal liver necrosis
5	In deep coma	Confluent multilobular focal necrosis
6	One day after recovery from hepatic coma	Multilobular focal necrosis
7	In deep coma	Confluent multilobular focal necrosis

of serum bilirubin (17.96 mg/100 ml) and serum GPT (903 U). His serum albumin fell to 3.17 g/100 ml and the prothrombin activity fell to less than 20%. The day after admission, the grade of coma progressed from grade II to grade V; liver biopsy was carried out immediately after infusion of prothrombin complex. The pathological diagnosis was confluent multilobular focal liver necrosis (Fig. 1), which demonstrated that the fulminant hepatic failure was due to severe hepatic necrosis. The patient received 10 mg thymic peptide per day, given by i.v. drip in combination with the other treatments as indicated above. Seven days after treatments, the patient recovered from unconsciousness and presented laboratory evidence of improvement in liver function. Repeated liver biopsy was performed during that period, revealing that the liver necrotic changes markedly lessened along with the decrease of portal inflammation (Fig. 2). Six months after treatment, the clinical symptoms of hepatitis had completely disappeared, the biochemical liver function test returned to normal, and the third liver biopsy performed at that time confirmed that the morphological findings of the liver were nearly normal (Fig. 3).

4.2. Case 2

A 25-year-old man was admitted to the hospital on August 15, 1982, with fatigue, loss of appetite, nausea, and jaundice for 7 days' duration. Physical examination of the patient showed extreme fatigue with deep yellow color of the skin and sclera. Liver function test showed serum bilirubin 8.8 mg/100 ml, GPT 576

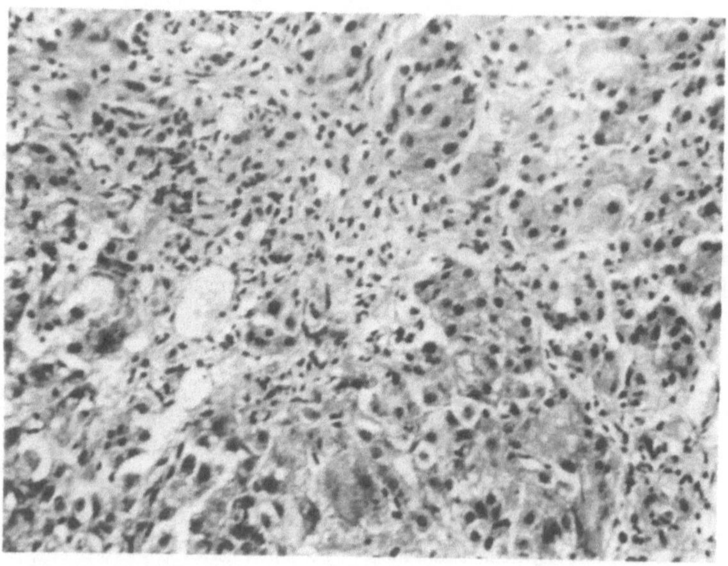

FIGURE 1. Case 1. Liver biopsy on the second day of coma. Histopathological examination reveals a confluent multilobular focal necrosis.

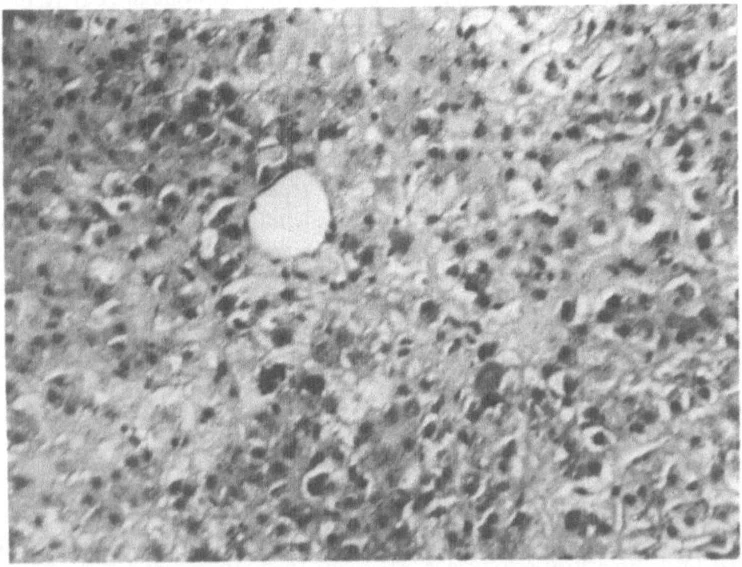

FIGURE 2. Case 1. Liver biopsy on the day after recovery from unconsciousness. Liver necrosis and inflammation have markedly decreased.

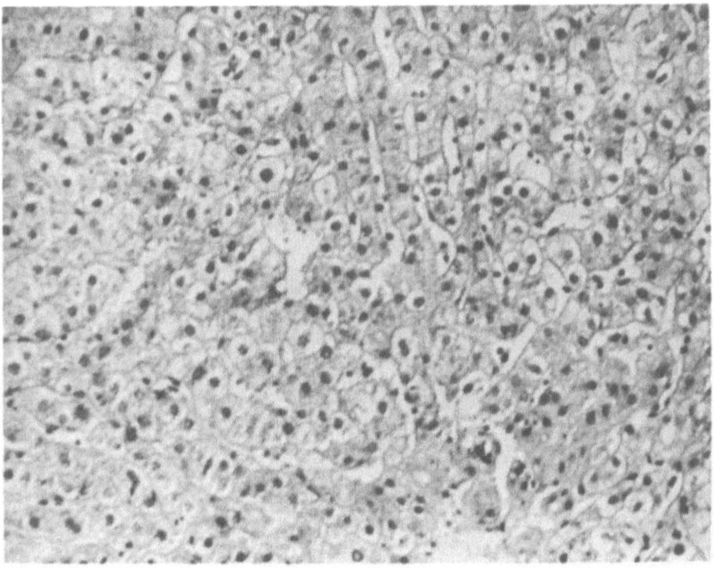

FIGURE 3. Case 1. Liver biopsy 6 months after fulminant illness. The hepatic morphological findings have become nearly normal.

U, serum albumin 3.06 g/100 ml, and decrease of prothrombin activity to 48%. On the second day of admission, the patient's condition rapidly exacerbated, encephalopathy developed to grade III and rapidly progressed within a few days to grade IV hepatic coma. Liver biopsy carried out on the fourth day after hepatic coma showed a confluent multilobular liver necrosis (Fig. 4). After being treated with combined therapy for 7 days, the patient recovered from coma. Ten days after recovery from unconsciousness, he received repeated liver biopsy. The biopsy findings revealed that hepatocyte necrosis and inflammation were markedly decreased (Fig. 5). Three months after treatment the patient recovered. Follow-up liver biopsy performed 6 months after discharge showed that the hepatic morphological findings became nearly normal (Fig. 6).

5. DISCUSSION

Thymic hormone treatment for fulminant viral hepatitis is still in its early stages of clinical study. The incorporation of this immune agent into the treatment of fulminant viral hepatitis is based on the following concepts which have recently been established according to the results of investigation in China, i.e., most of the patients with severe viral hepatitis are often complicated with failure of cell-mediate immunity and decreased host defense (Zhang *et al.*, 1979; Su and Cui, 1981). Thus, adding thymic hormone to the composite therapy of fulminant viral hepatitis is probably of benefit to the patient's resistance to infection caused by

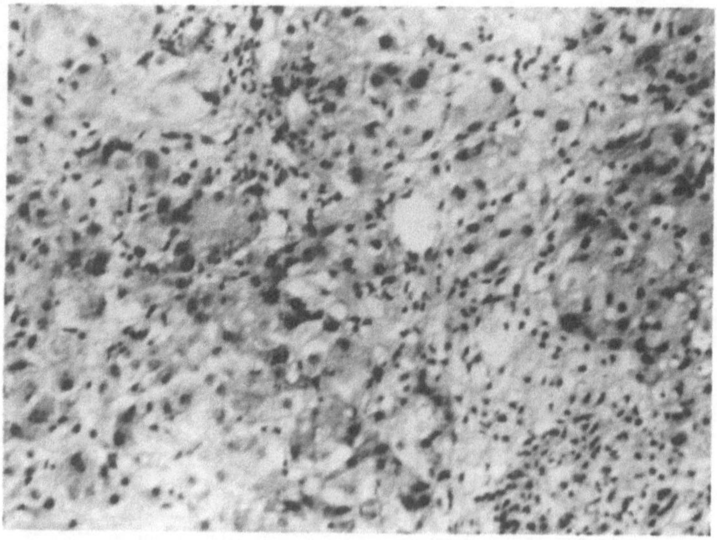

FIGURE 4. Case 2. Biopsy obtained on the fourth day of coma. A confluent multilobular focal necrosis is seen.

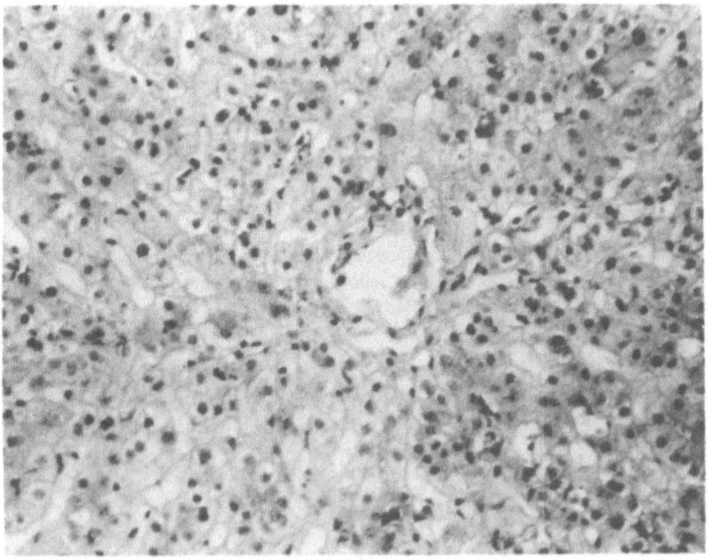

FIGURE 5. Case 2. Liver specimen obtained on the day after recovery from coma. The necrotic and inflammatory changes in the liver show marked remission.

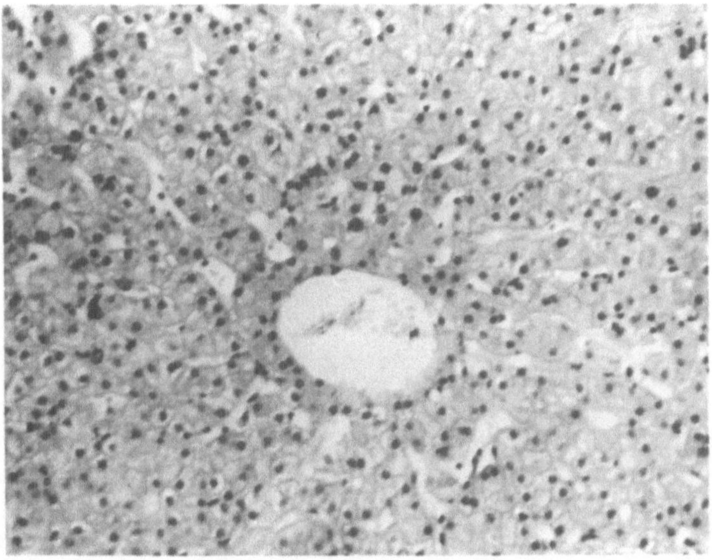

FIGURE 6. Case 2. Follow-up liver biopsy performed 6 months after discharge. The histological findings of the liver have nearly returned to normal.

bacteria, fungi, and other viruses and thereby may prevent the development of fatal complication of infection. But the effect of thymic hormone on the treatment of fulminant viral hepatitis remains to be studied carefully under controlled circumstances.

REFERENCES

Gimson, M. B., 1982, Fulminant hepatic failure and artificial liver support, *Gastroenterol. Jpn.* **17**:144.

Su, S., and Cui, Z. Y., 1981, A study of the immunological characteristics of 90 cases of fulminating chronic hepatitis B, *Chin. J. Intern. Med.* **20**:479.

Zhang, D., Liu, B. Y., and Liu, Z. C., 1979, Observations on the cellular immunity in hepatitis B, *Chin. J. Intern. Med.* **18**:163.

Use of Porcine Thymic Immunomodulator in the Treatment of Chronic Hepatitis B

SU SHENG and JIN YIFENG

1. INTRODUCTION

Studies of hepatitis B surface antigen (HBsAg) prevalence reveal that more than 100,000,000 healthy persons in the world are actually infected with hepatitis B virus and some of these infections may progress to chronic active hepatitis (CAH).

So far, the mechanism of CAH B has not been elucidated. The working hypothesis suggested by Eddleston and Williams (1974) was that HBsAg-positive and some of the HBsAg-negative cases of CAH were initiated by exposure to hepatitis B virus. The production of an autoimmune reaction during the course of CAH B is considered to be associated with the synthesis and release of the damaging autoantibody, in which the interactions between T and B lymphocytes may play an important role. According to the hypothesis suggested by Mackay (1975), there are two possible processes for the induction of autoantibody production. The first is the escape of T cells from tolerance with reactivity against self, which could result in activation of cytotoxic T cells and/or helper T cells to stimulate B cells to produce autoantibody. Second, deficiency in suppressor T-cell function could allow hyperproduction of autoreactive B cells and thereby also increase autoantibody production. Thus, enhancement of the function of suppressor T cells is considered to be of fundamental importance in the treatment of CAH B. For this purpose, it may be mentioned that Bach (1977) demonstrated that experimental thymic hormone treatment of animals could restore suppressor T-cell function or prevent its decline.

For the above-mentioned reason, the thymic hormones have been widely used for the treatment of CAH B in China since 1979 (Nanjing Associated Group for

SU SHENG • The First Hospital for Infectious Diseases, Beijing, China. JIN YIFENG • Biology Department, Nanjing University, Nanjing, China.

Study on Thymic Hormones, 1980; Chang, 1981; Li, 1980; The First Hospital for Infectious Diseases, 1980).

2. FIRST TRIAL

The first clinical trial was carried out in 1979 and 1980 by the Nanjing Associated Group for Study on Thymic Hormones. Thymic hormone, provided by one of us (J.Y.), was isolated from porcine thymocytes and termed porcine thymic immunomodulator (PTI). This extract consists of at least 40–50 polypeptides or proteins with molecular weight ranging from 7000 to 9600 and it has been shown to have E-rosette augmenting activity, to have no acute toxicity and no hypersensitive reaction, and to be free from pyrogen. During this clinical trial, 72 consecutive adult patients, including 33 CAH B, 18 HBsAg-positive chronic persistent hepatitis (CPH), and 21 CAH B with subacute hepatic necrosis, were observed. The clinical diagnosis of all patients was established according to criteria of the Prevention and Treatment Scheme of Viral Hepatitis formulated at the National Conference on Viral Hepatitis in 1978 in China. All of our CAH and CPH patients had severe symptoms and obvious biochemical evidence of liver dysfunction, and all cases with clinical diagnosis of CAH B with subacute hepatic necrosis had progressed to liver failure with greater than grade II hepatic coma before the initiation of the clinical observation. PTI 10 mg/day was given alone by i.m. injection for 3–6 months to patients with CAH and CPH, but for patients with CAH B with subacute hepatic necrosis, the dosage was raised to 20–40 mg/day in combination with infusion of fresh frozen plasma, albumin, glucose, and electrolytes, i.v. injection of prednisolone, supportive treatment, and relevant symptomatic management.

The results of PTI therapy as shown in Table I indicate that PTI treatment seems to be effective for CAH B and the therapeutic efficacy of PTI appears to be greater for CAHB than for both CPH B and CAH with subacute hepatic necrosis.

3. SECOND TRIAL

On the basis of the results of the first clinical trial, a control randomized study was carried out from 1980 to 1981 by the Nanjing Associated Group for Study on Thymic Hormones. Forty-eight consecutive adult patients with clinical diagnosis of CAH B were observed. The patients were randomly assigned to one of the following regimes of treatment: 24 patients were given glucuronate 800 mg/day for 6 month's duration plus vitamin C, vitamin B, and other drugs which are regularly used in the treatment of viral hepatitis in China. Findings on physical examination, results of liver function tests, and the persistence of serum HBsAg were recorded before the treatments and every 4 weeks afterwards. The results of this clinical trial are summarized in Table II. The data as shown in Table II reveal that in 16 of the 24 patients who were treated with PTI, remission was induced. Before the thera-

TABLE I. The Results of PTI Treatment in the Three Types of Chronic Hepatitis B

			Result			
Clinical diagnosis	No. of cases	Hepatic coma	Remission[a] (%)	No effect[b] (%)	Survival (%)	Death (%)
CAH B	33	0	24 (72.7)	9 (27.3)		
CPH B	18	0	7 (38.9)	11 (61.1)		
CAH B with subacute hepatic necrosis	21	21			8 (38.1)	13 (61.9)

[a] "Remission" means that all the symptoms have disappeared and the results of biochemical liver function tests have returned to normal.
[b] "No effect" indicates no significant changes in clinical symptoms and biochemical liver function tests after treatment.

peutic trial, all 16 patients exhibited varying degrees of general debility and loss of appetite and weight. Some of them had a slowly progressive jaundice. Liver function tests before the initiation of the clinical trial showed that all 16 patients had a high GPT (200–700) and TTT (7–26 U). After being treated with PTI for a period of 3–6 months, complete remission was seen and their serum GPT and TTT returned to normal. In contrast, improvements in both clinical symptoms and liver function tests were found in only four of the 24 control patients ($p < 0.05$). However, only a few cases (4/24) showed a conversion to HBsAg negative after PTI treatment, which was of no statistical significance as compared with the controls.

4. CONCLUSION

From this limited experience we may conclude that PTI treatment may have a beneficial effect on CAH B cases. Definite conclusions can only be made after a more extended randomized double-blind clinical study.

TABLE II. The Results of PTI Treatment in HBsAg-Positive CAH

	After treatment			
Group	GPT return to normal (%)	TTT return to normal (%)	HBsAg convert to negative (%)	Remission (%)
PTI (n = 24)	16 (66.7)	16 (66.7)	4 (16.7)	16 (66.7)
Control (n = 24)	4 (16.7)	4 (16.7)	0	4 (16.7)

REFERENCES

Bach, J.-F., 1977, Thymic hormones and autoimmunity, in: *Autoimmunity* (N. Talal, ed.), pp. 208–230, Academic Press, New York.

Chang, T. F., 1981, Experimental and clinical study of immunomodulatory agents in viral hepatitis B, in: *A Corpus of Thymic Hormone Data* (Biology Department of Nanjing University, ed.), pp. 53–63, Nanjing University, Nanjing, China.

Eddleston, A. L., and Williams, R., 1974, Inadequate antibody response to HBsAg or suppressor T-cell defect in development of active chronic hepatitis, *Lancet* **2:**1543.

The First Hospital for Infectious Diseases, 1980, Incorporation of thymic hormones into the treatment of severe viral hepatitis, in: *A Corpus of Thymic Hormone Data* (Biology Department of Nanjing University, ed.), pp. 35–39, Nanjing, China.

Li, F., 1980, Clinical investigation on thymic hormone treatment in viral hepatitis, in: *A Corpus of Thymic Hormone Data* (Biology Department of Nanjing University, ed.), pp. 45–48, Nanjing, China.

Mackay, I. R., 1975, Chronic active hepatitis, in: *Immune Disorders* (L. Van Der Reis, ed.), pp. 143–177, Freeman, San Francisco.

Nanjing Associated Group for Study of Thymic Hormones, 1980, A summary on the clinical application of thymic hormones, in: *A Corpus of Thymic Hormone Data* (Biology Department of Nanjing University, ed.), pp. 1–8, Nanjing, China.

Biological Properties and Clinical Use of Calf Thymus Extract TFX-Polfa

ALEKSANDER B. SKOTNICKI, BARBARA K. DABROWSKA-BERNSTEIN, MAREK P. DABROWSKI, ANDRZEJ GORSKI, JAN CZARNECKI, and JULIAN ALEKSANDROWICZ

1. INTRODUCTION

In the early 1970s, in searching for approaches which could reinforce immune responsiveness in our immunocompromised patients with aplastic or proliferative blood disorders, our attention became focused on the thymus gland. Previous studies on human fetal thymus transplantation in patients displaying various primary and acquired immunodeficiencies had proved to be partially successful in terms of immunorestoration and concomitant improvement of clinical status (August *et al.*, 1968; Cleveland *et al.*, 1968; Marcolongo and Di Paolo, 1973; Serrou, 1974; Stutzman *et al.*, 1971).

Between 1972 and 1974, we studied the clinical usefulness of myasthenic thymus transplantation in 30 patients suffering from acute and chronic leukemia and Hodgkin's disease (Aleksandrowicz *et al.*, 1973; Rzepecki *et al.*, 1973, 1974; Szmigiel *et al.*, 1975). As a result of this treatment, more than 50% of the patients showed improvement in immune and hematological parameters with only one case of GVH reaction.

As the clinical effects of this "transplantation thymotherapy" were transient, lasting only 4–8 weeks, one of us (J.A.) proposed maintenance treatment with thymus-derived humoral factors. Taking into consideration the potential advantages

ALEKSANDER B. SKOTNICKI, BARBARA K. DABROWSKA-BERNSTEIN, MAREK P. DABROWSKI, ANDRZEJ GORSKI, JAN CZARNECKI, and JULIAN ALEKSANDROWICZ • Hematology Department, Cracow Academy of Medicine, Institute of Infectious Diseases and Institute of Transplantation, Warsaw Academy of Medicine, Polfa Pharmaceuticals, Jelenia Gora, Poland

of "humoral thymotherapy," such as the possibility of repeated injections of known amounts of thymic hormone-like factors and their species nonspecificity, we collaborated with Polfa Pharmaceuticals in Jelenia Gora, Poland, in an attempt to obtain a biologically active extract from calf thymus.

Such a semipurified extract, designated thymus factor X (TFX-Polfa), was obtained in 1973 and extensive efforts were made by several Polish investigators to evaluate its biological activity and potential clinical usefulness in various experimental systems.

In this paper, we shall summarize the known characteristics of TFX and present the results of clinical studies with TFX in selected groups of patients.

2. PURIFICATION AND CHEMICAL CHARACTERIZATION OF TFX

TFX is an aqueous extract of the thymus gland from 5- to 7-week-old calves. The purification procedure employs the Polfa patented method involving ammonium sulfate fractionation, desalting through a G-25 porosity molecular sieve, and ion-exchange chromatography (Aleksandrowicz *et al.*, 1975; Czarnecki and Jaskolski, 1978). TFX is a nucleotide- and lipid-free polypeptide mixture, with a major-component molecular weight of 4200, accompanied by traces of several other fractions with molecular weight ranging from 2000 to about 18,000, which can be detected by 10% polyacrylamide gel electrophoresis at pH 6.8 (Skotnicki, 1978; Staroscik *et al.*, 1978). The detailed chemical characterization of TFX will be published elsewhere (Low et al., in preparation).

3. BIOLOGICAL PROPERTIES OF TFX

During the last 12 years, TFX was tested in several assays designed for studying thymus-derived products. Investigations included the effect of TFX on both T-cell markers and proliferative capacity, as well as the effect on T-cell effector functions.

The following list presents a summary of some of the biological properties of TFX.

3.1. Activities *in Vitro*

1. Restoration of azathioprine sensitivity of E-rosette-forming spleen cells from adult, thymectomized mice (Skotnicki, 1978; Gieldanowski and Kowalczyk-Bronisz, 1978).
2. Increase of steroid resistance of murine thymocytes (Gieldanowski and Kowalczyk-Bronisz, 1978).
3. Restoration of the ability of spleen cells from neonatally thymectomized mice preincubated with TFX to elicit GVH reaction (Slopek *et al.*, 1978).
4. Enhancement of murine MLR response (thymocytes against blocked allogeneic spleen cells) (Skotnicki *et al.*, manuscript in preparation).

5. Enhancement of allogeneic MLR response in patients with autoimmune disorders (Skotnicki *et al.*, manuscript in preparation).
6. Increase of number of E-rosette-forming cells in cord blood, and in patients with rheumatoid arthritis, chronic hepatitis, multiple sclerosis, and in some patients with solid tumors (Skotnicki, 1978; Dabrowski *et al.*, 1980; Skotnicki *et al.*, manuscript in preparation).
7. Increase of Con A-induced suppressor activity of lymphocytes from patients with chronic hepatitis (Dabrowski and Dabrowska-Bernstein, 1981; Dabrowski *et al.*, 1983).
8. Enhancement of spontaneous, as well as SEA, Con A, autologous or allogeneic MLR-induced γ-interferon production by lymphocytes from normal donors and patients with rheumatoid arthritis and multiple sclerosis (Skotnicki *et al.*, manuscript in preparation).
9. Enhancement of PHA-induced TCGF (IL-2) production by lymphocytes from normal donors and patients with multiple sclerosis and rheumatoid arthritis (Skotnicki *et al.*, manuscript in preparation).
10. Enhancement of colony-stimulating factor (CSF) production by human leukocytes (Gorski *et al.*, 1981).
11. Increase of PHA-induced human T-cell colony formation in soft agar culture (Gorski *et al.*, 1983).
12. Increase of PHA- and 2-mercaptoethanol-induced murine T and B lymphoid colony formation in soft agar culture (Gorski *et al.*, 1981).
13. Increae of myelopoiesis by murine spleen cells (Skotnicki, 1980; Gorski *et al.*, 1981).
14. Effects of chronic lymphocytic leukemia lymphocytes: increase of the number of E-rosette-forming cells, increase of the intracellular cAMP levels (with no effect on phosphodiesterase activity) and protein kinase activity; reduction of abnormal glycogen content, alteration of surface structure scanning microscope characteristics and lymphocyte lysosomal activity (Skotnicki *et al.*, 1976, 1978; Aleksandrowicz *et al.*, 1977; Thomson *et al.*, 1979).
15. Enhancement of T-cell-dependent immunoglobulin synthesis by human PBL from normal donors with no effect on direct B-cell response, suggesting influence on T-helper cells (Gorski *et al.*, 1983).
16. Decrease of antibody response of "high" responder to B-cell challenge, e.g., lymphocytes from renal allograft recipients with acute rejection episodes (Gorski *et al.*, 1982).

3.2. Activities *in Vivo*

1. Leukocytopoietic effect in rabbits and normal adult thymectomized mice (Jaszcz *et al.*, 1980; Turowski *et al.*, 1975).
2. Enlargement of thymus-dependent areas of lymph nodes and spleen in normal and thymectomized animals (Jaszcz *et al.*, 1980; Slopek *et al.*, 1980).

3. Dilatation of cortical zones of thymus with increased numbers of thymocytes (Slopek *et al.*, 1980).
4. Rejuvenation of bone marrow hemopoietic cell population (Slopek *et al.*, 1980).
5. Increase of survival of irradiated animals (Wazewska-Czyzewska *et al.*, 1977).
6. Enhancement of PHA and Con A response of lymph node lymphocytes derived from TFX-treated mice (Skotnicki, 1978).
7. Enhancement of T and B cells and granulocyte–macrophage colony-forming capacity of murine spleen cells after *in vivo* TFX administration (Gorski *et al.*, 1981).
8. Stimulation of spermatogenesis and fertility in mice and rats (Slopek *et al.*, 1980).
9. Antitumor effect on several models of transplantable tumors (Ehrlich cancer, Sa-180, polyoma, Sa-L_1, LLC) expressed by the delay or inhibition of tumor growth, the appearance of lymphocyte and macrophage infiltration of tumor-surrounding tissue, inhibition of development of lung metastases after i.v. injection of tumor cells, prolongation of survival time of tumor-bearing animals (Jaszcz *et al.*, 1981).

4. CLINICAL TRIALS WITH TFX

4.1. Patients with Primary Immune Deficiency Syndromes

Two patients with severe hypogammaglobulinemia (age 18 and 40) were the subjects on one of the first trials with TFX (Skotnicki *et al.*, 1975). They suffered from frequent upper and lower respiratory tract infections and were on almost constant antibiotic therapy. Daily injections of TFX for 3 months resulted in reversal of some of the abnormal laboratory parameters; however, the low levels of IgG remained (Table I). On the other hand, TFX therapy lowered the previous rate of infections, possibly due to the normalization of the coexisting decreased cellular immunity.

In another patient, with Louis–Bar syndrome (ataxia telangiectasia), age 10, injections of TFX caused an increase of lymphocyte count, especially with T-cell markers and subsequent clearing of the lung X-ray picture from long-lasting pulmonary infection. The severe neurological state remained unchanged; however, no improvement in this aspect was expected.

4.2. Patients with Bone Marrow Failure

The rationale for clinical trials with TFX in patients demonstrating bone marrow insufficiency was based on experimental data showing the relationship between thymic humoral function and hemopoiesis (Trainin and Resnitzky, 1969; Zipori and Trainin, 1975), as well as the effect of TFX on mouse and human myelopoiesis (Skotnicki *et al.*, 1980; Gorski *et al.*, 1981).

TABLE I. Effect of TFX on Some Immunological Parameters in Patients with Hypogammaglobulinemia[a]

Test	TFX[b]	Patient T.S.	Patient J.Z.
Absolute number of PBL/mm^3	Before	1428	785
	After	3016	2450
E-rosette-forming cells	Before	18%	48%
	After	60%	77%
IgG in mg/100 ml	Before	222	39
	After	330	330
Delayed hypersensitivity reaction	Before	−	−
against PPD	After	− +	+ + +
Lymphocyte glycogen content	Before	+ + +	+ +
	After	+	+
NBT test nonstimulated and *E. coli*	Before	ND	8% (16%)
stimulated (in parentheses)	After	ND	20% (37%)

[a] From Skotnicki *et al.* (1975).
[b] Daily s.c. TFX injections (10 mg) for 3 months.

Recent observation of hematological recovery in three out of six patients with idiopathic aplastic disorders treated with a bovine thymic factor, thymostimulin Tp-1 (Guglielmo *et al.*, 1983), confirmed our previous similar clinical results with TFX. Moreover, the administration of Tp-1 was followed in all patients by the restoration of normal T-cell helper/suppressor ratio due to a selective increase in absolute T-helper cell count. A similar finding was found in seven patients with aplastic anemia treated for 3 weeks with TFX (Gorski *et al.*, 1983).

To test the potential therapeutic value of TFX treatment in patients with bone marrow failure, four studies were performed during the last 7 years, which are summarized as follows.

I. In 7 out of 10 patients with primary aplastic anemia or dyserythropoietic syndrome refractory to conventional myelotrophic treatment, 14–30 days of TFX administration (10 mg/injection) resulted in increased reticulocyte, erythrocyte, granulocyte, and lymphocyte counts followed by a rise in platelet levels. Follow-up evaluation of the bone marrow picture showed increased cellularity and maturation of blood cell precursors (Aleksandrowicz and Skotnicki, 1976; Aleksandrowicz *et al.*, 1975).

II. Patients with malignant disorders showing secondary leukopenia (or pancytopenia) due to bone marrow toxic agents (cytostatic drugs, X-ray therapy) were divided into two groups (Aleksandrowicz *et al.*, 1975; Skotnicki, 1980):

1. Patients (N = 4) with severe leukopenia and septicemia as a result of prior therapy—TFX administration resulted in the prevention of further deterioration, increased WBC count, and improvement of clinical state.

2. Patients (N = 6) undergoing long-term chemo- or radiotherapy with pancytopenia tendency—TFX injections made the continuation of necessary antiproliferative treatment possible with the achievement of greater therapeutic antitumor effect.

III. Thirty patients with long-lasting secondary leukopenia caused by different drugs (i.e., antibiotics, mainly chloramphenicol, anticonvulsants, antiarthritics, antithyroids) ($N = 14$), professional exposure to X-ray or chemical agents ($N = 6$), or considered as "idiopathic" ($N = 10$) were studied. The patients were resistant to leukopoietic therapy during several months or years of ambulatory observation and most of them were suffering from frequent viral and bacterial infections. Their mean WBC counts in repeated tests performed at least 1 month prior to entry into the trial with TFX ranged between 2400 and 3100/mm^3. TFX was injected s.c. twice weekly in doses of 10 mg/injection for 12 weeks to otherwise untreated patients. More than half of the patients (53%) responded to TFX with a significant increase ($p < 0.01$ and $p < 0.05$) of their mean WBC in 8 of the 12 weekly assessments. The remaining patients did not respond, and in some a significant decrease of WBC count was observed; however, the drop of WBC never exceeded the potentially dangerous value of 2000/mm^3.

The different and even opposite effect of TFX administration in the above nonpreselected patients with secondary leukopenia could reflect not only irreversible damage of leukopoiesis in some of the nonresponders, but also different mechanisms leading to leukopenia, i.e., inadequate production or excessive immune destruction of cells which could be enhanced by thymic factors in some of the patients.

There were no significant changes in relative proportions of neutrophils, monocytes, and lymphocytes. However, TFX normalized the percentage of E-rosette-forming cells by increasing it from lower values (from 20–60%, mean 50%) in 13 patients and decreasing it from higher than normal values (71–91%, mean 80%) in 14 patients.

The simultaneously performed determination of erythrocyte and platelet counts, as well as liver enzyme and kidney function tests, protein and iron levels, and granulocyte lysosomal enzyme activity, did not show significant changes during the 3 months of TFX injections. Clinically, a decreased rate of infections was observed in the whole group of patients during and after TFX treatment.

It was concluded that TFX is a safe agent which is able to stimulate leukopoiesis in about 50% of patients with secondary leukopenia resistant to other forms of therapy (Skotnicki *et al.*, 1983b).

IV. A randomized open trial in patients with primary lung cancer showed that the combined treatment of radiotherapy plus TFX injections resulted in better hematological tolerance than in patients subjected only to radiotherapy (Table II). It is worthwhile to note that no decrease of WBC count below 3000/mm^3 occurred in the TFX group, as compared to a decrease of 24% in the control group (Zeromski *et al.*, 1976).

4.3. Patients with Autoimmunoaggressive Syndromes

The pathogenesis of these diseases seems to be related to the breakdown of immune tolerance due to abnormalities in endogenous immunoregulation.

The lack of autoreactivity in healthy subjects is thought to be the result of constant active control—regulation of the immune reactions against autologous tissues throughout life.

The physiological control which maintenances the autoimmune unresponsiveness is attributed to suppressor T-cell function and to the shifting balance between

TABLE II. Effect of TFX on Hematological Values of Patients with Primary Lung Cancer[a]

Therapy	Hb. erythrocytes			Leukocytes			
	Increase	No change	Decrease	Increase	No change	Decrease	Decrease below $3000/mm^3$
^{60}CO (N = 54)	0%	72%	28%	0%	24%	76%	24%
^{60}Co + TFX (N = 41)[b]	44%	39%	17%	47%	29%	24%	0%

[a] From Zeromski *et al*. (1976).
[b] 20–30 days' radiotherapy with simultaneous daily TFX injections.

helper and suppressor populations. Suppressor T lymphocytes create a kind of "suppressor umbrella" (Fig. 1). This suppressor barrier is thymus dependent as thymectomy or antithymocyte serum abolishes it, resulting in the derepression of autoreactivity with subsequent humoral and cellular aggression against self. This also occurs when autoantibody-inducing factors, i.e., some drugs, stress, or injection agents, operate on genetically susceptible individuals.

Four ostensibly nonrelated autoaggressive diseases, lupus erythematosus, rheumatoid arthritis, chronic active hepatitis, and multiple sclerosis, although so different in their clinical presentation, have many common characteristics such as chronic self-perpetuating course, the presence of polyclonal hypergammaglobulinemia with multiple autoantibodies, the accumulation of lymphocytes and plasma cells in damaged tissue, the pathogenic role of immune complexes with neutrophil-dependent tissue injury, and the presence of specific sensitized cytotoxic T cells. Disordered cellular immunity with low suppressor cell activity expressed predominantly during the exacerbation phase of disease, genetic predisposition in young female adults, and therapeutic effectiveness of corticotherapy are other common features (Fig. 2).

It seems important that the treatment of patients with autoimmunoaggressive syndromes be directed toward the restoration and maintenance of active immunoregulatory mechanisms and not only just to the effector arm of the autoimmune response as is the case with immunosuppressive or anti-inflammatory drugs.

The above approach can be realized by substitution of endocrine thymic function by thymic humoral factors, or thymomimetic drugs like levamisole. It has already been found that this type of immunotherapy results in serological and clinical improvement in several patients with different autoimmunopathies (Veys *et al.*, 1981; Skotnicki, 1983). Moreover, the maintenance of the immune balance by

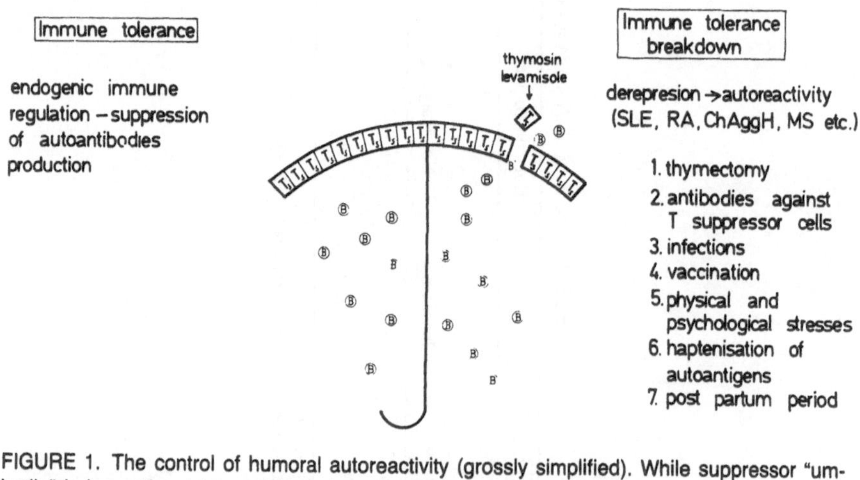

FIGURE 1. The control of humoral autoreactivity (grossly simplified). While suppressor "umbrella" is hermetic, autoreactivity is kept in check. Several indicated factors could break this barrier in genetically susceptible subjects making it permeable to self-reactive B cells (the same is related to autoreactive T cells). This results in autoimmune syndromes like systemic lupus erythematosus (SLE), rheumatoid arthritis (RA), chronic aggressive hepatitis (ChAggH), multiple sclerosis (MS), etc. Thymic extracts or thymomimetic drugs like levamisole could reinforce the endogenous immunoregulatory mechanisms and temper the disease activity.

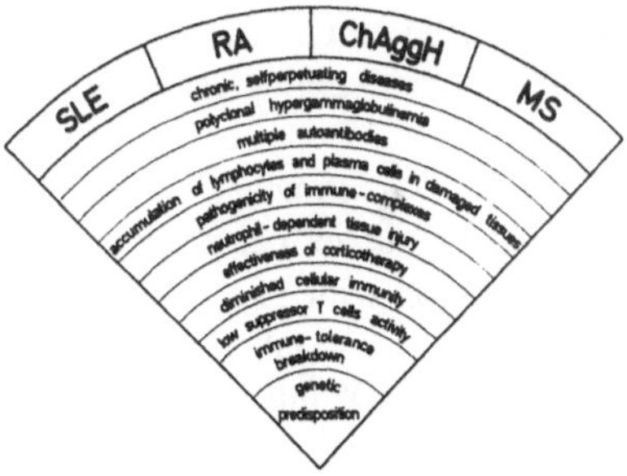

FIGURE 2. The similarity in pathogenesis of four symptomatically different human diseases.

prolonged immunoregulatory treatment is generally safe and inexpensive, decreases sensitivity to infections, and can be applied to young people in the early stages of disease with potentially reversible immune and tissue damage, thus halting the development of the disease.

4.3.1. Patients with Rheumatoid Arthritis

TFX was administered s.c. in 20 patients with rheumatoid arthritis (ages 18–45), 10 mg daily doses for 2 weeks, and then twice weekly for 6 months (Skotnicki *et al.*, 1983).

In 16 patients (80%), clinical evaluation as well as laboratory findings revealed the suppression of the clinical manifestation of the disease and reduction of inflammatory activity and hyperimmune phenomena.

The clinical improvement was manifested in both subjective (pain and morning stiffness) and objective (number of swollen and tender joints, muscle strength) parameters. Serological evaluation showed decreased rheumatoid factors levels, normalization of hypergammaglobulinemia and blood sedimentation rate, as well as increased hemogloblin and iron levels.

The clinical improvement preceded the change of laboratory findings, and was generally observed after 4 to 12 weeks of TFX administration.

The increased tolerance to weather changes and increased vigor and general mobility of the patients subjected to TFX treatment were also noted.

Apart from transient joint exacerbation during the first few weeks in one-third of the patients, no other side effects which could be related to TFX therapy were found.

Longer follow-up studies are necessary to evaluate the role of immunomodulation by TFX in rheumatoid arthritis patients, including the radiological progression of the disease.

4.3.2. Patients with Virus B Hepatitis

4.3.2a. Patients with Acute Hepatitis. In a placebo-controlled double-blind trial in 60 patients with acute hepatitis, 15 injections of TFX for 20 days beginning from the day of diagnosis, caused faster decrease of total bilirubin and iron levels with earlier clinical improvement and reappearance of appetite, as compared to the control placebo group. There was no significant difference between the two groups in terms of other serological parameters (Kiczka and Ozynski, 1983).

4.3.2b. Patients with Chronic Active Hepatitis (CAH). Selection of the patients was based on subnormal proportion of E-rosetting lymphocytes elevated after *in vitro* incubation with TFX, weak PHA and Con A responses, and in some cases, negative skin tests to tuberculin and DNCB. To date, the study includes 27 patients receiving s.c. TFX injections for 2 to 4½ years. TFX was administered daily for the first 30 days and every second or third day thereafter (Dabrowska-Bernstein *et al.*, 1980; Dabrowski *et al.*, 1980).

Immunological monitoring included (1) evaluation of E-rosette-forming capacity in terms of so-called instant rosettes (15 min incubation of PBL with SRBC at room temperature), early rosettes (1 hr incubation at 4°C), and late rosettes (18 hr incubation at 4°C); (2) proliferative response of PBL to T-cell mitogens PHA (5 μg/ml) and Con A (20 μg/ml); (3) Con A-induced suppression of PHA response by autologous PBL. After 3 weeks of treatment, TFX caused the normalization of percent E-rosetting cells in about 90% of the patients. The repeated assessments showed the maintenance of normal values of the three classes of E-rosettes during the prolonged TFX injections. This first signal of immunocorrection was followed during 2 to 8 months of immunotherapy by the improvement of *in vitro* proliferative Con A responses and suppressor activity. Normalization of PHA responses was found starting at 1 to 2½ years of TFX therapy.

The observed sequence of immunorestoration induced by TFX which is similar to the ontogenetic development of lymphocyte immunocompetence, suggests physiological effects of this putative thymic hormone on target cells.

Serological, clinical, and histological assessments were completed in 21 patients. They revealed an improvement or normalization of serum values of transaminase, bilirubin, albumin, γ-globulin, cholinesterase, γ-glutamyltransferase, VII and X clotting factors, and prothrombin time in about 60 to 80% of cases. The disappearance of HBsAg was observed in 13 patients (61%) about 16 months after the beginning of immunotherapy. An evident drop of the autoantibody levels, mainly antinuclear, anti-smooth muscles, and antimitochondrial, was also found.

The histological pictures in the repeated liver biopsies (in 11 to 18 months of TFX therapy) showed a resolution of mononuclear infiltrations in 50% of patients and in the remaining cases there was no significant change compared to the pretrial picture suggesting the inhibition of progressive liver tissue damage and fibrotic replacement.

Clinically, more than 50% of patients showed subjective and objective improvement with disappearance of dysgeusia, diminished liver size and compactness, decreased splenomegaly in two cases, and resolution of ascites and edemas in one patient in the third year of TFX injections.

In another study (Kiczka *et al.*, 1983) which included 14 patients with CAH, 12–16 months of TFX injections resulted in the rise of alanine transaminase levels during the first 3 months of immunotherapy (0.5- to 3-fold increase) with normalization in some of the cases thereafter. In 30% of the patients, transient pain and discomfort, under the right costal arch, appeared during the first few weeks of TFX therapy. No elimination of HBsAg was observed in seven seropositive cases during the time of reported evaluation. The histological picture in repeated liver biopsy revealed a regression from active hepatitis to persisting in three cases, and to minimal in eight, with no changes in two, and progression in one patient. TFX administration was connected with general clinical improvement, increased appetite, and return of libido in male subjects.

4.3.3. Patients with Multiple Sclerosis (MS)

The first clinical trial with TFX in 13 patients with MS was not promising. In eight patients, the natural course of the disease remained unchanged, two slowly improved, two markedly worsened, and one died (Cendrowski, 1980). However, this trial was of short duration (1–14 months), comprised mainly of patients with long-lasting disease (one-half of the group from 7 to 23 years) and with advanced disability (mean degree 6 in Kurtzke's scale).

In another study, two out of three patients with MS treated with TFX for 3 to 7 months, showed clinical improvement with amelioration of paresthesia, pyramidale syndrome, and sphincter disorders (Hausmanowa-Petrusewicz *et al.*, 1983). Parallel observations showing reconstitution of the T-cell-dependent immune compartment (E-rosettes, PHA and Con A response, Con A-induced suppression function) in MS patients treated with TFX (Dabrowski *et al.*, 1980) encouraged us to undertake a new trial in a selected group of these patients. Mainly patients with definite MS, but in early disease stage, with relapsing-remitting form and with potentially reversibel neurological and immunological disturbances, were chosen. We decided to combine TFX administration with nitrogen mustard (cyclophosphamide metabolite) alternating therapy, using low noncytotoxic doses of the drug. The results of this combined immunomodulatory treatment, which will be continued irrespective of the actual phase of the disease, even during remission (in an attempt to prevent or delay relapse), will be evaluated after at least 2 to 3 years. Similar therapy approach using nitrogen mustard and levamisole was found to be beneficial in about 80% of 50 MS patients with mainly relapsing-remitting course observed for 2 to 3 years (Aleksandrowicz *et al.*, 1981; Skotnicki *et al.*, 1983).

Interestingly, Gorski *et al.* (1983) recently showed that TFX could exert immunosuppressive effects in patients with B-cell-system hyperactivity. Moreover, they observed that TFX increased the sensitivity of immunoglobulin-synthesizing B cells to an inhibitory effect of cyclophosphamide. The rationale for our therapy approach was also supported by Nagai *et al.* (1982), who reported the intensive suppression of experimental allergic encephalomyelitis—the animal model for MS—by blood-derived facteur thymique sérique (FTS).

4.3.4. Patients with Recurrent Aphthous Stomatitis (RAS)

An open trial was performed in five patients suffering from RAS for 2 to 8 years. The pathogenesis of the disease—characterized by periodic, recurrent, oral ulcerations—seems to be related to different abnormalities of immune reactivity with local immune complex formation (Gorska, 1980; Lehner, 1977). As no drug is known to be effective in preventing recurrences of RAS, the immunoregulatory treatment with TFX was initiated. TFX was applicated twice weekly for 1 to 2 years. This small trial showed an overall beneficial effect of TFX in four out of five patients, expressed by decreased frequency and severity of attacks with reduction of duration and pain of the ulcerative lesions in two patients, and the complete disappearance of disease episodes in two others. In the latter patients, however, the discontinuation of TFX administration was followed after 7 and 12 months, by the return of the pretreatment picture of the disease (Skotnicki, unpublished).

4.4.1. Patients with Hodgkin's Disease

In 10 patients with advanced disease (stage III and IV), TFX administration resulted in an increased lymphocyte count (even in patients with histological lymphocyte depletion), increased cellular immunity (skin tests, E-rosettes, PHA response), and improvement of hematological tolerance in patients simultaneously receiving cytostatic drugs or radiotherapy (Marjanska-Radziszewska *et al.*, 1975). Hodgkin's disease, as a syndrome of "hypolymphocytic hypothymism" (Svet-Moldarsky, 1966), is in our opinion one of the most promising pathological situations as a candidate for thymotherapy, applied together with or after conventional treatment.

4.4.2. Patients with Chronic Lymphocytic Leukemia (CLL)

In 14 out of 20 untreated patients with early CLL, TFX caused a decrease of total lymphocyte count with a simultaneous increase of E-rosette-forming cells, lymphocytes with high lysosomal acid phosphatase activity, immunoglobulin levels, increase in reticulocyte and hemoglobin levels, and normalization of tuberculin skin sensitivity in previously anergic subjects (Aleksandrowicz and Skotnicki, 1976; Wazewska-Czyzewska *et al.*, 1976). The drop of total lymphocytosis in CLL patients after TFX treatment could be related to an enhancement of T-cell suppressor function with subsequent inhibition of pathological B-like cell proliferation in this disorder.

4.4.3. Patients with Primary Lung Cancer

The randomized prospective open trial (Zeromski *et al.*, 1976; Slowik-Gabryelska and Krzysko, 1980) included 135 patients with primary lung cancer. Histologically, the above population contained 96 cases with undifferentiated cell carcinoma, 37 with squamous cell carcinoma, and 2 with adenocarcinoma. Out of the total population, 53 patients were chosen randomly for the TFX study, and the remaining 82 served as controls.

Twelve patients from the TFX group with advanced bronchogenic cancer were excluded from conventional treatment because of the disseminated disease process, or the coexisting disease. TFX was administered twice weekly for 10 weeks. This

TABLE III. Effect of TFX on the Clinical Course of Patients with Primary Lung Cancer[a]

No. of treated patients	Therapy	Total remission[b]	Partial remission	Inhibition of neoplastic process	Progression of neoplastic process
54	^{60}CO	0%	24%	39%	37%
14	^{60}CO + TFX 100–300 mg[c]	14%	7%	46%	36%
27	^{60}Co + TFX 550 mg[d]	22%	37%	33%	7%

[a] From Zeromski *et al.* (1976).
[b] Based on X-ray examination.
[c] Patients accepted for radiotherapy (^{60}Co) received simultaneously s.c. TFX injections: 7 patients—5 mg daily for 20 days (total dose = 100 mg); 7 patients—10 mg daily for 30 days (total dose = 300 mg).
[d] Patients received radiotherapy + TFX (10 mg daily) for 30 days and 10 mg every 7 days for 5 months thereafter (total dose = 550 mg).

immunotherapy resulted in subjective and objective clinical improvement in 10 out of 12 patients, correlated with the inhibition of local growth and metastatic spreading to mediastinal lymph nodes or other organs. In three patients, partial regression of tumor mass was observed. The 6-month survival time in this group was 42%, comparing to 7% in the control group (N = 28).

The remaining patients from the TFX group (N = 41) with less advanced bronchogenic cancer were admitted to cobalt therapy. The association of ^{60}Co with TFX (10 mg daily for 30 days and thereafter every 7 days for 5 months) favorably influenced the remission rate and survival time compared to the control group subjected to radiotherapy alone (Tables III and IV).

Interestingly, the patients who received TFX (total dose 300 mg) during radiotherapy showed enhancement of the delayed hypersensitivity reaction (DHR) to PPD in 39% (including two patients with total positive conversion) compared to 12% (6/48) in the control group. This reinforcement of cell-mediated immunity was correlated with the mean survival time which increased from 13.2 months to

TABLE IV. Effect of TFX on the Survival Time of Patients with Primary Lung Cancer[a]

No. of treated patients	Therapy	Survival time	
		> 6 months	> 12 months
28	Symptomatic treatment[b]	7%	
12	Symptomatic treatment + TFX[b]	42%	
54	^{60}Co	31%	7%
14	^{60}Co + TFX 100—300 mg[c]	42%	21%
27	^{60}Co + TFX 550 mg[d]	96%	51%

[a] From Zeromski *et al.* (1976).
[b] Patients were disqualified for radiotherapy due to the disseminated disease process.
[c] Patients accepted for radiotherapy (^{60}Co) received simultaneously s.c. TFX injections: 7 patients—5 mg daily for 20 days (total dose = 100 mg); 7 patients—10 mg daily for 30 days (total dose = 300 mg).
[d] Patients received radiotherapy + TFX (10 mg daily) for 30 days and 10 mg every 7 days for 5 months thereafter (total dose = 550 mg).

TABLE V. Clinical Studies with TFX—Summary

Disease	No. of patients	Length of treatment	Effects on laboratory data	Clinical effects	Investigators
Primary immunodeficiency	3	1–3 months	↑ E-RFC, ↑ PHA response ↑ DHR to PPD ↑ NBT	↓ frequency of infections	Skotnicki *et al*. (1975)
Bone marrow failure	91	0.5–6 months	↑ Hb, ↑ RBC, ↑ WBC ↑ E-RFC ↑ platelet in 50–60%	Improvement of general state, resolution of infections	Aleksandrowicz *et al*. (1975), Aleksandrowicz and Skotnicki (1976), Skotnicki (1982), Skotnicki *et al*. (1976), Zeromski *et al*. (1976)
Acute B hepatitis	30	20 days	Faster ↓ bilirubin and iron level	Accelerated recovery compared to double-blind control	Kiczka and Ozynski (1983)
Chronic active hepatitis	35	1–5 years	↓ transaminase, IgG, and bilirubin levels ↑ albumin, factor VII + X levels ↑ cholinesterase activity HBsAg disappeared in 46% of 28 seropositive cases Histological regression of active hepatitis in 60% ↑ E-RFC, Con A, PHA response ↑ Con A-induced suppression	Subjective and objective improvement in 50–60%	Cianciara *et al*. (1983), Dabrowska-Bernstein *et al*. (1980), Dabrowski *et al*. (1980), Kiczka *et al*. (1983)
Rheumatoid arthritis	20	6–12 months	↑ E-RFC, Hb, Ht ↓ serum IgG, ESR ↓ rheumatoid factor in 45% of cases	Improvement decreased inflammatory symptoms and signs	Skotnicki *et al*. (1983)
Neurological diseases					Hausmanowa-Petrusewicz *et al*. (1983),

Amyotrophic lateral sclerosis	15	1–10 months	↑ E-RFC, ↑ Con A, ↑ PHA response ↑ Con A-induced suppression	Inhibition of progression or improvement in 60% of cases	Dabrowski *et al.* (1983), Cendrowski (1980)
Myasthenia gravis (partially resistant to thymectomy treatment)	4	0.3–2.5 years	Same as above	Improvement	
Polyneuropathy	5	0.1–2.5 years	Same as above	Improvement in 3 cases	
Multiple sclerosis	16	1–14 months	Same as above in 3 tested patients	Improvement in 4 cases	
Encephalomyelitis	1	15 months		Improvement	
Recurrent aphthous stomatitis	5	1–2 years		Improvement in 4 cases	Skotnicki (unpublished)
Nephrotic syndrome	16	2.5–12 months	Proteinuria—no change or worse	—	Gorski *et al.* (unpublished)
Chronic renal allograft rejection	5	2.5–4 months	↓ serum creatinine in 3 cases	Prolonged graft survival	Gorski *et al.* (1982)
Hodgkin's disease	10	3–6 months	↑ Hb and lymphocyte count ↑ E-RFC, ↑ DHR to PPD	Improvement	Marjanska-Radziszewska *et al.* (1975)
Chronic lymphocytic leukemia	20	3–6 months	↑ lymphocyte count in 60% ↑ Hb, ↑ E-RFC, ↑ serum IgG ↑ DHR to PPD	↑ frequency of infections	Aleksandrowicz and Skotnicki (1976), Wazewska-Czyzewska *et al.* (1976)
Colorectal cancer	50	6–24 months	↑ WBC, ↑ E-RFC ↑ PHA response Histological evidence for increased local immune response and tumor necrosis	Improvement Prolonged survival time	Aleksandrowicz *et al.* (1975), Turowski *et al.* (1975, 1976), Urban *et al.* (1977)
Primary lung cancer	53	6 months	↑ Hb, ↑ Ht, ↑ WBC ↑ DHR to PPD	Increased remission rate Prolonged survival time	Krzysko and Slowik-Gabryelska (1981), Slowik-Gabryelska and Krzysko (1980), Zeromski *et al.* (1976)
Total	379				

18.4 months in ^{60}Co plus TFX groups in patients who showed improvement in DHR. Similar data for the control ^{60}Co group were 7.2 and 8.8 months, respectively (Krzysko and Slowik-Gabryelska, 1981).

4.4.4. Patients with Colorectal Cancer

In 50 patients with inoperable diseases, repeated TFX injections caused an increase in granulocyte and lymphocyte counts with enhancement of cell-mediated immunity and clinical improvement (Turowski *et al.*, 1975, 1976). In 12 cases subjected to TFX immunotherapy, the evaluation of histological changes within the tumor tissue has been completed. These changes included inflammatory, granulomatous, and fibroblastic reactions with focal calcification and tumor necrosis (Urban *et al.*, 1977). Similar microscopic changes within the stromal tissue at the margin of the neoplastic growth are seen in cases of spontaneous tumor regression (Slausen *et al.*, 1975) and are considered as the expression of the natural host response to the invasion of neoplastic tissue. Thus, it is possible that the evident prolongation of survival time of the observed patients was due to the mobilization of TFX of local tissue defense reactions against tumor growth.

5. CONCLUDING REMARKS

Our clinical experience with TFX has showed that this preparation causes an increase of general and local immune responses, expressed by more rapid elimination of infectious agents (viral, bacterial, mycotic) and by clearing of organ pathology (e.g., improved liver function in cases of chronic hepatitis, or functional amelioration of peripheral nervous system in cases with polyneuropathies) (Table V).

Second, TFX suppresses autoimmune phenomena with improvement of target organ function (e.g., decreased symptoms and signs in patients with rheumatoid arthritis).

Third, TFX counteracts the myelo- and immunosuppressive properties of cytostatic agents by a stimulating influence on hemo- and lymphopoietic restoration after iatrogenic damage.

Finally, TFX regulates humoral immune response by increasing immunoglobulin levels in patients with primary or secondary hypogammaglobulinemia and by inhibiting immunoglobulin production in cases of hypergammaglobulinemia.

Thus, substitution of thymic function could create an effective obstacle against the development of symptomatically different, but pathogenetically similar human diseases. Taking into consideration the lack of side effects of this organ preparation, its relatively low price and facility even in self-application, TFX can be used both in immune prophylaxis as well as in immune intervention. As in every substitution therapy, TFX should be used constantly in repeated therapeutic cycles, especially in patients with genetically related chronic immune-mediated diseases.

However, further studies are still necessary to estimate the best immune monitoring system, indications, the optimal therapy protocol, and the mechanism of immunorestoration induced by TFX.

ACKNOWLEDGMENTS. We would like to express our sincere gratitude to Dr. A. E. R. Thomson, Imperial Cancer Research Fund, London; Dr. A. White, Syntex Research, Palo Alto, California; Dr. A. L. Goldstein, The George Washington University, Washington, D.C.; and Dr. M. M. Zatz, The George Washington University, Washington, D.C., for their interest, critical judgement, and help in experimental studies.

REFERENCES

Aleksandrowicz, J., 1977, Perspectives of thymotherapy and metalotherapy in hematology, *Haematologica* **62**:98–118.

Aleksandrowicz, J., and Skotnicki, A. B., 1976, The role of the thymus and thymic humoral factors in immunotherapy of aplastic and proliferative diseases of the haemopoietic system, *Acta Med Pol.* **17**:1–17.

Aleksandrowicz, J. and Skotnicki, A. B., 1982, Humoral thymotherapy in patients with primary and secondary immune deficiency states and bone marrow hypoplasia, in: *Leukemia Ecology* (J. Aleksandrowicz and A.B. Skotnicki, eds.), pp. 20–30, Natl. Cent. Sci. Tech. Econ. Inf., Warsaw.

Aleksandrowicz, J., Rzepecki, W. M., Szmigiel, Z., Lukasiewicz, M., Skotnicki, A. B., and Lisiewicz, J., 1973, Preliminary results of thymus transplantation from myasthenia gravis patients to patients with leukemias and Hodgkin's disease, *Przegl. Lek.* **30**:518–522.

Aleksandrowicz, J., Blicharski, J., Janicki, K., Lisiewicz, J., Skotnicki, A. B., Sliwczynska, B., Turowski, G., Szmigiel, Z., and Wazewska-Czyzewska, M., 1975a, Effect of the thymus extract on congenital hypogammaglobulinaemia and immunological deficiency accompanying the proliferative and aplastic haematological diseases, in: *Biological Activity of Thymus Hormones* (D.K. van Bekkum, ed.), pp. 37–39, Kooyker, Rotterdam.

Aleksandrowicz, J., Turowski, G., Cybulski, L., Szmigiel, Z., and Skotnicki, A. B., 1975b, Immunohaematological and clinical response in patients with malignant diseases after repeated administration of thymus extract, *Ann. Immunol.* **7**:97.

Aleksandrowicz, J., Turowski, G., Czarnecki, J., Szmigiel, Z., Cybulski, L., and Skotnicki, A. B., 1975c, Thymic factor X (TFX) as a biologically active extract, *Ann. Immunol.* **7**:97.

Aleksandrowicz, J., Blicharski, J., Cichocki, T., Dobrowolski, J., Kwiatkowski, A., Lisiewicz, J., Sasiadek, U., Skotnicki, A. B., and Wazewska-Czyzewska, M., 1977, The effect of calf thymus extract on surface structure and lysosomal apparatus in lymphocytes of patients with chronic lymphocytic leukemia, *Haematologica* **63**:267–276.

Aleksandrowicz, J., Blicharski, J., Retinger, M., Skotnicki, A. B., and Zdunczyk, Z., 1981, Some aspects of immunopathogenesis of multiple sclerosis and therapeutic perspectives, *Pol. Tyg. Lek.* **13**:1363–1365.

August, C. S., Rosen, F. S., Filler, R. M., Janeway, C. A., Markowski, B., and Kay, H. E. A., 1968, Implantation of fetal thymus restoring immunological competence in patient with thymic aplasia (Di George's syndrome), *Lancet* **2**:1210–1211.

Cianciara, J., Gorska, E., Babiuch, L., Dabrowska-Bernstein, B. K., and Dabrowski, M. P., 1983, The immunomodulation treatment with TFX in patients with chronic active hepatitis,

Cleveland, W. W., Fogel, B. J., Brown, W. T., and Kay, H. E. A., 1968, Fetal thymic transplant in a case of Di George's syndrome, *Lancet* **2**:1211–1212.

Czarnecki, J., and Jaskolski, B., 1978, Preparation TFX-Polfa Jelenia Gora, *Arch. Immunol. Ther. Exp.* **26**:475.

Dabrowska-Bernstein, B. K., Dabrowski, M. P., Cianciara, J., Brzosko, W. J., Babiuch, L., and Kassur, B., 1980, The treatment of chronic aggressive hepatitis with Polish thymus factor (TFX), *Erfahrungsheilkunde* **29**:71.

Dabrowski, M. P., and Dabrowska-Bernstein, B. K., 1981, Diagnostic significance of the involvement of E-rosetting lymphocytes and monocytes in suppressive regulation of lymphocyte proliferative response in man, International Congress of Immunology, "Immunoregulation 81," Urbino, Italy.

Dabrowski, M. P., Dabrowska-Bernstein, B. K., Babiuch, L., and Brzosko, W. J., 1980a, Calf thymic hormone/TFX/driven maturation of human T-lymphocytes, 4th Intern. Congress Immunol., Paris, July, Abstracts, 17,2,05.

Dabrowski, M. P., Dabrowska-Bernstein, B. K., Badmajew, W. P., Brzosko, W. J., 1980b, Quantitative estimation of human T-cell mediated suppressive activity, 4th International Congress of Immunology, Paris, Abstract 4.4.06.

Dabrowski, M. P., Dabrowski-Bernstein, B. K., Brzosko, W. J., Babiuch, L., and Kassur, B., 1980c, Immunotherapy of patients with chronic virus B hepatitis. I. Maturation of human T-lymphocytes under influence of calf thymic hormone, *Clin. Immunol. Immunopathol.* **16**:297–307.

Dabrowski, M. P., Dabrowska-Bernstein, B. K., Badmajew, W., Wasowska, B., Stasiak, A., Lawkowski, Z., and Brzosko, W. J., 1983, Concanavalin A induced cell activity in healthy donors and patients with calf thymus extract (TFX-Polfa): Mechanisms and simple method of estimation, *Thymus* (submitted).

Gieldanowski, J., 1981, Immunomodulators—Thymus factor X (TFX) and levamisole in immune reactions and inflammatory processes, *Arch. Immunol. Ther. Exp.* **29**:121–132.

Gieldanowski, J., and Kowalczyk-Bronisz, S. H., 1978, Investigations on immunological and pharmacological activities of thymus factor X (TFX), *Arch. Immunol. Ther. Exp.* **26**:485.

Gieldanowski, J., Slopek, S., and Kowalczyk-Bronisz, S. H., 1980, Immunological and pharmalogical properties of thymus factor X (TFX). II. Immunotropic activity, *Arch. Immunol. Ther. Exp.* **28**:853–857.

Gorska, R., 1980, Studies of cellular and humoral immunity in recurrent aphthous stomatitis, *Immunol. Pol.* **5**:291–295.

Gorski, A., Skotnicki, A. B., Gaciong, Z., and Korczak, G., 1981, The effect of calf thymus extract TFX on mouse and human hemopoiesis, *Thymus* **3**:129–141.

Gorski, A., Korczak-Kowalska, G., Nowaczyk, M., Gaciong, Z., and Skopinska-Rozewska, E., 1982, Thymosin, an immunomodulator of antibody production in man, *Immunology* **47**:497–501.

Gorski, A., Paczek, L., Rancewicz, A., Mokrzycka-Michalik, M., Korczak-Kowalska, G., and Dierzanowska, D., 1983a, Factors affecting immunoglobulin production of lymphocytes from renal allograft recipients, *Transplant. Proc.*

Gorski, A., Rancewicz, Z., Nowaczyk, M., Malejczyk, M., and Wasik, M., 1983b, Diminished synthesis of immunoglobulins by lymphocytes of patients treated with thymosin (TFX) and cyclophosphamide,

Guglielmo, P., Giustolisi, R., Cacciola, E., and Milone, G., 1983, Immunomodulating treatment in patients with aplastic anemia, *N. Engl. J. Med.* **308**:1362–1363.

Jaszcz, W., Chlap, Z., Turowski, G., and Piotrowska, K., 1975, Studies on biologic activity of thymic extract (TFX). I. Effect on the blood picture and lymphatic organs in mice, *Patol. Pol.* **26**:387–396.

Jaszcz, W., Rzepecki, W., Wojciechowski, Z., and Piotrowska, K., 1980, The effect of TFX on the pictures of the blood and lymphatic organs of thymectomized mice in the post-neonatal period, *Patol. Pol.* **31**:273–285.

Jaszcz, W., Chlap, Z., Rzepecki, W., Szot, W., and Piotrowska, K., 1982, Evaluation of the effect of calf thymus extract (TFX) administration to transplantable-tumor bearing animals, *Mater. Med. Pol.*

Kiczka, W., and Ozynski, R., 1983, Double blind control trial with calf thymus extract TFX in patients with acute B hepatitis,

Kiczka, W., Juszczyk, J., Adamek, J., and Zengteler, G., 1983, TFX in the treatment of chronic active hepatitis,

Korczak-Kowalska, G., 1983, Thymosin (TFX) and suppressor cells of immunoglobulin production in man, *Thymus*

Krzysko, R., and Slowik-Gabryelska, A., 1981, TFX in primary bronchial cancer and some indices of its activity, **6**:223–230.

Lehner, T., 1977, Progress report: Oral ulceration and Behcet's syndrome, *Gut* **18**:491–511.

Low, T. L. K., Mercer, R., Dahmberg, K., Czarnecki, J., and Skotnicki, A. B., Thymic factor X—TFX-Polfa: Biochemical characterization (in preparation).

Marcolongo, R., and Di Paolo, N., 1973, Fetal thymic transplant in patients with Hodgkin's disease, *Blood* **41**:625–630.

Marjanska-Radziszewska, J., Bicz-Ciecialowa, M., Szmigiel, Z., and Skotnicki, A. B., 1975, Application of the thymic extract in patients suffering from Hodgkin's disease, XIth Congr. Pol. Haematol. Soc., Gdansk, Abstracts, p. 118.

Rzepecki, W. M., Lukasiewicz, M., Aleksandrowicz, J., Szmigiel, Z., Skotnicki, A. B., and Lisiewicz, J., 1973, Thymus transplantation in leukemia and malignant lymphogranulomatosis, *Lancet* **2:**508.

Rzepecki, W. M., Lukasiewicz, M., Aleksandrowicz, J., Szmigiel, Z., Skotnicki, A. B., and Lisiewicz, J., 1974, Thymus transplantation in leukemia and malignant lymphogranulomatosis, *Lancet* **1:**990.

Serrou, B., 1974, Thymus transplantation in advanced cancer, *Lancet* **2:**1290–1291.

Skotnicki, A. B., 1978, The biological activity and physicochemical properties of the thymus extract TFX, *Pol. Tyg. Lek.* **33:**1119–1122.

Skotnicki, A. B., 1979, Thymus extract in the treatment of aplastic anemia, Vth meeting of Int. Soc. Haematol., Hamburg, Abstracts III, p. 73.

Skotnicki, A. B., 1980, Thymus extract—TFX—in the treatment of bone marrow hypoplasia, *Erfahrungsheilkunde* **29:**73.

Skotnicki, A. B., 1984, Immunoregulatory treatment in autoimmunoaggressive syndromes, *Postepy Hig. Med. Dosw.* (submitted).

Skotnicki, A. B., Aleksandrowicz, J., and Czarnecki, J., 1975a, Immunopotentiation by thymic extract in human leukemia, IIIrd meeting of Int. Soc. Haematol., London, Abstracts, p. 46.

Skotnicki, A. B., Aleksandrowicz, J., and Lisiewicz, J., 1975b, Effect of the calf thymus extract on immunologic reactivity in patients with hypogammaglobulinaemia, *Boll. Ist. Sieroter. Milan.* **54:**500–501.

Skotnicki, A. B., Thomson, A. E. R., and Tisdale, M. J., 1976, Influence of Polish TFX on levels of adenosine 3′: 5′ cyclic monophosphate (cAMP) in human lymphocytes, in: *Scientific Report*, p. 61, Imperial Cancer Research Fund, London.

Skotnicki, A. B., Thomson, A. E. R., and Tisdale, M. J., 1978, The influence of calf thymus extract on lymphocyte populations in chronic lymphocytic leukemia, in: *Immunotherapy of Malignant Diseases* (H. Rainer, ed.), pp. 27–30, Schattauer, Stuttgart.

Skotnicki, A. B., Gorski, A., Gaciong, Z., and Korczak, G., 1980, The effect of thymus extract on in vitro proliferation of hematopoietic progenitor cells, *Fed. Proc.* **39:**4696.

Skotnicki, A. B., Aleksandrowicz, J., Retinger-Grzesiula, M., Zdunczyk, A., Grochmal, S., and Huczynski, J., 1983a, Multiple sclerosis: Immune pathogeneis, up to date therapeutic strategy and the combined immunoregulatory treatment, *Postepy Hig. Med. Dosw.*

Skotnicki, A. B., Blicharski, J., Lisiewicz, J., Sasiadek, U., Janik, J., Wolska, T., Zdunczyk, A., and Aleksandrowicz, J., 1983b, Calf thymus extract TFX-Polfa in the treatment of patients with secondary leukopenia: Therapy and drugs, (submitted).

Skotnicki, A. B., Zatz, M. M., Gyorke, A., Seals, C., Cuddy, D., Low, T. L. K., Jacobs, R., Kind, P. H., and Molinari, G., Thymic factor X—TFX-Polfa: Biological properties, (manuscript in preparation).

Slausen, D. O., Osburn, B. I., Skifrine, M., and Dungworth, D. L., 1975, Regression of feline sarcoma virus-induced sarcomas in dogs. I. Morphologic investigation, *J. Natl. Cancer Inst.* **54:**361–370.

Slopek, S., Szymaniec, S., and Bromirska, J., 1978, Evaluation of biological activity of TFX-Polfa, a product obtained from calf thymus, *Arch. Immunol. Ther. Exp.* **26:**481–482.

Slopek, S., Gieldanowski, J., and Kowalczyk-Bronisz, S. H., 1980, Immunobiological and pharmacological properties of thymus factor X (TFX). I. Biological and pharmacological activity, *Arch. Immunol. Ther. Exp.* **28:**827–857.

Slowik-Gabryelska, A., and Krzysko, R., 1980, The relationship between the dose of TFX and the clinical course of primary lung cancer, *Pneumol. Pol.* **48:**187–195.

Staroscik, K., Boratynski, J., and Lisowski, J., 1978, Properties of TFX, a product obtained from calf thymus, *Arch. Immunol. Ther. Exp.* **26:**477–479.

Stutzman, L., Mittelman, A., Onkochi, T., and Ambrus, J., 1971, Fetal thymus transplantation in Hodgkin's disease, *Proc. Am. Assoc. Cancer Res.* **12:**101–102.

Svet-Moldavski, G. J., 1966, Is Hodgkin's disease a syndrome of hypothymism?, *Nature (London)* **209:**933–934.

Szmigiel, Z., Turowski, G., Rzepecki, W. M., Aleksandrowicz, J., and Skotnicki, A. B., 1975, The effect of myasthenic thymus fragment transplantation on the immune state of patients with proliferative and aplastic diseases of the haemopoietic system, *Acta Med. Pol.* **16:**61–82.

Thomson, A. E. R., Tisdale, M. J., and Skotnicki, A. B., 1979, The influence of Polish thymus extract "TFX" on lymphocytes in chronic lymphocytic leukemia, in: *Scientific Report*, pp. 85–86, Imperial Cancer Research Fund, London.

Turowski, G., Cybulski, L., and Urban, A., 1975a, The first thymus extract administration to a patient with advanced cancer in: *The Biological Activity of Thymic Hormones* (D.W. van Bekkum, ed.), p. 41, Kooyker, Rotterdam.

Turowski, G., Stachura, J., Kolodziej, J., Papla, B., and Wojcik, R., 1975b, Haematologic changes in rabbits under the influence of calf thymus extract (TFX), *Patol. Pol.* **26**:379–386.

Turowski, G., Cybulski, L., Politowski, M., Turaszwili, T., and Zubel, M., 1976, First trials of immunopotentiation by thymic extract (TFX) in surgical patients with malignant disease, *Acta Med. Pol.* **17**:18–32.

Urban, A., Turowski, G., and Cybulski, L., 1977, The histological changes of the stroma of colon and rectum cancers observed in patients after immunopotentiation by thymus extract (TFX), *Patol. Pol.* **28**:47–59.

Veys, E. M., Mielnats, H., Verbruggen, G., Dhondt, E., Goethals, L., Cheroutre, L., and Buelens, H., 1981, Levamisole as basic treatment of rheumatoid arthritis: Long term evaluation, *J. Rheumatol.* **8**:45–56.

Wazewska-Czyzewska, M., Aleksandrowicz, J., Turowski, G., Blicharski, J., Dobrowolski, J., and Dolezalowa, M., 1976, A trial of immunologic stimulation with thymic extract (TFX) in untreated chronic lymphocytic leukemia, *Acta Med. Pol.* **17**:97–110.

Wazewska-Czyzewska, M., Aleksandrowicz, J., Szybinski, Z., Kulig, D., Plonka, I., and Skotnicki, A. B., 1977, The effect of calf thymus extract (TFX) on the survival of mice exposed to ionizing radiation, *Pol. Przegl. Radiol. Med. Nukl.* **41**:51–55.

Zeromski, J., Slowik-Gabryelska, A., Krzysko, R., 1976, The preliminary evaluation of TFX administration in advanced bronchogenic carcinoma, Seminar on cellular and humoral immunity in lung diseases, Poznan, Abstracts, pp. 44–46.

Role of Thymostimulin (TP-1) as an Immunomodulator in Hodgkin's Disease

STEPHEN DAVIS, PIETRO RAMBOTTI, and FAUSTO GRIGNANI

1. INTRODUCTION

With the advent of effective chemotherapy and the introduction of megavoltage radiotherapy techniques, the relapse-free survival (RFS) in previously untreated Hodgkin's disease has been radically prolonged (Faguet and Davis, 1982). It would appear that the prolonged RFS is, to various degrees, dependent on age, histological subtype, presence of constitutional symptoms, and extent of disease at presentation (Young *et al.*, 1972).

The cellular immune defects in Hodgkin's disease have been recognized since 1902 (Reed, 1902). However, studies correlating immunocompetence in untreated Hodgkin's disease patients with prognosis have failed to demonstrate any prognostic value to pretreatment *in vitro* or *in vivo* immunological parameters (Young *et al.*, 1972). More recently, however, Bjorkholm *et al.* (1976) and Faguet and Davis (1982) suggested that pretreatment immunocompetence correlates positively with survival.

With an assumption that further improvements in survival and cure rates may not be obtainable with more aggressive staging or treatment but with immunomodulation of the patient prior to therapy, we undertook a systematic study of the *in vivo* effects of the bovine thymic extract thymostimulin (TP-1) on various *in vivo* and *in vitro* parameters of T-cell activity and T-cell-dependent specific immune system in untreated patients with Hodgkin's disease (Falchetti *et al.*, 1977).

STEPHEN DAVIS, PIETRO RAMBOTTI, and FAUSTO GRIGNANI • Istituto di Clinica Medica, Universita di Perugia, Italy.

2. RESULTS

Lymphocyte counts in patients prior to TP-1 therapy were similar to those of controls, and the mean percentage of E-rosetting lymphocytes in patients was reduced (47% vs. 59.7%, $p < 0.001$) (Table I). Following TP-1 therapy, total lymphocyte counts remained stable. The 11 Hodgkin's disease patients with depressed initial E-rosette levels (less than 50%) displayed as a group an increase in E-rosette-forming cells (36.8% to 51.9%; $p < 0.005$). TP-1 treatment brought E-rosette values to normal in seven and was without effect in four. TP-1 therapy did not result in significant change in the E-rosetting capacity in those patients whose initial E-rosette level was greater than 50% nor in the controls.

Lymphocyte PHA stimulation was depressed in patients compared to controls prior to TP-1 administration at all PHA doses tested ($p < 0.001$) (Table II). TP-1 administration induced an increase in PHA response in these 15 patients who had initially depressed SIs (< 50) ($p < 0.005$ at the two higher PHA concentrations and < 0.02 at the lowest). These 15 cases included 10 patients with depressed E-rosette levels plus five patients with normal E-rosetting cells. Most patients demonstrated an increased SI following TP-1 treatment and six reached control SI levels. TP-1 did not alter the patient's capacity to respond to mitogen when the initial SI was greater than 50. The slope of the regression line between dose of PHA SIs demonstrated that the effect of TP-1 on mitogen response is greater at the lower mitogen concentrations (1.9 and 5.0 μg/ml).

Serum leukocyte migration inhibitory activity (serum LIF) of patients before and after TP-1 treatment was evaluated. Before treatment, only one patient had a positive LIF test, the others showing no inhibitory activity. The mean LIF activity prior to treatment was 1.03 ± 0.23. Following TP-1 administration, mean LIF activity was 0.75 ± 0.17 ($p < 0.005$); nine patients displayed a positive LIF test.

The results of TP-1 administration upon skin test reactivity to recall antigens in patients and controls are shown in Table III. Skin tests were positive in 10 patients (52.6%) before treatment and in 18 after treatment (94.7%; $p < 0.05$).

The statistical evaluation of changes induced by TP-1 administration on patients grouped according to lymphocyte count, clinical stage, or histology showed that E-rosette values, mitogen response, and skin test reactivity tended to preferentially increase in the low lymphocyte count rather than the high lymphocyte group. When patients were grouped according to clinical stage or histology, a clear effect of TP-1 is apparent in all subgroupings.

Serum platelet aggregation titers of patients before and after TP-1 administra-

TABLE I. Effect of TP-1 Administration on E-Rosette-Forming Cells

	Normal controls	Patients
Lymphocytes/ml	2208 ± 700	2068 ± 900
% E-rosettes, pretreatment	58 ± 9	47 ± 14
% E-rosettes, posttreatment	58.9 ± 9	56 ± 12

TABLE II. Effect of TP-1 Administration on PHA Responses

	PHA stimulation index					
	Before TP-1			After TP-1		
	60 μg/ml	15 μg/ml	2 μg/ml	60 μg/ml	15 μg/ml	2 μg/ml
Patients	48.3 ± 21	42.2 ± 19	34.5 ± 22	66.4 ± 26	59 ± 27	51 ± 27
Controls	76.9 ± 15	86.2 ± 18	71 ± 16	77.2 ± 17	85.7 ± 16	70 ± 15

tion were performed. The titers in normal controls never exceeded 1 : 8, whereas titers were positive (≥ 1 : 16) in 12 of 32 patients before TP-1 treatment. Following TP-1 therapy, only 3 of 32 patients remained elevated ($p < 0.01$). Using the ClqB-ELISA test, normal controls had a CIC level of 16.35 ± 4.1 μg/ml. Before TP-1 therapy, eight Hodgkin's patients had increased CIC levels (≥ 24.5 μg/ml) with a mean value in this group of 42.33 ± 17.8. After TP-1 treatment, the levels remained raised in only three. The mean CIC level for all 32 patients was 20.13 ± 16.35 before TP-1 treatment and 14.9 ± 12.7 μg/ml after TP-1 ($p < 0.05$). CIC elevations by either assay were not correlated to histology or clinical staging of the patient.

The mean serum lysozyme in normal controls was 8.3 ± 1.4 μg/ml. Before TP-1 therapy, 11 Hodgkin's patients had elevated lysozyme levels (> 11.1 μg/ml) with the mean for the 32 patients being 10.7 ± 7.1 μg/ml. None of the patients or controls had renal insufficiency or peripheral monocytosis.

When patients were grouped according to elevated or normal CIC as evaluated by either test, a clear-cut polarity was evident (Table IV). The mean serum lysozyme level was increased with respect to the controls only in a group of 19 Hodgkin's

TABLE III. Effect of Treatment with TP-1 on Skin Test Reactivity of 19 Patients with Untreated Hodgkin's Disease and 10 Control Subjects

	No. positive/No. tested	
	Before TP-1 treatment	After TP-1 treatment
Patients (N = 19)		
Tuberculin (5 U)	5/19 (26%)	6/19 (31.5%)
Candida (1 : 10)	2/19 (10.5%)	4/19 (21%)
Streptokinase (40 U)	8/19 (42%)	17/19 (89%)**
Total[a]	10/19 (52.6%)	18/19 (94.7%)*
Controls (N = 10)		
Tuberculin (5 U)	2/10 (20%)	2/10 (20%)
Candida (1 : 10)	4/10 (40%)	4/10 (40%)
Streptokinase (40 U)	9/10 (90%)	9/10 (90%)
Total	10/10 (100%)	10/10 (100%)

[a] Represents number (percent) of patients with at least one positive skin test.
**$p < 0.01$, *$p < 0.05$ (McNemar's test).

TABLE IV. Mean Serum Lysozyme Levels (μg/ml) of Patients Grouped According to Presence (CIC-Patients) or Absence (CIC-Negative Patients) of Circulating Immune Complexes (CIC) or Evaluated by Platelet Aggregation and ClqB-ELISA Tests before and after TP-1 Treatment

	Before TP-1	After TP-1	Paired t test
Whole patient population ($N = 32$)	10.66 ± 7.09*	13.5 ± 11.73	$p < 0.05$
CIC-positive patients ($N = 13$)	7.32 ± 4.15	10.42 ± 5.2	$p < 0.05$
CIC-negative patients ($N = 19$)	12.95 ± 7.85**	15.61 ± 14.4	NS
t test[a]	$p < 0.02$	NS	

[a] t test: *NS and **$p < 0.05$ vs. serum lysozyme in 20 normal control subjects, 8.3 ± 1.42 (mean ± S.D.).

patients where no increases in CICs were detected ($p < 0.05$), whereas the 13 patients with elevated CICs had a lower mean lysozyme level than the control ($p < 0.02$). TP-1 increased serum lysozyme levels from 10.7 ± 7.1 μg/ml to 13.5 ± 11.7 μg/ml in Hodgkin's patients as a group ($p < 0.05$). This was mainly accounted for by the group of patients in whom the CICs were initially elevated (7.32 to 10.42 μg/ml; $p < 0.05$).

3. DISCUSSION

There is convincing evidence that patients with Hodgkin's disease have abnormalities in immune function (Young *et al.*, 1972). The pathogenesis of these defects remains unclear. While it seems that the percentage of cells positive for the T_3 antigen is normal in the majority of cases, E-rosetting cells are decreased and T-cell function is impaired. The abnormality in immune function may be due to an interaction of lymphocyte membrane receptors with various serum factors. These serum factors include a low-density lipoprotein, ferritin, antilymphocyte antibodies against T cells, and circulating immune complexes (Payne *et al.*, 1976; Siegel *et al.*, 1976). Schechtor and Soehnlen (1978) have shown excessive monocyte suppressor cell activity against T-cell function. Goodwin *et al.* (1977) suggest that suppressor cells secreting prostaglandins are involved in the impaired T-cell response.

Our data confirm previous observations that lymphocytes from patients with Hodgkin's disease frequently fail to exhibit normal circulating lymphocytes (Aisenberg, 1966). We similarly confirm that a variable proportion of Hodgkin's patients have elevated CICs and impaired skin test reactivity; however, there are no previous reports on impaired LIF release.

Following *in vitro* incubation with TP-1, rosetting capacity reached control levels. TP-1 also induces and increases in lymphocyte PHA blastogenic response. Unlike the E-rosette response, the PHA responses after TP-1 incubation did not reach control values in most cases. It is noteworthy that increases of E-rosetting cells of PHA responsiveness were more pronounced in patients with more depressed initial values, supporting the hypothesis that TP-1 reconstitutes rather than stimulates immune functions. Improvement in T-cell function was also greatest in those patients with a low lymphocyte count.

The *in vivo* administration of TP-1 similarly induced an increase in E-rosette-forming cells and PHA reactivity. In addition, LIF positively occurred in nine patients who prior to treatment had no detectable LIF activity. It appears that TP-1 has a broad effect on T-cell-mediated immunological parameters.

A 3-week course of TP-1 significantly reduced the CIC-like material in our patients. There were a greater number of reductions using the platelet aggregation test than with the ClqB-ELISA test, suggesting that TP-1 exerts a greater effect on non-complement-fixing complexes. Data on serum lysozyme levels obtained before and after TP-1 treatment offer some insight as to the significance of elevated CICs. High CIC levels tended to be associated with low-normal lysozyme values, whereas high lysozyme levels were associated with CIC levels that were not elevated. TP-1 seemed to convert a high-CIC low-normal-lysozyme state to a high-lysozyme normal-CIC state. As serum lysozyme has been correlated with monocyte function and/or mass, our data suggest that patients with elevated CICs in Hodgkin's disease have insufficient monocyte and/or neutrophil capacity to clear CICs. TP-1 may increased the T-cell interaction with nonspecific defense mechanism and thus reduce CICs.

Untreated patients with Hodgkin's disease clearly have defects in cellular immunity. Regardless of the pathogenesis of these defects, there appears to be a clear relationship between T-cell function and survival (Faguet and Davis, 1982). That immuological parameters remain abnormal in long-term "disease-free" survivors may connote an inherent prediagnosis characteristic of the Hodgkin's patient (Ficher *et al.*, 1980). Our data suggest that T-lymphocyte function and nonspecific immune mechanisms can be restored in a population of Hodgkin's disease patients prior to therapy. A controlled study will be needed to see if immunomodulation alters survival, infectious complications, or the incidence of second malignancies.

REFERENCES

Aisenberg, A. C., 1966, Immunologic status of Hodgkin's disease, *Cancer* **19**:385–390.

Bjorkholm, M., Holm, G., Mellstedt, H., and Johnson, B., 1976, Immunodeficiency and prognosis in Hodgkin's disease, *Acta Med. Scand.* **295**:1273.

Faguet, G. B., and Davis, H. C., 1982, Survival in Hodgkin's disease: The role of immunocompetence and other major risk factors, *Blood* **59**:938.

Falchetti, R., Bergesi, G., and Caprino, L., 1977, Isolation, partial characterization and biological effects of a calf thymus factor, 3rd European Immunology Meeting, Copenhagen.

Ficher, R., DeVita, V. T., Bostick, F., Vanhaelen, C., Hawsor, D. M., Hubbard, S., and Young, R., 1980, Persistent immunologic abnormalities in long-term survivors of advanced Hodgkin's disease, *Ann. Intern. Med.* **92**:595–599.

Goodwin, J. S., Messner, R. P., and Peake, G. T., 1977, Prostaglandin producing cells in Hodgkin's disease, *N. Engl. J. Med.* **297**:963–968.

Payne, S. V., Jones, D. B., Haegert, D. G., Smith, J. C., and Wright, D. H., 1976, T and B lymphocytes and Reed–Sternberg cells in Hodgkin's disease lymph nodes and spleen, *Clin. Exp. Immunol.* **24**:280–286.

Reed, D. M., 1902, On the pathological changes in Hodgkin's disease with respect to the problem of homotransplantation, *Johns Hopkins Hosp. Rep.* **10:**133.

Schechtor, G. P., and Soehnlen, F., 1978, Monocyte mediated inhibition of lymphocyte blastogenesis in Hodgkin's disease, *Blood* **52:**261–271.

Siegel, F. P., 1976, Inhibition of T-cell rosette formation by Hodgkin's disease serum, *N. Engl. J. Med.* **295:**1313–1316.

Young, R. C., Corder, M. P., Haynes, H. A., DeVito, V. T., 1972a, Delayed hyposensitivity in Hodgkin's disease, *Am. J. Med.* **52:**63.

Young, R. C., Corder, M. P., Haynes, H. A., and DeVita, V. T., 1972b, Delayed hypersensitivity in Hodgkin's disease: A study of 103 patients, *Am. J. Med.* **52:**63–68.

Thymosin Fraction 5 Treatment of Patients with Cellular Immunodeficiency Disorders

D. W. WARA, M. J. COWAN, and A. J. AMMANN

1. INTRODUCTION

Restoration of cellular immunity in children with primary immunodeficiency disorders is best achieved by bone marrow transplantation from a histocompatible donor. In the absence of such a donor, there is no consistent source of cells or factors to reconstitute these patients. Fetal thymus transplantation may result in partial restoration of cellular immunity (Ammann *et al.*, 1975). Steele *et al.* (1972) suggested that the fetal thymus produces a factor(s) capable of rapidly reconstituting cellular immunity in infants with Di George syndrome following transplantation in a cell-impermeable chamber.

We and others have documented that incubation of peripheral blood lymphocytes (PBL) from patients with cellular immunodeficiency with thymosin fraction 5 (TF5) enhances T-cell rosette formation (Wara *et al.*, 1975). We have further demonstrated that TF5 incubation increases lymphocyte response to allogeneic cells in a subgroup of these patients (Wara *et al.*, 1980).

Following the demonstration that TF5 incubation with lymphocytes isolated from patients with primary immunodeficiency disorders may enhance T-cell rosette formation and/or response to allogeneic cells, we initiated a clinical trial of TF5 in this patient group. Since 1973 we have treated 34 patients with TF5 and have achieved at least 1 year follow-up in 22 (see Tables I and II).

Duration of treatment varied form 2 weeks to 7 years. Diagnoses of patients treated include severe combined immunodeficiency disease (4), combined immunodeficiency disease (6), Di George syndrome (6), ataxia telangiectasia (4), chronic mucocutaneous candidiasis (4), Wiskott–Aldrich syndrome (3), histiocytosis X (1),

D. W. WARA, M. J. COWAN, and A. J. AMMANN • Pediatric Immunology, University of California, San Francisco, California 94143.

TABLE I. Thymosin Fraction 5 Therapy: Di George Syndrome

Disorder	T-cell number		T-cell function		Duration of treatment, maximum dose	Status
	% T-cell rosettes	Total T cells/mm³	PHA[a]	MLR[a]		
Di George-1 BD: 12/80	$9 \xrightarrow{\frac{1}{2}} 45$[b]	220 → 2747	$0 \xrightarrow{3} 100$	$46 \xrightarrow{3} 100$	4 mo, 2 mg/kg	Alive, well
Di George-2 BD: 9/76	$18 \xrightarrow{1\frac{1}{2}} 56$	398 → 2751	$56 \xrightarrow{1\frac{1}{2}} 100$	$0 \xrightarrow{1\frac{1}{2}} 100$	1½ mo, 2 mg/kg	Expired, congenital heart disease
Di George-3 BD: 7/78	$34 \xrightarrow{\frac{1}{2}} 57$	503 → 3184	$87 \xrightarrow{2} 100$	$100 \xrightarrow{2} 100$	2 mo, 6 mg/kg	Alive, well
Di George-4 BD: 6/77	$9 \xrightarrow{2} 44$	248 → 1850	$0 \xrightarrow{2} 30$	$0 \xrightarrow{2} 41$	2 mo, 2 mg/kg	Alive, well; S/P thymus tx
Di George-5 BD: 9/76	18 → 10	343 → 214	100 → 100	35 → 30	½ mo, 2 mg/kg	Expired, septicemia and congenital heart disease

[a] Percent of normal (highest level achieved) before and during therapy.
[b] Between pre- and maximum post-treatment value. Number above arrow is time in months on TF5 to achieve value.

TABLE II. Thymosin Fraction 5 Therapy: Combined Immunodeficiency Disease

Disorder	T-cell number: % T-cell rosettes	T-cell number: Total T cells/mm³	T-cell function: PHA	T-cell function: MLR	Duration of treatment, maximum dose	Status
CID-1 BD: 12/68	$12 \xrightarrow{2} 58$	208 → 1241	$8 \xrightarrow{12} 35$	$0 \xrightarrow{12} 0$	30 mo, 4 mg/kg	Alive, well; age 15 years
CID-2 BD: 2/70	$13 \xrightarrow{1} 69$	110 → 332	$0 \xrightarrow{9} 61$	$0 \xrightarrow{9} 25$	9 mo, 2 mg/kg (wheezing)	Expired, off treatment, varicella
CID-3 (SLD) BD: 3/79	$57 \xrightarrow{1/2} 75$	1282 → 1125	$10 \xrightarrow{5} 100$	$0 \xrightarrow{5} 100$	6 mo, 4 mg/kg	Alive, S/P bone marrow tx
CID-4 (SLD) BD: 8/67	$19 \xrightarrow{1} 19$	286 → 199	$13 \xrightarrow{4} 22$	0 → 0	4 mo, 2 mg/kg	Expired, chronic lung disease
CID-5 BD: 8/78	$73 \xrightarrow{4} 82$	451 → 841	$65 \xrightarrow{4} 55$	$20 \xrightarrow{4} 46$	7 mo, 2 mg/kg	Expired, chronic lung disease
CID-6 BD: 11/66	$86 \xrightarrow{1/2} 70$	578 → 700	$77 \xrightarrow{1/2} 69$	$38 \xrightarrow{1/2} 17$	2 wk, 2 mg/kg (wheezing)	Alive, chronic lung disease, *Candida*

hyper-IgE syndrome (4), acquired hypogammaglobulinemia (1), and acquired immunodeficiency syndrome (1).

2. METHODS

2.1. Patient Evaluation and Selection for Treatment

Patients were evaluated in the Pediatric Immunology/Rheumatology clinic at the University of California, San Francisco, between 1973 and 1982. At the time of initial evaluation, age ranged from 2 weeks to 43 years. Patients with evidence of significant malnutrition or active infection were vigorously treated and reevaluated following therapy.

Evaluation of cellular immune function before therapy with TF5 included total lymphocyte count, determination of percent T-cell rosettes, response of isolated PBL to allogeneic cells (MLR) and to phytohemagglutinin (PHA). Assessment of antibody-mediated immunity included quantitative immunoglobulin levels, and antibody response to pneumococcal polysaccharide and to keyhole limpet hemocyanin if patients were older than 2 years. Methods have previously been described in detail (Ammann *et al.*, 1973). In addition, enhancement of T-cell rosette formation (Wara *et al.*, 1975) and of lymphocyte response to allogeneic cells (Wara *et al.*, 1980) following incubation of isolated PBL with TF5 were determined. TF5 was obtained from Hoffmann–La Roche Inc. and was added to cell suspensions in concentrations varying from 50 to 500 μg/ml.

Patients were accepted for treatment with TF5 if cellular immunodeficiency was documented by decreased T-cell rosette formation (less than 30%) and decreased lymphocyte response to PHA and/or in MLR (less than 30% normal). However, patients with Di George syndrome diagnosed by standard criteria were accepted for treatment with low T-cell numbers but normal T-cell function. In addition, enhanced T-cell rosette formation (50%) and/or lymphocyte response to allogeneic cells (100%) with TF5 incubation was required. Patients whose lymphocytes did not respond to TF5 incubation *in vitro* were excluded from the study.

2.2. Thymosin Treatment Protocol

An initial test dose of 1 mg TF5 was given i.d. If no immediate hypersensitivity was observed, a regimen of daily s.c. injections (2–6 mg/kg) was begun for 14 consecutive days. Thereafter, TF5 was administered once weekly in a dose varying between 2 and 6 mg/kg and dependent upon patient response.

Each patient's antibody and cellular immunity were reevaluated at the completion of the 2-week induction period and at 3-month intervals thereafter until completion of therapy. Follow-up evaluations included total lymphocyte count, T-cell rosette formation, lymphocyte response to PHA and to allogeneic cells, and quantitative immunoglobulins. Drug toxicity studies, including evaluation of renal, liver, and hematopoietic function, were carried out at the same time intervals.

Therapy was terminated at 6 months if neither clinical nor laboratory improvement was observed. If an adverse reaction or clinical deterioration occurred, therapy was terminated earlier. Therapy was discontinued approximately 2 months following normalization of cellular immunity in patients with Di George syndrome.

2.3. Statistical Analysis

A 95% confidence limit was calculated for normal adult control lymphocyte responsivity to PHA and to allogeneic cells for each 6-month interval between 1973 and 1982. Twenty normal control values were included for each time interval. Patient results are expressed as the percentage of the 95% confidence limit for normal controls during the matched time interval.

3. RESULTS

3.1. Di George Syndrome

Four of six patients treated with TF5 were evaluable for longer than 2 weeks following initiation of therapy. Patient No. 5 expired after 2 weeks of daily therapy and a sixth patient expired after 1 week of therapy with congenital heart disease. Increased T-cell rosette formation and total T cells/mm^3 were found in patients No. 1 through 4 from 0.5 to 2 months after initiation of therapy. T-cell function improved from 1.5 to 3 months after therapy was begun. Unfortunately, patient No. 2 expired from cardiac disease. Patient No. 4 required a fetal thymus transplant for complete restoration of cellular immunity. However, patients No. 1 and 3 are alive and well, with normal immune function, 3 and 5 years after TF5 therapy. No deleterious effects were seen with TF5 therapy in this group of patients.

3.2. Combined Immunodeficiency Disease (CID)

Three of six patients with CID improved during treatment with TF5. An increase in T-cell numbers was seen between 2 weeks and 2 months after initiation of treatment while at least one measure of T-cell function (PHA or MLR) improved after 5 to 12 months of therapy. Of the remaining three patients, two expired with chronic lung disease while a third is alive with pulmonary disease.

Patient No. 1 has been reported in detail previously (Wara *et al.*, 1975). While receiving TF5, she improved clinically with weight gain (26 to 42 lbs) and decreased infection and diarrhea. Following a period of recurrent infection, TF5 was discontinued. A fetal thymus transplant was performed, but improvement in cellular immunity was not observed. Although she has received no specific immunotherapy for 3 years, she is alive and well at age 15.

Patient No. 3 had a transient improvement in cellular immunity. Because normal T-cell function was not sustained, TF5 was discontinued after 6 months. An allogeneic bone marrow transplantation, modified by soybean lectin agglutination, was performed.

Patients No. 2 and 6 experienced significant side effects following injections of TF5. Both developed wheezing, in addition to erythema at the injection site. TF5 was discontinued, in both patients. The two patients both had polyclonal hypergammaglobulinemia and documented capacity to form specific antibody, although no antibody to TF5 was found.

Thus, three of six patients with CID responded to treatment with TF5 by increase in T-cell number and/or enhanced function as well as apparent clinical improvement.

3.3. Primary Immunodeficiency—Other

Four patients with ataxia telangiectasia were treated with TF5 for 4 months to 7 years. Two of the four expired from lymphoma while receiving TF5. T-cell numbers were maintained and T-cell function improved but was not sustained in all four. However, patients with this disorder have fluctuating T-cell function without immunotherapy. Therefore, the results of treatment with TF5 in this group are difficult to interpret.

Three children with Wiskott–Aldrich syndrome received TF5. All three had increased T-cell numbers and enhanced T-cell function while receiving therapy. Unfortunately, all were lost to follow-up.

We did not observe a sustained improvement in T-cell numbers or function in any other patient treated with TF5. Although one patient with chronic mucocutaneous candidiasis had partial clearing of her *Candida*, this was not associated with altered immune function. Four patients with severe CID had neither clinical nor laboratory improvement.

4. DISCUSSION

The response by patients with primary immunodeficiency disorders to therapy with TF5 has been variable. A subgroup of patients with Di George syndrome or CID appear to respond to TF5 therapy with enhanced cellular immunity and clinical well-being. However, interpretation of clinical status must be cautious because patients often received additional treatment such as γ-globulin and/or antibiotics on a regular basis.

The results of TF5 treatment in patients with Di George syndrome are encouraging. However, those who responded most completely had mild defects in cellular immunity (Barrett *et al.*, 1980) and we are unable to exclude spontaneous resolution which is known to occur in these patients without immunotherapy.

Although three patients with CID had significant enhancement of cellular immunity while receiving TF5, the improvement was not always sustained (patient No. 3). In contrast, patient No. 1 has received no immunotherapy for 3 years and is clinically well. Although this is difficult to explain, it is possible that a component of TF5 stimulated her own thymus to produce factors which then maintained her immune function. Two patients in this group developed wheezing within 20 min after TF5 injections. Thus, significant side effects may occur with TF5 and appear to be more common in patients who are capable of antibody production.

Patients with other diagnoses of primary immunodeficiency did not appear to respond to TF5. However, many did not receive therapy for the time period of 3 to 6 months which appears necessary for quantitation of improved cellular immunity. In addition, treatment was initiated in some patients with active infection, making the assessment of clinical improvement difficult. The success of TF5 therapy in patients with primary immunodeficiency diseases is most likely related to several factors. The most important appears to be the underlying diagnosis. Patients with Di George syndrome or CID may respond to TF5 therapy while those with other disorders do not. Both patients with Di George syndrome and CID have normal bone marrow-derived stem cells but abnormal thymus glands. TF5 may mature the stem cells, replacing the function of the thymus. Alternatively, in patients with Di George syndrome, TF5 may act directly on the thymus and induce maturation, providing rapid and permanent reconstitution of cellular immunity.

The dosage of TF5 may be important. Our study did not define either the maximum tolerated dose, the optimal dose, or the schedule for efficacy. The duration of treatment is an additional variable. Patients treated for less than 3 months, with diagnoses other than Di George syndrome, did not have improved cellular immunity.

Although our clinical study suggests that TF5 therapy may be effective in some patients with Di George syndrome or CID, the results are not conclusive. However, in the absence of a histocompatible bone marrow donor, a trial of TF5 in patients with these disorders appears warranted. We propose a multicenter randomized trial of TF5 in patients with these two disorders.

Note added in proof: Patient CID-3 (SLD) expired with disseminated aspergillosis following a haploincompatible soybean lectin-separated bone marrow transplant.

ACKNOWLEDGMENTS. The authors' research reported herein was supported in part by the Division of Research Resources, National Institutes of Health–Pediatric Clinical Research Center (5M01-RR-00079-17), and in part by the American Cancer Society–Eleanor Roosevelt International Cancer Fellowship awarded by the International Union Against Cancer (to D.W.W.).

REFERENCES

Ammann, A. J., Wara, D. W., Salmon, S., and Perkins, H., 1973, Thymus transplantation: Permanent reconstitution of cellular immunity in a patient with sex-linked combined immunodeficiency, *N. Engl. J. Med.* **289**:5–9.

Ammann, A. J., Wara, D. W., Doyle, N. E., and Golbus, M. S., 1975, Thymus transplantation in patients with thymic hypoplasia and abnormal immunoglobin synthesis, *Transplantation* **20**:457–466.

Barrett, D. B., Wara, D. W., Ammann, A. J., and Cowan, M. J., 1980, Thymosin therapy in the Di George syndrome, *J. Pediatr.* **97**:66–71.

Steele, R. W., Lenias, C., Thurman, G. B., Schuelein, M., Bauer, H., and Bellanti, J. A., 1972, Familial thymic aplasia: Attempted reconstitution with fetal thymus in a Millipore diffusion chamber, *N. Engl. J. Med.* **287**:787–791.

Wara, D. W., Goldstein, A. L., Doyle, N. E., and Ammann, A. J., 1975, Thymosin activity in patients with cellular immunodeficiency, *N. Engl. J. Med.* **292**:70–74.

Wara, D. W., Barrett, D. J., Ammann, A. J., and Cowan, M. J., 1980, In vitro and in vivo enhancement of mixed lymphocyte culture reactivity by thymosin in patients with primary immunodeficiency disease, *Ann. N.Y. Acad. Sci.* **332**:128–135.

Effect of Thymosin on *in Vitro* Immune Function in Patients with Rheumatoid Arthritis

ROBERT P. JACOBS,
CHRISTINE S. E. RICHARDSON, SUSAN B. RILEY,
MARION ZATZ, JANELLE HATCHER,
and ALLAN L. GOLDSTEIN

1. INTRODUCTION

The pathogenesis of rheumatoid arthritis (RA) is not known, but an abnormality in immunoregulation is thought to play a role. Some of the immunological abnormalities associated with other autoimmune diseases have been reported to be normalized *in vitro* by exposure of peripheral blood lymphocytes (PBL) to thymosin fraction 5 (TF5) (Horowitz *et al.*, 1977; Scheinberg *et al.*, 1979; Lavastida *et al.*, 1981; Goldstein *et al.*, 1976), thus providing a rationale for the use of this immunomodulating agent in these disorders. The studies being presented today were undertaken to evaluate the effect of TF5 on *in vitro* immunoregulatory function in patients with RA and establish the relevance of clinical application of thymosin in the treatment of this common disease.

2. MATERIALS AND METHODS

Selection of Patients and Normal Controls. Two groups of patients with definite or classical RA (Ropes *et al.*, 1958) and age- and sex-matched normal controls were studied (Table I). Group I RA patients were selected randomly from the

ROBERT P. JACOBS, CHRISTINE S. E. RICHARDSON. amd SUSAN B. RILEY • Department of Medicine, The George Washington University School of Medicine and Health Sciences, Washington, D.C. 20037. MARION ZATZ, JANELLE HATCHER, and ALLAN L. GOLDSTEIN • Department of Biochemistry, The George Washington University School of Medicine and Health Sciences, Washington, D.C. 20037.

TABLE I. Clinical Features of RA Patients

	Group I	Group II
Number	22	14
Males/females	0/22	3/11
Mean age (years)	43.8	56.1
(age range)	(24–83)	(35–66)
Number on antirheumatics	20	0
Number seropositive	15	11

Rheumatology clinics at The George Washington University Medical Center where they were being treated with conventional anti-inflammatory and antirheumatic drug therapy. Group II RA patients included participants in a double-blind, placebo-controlled trial of TF5. Patients in the second group were studied a minimum of 3 months following discontinuation of antirheumatic drug therapy and prior to entry in the thymosin trial.

Thymosin. All thymosin incubations were performed with calf TF5, Lot C100496, provided by Hoffmann–La Roche Inc., Nutley, New Jersey. The thymosin was of clinical grade and endotoxin free.

Isolation of PBL. PBL from normal and patient volunteers were separated from heparinized blood on Lymphoprep gradients (Accurate Chemical and Scientific Co, Hicksville, N.Y.). The PBL were washed twice in HBSS and once with Hepes-buffered RPMI 1640 (HRPMI) (GIBCO, Grand Island, N.Y.), then resuspended in HRPMI.

T-Cell Subsets. PBL were adjusted to 5×10^6/ml, dispensed as 200-μl aliquots into 12 × 75-mm tubes, and incubated with 50 μl OKT3, OKT4, OKT8, or MO2 (Orthoclone, Ortho Pharmaceuticals) antibody or medium. The tubes were incubated for 30 min at 4°C and then washed twice in PBS buffer. The cells were then exposed to aggregate-free, fluorescein-labeled goat anti-mouse immunoglobulin, incubated for an additional 30 min at 4°C, washed twice, and resuspended in 1 ml filtered Isoton for quantitative analysis on a fluorescence-activated cell sorter by flow microfluorometry.

E-Rosette-Forming Cell Assay. PBL were adjusted to 5×10^6/ml and dispensed as 100-μl aliquots into triplicate round-bottom tubes. Medium or TF5 (100 μg/ml) was added as 100-μl aliquots and the tubes incubated for 10 min at 37°C in 5% CO_2/air. Sheep erythrocytes (SRBC) at 80×10^6/ml were added as 200-μl aliquots. The tubes were centrifuged at 180*g* for 5 min at 4°C and then incubated overnight at 4°C. Two hundred mononuclear cells per tube were counted in a standard hemocytometer chamber, cells binding $\geqslant$ 3 SRBC considered rosettes.

Active E-RFC were prepared by incubating the initial PBL–SRBC mixture for 5 min at 37°C, centrifuging at 180*g* for 5 min at room temperature, and then incubating in a 29°C water bath overnight.

Suppressor Cell Assay. The suppressor cell assay was modified from previously published methods (Shou *et al.*, 1976; Fineman *et al.*, 1979). PBL were adjusted to 5×10^6/ml, dispensed as 2-ml aliquots in 15-ml round-bottom tissue culture tubes, and incubated for 48 hr in the presence or absence of Con A at 20 μl/ml. Replicate tubes were set up in the presence or absence of TF5 at 100 μg/ml. At the end of the 48-hr preincubation, the cells were washed twice in HBSS containing α-methyl-D-mannoside at 5 mg/ml and once in HRPMI. The washed cells were then inactivated with mitomycin C, at 50 mg/ml for 30 min, washed, and added as stimulators to fresh normal responder PBL in MLC. Thus, the preincubated cells served as both potential suppressors and allogeneic stimulators in this assay. An index of suppression was calculated for each of the preincubation groups as:

$$\text{Percent change} = 100 \times \frac{\text{cpm (Con A and/or thymosin)} - \text{cpm (no Con A or thymosin)}}{\text{cpm (no Con A or thymosin)}}$$

Thus, induction of suppression resulted in negative indices.

3. RESULTS

T-Cell Subsets. The distribution of T-cell subsets detected by monoclonal antibodies OKT3, OKT4, and OKT8 was evaluated in PBL from group II RA patients and normal controls (Table II). Monocytes were detected with the monoclonal antibody MO2. No differences between RA patients and controls were observed with these reagents, although the small sample size must be noted. Preincubation of PBL with TF5 did not induce any significant changes in cell subsets in either study group (data not shown).

E-RFC. E-RFC and active E-RFC, expressed as a percent of total PBL, were the same in group I and group II RA patients and their corresponding normal controls (Table III). Furthermore, the effect of TF5 on these T-cell parameters was the same.

TABLE II. T-Cell Subsets in PBL from Group II RA Patients

	Controls	Patients
OKT3	57.2 ± 1.8[a]	57.5 ± 3.0
OKT4	40.6 ± 3.1	42.0 ± 5.0
OKT8	18.4 ± 1.4	18.5 ± 3.9
MO2	9.7 ± 0.9	9.0 ± 1.9
T4/T8	2.4 ± 0.2	2.7 ± 0.5

[a] Results expressed as a percent of total PBL. Differences between RA patients and controls are not significant.

TABLE III. Effect of Thymosin on E-RFC in PBL from RA Patients

	% E-RFC[a]			
	4°C		29°C	
	− thymosin	+ thymosin	− thymosin	+ thymosin
Group I				
Controls (11)	62.6[b]	66.8	ND	ND
Patients (12)	55.9	58.8	ND	ND
Group II				
Controls (28)	72.3[b]	ND	42.6	ND
Patients (13)	65.5	72.9	44.8	43.0

[a] E-RFC expressed as a percent of total PBL.
[b] Differences between RA patients and controls are not significant.

Suppressor Cell Assay. PBL from group I RA patients, matched normal controls, and a large group of unmatched controls were evaluated in the suppressor cell assay. Suppression induced by preincubation of PBL from RA patients with Con A (Table IV) was significantly greater than that seen in both the matched and unmatched control groups ($p < 0.05$ and $p < 0.02$, respectively). A comparison of Con A-induced suppression in seronegative and seropositive RA patients revealed similar responses (−59.4 and −54.0%, respectively).

Suppression induced by preincubation of PBL from RA patients with TF5 (Fig. 1) was significantly greater than that seen with either the matched normal controls

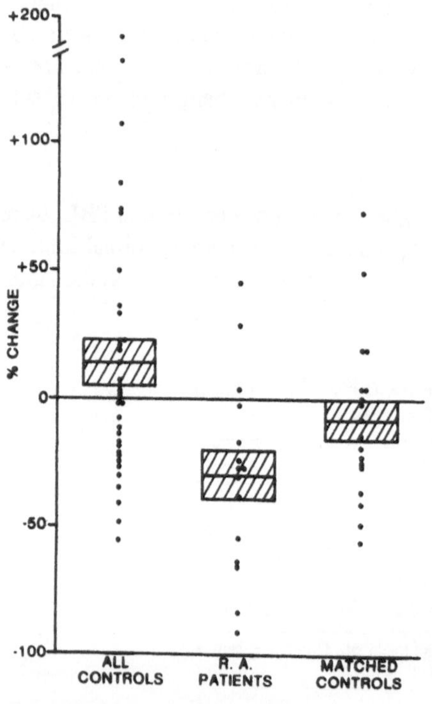

FIGURE 1. Suppressor cell assay: Thymosin-induced suppression in group I RA patients. The percent change in MLC response to PBL preincubated in the presence or absence of TF5 at 100 μg/ml is shown. The mean percent change is shown by the short horizontal line; the shaded area designates the standard error of the mean. Points below the solid horizontal line at 0% change indicate suppressed responses. The difference in the mean percent change between RA patients and all controls or matched normal controls was significant at $p < 0.005$ and $p < 0.05$, respectively.

TABLE IV. Suppressor Cell Assay: Con A-Induced Suppression in Group I RA Patients

	% Δ ± S.E.M.
Unmatched controls (38)	−18.4 ± 9.3[a]
Matched controls (13)	−24.4 ± 18.9[b]
Patients (13)	−57.5 ± 11.5

[a] Difference between RA patients and unmatched controls is significant at $p < 0.02$.
[b] Difference between RA patients and matched controls is significant at $p < 0.05$.

or an unmatched panel of normals ($p < 0.05$ and $p < 0.005$, respectively). A comparison of thymosin-induced suppression in seronegative and seropositive RA patients revealed similar responses (−34.2 and −28.3%, respectively).

When cells from RA patients and controls were preincubated with Con A plus TF5, the induced suppression was similar to that seen with Con A alone (Table V). These results suggest that thymosin is inducing a proportion of the same suppressor cell population that is induced by Con A.

4. DISCUSSION

One theory of autoimmunity suggests that a loss of normal immunoregulation contributes to the emergence of autoreactivity and the development of disease (Reinherz and Schlossman, 1980; Waldmann *et al.*, 1978). In an attempt to identify an immunoregulatory defect in RA patients, there have been multiple studies of T-cell subsets and suppressor cell function. Moutsopoulos *et al.* (1976) noted that E-RFC were normal in RA patients, but Espinoza *et al.* (1980) reported that a subpopulation of T cells detected by the active E-RFC assay was abnormal, particularly in active disease. Our data indicate that both E-RFC and active E-RFC are normal in patients with active RA.

There have been conflicting reports of T-cell subsets measured by monoclonal antibodies in PBL of RA patients. Veys *et al.* (1981) reported that patients with active disease had normal numbers of T cells reacting with OKT3, but a decrease

TABLE V. Suppressor Cell Assay: Combined Effect of Con A and Thymosin in Group I RA Patients

	Thymosin	Con A	% Δ
Controls (10)	−	+	−42.5
	+	−	−24.2
	+	+	−43.0
Patients (12)	−	+	−77.0
	+	−	−38.3
	+	+	−66.0

in OKT8-bearing lymphocytes and an increase in the OKT4/OKT8 ratio. Burmester *et al.* (1981), however, could identify no abnormality in the percentage of lymphocytes bearing OKT4 or OKT8 compared to normal controls. Our preliminary data are in agreement with this latter report.

Studies of suppressor cell function in PBL from RA patients have also been conflicting, with reports of increased, normal, and decreased suppressor cell activity (Zilko *et al.*, 1980; Doubloug *et al.*, 1981; Abdou *et al.*, 1981; Keystone *et al.*, 1980; Tosato *et al.*, 1981; Chattopadhyay *et al.*, 1979). Our results show an abnormality in the generation of suppressor cells in RA patients characterized by an excess of suppressor cell activity induced by both Con A and TF5. These data are in agreement with the observations of Zilko *et al.* (1980) who suggested that this immunoregulatory defect correlated with disease activity.

The induction of excess suppressor cell activity by both Con A and TF5 suggests that RA patients may have an excess of suppressor cell precursors. These observations are consistent both with the effect of TF5 on the induction of suppressor cell activity in SLE patients (Horowitz *et al.*, 1977) and the report that TP-5, a synthetic immune modulator, increases the proportion of OKT8 cells in PBL from RA patients (Veys *et al.*, 1981).

It is important to attempt to relate our observations regarding increased suppressor cell activity in RA patients to our current understanding of the immunopathogenesis of the disease. At first glance, an increase in suppressor cell activity in an autoimmune disease seems paradoxical. One would expect a loss of suppressor cells associated with the emergence of autoreactivity. Several possible explanations can be offered for this seemingly paradoxical observation. First, while an excess of inducible or precursor suppressor cell activity may be present in RA, specific suppressor activity directed toward relevant antigens might be depressed. There have been reports of depressed antigen-specific suppressor cell responses in RA patients to ovalbumin (Keystone *et al.*, 1980) and, recently, to Epstein–Barr virus (Tosato *et al.*, 1981). Second, an excess of inducible suppressor cells may indicate a block in the generation of functionally active suppressor cells in RA patients. This is suggested by reports of reduced numbers of OKT8 cells in PBL and synovium from RA patients (Veys *et al.*, 1981; Janossy *et al.*, 1981).

In view of current theories regarding the pathogenesis of autoimmune diseases and the results of *in vitro* studies, we have initiated a double-blind, placebo-controlled trial of TF5 in RA. The objectives of this study are to assess the efficacy of this immunomodulating agent in the treatment of RA and its effect on *in vitro* immunological function. The present observations may offer an *in vitro* tool for monitoring changes in cellular immunity in patients treated with immunomodulating agents, thereby providing information about the relevance of these immunoregulatory abnormalities to the clinical status of the patient.

ACKNOWLEDGMENTS. We wish to thank Hoffmann–La Roche Inc. for providing the TF5 used in these studies, Barbara Ekstrand for her contributions in establishing the Con A assay, and Marc Roskelley for his excellent secretarial assistance.

REFERENCES

Abdou, N. I., Lindsley, H. B., Racela, L. S., Pascual, E., and Hassanein, K. M., 1981, Suppressor T-cell dysfunction and antisuppressor cell antibody in active early rheumatoid arthritis, *J. Rheum.* **8**:9.

Burmester, G. R., Yu, D. T. Y., Irani, A. M., Kunkel, H. G., and Winchester, R. J., 1981, Ia + T-cells in synovial fluid and tissues of patients with rheumatoid arthritis, *Arthritis Rheum.* **24**:1370.

Chattopadhyay, C., Chattopadhyay, H., Natvig, J. B., and Melbye, O. J., 1979, Rheumatoid synovial lymphocytes lack concanavalin-A-activated suppressor cell activity, *Scand. J. Immunol.* **10**:479.

Doubloug, J. H., Chattopadhyay, C., Forre, O., Hoyeraal, H. M., and Natvig, J. B., 1981, Con A-induced suppressor cell activity and T-lymphocyte of patients with rheumatoid arthritis and juvenile rheumatoid arthritis, *Scand. J. Immunol.* **13**:367.

Espinoza, L. R., Gaylord, S. W., Bergen, L., Vasey, F. B., Germain, B. F., and Osterland, C. K., 1980, The "active" rosette test in rheumatoid arthritis: Correlation with disease activity, *Clin. Immunol. Immunopathol.* **17**:110.

Fineman, S. M., Mudawwar, F. B., and Geha, R. S., 1979, Characteristics and mechanisms of action of the concanavalin-A activated suppressor cell in man, *Cell. Immunol.* **45**:120.

Goldstein, A. L., Thurman, G. B., Cohen, G. H., and Rossio, H., 1976, The endocrine thymus: Role for thymosin in the treatment of autoimmune disease, *Ann. N.Y. Acad. Sci.* **274**:390.

Horowitz, S., Borcherding, W., Moorthy, A. V., Chesney, R., Schulte-Wisserman, H., Hong, R., and Goldstein, A. L., 1977, Induction of suppressor T-cells in systemic lupus erythematosus by thymosin and cultured thymic epithelium, *Science* **197**:999.

Janossy, G., Panayi, G., Duke, O., Bofill, M., Poulter, L. W., and Goldstein, G., 1981, Rheumatoid arthritis: A disease of T-lymphocyte/macrophage immunoregulation, *Lancet* **2**:839.

Keystone, E. C., Gladman, D. C., Buchanan, R., Cane, D., and Poplonski, L., 1980, Impaired antigen-specific suppressor cell activity in patients with rheumatoid arthritis, *Arthritis Rheum.* **23**:1246.

Lavastida, M. I., Goldstein, A. L., and Daniels, J. C., 1981, Thymus administration in autoimmune disorders, *Thymus* **2**:287.

Moutsopoulos, H., Fye, K. H., Sawada, S., Becker, M. J., Goldstein, A., and Talal, N., 1976, *In vitro* effect of thymosin on T-lymphocyte rosette formation in rheumatic diseases, *Clin. Exp. Immunol.* **26**:573.

Reinherz, E. L., and Schlossman, S. F., 1980, Regulation of the immune response-inducer and suppressor T-lymphocyte subsets in human beings, *N. Engl. J. Med.* **303**:370.

Ropes, M. W., Bennett, G. A., Cobb, S., Macox, R., and Jessar, R. A., 1958, Diagnostic criteria for rheumatoid arthritis, *Bull. Rheum. Dis.* **9**:175.

Scheinberg, M. A., Cathcart, E. S., and Goldstein, A. L., 1979, Thymosin induced reduction of null cells in peripheral blood lymphocytes of patients with systemic lupus erythematosus, *Lancet* **1**:424.

Shou, L., Schwartz, S. A., and Good, R. A., 1976, Suppressor cell activity after concanavalin A treatment of lymphocytes from normal donors, *J. Exp. Med.* **143**:1100.

Tosato, G., Steinberg, A. D., and Blaese, R. M., 1981, Defective EBV-specific suppressor T-cell function in rheumatoid arthritis, *N. Engl. J. Med.* **305**:1238.

Veys, E. M., Hermanns, P. L., Goldstein, G., Kung, P., Schindler, J., and Van Wauwe, J., 1981, T-cell subpopulations determined by monoclonal antibodies in RA—Influence of immunomodulating agents, *Adv. Inflammation Res.* **3**:155.

Waldmann, T. A., Blaese, R. M., Broder, S., and Krakauer, R. S., 1978, Disorders of suppressor immunoregulatory cells in the pathogenesis of immunodeficiency and autoimmunity, *Ann. Intern. Med.* **88**:225.

Zilko, P. J., Dawkins, R. L., and Carrano, J. A., 1980, Nonspecific suppression and genetic factors in autoimmunity, in: *Immunoregulation and Autoimmunity* (R. W. Krakauer and M. K. Cathcart, eds.), pp. 173–181, Elsevier/North-Holland, Amsterdam.

Phase I Trial Using Thymosin Fraction 5 in Renal Cancer

Northern California Oncology Group (NCOG) Report

WILLIAM M. WARA, MARGUERITE H. NEELY, LINDA J. FIPPIN, and DIANE W. WARA

1. INTRODUCTION

A number of laboratories have now isolated and purified thymic hormone-like factors from animal tissues and blood. These thymic hormones may have the potential to mature precursor lymphocytes to adult functioning cells. This effect could be advantageous for cancer patients to reverse the immunosuppressive effects of chemotherapy, radiotherapy, surgery, and of the tumor itself (W. Wara *et al.*, 1975).

Thymosin fraction 5 (Hoffman–LaRoche) has been evaluated in two large cancer trials to date. Chretien and associates have demonstrated a probable increase in survival in small cell carcinoma of the lung in 55 patients when given thymosin in addition to standard therapy. Median survival in patients with irradiation for all detectable tumor and adjuvant chemotherapy was 240 days for placebo patients versus 450 days for thymosin fraction 5-treated patients (Chretien *et al.*, 1978). The second trial, conducted at the University of California, San Francisco (UCSF), evaluated advanced head and neck and esophageal cancer patients receiving placebo or thymosin fraction 5 (60 mg/m^2 daily for 2 weeks, then twice a week for 50 weeks). The results of this trial demonstrated that thymosin could reverse the

WILLIAM M. WARA • Department of Radiation Oncology, University of California, San Francisco, California 94143, and Northern California Cancer Program, Palo Alto, California 94303. MARGUERITE H. NEELY and LINDA J. FIPPIN • Department of Radiation Oncology, University of California, San Francisco, California 94143. DIANE W. WARA • Pediatric Immunology, University of California, San Francisco, California 94143.

secondary immunosuppression from radiation therapy and a difference in disease-free survival of 61% versus 45% favoring the thymosin-treated patients (Wara *et al.*, 1981).

In both trials an arbitrary dose and schedule were selected to treat patients based on the experience in pediatric immunodeficient patients (D. Wara *et al.*, 1975). No dose or schedule modification was attempted and no drug toxicity was encountered. Because of these facts a new trial was designed to look at different doses of thymosin given to patients with all types of advanced malignant disease who had failed conventional treatment.

2. MATERIALS AND METHODS

Patients with advanced malignancies who had failed conventional chemotherapy and/or radiation therapy were referred for placement on a phase I trial of thymosin fraction 5 at UCSF. Patients had to have a biopsy-proven tumor, no possible alternative therapy, and informed consent.

Immunological evaluation was obtained on all patients prior to thymosin therapy—cell-mediated immunity to phytohemagglutinin (PHA), *in vitro* response to irradiated allogeneic cells (MLC) with and without thymosin incubation, and T-cell rosette formation with and without thymosin using methods previously described (Wara and Ammann, 1978). At completion of thymosin, all patients had their immunological tests repeated.

The proposed dose escalation schema appears in Table I and was administered after a 1-mg test dose.

3. RESULTS

Seventy-four patients were entered on study, with 63 fully evaluable patients. The escalation from 60 mg/m^2 to 210 mg/m^2 was achieved without difficulty. Twenty-three patients with renal cell carcinoma, eight patients with lung carcinoma, and eight patients with head and neck carcinoma comprised the largest patient group

TABLE I. Thymosin Fraction 5

Step	No. of patients	Dose (mg/m^2)	Route, days, frequency
1	10	60	s.c. qd × 15
2	3–10	90	s.c. qd × 15
3	3–10	120	s.c. qd × 15
4	3–10	150	s.c. qd × 15
5	3–10	180	s.c. qd × 15
6	3–10	210	s.c. qd × 15

TABLE II. Total T Cells—Thymosin Fraction 5

	Dose					
	60–90 mg/m^2		120–150 mg/m^2		180–210 mg/m^2	
	Pre	Post	Pre	Post	Pre	Post
Mean ± S.D. (cpm)	878 ±614	932 ±605	897 ±668	975 ±591	1100 ±250	1010 ±486
			NS			

with 35 other patients with various diseases. The toxicity to date was minimal and acceptable. Approximately one-third of the patients showed local toxicity consisting of erythema at the injection site. There was no major systemic toxicity noted in this group of patients. Of interest were the results of the immunological tests for these patients. There was no specific test which correlated with immunological reconstitution or clinical response. Immunological reconstitution also did not correlate with escalating doses as shown in Tables II, III, IV, and V.

There were 9 renal cell carcinoma patients treated with varying doses of thymosin fraction 5 who responded or stabilized out of a total of 23; 3 (13%) have shown partial responses (PR) and 6 (26%) have remained stable. Two patients with recurrent prostatic carcinoma have also shown PR as well as one patient with squamous cell carcinoma of the head and neck. The patients' characteristics, doses of thymosin, and duration of response are shown in Table VI. Of interest is that two of these patients developed brain metastases while on thymosin with no progression of their systemic disease. They continue on maintenance thymosin and still show no progression with stabilization of their brain metastases after irradiation.

4. DISCUSSION

The preliminary results of our phase I trial indicate that thymosin fraction 5 can correct secondary immunodeficiency *in vitro* and produce clinical responses in selected patients. These responses did not correlate with immunological data so we

TABLE III. PHA Stimulation—Thymosin Fraction 5

	Dose					
	60–90 mg/m^2		120–150 mg/m^2		180–210 mg/m^2	
	Pre	Post	Pre	Post	Pre	Post
Mean ± S.D. (cpm)	16,000 ±13,248	19,304 ±15,043	23,335 ±23,472	23,472 ±20,247	13,627 ±15,500	18,297 ±11,917
			NS			

TABLE IV. Mixed Lymphocyte Culture without Thymosin Fraction 5

	Dose					
	60–90 mg/m²		120–150 mg/m²		180–210 mg/m²	
	Pre	Post	Pre	Post	Pre	Post
Mean ± S.D. (cpm)	8646 ±7268	9271 ±5447	8551 ±7273	11,571 ±6631	9104 ±6431	13,148 ±11,785
				NS		

TABLE V. Mixed Lymphocyte Culture with Thymosin Fraction 5

	Dose					
	60–90 mg/m²		120–150 mg/m²		180–210 mg/m²	
	Pre	Post	Pre	Post	Pre	Post
Mean ± S.D. (cpm)	16,733 ±10,872	22,044 ±19,640	20,826 ±16,376	22,027 ±12,949	16,108 ±12,809	21,283 ±13,614
				NS		

TABLE VI. Thymosin Fraction 5, Phase I Patients, June 1983

Patient	Dose (mg/m²)	Diagnosis	Time (months)	Response
50/F	60	Renal	11	Partial
64/M	150	Renal	17+	Partial
67/M	150	Renal	17+	Partial
47/M	90	Renal	23	Stable
70/F	150	Renal	17+	Stable
63/F	120	Renal	11	Stable
46/F	180	Renal	15+[a]	Stable
47/M	180	Renal	13	Stable
46/M	150	Renal	29	Stable
62/M	60	Prostate	12+	Partial
64/M	150	Prostate	19+	Partial
49/F	210	Tongue	13+[a]	Partial
6/M	60	Pancreas	13+	Stable

[a] Subsequent brain metastases.

may be observing an indirect effect on the pituitary–hypothalamic axis in an endocrine-sensitive tumor from a thymic hormone. As essentially no toxicity has been produced to date and a subpopulation of patients has been identified who clinically respond, we have now initiated a phase II trial to look at the response of patients with metastatic renal cell carcinoma and prostatic carcinoma. If we corroborate our initial phase I trial with careful, systematic clinical observations, we will be able to pursue a randomized trial with renal carcinoma and prostatic carcinoma.

ACKNOWLEDGMENTS. The authors' research reported herein was supported by NCI-CM-07343-22 and performed in part during an American Cancer Society–Eleanor Roosevelt International Cancer Fellowship awarded by the International Union Against Cancer.

REFERENCES

Chretien, P. B., Lipson, S. D., Makuck, R., Kenady, D. E., Cohen, M. H., and Minna, J. D., 1978, Thymosin in cancer patients—In vitro effects: Correlations with clinical response to thymosin immunotherapy, *Cancer Treat. Rep.* **62:**1787.

Wara, D. W., and Ammann, A. J., 1978, Thymosin treatment of children with primary immunodeficiency disease, *Transplant. Proc.* **10:**203–208.

Wara, D. W., Goldstein, A. L., Doyle, N. E., and Ammann, A. J., 1975, Thymosin activity in patients with cellular immunodeficiency, *N. Engl. J. Med.* **292:**70–74.

Wara, W. M., Phillips, T. L., Wara, D. W., Ammann, A. J., and Smith, V., 1975, Immunosuppression following radiation therapy for carcinoma of nasopharynx, *Am. J. Roentgenol.* **123:**482–485.

Wara, W. M., Neely, M. H., Ammann, A. J., and Wara, D. W., 1981, Thymosin adjuvant therapy in advanced head and neck cancer, in: *Adjuvant Therapy of Cancer III* (S. E. Salmon and S. E. Jones, eds.), pp. 169–173, Grune & Stratton, New York.

Clinical Aspects of Thymulin (FTS)

JEAN-FRANÇOIS BACH and MIREILLE DARDENNE

Thymulin, formerly called FTS (facteur thymique sérique), is a peptide initially prepared from porcine serum but also isolated to purity from human serum and calf thymus.

1. UPDATE OF THYMULIN BIOLOGY

Thymulin is a nonapeptide (Glu-Ala-Lys-Ser-Gln-Gly-Gly-Ser-Asn). It binds zinc with a dissociation constant of 10^{-7} M (as evaluated by equilibrium chromatography) (Dardenne *et al.*, Chapter 2, this volume), and the presence of zinc is necessary for its biological activity (Dardenne *et al.*, 1982). More than 40 analogs of thymulin have been synthesized and immunologically evaluated (in bioassays and receptor assays) permitting the localization of the molecule's activity on the seven terminal amino acids (Bach, 1983). Some analogs have been shown to bind to the receptor and to inhibit thymulin activity, behaving like antihormones (Pléau *et al.*, 1979).

Thymulin is exclusively produced by the thymic epithelium. Recent studies using antithymulin monoclonal antibodies have shown that thymulin is only present in thymic epithelial cells (Savino *et al.*, 1982) and that the number of thymulin-containing cells augments after peripheral depletion of the hormone, either by active immunization against the peptide or by injecting an antithymulin monoclonal antibody. Conversely, the increase in thymulin-containing cells normally noted in cultured thymic epithelial cells is slowed down by the addition of synthetic thymulin. These data suggest the existence of a feedback regulation of thymulin secretion, thymulin itself being the inhibitory signal. Thymulin circulates in the blood, partly bound to a 40,000- to 60,000-dalton carrier molecule. It is subject to the action of several inhibitors (Bach, 1983).

JEAN-FRANÇOIS BACH and MIREILLE DARDENNE • INSERM U-25, Hôpital Necker, 75015 Paris, France.

Thymulin apparently acts exclusively on T cells. It binds to T-cell membrane receptors with a high affinity (there are two sites with respective K_d of 10^{-9} and 10^{-7} M), without negative cooperativity (Pléau *et al.*, 1980). Thymulin does not clearly stimulate cAMP synthesis but has been shown to enhance PGE_2 production by human lymphocytes (Gualde *et al.*, 1982). As far as its biological activity is concerned, thymulin has been shown to enhance the function of the various T-cell subsets (Table I). This does not mean that the hormone will be efficient in all systems considered. It will only be active in adequate recipients (with the putative postthymic target cells), when the simultaneous effects of suppressor and helper

TABLE I. Effects of Thymulin on T-Cell Functions

Proliferation	
PHA-induced	Mouse: mice Tx[a] at 3 weeks Rat: ATx rats Human: immunodeficiencies
Autologous mixed lymphocyte reaction (lupus[b])	
Cytotoxicity and delayed hypersensitivity	
Allogenic cytotoxicity	ATx mice
Anti-TNP cytotoxicity	Normal thymocytes[b]
Graft-versus-host reaction	Normal mice
Rejection of MSV-induced sarcoma (low dose)	B mice
Stimulation of delayed-type hypersensitivity	ATx mice
Helper T cells	
Antibody production (SRBC)	Aging mice
Induction of IgA (and IgE) synthesis	Ataxia telangiectasia and variable immunodeficiency
Production of interleukin-2	Normal thymocytes and nude mouse spleen cells
Increase in anti-DNA IgG autoantibodies	Young B/W mice (females)
Suppressor T cells	
Retardation of skin allograft rejection	Normal mice
Depression of antibody production	
SRBC	Normal mice
PVP	NZB mice
DNA	
Depression of T-cell mediated cytotoxicity	
Depression of delayed-type hypersensitivity	
Stimulation of Con A-induced suppression	
Other effects	
Increase of NK cell activity (*in vitro* and *in vivo*) in humans and in mice	
Stimulation of colony-forming units (CFU-S) entry into DNA synthesis in Tx mice after thymus-dependent antigen treatment	
Alteration of the migration capacity of fetal hemopoietic precursor cells[b]	
Increased resistance to *Salmonella typhimurium*	
Enhancement of *in vitro* LPS-induced polyclonal B-cell responses of CBA/N mice	
Increase in "early" null radiation-induced leukemias in Tx AKR mice	
Decrease in TdT expression in the mouse[b]	

[a] Tx = thymectomized.
[b] *In vitro* experiments.

(or effector) cells do not counterbalance each other (Bach, 1983). Interestingly, thymulin shows a preferential action on suppressor T cells at high doses, particularly in normal recipients. At low doses, or in immunodeficient patients, it is more difficult to predict which T-cell subset will be preferentially stimulated.

2. CLINICAL PHARMACOLOGY OF THYMULIN

Thymulin has a short half-life. When injected into mice i.v. or i.p. without coupling to a carrier protein or zinc, it is only detectable in serum for a few minutes. This short half-life may be augmented in several ways. Coupling to zinc is the simplest and most readily accessible method (Bach, 1983). In addition to increasing thymulin half-life, it enhances its activity. In fact, in the absence of zinc the peptide is biologically totally inactive, as may be the case in uncontrolled synthetic preparations. There is usually some zinc in synthetic thymulin (due to the metal's presence in synthesis reagents) but optimal zinc coupling needs calibrated addition of the metal (at best after previous chelation of the peptide, in zinc/peptide ratio of 1). Another method consists of preincubating the peptide with serum proteins (which contain carrier proteins). Lastly, one may use long-lived analogs, such as those obtained by substituting homoarginine to lysine (third residue). The two former techniques have been tested in man successfully. Interestingly, thymulin penetrates the central nervous system as is demonstrated by the appearance of rosette-dosable activity in the cerebrospinal fluid of multiple sclerosis patients injected s.c. with the hormone.

Thymulin circulates in the blood partly bound to carrier protein(s) (Dardenne and Bach, 1977). This carrier protein has been identified in mouse serum as a molecule with molecular weight close to 50,000, which is similar to that of prealbumin, a molecule known to be a carrier of several substances and shown to be active in the rosette assay used for thymulin characterization (Burton *et al.*, 1978). Thre is no direct proof, however, that prealbumin is indeed the thymulin carrier molecule.

Thymulin *is not toxic* in short- or long-term treatments even when used at high doses (up to 1 mg/kg) in mice, which are doses much higher than those utilized in man (between 2 and 15 μg/kg). Thymulin is immunogenic when coupled to a carrier protein (e.g., bovine serum albumin) but has not been shown to induce antibody production when used alone chronically in man. No sign of anaphylaxis has ever been noted in short- or long-term treatment.

Thymulin interaction with other drugs has not been studied thoroughly. Note, however, that indomethacin inhibits thymulin effects in the rosette assay (Bach, 1974), a finding in keeping with the demonstration of induction of prostaglandin synthesis by thymulin in human lymphocytes (Gualde *et al.*, 1982). It is not known whether this interaction is clinically relevant [e.g., rheumatoid arthritis (RA) patients, receiving nonsteroidal anti-inflammatory agents]. Preliminary data (to be discussed further) indicate that it might not be the case.

The selection of thymulin dosage is a matter of great difficulty. One may be guided by the follow-up of serum levels (as assessed by the rosette assay) or by the animal data. The latter indicate that in immunodeficient animals, doses of 0.1 ng/mouse to 10 ng/g, i.e., 1–10 μg/kg, can restore deficient cell-mediated immunity or stimulate helper T-cell function (Bach, 1983). In normal mice, there is little effect on helper or effector cells whatsoever. Conversely, one may stimulate suppressor T cells at high doses (1–10 μg/mouse, i.e., 10 mg/kg). The differential effect of variable doses of thymulin has also been observed on the rejection of MSV-induced sarcomas: low doses (0.1 μg/mouse) stimulating sarcoma rejection, higher doses (1–10 μg/kg) inhibiting (our unpublished results). Based on these considerations, we have selected two dose levels: 1–5 μg/kg for stimulation of helper or effector T cells (e.g., in immunodeficiency or in viral infections) and 15–20 μg/ml for stimulation of suppressor T cells (e.g., in RA). This approach has been justified retrospectively by the good effect of thymulin (at the low-dose schedule) in immunodeficient children and the much better effect obtained at the high- than at the low-dose protocol in RA.

3. POTENTIAL CLINICAL INDICATION OF THYMULIN

The selection of clinical indications of thymic hormones and more precisely of thymulin may be based on three sets of arguments (see Table II):

1. Determination of serum thymic hormone level (if one aims at a restoration of physiological levels).
2. Existence of T-cell anomalies, whether or not they are due to the thymus, or more precisely, thymic hormone deficiency (one may then undertake a physiological substitutive treatment or use pharmacological doses, as have been shown to work in animals with normal production of the hormone).
3. Results of therapeutic experiments in animal models of immunological diseases such as murine lupus and experimental allergic encephalomyelitis.

The evaluation of *circulating thymulin levels* has provided interesting results (reviewed by Dardenne, 1983). Thymulin serum level is depressed in a number of pathological conditions which can be, for most of them, classified in immunodeficiency syndromes or autoimmune diseases. Among the former, thymulin serum level is low in Di George syndrome, most cases of ataxia telangiectasia, common variable immunodeficiency with T-cell defect, zinc deficiency, Down's and Cockayne's syndromes, and some cases of severe combined immunodeficiency. It is also depressed (or nil) in aging animals and humans (after 30 years of age). As far as autoimmune diseases are concerned, thymulin serum level declines prematurely with age in systemic lupus (mice and humans) (Bach *et al.*, 1973, 1978) and autoimmune diabetes (Dardenne *et al.*, 1983). It has been verified in NZB and MRL/1 mice that this decline is indeed due to a diminution of thymulin synthesis by the thymic epithelium (as assessed by immunofluorescence on thymic sections using antithymulin monoclonal antibodies) rather than to a peripheral destruction by circulating antithymulin autoantibodies. Interestingly, thymulin levels are not depressed (and even relatively increased) in RA and myasthenia gravis (at least in

TABLE II. Potential Clinical Indications of Thymulin

Disease	Low thymulin level	T-cell anomaly (potential benefit of T-cell restoration)	Potential benefit of T-cell pharmacological stimulation	Experimental data in animal models
Immunodeficiency (ID)				
Di George syndrome	+	+	−	+
Common variable ID with T-cell defect	+	+	−	−
Ataxia telangiectasia	+	+	−	−
Aging	+	+	−	−
Drug- or radiation-induced immunosuppression (e.g., bone marrow graft recipient)	−	+	+	−
Autoimmunity				
SLE	+	+	+	+
Rheumatoid arthritis	−	+	+	−
Multiple sclerosis	−	+	+	−
Autoimmune diabetes	+ (mice)	+	+	+
Infections				
Lepromatous leprosy	−	+	+	−
Chronic bacterial infections	−	−	+	+
Herpes virus infection	−	+	+	−
AIDS	+	+	+	−
Other diseases				
Atopy	−	±	+	−
Tumors	NT	±	+	+
Organ transplantation	−	−	+	+
Histiocytosis X	−	+	−	−

older patients) but it is not proven in PR cases if the activity dosed by the rosette assay is indeed due to thymulin itself rather than to other factors such as the allogeneic factor also active in the rosette assay (Dardenne, 1983). Note also that thymulin levels are increased in patients with mycosis fungoides.

T-cell anomalies have been described in many diseases which are thus potential candidates to thymic hormone therapy. One may add to these diseases those without obvious primary (or secondary) T-cell defect but in which T-cell pharmacological stimulation could be beneficial. With these two concepts in mind, one may draw a long list of potential indications as shown in Table II. One may thus assume that stimulation of helper or effector T cells could be useful in patients with tumors or viral infections and that of suppressor T cells in patients with organ transplant, autoimmune disease, or severe atopy. One should realize, however, the fragility of the experimental evidence supporting the indication in some of these conditions, and in all cases the risk of unexpected aggravation under thymic hormone therapy, due to inappropriate stimulation of a T-cell subset whose function is detrimental to the course of the disease (as will now be discussed in B/W mice).

Experimental models of immunopathology can be used with benefit to evaluate the potential clinical indications of thymic hormones. NZB and (NZB × NZW)F_1 (B/W) mice have been treated with thymulin in several ways. Treatment of young female NZW mice from an early age prevents the onset of Sjögren's syndrome and reduces the autoimmune hemolytic anemia. Treatment of young female B/W mice also prevents Sjögren's syndrome but has no clear effect on anti-DNA antibody production and the glomerulonephritis (Bach *et al.*, 1980). It may even have an accelerating effect on these manifestations, probably by stimulation of helper T cells (overrunning the effect on deficient suppressor T cells perhaps altered by autoantibodies). Conversely, in aging male (NZB × NZW)F_1 mice, thymulin shows a favorable effect with decrease in anti-DNA antibody production and improvement of glomerulonephritis (Israel-Biet *et al.*, 1983). One should lastly mention here the report by Nagai *et al.* (1982) that thymulin prevents the development of experimental allergic encephalomyelitis in guinea pigs. Animal models also provide useful information in more conventional models of immunodeficiencies (mimicking the human immunodeficiency syndromes), organ transplantation, and atopy (Table II).

4. FIRST CLINICAL TRIALS OF THYMULIN

Thymulin has now been administered to more than 60 patients with immunodeficiency or immunological abnormalities. The most striking results were obtained in primary immunodeficiency syndromes of children (Bordigoni *et al.*, 1982; Griscelli, unpublished data; Faure *et al.*, 1983). In three cases of Di George syndrome, one noted a definite restoration of T-cell number and functions. Three patients with SCID received synthetic thymulin for several weeks at 2–5 μg/kg. In one case, thymulin induced reversible GVH reaction, probably secondary to the activation of the mother's cells persisting in the infant. In the two other cases, where B cells were present in excessive number, thymulin induced a normalization of B-cell level with corresponding increase in T-cell number and functions. IgA production, which was deficient in one of these children, increased. Most striking results have been obtained in ataxia telangiectasia. In four consecutive cases, Bordigoni *et al.* (1982) and Faure *et al.* (1983) showed altogether clinical improvement (cessation of infections), correction of T-cell defects (markers and functions), and unexpectedly rapid and major positive effects on IgA production which appeared in three cases. Interestingly, the effect depended on continuation of thymulin therapy. Thymulin was also used successfully in a case of common variable hypogammaglobulinemia for a 27-month period, with correction of T-cell defects, improvement in IgG and IgA production, and favorable clinical efficacy.

Preliminary clinical trials have also been initiated in RA and herpes virus infections. Treatment of eight chronic RA patients with 1 mg thymulin has been associated with clear improvement of the clinical signs (particularly those directly related to inflammation), but this result needs confirmation in a randomized double-blind study (Amor *et al.*, 1984). Similarly, data to be confirmed suggest that thymulin may very significantly improve the early phases of RA, where clear amelioration of clinical and immunological manifestations was noted (Faure, personal communication). As far as herpes virus infections are concerned, more limited data are available but two cases of disease cure (one case of generalized herpes, and one case of recurrent herpes) are promising, for complete cure of the herpes infection was obtained in a sustained fashion.

In conclusion, although bearing on a limited number of patients, these preliminary clinical trials are most encouraging. The synthetic peptide is available in large amounts and can be used at low doses (at least when stimulation of helper T cells is attempted). It is nontoxic and has been shown to be efficient in several indications. It is too early, however, to foresee the spectrum of its clinical usefulness.

REFERENCES

Amor, B., Dougados, M., Mery, C., De Gery, A., Choay, J., Dardenne, M., and Bach, J.-F., 1984, Thymulin (FTS) in rheumatoid arthritis, *Arthritis Rheum.* **27:**117.

Bach, J.-F., 1983, In: *Clinics in Immunology and Allergy*,Vol. 3, p. 133, Saunders, Philadelphia.

Bach, J.-F., Dardenne, M., and Salomon, J. C., 1973, Studies on thymus products. IV. Absence of serum thymic activity in adult NZB and (NZB × NZW)F_1 mice, *Clin. Exp. Immunol.* **14:**247.

Bach, J.-F., Bach, M. A., Blanot, D., Bricas, E., Bricas, D., Charreire, J., Dardenne, M., Fournier, C., and Pléau, J. M., 1978, Thymic serum factor (FTS), *Bull. Inst. Pasteur (Paris)* **76:**325.

Bach, M. A., 1974, Transient loss of theta antigen after indomethacin treatment, *Ann. Immunol.* **125c:**325.

Bach, M. A., Droz, D., Noël, L. H., Blanchard, D., Dardenne, M., and Pekin, A., 1980, Effect on murine lupus of long term treatment with circulating thymic factor, *Arthritis Rheum.* **23:**1351.

Bordigoni, P., Faure, G., Bene, M. C., Dardenne, M., Bach, J.-F., Duheille, J., and Olive, D., 1982, Improvement of cellular immunity and IgA production in immunodeficient children after treatment with synthetic thymic factor (FTS), *Lancet* **2:**293.

Burton, P., Iden, S., Mitchell, K., and White, A., 1978, Thymic hormone-like restoration by human prealbumin of azathioprine sensitivity of spleen cells from the thymectomized mice, *Proc. Natl. Acad. Sci. USA* **75:**823.

Dardenne, M., 1983, Evaluation of thymic hormone serum levels in health and disease, in: *Clinics in Immunology and Allergy*, Vol. 3, p. 157, Saunders, Philadelphia.

Dardenne, M., and Bach, J.-F., 1977, Demonstration and characterization of a serum factor produced by activated T cells, *Immunology* **33:**643.

Dardenne, M., Pléau, J. M., and Bach, J.-F., 1980, Evidence for the presence of a carrier of a serum thymic factor (FTS), *Eur. J. Immunol.* **10:**83.

Dardenne, M., Pléau, J. M., Nabarra, B., Lefrancier, P., Derrien, M., Choay, J., and Bach, J.-F., 1982, Contribution of zinc and other metals to the biological activity of the serum thymic factor (FTS), *Proc. Natl. Acad. Sci. USA* **79:**5370.

Dardenne, M., Savino, W., Gastinel, L. N., Nabarra, B., and Bach, J.-F., 1983, Thymic dysfunction in the mutant diabetic (*db/db*) mouse, *J. Immunol.* **130:**1195.

Faure, G., Bordigoni, P., Bene, M. C., Olive, D., and Duheille, J., 1983, Thymic factor therapy of primary T lymphocyte immunodeficiencies, in: *Serono Symposium*, Academic Press, New York.

Gualde, N., Rigaud, M., and Bach, J.-F., 1982, Stimulation of prostaglandin synthesis by the serum factor (FTS), *Cell. Immunol.* **70:**362.

Israel-Biet, N., Noël, L. H., Bach, M. A., Dardenne, M., and Bach, J.-F., 1983, Marked reduction of DNA antibody production and glomerulopathy in (FTS-Zn) and cyclosporin A-treated (NZB × NZW)F_1 mice, *Clin. Exp. Immunol.* **54:**359–365.

Nagai, Y., Osanai, T., and Sakakibara, K., 1982, Intensive suppression of experimental allergic encephalomyelitis (EAE) by serum thymic factor and therapeutic implication, *Jpn. J. Exp. Med.* **52:**213.

Pléau, J. M., Dardenne, M., Blanot, D., Bricas, E., and Bach, J.-F., 1979, Antagonistic analogue of serum thymic factor (FTS) interacting with the FTS cellular receptor, *Immunol. Lett.* **1**:179.

Pléau, J. M., Fuentes, V., Morgat, J. L., and Bach, J.-F., 1980, Specific receptor for the serum thymic factor (FTS) in lymphoblastoid cultured cell line, *Proc. Natl. Acad. Sci. USA* **77**:2861.

Savino, W., Dardenne, M., Papiernik, M., and Bach, J.-F., 1982, Thymic hormone-containing cells: Characterization and localization of serum thymic factor in young mouse thymus studied by monoclonal antibodies, *J. Exp. Med.* **156**:628.

An Evaluation of Two Different Schedules of Synthetic Thymosin α_1 Administration in Patients with Lung Cancer

Preliminary Results

RICHARD S. SCHULOF, MARGARET LLOYD, JOHN COX, SUSAN PALASZYNSKI, JOHN E. McCLURE, GENEVIEVE S. INCEFY, and ALLAN L. GOLDSTEIN

1. INTRODUCTION

Thymosin α_1 (molecular weight 3108) is one of the many active polypeptides isolated from thymosin fraction 5 (TF5) (Low and Goldstein, 1979; Goldstein *et al.*, 1982). α_1 was initially purified from extracts of calf thymus glands. Biologically active α_1 has now been successfully synthesized by classical solution (Birr and Stollenwerk, 1979; Wang *et al.*, 1978), solid-phase (Wang *et al.*, 1980; Folkers, *et al.*, 1980) and recombinant DNA procedures (Wetzel *et al.*, 1980).

In immunodeficient animal models, α_1 has been associated with the generation of functional helper T cells (Goldstein *et al.*, 1981). Synthetic α_1 has been the first thymosin polypeptide to enter phase I clinical trials in patients with advanced cancer (Dillman *et al.*, 1982). In man, α_1 has proved to be nontoxic when administered parenterally at doses up to 9.6 mg/m^2 (Dillman *et al.*, 1982). Preliminary immu-

RICHARD S. SCHULOF • Departments of Medicine and Biochemistry, The George Washington University School of Medicine and Health Sciences, Washington, D.C. 20037. MARGARET LLOYD, JOHN COX, and SUSAN PALASZYNSKI • Department of Medicine, The George Washington University School of Medicine and Health Sciences, Washington, D. C. 20037. JOHN E. McCLURE and ALLAN J. GOLDSTEIN • Department of Biochemistry, The George Washington University School of Medicine and Health Sciences, Washington, D.C. 20037. GENEVIEVE S. INCEFY • The Memorial Sloan-Kettering Cancer Center, New York, New York 10021.

nological screening of patients with advanced cancer treated with single i.m. injections of α_1 suggested that the maximal immunomodulatory dose of α_1 is 1.2 mg/m^2 (Dillman *et al.*, 1983).

Up until the present time, little attention has focused on defining an optimal schedule for administering α_1 or other partially purified or purified thymic factors. Previous randomized phase II studies of TF5 in cancer patients have utilized either a twice a week (BIW) (Cohen *et al.*, 1979) or a loading dose schedule of administration (i.e., daily × 14 days followed by twice weekly maintenance) (Wara *et al.*, 1981). However, in these prior studies, the schedules were empirically chosen.

In the present communication, we report the preliminary results of a randomized double-blind phase II trial of synthetic α_1 in patients with locally advanced non-small-cell lung cancer. Our study was designed specifically to contrast the pharmacokinetics and immunorestorative effects of α_1 administered by either a BIW or a loading dose schedule.

2. MATERIALS AND METHODS

2.1. Thymosin α_1

Synthetic α_1 was kindly provided by Hoffmann–La Roche Inc., Nutley, New Jersey. It was provided in powder form in 2-mg vials and stored at 4°C until use. Just prior to use, α_1 was reconstituted using 1.4% bicarbonate diluent.

2.2. Patients

All patients entered onto the trial had locally advanced, unresectable, but not distantly metastatic, non-small-cell lung cancer. At our institution, such patients receive primary radiation therapy (RT) as conventional treatment. The post-RT patients provided a uniformly immunosuppressed group of subjects with which to evaluate the immunorestorative properties of α_1 (Schulof *et al.*, 1982).

To be eligible for study, the following criteria had to be satisfied: (1) a diagnosis of non-small-cell lung cancer (large cell, adenocarcinoma, or squamous cell), (2) negative metastatic disease work-up, (3) age < 73 years, (4) no active chronic diseases, (5) normal hepatic and renal function, (6) no prior or concurrent chemotherapy, (7) off immunosuppressive drugs (e.g., steroids), and (8) the tumor had to be stabilized or show evidence of regression during RT (i.e., patients whose tumors progressed during RT were excluded).

2.3. Radiation Therapy

Megavoltage RT was administered with a linear accelerator over 6–8 weeks using a split course technique and 200-rad fractions, 5 days a week. Patients with prior surgical reductions received RT only to the mediastinum whereas all others received RT to both the primary lesion as well as the mediastinum.

2.4. Randomization

The trial was designed to compare BIW and loading dose schedules of subcutaneous (SQ) α_1 administration using the same dose (900 $\mu g/m^2$). As we could not predict the normal recovery pattern, if any, of thymus-dependent immunity after RT, we also included a group of patients treated with placebo to provide baseline serial immune data for comparison to the α_1-treatment groups.

Thus, in our study, patients were randomized to one of three groups: (1) α_1 loading dose, daily $\times$ 14 followed by twice weekly maintenace, (2) α_1 twice weekly (BIW), and (3) placebo twice weekly. α_1/placebo administration began within a week after completion of RT and continued for up to 1 year or until relapse. All patients began with 14 daily injections of α_1 or placebo, with placebo substituted when appropriate. α_1 administration is not associated with any significant side effects (Dillman *et al.*, 1982), thus enabling the study to be performed with a double-blind design.

2.5. Immune Profiles

Detailed analyses of peripheral blood T-cell numbers and function were preformed prior to RT (if possible), following RT (prior to thymosin/placebo), at 1 and 3 weeks after initiation of thymosin/placebo, then monthly $\times$ 5 and bimonthly $\times$ 3 until relapse. For comparative purposes, 97 sex- and age-matched normal donors (mean age 52 years, range 41–68) have been studied concurrently with patients.

2.6. Isolation of Lymphoid Cells/Immune Parameters

Monocyte-depleted peripheral blood lymphocytes (PBL) and purified T cells were isolated from fresh heparinized blood as previously described (Schulof *et al.*, 1981). PBL were utilized for determining the percentage of E-rosette-forming cells ($E_{4°}$-RFC) (Bentwich *et al.*, 1973) and high-affinity E-RFC ($E_{29°}$-RFC) (West *et al.*, 1976). Cryopreserved PBL were utilized to determine the percentage of cells expressing the OKT3 (pan T), OKT4 (helper/inducer), and OKT8 (cytotoxic/suppressor) antigens (Reinherz and Schlossman, 1980). OKT antigen-positive cells (OKT+) were assessed with appropriate monoclonal antibodies (Ortho Labs, Raritan, N.J.) using indirect immunofluorescence and flow cytometry (FACS IV, Becton–Dickinson, Sunnyvale, Calif.). Absolute T-cell numbers were determined by multiplying % E-RFC or % OKT+ $\times$ absolute lymphocyte count. Purified T cells were employed in functional assays which included: (1) phytohemagglutinin (PHA)-induced proliferative responses (Schulof *et al.*, 1981) and (2) proliferative responses to pooled allogeneic mononuclear (mitomycin C-treated) stimulator cells in 6-day mixed lymphocyte cultures (MLR). Preliminary immune results of the pre-and post-RT patient populations indicated that the post-RT patients exhibited marked depressions in total T-cell numbers as well as in T-cell function (Schulof *et al.*, 1982).

2.7. Pharmacokinetic Studies

Detailed pharmacokinetic studies were performed using a specific radioimmunoassay (RIA) for α_1 as previously reported (McClure *et al.*, 1982). In five patients who had received α_1 and one who had received placebo, plasma specimens were also assayed for FTS-like thymic hormone bioactivity using the murine azathioprine-rosette bioassay of Dardenne and Bach (1973). The methods for preparation of plasma specimens and for performance of the bioassay have previously been described (Iwata *et al.*, 1981). All specimens were drawn at the same time of day (8 a.m.) in order to control for the diurnal variation in circulating α_1 levels (McGillis *et al.*, 1983).

2.8. Statistical Methods

Immune comparisons were performed utilizing least-squares analysis of variance and pairwise *t* tests. All *p* values refer to two-sided tests (Snedecor and Cochran, 1980).

3. RESULTS

The results presented in this communication are preliminary and represent an interim analysis of the trial which is scheduled for completion and final analysis in 2–3 months. Patient accrual for the trial continued from November 1980 through January 1983 and the study will be completed by July 1983. In order to present our preliminary results, the codes for the three treatment groups have been broken. However, the codes for individual patients still on study have not been revealed.

3.1. Patient Groups

Forty-two patients were entered onto study and 41 are evaluable for follow-up study. One patient has been excluded from analysis because at the time of entry he had complained of abdominal pain. Subsequent work-up revealed the presence of metastatic disease and he was dropped from the study after having only received several injections. Currently, all patients have had a minimal follow-up period of at least 13 weeks (mean 38.5 weeks). The clinical characteristics of the three patient groups, I (placebo), II (α_1 BIW), and III (α_1 loading), are shown in Table I. As can be seen, the three groups are well matched with regard to most prognostic determinants including mean age, performance status, stage, and RT administered. However, several imbalances of prognostic factors are apparent which include a greater proportion of surgical reductions in the thymosin treatment groups (groups II and III) and a greater percentage of patients with pretreatment weight loss in the placebo group (group I).

TABLE I. Patient Group Characteristics (All Evaluable Patients)

	Patients entered	Male/female	Mean age (years)	Mean entry performance status (Karnofsky)	Histology			Stage (TNM)			History of weight loss > 10% in 6 months
					LC	AD	SQ	I	II	III	
Group I (placebo)	13	10/3	55.6	90	3	6	4	0	1	12	7
Group II (α_1 BIW)	15	8/7	57.3	91	1	6	8	2	1	12	4
Group III (α_1 loading)	13	8/5	52.8	89	2	9	2	0	2	11	4
Totals	41	26/15	55.4	90	6	21	14	2	4	35	15

	Tumor bulk[a] (pre-RT)			Number of surgical reductions	Mean RT dose (rads)		Response to RT			
	B	Non-B	CA[b]		Prim	Med	MR	PR	CR	Stable
Group I (placebo)	4	7	2	1	5709 (12)[c]	4418 (13)	5	3	0	5
Group II (α_1 BIW)	7	8	0	5	5728 (12)	4275 (15)	5	3	0	7
Group III (α_1 loading)	7	6	0	6	5825 (8)	4734 (13)	2	3	0	8
Totals	18	21	2	12	5745 (32)	4466 (41)	12	9	0	20

[a] Bulky defined as diameter > 5 cm.
[b] Cannot assess diameter.
[c] Number of patients indicated in parentheses.

3.2. Pharmacokinetic Studies

There were no significant differences between baseline plasma α_1 levels (in pg/ml) of the pre-RT (1019 ± 129 S.E., n = 14) and post-RT (846 ± 75, n = 26) patients compared to the healthy age- and sex-matched normal subjects (1017 ± 86, n = 47). Although there was a drop in baseline α_1 levels following RT, this was not statistically significant. By 1 hr following the SQ injection, plasma α_1 levels had increased 10-fold (Fig. 1). Peak levels achieved (25–30 ng/ml) were pharmacological and were 10–50 times greater than the highest physiological levels ever achieved either in infants or in patients with AIDS (Naylor *et al.*, Chapter 6, this volume). Peak levels persisted for approximately 6 hr and then returned to near baseline over the next 18 hr. Figure 2 illustrates baseline α_1 levels of the three treatment groups over 15 weeks of serial monitoring. Specimens for these analyses were drawn just prior to the injections of thymosin/placebo and thus represent the lowest circulating levels detected during prolonged administration of α_1. As seen in Fig. 2, both thymosin treatment groups exhibited a gradual increase in baseline

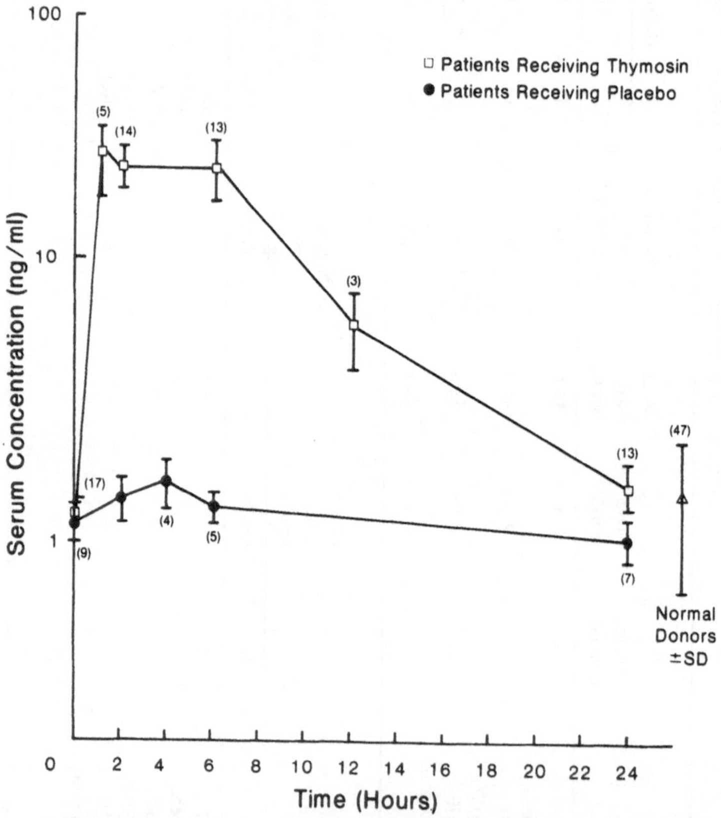

FIGURE 1. Plasma thymosin α_1 levels measured by radioimmunoassay following the first injection. Data from both the twice a week group and loading dose group of thymosin-treated patients have been pooled.

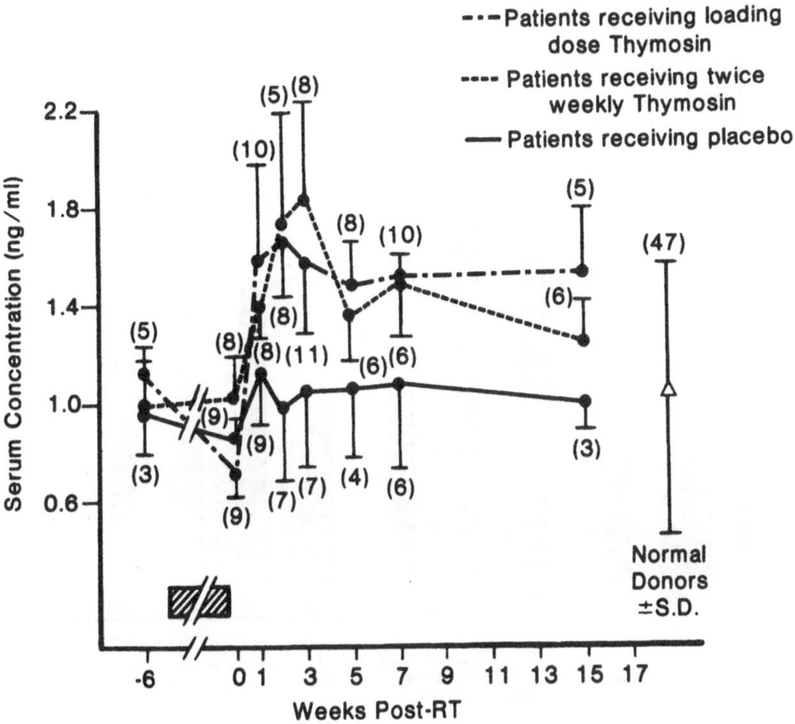

FIGURE 2. Baseline plasma thymosin α_1 levels over 11 weeks for the three treatment groups.

plasma α_1 levels over the initial 2 weeks to levels that were close to 2 S.D. above those of the normal donors and patients receiving placebo. Levels were then maintained at approximately 1 S.D. above the normal mean throughout 11 weeks of serial monitoring although the loading dose group exhibited a higher level than the BIW group at week 11.

Figure 3 illustrates a typical result of our pharmacokinetic study using the Dardenne–Bach bioassay. Five of the six patients studied exhibited pretreatment levels of $\leq 1/4$ (reciprocal of greatest dilution of plasma exhibiting activity) which is considered normal for their age. We did not observe any significant change in plasma FTS-like bioactivity following the administration of α_1 in five separate patients monitored over 11 weeks. There was no demonstrable plasma bioactivity even when levels of α_1, detected by RIA, were extremely high at 2 hr following administration.

3.3. Serial Immune Monitoring

Figures 4, 5, and 6 illustrate typical results of the serial immune follow-up studies for the three different patient groups. As seen in Fig. 4, the post-RT patients exhibited a significant depression in MLR compared to either the normal donors or the pre-RT patients ($p = 0.02$). There was no significant change in MLR for patients

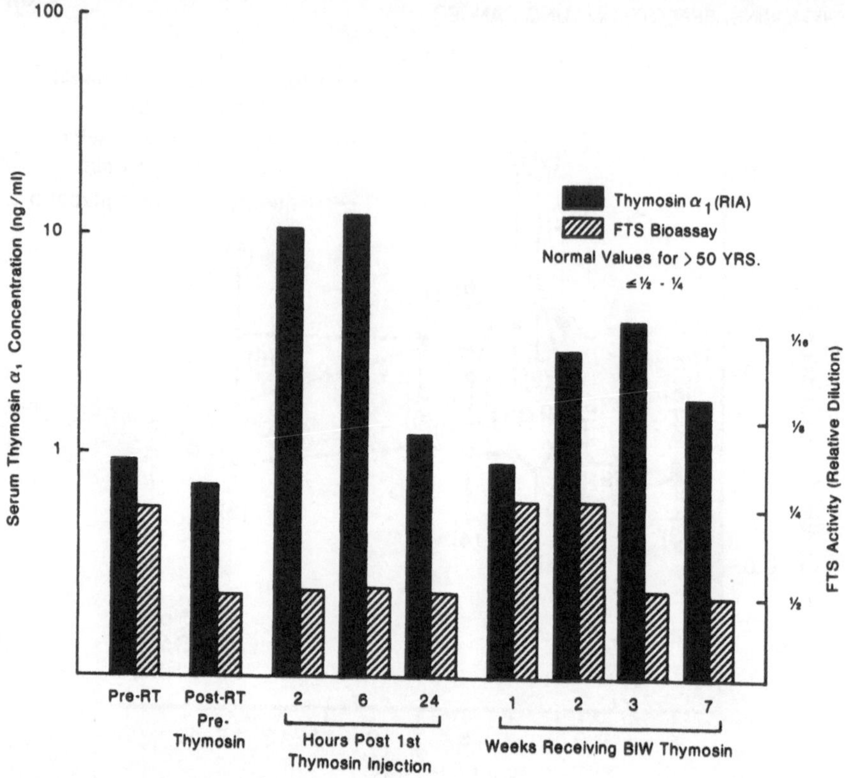

FIGURE 3. A comparison of plasma thymosin α_1 levels measured by radioimmunoassay and bioactivity assessed with the Dardenne–Bach bioassay in a patient treated with the twice a week regimen.

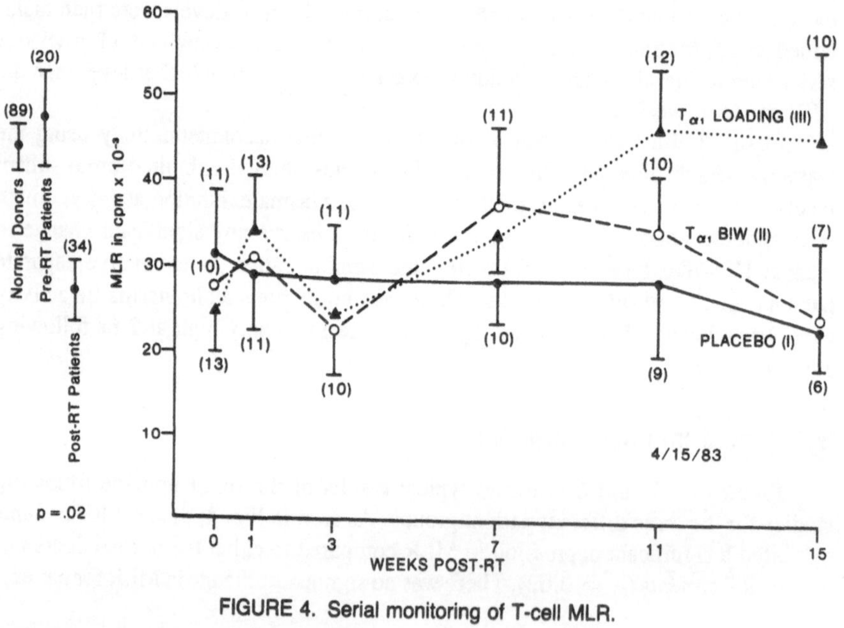

FIGURE 4. Serial monitoring of T-cell MLR.

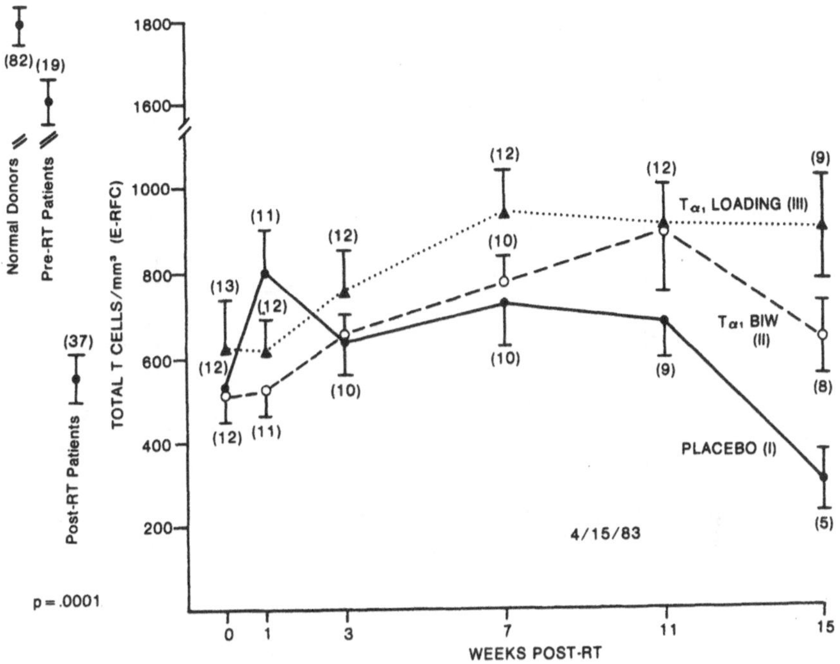

FIGURE 5. Serial monitoring of total T-cell (E-RFC) numbers.

receiving placebo (group I) over a 15-week follow-up period. In contrast, patients in the loading dose α_1 group (III) had complete restoration to normal of MLR which only became apparent beginning at 7 weeks after initiation of injections. Group II (α_1 BIW) exhibited a mild improvement in MLR at weeks 7–11 but this was only transient. Similarly, the post-RT patients exhibited a significant depression ($p = 0.0001$) in total T-cell numbers (absolute $E_{4°}$-RFC/mm^3) and the α_1 loading dose group was associated with the greatest restoration of total T cells (Fig. 5) although this only approached two-thirds of the mean for the pre-RT patients. Group II demonstrated the best maintenance of the T4/T8 ratio at week 11 compared to group I or group III (Fig. 6).

3.4. Patient Relapses

Although our study was not designed as an efficacy trial, its experimental design allowed us to assess the influence of α_1 administration on relapse rates and overall patient survival. The patient population under study has an extremely poor prognosis and it is expected that if no further treatment were administered following RT, less than 5% would survive for more than 2 years (Roswit *et al.*, 1968).

We have performed preliminary relapse-free survival (RFS) comparisons for the three treatment groups using the actuarial method of Kaplan and Meier (1958). These results must be interpreted cautiously as several prognostic imbalances (i.e.,

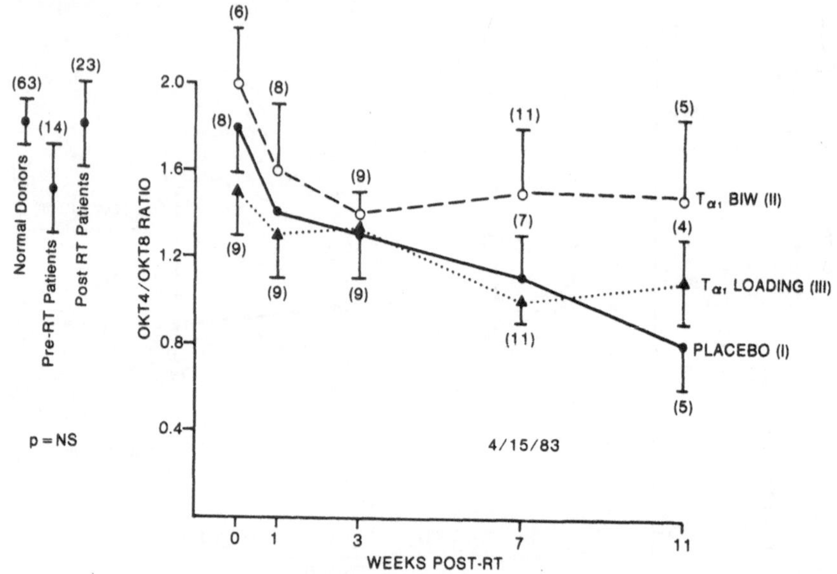

FIGURE 6. Serial monitoring of the T4/T8 ratio.

surgical reductions and pretreatment weight loss) favor the thymosin treatment groups, which, in a study with small patient numbers, could totally account for significant survival differences. Our preliminary analysis indicates that both thymosin treatment groups ($p = 0.03$) are showing a significant improvement in RFS compared to the placebo group. This same finding holds if the analysis is restricted to patients who did not have surgical reductions. Four of the five patients in remission for more than a year received α_1 by the loading dose schedule.

We have also performed a preliminary analysis to identify the patients most likely to benefit from parenteral α_1 administration. We could not identify any difference in relapse rates among the three groups according to tumor histology (i.e., squamous vs. nonsquamous) or location of relapse (i.e., local vs. distant). However, the improved RFS of groups II and III was limited to patients with nonbulky (pre-RT diameter < 5 cm) tumors. The relapse rates for patients with bulky tumors were approximately the same for patients in groups I, II, and III (71–86%). For patients with nonbulky tumors, the relapse rates for groups I, II, and II were 86, 50, and 33%, respectively.

4. DISCUSSION

In this communication we have presented preliminary results of a randomized double-blind phase II trial of synthetic α_1 in which we have contrasted two different schedules of administration. The trial was performed in patients with locally advanced non-small-cell lung cancer who had just completed primary radiation therapy and thus composed a uniformly immunosuppressed group of subjects with which to assess the immunorestorative effects of α_1.

Our preliminary serial immune monitoring results suggest that the loading dose α_1 regimen was superior to the twice a week schedule with regard to immune reconstitution. Only the loading dose schedule was capable of fully restoring T-cell function in MLR to normal. Although neither schedule of α_1 administration could restore total T-cell numbers back to pretreatment values, the loading dose schedule restored levels to approximately 1000/mm^3 and levels were sustained whereas this was not observed with the twice a week schedule. At present, the discrepancy between restoration of T-cell function and numbers is unexplained. It is possible that α_1 may act selectively on restoring T-cell function whereas the administration of other thymic peptides may be required in order to fully restore T-cell levels back to normal. Alternatively, it is possible that more intensive administration of α_1 (e.g., daily) may be required in order to fully restore T-cell levels to normal. However, the loading dose α_1 schedule was associated with a relative increase in recovery of OKT8$^+$ cells compared to OKT4$^+$ cells reflected in a drop in the OKT4/OKT8 ratio. The significance of this finding is currently unknown but ideally, thymic factors should be able to regenerate T-cell numbers while maintaining normal OKT4/OKT8 ratios. Thus, it remains for future trials in post-RT patients to determine whether other thymic factors are capable of completely restoring T-cell numbers while at the same time maintaining normal T-cell subset percentages.

Our preliminary results suggest that the administration of α_1 improves RFS and thus has potential as a therapeutic adjunct for lung cancer patients treated with radiotherapy. However, because our study is based on small patient numbers, and because prognostic imbalances favor the thymosin treatment groups, these results must be interpreted cautiously and require confirmation in a larger multicenter trial. In the final analysis of our trial, we plan to perform logistic regression analyses to balance prognostic factors so that we can definitively establish the impact on α_1 therapy on relapse-free and overall survival. Nevertheless, based on our initial findings, we have recommended that large-scale confirmatory trials be designed and two such studies will be initiated shortly by the Radiation Therapy Oncology Group (RTOG) and the Mid-Atlantic Oncology Program (MAOP). The latter study will be restricted to patients with nonbulky tumors.

ACKNOWLEDGMENTS. The authors' research reported herein was supported by NCI Contract NOI-CM-07446 and grants from the American Cancer Society (JFCF 630B) and Hoffmann–La Roche, Inc. We gratefully acknowledge the excellent technical assistance of Dao Mai, Karen Anderson, and Janelle Hatcher. We thank Sharon La Valle for performing our data computerization and Paddy Cleary for performing the statistical analyses. We thank Dr. Oliver Alabaster for performing monoclonal antibody analysis on a FACS IV and Connie Spooner, Cathy Carr, and Leony Leondaridis for technical assistance.

REFERENCES

Bentwich, Z., Douglas, S. D., Siegal, F. P., and Kunkel, H. G., 1973, Human lymphocyte–sheep erythrocyte rosette formation: some characteristics of the interaction, *Clin. Immunol. Immunopathol.* **1**:511–522.

Birr, C., and Stollenwerk, U., 1979, Synthesis of thymosin α_1, a polypeptide of the thymus, *Angew. Chem.* **91**:422.

Cohen, M. H., Chretien, P. A., Ihle, D. C., Fossieck, B. E., Makuch, R., Bunn, P. A., Johnston, A. V., Shackney, S. E., Mathews, M. J., Lipson, S. D., Kenady, D. E., and Minna, J. D., 1979, Thymosin fraction V and intensive combination chemotherapy prolonging the survival of patients with small-cell lung cancer, *J. Am. Med. Assoc.* **241**:1813–1815.

Dardenne, M., and Bach, J.-F., 1973, Studies on thymus products: Mofidifcation of rosette-forming cells by thymic extracts. I. Determination of the target RFC subpopulation, *Immunology* **25**:343–352.

Dillman, R. O., Beauregard, J. C., Medelsohn, J., Green, M. R., Howell, S. B., and Royston, I., 1982, Phase I trial of thymosin fraction 5 and thymosin α_1, *J. Biol. Resp. Mod.* **1**:35–42.

Dillman, R. O., Beauregard, J. C., Zavanelli, M. I., Halliburton, B. L., Wormsley, S., and Royston, I., 1983, *In vivo* immune restoration in advanced cancer patients after administration of thymosin fraction 5 or thymosin α_1, *J. Biol. Resp. Mod.* **2**:139–149.

Folkers, K., Lehan, J., Sakura, N., Rampold, G., Lundanes, E., Dahmen, J., Lebek, M., Ohta, J., and Bowers, C. Y., 1980, Current advances on biologically active synthetic peptides, in: *Polypeptide Hormones* (R. F. Beers and E. G. Bassett, eds.), p. 149, Raven Press, New York.

Goldstein, A. L., Low, T. L. K., Thurman, G. B., Zatz, M. M., Hall, N., Chen, J., Hu, S. K., Naylor, P. B., and McClure, J. E., 1980, Current status of thymosin and other hormones of the thymus gland, *Recent Prog. Horm. Res.* **37**:369–416.

Goldstein, A. L., Low. T. L. K., Thurman, G. B., Zatz, M. M., Hall, N. R., McClure, J. E., Hu, S. K., and Schulof, R. S., 1982, Thymosins and other hormone-like factors of the thymus gland, in: *Immunological Approaches to Cancer Therapeutics* (E. Mihich, ed.), pp. 137–190, John Wiley and Sons, New York.

Iwata, T., Incefy, G. S., Cunningham-Rundles, G., Smithwick, E., Geller, N., O'Reilly, R., and Good, R., 1981, Circulating thymic hormone activity in patients with primary and secondary immunodeficiency diseases, *Am. J. Med.* **71**:385–394.

Kaplan, E. L., and Meier, P., 1958, Non-parametric estimation from incomplete observations, *J. Am. Stat. Assoc.* **53**:457–481.

Low, T. L. K., and Goldstein, A. L., 1979, The chemistry and biology of thymosin. II. Amino acid sequence analysis of thymosin α_1 and polypeptide β_1, *J. Biol. Chem.* **254**:987–995.

McClure, J. E., Lameris, N., Wara, D. W., and Goldstein, A. L., 1982, Immunochemical studies on thymosin: Radioimmunoassay of thymosin α_1, *J. Immunol.* **128**:368–375.

McGillis, J., Hall, N., and Goldstein, A. L., 1983, Circadian rhythm of thymosin α_1 in normal and thymectomized mice, *J. Immunol.* **131**:148–151.

Reinherz, E. L., and Schlossman, S. F., 1980, The differentiation and function of human T lymphocytes, *Cell* **19**:821–827.

Roswit, B., Patno, M. E., Rapp, R., Veinbergs, A., Feder, B., Stuhlbarg, J., and Reid, C., 1968, The survival of patients with inoperable lung cancer: A large-scale randomized study of radiation therapy versus placebo, *Radiology* **90**:688.

Schulof, R. S., Lacher, M. J., and Gupta, S., 1981, Abnormal phytohemagglutinin-induced T-cell proliferative responses in Hodgkin's disease, *Blood* **57**:607–613.

Schulof, R. S., Lloyd, M., Cox, J., Palaszynski, S., Mai, D., McClure, J., and Goldstein, A. L., 1982, The immunopharmacology and pharmacokinetics of thymosin α_1 administration in man: A prototypic thymic hormone efficacy trial in patients with lung cancer, in: *Current Concepts in Human Immunology and Cancer Immunomodulation* (Serrou, C. Rosenfeld, J. C. Daniels, and J. P. Saunders, eds.), p. 545, Elsevier, Amsterdam.

Snedecor, G. W., and Cochran, W. G., 1980, *Statistical Methods*, 7th ed., pp. 83–102, 175–191, Iowa State University, Ames.

Wang, S. S., Kulesha, I. D., Winter, D. P., 1978, Synthesis of thymosin α_1, *Journal Amer. Chem. Soc.* **101**:253–250.

Wang, S. S., Mokofska, R., Bach, A. E., and Merrifield, R. B., 1980 Solid phase synthesis of thymosin α_1, *Int. J. Pept. Protein Res.* **15**:1–5.

Wara, W. M., Neely, M. H., Ammann, A. J., and Wara, D. W., 1981, Biologic modification of immunologic parameters in head and neck cancer patients treated with thymosin fraction V, in: *Lymphokines and Thymic Hormones: Their Potential Utilization in Cancer Therapeutics* (A. L. Goldstein and M. A. Chirigos, eds.), pp. 257–262, Raven Press, New York.

West, W. W., Sienknecht, C. W., Townes, A. S., and Herberman, R. B., 1976, Performance of a rosette assay between lymphocytes and sheep erythrocytes at elevated temperatures to study patients with cancer and other diseases, *Clin. Immunol. Immunopathol.* **5**:60–66.

Wetzel, R., Heyneker, H. L., Goeddel, D. V., Thurani, P., Shapiro, J., Crea, R., Low, T. L. K., McClure, J. E., and Goldstein, A. L., 1980, Production of biologically active Nα-desacetylthymosin α_1 in *Escherichia coli* through expression of a chemically synthesized gene, *Biochemistry* **19**:6096–6104.

Effect of Thymosin Fraction 5 on Purine Enzymes and Surface Markers in Lymphocytes from Normal Individuals and Homosexuals with Acquired Immune Deficiency Syndrome

J. L. MURRAY, J. M. REUBEN, C. G. MUNN, and E. M. HERSH

1. INTRODUCTION

Thymosin fraction 5 has numerous immunomodulatory effects, including the induction of Ly-123$^+$ phenotype expression from Thy-1$^+$ mouse thymocytes (Ahmed *et al.*, 1978) and augmentation of E-rosette-forming cells in both cancer patients (Schafer *et al.*, 1976) and children with primary immunodeficiencies (Wara *et al.*, 1975). In addition, thymosin has been shown to induce various enzyme changes in lymphocytes, such as the appearance of TdT in mouse bone marrow (Pazmino *et al.*, 1978) and elevations in 5′-nucleotidase (5′-NT) in human thymocytes (Cohen *et al.*, 1981).

Two other enzymes in addition to 5′-NT which play an important role in lymphocyte differentiation are adenosine deaminase (ADA) and purine nucleotide phosphorylase (PNP). In man, both enzymes are elevated in prothymocytes and subsequently fall as cortical thymocytes mature into medullary thymocytes. Mature T cells have very low levels of ADA and high levels of PNP and 5′-NT (Ma *et*

J. L. MURRAY, J. M. REUBEN, C. G. MUNN, and E. M. HERSH • Department of Clinical Immunology and Biological Therapy, The University of Texas System Cancer Center, M. D. Anderson Hospital and Tumor Institute, Houston, Texas 77030.

al., 1982). Several cases of immunodeficiency syndrome in children have been reported in which abnormally low levels of a specific enzyme (i.e., ADA or PNP) were present in circulating red cells (Giblett *et al.*, 1972, 1975).

We wanted to determine whether enzyme abnormalities occurred in lymphocytes from persons with acquired immune deficiency states such as homosexual males with acquired immune deficiency syndrome (AIDS) (Gottlieb *et al.*, 1981). In addition, we examined the effect of thymosin fraction 5 on purine enzyme changes in parallel with changes in surface antigen expression on null- and T-enriched lymphocytes from patients with AIDS compared to a healthy aged-matched control population.

2. MATERIALS AND METHODS

2.1. Patients

Twenty-two homosexual patients were studied. Six were clinically asymptomatic, five had documented Kaposi's sarcoma (KS), and 11 had opportunistic infections and/or had prodromal symptoms consisting of fever, weight loss, and lymphadenopathy. Twenty had reversal of the percentage of helper T cells to suppressor T cells with helper/suppressor ratios less than the normal mean of 1.8. All patients were being followed as outpatients at M. D. Anderson Hospital, and none had received chemotherapy or immunotherapy.

2.2. Lymphocyte Separation Techniques

Peripheral blood mononuclear cells (PBM) were purified using Ficoll–Hypaque gradients. PBM were then depleted of monocytes and B cells and enriched for null and T cells using two comparable techniques. In the first method, nylon wool-purified lymphocytes ($< 2 \pm 1.3\%$ Sig^+; $< 1\%$ $ANAE^+$) were incubated for 10 min with neuraminidase-treated SRBC (Galli and Schlessinger, 1974), pelleted, and placed on ice for 30 min. Rosettes were gently resuspended, layered on Ficoll–Hypaque, and centrifuged for 15 min at 2200 rpm. Interface cells (null enriched) contained $< 30\%$ lymphocytes staining positive by the monoclonal marker OKT3 (Pan T, Ortho Diagnostics, Raritan, N.J.). After lysing SRBC with Tris-NH_4Cl, pelleted cells were $> 90\%$ enriched for OKT3. In method 2, PBM were enriched for T cells by panning on anti-human immunoglobulin petri dishes (Mage *et al.*, 1977). Five micrograms of goat anti-mouse immunoglobulin (Tago, Burlingame, Calif.) was coupled to ox erythrocytes (OE) using chromic chloride. Panned lymphocytes were coated with anti-Leu-1 monoclonal antibody (Becton–Dickinson, Sunnyvale, Calif.) incubated with pretreated OE (0.5% suspension) for 10 min and pelleted on ice for 60 min. Rosettes were separated from nonrosetting cells by Ficoll–Hypaque (Gualde and Goodwin, 1982). Interface cells were $< 12 \pm 4\%$ Leu 1^+; pelleted cells were $> 95 \pm 30\%$ $OKT3^+$. Although purity was better using method 2, enzyme levels were similar in null- and T-enriched cells obtained by either method.

2.3. Lymphocyte Surface Markers

From 5 to 10 × 10^6 nonadherent lymphocytes were incubated overnight at 37°C with or without thymosin fraction 5 (150 μg/ml). Following incubation, cells were washed and 0.5 to 1 × 10^6 cells were aliquoted into glass tubes. Ten microliters (2 μg/ml) of fluorescein-labeled monoclonal antibodies OKT4, OKT8 (Ortho Diagnostics), Leu1 (Becton–Dickinson), and the unfluoresceinated markers OKT10, OKT9, Ia (Ortho), and anti-Tac (antibody reacting with the IL-2 receptor; generously donated by T. A. Waldmann) were added. Cells were incubated at 4°C for 20 min. In specific instances, cells were washed twice and indirect staining was performed by adding 50 μl of a 1:100 dilution of fluoresceinated goat anti-mouse immunoglobulin (Kallestad Laboratories, Dallas, Tex.). After another 20-min incubation at 4°C, cells were washed and the percent cells labeled with each antibody was examined, using an Ortho Spectrum III cytofluorograph.

2.4. Purine Enzyme Analysis

Purine enzymes (ADA, PNP, 5′-NT) were analyzed in T- and null-enriched lymphocytes using a modification of a previously published microtechnique (Van Laarhoven *et al.*, 1980). From 1000 to 3000 lymphocytes in normal saline containing 0.50% BSA were added to individual wells of Terasaki microtiter plates (Falcon Plastics, Oxnard, Calif.), frozen at −30°C, and lyophilized. Five-microliter aliquots of [^{14}C]adenosine (ADA substrate), [^{14}C]inosine (PNP substrate), and [^{14}C]-AMP (5′-NT substrate) in Tris buffer, 0.3% BSA, were added to each cell extract and incubated in high humidity at 37°C for 30 min (ADA) or for 2 hr (PNP, 5′-NT). Following incubation, the plates were placed on ice, and 1-μl aliquots were removed and spotted on glass α-cellulose-coated (Avicel) TLC plates (Analtech Inc., Newark, Del.). Plates were incubated for 3 hr in appropriate solvents, dried, and the products identified using UV fluorescence. The products were scraped off and added to 5-ml plastic vials, along with 3 ml of Aquasol (Packard Instruments, Downers Grove, Ill.). Radioactivity was measured using a beta scintillation counter (Packard) and the percent conversion of substrate to product calculated. Results were expressed in nanomoles per 10^6 cells per hour. In comparing the effects of thymosin on enzyme changes, individual null- and T-enriched fractions were incubated for 18 hr with or without 150 μg/ml thymosin, prior to enzyme analysis. In all individuals, enzyme changes and changes in lymphocyte surface antigens were studied in parallel.

3. RESULTS

Patients had a greater number (4 × 10^6 ± 0.74, mean ± S.E.M.) and percentage (17 ± 3) of null-enriched cells compared to normals (2.6 × 10^6 ± 0.48, 9 ± 1.6; $p < 0.10$ and <0.002, respectively). AIDS patients had a lower absolute mean number of T-enriched cells (7.1 × 10^6 ± 1.2) than normals (12 × 10^6 ± 2.1;

TABLE I. Purine Enzyme Activity in Null- and T-Enriched Lymphocytes from AIDS/KS vs. Normals

Enzyme	ADA[a]	PNP	5′-NT
Normals (16)			
Null	118 ± 8[b]	63 ± 6[c]	9 ± 0.8
T	100 ± 7	60 ± 5	18 ± 1[d]
AIDS/KS (22)			
Null	150 ± 11[b]	93 ± 7[c]	7 ± 0.5
T	115 ± 7	67 ± 5	10 ± 1.2[d]

[a] Enzyme activity expressed as nmoles/10^6 cells/hr.
[b] AIDS/KS null-ADA significantly elevated vs. normals ($p < 0.04$).
[c] AIDS/KS null-PNP vs. normal null-PNP ($p < 0.004$).
[d] AIDS/KS T-5′-NT vs. normal T-5′-NT ($p < 0.0002$).

$p < 0.04$). Mean T lymphocyte percentage was also decreased (AIDS = 22 ± 6, controls = 45 ± 8; $p < 0.006$). These values were calculated as a final absolute count and percentage of nonadherent B lymphocyte and monocyte-depleted cells remaining after either nylon wool purification or adherence to goat anti-human immunoglobulin-coated plates (see Materials and Methods).

Enzyme values were measured in cell extracts of null-and T-enriched lymphocytes in patients and normals (Table I). Patients had a significiant elevation of mean ADA levels in null-enriched cells, compared to normals. Likewise, mean PNP was elevated in patient vs. normal null. There were no significant differences in these enzymes between patient and normal T-enriched cells. In contrast, 5′-NT levels were higher in normal vs. patient T-enriched cells. Null 5′-NT was not significantly different between groups.

In Table II, nonadherent lymphocytes were analyzed with respect to surface marker expression. Symptomatic patients had a significantly higher percentage of $OKT10^+$ and Ia^+ cells, and a lower percentage of $OKT4^+$ cells than normals. Preliminary data on three controls and three patients suggested that the increases in OKT10 and Ia occurred in the null-enriched fraction rather than in the T fraction (data not shown).

Preincubation of both null- and T-enriched lymphocytes with 150 μg/ml thymosin fraction 5 (a dose found to be optimal in earlier dose–response experiments)

TABLE II. A Comparison of Lymphocyte Subpopulations in AIDS vs. Controls Using Monoclonal Antibodies

Monoclonal (% cells labeled)	OKT10	OKT9	Ia	Anti-Tac	Leu1	OKT4	OKT8
Normals (10)	11 ± 1.8[a]	2 ± 0.6	5 ± 0.8	3 ± 0.6	60 ± 3.7	35 ± 2.2	31 ± 2.0
AIDS/KS (15)	20 ± 3.3*	2 ± 0.7	13 ± 0.9*	3 ± 0.9	50 ± 7	22 ± 3.4*	39 ± 4.3

[a] Percent ± S.E.M. Markers performed on T-enriched lymphocytes depleted of monocytes (< 1% $ANAE^+$) and B cells (< 2% sIg^+).
*AIDS vs. normals, $p < 0.05$.

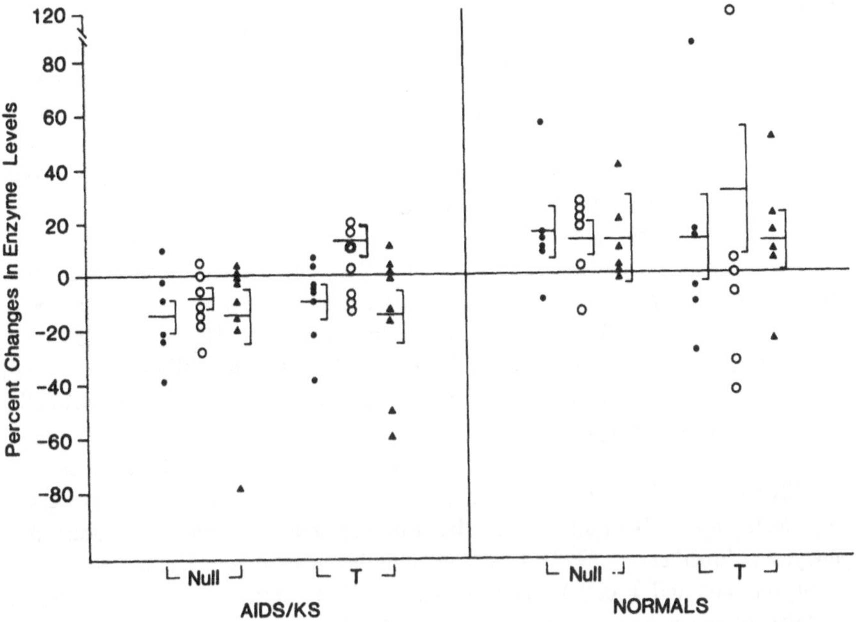

FIGURE 1. Percent change (+ or −) in lymphocyte purine enzymes ADA (●), PNP (○), and 5′ NT (▲) after preincubation with Thymosin (150 μg/ml). Each symbol represents one person. Horizontal bars denote mean; vertical bars denote S.E.M. Significant ($p < 0.05$) decreases in ADA and PNP occurred in AIDS/KS null cells in the presence of thymosin.

caused moderate enzyme changes (Fig. 1). Patients had a mean percent decrease in all null enzymes following thymosin incubation, whereas normals demonstrated an increase. Similar changes occurred with T-enriched cells, although results were variable for PNP. In general, the higher the initial enzyme level, the greater the decrease following thymosin exposure, and vice versa. As seen in Table III, similar changes occurred in lymphocyte surface markers. Leu1, OKT4, OKT8, OKT10, and Ia percentages in patients were suppressed by thymosin. In contrast, thymosin increased OKT10 and Ia, and OKT8 expression in normals.

TABLE III. Percent Change in Phenotypic Marker Expression following Thymosin Incubation: Normal vs. AIDS

	OKT10	Ia	Leu1	OKT4	OKT8
TF5, 150 μg/ml[a]					
Normals (6)	+8[b]	+20	−7	+15	+9
AIDS (11)	−14[b]	−20	−10	−11	−7

[a] B-cell- and monocyte-depleted lymphocytes preincubated with or without thymosin (150 μg/ml) and markers examined.
[b] (+) = mean percent increase in lymphocytes expressing OKT10; (−) = mean percent decrease. Changes in early markers (i.e., OKT10) correlated with similar changes in enzymes [if change in normals ADA (null) = +8%; change in AIDS ADA (null) = −14%]. See Fig. 1.

4. DISCUSSION

This study demonstrates several interesting findings. First, homosexuals with AIDS/KS had elevations of ADA and PNP in null-enriched lymphocytes, rather than low levels as has been observed in several immunodeficiency syndromes. Second, thymosin was capable of modulating enzyme levels and phenotypic marker expression in opposite directions, depending on whether each parameter was elevated or decreased initially.

The reasons for elevations in ADA and PNP levels in null lymphocytes from AIDS patients are unknown. However, ADA and PNP are increased in either very early prothymocytes (Ma *et al.*, 1982) or activated T cells (Hovi *et al.*, 1976). In four patients, over 30% of null cells did not mark with either OKT10, Ia, OKT8, or Leu7 (a monoclonal antibody defining a surface antigen expression on NK cells) Abo *et al.*, 1981), suggesting that a greater number of null cells might contain lymphocytes derived from early prothymocytes or pre-B cells. A decrease in ADA and PNP seen in AIDS following thymosin incubation could conceivably be due to a maturational effect such as has been demonstrated previously in mouse thymocytes (Ahmed *et al.*, 1978). It is likely that the elevation on OKT10 and Ia (both found on activated T cells) (Janossy *et al.*, 1980) could be related to increased enzyme levels, as these antigens were expressed largely on null-enriched cells (data not shown).

Thymosin has been shown to have variable effects on several immune parameters, depending on the initial immune status of the individual. For example, thymosin was shown to increase the percentage of E-rosette-forming cells in immunosuppressed cancer patients and in children with immunodeficiency syndromes. Significant increases often correlated with improved prognosis (Chretien *et al.*, 1978; Wara *et al.*, 1975). Schafer *et al.* (1977) demonstrated an improvement in lymphocyte PHA response in cancer patients upon incubation with thymosin. Decreased response occurred in patients in whom blastogenesis was elevated. In this study, positive or negative changes in enzymes and phenotypic markers correlated with clinical status and whether these parameters were initially elevated, normal, or suppressed. Interestingly, the increases noted in enzymes and in OKT10 and Ia, along with the decrease of the OKT4/OKT8 ratio following thymosin incubation in normals, are similar to the effects seen following lymphocyte activation (Burns *et al.*, 1982). Because AIDS patients had initial elevations in immature and/or "activation" antigens, as has been previously reported (Schroff *et al.*, 1983), perhaps thymosin induced a down-regulation of receptors along with lymphocyte maturation. Further studies examining the relationship of these changes to clinical status and other immune parameters such as lymphocyte blastogenesis are in progress.

REFERENCES

Abo, T., Cooper, M. D., and Balch, C. M., 1981, A differentiation antigen of human NK cells and K cells identified by a monoclonal antibody (HNK-1), *J. Immunol.* **127**:1024–1028.

Ahmed, A., Smith, A. H., and Sell, K. W., 1978, Maturation of thymus-derived (T) cells under the influence of thymosin, in: *Immune Modulation and Control of Neoplasia by Adjuvant Therapy* (M. A. Chirigos, ed.), Vol. 7, pp. 293–324, Raven Press, New York.

Burns, G. F., Battye, F. L., and Goldstein, G., 1982, Surface antigen changes occurring in short-term cultures of activated human T lymphocytes: Analysis by flow cytometry, *Cell. Immunol.* **71**:12–26.

Cohen, A., Dosch, H. M., and Gelfand, W. W., 1981, Induction of ecto-5'-nucleotidase activity in human thymocytes, *Clin. Immunol. Immunopathol.* **18**:287–290.

Chretien, P. B., Lipson, S. P., Makuch, R., Kenady, D. E., Cohen, M. H., and Minna, J. D., 1978, Thymosin in cancer patients: In vitro effects and correlations with clinical response to thymosin immunotherapy, *Cancer Treat. Rep.* **62**:1787.

Galli, V., and Schlessinger, M., 1974, The formation of stable rosettes after neuraminidase treatment of either human peripheral blood lymphocytes or sheep red blood cells, *J. Immunol.* **112**:1628–1634.

Giblett, E. R., Anderson, J. E., Cohen, F., Pollara, B., and Menwissen, J. H., 1972, Adenosine deaminase deficiency in two patients with severely impaired cellular immunity, *Lancet* **2**:1607–1609.

Giblett, E. R., Ammann, A. J., Sandman, R., Wara, D. W., and Diamond, L. K., 1975, Nucleoside-phosphorylase deficiency in a child with severely defective T-cell immunity and normal B cell immunity, *Lancet* **2**:1010–1013.

Gottlieb, M. S., Schroff, R., Schanker, H. M., Weisman, J. D., Fan, P. T., Wold, R. A., and Saxon, A., 1981, *Pneumocystis carinii* pneumonia and mucosal candidiasis in previously healthy homosexual men: Evidence of a new acquired cellular immunodeficiency, *N. Engl. J. Med.* **305**:1425–1430.

Gualde, N., and Goodwin, J. S., 1983, Effects of prostaglandin E_2 and preincubation of lectin-stimulated proliferation of human T cell subsets, *Cell. Immunol.* **70**:373–379.

Hovi, T., Smyth, J. F., Allison, A. C., and Williams, S. C., 1976, Role of adenosine deaminase in lymphocyte proliferation, *Clin. Exp. Immunol.* **23**:395–403.

Janossy, G., Tidman, N., Papageorgiou, E. S., Kung, P. C., and Goldstein, G., 1980, Distribution of T lymphocyte subsets in the human bone marrow and thymus: An analysis by monoclonal antibodies, *J. Immunol.* **126**:1608.

Ma, D. D. F., Sylwestrowicz, T. A., Granger, S., Massaia, M., Franks, R., Janossy, G., and Hoffbrand, V., 1982, Distribution of terminal deoxynucleotidyl transferase and purine degradative and synthetic enzymes in subpopulations of human thymocytes, *J. Immunol.* **129**:1430–1435.

Mage, M. G., McHugh, L. L., and Rothstein, T. L., 1977, Mouse lymphocytes with and without surface immunoglobulin: Preparative scale separation in polystyrene tissue culture dishes coated with specifically purified anti-immunoglobulin, *J. Immunol. Methods* **15**:47–55.

Pazmino, N. H., Ihle, J. N., and Goldstein, A. L., 1978, Induction in vivo and in vitro of terminal deoxynucleotidyl transferase by thymosin in bone marrow cells from athymic mice, *J. Exp. Med.* **147**:708.

Schafer, L. A., Gutterman, J. U., Hersh, E. M., Mavligit, G. M., Dandridge, K., Cohen, G., and Goldstein, A. L., 1976, Partial restoration by in vivo thymosin of E-rosettes and delayed type hypersensitivity reactions in immunodeficient cancer patients, *Cancer Immunol. Immunother.* **1**:259–264.

Schafer, L. A., Gutterman, J. U., Hersh, E. M., Mavligit, G. M., and Goldstein, A. L., 1977, In vitro and in vivo studies of thymosin activity in cancer patients, in: *Immune Modulation and Control of Neoplasia by Adjuvant Therapy* (M. A. Chirigos, ed.), Vol. 2, p. 329, Raven Press.

Schroff, R. W., Gottlieb, M. S., Prince, H. E., Chai, L., and Fahey, J. L., 1983, Immunological studies of homosexual men with immunodeficiency and Kaposi's sarcoma, *Clin. Immunol. Immunopathol.* **27**:300–314.

Van Laarhoven, J. P. R. M., Spierenburg, G. Tau, and deBruyn, C. H. M. M., 1980, Enzymes of purine nucleotide metabolism in human lymphocytes, *J. Immunol. Meth* **39**:47–58.

Wara, D. W., Goldstein, A. L., Dayle, N. E., and Ammann, A, 1975, Thymosin activity in patient with cellular immunodeficiency, *N. Engl. J. Med.* **292**:70–74.

Demonstration of a Block at an Early Stage of T-Cell Differentiation in Severe Combined Immunodeficiency Disease

RAJENDRA PAHWA, DAHLIA KIRKPATRICK, ANDRE VASU, and SAVITA PAHWA

1. INTRODUCTION

The syndrome of severe combined immunodeficiency disease (SCID) represents a group of congenital lethal disorders, characterized by defects of both T- and B-cell systems (Hitzig *et al.*, 1971; Giblett *et al.*, 1972; Gatti *et al.*, 1969; Pyke *et al.*, 1975; Hong *et al.*, 1978; Gelfand *et al.*, 1977; Pahwa *et al.*, 1979). Although the ultimate immunological expression of these disorders is quite similar, the defect of lymphoid development may vary. Studies using thymic extracts, thymic epithelial monolayers, and thymic hormones have revealed that the differentiation of precursor cells may be abnormal in SCID (Touraine *et al.*, 1974; Pahwa *et al.*, 1977, 1979). Most of the previously reported studies on T-lymphocyte differentiation have employed mainly rosette formation with SRBC (E-rosettes) as a differentiation marker for T cells. In this communication, we have investigated the differentiation of T-cell surface antigens using mouse monoclonal antibodies on two patients' blood and bone marrow lymphocytes using cultured thymic epithelial monolayers and thymosin fraction 5 (TF5) as inducing agents.

RAJENDRA PAHWA and DAHLIA KIRKPATRICK • The Memorial Sloan-Kettering Cancer Center, New York, New York 10021. ANDRE VASU and SAVITA PAHWA • North Shore University Hospital, Manhasset, New York 11030.

2. METHODS

Patients. The patients under study (under 1 year of age) were diagnosed as having SCID on clinical and laboratory bases. Both patients were lymphopenic with absence of T cells as shown in Table I. B cells were normal in number. Proliferative responses to mitogens, antigens, and allogeneic cells were severely depressed in both patients. Patient No. 1 had associated deficiency of the enzyme adenosine deaminase (ADA).

Cultures of Thymic Epithelium. Thymic epithelial monolayers were established as described previously (Pyke and Gelfand, 1974; Pahwa *et al.*, 1977). Briefly, normal thymic tissues obtained from children undergoing cardiac surgery were cut into small pieces, teased with forceps, and then cultured in small petri dishes in RPMI 1640 medium supplemented with 30% heat-inactivated fetal calf serum, gentamycin (4 μg/ml), and amphotericin B (1 μg/ml). The cultures were incubated in an atmosphere of 5% CO_2–95% air at 100% relative humidity. The culture media were changed weekly, and the supernatants from the cultures were collected, centrifuged, filtered through Millipores, and stored at −20°C. These were designated thymic epithelial culture supernatants (TCS). Cells from all cultures of epithelial monolayers were tested to ensure that they did not contain any E-rosetting cells before using these monolayers as inducers of differentiation. Satisfactory epithelial monolayers were established after a culture period of 3–5 weeks. Monolayers from fetal kidney or skin and their supernatants (also stored at −20°C at weekly intervals) were used as controls.

Cell Isolation. Small volumes (0.5–0.7 ml) of bone marrow from patients were aspirated from a few sites of the iliac crest into 10-ml heparinized glass syringes. Mononuclear cells from the marrow of patient No. 1 were separated on Ficoll–Hypaque gradients by centrifugation at 400*g* for 30 min at 22°C. Cells at the interface were washed thrice with RPMI 1640 medium, and adjusted to a concentration of 2×10^6/ml prior to induction studies. For remaining samples (marrow and blood from patient No. 2, blood from patient No. 1), 1-ml volumes of unfractionated samples were utilized for cell surface phenotype determination.

Induction of Markers. The unseparated blood from both patients, unseparated bone marrow from patient No. 2, and mononuclear cells from bone marrow cells

TABLE I. Features of Test Patients with SCID

		Lymphocytes (% positive cells)			Immunoglobulins (mg/dl)		
Patient	ADA	E-rosettes[a]	OKT11[b]	sIg[c]	IgG	IgA	IgM
1[d]	−	1.0	17.1	7.4	518	30	50
2	+	2.5	6.2	11.2	280	3	16

[a] Forming spontaneous rosettes with SRBC (normal ± 1 S.D., 72 ± 12).
[b] Characterization with fluorescein-conjugated monoclonal antibodies (normal ± 1 S.D., 70 ± 7).
[c] Surface immunoglobulin-bearing cells (normal ± 1 S.D., 17 ± 7) tested with fluorescein-conjugated polyvalent rabbit anti-human immunoglobulin antisera.
[d] Patient No. 1 was on plasma infusion therapy.

of patient No. 1 were cocultured with thymic epithelial monolayers, TF5 (50 and 100 μg/ml) and medium alone for 15 hr at 37°C in a humidified 5% CO_2–95% air incubator at 100% humidity. After incubation, cells were stained for T-cell surface antigen using the following mouse monoclonal antibodies: OKT3, T4, T6, T8, T9, T10, T11 (Ortho Diagnostics). The antibodies OKT3, T4, T6, T8, and T11 were fluorescein-conjugated antibodies, while the T9 and T10 antibodies were not. Cells treated for T9 and T10 antibodies were secondarily stained with goat anti-mouse fluorescein-conjugated antibodies. The test samples were analyzed by Ortho Spectrum III.

Phenotypic Characterization of Cells with Monoclonal Antibodies. Whole blood (0.1 ml) was mixed with 0.01 ml of appropriate fluorescein-conjugated mouse monoclonal antibodies (OKT3, T4, T6, T8, T11) for 30 min at 4°C. The red cells were then lysed with Ortho lysing reagent. The remaining cells were washed twice with PBS (Ca and Mg free), resuspended in 1 ml of PBS, and analyzed on Spectrum III. Cells treated with OKT9 and T10 antibodies were stained with goat anti-mouse fluorescein-conjugated antibodies before analysis. The trigger zone was focused on the lymphocyte population. Both percentages and absolute numbers of positively stained lymphocytes were noted in each experiment.

3. RESULTS

Table II depicts results of surface phenotype in blood and bone marrow cells of patient No. 1 with SCID. A marked deficit of cell surface phenotype characteristics of OKT3, T4, T6, and T8 was observed. Cell surface phenotype characteristics of OKT9 and T10 were present but OKT11 was markedly diminished (17.1% in blood and 0.0% in bone marrow cells). Following incubation with thymic epithelial monolayers and TF5 (50 μg/ml), only induction of $OKT10^+$ cells was observed as shown in Table II. In blood, increase was observed from 38.7% to 51.2% with thymic epithelial monolayers and 53.3% with TF5, while in the bone marrow

TABLE II. Induction of T-Cell Antigens in Blood and Bone Marrow Cells of Patient No. 1 with SCID[a]

		T-cell antigen (% positive cells)						
Sample	Induction	T3	T4	T6	T8	T9	T10	T11
Blood	Medium	2.2	2.6	1.1	1.8	18.1	38.7	17.1
	TEM	1.3	3.0	0.2	1.0	13.7	*51.2*	10.7
	TF5, 50 μg/ml	2.0	1.6	0.0	2.9	18.0	*53.3*	20.2
Bone marrow	Medium	0.0	0.0	0.0	0.0	23.5	38.3	0.0
	TEM	0.8	1.5	1.2	0.5	24.3	*55.4*	1.2
	TF5, 50 μg/ml	0.5	0.4	1.2	0.0	24.6	*53.5*	0.9

[a] $T10^+$ cells were present in bone marrow and blood. Further induction of $T10^+$ cells was noted following incubation of blood and bone marrow samples with thymic epithelial monolayers (TEM) or TF5.

TABLE III. Induction of T-Cell Antigens in Blood and Bone Marrow of Patient No. 2 with SCID[a]

Sample	Incubation	T-cell antigens (% positive cells)						
		T3	T4	T6	T8	T9	T10	T11
Blood	Medium	6.2	4.5	4.3	1.9	36.2	1.3	6.2
	TEM	4.4	4.4	1.5	2.0	37.7	5.3	6.7
	TF5, 50 μg/ml	5.5	6.8	3.4	1.1	32.3	1.5	5.1
Bone marrow	Medium	2.1	1.3	0.7	0.1	30.9	3.5	5.5
	TEM	2.6	1.1	0.2	1.2	27.8	*24.7*	7.4
	TF5, 50 μg/ml	2.4	2.0	1.1	0.2	32.5	6.7	6.6
	TF5, 100 μg/ml	1.9	1.4	0.7	0.3	36.4	*18.2*	7.2

[a] $T10^+$ cells were absent in bone marrow and blood. However, induction of $T10^+$ cells were noted in bone marrow following incubation with thymic epithelial monolayers (TEM) or TF5.

increase was from 38.3% to 55.4% with thymic epithelial monolayers and 53.5% with TF5.

Results of surface phenotype in blood and bone marrow cells of patient No. 2 with SCID are shown in Table III. In this patient a marked deficit of all cell surface phenotype characteristics of OKT3, T4, T6, T8, T10, and T11 were observed except OKT9 (36.2% in blood and 30.9% in bone marrow cells). Following incubation with thymic epithelial monolayers and TF5 (50 and 100 μg/ml), only induction of $OKT10^+$ cells was observed in bone marrow cells as shown in Table III. Increase was observed from 3.5% to 24.7% with thymic epithelial monolayers and 18.2% with TF5 (100 μg/ml).

4. DISCUSSION

Previous studies of T-cell differentiation on the marrow cells of patients with SCID revealed varying defects, ranging from a complete absence to partial differentiation of definable precursor cells (Touraine *et al.*, 1974; Pahwa *et al.*, 1977, 1979, 1980; Incefy *et al.*, 1976; Hong *et al.*, 1976, 1978; Pyke *et al.*, 1975; Gelfand *et al.*, 1977; Reinherz *et al.*, 1981). The present study, performed on blood from two patients with SCID, has revealed that the possible defect of the early stem cells might involve (1) absence of T10 antigen-bearing cells with absence of T3, T4, T6, and T11 antigens as seen in patient No. 2, and (2) presence of T10 antigen-bearing cells which are associated with $T11^+$ and $T11^-$ cells but with absence of T3, T4, T6, and T8 antigens as seen in patient No. 1.

In a larger series of five SCID patients, Reinherz *et al.* (1981) have described three subtypes of SCID based on cell surface phenotype using T-cell monoclonal antibodies. One subtype was associated with failure to develop lymphocytes that express any thymus-specific antigens, another with failure to differentiate beyond the early prothymocyte-thymocyte ($T9^+$, $T10^+$) stage, while a third subtype was associated with failure to differentiate beyond a late thymocyte ($T3^+$, $T4^+$, $T5^+$, $T8^+$, and $T10^+$) stage. According to this scheme, one of our patients (No. 2) would fit into the first subtype as he was lacking identifiable thymus-specific antigens. The second patient (No. 1) had T10 antigens but was lacking T9 and had in addition a small percentage of $T11^+$ lymphocytes. He could be considered as representing a subgroup within the second subtype according to the scheme of Reinherz *et al.* (1981).

The studies attempting to induce differentiation along the T-cell pathway in these patients indicate that their cells can be pushed at least partially from one subtype into the next. Expression of T10 antigen was induced in patient No. 2 and the population expressing this antigen was expanded in patient No. 1 following incubation of marrow cells from both patients with thymic influences employed herein, i.e., thymic epithelial monolayers and TF5. The T10 antigen-bearing population was expanded even in lymphocytes from peripheral blood in patient No. 1. No induction of other surface phenotypes, i.e., T3, T4, T8 or T11, was observed. Likewise, E-rosette-forming capacity was also not induced in either patient (data not shown).

Further in patient No. 1, only 1% of the lymphocytes formed E-rosettes with SRBC while expressing 17.1% cell surface phenotype characteristic of OKT11, thus indicating a dissociation between expression of OKT11 antigen and E-rosette formation. Such findings have also been observed in patients with childhood acute lymphoblastic leukemia (Narula *et al.*, 1982). The mechanism and significance of this dissociation between these two markers are not clear.

Our induction studies indicate that in both patients the true block in differentiation lies at an early stage, one presumably involving differentiation of prothymocytes to thymocytes. An assessment of the possible site of defect in SCID with the type of *in vitro* experiments described herein will help to provide a better understanding of the steps involved in normal T-lymphocyte differentiation.

ACKNOWLEDGMENTS. We thank Dr. R. J. O'Reilly and other members of the Bone Marrow Transplant Units of The Memorial Sloan-Kettering Cancer Center for providing us the bone marrow and blood specimens of their patients. The skilled technical assistance of Deepak Rajpoot and secretarial assistance of Miss Phyllis Hornstein are deeply appreciated. This work was aided by Grant CA-33050 from the National Institutes of Health.

REFERENCES

Gatti, R. A., Platt, N., Hong, R., Langer, L. O., Kay, H. E. M., and Good, R. A., 1969, Hereditary lymphopenic agammaglobulinemia associated with a distinctive form of shortlimbed dwarfism and ectodermal dysplasia, *J. Pediatr.* **75**:675–684.

Gelfand, E. W., Dosch, H. M., Huber, J., and Shore, A., 1977, In vitro and in vivo reconstitution of severe combined immunodeficiency disease with thymic epithelium, *Clin. Res.* **25**:358A.

Giblett, E. R., Anderson, J. E., Cohen, F., Pollara, B., and Meuwissen, H. J., 1972, Adenosine-deaminase deficiency in two patients with severely impaired cellular immunity, *Lancet* **2**:1067–1069.

Hitzig, W. H., Landolt, R., Muller, G., and Bodmer, P., 1971, Heterogeneity of phenotypic expression in a family with Swiss-type agammaglobulinemia: Observations on the acquisition of agammaglobulinemia, *J. Pediatr.* **78**:968–980.

Hong, R., Santosham, M., Schulte-Wissermann, H., Horowitz, S., Hsu, S. H., and Winkelstein, J. A., 1976, Reconstitution of B and T lymphocytes function in severe combined immunodeficiency disease after transplantation with thymic epithelium, *Lancet* **2**:1270–1272.

Hong, R., Schulte-Wissermann, H., Horowitz, S., Borzy, M., and Finlay, J., 1978, Cultured thymic epithelium in severe combined immunodeficiency, *Transplant. Proc.* **10**:201–202.

Incefy, G. S., Grimes, E., Kagan, W. A., Goldstein, G., Smithwick, E., O'Reilly, R. J., and Good, R. A., 1976, Heterogeneity of stem cells in severe combined immunodeficiency, *Clin. Exp. Immunol.* **25**:462–471.

Narula, N., Miller, D. R., Good, R. A., and Pahwa, R. N., 1982, Effect of thymosin fr. V on lymphoblasts from bone marrow and peripheral blood of newly diagnosed patients with acute lymphoblastic leukemia, *Fed. Proc.* **41**:407.

Pahwa, R., Pahwa, S., Good, R. A., Incefy, G. S., and O'Reilly, R. J., 1977, Rationale for combined uses of fetal liver and thymus for immunological reconstitution in patients with variants of severe combined immunodeficiency, *Proc. Natl. Acad. Sci. USA* **74**:3002–3005.

Pahwa, R. N., Pahwa, S. G., and Good, R. A., 1979, T lymphocyte differentiation in severe combined immunodeficiency: Defects of stem cells, *J. Clin. Invest.* **64**:1632–1641.

Pahwa, S., Pahwa, R., and Good, R. A., 1980, Heterogeneity of B lymphocyte differentiation in severe combined immunodeficiency, *J. Clin. Invest.* **66**:543–550.

Pyke, K. W., and Gelfand, E. W., 1974, Morphological and functional maturation of human thymic epithelium in culture, *Nature (London)* **251**:421–423.

Pyke, K. W., Dosch, H. M., Ipp, M. M., and Gelfand, E. W., 1975, Demonstration of an intrathymic defect in a case of severe combined immunodeficiency disease, *N. Engl. J. Med.* **293**:424–428.

Reinherz, E. L., Cooper, M. D., and Schlossman, S. F., 1981, Abnormalities of T cell maturation and regulation in human beings with immunodeficiency disorders, *J. Clin. Invest.* **68**:699–705.

Touraine, J. L., Incefy, G. S., Touraine, F., L'Esperance, P., Siegal, F. P., and Good, R. A., 1974, T lymphocyte differentiation in vitro in primary immunodeficiency diseases, *Clin. Immunol. Immunopathol.* **3**:228–235.

Part V Summary

Clinical Applications

ROBERT K. OLDHAM and PAUL B. CHRETIEN

BIOLOGICALS OTHER THAN THYMIC HORMONES

Dr. Chretien and I have divided our responsibilities for the Clinical Applications Session. I will summarize the presentations on biologicals other than thymosin and he will summarize the presentations on thymic factors. Since this transfers most of the work to my cochairman, I will follow with only a few comments. First, the presentation by Dr. Dumonde on the use of crude extracts of the cell line RPMI (1788) was of some interest. I believe although we would prefer to study purified molecules and combinations of purified molecules instead of extracts of poorly defined cell supernatants, these kinds of preliminary clinical studies are of use. They do give us some appreciation of the kinds of clinical activities that can be seen with reagents that are mixtures and contain a variety of biological activities as measured by our biological assays. Although the time will come when we will put the components of such extracts together more precisely, these kinds of studies do give us some preliminary information. As was well pointed out by Dr. Dumonde, such studies are very limited in what we can learn from them in terms of precise molecular interrelationships. The interlesional injection with subsequent tumor necrosis was interesting, particularly with the data on cell infiltrates which made it appear as though an immunological response was stimulated when this extract was injected into the lesions. In the Biological Response Modifiers Program (BRMP), we are conducting a study of genetically engineered interferon interlesionally, and although we see necrosis in some injected nodules, this kind of infiltrate has not been seen. It may well be that the lymphokine mixture draws forth an inflammatory response not stimulated by a single purified recombinant α-interferon molecule. It will be interesting to investigate this in more detail.

ROBERT K. OLDHAM • Biological Response Modifiers Program, Division of Cancer Treatment, NCI-Frederick Cancer Research Facility, Frederick, Maryland 21701. PAUL B. CHRETIEN • Department of Surgery, University of Maryland, Baltimore, Maryland 21201.

I am not an endocrinologist so I'm very unsure about the hormonal relationships described by Dr. Dumonde, but I think they bear study and critical review. Dr. Joshua presented some interesting information with another clinical study on crude lymphokine-containing material. Measurable responses were seen in a select group of those patients with interlesional injections of this extract. Both Dr. Dumonde and Dr. Joshua need to pursue these studies applying placebos to analogous lesions in the skin simultaneously. We have done that in our interlesional interferon study and have not seen saline giving responses in injected nodules, whereas the interferon has. When one does these studies, it is important to use some kind of placebo control and also to describe whether distant lesions respond simultaneously with the injected lesion. It is also important to measure the injected biological in the blood and to understand the pharmacokinetics and effects of the systemically distributed materials after intralesional injection.

Dr. Palladino gave some interesting information on IL-2 production in patients with acquired immunodeficiency disease syndrome (AIDS). It is curious that a disease appears to have been born at about the same time the cures are being discovered. I do not know why that is, and I do not know why AIDS did not show up 10 to 15 years ago before we were genetically engineering lymphokines. Why AIDS showed up in the early 1980s at precisely the time our techniques are allowing us to produce, in high quantity and high quality, interferons and other lymphokines such as IL-2, which may ultimately be of use in curing the disorder, is an interesting philosophical point to consider. The data on IL-2 production and effects of this lymphokine in AIDS, in other immunodeficiency states, and in cancer patients are of great interest.

In the poster sessions, the poster by Dr. Basham stuck me as particularly interesting. It illustrates the effect of interferon on cell surface molecules. They reported an increase in antigen display with respect to Ia antigens and the more powerful action, at least in one system, of γ-interferon as compared to α- and β-interferon. This area deserves considerable study because it is clear that as biologicals increase antigenic display, this may be of use in integrated studies with active specific immunotherapy.

A final comment has to do with the question of endotoxin which, because of time restraints, was not discussed in great detail today. We have noted that endotoxin can be seen by virtue of its biological effect at 2–3 logs below detectable levels by the limulus assay, if its presence is detected by nature of endotoxin's synergistic effect on other lymphokine activities. I believe this indicates that the limulus assay is a very limited tool in determining that preparation is endotoxin negative. The analogy is the common statement that all cell lines were tested for mycoplasma and were negative. The tests used are crucial in defining the absence of a contaminant from any biological substance. I do not believe that any of us can confidently rely on the current techniques for endotoxin measurement, and we must assume that small quantities are often there in the preparation of crude lymphokine extracts. The comment that Dr. Oppenheim made about antibodies is very important. Methods to rid preparations of endotoxin activity, such as polymyxin B or perhaps antibody to a particular endotoxin, would be very useful in studying these crude extracts.

This is a time of great excitement in biology. With monoclonal antibodies, genetically engineered lymphokines/cytokines, improved instrumentation and fermentation techniques, and recent advances in oncogene research, we can all look forward to the certainty of biologicals as the fourth modality of cancer therapy.

R. K. O.

THYMIC HORMONES

The privilege afforded me by our conference host, Dr. Allan Goldstein, and my session cochairman, Dr. Robert Oldham, to comment on the clinical studies with thymic hormone is opportune, because these reports, combined with other recent ones, appear to mark a crossroads in clinical immunology. These studies firmly document that extracts and peptides of thymus gland origin have immunological activity which is reflected in clinical benefit. The results give legitimacy to the use of these agents in the treatment of disease and move the challenge from demonstration of efficacy in pilot studies to controlled trials in well-defined patient populations to establish activity of individual thymic hormones in distinct clinical diseases. My comments are limited to the clinical aspects of the reports in this session and the relevance to future clinical research, so as not to overlap discussions of preclinical aspects of the hormones presented in the previous sections.

In Diane Wara's summary of her experience since her pioneering utilization of thymosin fraction 5 in the treatment of congenital immune deficiencies, she prudently advises caution in the interpretation of her uncontrolled observations, because of the possibility of spontaneous improvement in these diseases and the potential benefit from other therapies that were employed. Since her initial report, however, other thymic hormones have been evaluated and the accumulated data leave no doubt of their efficacy in these disorders. Her results, combined with the reports by Skotnicki and Bach in this session and the successful use of TP-5 previously reported, justify thymic hormones as a treatment of choice or as a therapeutic alternative for certain congenital immune deficiencies.

In the course of her initial studies, Dr. Wara observed that thymosin improved peripheral blood *in vitro* lymphocyte reactivity in the mitogen-induced cellular proliferation and mixed leukocyte reaction assays, and that this response correlated with the clinical benefit after a course of thymosin. She also found that these immune parameters progressed toward normal levels in the children who improved clinically after administration of thymosin. Thus, utilization of this *in vitro* effect of thymic hormones on assays of cellular immunity as a predictor of the clinical response may have great value in delineating patient populations potentially amenable to treatment with thymic hormone and also provide a means of comparing the relative clinical efficacy of individual thymic hormones.

In a study of patients with untreated Hodgkin's disease, Davis observed improvement in assays of cellular immunity after *in vitro* incubation with TP-1 of peripheral blood lymphocytes from the patients, similar to the results obtained with thymosin by Schaeffer and Kenady in patients with malignancies. Davis also confirmed their finding of the greatest improvement in the immune assays with specimens from patients who had the most severely impaired cellular immunity. In contrast to the observations by Wara in congenital immune deficiency, however, in which these results were predictive of the clinical response with thymosin as the sole therapy, the use of *in vitro* response to thymosin as a predictor of eventual clinical course in malignancy is dependent on an effective primary tumoricidal treatment with which the thymic hormone is used as an adjuvant. Results from several studies illustrate this relationship. In trials with thymosin, Lipson found a correlation between low T-cell levels before treatment and duration of survival in patients with oat cell carcinoma of the lung treated with chemotherapy and Patt showed a relationship between tumor stage (which correlates with cellular immunity) in melanoma and duration of the clinical response to chemotherapy in melanoma. Thus, in the patients Davis studied, the clinical courses after primary therapy and adjuvant TP-1 offer an opportunity to further investigate this relationship.

Schulof's clinical trial in bronchogenic carcinoma, in which thymosin α_1 was employed in conjunction with palliative radiation therapy, is exemplary in its rationale based on preclinical and *in vitro* studies of the effects of thymic hormones, in the battery of assays used to monitor immunological responses and in its highly scientific utilization of a control population which was randomized double-blind to receive either thymosin α_1 or placebo. The rationale for thymic hormone as an adjunct to radiotherapy to the chest is derived from studies that document profound declines in lymphocyte and T-cell populations during irradiation to the chest and a significant increase in T-cell levels *in vitro* after incubation with thymosin of the peripheral blood lymphocytes obtained from the same patient population. Although administration of α_1 was associated with a modest increase in helper T-cell levels and mixed leukocyte responsiveness to alloantigens that did not return to pretreatment levels, nevertheless, the results are extremely gratifying when the clinical setting is considered. In patients with congenital immune deficiencies who benefited from thymic hormones, among whom Wara and others observed a return to normal levels of the immune parameters after administration of thymic hormones, there was not an ongoing pathologic process which diminishes immune reactivity. By contrast, in patients with cancer and infectious diseases, the effects of the microbial infection or the malignant cells exert a continuous immunosuppressive effect, and reversal of this effect must be sought through effective primary therapies. In these latter settings, mere stabilization of the progressive decline in immune reactivity is indicative of a marked immunological effect of the hormone, and the associated therapeutic benefit should be sought through comparisons of the clinical course with concomitant control populations.

In Schulof's trial, the serial determination of peripheral blood levels of thymosin α_1 by radioimmunoassay is an important advance in immunotherapy with thymic hormones. The determination of circulating levels of the hormone prior to treatment has obvious usefulness in delineating patient populations and disease entities that are appropriate for treatment with the agents. During administration of the hormones, serial determination of the levels should provide useful correlations with clinical course that give insight into the effects of increasing circulating levels of the hormones to the normal range in patients with diminished levels and whether additional benefit or deleterious effect accrues with increase in the levels to even higher ranges. It is theoretically possible that patients with cellular immune deficiency and high levels of thymic hormones may nevertheless benefit from treatment with the agents, and studies which incorporate monitoring of hormone levels will be useful in exploring this hypothesis.

Murray considered the potential usefulness of lymphocyte enzymes as predictors of the maturational effects of thymic hormones and found elevated levels of the purine enzymes adenosine deaminase (ADA) and purine nucleotide phosphorylase (PNP) in null cells in AIDS that decreased after *in vitro* incubation of the lymphocyte preparations with thymosin. The data suggest maturation of null cells by the hormone, but regression to the T-cell lineage was not detected. One interpretation of the phenomenon observed is that a cell population not presently measurable was generated. The potential importance of these observations merits further investigation, especially in diseases associated with low blood levels of thymic hormones.

William Wara's observation of tumor regression in patients with advanced malignancies of the kidney and prostate who received relatively large doses of thymosin fraction 5 provokes speculation concerning effects of thymic hormone other than immune modulation. Although augmentation of immunological processes that lead to tumor regression cannot be excluded, the high-dosage schedule requires consideration of the corticogenic effects of thymosin. In recent studies, thymosin α_1 introduced into the CNS provoked elaboration of adrenal corticosteroids, thus implicating neuroendocrine pathways for this effect of the hormone. Since corticosteroids have induced regression of renal and prostate tumors, this explanation for the tumor responses obtained with thymosin in this trial must be considered. The possibility that thymic hormones induce immunological alterations in conventional dosages, but stimulate endocrine changes in higher dose regimens should also be explored for possible clinical usefulness in the treatment of other diseases.

The use of a thymic extract for treatment of fulminant viral hepatitis B in China reported by Su addresses the formidable challenge of immunotherapy in infectious diseases. Most infections either resolve or become indolent after a short clinical course or are fatal after a short interval, and thus do not provide sufficient time for the pathologic process to be altered by improved or stabilized cellular immunity induced by thymic hormones. Of the infectious diseases with courses advantageous for evaluation of the clinical efficacy of thymic hormones, infectious hepatitis B is relatively ideal because of an associated impairment of cellular immunity and a morbid course with current therapies. The lower mortality in fulminant hepatitis B infection in this pilot provides adequate rationale for the institution of controlled clinical trials in the disease. This conclusion is further supported by similarly encouraging results obtained in a controlled trial with TFX reported by Skotnicki. Since patients who recover from acute hepatitis B infection may enter a carrier state with persistence of infectious virus that also is associated with depressed cellular immunity, similar studies should be considered in this patient population.

An increase in suppressor activity in patients with rheumatoid arthritis found by Jacobs introduces the current problem of the extreme variation and conflicting results of immunological studies in patients with autoimmune diseases thus far reported. Most investigators found decreased suppressor activity in rheumatoid arthritis, systemic lupus erythematosus, asthma, chronic autoimmune hepatitis, and similar diseases. However, almost invariably, the populations studied were on one or more medications, such as corticosteroids or nonsteroidal anti-inflammatory drugs, and in virtually all the diagnosis had been made a number of years beforehand. Studies on the effect of disease activity, duration since first onset, age of the patient, and current or prior medications usually found that these variations affect results of the immune assays. Before full-scale clinical trials of thymic hormones in autoimmune diseases are initiated, the effects of these variables should be determined with certainty. Equally important is the determination of the immunological status during periods of inactive disease. Relatively normal immunologic activity during these intervals advises caution in the use of thymic hormones and other immune modulators for prophylaxis against acute exacerbations because of the theoretical possibility of activating the disease.

The report by Sheng of clinical improvement in patients with chronic autoimmune hepatitis with thymus extract in China gives great impetus to evaluation of thymic hormones in autoimmune diseases. These clinical results are important extensions of experimental animal studies and *in vitro* studies of the effects of thymic hormones utilizing blood specimens from patients with chronic autoimmune hepatitis and other autoimmune diseases, which predict that the hormones should be clinically beneficial in these disorders. Future therapy trials in autoimmune diseases should include randomized controls as utilized by Dr. Sheng, and also begin to concentrate on the derivation of maximally effective dose-regimens and treatment schedules of the hormones.

The review by Skotnicki of the immunological and clinical studies that have been conducted in Poland with TFX shows that this thymus extract has biological activity equivalent to other thymic hormones. The double-blind, placebo controlled trial in hepatitis B enjoys the distinction of the first study in infectious diseases of thymic hormones thus reported with this scientific level of clinical design. The positive clinical results support the observations by Su in the same disease and should encourage trials in othr viral diseases associated with extreme morbidity and resistance to other therapeutic modalities.

In the progress report by Bach on the studies which have been conducted with thymulin, the low circulating blood levels of this nonapeptide in children with congenital immune deficiency, combined with the improvement in the patients after administration of the peptide, are evidence that thymic hormones of this small size can exert immune modulating properties and corresponding clinical benefits equal to that achieved with larger peptides and thymic extracts. The short half-life of thymulin, and of TP-5, which also is effective in congenital immune deficiencies, further show that clinical benefit is not necessarily dependent on persisting elevations of blood levels of the hormones. It is more than possible that Bach's correlations of immunological activity with dose regimen of thymulin, i.e., that augmentation of helper function is only achieved with lower doses while higher doses lead to intensification of suppressor activity, apply to other thymic hormones. Therefore, it is imperative that investigations which explore this relationship be conducted with the other preparations to determine not only the efficacy but the potential hazards associated with clinical use of thymic hormones.

CONCLUSIONS

A major conclusion from these clinical trials with thymic hormones, along with studies reported earlier, is that both thymic extracts and individual peptides thus far evaluated exerted immunological effects and corresponding clinical effects in several major categories of diseases. In cancer and infectious diseases, subpopulations associated with impaired immune reactivity have manifested elevations of these parameters toward normal and improvement in disease course. In autoimmune diseases, significant proportions of patients have displayed clinical improvement and the preliminary results indicate that this response depended on the presence of impaired suppressor activity. These immunological effects of thymic hormones define immune modulation and distinguish these preparations from agents whose beneficial effect is related to augmentation of immune reactivity and thus are more appropriately termed immune stimulants.

An important area for future investigation is the relative clinical advantages, and disadvantages, of thymic extracts, such as thymosin fraction 5, thymic humoral factor, TFX, TP-1, etc. compared with single peptides, such as thymosin α_1, TP-5, and thymulin. A potential disadvantage of thymus extracts is the variation in number and concentrations of biologically active peptides present in each batch and the corresponding variations in immunological effects. The variation may be due to strain, age, and sex of the animals, and diet, season, etc., and the procedures used for extraction of the polypeptide mixtures. By contrast, most of the individual peptides currently being used in clinical trials are synthesized, and thus have reproducible chemical composition and biological activity. The single peptides also offer the opportunity for assessment of the blood levels of the naturally occurring peptides before administration of the synthetic or purified preparation, and permit correlations of blood levels before and during administration of the peptide with the clinical course. These considerations give priority to comparisons of the relative effects of thymic extracts and individual peptides on lymphocyte maturation and other immunological changes. The comparisons should be extended to include evaluation of combinations of single peptides that may yield immunological effects which are equal to or greater than single peptides or extracts. This proposal is derived from several studies in which thymic hormone had differing immunological effects on lymphocyte precursors and other cells. For example, thymosin α_1 and thymulin have been shown to exert maturational effects beginning with terminal deoxynucleotidyl transferase positive (TDT^+) prothymocytes, but thymosin fraction 5 and one of its constituents, β_4, exert expression of the enzyme in TDT^- bone marrow cells. Thus, these data support investigations for complementary activities of the individual peptides in an attempt to derive peptide combinations that have greater immunological effects than obtained with the peptides used singly.

The relation of the immunological and clinical effect of the hormones on the magnitude of immunological abnormality associated with the disease serves to delineate patient subpopulations that should be evaluated in future trials. In cancer, the ideal patient groups have stages of the disease associated with significant impairment of immune reactivity and are treated with an effective but immunosuppressive therapy. In infectious diseases, it appears advantageous to study highly morbid and immunosuppressive processes that have clinical courses which extend at least several weeks. In autoimmune diseases, there are several ideal populations, e.g., chronic hepatitis B, autoimmune hepatitis, rheumatoid arthritis, asthma, etc. In this category, however, relatively much more data are needed concerning correlations of immune status and response to thymic hormone with disease activity, duration of disease since onset, and effects of concomitant and previously administered drugs. The favorable results thus far reported with autoimmune diseases encourage the initiation of these admittedly tedious studies. Additional impetus arises from the evidence that in properly selected patient groups with autoimmune diseases, and potentially autoimmune diseases such as multiple sclerosis and recurrent polyserositis, thymic hormones offer promise of being effective as a primary treatment modality. Also, the symptom complex of the patients with these diseases should lead to the most dramatic and most easily discerned clinical benefit.

In all of the diseases discussed, prior to the clinical studies, the effects of thymic hormones *in vitro* on assays of cellular immunity, using peripheral blood lymphocytes from well-defined patient groups, should be employed to identify the appropriate populations for the trials and the ideal interval for introduction of the hormones and the duration of treatment.

In summary, the current challenge in the study of thymic hormones is not defense of the clinical applications being explored, but the meticulous design and conduct of clinical trials that include a range of dose regimens and durations of treatment in an attempt to achieve optimum immune modulation and avoid adverse effects which may arise through administration of thymic hormones to inappropriate populations. Future trials, except for pilot studies, must include concomitant control groups treated with placebo that are derived through double-blind randomization, and should employ a battery of cellular immune assays which appear most likely to correlate with the clinical effects of the hormones during the study. Hopefully, the great potential benefit from the use of thymic hormones in a number of diseases for which there is no satisfactory treatment at present will accelerate the widespread initiation of these studies.

P. B. C.

Participants

PAUL G. ABRAMS
Biological Response Modifiers Program
Division of Cancer Treatment
NCI-Frederick, Cancer Research Facility
Frederick, Maryland 21701

BHARAT B. AGGARWAL
Department of Protein Biochemistry
Genentech Inc.
South San Francisco, California 94080

MUSHTAQ AHMAD
Bio-Organic Chemistry Department
Roche Research Center
Hoffman–La Roche Inc.
Nutley, New Jersey 07110

JULIAN ALEKSANDROWICZ
Hematology Department
Cracow Academy of Medicine
Institute of Infectious Diseases and
Institute of Transplantation
Warsaw Academy of Medicine
Polfa Pharmaceuticals
Jelenia Gora, Poland

A. J. AMMANN
Pediatric Immunology
University of California
San Francisco, California 94143

EMMANUEL ARNOUX
Groupe de recherche sur les Maladies
Immunitaries en Haiti
Port-au-Prince, Haiti

SURESH K. ARYA
Laboratory of Tumor Cell Biology
Division of Cancer Treatment
National Cancer Institute
National Institutes of Health
Bethesda, Maryland 20205

JEAN-FRANCOIS BACH
INSERM U-25
Hôpital Necker
75015 Paris, France

FRANCES G. BEACH
Poultry Disease Research Center
Department of Avian Medicine
College of Veterinary Medicine
University of Georgia
Athens, Georgia 30605

F. BETTENS
Institute for Clinical Immunology
Inselspital
University of Bern
Bern, Switzerland

CHRISTIAN BIRR
Max-Planck-Institut für
Medizinische Forschung
and Organogen, Medizinisch-
Molekularbiologische
Forschungsgesellschaft m.b.h.,
D-6900
Heidelberg, Federal Republic of Germany

I. BLAZSEK
Institut de Cancérologie et d'Immunogénétique
(INSERUM U-50)
Hôpital Paul-Brousse
94804 Villejuif, France

G. D. BONNARD
Institute for Clinical Immunology
Inselspital
University of Bern
Bern, Switzerland

TERRY BOWLIN
LBI-Basic Research Program
NCI-Frederick Cancer Research Facility
Frederick, Maryland 21701

CAROL J. BURGER
Department of Biology
Microbiology Section
Virginia Polytechnic Institute and
State University
Blacksburg, Virginia 24061

R. CHRISTOPHER BUTLER
Departments of Medical Microbiology and
Immunology
Arlington Hospital
Arlington, Virginia 22205

S. Y. CHANG
The First Hospital for
Infectious Diseases
Beijing, China

JIEPING CHEN
Department of Biochemistry
The George Washington University
School of Medicine and
Health Sciences
Washington, D.C. 20037

MICHAEL A. CHIRIGOS
Immunopharmacology Section
Biological Therapeutics Branch
Biological Response Modifiers Program
Division of Cancer Treatment
NCI-Frederick Cancer Research Facility
Frederick, Maryland 21701

PAUL S. CHRETIEN
Department of Surgery
University of Maryland
Baltimore, Maryland 21201

GEORGE P. CHROUSOS
Developmental Endocrinology Branch
NICHD
National Institutes of Health
Bethesda, Maryland 20205

NICULAE CIOBANU
The Memorial Sloan-Kettering
Cancer Center
New York, New York 10021

ANGELO CONSTANTINO
Department of Molecular Immunology
Roswell Park Memorial Institute
Buffalo, New York 14263

TERRY D. COPELAND
Laboratory of Molecular Virology
and Carcinogenesis
LBI-Basic Research Program
NCI-Frederick Cancer Research Facility
Frederick, Maryland 21701

JEFFREY COSSMAN
Laboratory of Pathology
National Cancer Institute
National Institutes of Health
Bethesda, Maryland 20205

M. J. COWAN
Pediatric Immunology
University of California
San Francisco, California 94143

JOHN COX
Department of Medicine
The George Washington University School of
Medicine and Health Sciences
Washington, D.C. 20037

L. X. CUI
Chinese Academy of Medical Sciences
Beijing, China

Z. Y. CUI
The First Hospital for
Infectious Diseases
Beijing, China

GUY CUNNINGHAM
National Institute of Neurological and
Communicative Disorders and Stroke
National Institutes of Health
Bethesda, Maryland 20205

JAN CZARNECKI
Hematology Department
Cracow Academy of Medicine
Institute of Infectious Diseases and Institute of
Transplantation
Warsaw Academy of Medicine
Polfa Pharmaceuticals
Jelenia Gora, Poland

BARBARA K. DABROWSKA-BERNSTEIN
Hematology Department
Cracow Academy of Medicine
Institute of Infectious Diseases and Institute of
Transplantation
Warsaw Academy of Medicine
Polfa Pharmaceuticals
Jelenia Gora, Poland

MAREK P. DABROWSKI
Hematology Department
Cracow Academy of Medicine
Institute of Infectious Diseases and Institute of
Transplantation
Warsaw Academy of Medicine
Polfa Pharmaceuticals
Jelenia Gora, Poland

MARINOS C. DALAKAS
National Institute of Neurological and
Communicative Disorders and Stroke
National Institutes of Health
Bethesda, Maryland 20205

W. P. DAO
The First Hospital for
Infectious Diseases
Beijing, China

MIREILLE DARDENNE
INSERM U-25
Hôpital Necker
75015 Paris, France

ARTHUR E. DAVIS, Jr.
Roche Biomedical Laboratories, Inc.
Burlington, North Carolina 27215

STEPHEN DAVIS
Istituto di Clinica Medica
Universita di Perugia
Italy

ISAAC DJERASSI
Mercy Catholic Medical Center
Philadelphia, Pennsylvania 19143

GINO DORIA
ENEA-EURATOM
Immunogenetics Group
Laboratory of Pathology
C.R.E. Casaccia
Rome, Italy

MARY K. DOYLE
Cancer Research Center and
Ellis Fischel State Cancer Center
Columbia, Missouri 65205

D. C. DUMONDE
Department of Immunology
St. Thomas' Hospital
London SE1 7EH, England

PAMELA A. DUNN†
Cancer Research Center and
Ellis Fischel State Cancer Center
Columbia, Missouri 65205

JEAN-MARIE DUPUY
Immunology Research Center
Institut Armand-Frappier
Université du Québec
Laval-des-Rapides
Quebec, Canada

WILLIAM EATON
Cancer Research Center and
Ellis Fischel State Cancer Center
Columbia, Missouri 65205

KLAUS D. ELGERT
Department of Biology
Microbiology Section
Virginia Polytechnic Institute and State University
Blacksburg, Virginia 24061

ROBERT ELIE
Groupe de recherche sur les Maladies Immunitaries en Haiti
Port-au-Prince, Haiti

MICHAEL R. ERDOS
Department of Biochemistry
The George Washington University School of Medicine and Health Sciences
Washington, D.C. 20037

WILLIAM B. ERSHLER
Department of Medicine
University of Vermont
Burlington, Vermont 05405

CHARLES H. EVANS
Tumor Biology Section
Laboratory of Biology
National Cancer Institute
National Institutes of Health
Bethesda, Maryland 20205

CARTESIO FAVALLI
Department of Experimental Medicine
2° Rome University
00173 Rome, Italy

ARTHUR M. FELIX
Bio-Organic Chemistry Department
Roche Research Center
Hoffmann–La Roche Inc.
Nutley, New Jersey 07110

ROSEMARY P. FIORE
Department of Zoology and Physiology
Rutgers University
Newark, New Jersey 07102

LINDA J. FIPPIN
Department of Radiation Oncology
University of California
San Francisco, California 94143

KARL FOLKERS
Institute for Biomedical Research
The University of Texas
Austin, Texas 78712

KENNETH A. FOON
Biological Response Modifiers Program
Division of Cancer Treatment
NCI-Frederick Cancer Research Facility
Frederick, Maryland 21701

DANIELA FRASCA
ENEA-EURATOM
Immunogenetics Group
Laboratory of Pathology
C.R.E. Casaccia
Rome, Italy

MAYA FREUND
Laboratory of Immunopharmacology
Department of Human Microbiology
Sackler Faculty of Medicine
Tel-Aviv University
Tel-Aviv 69978, Israel

HERMAN FRIEDMAN
Department of Medical Microbiology and Immunology
University of South Florida College of Medicine
Tampa, Florida 33612

JERI M. FRIER
Departments of Medical Microbiology and Immunology
Arlington Hospital
Arlington, Virginia 22205

ROBERT C. GALLO
Laboratory of Tumor Cell Biology
Division of Cancer Treatment
National Cancer Institute
National Institutes of Health
Bethesda, Maryland 20205

ENRICO GARACI
Department of Experimental Medicine
2° Rome University
00173 Rome, Italy

JOHN G. GARTNER
Department of Physiology
McGill University Faculty of Medicine
Montreal, Quebec, Canada

L. GASTINEL
INSERM U-25
Hôpital Necker
75015 Paris, France

IGAL GERY
Laboratory of Vision Research
National Eye Institute
National Institutes of Health
Bethesda, Maryland 20205

DIETER GILLESSEN
F. Hoffmann–La Roche and Co., Ltd.
Basel, Switzerland

HY GOLDMAN
Department of Pediatrics
McGill University Faculty
of Medicine
Montreal, Quebec
Canada

ALLAN L. GOLDSTEIN
Department of Biochemistry
The George Washington University
School of Medicine and
Health Sciences
Washington, D.C. 20037

ANDRZEJ GORSKI
Hematology Department
Cracow Academy of Medicine
Institute of Infectious Diseases and Institute of
Transplantation
Warsaw Academy of Medicine
Polfa Pharmaceuticals
Jelenia Gora, Poland

GALE A. GRANGER
Department of Molecular Biology
and Biochemistry
University of California
Irvine, California 92717

FAUSTO GRIGNANI
Istituto di Clinica Medica
Universita di Perugia
Italy

ELIZABETH GRIMM
Surgery Branch
Division of Cancer Treatment
National Cancer Institute
National Institutes of Health
Bethesda, Maryland 20205

JEAN-MICHAEL GUÉRIN
Groupe de recherche sur les Maladies
Immunitaries en Haiti
Port-au-Prince, Haiti

JOHN W. HADDEN
Department of Immunopharmacology
University of South Florida
College of Medicine
Tampa, Florida 33612

NICHOLAS R. HALL
Department of Biochemistry
The George Washington University
School of Medicine and
Health Sciences
Washington, D.C. 20037

ANNE S. HAMBLIN
Department of Immunology
St. Thomas' Hospital
London SE1 7EH, England

RICHARD N. HARKINS
Department of Protein Biochemistry
Genentech Inc.
South San Francisco, California 94080

JANELLE HATCHER
Department of Biochemistry
The George Washington University
School of Medicine and
Health Sciences
Washington, D.C. 20037

BARTON F. HAYNES
Department of Medicine
Division of Rheumatic and
Genetic Diseases
Duke University School of Medicine
Durham, North Carolina 27710

DAVID L. HEALY
Pregnancy Research Branch
NICHD
National Institutes of Health
Bethesda, Maryland 20205

EDGAR P. HEIMER
Bio-Organic Chemistry Department
Roche Research Center
Hoffmann–La Roche Inc.
Nutley, New Jersey 07110

E. M. HERSH
Department of Clinical Immunology and Biological Therapy
The University of Texas System Cancer Center
M. D. Anderson Hospital and Tumor Institute
Houston, Texas 77030

HOWARD R. HIGLEY
Department of Pathology
Loyola University Medical School
Maywood, Illinois 60153

GARY D. HODGEN
Pregnancy Research Branch
NICHD
National Institutes of Health
Bethesda, Maryland 20205

F. C. DEN HOLLANDER
Organon International BV
Oss, The Netherlands

BERNARD L. HORECKER
Roche Institute of Molecular Biology
Roche Research Center
Nutley, New Jersey 07110

C. S. HSU
Chinese Academy of Medical Sciences
Beijing, China

RAYMOND HUBBARD
National Institute of Neurological and Communicative Disorders and Stroke
National Institutes of Health
Bethesda, Maryland 20205

JAMES N. IHLE
LBI-Basic Research Program
NCI-Frederick Cancer Research Facility
Frederick, Maryland 21701

GENEVIEVE S. INCEFY
The Memorial Sloan-Kettering Cancer Center
New York, New York 10021

HIDEO ISHITSUKA
Nippon Roche Research Center
Kamakura City
Kanagawa 247, Japan

ROBERT P. JACOBS
Department of Medicine
The George Washington University School of Medicine and Health Sciences
Washington, D.C. 20037

DIANE L. JOHNSON
Department of Molecular Biology and Biochemistry
University of California
Irvine, California 92717

F. JONCOURT
Institute for Clinical Immunology
Inselspital
University of Bern
Bern, Switzerland

HENRY JOSHUA
Mercy Catholic Medical Center
Philadelphia, Pennsylvania 19143

JONATHAN KELLER
LBI-Basic Research Program
NCI-Frederick Cancer Research Facility
Frederick, Maryland 21701

N. KIGER
Institut de Cancérologie et d'Immunogénétique (INSERM U-50)
Hôpital Paul-Brousse
94804 Villejuif, France

DAHLIA KIRKPATRICK
The Memorial Sloan-Kettering
Cancer Center
New York, New York 10021

AURELLA KREZLEWICZ
National Institute of Neurological and
Communicative Disorders and Stroke
National Institutes of Health
Bethesda, Maryland 20205

F. KRISTENSEN
Institute for Clinical Immunology
Inselspital
University of Bern
Bern, Switzerland

STEPHEN LAHEY
Department of Surgery
Brigham & Women's Hospital and Dana-Farber
Cancer Institute
Harvard Medical School
Boston, Massachusetts 02115

THEODORE J. LAMBROS
Bio-Organic Chemistry Department
Roche Research Center
Hoffmann–La Roche Inc.
Nutley, New Jersey 07110

WAYNE S. LAPP
Department of Physiology
McGill University Faculty of Medicine
Montreal, Quebec, Canada

A. CLAUDE LAROCHE
Groupe de recherche sur les Maladies
Immunitaries en Haiti
Port-au-Prince, Haiti

SANG HE LEE
Department of Pharmacological Sciences
Genentech Inc.
South San Francisco, California 94080

M. LENFANT
Institut de Cancérologie et d'Immunogénétique
(INSERM U-50)
Hôpital Paul-Brousse
94804 Villejuif, France

JOSE L. LEPE-ZUNIGA
Laboratory of Vision Research
National Eye Institute
National Institutes of Health
Bethesda, Maryland 20205

YUAN LIN
Central Research and
Development Department
E. I. du Pont de Nemours
and Company
Glenolden Laboratory
Glenolden, Pennsylvania 19036

STEPHEN G. LINDNER
Laboratory of Tumor Cell Biology
Division of Cancer Treatment
National Cancer Institute
National Institutes of Health
Bethesda, Maryland 20205

S. L. LIU
Chinese Academy of Medical Sciences
Beijing, China

MARGARET LLOYD
Department of Medicine
The George Washington University
School of Medicine and
Health Sciences
Washington, D.C. 20037

CEDRIC W. LONG
Biological Response Modifiers Program
Division of Cancer Treatment
NCI-Frederick Cancer Research Facility
Frederick, Maryland 21701

D. LYNN LORIAUX
Developmental Endocrinology Branch
NICHD
National Institutes of Health
Bethesda, Maryland 20205

MICHAEL LOTZE
Surgery Branch
Division of Cancer Treatment
National Cancer Institute
National Institutes of Health
Bethesda, Maryland 20205

THERESA L. K. LOW
Department of Biochemistry
The George Washington University
School of Medicine and
Health Sciences
Washington, D.C. 20037

DAVID L. MADDEN
National Institute of Neurological and
Communicative Disorders and Stroke
National Institutes of Health
Bethesda, Maryland 20205

MICHIYUKI MAEDA
Chest Disease Research Institute
Kyoto University
Kyoto, Japan

RODOLPHE MALEBRANCE
Groupe de recherche sur les Maladies
Immunitaries en Haiti
Port-au-Prince, Haiti

URSULA MALORNY
Department of
Experimental Dermatology
Universitäts-Hautklinik
D-4400 Münster
Federal Republic of Germany

ANNETTE MALUISH
Biological Response Modifiers Program
Division of Cancer Treatment
NCI-Frederick Cancer Research Facility
Frederick, Maryland 21701

GEORGE L. MANDERINO
Abbott Laboratories
North Chicago, Illinois 60064

JEANNE MARRAZO
Department of Surgery
Brigham & Women's Hospital and Dana-Farber
Cancer Institute
Harvard Medical School
Boston, Massachusetts 02115

IRENE K. MASUNAKA
Department of Molecular Biology
and Biochemistry
University of California
Irvine, California 92717

G. MATHÉ
Institut de Cancérologie et d'Immunogénétique
(INSERM U-50)
Hôpital Paul-Brousse
94804 Villejuif, France

AMITABHA MAZUMDER
Surgery Branch
Division of Cancer Treatment
National Cancer Institute
National Institutes of Health
Bethesda, Maryland 20205

JOHN E. McCLURE
Allergy Immunology Service
Department of Pediatrics
Texas Children's Hospital
Houston, Texas 77030

JOHN E. McENTIRE
Cancer Research Center and
Ellis Fischel State Cancer Center
Columbia, Missouri 65205

JOSEPH P. McGILLIS
Department of Biochemistry
The George Washington University
School of Medicine and
Health Sciences
Washington, D.C. 20037

RUDOLPH MEDICUS
LBI-Basic Research Program
NCI-Frederick Cancer Research Facility
Frederick, Maryland 21701

JOHANNES MEIENHOFER
Bio-Organic Chemistry Department
Roche Research Center
Hoffmann–La Roche Inc.
Nutley, New Jersey 07110

ROLAND MERTELSMANN
The Memorial Sloan-Kettering
Cancer Center
New York, New York 10021

ELMAR MICHELS
Department of
Experimental Dermatology
Universitäts-Hautklinik
D-4400 Münster
Federal Republic of Germany

BARBARA MOFFAT
Department of Protein Biochemistry
Genentech Inc.
South San Francisco, California 94080

HERBERT C. MORSE, III
National Institute of Allergy and Infectious Diseases
National Institutes of Health
Bethesda, Maryland 20205

C. G. MUNN
Department of Clinical Immunology and Biological Therapy
The University of Texas System Cancer Center
M. D. Anderson Hospital and Tumor Institute
Houston, Texas 77030

RICHARD MURAL
LBI-Basic Research Program
NCI-Frederick Cancer Research Facility
Frederick, Maryland 21701

J. L. MURRAY
Department of Clinical Immunology and Biological Therapy
The University of Texas System Cancer Center
M. D. Anderson Hospital and Tumor Institute
Houston, Texas 77030

KRISHNA K. MURTHY
Poultry Disease Research Center
Department of Avian Medicine
College of Veterinary Medicine
University of Georgia
Athens, Georgia 30605

PAUL H. NAYLOR
Department of Biochemistry
The George Washington University School of Medicine and Health Sciences
Washington, D.C. 20037

LEONARD M. NECKERS
Laboratory of Pathology
National Cancer Institute
National Institutes of Health
Bethesda, Maryland 20205

MARGUERITE H. NEELY
Department of Radiation Oncology
University of California
San Francisco, California 94143

RUTH NETA
Armed Forces Radiobiology
Research Institute
Bethesda, Maryland 20814

CHRISTINE NEUMANN
Department of
Experimental Dermatology
Universitäts-Hautklinik
D-4400 Münster
Federal Republic of Germany

ANITA D. NOVITT
Department of Zoology and Physiology
Rutgers University
Newark, New Jersey 07102

HERBERT F. OETTGEN
The Memorial Sloan-Kettering Cancer Center
New York, New York 10021

YUMIKO OHTA
Nippon Roche Research Center
Kamakura City
Kanagawa 247, Japan

ROBERT K. OLDHAM
Biological Response Modifiers Program
Division of Cancer Treatment
NCI-Frederick Cancer Research Facility
Frederick, Maryland 21701

JOOST J. OPPENHEIM
Laboratory of Molecular Immunoregulation
Biological Response Modifiers Program
Division of Cancer Treatment
NCI-Frederick Cancer Research Facility
Frederick, Maryland 21701

STEPHEN OROSZLAN
Laboratory of Molecular Virology and Carcinogenesis
LBI-Basic Research Program
NCI-Frederick Cancer Research Facility
Frederick, Maryland 21701

SALLY L. ORR
Department of Molecular Biology
and Biochemistry
University of California
Irvine, California 92717

GERALDINE M. OVAK
Department of Molecular Immunology
Roswell Park Memorial Institute
Buffalo, New York 14263

RAJENDRA PAHWA
The Memorial Sloan-Kettering
Cancer Center
New York, New York 10021

SAVITA PAHWA
North Shore University Hospital
Manhasset, New York 10030

EDMUND PALASZYNSKI
LBI-Basic Research Program
NCI-Frederick Cancer Research Facility
Frederick, Maryland 21701

SUSAN PALASZYNSKI
Department of Medicine
The George Washington University
School of Medicine and
Health Sciences
Washington, D.C. 20037

MICHAEL A. PALLADINO
The Memorial Sloan-Kettering
Cancer Center
New York, New York 10021
Present Address:
Genentech, Inc.
South San Francisco, California 94080

BEN W. PAPERMASTER
Cancer Research Center and
Ellis Fischel State Cancer Center
Columbia, Missouri 65205

JOHN L. PAULY
Department of Molecular Immunology
Roswell Park Memorial Institute
Buffalo, New York 14263

THOMAS M. PETRO
Department of Food and Nutrition
Purdue University
West Lafayette, Indiana 47907
Present Address:
Microbial Biochemistry Branch
Food and Drug Administration
Cincinnati, Ohio 45226

EDGAR PICK
Laboratory of Immunopharmacology
Department of Human Microbiology
Sackler Faculty of Medicine
Tel-Aviv University
Tel-Aviv 69978, Israel

GÉRARD PIERRE
Department of Pathology
Montreal Children's Hospital and
McGill University
Montreal Children's Hospital
Research Institute
Montreal, Quebec, Canada

ROSS PITCHER
Bio-Organic Chemistry Department
Roche Research Center
Hoffmann–La Roche Inc.
Nutley, New Jersey 07110

J. MICHAEL PLUNKETT
Department of Molecular Biology
and Biochemistry
University of California
Irvine, California 92717

MELANIE S. PULLEY
Department of Immunology
St. Thomas' Hospital
London SE1 7EH, England

WILLIAM L. RAGLAND
Poultry Disease Research Center
Department of Avian Medicine
College of Veterinary Medicine
University of Georgia
Athens, Georgia 30605

PIETRO RAMBOTTI
Istituto di Clinica Medica
Universita di Perugia
Italy

THANJAVUR RAVIKUMAR
Department of Surgery
Brigham & Women's Hospital and Dana-Farber Cancer Institute
Harvard Medical School
Boston, Massachusetts 02115

JANET H. RANSOM
Tumor Biology Section
Laboratory of Biology
National Cancer Institute
National Institutes of Health
Bethesda, Maryland 20205

ROBERT W. REBAR
Department of Obstetrics and Gynecology
Northwestern Memorial Hospital
Northwestern University Medical School
Chicago, Illinois 60611

J. M. REUBEN
Department of Clinical Immunology and Biological Therapy
The University of Texas System Cancer Center
M. D. Anderson Hospital and Tumor Institute
Houston, Texas 77030

RALPH D. REYNOLDS
Cancer Research Center and Ellis Fischel State Cancer Center
Columbia, Missouri 65205

CHRISTINE S. E. RICHARDSON
Department of Medicine
The George Washington University School of Medicine and Health Sciences
Washington, D.C. 20037

SUSAN B. RILEY
The George Washington University
Department of Medicine
School of Medicine and Health Sciences
Washington, D.C. 20037

CRISTINA RINALDI-GARACI
Institute of General Pathology
1° Rome University
00173 Rome, Italy

RICHARD J. ROBB
Central Research and Developmental Department
E. I. du Pont de Nemours and Company
Glenolden Laboratory
Glenolden, Pennsylvania 19036

MARY RODRICK
Department of Surgery
Brigham & Women's Hospital and Dana-Farber Cancer Institute
Harvard Medical School
Boston, Massachusetts 02115

STEVEN A. ROSENBERG
Surgery Branch
Division of Cancer Treatment
National Cancer Institute
National Institutes of Health
Bethesda, Maryland 20205

MAURY ROSENSTEIN
Surgery Branch
Division of Cancer Treatment
National Cancer Institute
National Institutes of Health
Bethesda, Maryland 20205

DONALD ROSS
Department of Surgery
Brigham & Women's Hospital and Dana-Farber Cancer Institute
Harvard Medical School
Boston, Massachusetts 02115

MARTIN ROSZKOWSKI
Bio-Organic Chemistry Department
Roche Research Center
Hoffmann–La Roche Inc.
Nutley, New Jersey 07110

GEOFFREY ROWDEN
Department of Pathology
Loyola University Medical School
Maywood, Illinois 60153

CYNTHIA W. RUSSELL
Department of Molecular Immunology
Roswell Park Memorial Institute
Buffalo, New York 14263

PIERRE RUSSO
Department of Pathology
Montreal Children's Hospital and
McGill University—
Montreal Children's Hospital
Research Institute
Montreal, Quebec, Canada

S. B. SALVIN
Department of Microbiology
University of Pittsburgh
School of Medicine
Pittsburgh, Pennsylvania 15261

MANGALASSERIL G. SARNGADHARAN
Laboratory of Tumor Cell Biology
Division of Cancer Treatment
National Cancer Institute
National Institutes of Health
Bethesda, Maryland 20205

W. SAVINO
INSERM U-25
Hôpital Necker
75015 Paris, France

RICHARD S. SCHULOF
Departments of Medicine
and Biochemistry
The George Washington University
School of Medicine and
Health Sciences
Washington, D.C. 20037

HEINRICH M. SCHULTE
Developmental Endocrinology Branch
NICHD
National Institutes of Health
Bethesda, Maryland 20205

THOMAS A. SEEMAYER
Department of Pathology
Montreal Children's Hospital and
McGill University—
Montreal Children's Hospital
Research Institute
Montreal, Quebec, Canada

JOHN L. SEVER
National Institute of Neurological and
Communicative Disorders and Stroke
National Institutes of Health
Bethesda, Maryland 20205

SU SHENG
The First Hospital for
Infectious Diseases
Beijing, China

STEPHEN A. SHERWIN
Biological Response Modifiers Program
Division of Cancer Treatment
NCI-Frederick Cancer Research Facility
Frederick, Maryland 21701

J. Y. SHI
The First Hospital for
Infectious Diseases
Beijing, China

HONG-MING SHIEH
Institute for Biomedical Research
The University of Texas at Austin
Austin, Texas 78712

ALEXANDER B. SKOTNICKI
Hematology Department
Cracow Academy of Medicine
Institute of Infectious Diseases and Institute of
Transplantation
Warsaw Academy of Medicine
Polfa Pharmaceuticals
Jalenia Gora, Poland

KENDALL A. SMITH
Department of Medicine
Dartmouth Medical School
Hanover, New Hampshire 03756

CLEMENS SORG
Department of
Experimental Dermatology
Universitäts-Hautklinik
D-4400 Münster
Federal Republic of Germany

BARBARA M. SOUTHCOTT
Department of Radiotherapy
Charing Cross Hospital
London W6, England

BRYAN L. SPANGELO
Department of Biochemistry
The George Washington University
School of Medicine and
Health Sciences
Washington, D.C. 20037

GLENN STEELE, Jr.
Department of Surgery
Brigham & Women's Hospital and Dana-Farber
Cancer Institute
Harvard Medical School
Boston, Massachusetts 02115

CORA STERNBERG
The Memorial Sloan-Kettering
Cancer Center
New York, New York 10021

HELEN R. STRAUSSER
Department of Zoology and Physiology
Rutgers University
Newark, New Jersey 07102

S. SU
The First Hospital for
Infectious Diseases
Beijing, China

JOSEPH SWISTOK
Bio-Organic Chemistry Department
Roche Research Center
Hoffmann–La Roche Inc.
Nutley, New Jersey 07110

K. TAO
Chinese Academy of Medical Sciences
Beijing, China

KEISUKE TESHIGAWARA
Institute for Immunology
Faculty of Medicine
Kyoto University
Kyoto, Japan

EMIKO TEZUKA
Nippon Roche Research Center
Kamakura City
Kanagawa 247, Japan

VOLDEMAR TOOME
Bio-Organic Chemistry Department
Roche Research Center
Hoffmann–La Roche
Nutley, New Jersey 07110

BRUCE TRAPP
National Institute of Neurological and
Communicative Disorders and Stroke
National Institutes of Health
Bethesda, Maryland 20205

ARNOLD TRZECIAK
F. Hoffmann–La Roche and Co., Ltd.
Basel, Switzerland

MITSURU TSUDO
First Department of Internal Medicine
Kyoto University
Kyoto, Japan

CAROL J. TWIST
Department of Molecular Immunology
Roswell Park Memorial Institute
Buffalo, New York 14263

TAKASHI UCHIYAMA
First Department of Internal Medicine
Kyoto University
Kyoto, Japan

YUKIO UMEDA
Nippon Roche Research Center
Kamakura City
Kanagawa 247, Japan

GEORGE V. VAHOUNY
Biochemistry Department
The George Washington University
School of Medicine and
Health Sciences
Washington, D.C. 20037

ANDRE VASU
North Shore University Hospital
Manhasset, New York 11030

JEANNE WALTER
Cancer Research Center and
Ellis Fischel State Cancer Center
Columbia, Missouri 65205

CHING-TSO WANG
Bio-Organic Chemistry Department
Roche Research Center
Hoffmann–La Roche Inc.
Nutley, New Jersey 07110

Y. WANG
Institute for Clinical Immunology
Inselspital
University of Bern
Bern, Switzerland

YUJI WANO
First Department of Internal Medicine
Kyoto University
Kyoto, Japan

DIANE W. WARA
Pediatric Immunology
University of California
San Francisco, California 94143

WILLIAM M. WARA
Department of Radiation Oncology
University of California
San Francisco, California 94143, and
Northern California Cancer Program
Palo Alto, California 94303

RONALD ROSS WATSON
Department of Family and
Community Medicine
University of Arizona Medical School
Tucson, Arizona 85724

THELMA WATSON
Biological Response Modifiers Program
Division of Cancer Treatment
NCI-Frederick Cancer Research Facility
Frederick, Maryland 21701

A. L. DE WECK
Institute for Clinical Immunology
Inselspital
University of Bern
Bern, Switzerland

BOGDA WEGRZYNSKI
Bio-Organic Chemistry Department
Roche Research Center
Hoffmann–La Roche Inc.
Nutley, New Jersey 07110

KARL WELTE
The Memorial Sloan-Kettering
Cancer Center
New York, New York 10021

FLOSSIE WONG-STAAL
Laboratory of Tumor Cell Biology
Division of Cancer Treatment
National Cancer Institute
National Institutes of Health
Bethesda, Maryland 20205

YASUO YAGI
Nippon Roche Research Center
Kamakura City
Kanagawa 247, Japan

ROBERT S. YAMAMOTO
Department of Molecular Biology
and Biochemistry
University of California
Irvine, California 92717

G. C. YANG
Chinese Academy of Medical Sciences
Beijing, China

JIN YIFENG
Biology Department
Nanjing University
Nanjing, China

JUNJI YODOI
Institute for Immunology
Faculty of Medicine
Kyoto University
Kyoto, Japan

MARION ZATZ
Department of Biochemistry
The George Washington University
School of Medicine and
Health Sciences
Washington, D.C. 20037

C. X. ZHENG
Beijing University
Beijing, China

J. SAMUEL ZIGLER, Jr.
Laboratory of Vision Research
National Eye Institute
National Institutes of Health
Bethesda, Maryland 20205

MARY LOUISE ZIMMERMAN
Laboratory of Vision Research
National Eye Institute
National Institutes of Health
Bethesda, Maryland 20205

Index